Hydraulic Structures

Also available from Taylor & Francis

Hydraulics in Civil and Environmental Engineering 4th edition
A. Chadwick *et al.*
Hardback: ISBN 978-041-530608-9
Paperback: ISBN 978-041-530609-6

Mechanics of Fluids 8th edition
B. Massey, by J. Ward Smith
Hardback: ISBN 978-0-415-36205-4
Paperback: ISBN 978-0-415-36206-1

Practical Hydraulics 2nd edition
M. Kay
Hardback: ISBN 978-0-415-35114-0
Paperback: ISBN 978-0-415-35115-7

Hydraulic Canals
J. Liria
Hardback: ISBN 978-0-415-36211-5

Information and ordering details
For price availability and ordering visit our website **www.sponpress.com**
Alternatively our books are available from all good bookshops.

Hydraulic Structures

Fourth Edition

P. Novak, A.I.B. Moffat and C. Nalluri

School of Civil Engineering and Geosciences,
University of Newcastle upon Tyne, UK

and

R. Narayanan

Formerly Department of Civil and Structural Engineering, UMIST,
University of Manchester, UK

Spon Press
an imprint of Taylor & Francis
LONDON AND NEW YORK

Fourth edition published 2007 by E & FN Spon
2 Park Square, Milton Park, Abingdon, Oxon, OX14 4RN

Simultaneously published in the USA and Canada
by E & FN Spon
605 Third Avenue, New York, NY 10017

Published 2007, 2008, 2010

E & FN Sopn is an imprint of the Taylor & Francis Group, an informa business

© 1990, 1996, 2001, 2007 Pavel Novak, Iain Moffat, the estate of Chandra Nalluri and Rangaswami Narayanan

The right of Pavel Novak, Iain Moffat, Chandra Nalluri and Rangaswami Narayanan to be identified as the Authors of this Work has been asserted by them in accordance with the Copyright, Designs and Patents Act 1988

Typeset in Times by Wearset Ltd, Boldon, Tyne and Wear

All rights reserved. No part of this book may be reprinted or reproduced or utilized in any form or by any electronic, mechanical, or other means, now known or hereafter invented, including photocopying and recording, or in any information storage or retrieval system, without permission in writing from the publishers.

The publisher makes no representation, express or implied, with regard to the accuracy of the information contained in this book and cannot accept any legal responsibility or liability for any efforts or omissions that may be made.

Notice:
Product or corporate names may be trademarks or registered trademarks, and are used only for identification and explanation without intent to infringe.

British Library Cataloguing in Publication Data
A catalogue record for this book is available from the British Library

Library of Congress Cataloging in Publication Data
Hydraulic structures / P. Novak . . . [et al.]. — 4th ed.
p. cm.
Includes bibliographical references and index.
ISBN-13: 978-0-415-38625-8 (alk. paper)
ISBN-13: 978-0-415-38626-5 (pbk. : alk. paper)
1. Hydraulic structures. I. Novák, Pavel.
TC180.H95 2007
627--dc22

ISBN13: 978-0-415-38625-8 Hardback
ISBN13: 978-0-415-38626-5 Paperback
ISBN13: 978-0-203-96463-7 e-book

Contents

Preface

The aim of the book, to provide a text for final year undergraduate and postgraduate students, remains the same as in the previous editions; we also trust that researchers, designers and operators of hydraulic structures will continue to find the text of interest and a stimulating up-to-date reference source.

This new edition enabled us to *update the text and references throughout*, and to introduce some important changes and additions reacting to new developments in the field. We have also taken note of some comments received on the previous edition; particular thanks for the constructive comments and help provided by Professor J. Lewin in redrafting Chapter 6 (Gates and valves).

The authorship of individual chapters remains the same as in previous editions; (Dr Narayanan carried out the work on this edition during his stay in the Faculty of Civil Engineering, Universiti Teknologi Malaysia, Johor Bahru, Malaysia). However, as our colleague Dr C. Nalluri unfortunately died in December 2003 'his' text was reviewed by Dr Narayanan (Chapter 13) and Professor Novak (Chapters 9, 10 and 12) who also again edited the whole text.

Readers of the previous (2001) edition may note the following *major* changes:

Chapter 1. Enhanced discussion of environmental issues including the World Commission on Dams report.

Chapter 2. New sections on partially saturated soils, small farm and amenity dams, tailing dams and lagoons and upgrading and rehabilitation of embankment dams; extended treatment of upstream face protection/rock armouring.

Chapter 3. Extended discussion of roller-compacted concrete dams and a new section on upgrading of masonry and concrete dams.

Chapter 4.	Substantially enhanced discussion of flow over stepped spillways.
Chapter 5.	Extended treatment of scour in plunge pools.
Chapter 6.	Enlarged treatment of hydrodynamic forces acting on low and high-head gates and new sections dealing with cavitation, aeration and vibrations of gates and automation, control and reliability.
Chapter 7.	Increased coverage of integrated risk analysis/management and contingency/emergency planning in dam safety.
Chapter 9.	Inclusion of barrages with raised sill.
Chapter 12.	New text on small hydraulic power development and tidal and wave power.
Chapter 14.	More detailed treatment of wave breaking, wave statistics and pipeline stability.
Chapter 15.	Enhanced treatment of beach profile and wave/structure interaction and a new section on coastal modelling.
Chapter 16.	Enlarged discussion of mathematical, numerical and computational models in hydraulic engineering.

In order not to increase the size of the book unduly some less relevant material has been omitted (particularly in Ch. 12).

P. Novak, A.I.B. Moffat and R. Narayanan
Newcastle upon Tyne, June 2006

Preface to the third edition

The main aim of the book, i.e. to provide a text for final year undergraduate and for postgraduate students, remains the same as for the previous two editions; we also hope that researchers, designers and operators of the many types of structures covered in the book will continue to find the text of interest and a stimulating, up-to-date reference source.

It is now almost six years since the manuscript of the second edition was completed and this new edition gave us the opportunity to correct the few remaining errors and to update the text and references throughout. At the same time, as a reaction to some important developments in the field, certain parts of the text have been rewritten, enlarged or reorganized. Readers of the second edition may wish to note the following *major* changes:

Chapter 1. The environmental and social issues associated with major reservoir projects are addressed in greater depth.

Chapter 2. New section on small embankments and flood banks and expanded discussion of seismicity and seismic analysis.

Chapter 4. Enlarged text on design flood selection and reservoir flood standards, aeration on spillways and in free flowing tunnels; extended treatment of stepped spillways.

Chapter 6. A new section on tidal barrage and surge protection gates and enlarged text on forces acting on gates; a new worked example.

Chapter 7. Enhanced text on reservoir hazard analysis and dam break floods.

Chapter 9. New paragraph on pressure distribution under piled foundation floors of weirs with a new worked example.

Chapter 14. This chapter – Coastal and offshore engineering in previous edition – has been divided into:
Chapter 14 'Waves and offshore engineering' and

Chapter 15 'Coastal engineering'
Consequently the whole material has been reorganized. The treatment of forces on cylindrical bodies in waves and currents has been significantly extended in Chapter 14. Chapter 15 now includes an extended treatment of wave overtopping and stability of breakwaters as well as a brief discussion of coastal management.

Chapter 16. *(formerly ch. 15).* Extended discussion of computational modelling of hydraulic structures.

P. Novak, A.I.B. Moffat, C. Nalluri and R. Narayanan
Newcastle upon Tyne, August 2000

Preface to the second edition

The main aim of the book, i.e. to provide a text for final year undergraduate and for postgraduate students, remains the same as for the first edition; equally we hope that researchers, designers and operators of the many types of hydraulic structures covered in the book will find the text of interest and a useful reference source.

We took the opportunity of a new edition to correct all (known) errors and to thoroughly update the text and references throughout. At the same time as a response to received comments and reviews as well as a reaction to some new developments in the field, certain parts of the text were rewritten or enlarged. Readers of the first edition may wish to note the following major changes.

Chapter 1. Extended text on site assessment for dams.

Chapter 2. Expanded treatment of geotechnical aspects, e.g. a new paragraph (2.8.3) on performance indices for earthfill cores, and a new brief section (2.10) on geosynthetics.

Chapter 3. Extended coverage of RCC dams with a new paragraph (3.7.3) dealing with developments in RCC construction.

Chapter 4. Enlarged text dealing with design flood estimation, reservoir sedimentation, interference waves and aeration on spillways and a new paragraph (4.7.6) on stepped spillways.

Chapter 5. Enlarged section on scour below spillways.

Chapter 6. A new paragraph (6.2.8) on overspill fusegates.

Chapter 7. Enlarged text on reservoir downstream hazard assessment.

Chapter 8. Enlarged text on multistage channels, geotextiles, Crump weir computation and a new section (8.6) on river flood routing.

Chapter 9. Extended text on fish passes and a new paragraph (9.1.6) on the effect of the operation of barrages on river water quality.

Chapter 10. Enlarged text on canal inlets and scour at bridges and below culvert outlets.
Chapter 13. A new short section (13.7) on benching.
Chapter 14. Change of title (from Coastal engineering) to Coastal and offshore engineering incorporating a substantial new section (14.7) on sea outfalls and the treatment of wave forces on pipelines in the shoaling region.
Chapter 15. Change of title (from Scale models in hydraulic engineering) to Models in hydraulic engineering to include in the general discussion of hydraulic models (15.1.1) a typology of mathematical models; also included a short paragraph (15.2.4) on modelling of seismic response.

The authors would like to thank the reviewers for their constructive comments and the publisher for providing the opportunity for this second edition.

P. Novak, A.I.B. Moffat, C. Nalluri and R. Narayanan
Newcastle upon Tyne, December 1994

Preface to the first edition

This text is loosely based on a course on 'Hydraulic Structures' which evolved over the years in the Department of Civil Engineering at the University of Newcastle upon Tyne. The final-year undergraduate and Diploma/MSc postgraduate courses in hydraulic structures assume a good foundation in hydraulics, soil mechanics, and engineering materials, and are given in parallel with the more advanced treatment of these subjects, and of hydrology, in separate courses.

It soon became apparent that, although a number of good books may be available on specific parts of the course, no text covered the required breadth and depth of the subject, and thus the idea of a hydraulic structures textbook based on the course lecture notes came about. The hydraulic structures course has always been treated as the product of team-work. Although Professor Novak coordinated the course for many years, he and his colleagues each covered those parts where they could make a personal input based on their own professional experience. Mr Moffat, in particular, in his substantial part of the course, covered all geotechnical engineering aspects. In the actual teaching some parts of the presented text may, of course, have been omitted, while others, particularly case studies (including the discussion of their environmental, social, and economic impact), may have been enlarged, with the subject matter being continuously updated.

We are fully aware that a project of this kind creates the danger of presenting the subject matter in too broad and shallow a fashion; we hope that we have avoided this trap and got it 'about right', with worked examples supplementing the main text and extensive lists of references concluding each chapter of the book.

This text is not meant to be a research monograph, nor a design manual. The aim of the book is to provide a textbook for final-year undergraduate and postgraduate students, although we hope that researchers, designers, and operators of the many types of hydraulic structures will also find it of interest and a useful reference source.

PREFACE TO THE FIRST EDITION

The text is in two parts; Part One covers dam engineering, and Part Two other hydraulic structures. Mr A.I.B. Moffat is the author of Chapters 1, 2, 3 and 7, and of section 15.2. Dr C. Nalluri wrote Chapters 9, 10, 12 and 13, and sections 8.4 and 8.5. Dr R. Narayanan of UMIST was invited to lecture at Newcastle for two years, on coastal engineering, and is the author of Chapter 14. The rest of the book was written by Professor P. Novak (Chapters 4, 5, 6 and 8, except for sections 8.4 and 8.5, Chapter 11 and section 15.1), who also edited the whole text.

P. Novak, A.I.B. Moffat, C. Nalluri and R. Narayanan
Newcastle upon Tyne, 1989

Acknowledgements

We are grateful to the following individuals and organizations who have kindly given permission for the reproduction of copyright material (figure numbers in parentheses):

Thomas Telford Ltd (4.1, 4.2); US Bureau of Reclamation (4.3, 4.7, 4.15, 4.16, 5.6, 5.7); Elsevier Science Publishers (4.5, 4.12, 4.13, 5.5, 5.8, 5.10. 11.1, 11.2, 11.10, 11.11, 11.16, 11.17, 11.18, 12.17); British Hydromechanics Research Association (4.11, 13.6, 13.9); Institution of Water and Environmental Management (4.18); ICOLD (4.19, 4.20); Figures 4.21, 6.2, 6.3, 6.4 reproduced by permission of John Wiley & Sons Ltd, from H.H. Thomas, *The Engineering of Large Dams*, © 1976; C.D. Smith (6.6, 6.7); MMG Civil Engineering Systems Ltd (8.20); E. Mosonyi (9.12, 9.13, 12.17); International Institute for Land Reclamation and Improvement, the Netherlands (10.14, 10.15); Morgan-Grampian Book Publishing (11.1, 11.5); Delft Hydraulics (11.7); Macmillan (14.12); C.A.M. King (14.13); C. Sharpe (11.2); J. Lewin (6.1, 6.2).

Cover image courtesy of Ingetec S.A. Colombia (Dr A. Marulanda)

List of tables

Main symbols

a	constant, gate opening, pressure wave celerity, wave amplitude
A	cross-sectional area
b	breadth, channel width, constant, length of wave crest
B	water surface width
$\bar{B}$	porewater pressure coefficient
c	apparent cohesion, coefficient, constant, unit shearing strength, wave celerity
C	Chezy coefficient, coefficient, concentration
C_d	coefficient of discharge
C_D	drag coefficient
C_v	coefficient of consolidation, coefficient of velocity
d	depth, diameter, sediment grain size
D	diameter, displacement of vessels
E	cut-off (core) efficiency, energy, Young's modulus
e	energy loss, pipe wall thickness
f	correction factor, frequency, function, Lacey's silt factor
F	factor of safety, fetch, force, function
F_D	drag force
Fr	Froude number
FSL	full supply level
g	gravitational acceleration
GWL	ground water level
h'	uplift pressure head
h	head, pump submergence, rise of water level above SWL, stage
H	total energy (head), head (on spillway etc.), wave (embankment) height
H_s	seepage head, significant wave height, static lift
HFL	high flood level
i	hydraulic gradient
I	inflow, influence factor, moment of inertia

k	coefficient (of permeability), effective pipe roughness, wave number
K	bulk modulus, channel conveyance, coefficient
Kc	Keulegan–Carpenter number
l	length
L	length, wavelength
m	mass
m_v	coefficient of volume compressibility
M	moment
n	Manning roughness coefficient
N	hydraulic exponent speed in rev/min
N_d	number of increments of potential in flownet
N_f	number of flow channels in flownet
N_s	specific speed
NWL	normal water level
O	outflow
p	number of poles, pressure intensity
p_v	vapour pressure
P	force, power, wetted perimeter
q	specific discharge
Q	discharge
Q_s	discharge of sediment
r	factor, radius
r_u	pore pressure ratio
R	hydraulic radius, resistance, resultant, radius
Re	Reynolds number
R_s	régime scour depth
S	maximum shearing resistance, slope
S_c	critical slope
S_f	friction slope
S_0	bed slope
Sh	Strouhal number
SWL	still water level
t	thickness, time
T	draught, time, wave period
u	local velocity (x direction)
u_w	porewater pressure
U	wind speed
U_*	shear velocity
v	velocity (general), velocity (y direction)
V	mean cross-sectional velocity, storage, volume
V_c	critical velocity
w	moisture content, velocity (z direction)
w_s	sediment fall velocity
W	régime width, weight

MAIN SYMBOLS

x	distance, x coordinate
y	flow depth, y coordinate
y'	stilling basin depth
y^+	depth of centroid of section A
y_c	critical depth
y_m	mean depth ($=A/B$)
y_s	maximum scour (local) depth, turbine setting
z	depth, elevation relative to datum, z coordinate
α	angle, constant, energy (Coriolis) coefficient, (seismic) coefficient, wave crest angle
β	angle, momentum (Boussinesq) coefficient, slope, angle
γ	specific (unit) weight ($=pg$)
δ	boundary layer thickness, deflection settlement
δ'	laminar sublayer thickness
Δ	relative density of sediment in water ($(\rho_s - \rho)/\rho$)
ϵ	strain
η	area reduction coefficient, efficiency
θ	angle, velocity coefficient
λ	Darcy–Weisbach friction factor, flownet scale transform factor
μ	dynamic viscosity of water
ν	kinematic viscosity of water, Poisson ratio
ξ	coefficient (head loss), parameter
ρ	density of water
ρ_s	density of sediment particle
σ	cavitation number, conveyance ratio, safety coefficient, stress, surface tension
$\sigma_{1,2,3}$	major, intermediate and minor principal stresses
σ'	effective stress, safety coefficient
τ	shear stress, time interval
τ_c	critical shear stress
τ_0	boundary shear stress
ϕ	angle of shearing resistance or internal friction, function, sediment transport parameter, speed factor
ψ	flow parameter
ω	angular velocity (radians s^{-1})

Part One

Dam engineering

Chapter 1

Elements of dam engineering

1.1 General

The construction of dams ranks with the earliest and most fundamental of civil engineering activities. All great civilizations have been identified with the construction of storage reservoirs appropriate to their needs, in the earliest instances to satisfy irrigation demands arising through the development and expansion of organized agriculture. Operating within constraints imposed by local circumstance, notably climate and terrain, the economic power of successive civilizations was related to proficiency in water engineering. Prosperity, health and material progress became increasingly linked to the ability to store and direct water.

In an international context, the proper and timely utilization of water resources remains one of the most vital contributions made to society by the civil engineer. Dam construction represents a major investment in basic infrastructure within all nations. The annual completion rate for dams of all sizes continues at a very high level in many countries, e.g. China, Turkey and India, and to a lesser degree in some more heavily industrialized nations including the United States.

Dams are individually unique structures. Irrespective of size and type they demonstrate great complexity in their load response and in their interactive relationship with site hydrology and geology. In recognition of this, and reflecting the relatively indeterminate nature of many major design inputs, dam engineering is not a stylized and formal science. As practised, it is a highly specialist activity which draws upon many scientific disciplines and balances them with a large element of engineering judgement; dam engineering is thus a uniquely challenging and stimulating field of endeavour.

1.2 Introductory perspectives

1.2.1 Structural philosophy and generic types of dams

The primary purpose of a dam may be defined as to provide for the safe retention and storage of water. As a corollary to this every dam must represent a design solution specific to its site circumstances. The design therefore also represents an optimum balance of local technical and economic considerations at the time of construction.

Reservoirs are readily classified in accordance with their primary purpose, e.g. irrigation, water supply, hydroelectric power generation, river regulation, flood control, etc. Dams are of numerous types, and type classification is sometimes less clearly defined. An initial broad classification into two generic groups can be made in terms of the principal construction material employed.

1. Embankment dams are constructed of earthfill and/or rockfill. Upstream and downstream face slopes are similar and of moderate angle, giving a wide section and a high construction volume relative to height.
2. Concrete dams are constructed of mass concrete. Face slopes are dissimilar, generally steep downstream and near vertical upstream, and dams have relatively slender profiles dependent upon the type.

The second group can be considered to include also older dams of appropriate structural type constructed in masonry. The principal types of dams within the two generic groups are identified in Table 1.1. Essential characteristics of each group and structural type are detailed further in Sections 1.3 and 1.4.

Embankment dams are numerically dominant for technical and economic reasons, and account for an estimated 85–90% of *all* dams built. Older and simpler in structural concept than the early masonry dam, the

Table 1.1 Large dams: World Register statistics (ICOLD, 1998)

Group	*Type*	*ICOLD code*		*%*
Embankment dams	Earthfill	TE		82.9
	Rockfill	ER		
Concrete dams (including masonry dams)	Gravity	PG		11.3
	Arch	VA		4.4
	Buttress	CB		1.0
	Multiple arch	MV		0.4
Total large dams			41 413	

embankment utilized locally available and untreated materials. As the embankment dam evolved it has proved to be increasingly adaptable to a wide range of site circumstances. In contrast, concrete dams and their masonry predecessors are more demanding in relation to foundation conditions. Historically, they have also proved to be dependent upon relatively advanced and expensive construction skills and plant.

1.2.2 Statistical perspective

Statistics are not available to confirm the total number of dams in service worldwide. Accurate statistical data are confined to 'large' dams entered under national listings in the World Register of Dams, published by the International Commission on Large Dams.

ICOLD is a non-governmental but influential organization representative of some 80 major dam-building nations. It exists to promote the interchange of ideas and experience in all areas of dam design, construction, and operation, including related environmental issues. Large dams are defined by ICOLD as dams exceeding 15 m in height or, in the case of dams of 10–15 m height, satisfying one of certain other criteria, e.g. a storage volume in excess of $1 \times 10^6 m^3$ or a flood discharge capacity of over $2000 m^3 s^{-1}$ etc. The World Register of 1998 (ICOLD, 1998) reported 41 413 large dams completed or under construction. Of this total, which excluded separately registered industrial tailings dams, over 19 000 were claimed by China and over 6000 by the US. These figures may be compared with a worldwide total of 5196 large dams recorded in 1950.

The 1998 edition of the World Register restricted the number of entries for certain countries, notably China, in the interests of saving space. This was achieved by listing only dams of 30 m height and above, a total of 25 410 dams.

Few reliable estimates of national totals of dams of all sizes have been published. Estimated total numbers for the UK and for the US are available, however, following national surveys. They are presented alongside the corresponding national figures for large dams in Table 1.2. From these statistics it may reasonably be inferred that the total number of dams in existence worldwide exceeds 300 000.

Table 1.2 Summary of numbers of British, US and Chinese dams (1998)

	Large dams	*Estimated total dams (national surveys)*	*Dams subject to national safety legislation*
UK	535	>5500	2650
USA	6375	75000	N.K.
China	*c.* 19100	>90000	N.K.

Rapid growth in the number of large dams has been accompanied by a progressive increase in the size of the largest dams and reservoirs. The physical scale of the largest projects is demonstrated by the statistics of height, volume, and storage capacity given in Tables 1.3, 1.4 and 1.5 respectively. Industrial tailings dams are excluded from Table 1.4.

In appreciating the progressive increase in the number of large dams and in the size of the largest, it must be recognized that the vast majority of new dams continue to be relatively small structures. They lie most

Table 1.3 Highest dams

Dam	*Country*	*Type*	*Completed*	*Height (m)*
Rogun	Tadjikistan	TE–ER	1985	335
Nurek	Tadjikistan	TE	1980	300
Xiaowan	China	TE	In progress	292
Grand Dixence	Switzerland	PG	1962	285
Inguri	Georgia	VA	1980	272
Manuel M Torres	Mexico	TE–ER	1980	261

38 dams greater than 200 m in height.

Table 1.4 Largest-volume dams

Dam	*Country*	*Type*	*Height (m)*	*Completed*	*Fill volume ($\times 10^6 m^3$)*
Tarbela	Pakistan	TE–ER	143	1976	105.9
Fort Peck	USA	TE	76	1937	96.1
Lower Usuma	Nigeria	TE	49	1990	93.0
Tucurui	Brazil	TE–ER–PG	106	1984	85.2
Ataturk	Turkey	TE–ER	184	1990	84.5
Guri (Raul Leoni)	Venezuela	TE–ER–PG	162	1986	78.0

Tailings dams excluded.

Table 1.5 Dams with largest-capacity reservoirs

Dam	*Country*	*Type*	*Height (m)*	*Completed*	*Reservoir capacity ($\times 10^9 m^3$)*
Kakhovskaya	Ukraine	TE–PG	37	1955	182.0
Kariba	Zimbabwe–Zambia	VA	128	1959	180.6
Bratsk	Russian Fedn.	TE–PG	125	1964	169.3
Aswan (High)	Egypt	TE–ER	111	1970	168.9
Akosombo	Ghana	TE–ER	134	1965	153.0
Daniel Johnson	Canada	VA	214	1968	141.8

commonly in the 5–10m height range. Earthfill embankments remain dominant, but rockfill is to some extent displacing earthfill for larger structures as it offers several advantages.

It is also important to recognize that many major dams are now necessarily built on less favourable and more difficult sites. For obvious reasons, the most attractive sites have generally been among the first to be exploited. A proportion of sites developed today would, in the past, have been rejected as uneconomic or even as quite unsuitable for a dam. The ability to build successfully on less desirable foundations is a reflection of advances in geotechnical understanding and of confidence in modern ground-improvement processes.

1.2.3 Historical perspective

The history of dam building dates back to antiquity, and is bound up with the earlier civilizations of the Middle East and the Far East. Countless small dams, invariably simple embankment structures, were constructed for irrigation purposes in, for example, China, Japan, India and Sri Lanka. Certain of these early dams remain in existence.

The dam built at Sadd-el-Kafara, Egypt, around 2600 BC, is generally accepted as the oldest known dam of real significance. Constructed with an earthfill central zone flanked by rock shoulders and with rubble masonry face protection, Sadd-el-Kafara was completed to a height of 14 m. The dam breached, probably in consequence of flood overtopping, after a relatively short period of service.

Numerous other significant dams were constructed in the Middle East by early civilizations, notably in modern Iraq, Iran and Saudi Arabia. The Marib embankment dam, completed in the Yemen around 750 BC to service a major irrigation project, was an example of particular note, as this dam was raised to a final height of 20 m. The first significant masonry dam, the 10 m high Kesis Gölü (North) dam in Turkey, dates from the same period.

The Romans made a significant later contribution in the Middle East and in countries bordering the Mediterranean. A number of Roman dams remain in service, and to the Romans probably falls the credit for first adopting the arch principle in dam construction. The 12 m high and 18 m long Baume arch dam, in France, was completed by the Romans in the second century AD.

In the Far East the construction of significant dams can be dated to the period commencing *c.* 380 BC. Activity initially centred upon Sri Lanka, where a remarkable period of dam building commenced with the 10 m high Bassawak embankment and culminated in the Giritale and Kantalai embankments (23 m and 20 m high respectively), completed in

AD 610. The Japanese and Indian entry into major dam building commenced *c.* AD 750, and both nations made a notable contribution to the early development of the embankment.

The period from AD 1000 onwards saw a spread of dam-building activity, with quite rapid growth in the height of dams and in the boldness of their concept. Of particular note was the construction of a series of masonry gravity dams in Iran where the first true arch dam, i.e. a masonry dam too slender to be stable as a gravity structure, was also built. The latter dam, at Kebar, 26 m high and of 55 m crest length with a base thickness of 6 m, was completed *c.* AD 1300. The remarkable 31 m high Sultan Mahmud dam in Afghanistan also dates from this time. This era also saw the commencement of serious dam building activity in many parts of Europe, e.g. the 6 m high embankment at Alresford, in Britain (*c.* 1195) or the 10 m high embankments at Mittlerer Pfauen, Germany (*c.* 1298) and at Dvořiště, Czech Republic (*c.* 1367) and many others.

The dam-builders of 16th-century Spain advanced masonry dam construction very considerably. The magnificent Tibi gravity dam, 42 m in height, was completed in 1594 and followed by a series of other outstanding masonry structures. The Elche masonry arch dam, 23 m high and 120 m in length, was completed in 1640 and is also of particular merit. With the rapid expansion of the Spanish Empire the expertise of the Spanish dam-builders was also exported to Central and South America. Representative of their breadth of vision and their ability to plan and to mobilize resources, the intensive metalliferous mining activity centred on Potosí (Bolivia) was, by the mid-17th century, served by a group of 32 reservoirs.

In the period from 1700 to 1800 the science of dam building advanced relatively slowly. The dawn of the first Industrial Revolution and the canal age gave considerable impetus to embankment dam construction in Britain and in Western Europe in the period from about 1780. Design continued to be based on a combination of empirical rules and proven experience. Despite the lack of rational design methods, dams steadily increased in size. As an example, the Entwistle embankment dam was completed in England in 1838 as the first of its type to exceed 30 m in height. In the 19th century British engineers advanced and developed embankment design and construction very successfully, notable projects in the UK including the magnificent Longdendale series of five principal dams, completed between 1854 and 1877, and many similar large structures constructed in India and elsewhere overseas.

Rational methods of analysis for masonry dams were developed and refined in various countries, notably France, Britain and the US, from about 1865. The design of embankment dams continued to be very empirical until much later. Advances in embankment construction were dependent upon the emergence of modern soil mechanics theory in the period from 1930. Subsequent progress has been relatively rapid, and major advances have been made in consequence of improvements in understanding of the behaviour of compacted earthfill and rockfill and with the introduction of modern high-capacity earthmoving plant. In the same period, partly in consequence of several major disasters, the vital importance of the interrelated disciplines of soil mechanics, rock mechanics and engineering geology to dam engineering was finally established.

Analytical techniques have also progressed rapidly in recent years, most specifically with the development of the elegant and extremely powerful finite element analyses (FEA), now widely employed for the most advanced analysis of all types of dam. The application of sophisticated FEA techniques has, in turn, been dependent upon the ready availability and power of the modern computer. However, limitations on the applicability of FEA remain, and they arise essentially from the complex load response of all construction materials utilized in dams. These limitations will be referred to further in Chapters 2 and 3 (Sections 2.7.2 and 3.2.8).

A comprehensive review of the history of dams lies beyond the scope of this text. Reference should be made to the international and comprehensive historical review of dams from earliest times published in Smith (1971) or to Schnitter (1994). The history prepared for the International Commission on Irrigation and Drainage (Garbrecht, 1987) gives particularly detailed descriptions of the earliest dams in parts of the Middle East and of Central Europe; the text also includes a useful review of the development of dams in Britain. More detailed and comprehensive accounts of early British dams, and of 19th-century dams built by prominent engineers of the period, are published in Binnie (1987a) and Binnie (1981) respectively. The latter provides a valuable insight into the reasoning underlying some design features of many older embankment dams.

1.2.4 Environmental and related issues

The environmental, economic and other socio-political issues associated with reservoir development must in all instances be acknowledged at the outset and fully addressed thereafter. This is especially important in the case of the larger high-profile projects and all others, large or lesser, sited in environmentally or politically sensitive locations.

Political and public consciousness with regard to environmental issues, compounded by a heightened awareness of issues associated with climate change and interest in promoting sustainable development, has led to growing international debate over the benefit derived from major dam projects. This resulted in the setting-up of a 12-man 'World Commission on Dams' (WCD; not to be confused with the International Commission on Large Dams, ICOLD) under the auspices of the World Bank and the World Conservation Union in 1998. WCD was charged with reviewing international experience in context with the emergent social and environmental controversies over large dam projects and reporting upon the role of such projects in development strategies. Looking to the future, the Commission was also tasked with identifying best practice in addressing critical policy and decision-making issues.

WCD reported in late 2000, stating that dams deliver significant development services in some 140 countries, with dam projects responsible for 19% of global electrical output, 12–15% of food production, and 12% of domestic and industrial water. It was also stated that dams provide for large-scale flood control and mitigation in at least 70 countries. The Commission examined alternatives for meeting water, energy and food needs, and identified a number of palliative organizational measures.

In terms of decision-making practice, the Commission's guidelines recommend outcomes based on multi-criteria analysis of technical, social, environmental, economic and financial parameters. The recommendations for future decision-making also included:

- Five core values: equity; sustainability; efficiency; participatory decision-making; accountability.
- A 'rights and risk' approach in negotiating development options.
- Seven strategy priorities for water resource development:

  Gain public acceptance
  Assess options
  Address existing dams
  Sustain rivers and livelihoods
  Recognize entitlements and share benefits
  Ensure compliance
  Share rivers for peace, development and security

- Clear criteria for assessing compliance, with 26 guidelines for reviewing and approving projects at five key stages in the decision-making process.

The WCD report has been criticized for not having given sufficient recognition to the positive dimension of major dam projects. The report has, however, made a significant contribution by stimulating considerable debate. Issues associated with future decision-making for development and sustainability are further examined and discussed in Pritchard (2000), Morrison and Sims (2001), Workman (2001), Bridle (2006), Collier (2006) and UNEP (2006).

Environmental impact and associated socio-political considerations can extend across a diverse spectrum of issues. The latter may range from population displacement, with consequent economic impacts, to the preservation of cultural or heritage sites; from the consequences of sedimentation and/or of changing flood regimes to altered patterns of disease.

The discussion of such an extensive and varied range of issues goes well beyond the scope of this textbook. Some general reference to selected issues is, however, dispersed through the text, e.g. Section 4.5 on sedimentation, or Section 9.1.7 on the effects of river barrages on water quality.

The broader issues are examined and discussed within Golzé (1977), in ICOLD (1988, 1992, 1994) and in specialist texts. Hendry (1994) examines legislative issues in the European context. The paper discusses the role of environmental assessment in terms of the appropriate European Directive (CEC, 1985), and discusses the provisions of the latter in relation to relevant UK provisions, e.g. DoE (1989). General questions of environmental evaluation, impact assessment and benefit appraisal are addressed in Clifton (2000), Thomas, Kemm and McMullan (2000), and in Gosschalk and Rao (2000). The latter reference includes a concise summary of the issues arising on three major high-profile dam projects, i.e. Aswan High (Lake Nasser, Egypt) completed in 1968, and projects currently completing at Sardar Sarovar (Narmada River, India) and Three Gorges (Yangtze River, China). The scale, and thus the overall impact, of the latter two multi-purpose projects is of particular note.

Sardar Sarovar, the principal component of the inter-state Narmada River development, is intended to irrigate some 1.9 million ha of land in the states of Gujarat and Rajasthan and provide 2450 MW of hydro-electric generating capacity. The concrete gravity dam is intended to reach a height of 138 m, and has a designed overflow capacity of $79 \times 10^3 \, m^3/s$. Construction commenced in the late 1980s, but opposition in the courts centred upon the displacement of an estimated 300 000 people from the very many village communities scheduled for inundation has delayed completion of the dam beyond an interim height of 110 m.

The Three Gorges project centres upon a 2331 m long and 184 m high concrete gravity dam impounding the Yangtze River. Design discharge capacity of the overflow system is $110 \times 10^3 \, m^3/s$. The immediate benefits associated with Three Gorges on project completion in 2008/2009 will be the availability of up to $22 \times 10^9 \, m^3$ of storage capacity for flood control on the notoriously difficult Yangtze and 18 200 MW of hydro-electric generating capacity from 26 turbines (see also Section 12.2). Three Gorges is also central to future development along some 600 km length of the upper Yangtze, the lock system which bypasses the dam (see also Sections 11.8.3 and 11.10) providing direct access to the heart of China for ships of up to 10 000 tonnes. The project has engendered considerable controversy however, since creation of the reservoir is estimated to displace at least 1.3 million people and submerge some 1300 known archaeological sites. Overall cost is officially stated as $14 billion, but it has been suggested that the true final figure will be considerably higher, with the most extreme estimates ranging up to $90–100 billion. An outline perspective on Three Gorges which makes plain the enormous scale and societal/environmental impact of this regional development project is presented in Freer (2000).

1.2.5 Dams: focus points

Dams differ from all other major civil engineering structures in a number of important regards:

- every dam, large or small, is quite unique; foundation geology, material characteristics, catchment flood hydrology etc. are each site-specific.
- dams are required to function at or close to their design loading for extended periods.
- dams do not have a structural lifespan; they may, however, have a notional life for accounting purposes, or a functional lifespan dictated by reservoir sedimentation.
- the overwhelming majority of dams are of earthfill, constructed from a range of natural soils; these are the least consistent of construction materials.
- dam engineering draws together a range of disciplines, e.g. structural and fluid mechanics, geology and geotechnics, flood hydrology and hydraulics, to a quite unique degree.
- the engineering of dams is critically dependent upon the application of informed engineering judgement.

In summary, dam engineering is a distinctive, broadly based and specialist discipline. The dam engineer is required to synthesize design solutions which, without compromise on safety, represent the optimal balance between technical, economic and environmental considerations.

1.3 Embankment dam types and characteristics

The embankment dam can be defined as a dam constructed from natural materials excavated or obtained close by. The materials available are utilized to the best advantage in relation to their characteristics as an engineered bulk fill in defined zones within the dam section. The natural fill materials are placed and compacted without the addition of any binding agent, using high-capacity mechanical plant. Embankment construction is consequently now an almost continuous and highly mechanized process, weather and soil conditions permitting, and is thus plant intensive rather than labour intensive.

As indicated in Section 1.2.1, embankment dams can be classified in broad terms as being earthfill or rockfill dams. The division between the two embankment variants is not absolute, many dams utilizing fill materials of both types within appropriately designated internal zones. The conceptual relationship between earthfill and rockfill materials as employed in

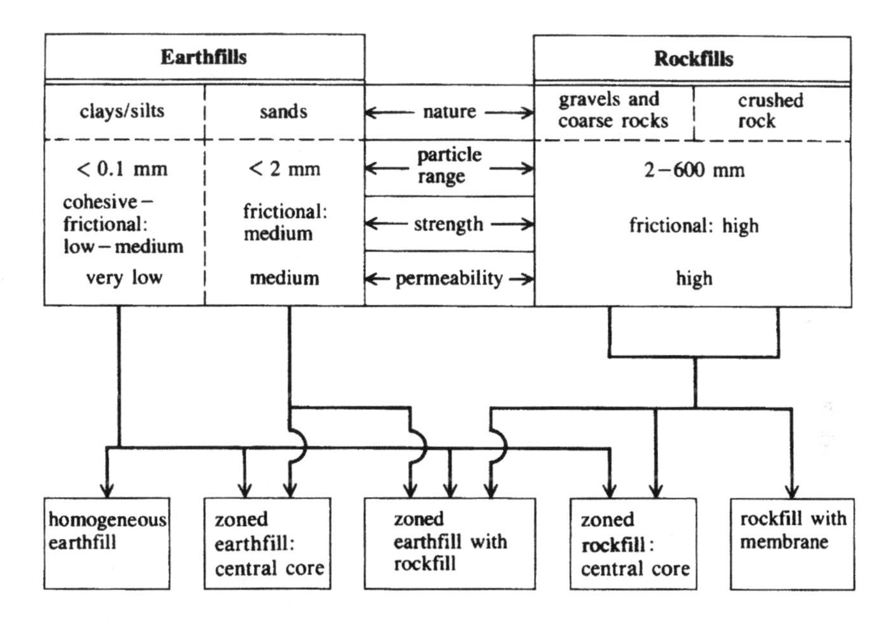

Fig. 1.1 Earthfills and rockfills in dam construction

embankment dams is illustrated in Fig. 1.1. Secondary embankment dams and a small minority of larger embankments may employ a homogeneous section, but in the majority of instances embankments employ an impervious zone or core combined with supporting shoulders which may be of relatively pervious material. The purpose of the latter is entirely structural, providing stability to the impervious element and to the section as a whole.

Embankment dams can be of many types, depending upon how they utilize the available materials. The initial classification into earthfill or rockfill embankments provides a convenient basis for considering the principal variants employed.

1. *Earthfill embankments.* An embankment may be categorized as an earthfill dam if compacted soils account for over 50% of the placed volume of material. An earthfill dam is constructed primarily of selected engineering soils compacted uniformly and intensively in relatively thin layers and at a controlled moisture content. Outline sections of some common variants of the earthfill embankment are illustrated in Fig. 1.2.
2. *Rockfill embankments.* In the rockfill embankment the section includes a discrete impervious element of compacted earthfill or a slender concrete or bituminous membrane. The designation 'rockfill embankment' is appropriate where over 50% of the fill material may be classified as rockfill, i.e. coarse-grained frictional material.

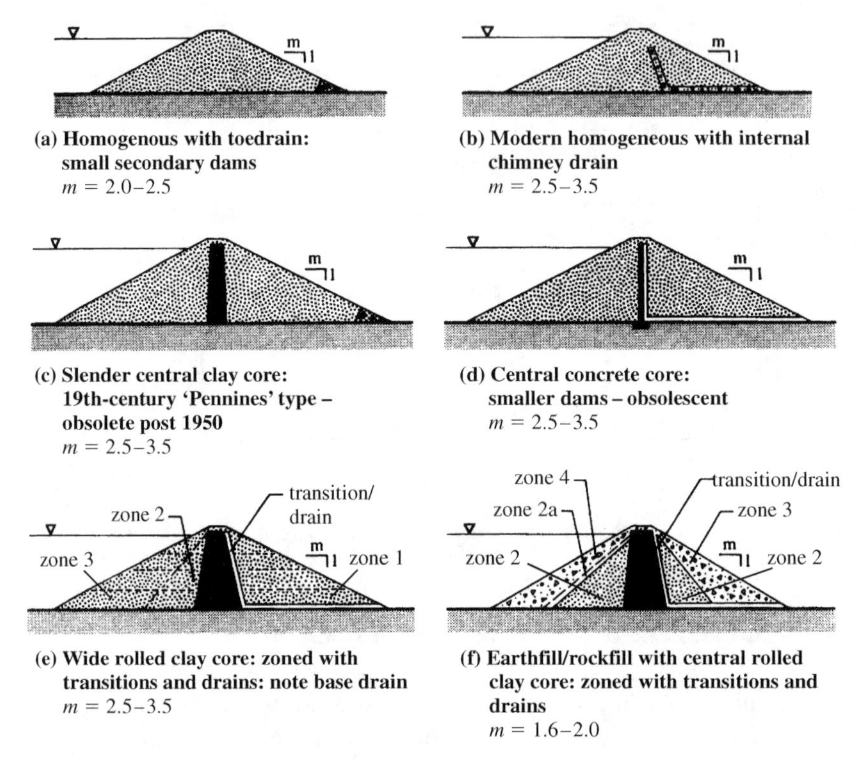

(a) Homogenous with toedrain: small secondary dams
$m = 2.0–2.5$

(b) Modern homogeneous with internal chimney drain
$m = 2.5–3.5$

(c) Slender central clay core: 19th-century 'Pennines' type – obsolete post 1950
$m = 2.5–3.5$

(d) Central concrete core: smaller dams – obsolescent
$m = 2.5–3.5$

(e) Wide rolled clay core: zoned with transitions and drains: note base drain
$m = 2.5–3.5$

(f) Earthfill/rockfill with central rolled clay core: zoned with transitions and drains
$m = 1.6–2.0$

Fig. 1.2 Principal variants of earthfill and earthfill–rockfill embankment dams (values of *m* are indicative only)

Modern practice is to specify a graded rockfill, heavily compacted in relatively thin layers by heavy plant. The construction method is therefore essentially similar to that for the earthfill embankment.

The terms 'zoned rockfill dam' or 'earthfill–rockfill dam' are used to describe rockfill embankments incorporating relatively wide impervious zones of compacted earthfill. Rockfill embankments employing a thin upstream membrane of asphaltic concrete, reinforced concrete or other manufactured material are referred to as 'decked rockfill dams'.

Representative sections for rockfill embankments of different types are illustrated in Fig. 1.3. Comparison should be made between the representative profile geometries indicated on the sections of Figs 1.2 and 1.3. The saving in fill quantity arising from the use of rockfill for a dam of given height is very considerable. It arises from the frictional nature of rockfill, which gives relatively high shear strength, and from high permeability, resulting in the virtual elimination of porewater pressure problems and permitting steeper slopes. Further savings arise from the reduced foundation footprint and the reduction in length of outlet works etc.

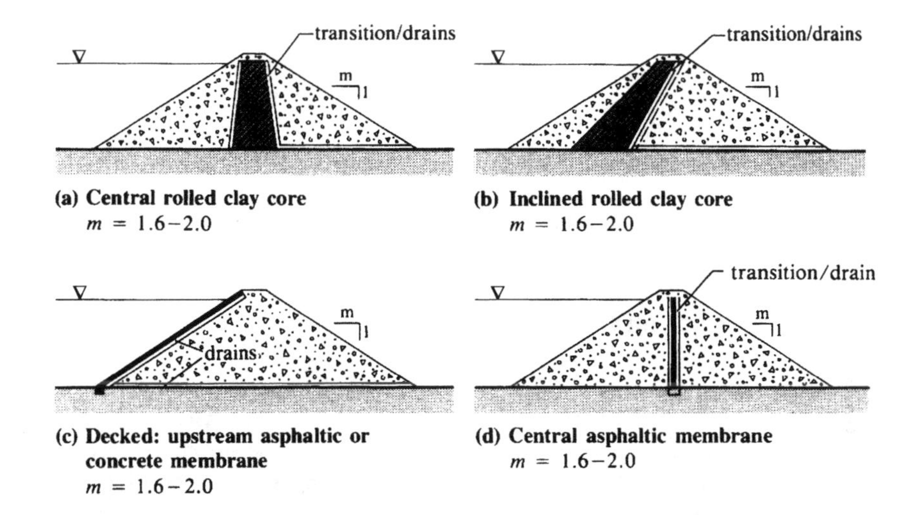

Fig. 1.3 Principal variants of rockfill embankment dams (values of *m* are indicative only)

The variants of earthfill and rockfill embankments employed in practice are too numerous to identify all individually. The more important are discussed further in appropriate sections of Chapter 2.

The embankment dam possesses many outstanding merits which combine to ensure its continued dominance as a generic type. The more important can be summarized as follows:

1. the suitability of the type to sites in wide valleys and relatively steep-sided gorges alike;
2. adaptability to a broad range of foundation conditions, ranging from competent rock to soft and compressible or relatively pervious soil formations;
3. the use of natural materials, minimizing the need to import or transport large quantities of processed materials or cement to the site;
4. subject to satisfying essential design criteria, the embankment design is extremely flexible in its ability to accommodate different fill materials, e.g. earthfills and/or rockfills, if suitably zoned internally;
5. the construction process is highly mechanized and is effectively continuous;
6. largely in consequence of 5, the unit costs of earthfill and rockfill have risen much more slowly in real terms than those for mass concrete;
7. properly designed, the embankment can safely accommodate an appreciable degree of deformation and settlement without risk of serious cracking and possible failure.

The relative disadvantages of the embankment dam are few. The most important include an inherently greater susceptibility to damage or destruction by overtopping, with a consequent need to ensure adequate flood relief and a separate spillway, and vulnerability to concealed leakage and internal erosion in dam or foundation. Examples of alternative types of embankment dam are illustrated and described in Thomas (1976), Golzé (1977) and Fell, MacGregor and Stapledon (1992).

1.4 Concrete dam types and characteristics

Rubble masonry or random masonry was successfully employed for many early dams. In the latter half of the 19th century masonry was used for high dams constructed in accordance with the first rational design criteria. Cyclopean masonry (i.e. stones of up to *c.* 10 t mass individually bedded in a dry mortar) was generally used, with a dressed masonry outer facing for durability and appearance (Binnie, 1987b).

Mass concrete, initially without the formed transverse contraction joints shown on Fig. 1.4(a), began to displace masonry for the construction of large non-embankment dams from about 1900 for economic reasons and also for ease of construction for more complex dam profiles, e.g. the arch. Early mass concrete commonly employed large stone 'displacers' (cf. cyclopean masonry). From about 1950 mass concrete increasingly incorporated bulk mineral additives, e.g. slags or pulverized fuel ash (PFA), in attempts to reduce thermal problems and cracking and to contain escalating costs.

The principal variants of the modern concrete dam are defined below.

1. *Gravity dams.* A concrete gravity dam is entirely dependent upon its own mass for stability. The gravity profile is essentially triangular, with the outline geometry indicated on Fig. 1.4(a), to ensure stability and to avoid overstressing of the dam or its foundation. Some gravity dams are gently curved in plan for aesthetic or other reasons, and without placing any reliance upon arch action for stability. Where a limited degree of arch action is deliberately introduced in design, allowing a rather slimmer profile, the term arched or arch-gravity dam may be employed.
2. *Buttress dams.* In structural concept the buttress dam consists of a continuous upstream face supported at regular intervals by downstream buttresses. The solid head or massive buttress dam, as illustrated by Figs 1.4(b) and 1.4(c), is the most prominent modern variant of the type, and may be considered for conceptual purposes as a lightened version of the gravity dam.
3. *Arch dams.* The arch dam has a considerable upstream curvature. Structurally it functions primarily as a horizontal arch, transmitting

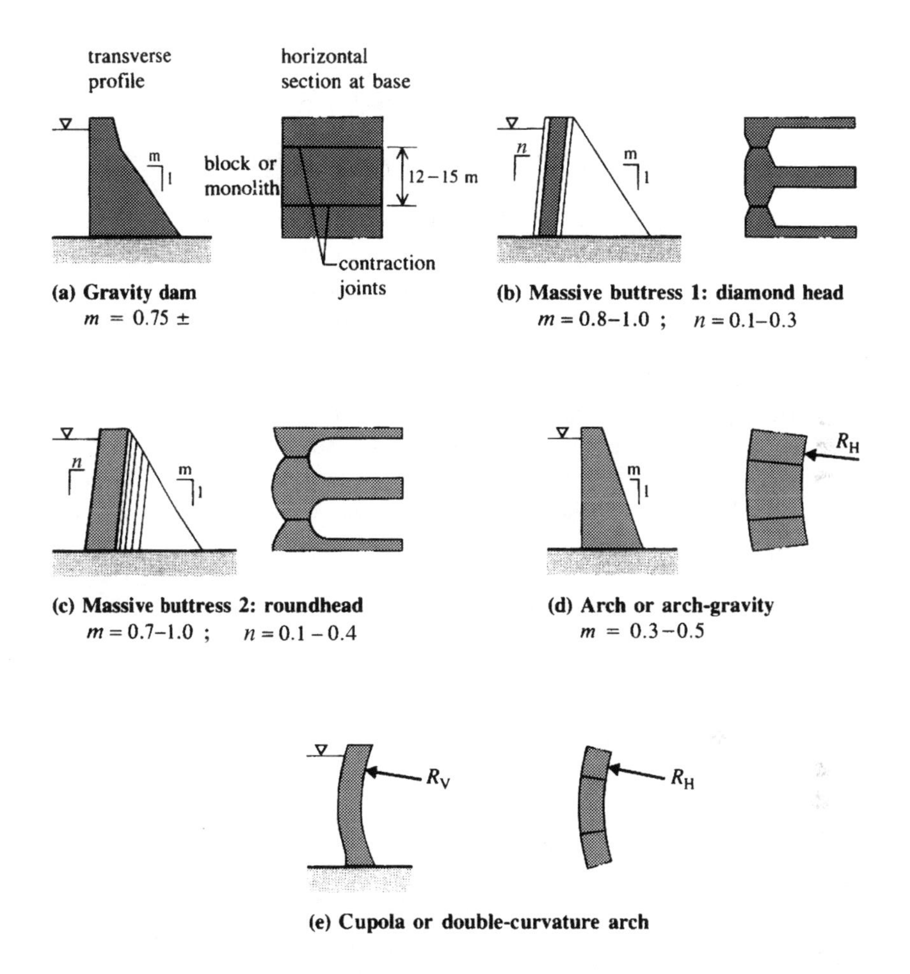

Fig. 1.4 Principal variants of concrete dams (values of *m* and *n* indicative only; in (e) R_H and R_V generally vary over dam faces)

the major portion of the water load to the abutments or valley sides rather than to the floor of the valley. A relatively simple arch, i.e. with horizontal curvature only and a constant upstream radius, is shown in Fig. 1.4(d). It is structurally more efficient than the gravity or buttress dam, greatly reducing the volume of concrete required. A particular derivative of the simple arch dam is the cupola or double-curvature arch dam (Fig. 1.4(e)). The cupola dam introduces complex curvatures in the vertical as well as the horizontal plane. It is the most sophisticated of concrete dams, being essentially a dome or shell structure, and is extremely economical in concrete. Abutment stability is critical to the structural integrity and safety of both the cupola and the simple arch.

4. *Other concrete dams.* A number of less common variants of the major types of concrete dams illustrated in Fig. 1.4 can also be identified. They include hollow gravity, decked buttress, flat slab (Ambursen) buttress, multiple arch, and multiple cupola dams, as illustrated in Fig. 1.5. The type names are self-explanatory, and the structural parentage of each as a derivative of one or other of the principal types is apparent from the figures. In view of this and the relative rarity of these variants they are not considered further in this text, but the comparative vulnerability of the slender flat slab and similar types to seismic disturbance etc. may be noted.

The characteristics of concrete dams are outlined below with respect to the major types, i.e. gravity, massive buttress and arch or cupola dams. Certain characteristics are shared by all or most of these types; many are, however, specific to particular variants. Merits shared by most concrete dams include the following.

1. Arch and cupola dams excepted, concrete dams are suitable to the site topography of wide or narrow valleys alike, provided that a competent rock foundation is accessible at moderate depth (<5 m).
2. Concrete dams are not sensitive to overtopping under extreme flood conditions (cf. the embankment dam).

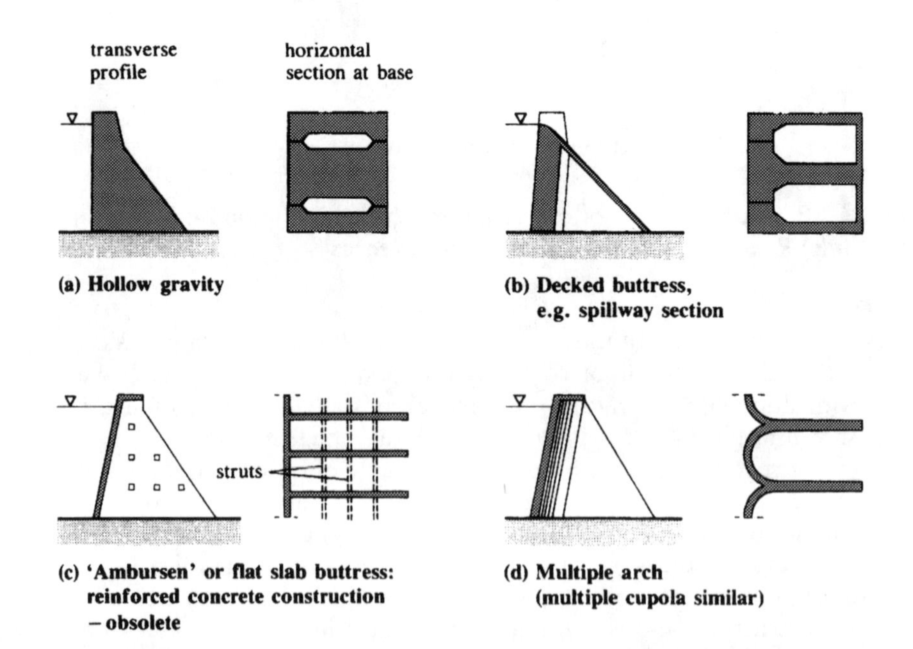

(a) Hollow gravity

(b) Decked buttress, e.g. spillway section

(c) 'Ambursen' or flat slab buttress: reinforced concrete construction – obsolete

(d) Multiple arch (multiple cupola similar)

Fig. 1.5 Further variants of concrete dams

3. As a corollary to 2, all concrete dams can accommodate a crest spillway, if necessary over their entire length, provided that steps are taken to control downstream erosion and possible undermining of the dam. The cost of a separate spillway and channel are therefore avoided.
4. Outlet pipework, valves and other ancillary works are readily and safely housed in chambers or galleries within the dam.
5. The inherent ability to withstand seismic disturbance without catastrophic collapse is generally high.
6. The cupola or double-curvature arch dam is an extremely strong and efficient structure, given a narrow valley with competent abutments.

Type-specific characteristics are largely determined through the differing structural *modus operandi* associated with variants of the concrete dam. In the case of gravity and buttress dams, for example, the dominant structural response is in terms of vertical cantilever action. The reduced downstream contact area of the buttress dam imposes significantly higher local foundation stresses than for the equivalent gravity structure. It is therefore a characteristic of the former to be more demanding in terms of the quality required of the underlying rock foundation.

The structural behaviour of the more sophisticated arch and cupola variants of the concrete dam is predominantly arch action, with vertical cantilever action secondary. Such dams are totally dependent upon the integrity of the rock abutments and their ability to withstand arch thrust without excessive yielding. It is consequently characteristic of arch and cupola dams that consideration of their suitability is confined to a minority of sites in relatively narrow steep-sided valleys or gorges, i.e. to sites with a width:height ratio at the dam crest level generally not exceeding 4–5.

A comparison of the general characteristics of concrete dams with those of the embankment dam suggests the following inherent disadvantages for the former.

1. Concrete dams are relatively demanding with respect to foundation conditions, requiring sound and stable rock.
2. Concrete dams require processed natural materials of suitable quality and quantity for aggregate, and the importation to site and storage of bulk cement and other materials.
3. Traditional mass concrete construction is relatively slow, being labour intensive and discontinuous, and requires certain skills, e.g. for formwork, concreting, etc.
4. Completed unit costs for mass concrete, i.e. cost per cubic metre, are very much higher than for embankment fills, typically by an order of magnitude or more. This is seldom counterbalanced by the much lower volumes of concrete required in a dam of given height.

A considered evaluation of the generalized characteristics in conjunction with Figs 1.3 and 1.4 will suggest further conclusions as to the corresponding advantages of embankment and concrete dams. However, the limitations of generalizations on the merits of either type must be appreciated. An open mind must be maintained when considering possible dam types in relation to a specific site, and evaluation must attach proper weight to local circumstances. Economic comparisons apart, other non-engineering factors may be of importance: this is referred to further in Section 1.6.

The variants of the concrete dam illustrated and their merits are further compared with those for the embankment dam in Thomas (1976), Golzé (1977) and USBR (1987).

1.5 Spillways, outlets and ancillary works

Dams require certain ancillary structures and facilities to enable them to discharge their operational function safely and effectively. In particular, adequate provision must be made for the safe passage of extreme floods and for the controlled draw-off and discharge of water in fulfilment of the purpose of the reservoir. Spillways and outlet works are therefore essential features. Other ancillary facilities are incorporated as necessary for the purpose of the dam and appropriate to its type. Provision for permanent flood discharge and outlet works and for river diversion during construction can prove to be technically difficult and therefore costly.

In this section, the more important structures and ancillary works associated with impounding dams are identified and briefly described. As such, it is intended as an introduction to subsequent chapters dealing with the design of dams (Chapters 2 and 3), spillways and outlets (Chapter 4), energy dissipators (Chapter 5) and gates and valves (Chapter 6). For a review of hydraulics of spillways and energy dissipators see also Khatsuria (2005)

1.5.1 Spillways

The purpose of the spillway is to pass flood water, and in particular the design flood, safely downstream when the reservoir is overflowing. It has two principal components: the controlling spillweir and the spillway channel, the purpose of the latter being to conduct flood flows safely downstream of the dam. The latter may incorporate a stilling basin or other energy-dissipating devices. The spillway capacity must safely accommodate the maximum design flood, the spillweir level dictating the maximum retention level of the dam, i.e. the normal maximum water level (NWL).

Spillways are normally uncontrolled, i.e. they function automatically as the water level rises above NWL, but they may be controlled by gates. In some instances additional emergency spillway capacity is provided by a fuse plug (see Section 4.7.7), i.e. an erodible subsidiary bank designed to wash out if a predetermined extreme flood level is attained. Alternative emergency provision can be made by reinforced concrete flap-gates designed to tip over by hydrostatic pressure under extreme flood conditions or by the use of crest-mounted fusegates (see Section 6.2.8). Concrete dams normally incorporate an overfall or crest spillway, but embankments generally require a separate side-channel or shaft spillway structure located adjacent to the dam.

1.5.2 Outlet works

Controlled outlet facilities are required to permit water to be drawn off as is operationally necessary. Provision must be made to accommodate the required penstocks and pipework with their associated control gates or valves. Such features are readily accommodated within a concrete dam, as noted in Section 1.4. For embankment dams it is normal practice to provide an external control structure or valve tower, which may be quite separate from the dam, controlling entry to an outlet tunnel or culvert.

A bottom discharge facility is provided in most dams to provide an additional measure of drawdown control and, where reasonable, to allow emptying of the reservoir. The bottom outlet must be of as high a capacity as economically feasible and consistent with the reservoir management plan. In most cases it is necessary to use special outlet valves (Section 6.3) and/or structures to avoid scouring and damage to the stream bed and banks downstream of the dam.

1.5.3 River diversion

This provision is necessary to permit construction to proceed in dry conditions. An outlet tunnel or culvert may be temporarily adapted to this purpose during construction, and subsequently employed as a discharge facility for the completed dam. In the absence of such a tunnel of adequate capacity alternative steps will be necessary, involving the construction of temporary upstream and downstream cofferdams or, in the case of concrete dams, by programming construction of one monolith or block to leave a temporary gap or formed tunnel through the structure.

The hydraulic aspects of river diversion are dealt with in detail in Vischer and Hager (1998).

1.5.4 Cut-offs

Seepage under and round the flank of a dam must be controlled. This is achieved by the construction of a cut-off below the structure, continued as necessary on either flank. Modern embankment cut-offs are generally formed from wide trenches backfilled with rolled clay, if impervious strata lie at moderate depths, and/or by drilling and grouting to form a cut-off screen or barrier to greater depths. Grout screen cut-offs are also customarily formed in the rock foundation under a concrete dam.

1.5.5 Internal drainage

Seepage is always present within the body of any dam. Seepage flows and their resultant internal pressures must be directed and controlled. Internal drainage systems for this purpose are therefore an essential and critical feature of all modern dams. In embankments drainage is effected by suitably located pervious zones leading to horizontal blanket drains or outlets at base level. In concrete dams vertical drains are formed inside the upstream face, and seepage pressure is relieved into an internal gallery or outlet drain. In the case of arch dams, seepage pressures in the rock abutments are frequently relieved by systems of bored drains and/or drainage adits or tunnels.

1.5.6 Internal galleries and shafts

In addition to their function alongside drains in effecting local control of seepage, galleries and shafts are provided as a means of allowing internal inspection, particularly in concrete dams. The galleries, shafts and any associated chambers to accommodate discharge valves or gates can also be used to accommodate instrumentation for structural monitoring and surveillance purposes (Chapter 7).

The ancillary structures and design features referred to are further described in subsequent chapters. Additional illustrations of these and other ancillary works are also contained in Thomas (1976), USBR (1987), Fell, MacGregor and Stapledon (1992) and Kennard, Owens and Reader (1996).

1.6 Site assessment and selection of type of dam

1.6.1 General site appraisal

A satisfactory site for a reservoir must fulfil certain functional and technical requirements. Functional suitability of a site is governed by the balance between its natural physical characteristics and the purpose of the reservoir. Catchment hydrology, available head and storage volume etc. must be matched to operational parameters set by the nature and scale of the project served. Technical acceptability is dictated by the presence of a satisfactory site (or sites) for a dam, the availability of materials suitable for dam construction, and by the integrity of the reservoir basin with respect to leakage. The hydrological and geological or geotechnical characteristics of catchment and site are the principal determinants establishing the technical suitability of a reservoir site. To these must be added an assessment of the anticipated environmental consequences of construction and operation of the dam, alluded to in Section 1.2.4. They are not considered further here.

The principal stages involved in site appraisal and leading to selection of the optimum dam site and type of dam for a major project are as indicated schematically in Fig. 1.6.

The considerable time which can elapse between initial strategic planning, with identification of the project requirement, and commencement of construction on site will be noted. A significant proportion of that time may be attributable to the 'political' decision-making processes and to arranging project funding.

In the reconnaissance phase, which may extend over a substantial period, the principal objective is to collect extensive topographical, geological and hydrological survey data. Large-scale maps and any records already available provide the starting point, but much more detailed surveys will inevitably be required. Aerial reconnaissance, employing modern sensors in addition to the traditional photogrammetric survey techniques, has a particular rôle to play in the preparation of accurate and large-scale site plans (e.g. 1:5000 and larger). In the hands of an experienced engineering geologist as interpreter, aerial surveys also provide valuable information on geology, on possible dam sites and on the likely availability of construction materials. Hydrological catchment and river surveys are directed to determining rainfall and run-off characteristics, and assessing historical evidence of floods etc.

The feasibility report prepared at the conclusion of the reconnaissance phase assembles and interprets all available information, data and records, and makes initial recommendations with respect to the technical and economic viability of the reservoir. Options with regard to the location, height and type of dam are set out, and comparisons drawn in terms

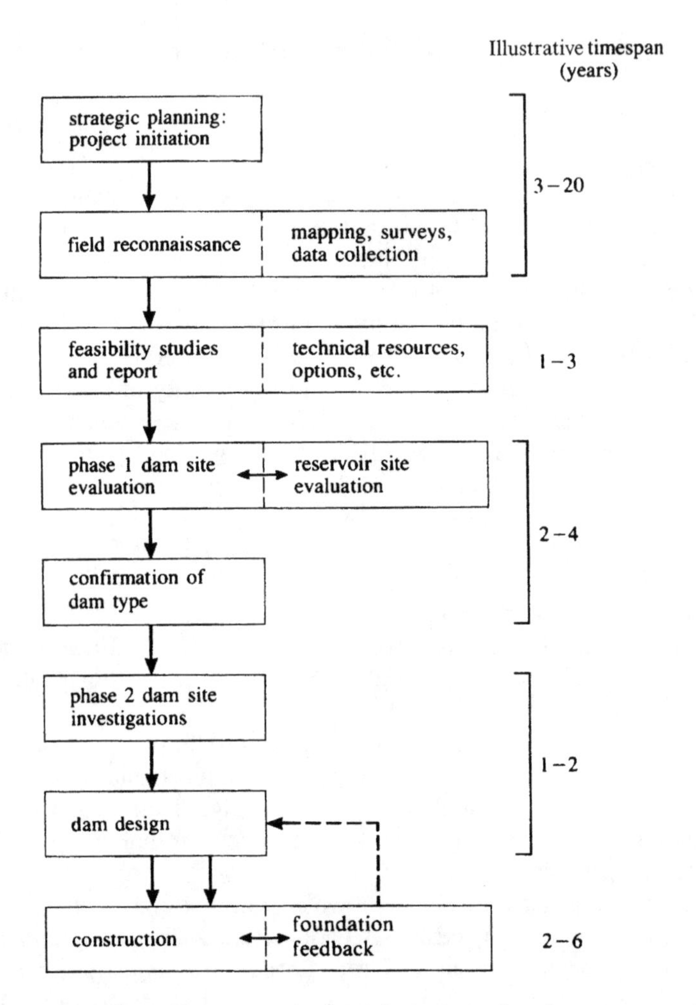

Fig. 1.6 Stages in dam site appraisal and project development: major projects

of estimated costs and construction programmes. Within the latter, account must be taken of the resource implications of each, i.e. financial outlay, labour and plant requirements etc. On the strength of this report a decision can be made with respect to the further detailed investigations required to confirm the suitability of the reservoir basin and preferred dam site (or sites).

Further investigation of the reservoir basin is principally directed to confirming its integrity with respect to water retention. A thorough geological assessment is necessary for this purpose, particularly in karstic and similarly difficult formations and in areas with a history of mining activity. The issue of less favourable sites for reservoirs and solution of the associated problems is addressed in ICOLD (1970) and in Fell, MacGregor and Stapledon (1992). As specific examples, investigations and conclusions drawn for Cow Green reservoir (UK) are described by Kennard and Knill (1969), and the initial leakage losses at May reservoir (Turkey) are discussed by Alpsü (1967).

Investigation of the reservoir margins to confirm the stability of potentially vulnerable areas, e.g. adjacent to the intended dam, is conducted as required. The availability of possible construction materials, e.g. suitable fills, sources of aggregates etc., is also assessed in considerable depth.

Hydrological studies are continued as necessary to confirm and extend the results of the initial investigations. In view of their very specialist nature they are not considered further here; reference may be made to Thomas (1976) and to Chapters 4 and 8 for details.

1.6.2 Dam site evaluation – general

The viability of the preferred dam site identified in a reservoir feasibility study must be positively established. Extensive investigations are conducted to confirm that the site can be developed on the desired scale and at acceptable cost. The nature of the soil and rock formations present, critical to foundation integrity, must be proved by subsurface exploration. Emphasis is placed upon confirmation of site geology and geotechnical characteristics, and on the evaluation of sources of construction materials (Sections 1.6.3–1.6.5).

Foundation competence is determined by stability, load-carrying capacity, deformability, and effective impermeability. All are assessed in relation to the type and size of dam proposed (Section 1.6.4).

In the case of a difficult site, the site evaluation programme can be protracted and expensive. Expenditure may be of the order of 1% up to, exceptionally, 2.5 or 3% of the anticipated cost of the dam. The scope of individual aspects of an investigation reflects circumstances unique to the site. The investigation may also relate to a specific type of dam if site conditions are such that options are restricted, e.g. by depth of overburden (Section 1.6.6).

Only the general principles underlying dam site evaluation can be presented here. A comprehensive review is provided in Thomas (1976), with outline summaries of example cases. An indication of the interaction which develops between site evaluation, local circumstance and type of dam is given in Bridle, Vaughan and Jones (1985), Coats and Rocke (1983) and Collins and Humphreys (1974) for embankment dams, or Bass and Isherwood (1978) for a concrete dam and Kennard and Reader (1975) for a composite dam, part concrete and part embankment. Walters (1974) presents simplified but informative summaries of site geology in relation to an international selection of dams.

In parallel with these investigations, extensive and detailed surveys are required to establish the location and extent of potential sources of construction materials in reasonable proximity to the site. The materials of interest may range from low-permeability cohesive soils and glacial tills for embankment cores through to sands and gravels suitable for shoulder fill or as concrete aggregates. Crushed rock may also be obtainable from excavations for underground works associated with the project.

Overall site viability is additionally subject to economic considerations, notably site preparation and construction material costs. It may also be influenced by seismicity, access development cost or other local constraints, including environmental and socio-political considerations.

In summary, dam site investigations require careful planning and the investment of sufficient time and resources. Wherever possible, *in situ* and field test techniques should be employed to supplement laboratory testing programmes. Proper and meticulous interpretation of geological and geotechnical data demands the closest cooperation between the engineering geologist, the geotechnical specialist and the dam engineer. Underinvestment in reservoir site appraisal and in the investigation and assessment of the site for a dam can have grave consequences, both technical and economic.

1.6.3 Geological and geotechnical investigations

Geological and geotechnical investigation of a dam site selected for detailed evaluation is directed to determination of geological structure, stratigraphy, faulting, foliation and jointing, and to establishing ground and groundwater conditions adjacent to the dam site, including the abutments.

The general objectives of these and allied investigations are

(a) to determine engineering parameters which can reliably be used to evaluate stability of the dam foundation and, on compressible foundations, i.e. soils, to estimate probable settlement and deformation,
(b) the determination of seepage patterns and parameters enabling assessment of the probable seepage regime, including quantities and pressures, and
(c) to confirm the containment integrity of the reservoir basin and the stability of its margins.

The relative importance of (a), (b) or (c) is dependent upon the site and the type of dam proposed. A fourth general objective is

(d) confirmation of the nature, suitability and availability of natural construction materials, including the determination of design parameters for fill materials etc.

General features to be identified and defined in the course of the site investigation include the interface between soil and rock, groundwater conditions, unstable and caving ground, e.g. karstic formations etc., and all significant discontinuities, i.e. rock faults, shatter zones, fissured or heavily fractured rock and the spacing and other characteristics of jointing and bedding surfaces etc. within the rock mass. Reference should be made to Attewell and Farmer (1976) and/or to Bell (1993) for a comprehensive perspective on engineering geology in relation to dam and reservoir sites.

Key features of this phase of the investigation include

(a) meticulous logging of all natural and excavated exposures and borehole records, etc.,
(b) careful correlation between all exposures, boreholes and other data, and
(c) excavation of additional trial pits, boreholes, shafts and exploratory adits as considered necessary.

It is at this stage that more extensive geophysical and *in situ* testing programmes may also be conducted, with the primary intention of extending and validating borehole and laboratory data. A further purpose of field testing at this time is confirmation of the natural groundwater regime, e.g. through installation of piezometers, pumping tests, etc.

Extensive use is made of rotary drilling and coring techniques to establish the rock structure at depth and to confirm its competence. Core recovery is a crude but useful index of rock quality, e.g. in terms of rock quality designation (RQD) (i.e. total recovered core in lengths of over 10 cm as a percentage of total borehole depth; RQD > 70 is generally indicative of sound rock). *In situ* tests, e.g. for permeability, strength and deformability, are used to estimate rock mass characteristics in preference to small-scale laboratory sample testing wherever possible. All cores are systematically logged and should ideally be retained indefinitely. Drilling, sampling and testing techniques are essentially those employed in conventional site investigation practice. A comprehensive review of the latter is presented in Clayton, Simons and Matthews (1995) and in the CIRIA site investigation manual (Weltman and Head, 1983). More specialist techniques, e.g. for large-scale *in situ* tests, are illustrated in Thomas (1976) and in Fell, MacGregor and Stapledon (1992). The applicability of different equipment and exploratory methods in the context of site investigation for dams are reviewed concisely in Wakeling and Manby (1989).

Evaluation of seismic risk for an important dam requires identification of the regional geological structure, with particular attention being paid to fault complexes. Activity or inactivity within recent geological history will require to be established from study of historical records and field reconnaissance. If historical records of apparent epicentres can be matched to key geological structures it is possible to make a probabilistic assessment of seismic risk in terms of specific intensities of seismic event. In the absence of reliable historical information it will be necessary to monitor microseismic activity as a basis for the probabilistic prediction of major seismic events. Either process is imprecise and will at best provide only an estimate of the order of seismic risk. As a measure of reassurance over seismicity it has been suggested that most well-engineered dams on a competent foundation can accept a moderate seismic event, with peak accelerations in excess of $0.2g$, without fatal damage. Dams constructed with or on low-density saturated cohesionless soils, i.e. silts or sands, are, however, at some risk of failure in the event of seismic disturbance due to porewater pressure buildup and liquefaction with consequent loss of stability.

Seismicity is discussed further in Sections 2.7 and 3.1, with a brief introduction to the application of pseudo-static seismic analysis.

1.6.4 Foundation investigations

Foundation competence of the dam site must be assessed in terms of stability, load-carrying capacity, compressibility (soils) or deformability (rocks), and effective mass permeability. The investigative techniques to be adopted will depend upon the geomorphology and geology of the specific site.

(a) Dams on competent stiff clays and weathered rocks

Serious underseepage is unlikely to be a problem in extensive and uniform deposits of competent clay. It is important, however, to identify and consider the influence of interbedded thin and more permeable horizons which may be present, e.g. silt lenses, fine laminations, etc. Considerable care is required in the examination of recovered samples to detect all such features. The determination of appropriate shear strength parameters for evaluating foundation stability is of major importance.

For a foundation on rock positive identification of the weathered rock profile may prove difficult. *In situ* determination of shear strength parameters may also be necessary, using plate loading tests in trial pits or adits, or dilatometer or pressuremeter testing conducted within boreholes. The latter techniques are particularly suitable in softer rocks containing very fine and closely spaced fissures.

(b) Dams on soft cohesive foundations

The presence of superficial soft and compressible clay deposits normally ensures that seepage is not a major consideration. The nature of such formations also ensures that investigations are, in principle, relatively straightforward.

The soft consistency of the clays may necessitate the use of special sampling techniques. In such situations continuous sampling or *in situ* cone penetrometer testing techniques offer advantages. Stability and settlement considerations will require the determination of drained shear strength and consolidation parameters for the clay.

(c) Dams on pervious foundations

Seepage-associated problems are normally dominant where a dam is to be founded on a relatively pervious foundation. In a high proportion of such instances the soil conditions are very complex, with permeable and much less permeable horizons present and closely interbedded.

(d) Dams on rock foundations

The nature of the investigation is dependent upon whether an embankment or a concrete dam is proposed. Where the decision is still open, the investigation must cover either option; both require a full understanding of the site geology.

CONCRETE DAMS

Concrete dam foundation stability requires careful assessment of the frequency, orientation and nature of the rock discontinuities, including the characteristics of infill material, e.g. clays etc. Foundation deformability will be largely dependent upon rock load response characteristics and on discontinuity structure. Rotary coring is widely employed, but to assess the rock structure reliably on the macroscale it is also advisable to expose and examine it in trial excavations and, wherever justifiable, by driving exploratory adits. The latter can be used subsequently for grouting or as permanent drainage galleries. Abutment stability and deformability are very important to all types of concrete dam in narrow steep-sided valleys, and most particularly if the design relies on some degree of arch action. Detailed investigations should, therefore, extend to the abutments, with particular regard to the possibility of large-scale wedge or block instability or excessive deformation and yielding. Large-scale *in situ* loading tests to evaluate the strength and load-response characteristics of the rock, while costly, should be conducted in parallel with laboratory testing whenever practicable. *In situ* tests of this nature can be carried out in exploratory or drainage adits, or at suitably prepared exposures, e.g. in excavations.

In situ permeability testing is generally conducted through use of borehole packer tests, but the proper interpretation of field permeability data can prove difficult.

EMBANKMENT DAMS

Foundation seepage is less critical than for the concrete dam, as seepage paths are much longer. Discontinuity shear strength is generally of less importance, but deformability and settlement involving determination of elastic moduli etc. may be a significant consideration if a decked, i.e. upstream membrane, embankment is contemplated.

As with the concrete dam, discontinuity shear strength, elastic moduli and related rock parameters are best determined from large-scale *in situ* tests wherever practicable and supplemented by appropriate laboratory tests. Moduli can be determined in the laboratory using cylindrical specimens of intact rock recovered from boreholes, as can uniaxial compressive strength, e.g. by direct compression or by point load tests etc. Intact moduli obtained from such small-scale specimens must be adjusted to values appropriate to the prototype rock mass, as governed by the discontinuity pattern, by application of a reduction factor j_c, thus:

$$E_{mass} = j_c \, E_{lab}.$$

The reduction factor is related to fracture spacing and other geological and physical characteristics, including degree of weathering. It may be as low as 0.1 in some circumstances, ranging up to 0.5 or more in others.

Durability of rock for use in rockfill and/or as upstream face protection may also require to be assessed where an embankment dam is contemplated. This will necessitate chemical, attrition and accelerated weathering tests to study longer-term degradation.

(e) Dams on karst foundations (carbonate rocks, e.g. limestones etc.)

The presence of extensive solution cavities and fissures renders all such sites peculiarly difficult. It is essential that the extent of the karstic features and, most importantly, their configuration in terms of void continuity be established. Geological studies can be useful for initial interpretation of the karstic landforms as a guide to the planning of detailed investigation. Aerial survey often reveals shallow karstic cavities, and geophysical methods can also be of value. It will be necessary to confirm the size and nature of all features initially identified by geophysical or other indirect techniques through drilling and other direct methods of investigation.

1.6.5 Materials for dam construction

Initial exploration for sources of materials is conducted by a combination of surface and aerial reconnaissance in conjunction with extensive geological surveys. Potential borrow areas or aggregate sources so revealed must be thoroughly evaluated in terms of the suitability of the materials they contain for different purposes, e.g. core material, material for shoulder zones and filters etc. or for aggregates. The quantities which can be realized from a source must also be estimated. Geophysical methods can play a useful part in this latter process, but they must always be correlated with hard evidence from natural or trial exposures or from conventional drilling programmes. Investigation of the suitability of an earthfill or rockfill material will invariably justify an instrumented trial fill, which can provide much invaluable data on physical characteristics, compaction characteristics and plant performance, and on geotechnical design parameters.

The evaluation of earthfill borrow sources is notoriously prone to overestimation of the available yield of suitable materials owing to undetected variations in soil type or quality. It is therefore essential to prove quantities of individual fill materials very substantially in excess of the estimated requirement for each, and figures as high as three and four times the nominal requirement have been suggested as prudent. The percentage of each material likely to prove unusable or anticipated extraction difficulties may be critical factors in relation to relative costs for alternative sources.

The proving of sources for rockfill is superficially more straightforward than for earthfill. The essential requirement is a source of sound durable rock, the location of which is generally apparent from the initial geological appraisal. Investigation of the suitability of the rockfill will normally require a trial fill and, in the case of excavated or quarried rocks, it will also be necessary to conduct blasting or ripping trials to determine rock fragment size, grading and shape, etc. The percentage of oversize or fine material or the excavation cost for a very dense hard rock will be critical economic factors.

Aggregate sources for concrete dams include natural borrow areas and the use of crushed aggregates derived from quarries or excavations. In checking the quality of aggregates, physical and mechanical properties and long-term chemical stability, e.g. with regard to risk of alkali–silica reaction (ASR), will require investigation.

1.6.6 Selection of type of dam

The optimum type of dam for a specific site is determined by estimates of cost and construction programme for all design solutions which are technically valid. Where site circumstances are such that viable alternatives exist it is important that options are kept open, assessing the implications of each with respect to resources, programme and cost, until a preferred solution is apparent. It may also be necessary to take account of less tangible socio-political and environmental considerations in the determination of that solution.

Four considerations of cardinal importance are detailed below.

1. *Hydraulic gradient*: the nominal value of hydraulic gradient, i, for seepage under, around or through a dam varies by at least one order of magnitude according to type.
2. *Foundation stress*: nominal stresses transmitted to the foundation vary greatly with dam type.
3. *Foundation deformability*: certain types of dams are better able to accommodate appreciable foundation deformation and/or settlement without serious damage.
4. *Foundation excavation*: economic considerations dictate that the excavation volume and foundation preparation should be minimized.

The first consideration is illustrated by reference to Figs 1.2–1.5 inclusive. Notional values of gradient range from about 0.5 for a homogeneous embankment (Fig. 1.2(a)), to 10 or more for a buttress or cupola dam (Figs 1.4(b) and 1.4(e)). The ability of softer and weaker or more erodible foundations to resist high hydraulic gradients safely is very limited.

In illustration of the second point, notional stress values for 100 m high dams of different types are shown in Table 1.6.

The significance of excessive or non-uniform foundation deformability, point 3 above, arises in relation to cracking and stress redistribution within the dam. The relative structural flexibility of a well-designed embankment dam may be advantageous.

With regard to the final consideration, the economic disincentive of excessive excavation, particularly in relation to a concrete dam, is self-evident.

It is inappropriate to generalize over considerations controlling the choice of dam type beyond the four major points referred to. Their

Table 1.6 Notional foundation stresses: dams 100 m in height

Dam type	*Notional maximum stress* ($MN\,m^{-2}$)
Embankment	1.8–2.1
Gravity	3.2–4.0
Buttress	5.5–7.5
Arch	7.5–10.0

collective importance is such, however, that Fig. 1.7 and Table 1.7 are provided in illustration. Selection of type of dam is discussed, with examples, in Fell, MacGregor and Stapledon (1992), USBR (1987), Thomas (1976) and Walters (1974).

The situation of a wide valley floored with deep deposits of fine-grained soils, e.g. glacial tills etc., is illustrated in Fig. 1.7(a). Considerations of foundation deformation and the depth of excavation required favour an earthfill embankment, given the ready availability of suitable fill, as at Kielder dam (Fig. 2.10 and Coats and Rocke, 1983). The availability of competent rock at shallow depth, however, as shown in Fig. 1.7(b), favours either a rockfill embankment or, alternatively, a concrete gravity or buttress dam. Availability of rockfill, and thus relative cost, would dictate the final choice. A narrow and steep-sided valley in sound rock, as illustrated in Fig. 1.7(c), may be suited to an arch or cupola dam given competent abutments. Economic considerations may, however, suggest the rockfill embankment as a viable alternative.

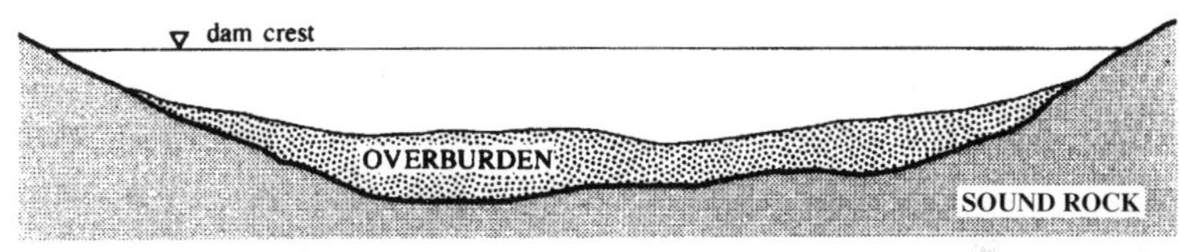

(a) Wide valley with deep overburden:
fine-grained deposits, e.g. Fine glacial soils, tills etc., over 5m deep favour earthfill embankment dam

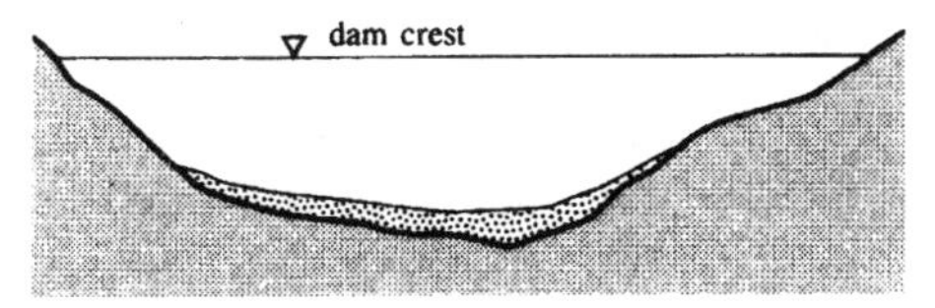

(b) Valley with little overburden:
suitable for embankment, gravity, or buttress dam

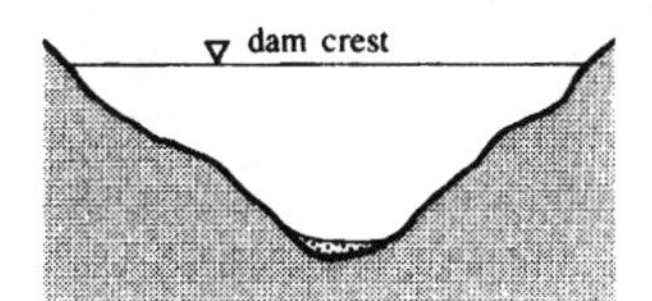

(c) Narrow valley, steep sides, little overburden:
suitable for arch, cupola, or rockfill embankment dam

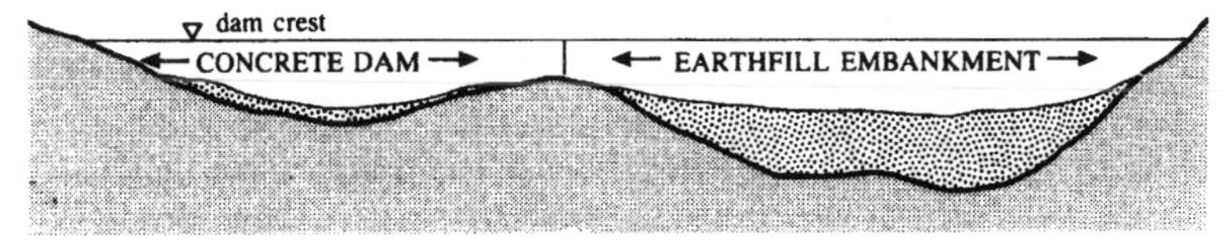

(d) Valley with irregular depth of overburden:
possible composite solution as shown; spillway on concrete dam

Fig. 1.7 Illustrative examples of dam type in relation to valley profile

The situation shown in Fig. 1.7(d), with deep overburden under one-half of the site, could well suggest the composite solution shown. An earth-fill embankment is constructed where overburden depth is considerable, the spillway being conveniently accommodated on a concrete gravity section where the required excavation depth is reasonable, e.g. at Cow Green dam (Kennard and Reader, 1975).

It will be appreciated that in all four illustrative instances the availability of alternative construction materials, and their relative cost, will impact heavily upon the final decision as to type of dam.

Figure 1.7 should be studied in conjunction with Table 1.7, which summarizes dam type characteristics in the context of selection of type.

The choice of type of dam may be influenced in part by factors which assume a particular importance in relation to a specific site. An example would lie in a situation where site conditions in a steep-sided valley might favour an earthfill embankment, but a spillweir and channel of the required size might prove disproportionately expensive to construct in the hillside round the flank of the dam. In such a case the economic balance might then tilt in favour of the gravity dam with an overflow crest, if the overburden depth were not excessive.

It may also be noted that local factors may assume greater relative importance for a specific site. Examples might include access, seismic risk,

Table 1.7 Dam selection: type characteristics

Type	*Notes and characteristics*
Embankment	
Earthfill	Suited to either rock or compressible soil foundation and wide valleys; can accept limited differential settlement given relatively broad and plastic core. Cut-off to sound, i.e. less permeable, horizons required. Low contact stresses. Requires range of materials, e.g. for core, shoulder zones, internal filters etc.
Rockfill	Rock foundation preferable; can accept variable quality and limited weathering. Cut-off to sound horizons required. Rockfill suitable for all-weather placing. Requires material for core, filters etc.
Concrete	
Gravity	Suited to wide valleys, provided that excavation to rock is less than *c.* 5 m. Limited weathering of rock acceptable. Check discontinuities in rock with regard to sliding. Moderate contact stress. Requires imported cement.
Buttress	As gravity dam, but higher contact stresses require sound rock. Concrete saved relative to gravity dam 30–60%.
Arch and Cupola	Suited to narrow gorges, subject to uniform sound rock of high strength and limited deformability in foundation and most particularly in abutments. High abutment loading. Concrete saving relative to gravity dam is 50–85%.

material processing and transportation costs, availability of plant and trained labour, ease of river diversion, risk of flood inundation during construction and construction period relative to the desired project commissioning date.

Design features of dams which can have major implications with regard to construction programming and costs include

- cut-offs,
- spillway systems, including channels and stilling basins,
- internal drainage systems,
- internal culverts, galleries, etc.,
- foundation preparation, including excavation and grouting, etc.,
- construction details, e.g. transitions or filters in embankments or contraction joint details in concrete dams,
- gates, valves and bottom outlet works, and
- river diversion works.

1.7 Loads on dams

1.7.1 General

The structural integrity of a dam must be maintained across the range of circumstances or events likely to arise in service. The design is therefore determined through consideration of the corresponding spectrum of loading conditions. In all foreseeable circumstances the stability of the dam and foundation must be ensured, with stresses contained at acceptable levels and watertight integrity essentially unimpaired.

Dams display a sophistication in their structural response which stands in sharp contrast to their apparent simplicity of structural concept and form. They are asymmetrical and three-dimensional structures, constructed from materials with complex physical properties and founded upon non-uniform and anisotropic natural formations. This is reflected in the interaction of the dam with its foundation and in the complex structural response of both when subjected to fluctuations in major loads and to the effects of progressive saturation. In comparison with most other engineering structures, dams are also required to function at or near their specified design load for a high proportion of their service life as noted earlier.

Certain loads are accurately predeterminate with regard to their distribution, magnitude and mode of action. Obvious examples are external water loads and structure self-weight. Other major loads, some equally important, are less reliably predeterminate and may also be time dependent. Examples are provided by internal seepage pressures, by the load redistribution effected through foundation deformations or, in concrete dams, by thermal effects associated with cement hydration etc.

It is convenient to classify individual loads as primary, secondary, or exceptional. Classification in this manner assists in the proper appreciation of load combinations to be considered in analysis. The classification is made in terms of the applicability and/or the relative importance of the load.

1. *Primary loads* are identified as universally applicable and of prime importance to all dams, irrespective of type, e.g. water and related seepage loads, and self-weight loads.
2. *Secondary loads* are generally discretionary and of lesser magnitude (e.g. sediment load) or, alternatively, are of major importance only to certain types of dams (e.g. thermal effects within concrete dams).
3. *Exceptional loads* are so designated on the basis of limited general applicability or having a low probability of occurrence (e.g. tectonic effects, or the inertia loads associated with seismic activity).

1.7.2 Schedule of loads

The primary loads and the more important secondary and exceptional sources of loading are identified schematically on Fig. 1.8, a gravity dam section being used for this purpose as a matter of illustrative convenience. Quantification of loads is left to Chapters 2 and 3.

Not all the loads identified may be applicable to a specific dam; an element of discretion is left in formulating combinations of loading for analysis.

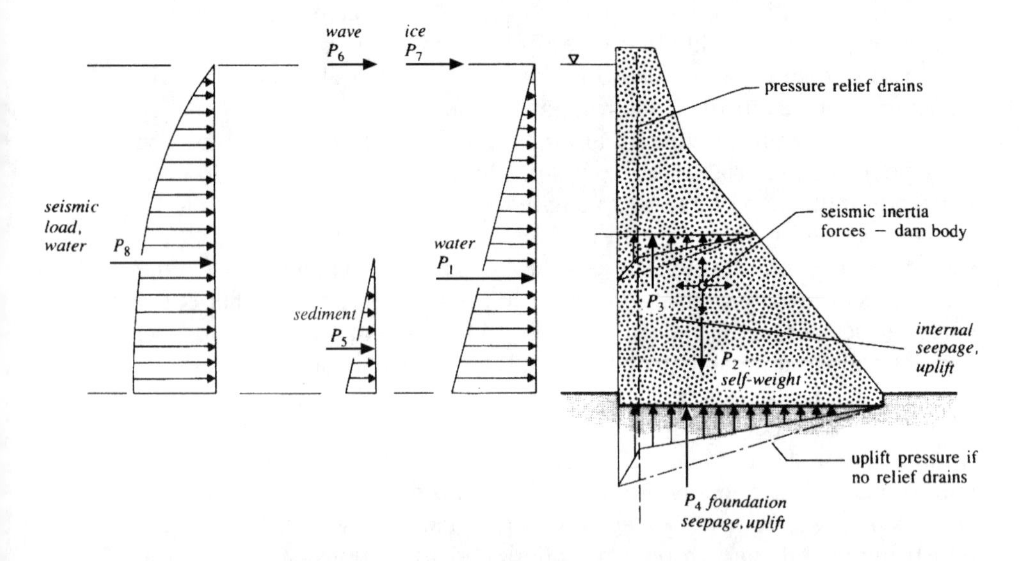

Fig. 1.8 Schematic of principal loads: gravity dam profile

(a) Primary loads

1. *Water load.* This is a hydrostatic distribution of pressure with horizontal resultant force P_1. (Note that a vertical component of load will also exist in the case of an upstream face batter, and that equivalent tailwater loads may operate on the downstream face.)
2. *Self-weight load.* This is determined with respect to an appropriate unit weight for the material. For simple elastic analysis the resultant, P_2, is considered to operate through the centroid of the section.
3. *Seepage loads.* Equilibrium seepage patterns will develop within and under a dam, e.g. in pores and discontinuities, with resultant vertical loads identified as internal and external uplift, P_3 and P_4, respectively. (Note that the seepage process will generate porewater pressures in pervious materials, and is considered in this light as a derivative of the water load for the embankment dam (Section 1.7.3).)

(b) Secondary loads

1. *Sediment load.* Accumulated silt etc. generates a horizontal thrust, considered as an equivalent additional hydrostatic load with horizontal resultant P_5.
2. *Hydrodynamic wave load.* This is a transient and random local load, P_6, generated by wave action against the dam (not normally significant).
3. *Ice load.* Ice thrust, P_7, from thermal effects and wind drag, may develop in more extreme climatic conditions (not normally significant).
4. *Thermal load (concrete dams).* This is an internal load generated by temperature differentials associated with changes in ambient conditions and with cement hydration and cooling (not shown).
5. *Interactive effects.* These are internal, arising from differential deformations of dam and foundation attributable to local variations in foundation stiffness and other factors, e.g. tectonic movement (not shown).
6. *Abutment hydrostatic load.* This is an internal seepage load in the abutment rock mass, not illustrated. (It is of particular concern to arch or cupola dams.)

(c) Exceptional loads

1. *Seismic load.* Oscillatory horizontal and vertical inertia loads are generated with respect to the dam and the retained water by seismic disturbance. For the dam they are shown symbolically to act through the section centroid. For the water inertia forces the simplified

equivalent static thrust, P_8, is shown. Seismic load is considered further within Section 2.7 in relation to embankment dams, and Section 3.1 with respect to concrete dams.

2. *Tectonic effects.* Saturation, or disturbance following deep excavation in rock, may generate loading as a result of slow tectonic movements. This lies beyond the scope of this text and is not considered further, but reference may be made to Golzé (1977).

1.7.3 Concepts of loading

The concurrent and persistent presence of all primary loads acting on an operational dam is self-evident. Some secondary loads, e.g. sediment load, will develop gradually and persist in concert with the primary loads. Other secondary loads will provide an infrequent and temporary additional load, e.g. ice thrust. Of the exceptional loads, tectonic action can generate a permanent addition to the overall loading on a dam. Seismic activity, however, is essentially random with a low probability of occurrence, generating an extreme but transient dynamic loading.

Study of the nature of the various individual loads and loading patterns leads to the logic of expressly defined load combinations being employed in analysis. Each load combination is related to a particular service condition and defines those loads considered to be concurrently operative on the dam. This logic can be extended to effecting a considered reduction in design safety margins to parallel the diminishing probability of occurrence of more rigorous load combinations.

The important elements of embankment dam and concrete dam design are introduced in Chapters 2 and 3 respectively. In anticipation of the analytical methods presented there a historical conceptual difference in the assumed mode of action of the primary loads should be identified. In embankment dam analyses water, seepage and self-weight loads are considered, correctly, to operate through distributed internal or body pressures and forces, as shown schematically on Fig. 1.9. The classical stability analysis of concrete dams, however, was developed on the basis of

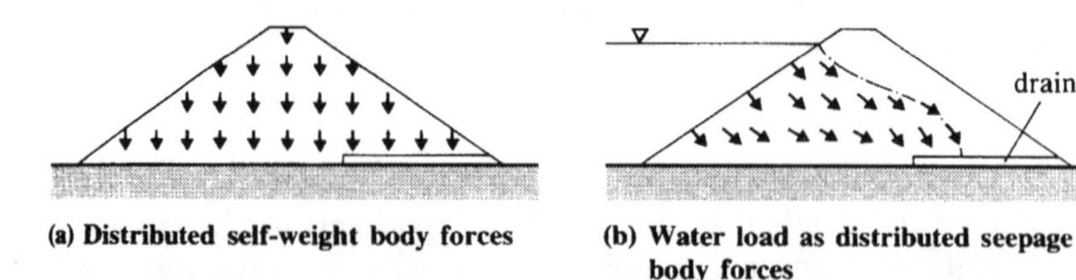

(a) Distributed self-weight body forces

(b) Water load as distributed seepage body forces

Fig. 1.9 Schematic of internal body forces: embankment dam

19th-century classical mechanics, and assumes water and seepage loads to act as surface pressures, with their resultant forces considered to act in conjunction with the resultant self-weight load as illustrated on Fig. 1.8.

This conceptual discrepancy is largely eliminated by the application of sophisticated modern techniques of mathematical modelling, most notably the extremely powerful and elegant finite element model. These techniques present their own problems, however, not least with regard to mathematical characterization of the prototype materials, and such is their specialist nature that beyond the brief appreciation given in Sections 2.7 and 3.2 their use is considered to lie outside the scope of this text.

References

Alpsü, I. (1967) Investigation of water losses at May reservoir, in *Transactions of the 9th International Congress on Large Dams*, Istanbul, International Commission on Large Dams, Paris, Q34 R27.

Attewell, P.B. and Farmer, I.W. (1976) *Principles of Engineering Geology*, Chapman and Hall, London.

Bass, K.T. and Isherwood, C.W. (1978) The foundations of Wimbleball dam. *Journal of the Institution of Water Engineers and Scientists*, 32: 187–97.

Bell, F.G. (1993) *Engineering Geology*, Blackwell Science, Oxford.

Binnie, G.M. (1981) *Early Victorian Water Engineers*, Thomas Telford, London.

—— (1987a) *Early Dam Builders in Britain*, Thomas Telford, London.

—— (1987b) Masonry and concrete gravity dams. *Industrial Archaeology Review*, 10 (1): 41–58.

Bridle, R.C. (2006) Dams: setting a new standard for sustainability, in *Proceedings of the Institution of Civil Engineers, 'Civil Engineering'*, 159 (May): 21–5.

Bridle, R.C., Vaughan, P.R. and Jones, H.N. (1985) Empingham dam – design, construction and performance. *Proceedings of the Institution of Civil Engineers*, 78: 247–89.

CEC (1985) Directive on the assessment of the effects of certain public and private projects on the environment (85/337/EEC), in *Official Journal of the European Communities,* L175: 40–8.

Clayton, C.R.I., Simons, N.E. and Matthews, M.C. (1995) *Site Investigation – a Handbook for Engineers*, 2nd edn. Blackwell Science, Oxford.

Clifton, S. (2000) Environmental assessment of reservoirs as a means of reducing the disbenefit/benefit ratio, in *Proceedings of Conference Dams 2000, Bath*, British Dam Society, Thomas Telford, London, 190–8.

Coats, D.J. and Rocke, G. (1983) The Kielder headworks. *Proceedings of the Institution of Civil Engineers*, 72: 149–76.

Collier, U. (2006) Dams are still not being properly planned, in *Proceedings of the Institution of Civil Engineers, 'Civil Engineering'*, 159 (1): 4.

Collins, P.C.M. and Humphreys, J.D. (1974) Winscar reservoir. *Journal of the Institution of Water Engineers and Scientists*, 28: 17–46.

DoE (1989) *Environmental assessment: a guide to the procedures.* Her Majesty's Stationery Office, London.

Fell, R., MacGregor, P. and Stapledon, D. (1992) *Geotechnical Engineering of Embankment Dams*, Balkema, Rotterdam.
Freer, R. (2000) The Three Gorges project. A progress report on a major new dam on the Yangtze River in China, in *'Dams & Reservoirs'*, 10 (2): 32–5.
Garbrecht, G. (ed.) (1987) *Historische Talsperren*, Konrad Wittwer, Stuttgart.
Golzé, A.R. (ed.) (1977) *Handbook of Dam Engineering*, Van Nostrand Reinhold, New York.
Gosschalk, E.M. and Rao, K.V. (2000) Environmental implications – benefits and disbenefits of new reservoir projects, in *Proceedings of Conference Dams 2000, Bath*, British Dam Society, Thomas Telford, London: 199–211.
Hendry, M. (1994) Environmental assessment – legislation and planning within the European framework, in *Proceedings of Conference on 'Reservoir safety and the environment'*, British Dam Society, London: 100–8.
ICOLD (1970) Recent developments in the design and construction of dams and reservoirs on deep alluvial, karstic or other unfavourable foundations, in *Transactions of the 10th International Congress on Large Dams, Montreal*, International Commission on Large Dams, Paris, Q37.
—— (1988) Reservoirs and the environment – experiences in managing and monitoring. *Transactions of the 16th International Congress on Large Dams*, San Francisco, International Commission on Large Dams, Paris, Q60.
—— (1992) *Dams and Environment – Socio-Economic Impacts*. Bulletin 86, International Commission on Large Dams, Paris.
—— (1994) *Dams and Environment – Water Quality and Climate*. Bulletin 96, International Commission on Large Dams, Paris.
—— (1998) *World Register of Dams*, International Commission on Large Dams, Paris.
Kennard, M.F. and Knill, J.L. (1969) Reservoirs on limestone, with particular reference to the Cow Green scheme. *Journal of the Institution of Water Engineers*, 23: 87–136.
Kennard, M.F. and Reader, R.A. (1975) Cow Green dam and reservoir. *Proceedings of the Institution of Civil Engineers*, 58: 147–75.
Kennard, M.F., Owens, C.L. and Reader, R.A. (1996) *Engineering Guide to the Safety of Concrete and Masonry Dam Structures in the UK*, Report 148, CIRIA, London.
Khatsuria, R.M. (2005) *Review of Hydraulics of Spillways and Energy Dissipators*, Marcel Dekker, New York.
Morrison, T. and Sims, G.P. (2001) Dams and development – final report of the World Commission on Dams, in *'Dams & Reservoirs'*, 11 (1): 32–5.
Pritchard, S. (2000) A milestone … or a millstone?, in *International Water Power and Dam Construction*, 53 (12): 18–21.
Schnitter, N.J. (1994) *A History of Dams – the Useful Pyramids*, Balkema, Rotterdam.
Smith, N.A. (1971) *A History of Dams*, Peter Davies, London.
Thomas, C., Kemm, H. and McMullan, M. (2000) Environmental evaluation of reservoir sites, in *Proceedings of Conference Dams 2000, Bath*, British Dam Society, Thomas Telford, London: 240–50.
Thomas, H.H. (1976) *The Engineering of Large Dams*, 2 vols, Wiley, Chichester.
UNEP (2006) Promoting sustainable development, in *International Water Power and Dam Construction* 58 (4): 12–15.

USBR (1987) *Design of Small Dams*, 3rd edn, US Government Printing Office, Washington DC.

Vischer, D.L. and Hager, W.H. (1998) *Dam Hydraulics*, Wiley, Chichester.

Wakeling, T.R.M. and Manby, C.N.D. (1989) Site investigations, field trials and laboratory testing, in *Proceedings of Conference on Clay Barriers for Embankment Dams*, Thomas Telford, London.

Walters, R.C.S. (1974) *Dam Geology*, 2nd edn, Butterworth, London.

Weltman, A.J. and Head, J.M. (1983) *Site Investigation Manual*, CIRIA Special Publication 25 (PSA Technical Guide 35), Construction Industry Research and Information Association, London.

Workman, J. (2001) Dams and development – a new framework for decision making, in *Proceedings of the Institution of Civil Engineers, 'Civil Engineering'*, 144 (1): 8–9.

Chapter 2

Embankment dam engineering

2.1 Introduction

An introductory presentation on the position of embankments in the history of dam engineering, as well as of the principal variants and their key components, was included in Chapter 1. The structure and contents of this chapter, which is necessarily concise, are dictated by the need to introduce basic elements of soil mechanics and applied geology in sections dealing with the nature and classification of engineering soils and with their characteristics. The text is also influenced by the design approach to embankment dams being in many respects less formalized than is the case for concrete dams (Chapter 3). After briefly reviewing embankment dam design principles and construction methods this chapter concentrates on the discussion of seepage, stability and settlement as the key factors in design. It concludes with a brief section dealing with rockfill and rockfill embankments.

For comprehensive background texts on soil mechanics, reference should be made to Craig (2004), or to Das (1997). Comprehensive discussion of earth and rockfill dam engineering is provided in texts by Fell, MacGregor and Stapledon (1992), Jansen (1988), Thomas (1976), and in Hirschfeld and Poulos (1973). Selected geotechnical issues are addressed in Penman (1986).

2.2 Nature and classification of engineering soils

2.2.1 The nature of soils

Soil is defined for engineering purposes as a natural aggregate of mineral grains separable by gentle mechanical means, e.g. agitation in water. Rock, in contrast, is a natural aggregate of minerals connected by strong and permanent cohesive bonds. The boundary between soil and rock is to some degree arbitrary, as exemplified by soft or weathered rocks, e.g. weathered limestones and shales, or weakly cemented sandstones.

All engineering soils of non-organic origin (i.e. excluding peats etc.) are formed by rock weathering and degradation processes. These may occur *in situ* forming *residual* soils. Alternatively, if the rock particles are removed and deposited elsewhere by natural agents, e.g. glaciation or fluvial action, they will form *transported* soils. Soft or weathered rocks form part of the range of residual soils. Transportation results in progressive changes in the size and shape of mineral particles and a degree of sorting, with the finest particles being carried furthest. All engineering soils are particulate in nature, and this is reflected in their behaviour.

An important distinction must be drawn between two generic inorganic soil groups which result from different weathering processes. The larger, more regularly shaped mineral particles which make up silts, sands and gravels are formed from the breakdown of relatively stable rocks by purely physical processes, e.g. erosion by water or glacier, or disintegration by freeze–thaw action.

Certain rock minerals are chemically less stable, e.g. feldspar, and undergo changes in their mineral form during weathering, ultimately producing colloidal-sized 'two-dimensional' clay mineral platelets. These form clay particles, the high specific surface and hence surface energy of which are manifested in a strong affinity for water and are responsible for the properties which particularly characterize clay soils, i.e. cohesion, plasticity and susceptibility to volume change with variation in water content. Differences in platelet mineralogy mean that clay particles of similar size may behave differently when in contact with water, and hence differ significantly in their engineering characteristics.

2.2.2 Description and classification of natural soils

Soil particles vary in size from over 100 mm (cobbles) down through gravels, sands and silts to clays of less than 0.002 mm size. Naturally occurring soils commonly contain mixtures of particle sizes, but are named according to the particle type the behaviour of which characterizes that of the soil as a whole. Thus a clay soil is so named because it exhibits the plasticity and cohesion associated with clay-mineral-based particles, but the mineral matrix invariably contains a range of particle sizes, and only a minor proportion of the fine material in the matrix may be clay sized, i.e. <0.002 mm (2 μm).

A comparison of two major systems used for defining and classifying the particle size ranges for soils is provided in Fig. 2.1. The divisions

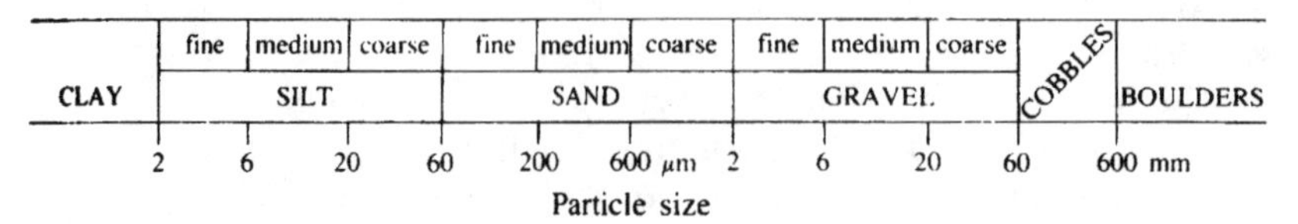

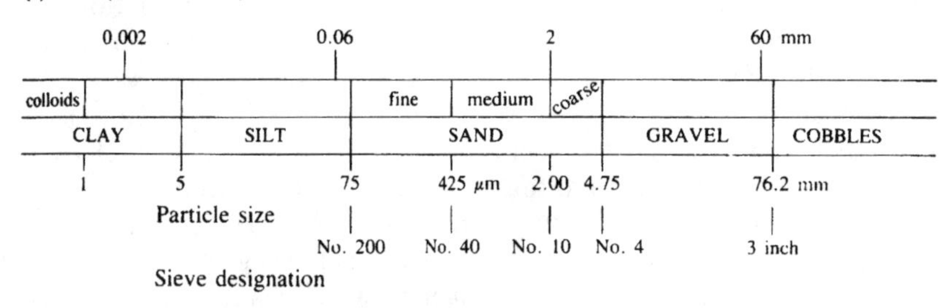

Fig. 2.1 Soil particle size classification systems (after Head, 1980)

between the named soil types correspond broadly to significant and identifiable changes in engineering characteristics. Particle size analysis is therefore employed for primary classification, to distinguish between gravels, sands and fine-grained silts and clays. A triangular chart for initial descriptive comparison and classification of soils from their particle size distribution is shown in Fig. 2.2.

Particle size analysis is insufficient for the complete classification of fine-grained soils or coarser soils where the matrix includes a proportion of plastic fines, i.e. clays. Secondary classification by degree of plasticity is then necessary, using consistency limits expressed in terms of percentage water content by mass, w.

The *liquid limit*, w_L, is the water content defining the change in soil state, i.e. consistency, from plastic to liquid; the *plastic limit*, w_P, defines the change-point below which a soil is too dry to exhibit plasticity. The range of water content over which a soil displays plastic behaviour is expressed by the *plasticity index*, I_P, with $I_P = w_L - w_P$. Secondary classification is determined through I_P and w_L using classification charts.

Common classification systems include the Unified Soil Classification System, used in the USA, and the British Soil Classification System for Engineering Purposes. In the latter system soils are divided into groups, each of which is denoted by a symbol, usually comprising two letters. The first letter refers to the dominant soil constituent, i.e. G, S, M and C for gravels, sands, silts and clays respectively. The second or qualifying letter provides descriptive detail based on, for example, particle size distribution for coarser soils, e.g. SW for well graded sand, or on degree of plasticity

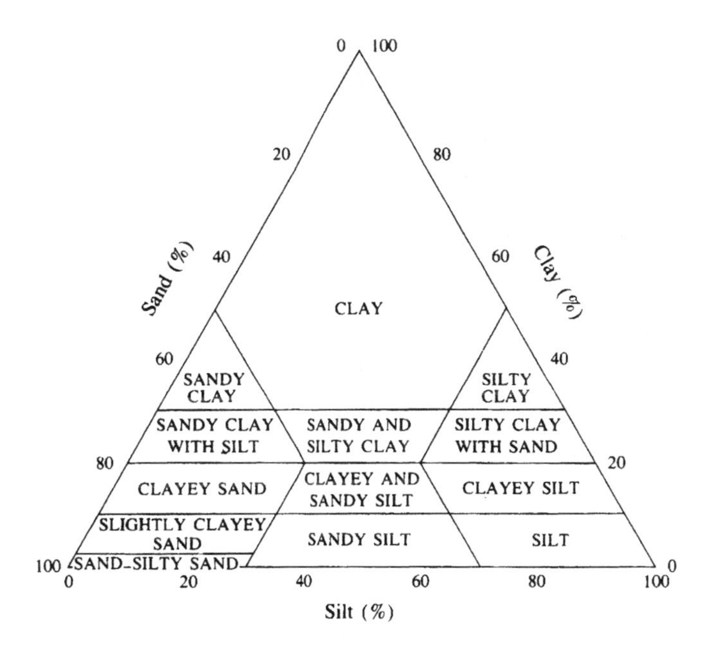

Fig. 2.2 Descriptive soil identification and classification chart (after Head, 1980)

where clay fines are present, e.g. I for intermediate, H for high etc. (BSI, 1999).

2.2.3 Phases in soil: soil porewater: effective stress

A soil may constitute a two- or three-phase system comprising solid soil matrix or skeleton and fluid, either water or gas or both. Water may exist in the soil in a variety of forms. Apart from being the main constituent of the liquid phase, water may also exist in a gaseous phase as water vapour and in the solid phase as adsorbed water. All mineral particles tend to form physico-chemical bonds with water, resulting in a surface film of adsorbed or fixed molecular water. This is most significant with the finest-grained soils as a result of their relatively high specific surface and the mineralogical composition of clay particles. The associated electrical phenomena occurring at the clay particle–water interface are primarily responsible for the cohesion and plasticity identified with clay soils.

Free water is the term used to describe that portion of the total porewater which obeys the normal laws of hydraulics. Provided that this water is present in the soil pores as a continuous liquid phase, Bernoulli's law applies. The *phreatic surface* is defined as the datum level at which the porewater pressure within a soil mass is zero, i.e. atmospheric. The stable water level attained in a standpipe is termed the *piezometric level.* The level of the water table (WT) or groundwater level (GWL) undergoes seasonal fluctuations and may also change as a direct result of construction operations. Below GWL the soil is assumed to be fully saturated, but it may contain small volumes of entrapped air. Above GWL, soil moisture may be held by capillary forces. The silty soils and clays frequently employed in embankment fills are generally non-saturated when first compacted, i.e. some pore space is filled with compressible pore air. The fill will progress to a saturated state, on which much of soil mechanics practice is predicated, as the seepage front advances through the embankment.

The vertical section through a soil mass generating a vertical total stress, σ, and static porewater pressure, u_w, on the horizontal plane X–X at depth z is shown in Fig. 2.3(a). Positive porewater pressure below the water table decreases interparticle contact pressure and the intergranular, i.e. effective, stress, σ', transmitted through the soil particles is less than the total stress σ by an amount equivalent to the porewater pressure, i.e. the effective pressure or stress is given by

$$\sigma' = \sigma - u_w \tag{2.1}$$

as illustrated in Fig. 2.3(b).

This, the effective stress relationship, is at the core of much of geotechnical practice since it is effective stress level which determines the

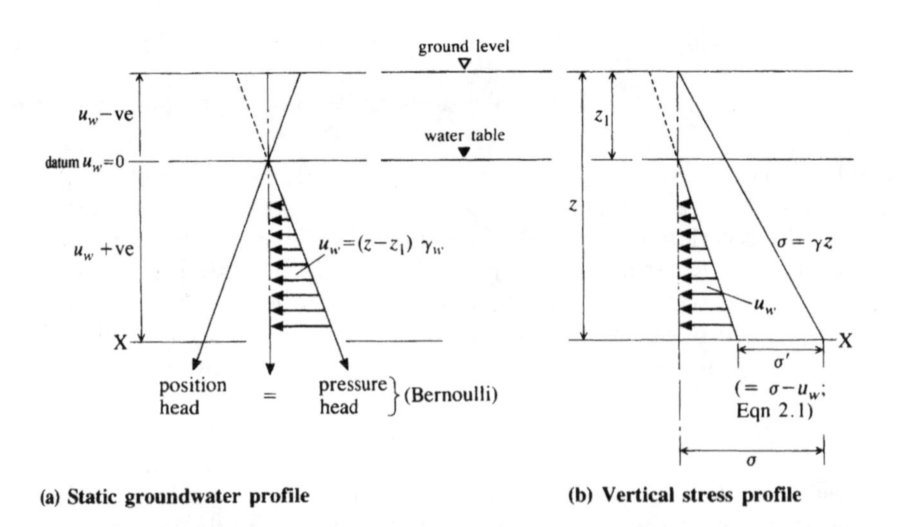

Fig. 2.3 Porewater pressures and vertical geostatic stresses: static groundwater case

shearing resistance which a soil can mobilize and the compressibility of clay soils (equations (2.6) and (2.7)).

Representative physical data from samples of common soil types are presented in Table 2.1. Water contents are expressed as percentages of the dry mass; the void ratio, e, and degree of saturation, S_r, are defined in accordance with standard soil mechanics terminology and expressed in volume terms.

2.3 Engineering characteristics of soils

2.3.1 Soil load response

Soil response is important in embankment dam construction with respect to the performance of engineered compacted soils in earthfills and that of the natural underlying foundation soils.

In earthfill construction it is necessary to consider the load-bearing characteristics of the compacted fill and also the behaviour of the soil as construction proceeds. It is convenient to categorize problems concerning the response of soils to specific loading conditions as problems of stability or of deformation. Problems of *stability* concern the equilibrium between forces and moments and the mobilized soil strength. When the former, arising from loading (or from the removal of support as in a trench excavation), exceed the shearing resistance which the soil can mobilize, failure will occur. This is generally manifested by progressive and, in the final phase, large and relatively rapid mass displacements, e.g. of a soil slope. Stability problems involve concepts of soil shear strength and stress–strain response.

Table 2.1 Representative physical characteristics of soils

Soil type	*Natural water content, w (%)*	*In situ unit weight, $\gamma (kN m^{-3})$*	*Void ratio, $e^{(see\ note)}$*	*Degree of saturation S_r (%)*
Dry uniform sand, loose	0	13.5	0.95	0
Well-graded sand	5	19.5	0.45	40
Soft clay	55	17.0	1.50	>95
Firm clay	20	19.5	0.70	>90
Stiff glacial till	10	23.0	0.30	>95
Peat and organic soils	>250	10.0	>3.50	>90

The specific gravity of soil mineral particles, $G_s \approx 2.65$–2.75.
e = Pore void volume relative to volume of matrix solids.

While a soil mass may be stable in the sense described above it may nevertheless undergo *deformation* as a result of changes in loading or drainage conditions. A limited amount of deformation occurs with no net volume change, and is thus comparable with the elastoplastic behaviour of many non-particulate materials. The most significant soil deformations, however, usually involve volume changes arising from alterations in the geometric configuration of the soil particle assemblage, e.g. a loosely packed arrangement of soil particles will on loading adopt a more compact and denser structure. Where the soil particles are relatively coarse, as with sands, such a change occurs almost immediately on load application. In saturated clayey soils, however, volume changes and settlement due to external loading take place slowly through the complex hydrodynamic process known as consolidation (Section 2.3.3).

The effective stress, σ', can be calculated from equation (2.1) if the total stress, σ, and porewater pressure, u_w, are known. While the total stress at a point may be readily determined by statics, the local porewater pressure is a more complex variable. In fine-grained clay-type soils the value of u_w for applied increments of total stress will depend upon the properties of the soil mineral skeleton and the pore fluid and will be strongly time dependent. The immediate ($t = 0$) response of porewater pressure in a particular soil to various combinations of applied total stresses is described through the concept of pore pressure coefficients.

From consideration of volume changes in a soil element under applied total stress (Fig. 2.4), the change in pore pressure Δu_3 due to an applied change in minor principal stress of $\Delta\sigma_3$ can be expressed as

$$\Delta u_3 = B\Delta\sigma_3 \tag{2.2}$$

where B is an empirical pore pressure coefficient.

If the major principal total stress, σ_1, is also then changed, by an increment $\Delta\sigma_1$, the corresponding change in pore pressure, Δu_1, is given by

$$\Delta u_1 = AB(\Delta\sigma_1 - \Delta\sigma_3) \tag{2.3}$$

where A is a further empirical coefficient.

The overall change in pore pressure, Δu_w, due to changes in both σ_3 and σ_1 is then given by

$$\Delta u_w = \Delta u_3 + \Delta u_1 = B[\Delta\sigma_3 + A(\Delta\sigma_1 - \Delta\sigma_3)]. \tag{2.4}$$

Pore pressure coefficients A and B permit estimation of effective stresses resulting from predicted or known changes in applied stress. In view of the importance of effective stresses in controlling soil behaviour, the coefficients are essential predictive tools in the solution of many soil-engineering problems. They are determined in special laboratory triaxial shear strength tests (Section 2.3.2).

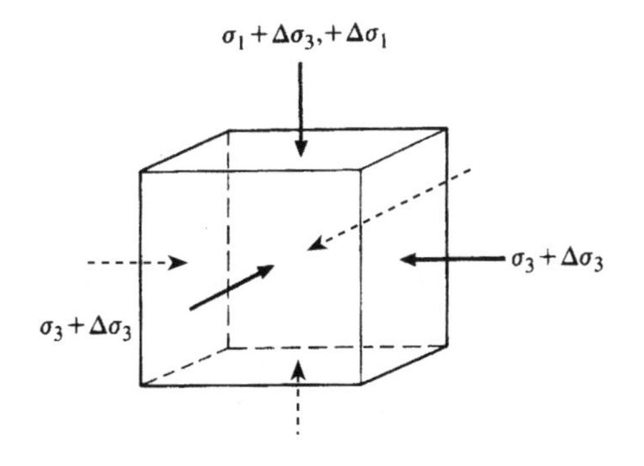

Fig. 2.4 Principal stress increments and porewater pressures

Equation (2.4) may be divided by $\Delta\sigma_1$ and rearranged as

$$\frac{\Delta u_w}{\Delta\sigma_1} = \bar{B} = B\left[1 - (1 - A)\left(1 - \frac{\Delta\sigma_3}{\Delta\sigma_1}\right)\right]. \tag{2.5}$$

Parameter $\bar{B}$ is an overall coefficient of particular relevance in the prediction of porewater pressures generated in the course of embankment dam construction. It is obtained from tests in which the sample is subjected to stress changes corresponding to those anticipated within the prototype embankment.

2.3.2 Shear strength

The *shear strength* of a soil is defined as the maximum resistance to shearing stresses which can be mobilized. When this is exceeded failure occurs, usually along identifiable failure surfaces. Shear strength is usually quantified through two component parameters:

1. apparent cohesion, c, essentially arising from the complex electrical forces binding clay-size particles together;
2. angle of shearing resistance, ϕ, developed by interparticle frictional resistance and particle interlocking.

The shear strength of a soil at a point on a particular plane can be expressed as a linear function of the normal stress, σ_n, at that same point using the Mohr–Coulomb failure criterion:

$$\tau_f = c + \sigma_n \tan\phi \tag{2.6a}$$

where τ_f is the shear strength at failure.

As noted previously, shearing resistance is determined by effective, i.e. interparticulate rather than total, stress level. A more appropriate form of equation (2.6a) is therefore

$$\tau_f = c' + \sigma'_n \tan\phi' \tag{2.6b}$$

where c' and ϕ' are the shear strength parameters expressed in terms of effective stresses, and σ'_n is the effective normal stress (equation (2.1)).

Equations (2.6a) and (2.6b) are represented by the failure envelopes AB and CD respectively on Fig. 2.5(a). Any combination of normal and shearing stresses represented by a point above the appropriate Mohr–Coulomb envelope represents a failure state for the soil; a point below represents a sustainable stress condition.

The envelopes AB and CD may be determined from the results of laboratory shear tests, e.g. by a triaxial shear test (Fig. 2.5(b)), in which the applied principal stresses σ_1 and σ_2 ($=\sigma_3$) acting on a cylindrical specimen enveloped in a rubber membrane within a water-filled cell (cell pressure, σ_3 ($=\sigma_2$)), are controlled in three orthogonal directions. It is an indirect shear test in that the inclination of the failure plane is not predetermined. The vertical (σ_1) and horizontal (σ_3 ($=\sigma_2$)) principal stresses do not corres-

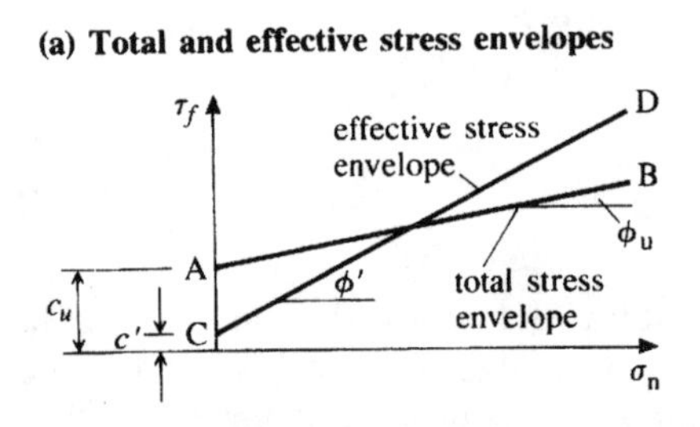

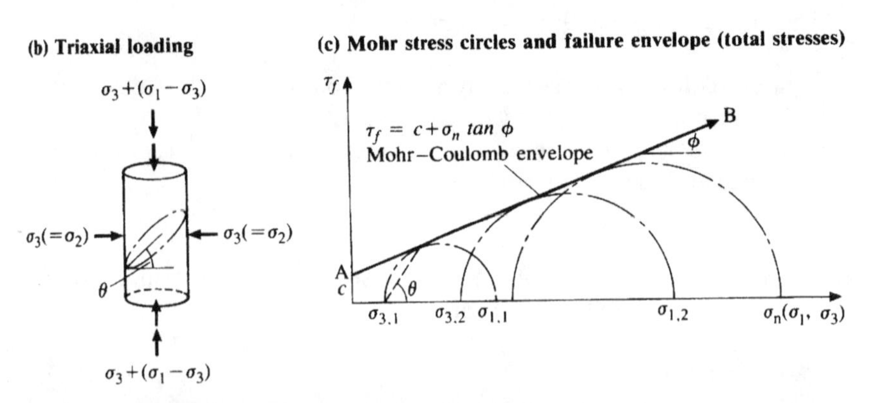

Fig. 2.5 Failure envelopes and the Mohr–Coulomb criterion

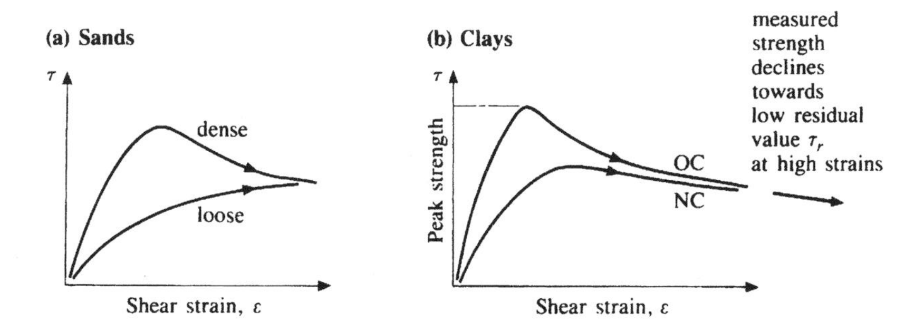

Fig. 2.6 Soil stress–strain response curves

pond to the normal and shear stresses on the failure plane, which must therefore be obtained indirectly, e.g. by the construction of a Mohr circle plot (Fig. 2.5(c)) or '*p–q*' plot (not shown). The stress–strain response of soils plotted from triaxial tests is essentially curvilinear, i.e. there is little elastic response to load, as shown in Fig. 2.6.

Coarse soils such as sands derive their shear strength essentially from particle interlock and internal friction, and are therefore termed cohesionless ($c = 0$) or frictional soils. When loaded under conditions of no drainage, saturated clays may appear to possess cohesion only. Clays are frequently identified in generic terms as cohesive soils ($c > 0$, $\phi = 0$). Soils of intermediate type, including the majority of 'cohesive' soils, will exhibit both cohesion and internal friction (a 'c–ϕ' soil). The shearing behaviour of each type can be represented through the Mohr–Coulomb relationships (equations 2.6(a) and 2.6(b)), giving the example failure envelopes illustrated in Fig. 2.5.

The measured shear strength of frictional soils is controlled largely by density, a higher density giving a greater angle of shearing resistance, ϕ (Fig. 2.6(a)).

Most engineering problems occur with fine cohesive soils, and arise from the nature of the clay particles. Because of their low permeability and strong affinity for water, natural clay soils usually occur in a saturated or near-saturated state. High porewater pressures are generated by changes in external loading conditions, including construction operations, and are very slow to dissipate. There is a clear relationship between shear strength and increasing water content: at high water contents the cohesive forces between clay particles rapidly weaken, resulting in very much reduced shear strengths.

The important factor influencing the shear strength and consolidation characteristics (Section 2.3.3) of a saturated clay is stress history rather than density. If the present *in situ* effective stresses are the greatest to which the clay has historically been subjected the clay is described as *normally consolidated* (NC). If, on the other hand, previous effective stress levels have been relieved, e.g. as a result of glaciation, the clay is described as *overconsolidated* (OC). The ratio of previous maximum effective stress to present *in situ* effective stress is the *overconsolidation ratio* (OCR). NC clays are relatively soft and compressible. Their undrained shear strength, c_u, developed where there is no relief of porewater pressure by drainage, is proportional to the pressure under which they have consolidated, and therefore increases with depth. OC clays, such as the glacial tills (OCR = 1–3) frequently used in UK embankment earthfills, are relatively stiff.

If the structure of certain cohesive soils is disturbed or remoulded, as in the process of compaction of earthfill in an embankment, a significant loss of shear strength may result. The ratio of undisturbed to remoulded undrained strength at the same moisture content is defined as the *sensitivity*, *S*. The sensitivity of most UK clays and tills lies between 1 and 3. Clays with values above 4 are referred to as sensitive.

Clay consistency may be loosely classified on the basis of undrained cohesive shear strength, c_u, as in Table 2.2.

2.3.3 Compressibility and consolidation

When load is applied to a soil mass compression and settlement may occur in consequence of one or more of three mechanisms:

1. elastic deformation of the soil particles;
2. compression of the pore fluid;
3. expulsion of the pore fluid from the stressed zone, with rearrangement of the soil particles.

Soil particles and water are sensibly incompressible, and compression or volume decrease in a saturated fine-grained soil due to applied stress or

Table 2.2 Descriptive consistency of clay soils (BSI, 1999)

Consistency	*Undrained strength, $c_u (kN m^{-2})$*
Very stiff or hard	>150
Stiff	100–150
Firm to stiff	75–100
Firm	50–75
Soft to firm	40–50
Soft	20–40
Very soft	<20

load is therefore accounted for almost entirely by mechanism 3 (expulsion of porewater) as excess porewater pressures dissipate. This hydrodynamic process is termed *consolidation* and is mainly relevant to clays and organic soils in which the volume change process is comparatively slow because of their very low permeability (Section 2.3.4). The consolidation process is partly reversible, i.e. compressible soils can swell on removal of load. One-dimensional vertical consolidation characteristics, determined in laboratory oedometer tests, are expressed in terms of two principal coefficients.

1. The coefficient of volume compressibility, m_v, is required to determine the magnitude of time-dependent primary consolidation settlement:

$$m_v = \Delta\epsilon_v / \Delta\sigma'_v \tag{2.7}$$

   where $\Delta\epsilon_v$ is the vertical strain increment produced by increment of vertical stress $\Delta\sigma'_v$, if no lateral yielding is permitted.
2. The coefficient of consolidation, c_v, is used to establish rates of settlement:

$$c_v = k / m_v \gamma_w \tag{2.8}$$

   where k is a coefficient of permeability.

The stress-dependency of the principal coefficients will be noted.

The coefficient of secondary consolidation, C_α, is determined in the same test and is used to describe subsequent and very slow continuing settlement due to creep of the soil structure under constant effective stress.

2.3.4 Soil permeability

Soil permeability is important to problems of seepage, stability of slopes and consolidation. It is also important in ground engineering processes, e.g. in grouting, and in dewatering. Relative permeability of saturated soils is assessed by the coefficient of permeability, k, expressed in units of velocity ($\mathrm{m\,s^{-1}}$). It is the most variable of soil properties between the extremes of a coarse gravel and an intact clay, and even within the confines of a notionally uniform soil. Illustrative values of k are given in Table 2.3.

The flow of water in a saturated soil can be represented by

$$v = -k \frac{\mathrm{d}h}{\mathrm{d}l} \tag{2.9}$$

where v is velocity and $\mathrm{d}h/\mathrm{d}l = i$, the hydraulic gradient.

The above relationship applies only if the soil is at or near full saturation and if laminar flow conditions prevail, and may be rewritten in the form of the familiar Darcy equation:

$$Q = kiA_s \tag{2.10}$$

where Q is the flow, and A_s is the gross cross-sectional area subject to flow.

Soil permeabilities are markedly anisotropic, with k_h, the coefficient of horizontal permeability, several times larger than k_v, the coefficient of vertical permeability. In compacted earthfills the ratio k_h/k_v may exceed 20. The coefficient of horizontal permeability is most reliably determined *in situ*, e.g. via field pumping tests in boreholes. Direct and indirect laboratory techniques are also available, but reproducibility of the results is poor and they are best regarded as indicative of relative orders of magnitude for permeability rather than as absolute values.

The permeability of non-saturated soils, e.g. embankment earthfills prior to impounding and saturation, is very much more complex (see Section 2.3.7). It is not considered in depth in this text, but reference may be made to Das (1997).

2.3.5 Compaction

Compaction is the process of densification by expulsion of air from the soil void space, and results in closer particle packing, improved strength, reduced permeability and reduced settlement. (The process must not be confused with consolidation, in which volume decrease is a result of the gradual expulsion of water under applied load – Section 2.3.3.) Field compaction of embankment fills is normally achieved by rolling the earthfill in thin layers, often assisted by vibratory excitation of the plant. The process may also be applied to *in situ* soils, and it is the most common and cheapest of large-scale ground improvement techniques.

The degree of compaction of a soil is measured in terms of dry density, ρ_d (or dry unit weight γ_d), i.e. the mass (or weight) of solids per unit volume of soil exclusive of moisture:

$$\rho_d = \rho/(1 + w) \tag{2.11}$$

where ρ is the bulk or *in situ* density, and w is the water content.

The dry density achieved during compaction varies with the water content of a soil and the compactive effort applied. The effects of these variables are apparent in the plots of dry density–moisture content relationships shown in Fig. 2.7. The application of a specified compactive effort to samples of soil prepared at different water contents yields curves with a characteristic inverted V-shape. At low water contents the soil is stiff and difficult to compact, producing low dry densities: as the water content is increased compaction is made easier and higher dry densities are obtained. At high water contents the water occupies an increasing volume within the soil void space, and the dry density is diminished. For a given compactive effort there is therefore an optimum water content, w_{opt}, at which a maximum value of dry density, $\rho_{d\ max}$, is achieved. If all the air in the soil could be expelled by compaction, the soil would be in a state of full saturation ($S_r = 100\%$) and ρ_d would attain the maximum possible value for the given moisture content. In practice this degree of compaction, analogous to 100% 'efficiency', cannot ever be achieved.

Increased compactive effort displaces the dry density–moisture content curve to give a higher maximum dry density at a lower optimum water content (Fig. 2.7). For a constant compactive effort, different soil types yield different dry density–water content curves, and in general coarse soils can be compacted to higher dry densities. For effective compaction, the soil layer thickness should be as low as is economically viable. Specified maximum layer thicknesses for effective field compaction generally lie in the range 150–250 mm.

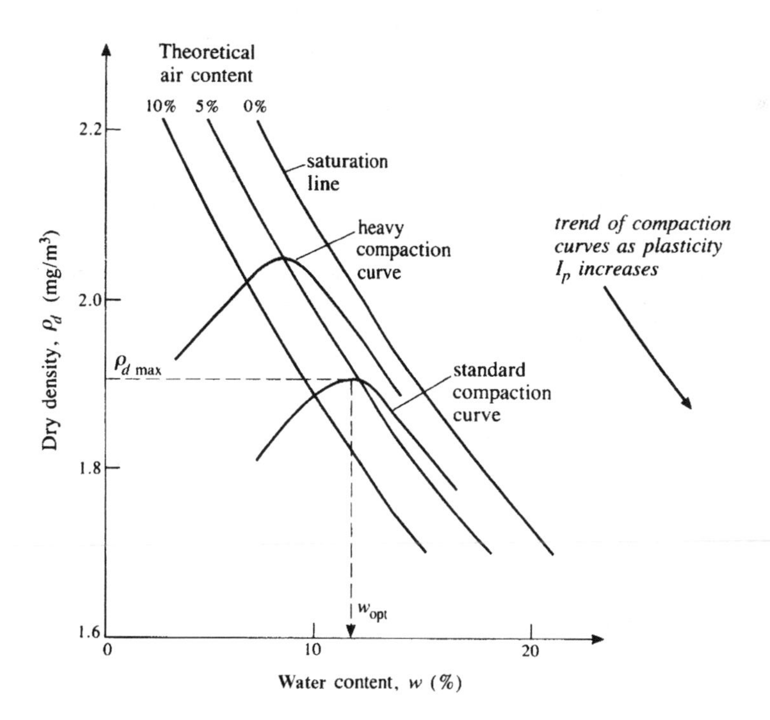

Fig. 2.7 Soil compaction relationships

A specification for field compaction of earthfill must ensure that a dry density is attained which will ensure strength and other characteristics adequate to fulfil stability, settlement or other criteria. The dry density achieved in the field is usually less than $\rho_{d\,max}$ obtained in the laboratory. The ratio of field dry density to $\rho_{d\,max}$, expressed as a percentage, is defined as the *relative compaction*. This ratio may be used to specify the required degree of site compaction, with limits placed on the water content of the soil. For major fill projects, such as embankments, it is now increasingly common to use a 'method'-type specification, based on comprehensive field trials. In broad terms this type of specification defines the layer thickness and water content limits and requires that a specified minimum number of passes of a nominated roller be applied to each layer of soil.

Compaction of a soil modifies the major engineering characteristics as summarised below.

1. *Shear strength.* Although shear strength is increased by compaction, the shear strength obtained under a given compactive effort varies with the water content. The maximum shear strength generally occurs at a water content less than w_{opt} and so at a dry density less than $\rho_{d\,max}$. In practice, the greatest long-term shear strength is likely to be achieved by compaction at a water content slightly greater than w_{opt}. In earthfills, particularly in the core of a dam, an increase in plasticity is also an important consideration in relation to reducing the risk of internal cracking or hydraulic fracturing (Section 2.7), and may encourage compaction at a few percent above w_{opt}.
2. *Compressibility.* A higher degree of compaction will reduce settlement due to subsequent compression and consolidation.
3. *Volume changes (due to changes in water content).* For a compressible soil compaction at a high water content generally tends to reduce subsequent swelling and increase potential shrinkage. Compaction at a low water content has the reverse effect.
4. *Permeability.* Greater compaction results in reduced permeability. The effects are rather less apparent with cohesive soils, where the very low permeability values are dictated by, *inter alia*, pore nature and structure and the presence of the molecular adsorbed water layer.

2.3.6 Representative engineering properties for soils

Representative ranges of values for the principal engineering properties of a number of generic soil types in their natural state are presented in Table 2.3.

Table 2.3 Illustrative engineering properties for selected soil types (compare with Table 2.6)

Descriptive	*Saturated unit weight,* $\gamma (kN m^{-3})$	*Shear strength (effective stress)*		*Coefficient of compressibility,* m_v $(\times 10^{-4} m^2 kN^{-1})$	*Coefficient of horizontal permeability* $k_h (m s^{-1})$
		Cohesion $c' (kN m^{-2})$	*Friction,* ϕ' *(degrees)*		
Gravels	17–22	0	30–45	(0.1–1.0)	10^{-1}–10^{-2}
Sands		0	30–45		10^{-2}–10^{-5}
Silts		<5	20–35		10^{-4}–10^{-6}
Clays (soft–medium)	15–21	0	20–30	(1.0–10.0)	Intact clay, $<10^{-8}$ If weathered, fissured, or with silt lenses 10^{-3}–10^{-8}
Clays (sensitive, silty)		<10	<30		
Clays (medium–stiff)		<50	<20		

1. Values of m_v and k_h are subject to wide variation; the figures quoted are a guide to the order of magnitude only.
2. $k_h > 10^{-4} m s^{-1}$ is necessary for good drainage, i.e. for dissipation of excess porewater pressures; $k_h < 10^{-8} m s^{-1}$ corresponds to being virtually impervious.
3. The properties of coarse-grained soils are controlled by relative density and particle shape; those of clay-type soils are influenced by their nature and stress history etc.

Care must be taken in the interpretation of Table 2.3, as the measured properties of notionally homogeneous and uniform soils are subject to very considerable variation. The ranges of values quoted should therefore be regarded as purely illustrative. They may be compared with the corresponding figures for compacted fills given in Table 2.6 (Section 2.5.1 should also be referred to).

2.3.7 Partially-saturated soils and the embankment dam

Engineering soils have to this point been considered as existing in the fully saturated state, i.e. all inter-particle void space is completely filled with pore water. An appreciation of the influence of degree of saturation of a soil, S_r, upon engineering behaviour is of considerable importance to the dam engineer, as compacted fine-grained soils are initially in a partially-saturated state. The soil is then a three-phase particulate material, with the solid particles forming the soil skeleton in point contact and the interstitial void spaces containing both pore water and compressible pore air.

It should be appreciated that the pore water within a fine-grained soil is a dilute electrolyte, the engineering characteristics of clayey soils reflecting their mineralogy and the concentration and nature of the 'free' ions present in the pore water. By the same token pore air should be correctly seen as embracing air, gases, and/or water vapour.

In the unsaturated state a meniscus is formed at the pore water/pore air interfaces. These menisci are responsible for the local pore water pressure, u_w, always being less than pore air pressure, u_a. The difference between pore air and pore water pressures is responsible for generating considerable intergranular contact forces between the soil particles. In fine-grained soils, especially those with an appreciable clay content, the matric suction, defined as $(u_a - u_w)$, diminishes dramatically with increasing degree of saturation, S_r, and plays a very significant role in determining important aspects of mechanical behaviour.

Specific soils display a quite unique and therefore characteristic relationship between matric suction and degree of saturation; the shape of the $(u_a - u_w)$ *v.* S_r curve is therefore a characteristic related to soil type, well-graded soils displaying a smoothly curving relationship with matric suction diminishing at a reducing rate as saturation level increases. More uniformly graded soils, on the other hand, display a much flatter curve, showing comparatively little diminution in matric suction over an appreciable range of increasing degree of saturation.

The increased matric suction associated with diminishing degree of saturation for a fine-grained soil will effect significant changes in several engineering characteristics of the soil of primary concern to the dam engineer, i.e. shear strength, collapse behaviour and permeability.

Shear strength is enhanced by increased suction, i.e. a lower degree of saturation, on account of the stabilizing influence of the increased intergranular forces which are generated. Apparent cohesion is increased, but there is often little change in angle of shearing resistance. In this lies the explanation for the seemingly greater stability of steep slopes in unsaturated soils. It will be evident that a subsequent increase in water content towards saturation will alter the suction, and hence reduce stability of the slope to the point where failure can occur. It may be noted that the effective stress relationship remains valid for partially saturated soils at high values of S_r

Collapse compression, which occurs quickly when the water content is increased, is a characteristic feature of unsaturated soils. The intergranular forces produced by soil suction create a metastable structure with a high void ratio; as saturation is approached the reduction in suction diminishes the intergranular forces, causing compression of the soil. If saturation is rapid there will be a correspondingly rapid reduction in soil volume, i.e. collapse compression of the soil will occur, which can give rise to serious problems. With certain clays, if saturation occurs at a reduced stress level undesirably large expansive strains may be developed, with denser soils displaying greater volume instability as swelling is highly dependent upon density.

The permeability of unsaturated fine-grained soils is very much dependent upon suction level. Reduction in the degree of saturation, particularly at values of $S_r < 90\%$, can significantly influence the matric suction and hence the intergranular contact forces.

The apparent permeability of the unsaturated soil may then change by several orders of magnitude. The characteristic curve descriptive of the suction *v.* degree of saturation relationship for a soil is therefore basic to a comprehensive description of the fluid flow processes within an unsaturated soil. Characterisation of the permeability of the latter is correspondingly of much greater complexity than for the soil in a fully saturated state. Considered in context with the embankment dam, fine-grained cohesive earthfills will invariably be in a partially saturated state at time of placing.

One illustrative instance of the significance of partial saturation arises with respect to compaction of a rolled core, i.e. whether it is advantageous to compact on the dry side of optimum (see Section 2.3.5). Opinion on this remains divided, but the risk of so doing is that the compacted soil will have a granular structure with larger pores and so be more prone to subsequent collapse compression and settlement. Analytical studies made for the Limonero dam (Spain) included comparison of the predicted effects of compacting the core wet and dry of optimum and a study of the influence of the soil microfabric. It was determined that the wet core showed less collapse compression and more effective dissipation of pore water pressure than the dry core, the latter showing a sharp wetting front and increased collapse compression accompanied by little dissipation of pore water pressures (Gens, 1998).

2.4 Principles of embankment dam design

2.4.1 Types and key elements

In its simplest and oldest form the embankment dam was constructed with low-permeability soils to a nominally homogeneous profile. The section featured neither internal drainage nor a cut-off. Dams of this type proved vulnerable to problems associated with uncontrolled seepage, but there was little progress in design prior to the 19th century. It was then increasingly recognized that, in principle, larger embankment dams required two component elements (Section 1.3 should also be referred to):

1. an impervious water-retaining element or core of very low permeability soil, e.g. soft clay or a heavily remoulded 'puddle' clay and
2. supporting shoulders of coarser earthfill (or of rockfill), to provide structural stability.

As a further design principle, from *c.* 1860 the shoulders were frequently subject to a degree of simple 'zoning', with finer more cohesive soils placed adjacent to the core element and coarser fill material towards either face.

Present embankment dam design practice retains both principles. Compacted fine-grained silty or clayey earthfills, or in some instances manufactured materials, e.g. asphalt or concrete, are employed for the impervious core element. Subject to their availability, coarser fills of different types ranging up to coarse rockfill are compacted into designated zones within either shoulder, where the characteristics of each can best be deployed within an effective and stable profile.

The principal advantages of the embankment dam which explain its continuing predominance were outlined in Section 1.3. Figures 1.2 and 1.3 illustrated the more important variants of embankment dam, and brief supplementary notes on each are given below.

Homogeneous embankments (Figs 1.2(a) and (b)) are now generally confined to smaller, less important dams and to dykes in river engineering (see Section 2.10 and Chapter 8). They require ready availability of sufficient low permeability soil, and careful design and internal detailing is necessary to control seepage and porewater pressures.

The central core earthfill profile, illustrated in Figs 1.2(c)–1.2(e), is the most common for larger embankment dams. Narrow cores of soft compressible 'puddle' clay or of concrete, as in the profiles of Figs 1.2(c) and 1.2(d), have been displaced since 1940–1950 by the technically superior wide rolled clay core profile of Fig. 1.2(e). The characteristics of soft puddle clays, as used in the obsolete profile of Fig 1.2(c), are reviewed in

Moffat (2002), as there are many such dams in service. The slender core can prove vulnerable to fracturing and internal erosion (Section 2.7.2), the wide core offering lower internal hydraulic gradients. The change to the wide core was coincident with the development of soil mechanics theory and with the introduction of high-capacity earthmoving and compaction plant. Core base width is now generally 20–40% of the height of the embankment (refer also to Figs 2.10(a) and (b)). Central and inclined core dams with shoulders of graded compacted rockfill are shown in Figs 1.2(f), 1.3(a) and 1.3(b).

The inclined core profile of Fig. 1.3(b) is sometimes considered advantageous in moderating the risk of core cracking as a result of load transfer between compressible core and stiffer rockfill shoulder (Section 2.7).

The decked rockfill embankment is illustrated in Fig. 1.3(c) and depicts an asphaltic or concrete impermeable upstream membrane. Thin asphaltic membranes (0.15–0.30 m thick) are now widely employed where soil suitable for core construction is either not available or uneconomic. An asphaltic membrane can accept a degree of deformation without rupture. Thicker (0.6–1.2 m) asphaltic membranes are also widely employed in the less vulnerable central position indicated in Fig. 1.3(d) (and Fig. 2.17).

Selection of the optimum type of embankment for a specific location is determined largely by the nature and availability of different fill materials in sufficient quantity. The much steeper face slopes possible with compacted rockfill shoulders (Figs 1.2(f) and 1.3(a)–(d)) can reduce the quantities of fill required for a given height of dam by 30–50%.

The primary loads acting on an embankment do not differ in principle from those applicable to gravity dams and outlined in Section 1.7. There are, however, the conceptual differences there referred to with regard to the water load which, in the case of all but decked embankments, is exerted inside the upstream shoulder fill. Self-weight load, similarly a distributed internal body load, is significant with respect to stability and internal stress for the embankment and for a compressible soil foundation. Because of such differences, embankment dam analysis is less formalized and is carried out quite differently from concrete dam analysis (Chapter 3). This will be developed further following consideration of the defects and failure modes which may affect embankment dams.

Charles (1998) presents a thoughtful discourse on the embankment dam.

2.4.2 Defect mechanisms, failure modes and design principles

The principal defect mechanisms and failure modes identifiable with embankment dams are illustrated in schematic form in Fig. 2.8. Certain mechanisms are interrelated, e.g. overtopping may result from inadequate spillway capacity or from a lack of freeboard which may, in turn, be the

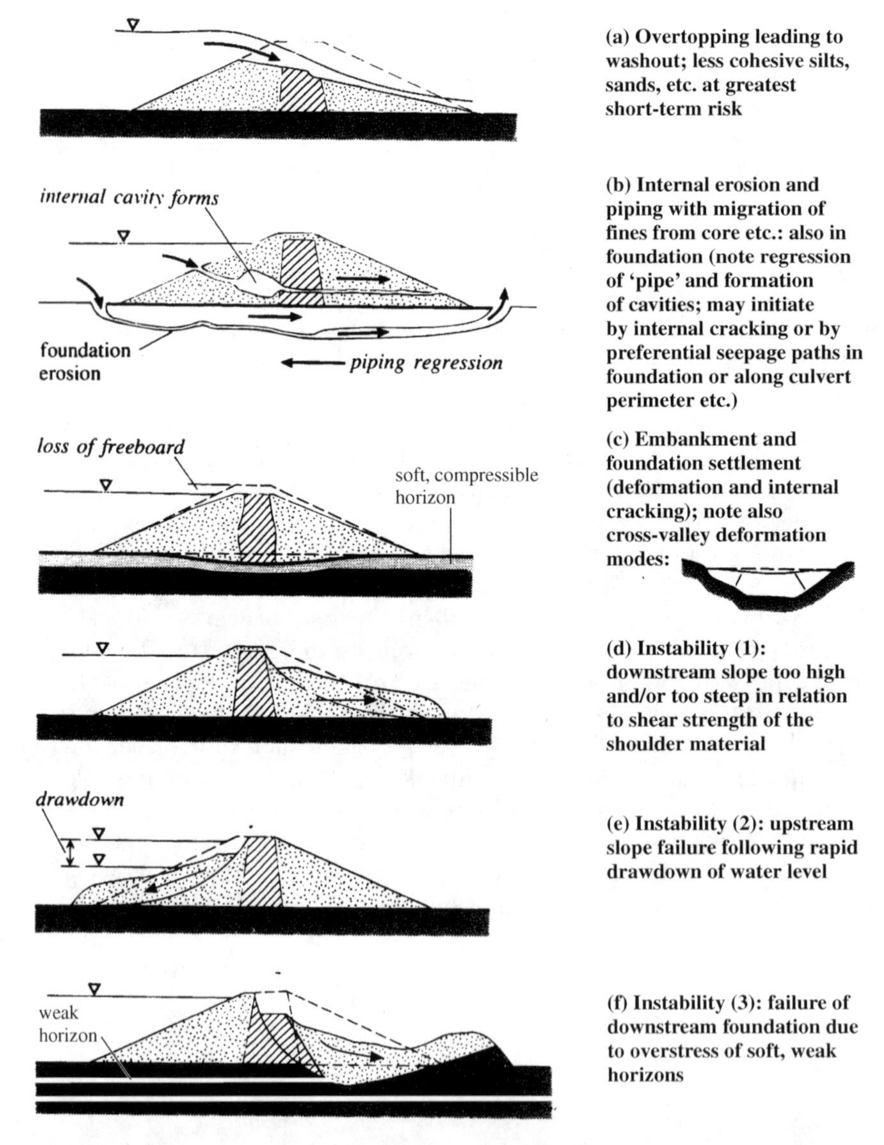

Fig. 2.8 Illustrative embankment defect mechanisms and failure modes

result of long-term deformation and settlement. The internal erosion and overtopping mechanisms are of particular concern, each being responsible for *c.* 30–35% of serious incidents and failures.

In drawing attention to the principal risks to be guarded against the schematic diagrams of Fig. 2.8 also highlight the principal design considerations:

1. *Overtopping and freeboard.* Spillway and outlet capacity must be sufficient to pass the design maximum flood (Section 4.2) without overstopping and risk of serious erosion and possible washout of the embankment. Freeboard, i.e. the difference between maximum watertight (i.e. core) level and minimum crest level of the dam, must also be sufficient to accept the design flood plus wave action without overtopping, and must include an allowance for the predicted long-term settlement of the embankment and foundation (Section 4.4).
2. *Stability.* The embankment, including its foundation, must be stable under construction and under all conditions of reservoir operation. The face slopes must therefore be sufficiently flat to ensure that internal and foundation stresses remain within acceptable limits.
3. *Control of seepage.* Seepage within and under the embankment must be controlled to prevent concealed internal erosion and migration of fine materials, e.g. from the core, or external erosion and sloughing. Hydraulic gradients, seepage pressures and seepage velocities within and under the dam must therefore be contained at levels acceptable for the materials concerned.
4. *Upstream face protection.* The upstream face must be protected against local erosion as a result of wave action, ice movement, etc.
5. *Outlet and ancillary works.* Care must be taken to ensure that outlet or other facilities constructed through the dam do not permit unobstructed passage of seepage water along their perimeter with risk of soil migration and piping.

Details of the more important defect mechanisms in relation to illustrative examples of causes and preventive measures are set out in Table 2.4.

Table 2.4 Embankment dam defect mechanisms and preventive measures

Defect	*Characteristics*	*Causes*	*Preventive–corrective measures*
External			
Overtopping	Flow over dam and possible washout; less cohesive soils most at risk; most serious if localized	Inadequate spillway and/or freeboard	Adequate spillway capacity and initial freeboard, and/or reinforced grass surface to slope
		Settlement reducing freeboard; spillway obstructed	Restoration of settlement; crest protection; good maintenance
Wave erosion	Damage to upstream face and shoulder	Face protection disturbed or damaged	Proper design and maintenance
Toe erosion	Flood discharge damaging toe	Spillway channel badly designed and/or located	Good hydraulic design; training walls
Gullying	Local concentrated erosion of downstream face by precipitation	Poor surface drainage	Vegetation, surface reinforcement and/or drainage
Internal seepage			
Loss of water	Increased seepage loss and/or irregularities in phreatic surface; soft spots on slopes or downstream	Pervious dam and/or foundation; cut-off inadequate	Cut-off and core grouting
		Internal cracking	Careful design; grouting
Seepage erosion (concealed internal erosion)	Turbid seepage through drainage system	Internal cracking	Internal drainage; filters; careful zoning of fill
		Leakage along perimeter of culverts, tunnels, pipework etc.	Detail design; use of collars; grouting

Instability			
Foundation slip	Surface displacement of ground near toe of slope	Soft or weak foundation and/or high porewater pressures	Consolidate soil; drainage; ground improvement
Face slopes	Change in morphology; bulging and deformation, leading to rotational or translational slip	High porewater pressure; slopes too steep; rapid draw-down on upstream slope	Drainage; flatten slope or construct stabilizing berms
Flowslide	Sudden liquefaction, rapid flow mechanism	Triggered by shock or movement; silty soils at risk	Adequate compaction/ consolidation or toe berm added
Deformation			
Settlement	Loss of freeboard; local low spots	Deformation and consolidation of dam and/or foundation; result of internal erosion etc.	Restoration of freeboard; good internal detailing to reduce risk of cracking, e.g. protective filters
Internal	External profile deformation; internal cracking	Relative deformation of zones or materials	Good detailing, with wide transition zones etc.

2.4.3 Design features and practice

The considerations summarized in Section 2.4.2 have major implications with regard to certain design features and good construction practice. Some of the more important points are outlined below, and representative examples of embankment dams illustrating modern practice are shown in Figs 2.10(a) and (b) and Fig. 2.17.

(a) Zoning of shoulder fills

The careful and correct zoning of the available materials is an important aspect of embankment design. The principles are as follows.

1. The core width should be as great as is economically viable.
2. The downstream shoulder should be underlain by a drainage blanket, or base drain, of free-draining material.
3. Finer shoulder material should be zoned closest to the core, with an intervening vertical drainage zone connecting to the base drain (see Fig. 2.10(b)).
4. Shoulder zones should be of progressively coarser material as the face slopes are approached.
5. Where a major change in the characteristics of material in adjacent zones is unavoidable, interface effects should be eased by the insertion of an intermediate or transition zone.

The permeability of successive zones should increase toward the outer slopes, materials with a high degree of inherent stability being used to enclose and support the less stable impervious core and filter. Pervious materials, if available, are generally placed in upstream sections to permit rapid porewater pressure dissipation on rapid drawdown (Section 2.7.1(a) and Fig. 2.10(b)).

The stability of an embankment and its foundation is determined by their collective ability to resist shear stresses. Embankments constructed with cohesive materials of low permeability generally have slopes flatter than those used for zoned embankments, which have free-draining outer zones supporting inner zones of less pervious fill material.

Lower quality random fill materials may be satisfactorily employed in areas within the dam profile where neither permeability nor shear strength is critical and bulk and weight are the primary requirements. Examples include the placing of stabilizing fill at the toes of embankments on low-strength foundations or so-called 'random zones' within the heart of either shoulder.

General points regarding zoning and core profile include the following.

1. A nominal hydraulic gradient through the core of the order of 1.5–2.5 is satisfactory; a value greater than 3.5–4 is undesirable.
2. Core geometry is not critical provided that the upstream core slope is not such as to control overall slope stability. It is preferable that the core be approximately central and it can ease placing of the downstream drain and transition/filter zones if that face of the core is kept vertical.
3. The dimensioning of structural zones or intermediate transition zones is governed by stability and deformation considerations. It is sometimes assisted by data from construction of special test fills.
4. A compromise must be made between sophistication of design and zoning and ease of construction. Internal zoning and associated specification requirements should be kept as simple as possible.

(b) Spillway location

Geotechnical and hydraulic design considerations require that to minimize the risk of damage to the dam under flood conditions the spillway and discharge channel are kept clear of the embankment. Spillways are therefore generally built on natural ground with the channel bypassing the flank of the dam and discharging to a stilling basin clear of the downstream toe. The alternative is to use a dropshaft-type spillway located within the reservoir and discharging via an outlet tunnel or culvert. In the latter case it is preferable to tunnel the outlet through the natural ground of the abutment wherever possible. The alternative is a concrete culvert if founded on incompressible rock. The hydraulic aspects of spillways are discussed in Chapter 4.

In a number of instances auxiliary or emergency spillways have been constructed on embankment crests using suitable proprietary grass reinforcement techniques (see Section 4.4).

(c) Freeboard

The provision necessary for long-term settlement within the overall minimum freeboard is determined by the height of dam and the depth of compressible foundation at any section. It is therefore customary to construct the crest of the dam to a longitudinal camber to accommodate the predicted consolidation settlement.

A proportion of the design freeboard is sometimes provided by construction of a continuous concrete wavewall along the upstream edge of the dam crest. This can also be done when it is necessary to uprate the freeboard of older dams following reassessment of the design flood. The overall minimum freeboard from spillway sill to dam crest (or top of a structural wavewall) should be at least 1.0m on the smallest reservoir embankment, and it will be very much greater for larger embankments and/or reservoirs (see Section 4.4).

(d) Foundation seepage control

Seepage flows and pressure within the foundation are controlled by cut-offs and by drainage. Cut-offs are impervious barriers which function as extensions of the embankment core into the foundation. They are generally located directly under the core, but can also be located a short distance upstream and connected to the core by an impervious horizontal blanket under the shoulder. The cut-off may penetrate to impervious strata (a 'fully penetrating' cut-off) or, if pervious material occurs to considerable depths, may terminate where the head loss across the cut-off is sufficient to effect the required degree of control (a 'partially penetrating' cut-off). Older cut-offs were frequently constructed as very narrow clay-filled 'puddle trenches', many proving vulnerable to hydraulic fracture (see Section 2.7.2), seepage damage and erosion. The principal variants of cut-off now employed are illustrated schematically in Fig. 2.9.

The relatively wide and shallow trench cut-off (Fig. 2.9(a)) is filled with rolled clay and forms the base to the core above. It is very effective, particularly if supplemented by a deep grout curtain, but excavation costs limit it to maximum trench depths of the order of 10–20 m even for the largest dams.

The grouted zone type cut-off, shown in Fig. 2.9(b), is now applicable to a wide range of foundation conditions due to developments in grouting technology, e.g. alluvial grouting techniques. The cut-off is formed by several parallel lines of staggered grout holes spaced at 2–3 m centres. Cement-based grouts are generally employed, but more sophisticated and costly chemical grouts are available for particularly difficult conditions. Grout cut-offs are most effective in fractured rock and in coarser-grained soils, where they may reduce permeability by one to three orders of magnitude. Cut-offs of this type have been constructed to depths in excess of 100 m. They can be installed, or improved retrospectively, by drilling through the body of a dam, but are generally relatively expensive.

The thin diaphragm-type cut-off, the result of advances made in geotechnical processes, is illustrated in Fig. 2.9(c). The cut-off is formed by excavation, in panel lengths, of a narrow slurry-stabilized trench which is subsequently backfilled with a permanent clay–sand–bentonite mix. A relatively weak and deformable, 'plastic' concrete backfill may alternatively be employed to form the impervious element. The diaphragm cut-off is very effective in alluvium and finer-grained soils, and can be constructed economically to depths exceeding 30–40 m.

Diaphragm walls of sheet piling may be driven to depths of up to 20–25 m to form a cut-off under low-head structures. The cost of this type of cut-off is moderate, but its efficiency is low unless supplemented by upstream grouting, e.g. with bentonite slurry.

Control of seepage downstream of the cut-off is also assisted by the almost universal provision of a horizontal blanket drain at ground level

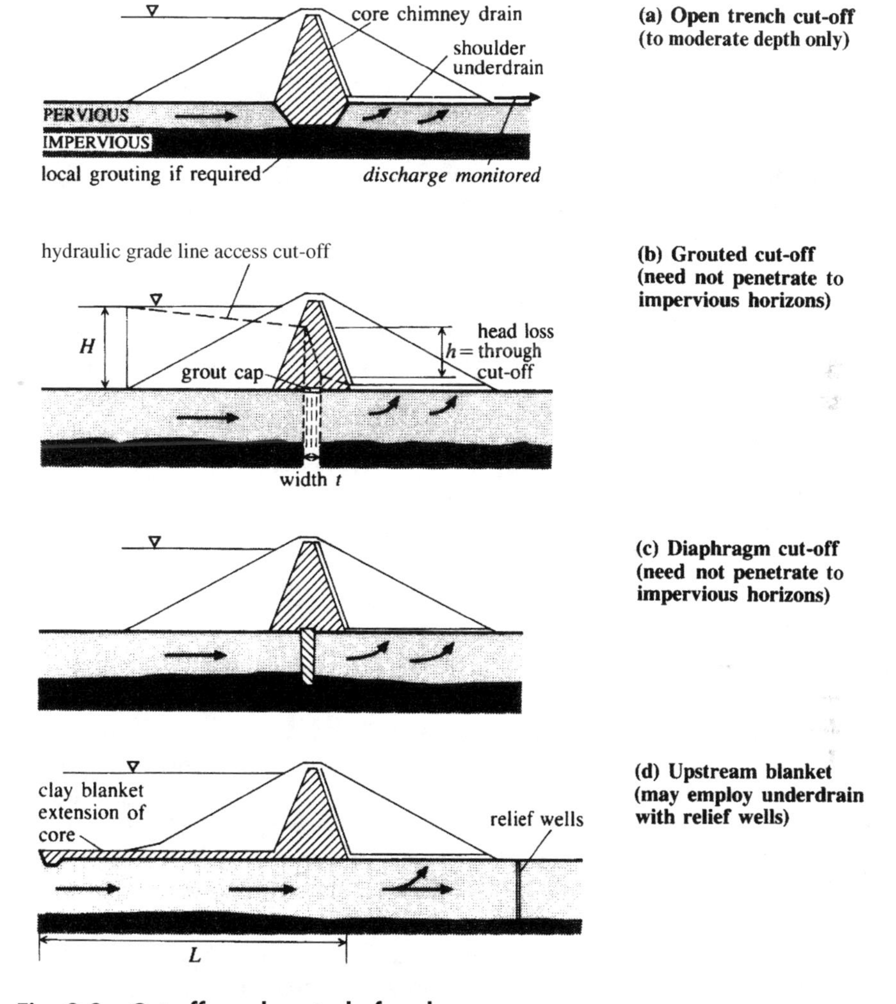

Fig. 2.9 Cut-offs and control of underseepage

under the downstream shoulder. This is frequently supplemented by deep relief wells under or near the toe. Both drainage features are identified in Figs 2.9(a)–2.9(d).

Seepage can also be moderated by a continuation of the core upstream as a horizontal impervious blanket extending over the bed of the reservoir (Fig. 2.9(d)). The blanket is carried upstream for a distance, L, sufficient to lengthen the seepage path, and hence reduce flow, to the required degree. The thickness or depth of compacted fill required may be taken as $c.\,1.0 + 0.1H$, where H (m) is the embankment height. The efficiency of an upstream blanket can be low relative to the considerable construction costs involved.

(e) Outlet works: tunnels and culverts

Outlet works (see also Section 4.8) should where practicable be constructed as a tunnel driven through the natural ground of the dam abutment. Where this is difficult or uneconomic a concrete culvert founded on rock is a satisfactory alternative, provided that care is taken to ensure that, if not founded in a shallow rock excavation (Fig. 2.17), a relatively rigid projecting culvert does not promote shear cracking of the embankment fill as the latter settles. The culvert cross-section should therefore be ogival, with transverse joints at intervals of 10–15 m and an external slip coating to assist settlement of the fill relative to the culvert. It is potentially dangerous to construct an outlet culvert on a compressible foundation within the fill itself owing to the effects of differential deformations and possible cracking.

Concrete culverts sometimes incorporate a number of external transverse seepage collars or plates at intervals along their length to inhibit preferential seepage and possible erosion at the culvert-fill interface (USBR, 1987). The effectiveness of such features can be questioned.

(f) Upstream face protection

Methods of wave prediction and the issue of possible wave overtopping, with its potential to degrade downstream slope stability, are discussed in Yarde, Banyard and Allsop (1996) (see also Section 4.4).

Several options are available for protection of the upstream face against wave erosion, ranging from traditional stone pitching with grouted joints through concrete facing slabs to the use of concrete blockwork and rock armouring. A heavy and thick protective layer is necessary between crest level and the minimum operating or drawdown level, with reduced protection typically in the form of beaching, i.e. smaller uniform rocks, provided down to the reservoir bed. Pitching is expensive and, while durable, it is not an efficient dissipator of wave energy. Concrete slabs have similar limitations, allowing considerable wave run-up which has to be included within the design freeboard. Open-jointed heavy concrete blockwork bedded on gravel and a granular filter is preferable, and is now widely employed. Where durable rock is available in large, angular sizes, rock armouring and dumped rock riprap provide the most effective protection.

The wave protection facing plus under-layer interacts with the immediately underlying embankment fill to perform as a composite system. The performance of the overall protection is particularly dependent upon the under-layer fulfilling the following functions:

- filtration, to resist movement of embankment fill by wave-induced under-layer water flows;
- drainage, to relieve wave-induced uplift forces under the facing;

- erosion protection at the interface between under-layer and embankment fill;
- separation, to prevent the protection layer penetrating the fill.

The principal characteristics of the outer face protection system which influence the under-layer are:

- flexibility, controlling the degree to which inter-layer contact can be maintained;
- permeability, determining the extent to which hydraulic conditions are transferred into the under-layer.

The size of stone required for rock armouring protection is related to the size and number of waves which the armouring is intended to withstand. These are in turn a function of sustained wind speed and time. The underlying criteria adopted for design purposes are those which give an appropriate combination of return period and possible damage to the armouring having regard to the risk category of the dam (see Section 4.2 and Table 4.1). In most cases only generalized wind data will be available (as presented for the UK in ICE (1996)); allowance should be made for enhanced wind speeds over an open unobstructed body of water, and it may be prudent to add a further margin to allow for uncertainty.

The principal parameters used in the design of rock armouring are the significant wave height, H_s, i.e. the mean height of the highest 33% of waves in a suitably extended wave train. H_s is thus a function of effective fetch, F, i.e. the unobstructed wind path determined across an arc normal to the axis of the dam, and the sustained wind speed, U, necessary to generate the wave train. (Reference should be made to Section 4.4 and Fig. 4.1 and to ICE (1996) with regard to estimation of fetch, F, and for design curves of the relationship linking predicted significant wave height to fetch and sustained wind speed, and also to Chapter 14.)

If the adjacent landforms and reservoir shape are such that a greater effective fetch oblique to the dam axis may reasonably be predicated consideration should be given to this in estimating H_s. In this event consideration should then be given to the consequent obliquity of waves breaking against the face of the dam. There is limited evidence that for a given significant wave height damage is reduced when waves impinge obliquely to the face.

The stone size required for rock armouring is normally expressed in terms of D_{50}, i.e. the mass of the median stone size. An empirical rule for the size of stone necessary for stability under wave action is given by:

$$M = 10^3 H_s^3 \qquad (2.12)$$

where M (kg) is the mass of stone required and H_s the significant wave height (m).

The overall minimum thickness specified for the rock armour layer is determined by two requirements:

1. layer thickness, normal to the face, of not less than 1.5 D_{50};
2. layer thickness to be sufficient to contain the largest rock.

An additional important requirement is that the rock armouring should be well graded to maximize stone interlocking and ensure stability. The rock armouring is protected from undermining and subsidence by the introduction of one or more underlying filter layers designed to allow free drainage beneath the armouring while preventing removal of fill or bedding material. Filter design criteria and recommended grading envelope limits for rock armouring are set out in Thompson and Shuttler (1976).

The thickness of a layer of rock armouring is generally of the order of 0.7–0.9 m for fetches up to *c*. 5 km, with maximum-size stones of *c*. 2000 kg (USBR (1987)).

Concrete blockwork face protection is generally designed by established empirical relationships. The conventional rule is that given the provision of an appropriate underlayer system and properly executed joints, a block thickness of D will be satisfactory for a significant wave height $H_s = 6D$. Later research has indicated that the permissible H_s/D ratio is dependent upon many factors, and it has been suggested (Hydraulics Research (1988)) that the H_s/D relationship is expressed by a function of the form $H_s/\delta D = \Phi$, where δ is the density of the blockwork relative to water (approximately 2.5 for concrete blocks) and Φ has a value dependent upon a number of factors including slope, wave characteristics, joint width and drainage (permeability) of the under-layer system. Stability of the blockwork is improved by narrow joint widths ($\leq$10 mm) to enhance interlocking.

Concrete slab face protection is constructed in large panels of the order of 6 m × 6 m, the narrow gaps between panels providing necessary articulation and allowing some transfer of hydraulic pressures into the under-layer. The smooth surface of such panels allows significant wave run-up (see Section 4.4), requiring an increase in overall design freeboard.

A monolithic asphaltic concrete facing, generally in the form of an upstream deck serving a dual purpose as both watertight element and face protection, offers a further alternative. It has the advantage that asphaltic concrete has the flexibility to accept local deformation without distress and therefore no articulation is necessary.

A soil-cement protective facing has been used on some dams in the USA, but long-term performance has not always proved satisfactory.

For further information on face protection see also Chapter 14 and Section 15.6

Face protection is discussed in depth in Thomas (1976), Thompson and Shuttler (1976), ICOLD (1993a) and in Besley *et al.* (1999). A comprehensive treatment of rock armouring and rip-rap is given in CIRIA, CUR, CETMEF (2005).

(g) Embankment crest

The crest should have a width of not less than 5 m, and should carry a surfaced and well-drained access road. (With old dams the latter provides valuable resistance to the erosive effect of occasional limited and short-term overtopping.)

2.5 Materials and construction

2.5.1 Earthfill materials

Three principal categories of fill material are necessary for earthfill embankment dams to fulfil the requirements for core, shoulder and drainage blankets, filters, etc.

Core fill should have low permeability and ideally be of intermediate to high plasticity to accommodate a limited degree of deformation without risk of cracking (Section 2.7.5). It is not necessary, and possibly disadvantageous, to have high shear strength. The most suitable soils have clay contents in excess of 25–30%, e.g. glacial tills etc., although clayey sands and silts can also be utilized. The core is the key element in an embankment and the most demanding in terms of material characteristics and uniformity, the properties of the compacted clay core being critical to long-term watertight integrity. The principal performance characteristics of the more important groups of soils suitable for rolled cores are summarized in Table 2.5.

Representative values for the more important engineering parameters of compacted earthfills, to be utilized in core or shoulders as appropriate, are summarized in Table 2.6. Note that the comparable parameters for compacted rockfill are also tabulated.

Shoulder fill requires sufficiently high shear strength to permit the economic construction of stable slopes of the steepest possible slope angle. It is preferable that the fill has relatively high permeability to assist in dissipating porewater pressures. Suitable materials range across the spectrum from coarse granular material to fills which may differ little from the core materials. The shoulder need not be homogeneous; it is customary to utilize different fills which are available within predetermined zones within the shoulders (rockfill for shoulders is discussed briefly in Section 2.9).

Drain/filter material must be clean, free draining and not liable to chemical degradation. Processed fine natural gravels, crushed rock and coarse to medium sands are suitable, and are used in sequences and gradings determined by the nature of the adjacent core and/or shoulder fills. The cost of processed filter materials is relatively high, and the requirement is therefore kept to a minimum. The properties of core materials are further discussed in McKenna (1989).

Table 2.5 Characteristics of core soils

Soil description (BS 5930)	*Cracking resistance*	*Erosion–piping resistance*	*Optimum compaction roller*	*Sensitivity to compaction water content control*
Very silty sands or gravels; 6% clay (GM–SM)	Low; increases with $<\rho_d$ and $>I_p$	Low; increases with $<\rho_d$ and $>I_p$	Pneumatic tyred (20–80 t)	High to avoid brittleness
Very clayey sands or gravels; 20% clay (GC–SC)	Intermediate at representative ρ_d	Intermediate	Pneumatic tyred (20–80 t)	Low to control u_w
Low-plasticity clays (CL)	Relatively flexible	High; increase with $>\rho_d$	Pneumatic tyred or sheepsfoot	Intermediate to high to control u_w
High-plasticity clays (CH)	Flexible; can resist large deformations	High; increase with $>\rho_d$	Sheepsfoot	High to control u_w

Table 2.6 Indicative engineering properties for compacted earthfills (compare with Table 2.3)

Fill type (BS 5930)	*Compaction characteristics*		*Shear strength (effective stress)*		*Coefficient of compressibility, m_v ($\times 10^{-4} m^2 kN^{-1}$)*	*Coefficient of horizontal permeability, $k_h(ms^{-1})$*	*Drainage characteristics (relief of u_w)*
	Unit weight, $\gamma_{d\,max}$ (kNm^{-3})	*Water w_{opt} (%)*	*Cohesion, c' (kNm^{-2})*	*Friction, ϕ' (degrees)*			
Gravels (GW–GC)	18–22	5–10	0	35–40	0.1–1.0	10^{-3}–10^{-5}	excellent
Sands (SW–SP)	16–20	10–20	0	35–40	0.5–1.5	10^{-4}–10^{-6}	good → fair
Silts (ML–MH)	16–20	15–30	<10	25–35	0.5–2.5	10^{-5}–10^{-8}	fair → poor
Clays (CL–CH)	16–21	15–30	<20	20–30	0.5–3.0	10^{-7}–10^{-10}	very poor → impervious
Crushed rockfill (2–600 mm size range)	17–21	N/A	0	40–55	N/A	10^{-1}–10^{-2}	free-draining: excellent

Crushed rockfill is shown here for comparative purposes only – refer to Section 2.9.

2.5.2 Construction

The construction operations which follow initial site development fall into four principal groups of activities, relating to (1) material source development, (2) foundation preparation and construction, (3) fill construction and control and (4) ancillary works construction.

Material source development activities involve the opening out of borrow areas or quarries, including the installation of fixed plant, e.g. crushers, conveyors, etc. Access and haulage roads are also constructed between the various borrow areas and the embankment site, and excavation and haulage plant is mobilized.

Foundation preparation activities, including river diversion, can proceed concurrently with the development of the fill sources. Temporary river diversion is commonly effected by driving a flanking tunnel, which in most cases subsequently houses the outlet works. (Where an outlet culvert through or under the embankment is planned rather than a flanking tunnel the culvert may be used temporarily for river diversion purposes.)

Topsoil and weathered surface drift deposits etc. are removed. In the case of a soft, compressible foundation, strength can be enhanced and construction accelerated by preconsolidation and/or the installation of sand drains, as for the Derwent dam, constructed on a difficult site in the UK (Ruffle, 1970). Foundation instrumentation is also installed at this stage to monitor pore pressures and cut-off performance (Chapter 7). Foundation construction is completed with the laying of the drainage blankets which will underlie the downstream shoulder.

Fill construction is an exercise in efficient plant utilization within the terms of the specification requirements as to materials compliance and compaction technique. Placing operations may be subject to the influence of weather conditions and to subtle changes in material characteristics. Control of placing is centred upon supervision of water content, layer thickness and compaction procedure. The quality and uniformity of the compacted core fill are critical. Recent UK practice is to employ a statistical approach to test-ing and quality control and to require that the undrained shear strength, c_u, should lie within specific limits. A specification requirement of $c_u = 55–100\,kN\,m^{-2}$ is typical for UK clays (Kennard *et al.*, 1979). Control of construction porewater pressures and acceleration of consolidation in cohesive fill materials of low permeability can require the installation of horizontal blanket drains in both shoulders, at vertical intervals of 3–6 m (Fig. 2.10(a) and (b); Gibson and Shefford (1968)). Knight (1989) reviews aspects of fill construction and control in terms of recent practice on a number of projects.

The installation of instrumentation in the core and shoulders proceeds in parallel with the placing of fill. Fill construction is concluded with the completion of upstream rock armouring or other face protection works.

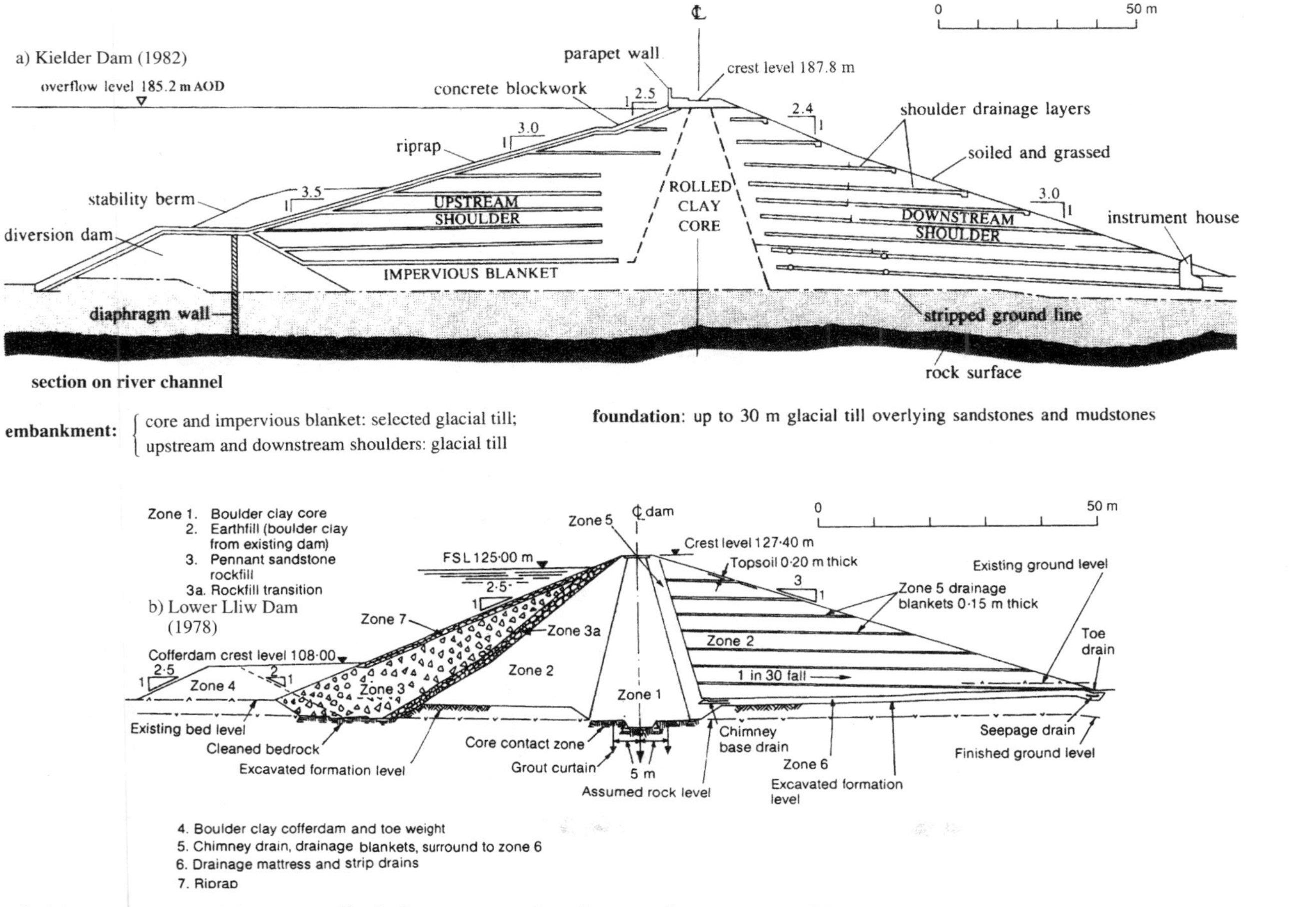

Fig. 2.10 Representative UK rolled clay core embankment dams: (a) Kielder (1982); (b) Lower Lliw (1978) (after Binnie, 1981)

Ancillary works construction embraces the construction of spillway and stilling basins, culverts or tunnels for outlet works etc., valve towers and similar control works. It also includes completion of crest details, e.g. roadway, drainage works, wavewall etc. and, where climatic conditions allow, grassing of the downstream face slope (Section 4.4).

Embankment construction practice, including planning and control, is further discussed in Thomas (1976), Wilson and Marsal (1979), USBR (1987), Jansen (1988) and Fell, MacGregor and Stapledon (1992). Two recent projects in the UK are described in greater depth in Coats and Rocke (1982) (Fig. 2.10a); for geotechnical details see Millmore and McNicol (1983)) and Bridle, Vaughan and Jones (1985). Recent construction practice is also described in Banyard, Coxon and Johnston (1992.)

2.6 Seepage analysis

2.6.1 Seepage

The phreatic surface of the seepage régime, i.e. the free surface, must be kept well clear of the downstream face to avoid high porewater pressures which may promote slope instability (Section 2.7). In the extreme case of the seepage line emerging on the face, local softening and erosion will occur and may initiate sloughing as a prelude to instability. Seepage pressures and velocities must also be controlled to prevent internal erosion and particle migration, particularly from the core. Seepage control is effected by the incorporation of vertical chimney drains and horizontal drainage layers, protected by suitable filters and transition layers.

In this section a basic knowledge of seepage theory and flownet construction is assumed, including entry and exit conditions, as provided in the soil mechanics texts referred to in Section 2.1. A more exhaustive general discussion of flownets and seepage analysis is provided in Cedergren (1977). Embankment flownets in particular are considered in depth, with numerous illustrations in Cedergren (1973). Embankment seepage control is also discussed in Volpe and Kelly (1985).

The fundamental relationships applicable to flownets for two-dimensional seepage are summarized below.

For anisotropic soils, with the coefficient of horizontal permeability $k_h > k_v$, the coefficient of vertical permeability, the horizontal scale transform factor, λ, and the effective permeability, k', are respectively given by

$$\lambda = (k_v/k_h)^{1/2} \tag{2.13}$$

and

$$k' = (k_v k_h)^{1/2}. \qquad (2.14)$$

The seepage flow, q (equation (2.10) should be referred to), is defined by

$$q = k'H\frac{N_f}{N_d} \qquad (2.15)$$

where H is the head differential and the ratio N_f/N_d is the flownet shape factor, i.e. the number of flow channels, N_f, in relation to the number of decrements in potential, N_d (N_f and N_d need not be integers).

For the unconfined flow situation applicable to seepage *through* a homogeneous embankment the phreatic surface is essentially parabolic. The curve can be constructed by the Casagrande–Kozeny approximation, defined in the references previously detailed, or from interpretation of piezometric data (Casagrande, 1961). In the case of a central core and/or zoned embankment, construction of the flownet is based upon consideration of the relative permeability of each element and application of the continuity equation:

$$q_{\text{upstream shoulder}} = q_{\text{core zone}} = q_{\text{downstream shoulder}} + q_{\text{drains}}. \qquad (2.16)$$

An illustrative flownet for seepage *under* an embankment is given in Fig. 2.11. (The figure forms the transform scale solution to worked example 2.1.) In Fig. 2.12 is shown the flownet for a simple upstream core two-zone profile, where piezometric data have been interpreted to define the phreatic surface within the core (Fig. 2.12 is the flownet solution to worked example 2.2).

The thickness of horizontal blanket drain, t_d, required to discharge the seepage flow and shown in Fig. 2.12 can be estimated from

$$t_d = (qL/k_d)^{1/2} \simeq 1.5\,H(k_c/k_d)^{1/2} \qquad (2.17)$$

where L is the downstream shoulder width at drain level and k_d and k_c are the drain and core permeabilities respectively (the factor 1.5 in equation (2.17) is derived from a representative embankment geometry).

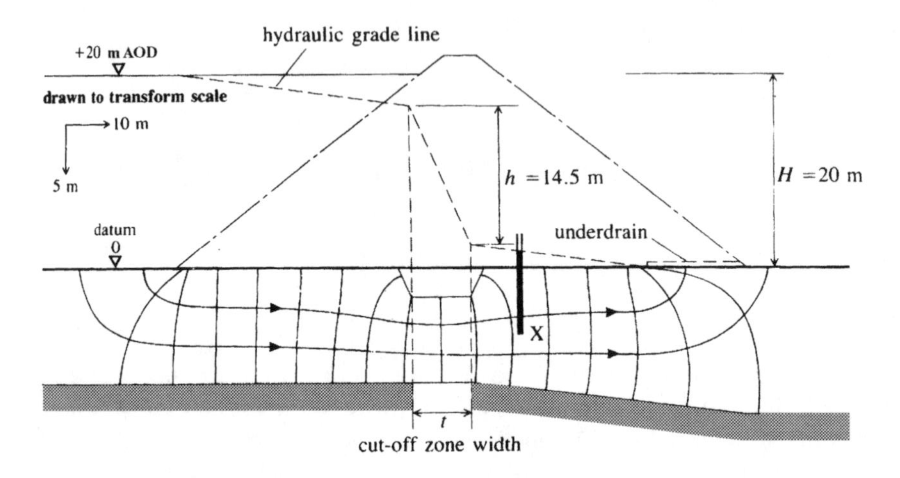

Fig. 2.11 Idealized flownet for foundation seepage (Worked example 2.1 should also be referred to)

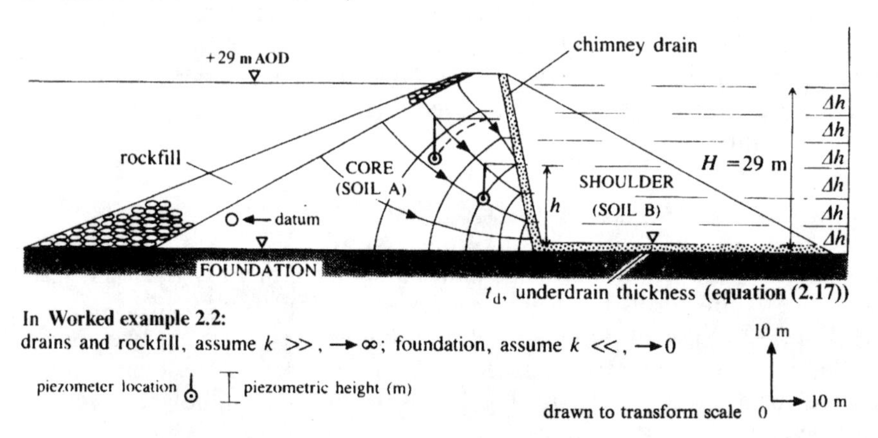

Fig. 2.12 Flownet for internal seepage in embankment core (Worked example 2.2 should also be referred to)

2.6.2 Core and cut-off efficiency

The effectiveness of a core or cut-off may be empirically defined in terms of two criteria (Telling, Menzies and Simons, 1978):

1. *head efficiency*,

$$E_H = h/H \tag{2.18}$$

where h is the head loss across the core or cut-off and H is the overall differential head, as shown in Figs 2.9(b) and 2.11;

2. *flow efficiency*,

$$E_Q = 1 - Q/Q_0 \tag{2.19}$$

where Q and Q_0 are, respectively, the seepage flows with and without the core or cut-off in position.

E_H may be determined from the piezometric levels upstream and downstream of core or cut-off, and E_Q from flow measurements. Both may be approximated from flownet studies (Section 2.6.1).

For the special case of a fully penetrating cut-off and a parallel boundary flow régime efficiencies E_H and E_Q are related thus:

$$\frac{E_H}{E_Q} = \frac{k_1}{k_1 - k_2} \tag{2.20}$$

where k_1 and k_2 are the coefficients of permeability of natural foundation and cut-off zone respectively. Efficiencies E_H and E_Q recorded for well-constructed and effective cut-offs are normally both in excess of 50–60%.

2.6.3 Filter design

The design of filters and transition layers to prevent seepage-induced migration of fines is discussed in soil mechanics texts and, in greater detail, in Mitchell (1983) and Sherard and Dunnigan (1985). Reference may also be made to ICOLD Bulletin 95 (ICOLD, 1994). They are required to be sufficiently fine to prevent migration of the protected soil (piping criterion) while being sufficiently permeable to freely discharge seepage (permeability criterion).

The essential principle of design is that any change from fine to coarse material must be effected gradually in staged filter or transition zones, e.g. clay core → sand → coarse sand → pea gravel → coarse shoulder, i.e.

[protected soil → $\underbrace{\text{transition or filters}\rightarrow}_{\text{multistage if required}}$ drain]

A widely employed empirical approach to defining appropriate filter material grading envelopes is given in the form of ratios for specified particle passing sizes as typified by the expressions

$$\frac{D_{15}(\text{filter})}{D_{85}(\text{soil})} \leqslant 5, \tag{2.21a}$$

$$\frac{D_{15}(\text{filter})}{D_{15}(\text{soil})} \geqslant 5 \tag{2.21b}$$

and

$$\frac{D_{50}(\text{filter})}{D_{50}(\text{soil})} \leqslant 25 \tag{2.21c}$$

where D_{15} etc. refer to the 15% passing size etc. as determined from particle size analysis. Expressions (2.21a) and (2.21b) set out piping and permeability criteria respectively; expression (2.21c) further defines permeability ratio.

A recent development is the suggestion that, in view of the potential problem of hydraulic fracturing and cracking etc., with risk of progressive erosion (Section 2.7.2), rational filter and transition design should be based on considerations of relative permeability (Vaughan and Soares, 1982). This approach introduces the concept of basing specification of the filter to protect a clay core on considerations of seepage water chemistry and clay floc or particle size in relation to filter void size and effectiveness. It is considered to define the way ahead in filter design for the protection of very fine soils such as clays, the empirical approach based on particle size ratios having proved unsafe in a number of instances (Vaughan *et al.*, 1970).

2.7 Stability and stress

2.7.1 Stability analyses

Embankment dam stability must be assessed in relation to the changing conditions of loading and seepage régime which develop from construction through first impounding into operational service, including reservoir drawdown. The slope stability analyses generally employed are detailed in the soil mechanics texts referred to in Section 2.1. In this section a basic understanding of established limit-equilibrium methods of two-dimensional stability analysis is assumed, and only a brief appreciation of certain fundamental points is given below. (Three-dimensional limit-equilibrium techniques have been developed, but they have not won wide acceptance for a number of reasons and are not considered further.)

Two-dimensional limit-equilibrium analysis is based on consideration of the static equilibrium of the potentially unstable and 'active' mass of soil overlying a conjectural failure surface. The factor of safety, F, is defined by

$$F = \Sigma\tau_f / \Sigma\tau \tag{2.22}$$

where τ_f and τ are, respectively, the unit shear resistance which can be mobilized and unit shear stress generated on the failure surface. The analysis is applied to all conceivable failure surfaces, and the supposed minimum factor of safety F_{min} is sought.

Stability is very sensitive to u_w, which must be estimated from a flownet or predicted on the basis of the pore pressure coefficients (Section 2.3.1) in the absence of field data. It is therefore sometimes more convenient in analysis to consider porewater pressures in terms of the dimensionless pore pressure ratio, r_u:

$$r_u = u_w/\gamma z \tag{2.23}$$

where z is the depth below ground surface and γz the local vertical geostatic stress.

Parameter r_u may effectively equate to $\bar{B}$ (equation (2.5)) in the case of a saturated fill. The value of r_u can often be taken as sensibly uniform within a cohesive downstream shoulder, and equilibrium reservoir full values will typically lie in the range 0.10–0.30. The initial porewater pressures generated in a cohesive fill develop as a result of the construction process itself, i.e. overburden load and plant loads. Dissipation rate is a time-dependent function of permeability and drainage path length (i.e. slope geometry). Construction porewater pressures are partially dissipated prior to impounding, after which they progressively stabilize to correspond to the advancing seepage front and, ultimately, the reservoir full steady-state seepage or other, varying, operational conditions (illustrated in Fig. 2.14). Abdul Hussain, Hari Prasad and Kashyap (2004) presents the results of a study on modelling pore water pressures and their influence on stability.

The form of the critical failure surface for F_{min} is controlled by many factors, including soil type and the presence of discontinuities or interfaces, e.g. between soft soil and rock. A number of failure surfaces representative of different embankment and/or foundation situations are illustrated schematically in Fig. 2.13. For most initial analyses involving relatively homogeneous and uniform cohesive soils, two-dimensional circular arc failure surfaces are assumed. The probable locus of the centre of the critical circle in such cases, with $r_u < 0.3$, can be approximated by

$$z_c = H \cot \beta (0.6 + 2 \tan \phi') \tag{2.24a}$$

and

$$y_c = H \cot \beta (0.6 - \tan \phi') \tag{2.24b}$$

where z_c and y_c are coordinates with respect to the toe, measured positive upwards and into the slope respectively, H is the height and β is the slope angle.

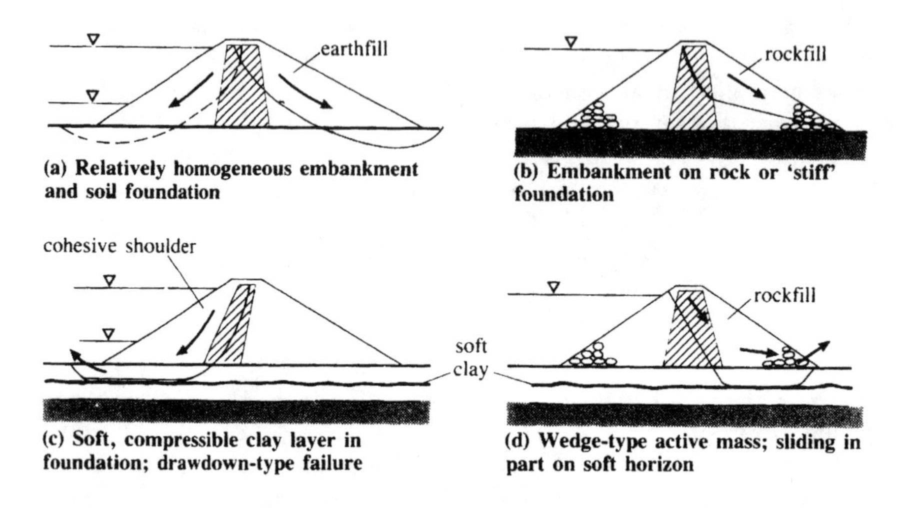

Fig. 2.13 Stability analysis: failure surface schematics

The following critical conditions must be analysed (see Fig. 2.14):

1. end of construction (both slopes);
2. steady state, reservoir full (downstream slope critical);
3. rapid drawdown (upstream slope critical);
4. seismic loading additional to 1, 2 and 3, if appropriate to the location.

Analysis is in terms of effective stress shear strength parameters c' and ϕ', with porewater pressures, u_w, or the pore-pressure ratio, r_u. Total stress parameters, c and ϕ, are suitable only for a short-term and approximate analysis, e.g. stability at an intermediate stage during construction.

Design parameters to be employed in stability analysis may be summarized as follows.

- *During and at end of construction.* Earthfill is compacted in a partially saturated state, i.e. initial values of u_w are negative. For an upper bound, and assuming no dissipation of porewater pressure, pore-pressure ratio $r_u = \bar{B}$, with pore pressure coefficient $\bar{B}$ (Section 2.3.1) set by the stress state within the dam. If high values of r_u are anticipated excess porewater pressures may be relieved by horizontal drainage layers (Fig. 2.10). The design and spacing of such layers, which should have a permeability of 10^5–10^6 times that of the fill, is discussed in Gibson and Shefford (1968).
- *Steady-state seepage, reservoir full.* Effective stress analysis should always be employed. Values of r_u in excess of 0.45 may occur in old homogeneous clay dams; effective internal drainage blankets may reduce r_u values to 0.10–0.20.

- *Rapid drawdown.* Values of r_u in the range of 0.30–0.60 may occur immediately after initial drawdown. Transient values of u_w may be estimated from flownets drawn for intermediate positions of the drawdown phreatic surface. (The complex response of an old embankment dam during drawdown is described in Holton, Tedd and Charles (1989))

Values of F_{min} determined in a comprehensive stability analysis must always be regarded as relative and not as absolute. The expressions determining F are of varying rigour and are inexact, a reflection of the complexity of the stability problem, where measured shear strength parameters can be subject to a variance of up to 30–40%. Economic considerations require acceptance of relatively low values of F for embankment slopes.

The factors of safety for embankment dam slopes are time dependent, varying significantly in accordance with changes in loading corresponding to construction and the subsequent operational cycle. This is illustrated schematically in Fig. 2.14. Representative guideline values of F_{min} corresponding to the major loading conditions are presented in Table 2.7. Values considered acceptable must always reflect the confidence attaching to geotechnical data and design parameters.

The expressions for F corresponding to the analytical methods most commonly employed are as follows.

(a) Swedish circle (Fellenius) solution: circular arc surface

$$F = \frac{c' L_a + \tan \phi' \Sigma (W \cos \alpha - u_w l)}{\Sigma W \sin \alpha} \tag{2.25}$$

where L_a is the overall length of the failure surface, W and l are, respectively, the weight and base length of the slices into which the active mass is subdivided for analysis and α is the angle of inclination of the slice base to

Table 2.7 Guideline factors of safety: effective stress stability analysis

Design loading condition	*Factor of safety, F_{min}*	
	Downstream slope	*Upstream slope*
(1) Under construction; end of construction	1.25	1.25
(2) Long-term operational; reservoir full	1.5	1.5
(3) Rapid drawdown	–	1.2
(4) Seismic loading with 1, 2 or 3 above	1.1	1.1

1. The above values **must** be interpreted in the context of the particular case considered, e.g. uncertainties regarding any of the principal parameters, u_w, c' and ϕ'.
2. Higher design minima are appropriate for analyses based on total stress parameters or on peak strengths in the case of more brittle soils.

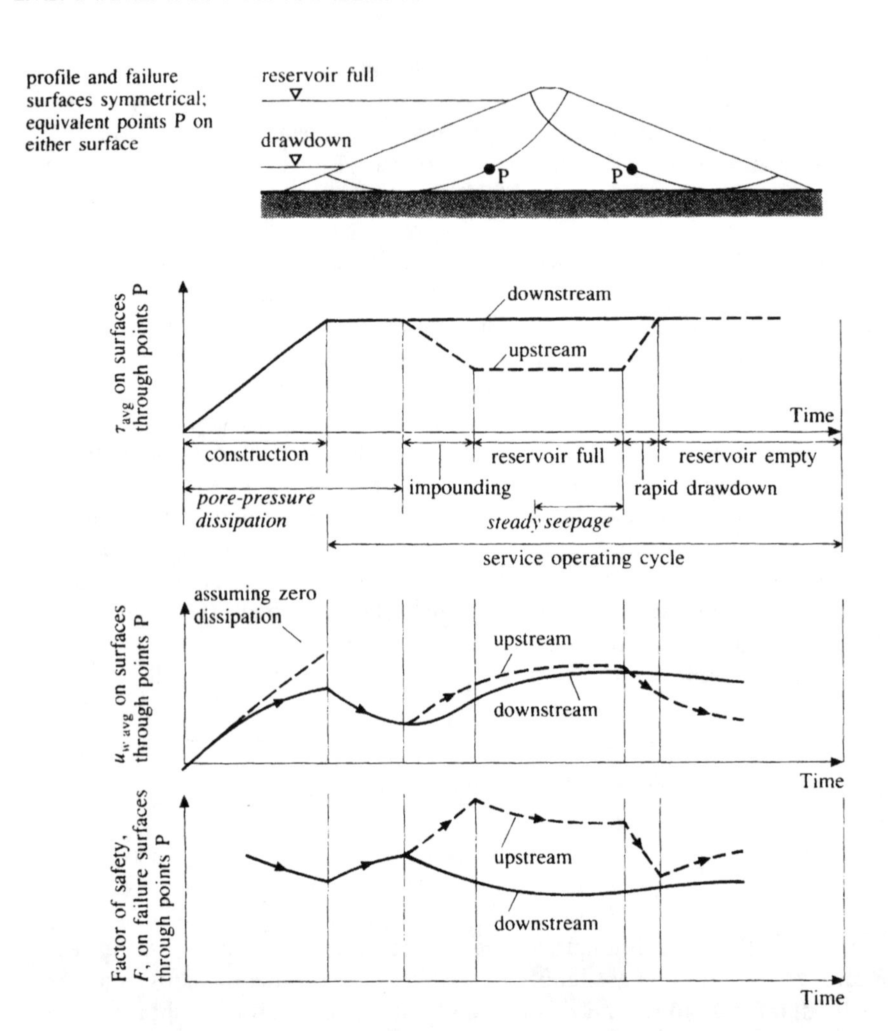

Fig. 2.14 Schematic of the variation of embankment stability parameters during construction and operation (after Bishop and Bjerrum, 1960)

the horizontal. The latter is considered positive if upslope from the lowest point on the failure arc. The Fellenius solution is, in practice, conservative by comparison with more rigorous analyses, and may underestimate F by 5–15%. This margin is generally unacceptable in terms of the cost implications.

(b) Bishop semi-rigorous solution: circular arc surface

The Bishop semi-rigorous solution (Bishop, 1955) differs from the Fellenius solution in the assumptions made with regard to the interslice forces required for static equilibrium:

$$F = \frac{1}{\Sigma W \sin \alpha} \Sigma \left\{ [c'b + (W - u_w b) \tan \phi'] \frac{\sec \alpha}{1 + (\tan \alpha \tan \phi')/F} \right\}. \quad (2.26a)$$

In the above iterative expression b is the width of any slice. Alternatively, expressing porewater pressure u_w in terms of predicted pore pressure ratio, r_u, for convenience in initial analysis, with $r_u = u_w/\gamma z = u_w b/W$ for any slice:

$$F = \frac{1}{\Sigma W \sin \alpha} \Sigma \left\{ [c'b + W(1 - r_u) \tan \phi'] \frac{\sec \alpha}{1 + (\tan \alpha \tan \phi')/F} \right\}. \quad (2.26b)$$

On the assumption of a saturated fill, the further substitution of $\bar{B}$ for r_u may be made in equation (2.26b).

In applying this method an appropriate trial value of F is first selected, subsequent iteration resulting in the expression converging rapidly to a solution. The Bishop expression may, with discretion, be applied to a non-circular arc failure surface, as shown in Fig. 2.15, which refers also to worked example 2.3. Charts of $m_\alpha = \cos \alpha [1 + (\tan \alpha \tan \phi')/F]$ for use with equation (2.26) are presented in Fig. 2.16.

More exhaustive circular arc analyses include the Bishop rigorous solution (Bishop, 1955): solutions to the analysis of non-circular and irregular failure surfaces are provided in Janbu (1973) and by Morgenstern and Price (1965). Parametric initial studies of the stability of homogeneous shoulders can be made using stability charts (Bishop and Morgenstern, 1960; O'Connor and Mitchell, 1977). Stability charts for rapid drawdown analysis are presented in Morgenstern (1963).

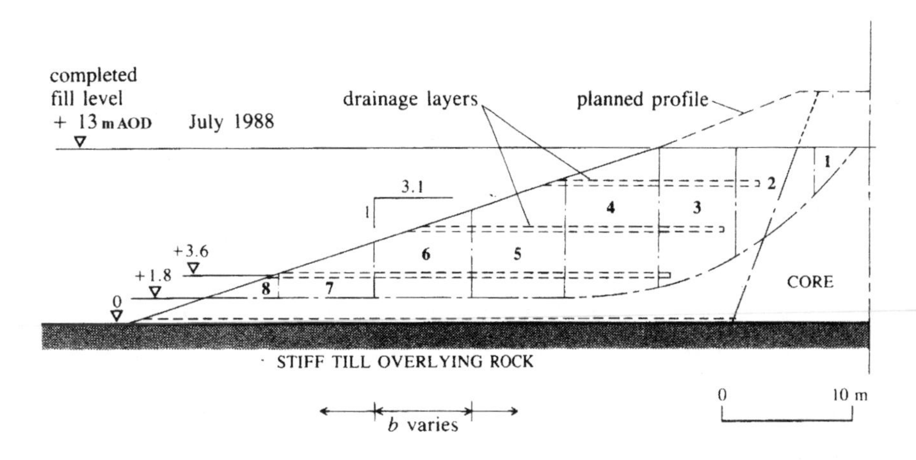

Fig. 2.15 Stability analysis: non-circular arc failure surface (Worked example 2.3 should also be referred to)

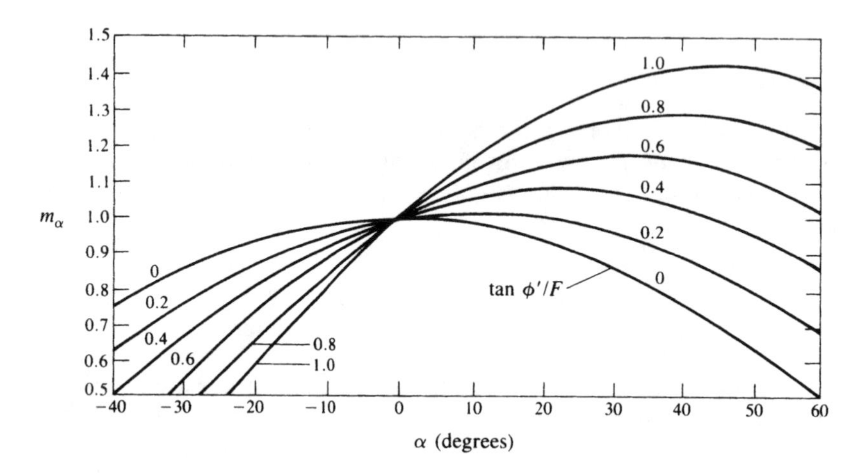

Fig. 2.16 Stability analysis (Bishop semi-rigorous method): curves of function $m_\alpha = \cos\alpha[1 + (\tan\alpha \tan\phi')/F]$

Kennard (1994) suggests that limit-state stability analysis using effective stresses predicted on pore pressure dissipation and consolidation requires the discipline of:

- conservative shear strength parameters, c' and ϕ';
- appropriate assumptions with respect to pore pressure;
- rigorous analysis of realistic slip surfaces, both circular and non-circular;
- design revision when performance data from the construction stage is available;
- realistic target factors of safety;
- the use of appropriate instrumentation (see Chapter 7).

An interesting comparison of conventional limit-equilibrium analysis and numerical analysis for a simplified enbankment profile has been published in Nandi, Dutta and Roychowdhury (2005). Vaughan, Kovacevic and Ridley (2002) have published an interesting perspective on the long-term issue of the effect of climate change on embankment stability.

2.7.2 Stress analysis, hydraulic fracturing and cracking

(a) Stress analysis

The application of sophisticated mathematical modelling techniques to embankment dam stress analysis is inhibited by difficulties occasioned by non-uniformity of the fill materials and by representation of their non-linear load response. Modelling techniques well established in other applications, e.g. finite element analysis (FEA), are therefore not widely employed other than for more rigorous and specific design studies. The principles of finite element modelling are referred to very briefly in Section 3.2.8, in the context of concrete dam stress analysis.

The application of mathematical modelling to the study of specific areas of behaviour, notably internal cracking and progressive failure in embankments, is addressed comprehensively in Vaughan, Dounias and Potts (1989), in Potts, Dounias and Vaughan (1990), and in Dounias, Potts and Vaughan (1996). The first also addresses the influence of core geometry, e.g. vertical or inclined, slender or wide, on embankment dam behaviour. All are also relevant to more general considerations of hydraulic fracturing and cracking, leading to potentially catastrophic internal erosion, as exemplified by problems which occurred at recently commissioned dams at Hyttejuvet (Finland) and Balderhead (UK). The latter instance is discussed in Vaughan *et al.* (1970).

(b) Hydraulic fracturing

The analysis of internal stresses is generally restricted to an assessment of the risk of hydraulic fracturing or of internal cracking, e.g. as a result of interface effects attributable to load transfer and strain incompatibility at zone boundaries. This is addressed in some depth in Kulhawy and Gurtowski (1976).

Hydraulic fracturing, with the risk of consequent internal erosion and migration of fines, may initiate if the total stress, σ_n, normal to any plane within a soil mass is less than the local porewater pressure u_w, making allowance for the limited tensile strength, σ_t, of the soil. The condition for fracturing to occur is thus $u_w > \sigma_n + \sigma_t$ on any internal plane. Given that such a fracture initiates in the core the factors critical to integrity are firstly whether the fracture propagates through the core and, secondly, whether seepage velocities etc. are such that ongoing erosion takes place. Development of the erosion process will depend upon whether the fracture self-heals and/or the ability of the material immediately downstream to 'trap' fines migrating from the core. Erodibility of soils is discussed in Atkinson, Charles and Mhach (1990) and in Sherard and Dunnigan (1985).

A transverse vertical plane can be shown to be the critical orientation within the core, and transverse fracturing on such a vertical plane will take place if, neglecting σ_t, $u_w > \sigma_{ha}$, where σ_{ha} is the axial horizontal total stress, i.e. parallel to the dam axis. The other critical plane is the transverse horizontal plane. Fracturing will develop on the latter orientation if vertical total stress, σ_v, falls below u_w as a result of arching of a compressible core as a result of load transfer to relatively incompressible (e.g. rockfill) shoulders. Mobilization of core shear strength and cracking is associated with consolidation of a clay core supported by relatively incompressible granular shoulders. Assuming that the full undrained shear strength of the core is mobilized in load transfer, it can be shown that the nominal vertical total stress σ_v at depth z below crest in a core of width $2a$ is given by

$$\sigma_v = z\left(\gamma - \frac{c_u}{a}\right). \tag{2.27}$$

The corresponding horizontal total stress, σ_h, is given by

$$\sigma_h = K_0(\sigma_v - u_w) + u_w \tag{2.28}$$

where K_0 is the 'at rest' coefficient of lateral earth pressure for zero lateral strain.

Most of the many recorded instances of hydraulic fracturing have occurred on first impounding or shortly thereafter. The phenomenon has been discussed by Vaughan *et al.* (1970), Sherard (1973, 1985) and Lun (1985). The risk of fracturing is moderated if a wide and relatively plastic core zone of low to intermediate shear strength is employed, and if core and shoulder are separated by a transition or filter zone. Evidence as to the influence of core water content and other factors on susceptibility to fracturing is contradictory. A laboratory simulation of fracturing and study of the influence of certain parameters for a specific soil type is presented in Medeiros and Moffat (1995).

(c) Cracking

Cracking other than by hydraulic fracturing is generally associated with strain incompatibilities, e.g. at interfaces. Potentially dangerous transverse or longitudinal cracking modes can develop from a number of causes:

1. shear displacements relative to very steep rock abutments or to badly detailed culverts;
2. differential strains and/or local arching across irregular foundations;
3. vertical steps along the axis of a deep cut-off trench;
4. progressive slope deformation and distress;
5. strain incompatibility at zone interfaces within the dam.

It has been suggested that the risk of serious internal cracking due to 5 will be of increased significance for earthfill embankments if the ratio of maximum post-construction settlement to $H \cot \beta$, where H is the height of embankment and β is the slope angle, exceeds 0.003–0.005 (Justo, 1973).

The risk of such cracking modes arising can be moderated by careful detail design, e.g. avoidance of major irregularities in foundation excavations, use of wetter and more plastic core materials, and careful zoning, with wide transitions adjacent to the core.

2.7.3 Seismicity and seismic load effects

SEISMICITY

Dynamic loads generated by seismic disturbances must be considered in the design of all major dams situated in recognized seismic 'high-risk' regions. The possibility of seismic activity should also be considered for dams located outside those regions, particularly where sited in close proximity to potentially active geological fault complexes.

The sites for major dams are normally subjected to rigorous seismological appraisal. Seismicity is assessed through a specialist review of regional and local geology in conjunction with historical evidence. Where a risk of seismic activity is confirmed, estimates of probable maximum intensity provide the basis for selecting seismic design parameters. In the case of smaller concrete dams, or dams on 'low-risk' sites, it is generally sufficient to specify a nominal level of disturbance for design purposes. Seismic risk to UK dams is reviewed in Charles *et al.* (1991) and ICE (1998).

Seismic activity is associated with complex oscillating patterns of accelerations and ground motions, which generate transient dynamic loads due to the inertia of the dam and the retained body of water. Horizontal and vertical accelerations are not equal, the former being of greater intensity. For design purposes both should be considered operative in the sense least favourable to stability of the dam. Horizontal accelerations are therefore assumed to operate normal to the axis of the dam. Under reservoir full conditions the most adverse seismic loading will then occur when a ground shock is associated with:

1. horizontal foundation acceleration operating upstream, and
2. vertical foundation acceleration operating downwards.

As a result of 1, inertia effects will generate an additional hydrodynamic water load acting downstream, plus a further inertia load attributable to the mass of the dam and also acting in a downstream sense. Foundation acceleration downwards, 2 above, will effectively reduce the mass of the structure. The more important recurring seismic shock waves have a frequency in the range 1–10 Hz. Seismic loads consequently oscillate very rapidly and are transient in their effect.

The strength of a seismic event can be characterized by its *magnitude* and its *intensity*, defined thus:

magnitude: a measure of the energy released; it therefore has a single value for a specific seismic event. It is categorized on the Richter scale, ranging upwards from M = 1.0 to M = 9.0.

intensity: a measure of the violence of seismic shaking attaching to an event, and hence of its destructiveness, at a specific location. Intensity thus varies with position and distance from the epicentre, and is commonly expressed on the modified Mercalli scale of MM = I to MM = XII.

The terminology associated with seismic safety evaluation includes a range of definitions, some of which are especially significant in the context of dams, thus:

Maximum Credible Earthquake (MCE): the event predicted to produce the most severe level of ground motion possible for the geological circumstances of a specific site.

Safety Evaluation Earthquake (SEE): the event predicted to produce the most severe level of ground motion against which the safety of a dam from catastrophic failure must be assured.

SEE may be defined as a proportion of the MCE or equal to it; an alternative is to specify SEE on the basis of a notional return period (cf. flood categorization in spillway design, Section 4.2). In recommended UK practice (Charles *et al.*, 1991) SEE takes the place of MCE as employed in US practice, e.g. USBR (1987).

Other terms employed in seismic design include Controlling Maximum Earthquake (CME), Maximum Design Earthquake (DBE) and Operating Basis Earthquake (OBE). These are not referred to further in this text.

Ground motions associated with earthquakes can be characterized in terms of acceleration, velocity or displacement. Only *peak ground acceleration*, PGA, generally expressed as a proportion of gravitational acceleration, g, is considered here. PGA can be rather imprecisely correlated with intensity, as indicated by Table 2.7. It has been suggested that in general terms seismic events with a high PGA of short duration are less destructive than events of lower PGA and greater duration.

Most energy released as ground motion is transmitted in the sub-10 Hz frequency band. An elastic structure having a natural frequency, f_n, within that range is therefore potentially at risk. Embankment dams will generally be in this category, since the theoretical natural frequency for an elastic long and uniform triangular embankment section is a function of height, H, and shear velocity, V_s. For larger embankments it can be calculated that natural frequency will be of the order of $f_n = 1$ Hz and above, thus registering in the frequency band of most seismic energy release. This has been demonstrated in field trials. (A concrete dam, by comparison, will typically have $f_n > 10$ Hz and will thus respond as a 'rigid' body.)

Certain ancillary structures, e.g. free-standing valve towers, gate structures and critical items of operating equipment such as valves etc., may be particularly at risk, irrespective of type of dam.

The seismic events of primary interest in dam design are essentially the outcome of tectonic and other natural ground movement, e.g. displacement on a fault system etc. There are, however, a number of instances where reservoir-induced seismic activity has been recorded, as at Koyna (India) and Kariba (Zimbabwe/Zambia). Seismicity of this nature is generally associated with high dams (H > 100 m) and reservoirs of high capacity (storage in excess of 100×10^6 m^3) located on tectonically sensitive sites.

Seismic categorization of a dam site can present difficulties, particularly in parts of the world which do not have a significant history of seismicity. Many countries faced with this problem resort to zoning, e.g. the USA has four zones defined by recorded levels of seismicity (see Table 2.7). The UK is similarly divided, into zones designated A, B and C, Zone A being the most severe (Charles *et al.*, 1991).

In categorizing a dam in the UK an empirical system of weighted classification parameters embracing dam height, reservoir capacity, evacuation requirements and damage potential is evaluated, the aggregated value placing the site in one of four categories. Categories I to IV are in turn linked by zone to suggested PGA values and notional SEE return periods. For Category IV, the most severe situation and applicable to only a very small number of UK dams, the SEE is defined as equivalent to MCE *or* to an upper-bound return period of 3×10^4 years. The associated PGA values range from 0.25 g (Zone C) up to 0.375 g (Zone A). At the other end of the scale, seismic safety evaluation is not normally considered necessary for Category I dams, but if deemed appropriate the SEE return period is 1×10^3 years, with PGAs of 0.05 g (Zone C) and 0.10 g (Zone A) (Charles *et al.*, 1991).

SEISMIC LOADING

Seismic loads can be approximated in the first instance by using the simplistic approach of *pseudostatic*, or seismic coefficient, analysis. Inertia forces are calculated in terms of the acceleration maxima selected for design and considered as equivalent to additional static loads. This approach, sometimes referred to as the *equivalent static load method*, is generally conservative. It is therefore now applied only to smaller and less vulnerable dams, or for purposes of preliminary analysis. For high dams, or dams in situations where seismicity is considered critical, more sophisticated procedures are necessary. In such circumstances seismological appraisal of the dam site should be followed by a comprehensive evaluation on the basis of *dynamic response analysis*. Pseudostatic and dynamic analyses are briefly introduced below.

Kaptan (1999) presents a review of the approach to determining seismic design criteria for dams in Turkey, a region noted for its high level of seismic activity. Fundamental criteria for dynamic analysis of embankments and of concrete dams are discussed. They are placed in perspective by notes on the evaluation of seismic criteria appropriate to the 184 m high Ataturk earthfill–rockfill embankment and on dynamic response analysis for the Sürgü rockfill dam of 66 m height.

PSEUDOSTATIC ANALYSIS

The intensity of a shock is expressed by acceleration coefficients α_h (horizontal) and α_v (vertical), each representing the ratio of PGA to gravitational acceleration, g. It is frequently assumed that $\alpha_h = (1.5\text{–}2.0)\alpha_v$ for the purposes of initial analysis. Representative seismic coefficients, α_h, as applied in design are listed in Table 2.8. Corresponding earthquake intensities on the modified Mercalli scale and a qualitative scale of damage are also shown, together with the equivalent US seismic zone designations.

Values of seismic coefficient greater than those tabulated are appropriate to more extreme circumstances, e.g. $\alpha_h = 0.4$ has been employed for dams on high-risk sites in Japan. (Ground accelerations equivalent to $\alpha_h = 0.5$ severely damaged the Koyna gravity dam, India, in 1967. Pacoima arch dam in the USA was similarly severely damaged in 1971 by an event estimated as equating to a base level horizontal seismic coefficient of $\alpha_h = 0.6\text{–}0.8$.)

Table 2.8 Seismic acceleration coefficients, α_h, and earthquake intensity levels

Coefficient α_h	*Modified Mercalli scale*	*General damage level*	*US seismic zone*	
0.0	–	Nil	0	(Note: there is no direct equivalence with UK zones A, B, C)
0.05	VI	Minor	1	
0.10	VII	Moderate	2	
0.15	*c.* VIII–IX	Major	3	
0.20		Great	4	
0.30	IX	Widespread destruction	–	–
0.40	IX		–	–

Inertia forces associated with the mass of a dam structure are then expressed in terms of the horizontal and vertical seismic acceleration coefficients, α_h and α_v respectively, and gravitational acceleration, g, thus:

horizontal forces: $\pm\alpha_h$. (static mass)
vertical forces: $\pm\alpha_v$. (static mass)

The reversible character of these forces will be noted: for initial analysis they may be assumed to operate through the centroid of the mass to which they are considered to apply.

It may be considered unrealistic to assume that peak downward vertical acceleration will be coincident with peak upstream horizontal acceleration, hypothetically the most adverse combination of seismic loads. Three load cases may then be defined as being more appropriate:

i) peak *horizontal* ground acceleration with zero *vertical* ground acceleration,
ii) peak *vertical* ground acceleration with zero *horizontal* ground acceleration, and
iii) an appropriate combination of peak horizontal and vertical ground accelerations, e.g. one component (normally the horizontal) at its peak value with the other at some 40–50% of its peak.

The application of pseudo-static analysis, together with the associated determination of inertia forces attributable to the reservoir water, is addressed in Section 3.1.1 with respect to concrete gravity dam analysis.

For initial seismic assessment of embankment dams, a rapid parametric study of the influence of different values of pseudostatic coefficient can be carried out using stability charts and considering the seismic effect as approximating to tilting the slope through angle $\tan^{-1}\alpha_h$. The factor of safety of a flatter slope will be much more affected by varying the assumed value of α_h than will that of a steeper slope.

More complex variants of pseudostatic analyses may, *inter alia*, employ a non-uniform value of α_h, increasing towards the crest, derived from a more rigorous analysis of the anticipated seismic event.

DYNAMIC RESPONSE ANALYSIS

The simplifications inherent in pseudostatic analysis are considerable. Complex problems of dam-foundation and dam-reservoir interaction are not addressed, and the load response of the dam itself is neglected. The interactions referred to are of great importance, as they collectively modify the dynamic properties of the dam and consequently may significantly affect its load response. They are accounted for in dynamic response analysis, where the coupled effects of the dam-foundation-reservoir system are addressed by a substructure analysis procedure.

In this approach the dam is idealized as a two-dimensional plane-strain or plane-stress finite element system, the reservoir being regarded as a continuum. The foundation zone is generally idealized as a finite element system equivalent to a viscoelastic half-space. The complexities of such an approach are evident, and take it outside the scope of this text.

It has been suggested that the application of more rigorous methods, such as dynamic analysis, in preference to the simplistic pseudostatic approach is appropriate if the anticipated PGA will exceed 0.15 g (Charles *et al.*, 1991). This may be compared with the suggestion of an anticipated PGA of over 0.25 g made in ICOLD (1989).

Dynamic response analysis is fully discussed by Chopra and also by Idriss and Duncan, both within Jansen (ed.) (1988). The seismic assessment of two dams in the UK, one an earthfill embankment, the other a concrete gravity dam, is presented in Taylor *et al.* (1985).

SEISMIC DAMAGE TO EMBANKMENT DAMS

Earthquake damage to embankments, which could potentially lead to breachings and catastrophic failure, can take the following forms:

- shear displacement and disruption by fault reactivation;
- instability of the face slopes;
- deep instability involving the foundation;
- liquefaction and flow slides;
- volume reduction and slumping of the embankment fill, leading to differential displacements and internal cracking (a vulnerable foundation may similarly be at risk);
- damage to ancillary structures, including spillways, outlet tunnels and culverts;
- instability of valley slopes leading to displacement, overtopping and instability, or to obstruction of the overflow.

The risk of liquefaction and a flow slide as a result of pore pressure generation and soil densification is effectively confined to loose, coarser soils such as sands, typically composed of particles in the 0.06 mm to 0.6 mm range.

Pseudostatic seismic analysis, described in Section 3.1.1, involves a major simplification. It assumes that the dynamic effects of seismic disturbance can be substituted by static forces derived from assumed peak accelerations, α, expressed as a proportion of the acceleration due to gravity, g. Applied to an embankment dam, pseudostatic analysis in its simplest form introduces an additional horizontal inertia load. This is similar in its influence to conducting a conventional stability analysis with the embankment slope tilted by $\tan^{-1}\alpha$. The limit equilibrium analysis developed in Sarma (1975) is frequently employed in this context as it is readily adapted to include horizontal interslice forces. A simplified approach to earthquake resistant design of embankment dams is presented in Sarma (1980).

The seismic risk to UK dams, including selection of design parameters, is addressed in Charles *et al.* (1991) and ICE (1998). Selection of seismic design parameters has also been reviewed in ICOLD Bulletin 72 (ICOLD, 1989). Reference should be made to Seed (1981) and to Jansen *et al.* (1988) for further detailed discussion of seismic analysis and in particular of dynamic analysis and its application in design.

2.8 Settlement and deformation

2.8.1 Settlement

The primary consolidation settlements, δ_1, which develop as excess porewater pressures are dissipated, can be estimated in terms of m_v, the coefficient of compressibility (Section 2.3.3), the depth of compressible soil, and the mean vertical effective stress increases, $\Delta\sigma'$. Subscripts 'e' and 'f' in the following equations refer to embankment and foundation respectively:

$$\delta_1 = f_n m_v \Delta\sigma' \qquad (2.29)$$

$$\delta_{1e} = m_{ve}\gamma H^2/2, \qquad (2.30)$$

where H is the embankment height, and

$$\delta_{if} = m_{vf} D_f \Delta\sigma'_f \qquad (2.31)$$

where D_f is the depth of the compressible foundation. $\Delta\sigma'_f$ is given by the relationship

$$\Delta\sigma'_f = I\gamma_f z_e \qquad (2.32)$$

where I is an influence factor determined by the foundation elasticity and the width:depth ratio. Curves for I under the centre of a symmetrical embankment are given in Mitchell (1983). For representative embankment dam–foundation geometries with $D_f < 0.5H$, $I = 0.90–0.99$.

The accuracy of settlement predictions is enhanced by subdividing the embankment and/or foundation into a number of layers, and analysing the incremental settlement in each in turn.

The secondary consolidation settlement, δ_2, can be estimated from the coefficient of secondary consolidation, C_α (Section 2.3.3). The general equation for δ_2, applied to embankment and foundation in turn, is given by

$$\delta_2 = C_\alpha z \log(t_2/t_1) \tag{2.33}$$

where z is the height H or depth D_f as appropriate, and times t_2 and t_1 are determined relative to the completion of primary consolidation.

Values for C_α are generally below 0.002 for overconsolidated clay fills etc., rising to between 0.005 and 0.5 for softer normally consolidated clays.

Anticipated maximum settlements within an earthfill embankment at end of construction for $H > 13$ m can initially be approximated from the relationship

$$\delta_{ec} = 0.035(H - 13) \text{ (m)}. \tag{2.34}$$

2.8.2 Deformation

The internal deformations which develop are complex and are not precisely determinate. The average transverse base strain can be approximated if it is assumed that on completion of foundation settlement the base deforms to a circular arc. For moderate settlements the average positive, i.e. tensile, base strain ϵ_h is then given by

$$\epsilon_h = -2\frac{[(L'/2)^2 + (\delta_f)^2]^{1/2}}{L'} + 1 \tag{2.35}$$

where L' is the base width of the embankment.

The corresponding mean vertical strain, ϵ_v, is expressed by

$$\epsilon_v = \delta_e/H = m_{ve}\gamma H/2. \tag{2.36}$$

Further aspects of internal deformation, notably at the critical core–filter–shoulder interfaces, are addressed in Mitchell (1983).

2.8.3 Performance indices for earthfill cores

Progressive deterioration or inadequate performance of an earthfill core may be indicated by excessive and possibly turbid seepage and leakage. Other indicators include localized crest or upstream face depressions, excessive general settlement or a high phreatic surface in the downstream shoulder.

Empirical performance indices have been suggested for certain critical parameters.

(a) Hydraulic fracturing index, HFI

Hydraulic fracturing index is a measure of susceptibility to fracturing, and is given by

$$\mathrm{HFI} = \frac{\sigma}{\gamma_w z_1} \qquad (2.37)$$

where σ is the total stress and $\gamma_w z_1$ the reservoir head, determined for a plane crossing the core. The horizontal total stress, σ_h, normal to a transverse vertical plane can be demonstrated to be critical, and $\mathrm{HFI} < 1$ indicates a risk of hydraulic fracturing (σ_h must be estimated or determined by *in situ* tests (Charles and Watts, 1987)).

(b) Settlement index, SI

Settlement index is defined as:

$$\mathrm{SI} = \frac{\Delta_S}{1000H \log t_2/t_1} \qquad (2.38)$$

where Δ_S is crest settlement in millimetres occurring over time interval t_1 to t_2 (years) after completion of an embankment of height H (metres). A value of $\mathrm{SI} \leqslant 0.02$ is considered to represent acceptable long-term settlement behaviour in terms of secondary consolidation of fill and/or foundation, or creep of a granular fill (Charles, 1986).

(c) Seepage index, QI

A seepage and leakage index may be defined as

$$\mathrm{QI} = \frac{q}{1000AKi} \qquad (2.39)$$

where q in litres per second is the flow through the core or water-retaining element and A the area of element in square metres. The coefficient k is the maximum acceptable permeability of the core in metres per second and i the mean hydraulic gradient across the core. A value of $\mathrm{QI} < 1$ might be expected if all flow is seepage, as opposed to leakage through imperfections etc. In view of the problems associated with determining q and k, the value of QI lies principally in its use as a long-term comparative index.

2.9 Rockfill embankments and rockfill

2.9.1 Rockfill embankments

Rockfill embankments may either have an internal watertight element or an upstream deck, as illustrated on Fig. 1.3(a) to (d). In the case of the former, the element is generally central and may be of earthfill or, in the absence of suitable soils, of asphaltic concrete. The latter type of slender bituminous element is increasingly common in Europe. The alternative upstream face membrane or deck can be constructed of reinforced concrete or of asphaltic concrete; in either event the deck is relatively thin. The concrete decked rockfill dam, commonly referred to as the CFRD type, has been widely used for very large dams in South America and Australia. Deck thickness generally increases with depth below crest level, and is typically of the order of $(0.3 + 0.005H)$ m for high dams. The concrete deck must, however, be subdivided into rectangular strips running normal to the dam axis, or into rectangular panels, to accommodate deformation of the rockfill. This requires an expensive and sometimes troublesome joint detail incorporating a waterstop. An asphaltic concrete deck, which is generally somewhat thinner, has the advantage of a degree of flexibility, and no joint system is therefore necessary.

A central asphaltic or bituminous core is typically of the order of 1.0–1.5 m maximum width, with a narrow transition zone on either side. A rockfill embankment with a central asphaltic core is shown in Fig. 2.17 and described in Gallacher (1988). The central membrane and the upstream deck each have advantages, e.g. a dam with an upstream deck will have greater resistance to horizontal sliding and is more easily repaired. The central membrane, on the other hand, is less vulnerable to deterioration and accidental or deliberate damage.

Rockfill embankments, if protected and/or reinforced, have the capability to accept throughflow or overflow of flood water without serious damage. Suitable protective measures are discussed in Maranha das Neves and ICOLD (1993c).

2.9.2 Rockfill

Compacted rockfills have displaced the dumped or sluiced rockfills formerly employed for embankment dam construction. The concept of well-graded and intensively compacted rockfill has developed from appreciation of the importance of grading and the introduction of heavy compaction plant. Compacted density is the principal factor governing rockfill shear strength and settlement.

Representative characteristics for graded compacted rockfill were

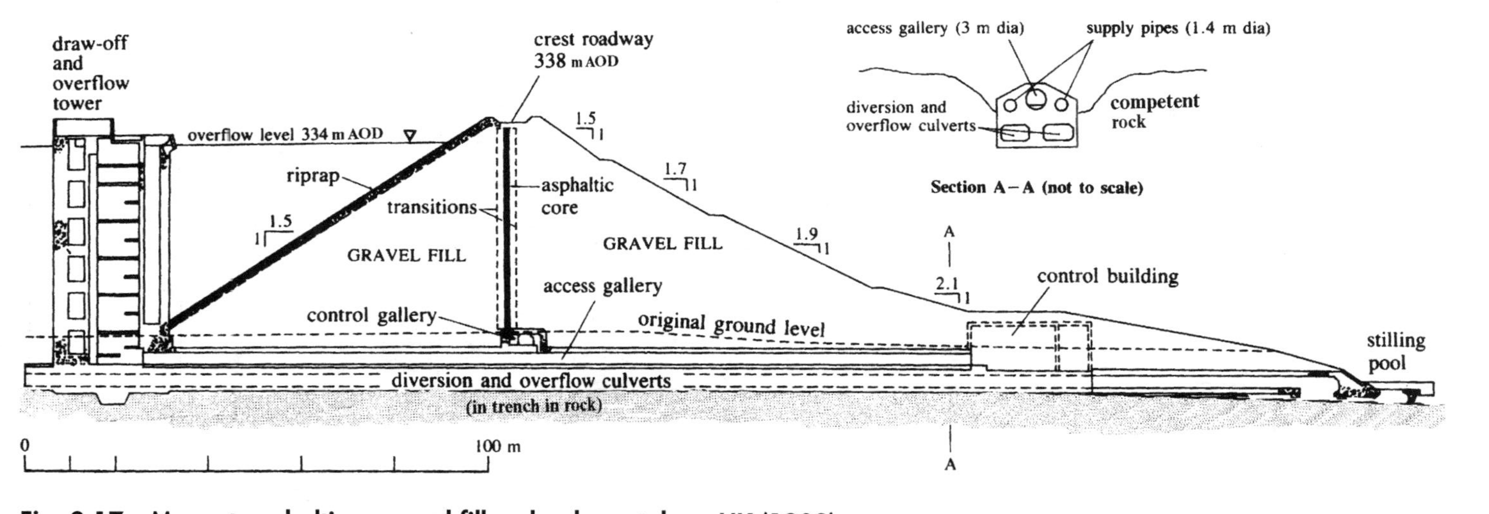

Fig. 2.17 Megget asphaltic core rockfill embankment dam, UK (1982)

indicated in Table 2.5. The major advantages of rockfill as an embankment construction material are high frictional shear strength, allowing the construction of much steeper face slopes than earthfill, and relatively high permeability, eliminating problems associated with construction or seepage porewater pressures.

The disadvantages of rockfill lie in the difficulties of controlling the grading of crushed rock, e.g. from excavations and tunnels, and in the construction and post-construction settlements, which are relatively high. This can result in interface problems where rockfill shoulders are adjacent to a compressible clay core (Section 2.7.2).

The quality and suitability of rockfill is discussed in Penman and Charles (1975) and ICOLD (1993b). An exhaustive study of the engineering characteristics of compacted rockfill and of the special large-scale test techniques required is reported in Marsal (1973).

The shear strength of compacted rockfills is defined by a curved failure envelope with the form (De Mello, 1977)

$$\tau_f = A(\sigma')^b. \tag{2.40}$$

Illustrative values of parameters A and b, from data presented in Charles and Watts (1980), and with τ_f and σ' in kN/m^2, are

$$A = 3.0 \text{ (poor quality slate)} \rightarrow 6.8 \text{ (sandstone)},$$

$$b = 0.67 \text{ (sandstone)} \rightarrow 0.81 \text{ (basalt)}.$$

The stability of slopes in compacted rockfill can be determined using limit-equilibrium methods, including those introduced in Section 2.7, and comparable analyses for wedge-type failure surfaces. Rapid parametric stability studies can be conducted using the dimensionless stability numbers, Γ_F (Fellenius analysis) and Γ_B (Bishop analysis) developed in Charles and Soares (1984) and based on the relationship of equation (2.40).

Rockfill settlement is associated with particle contact crushing and is greatly increased by saturation; it can therefore be accelerated during construction operations. The construction settlement occurring at crest level, δ_{r1}, can be estimated from the approximate relationships

$$\delta_{r1} = 0.001H^{3/2} \text{ (m)} \tag{2.41}$$

and

$$\delta_{r1} = \rho_r g H^2 / E_r \tag{2.42}$$

where E_r is the modulus of deformation and ρ_r is the density of the rockfill. Values of E_r are typically of the order of 20–50 MN m^{-2}.

Long-term post-construction settlement, δ_{r2}, is governed by a relationship of the form

$$\delta_{r2} = \alpha' (\log t_2 - \log t_1). \qquad (2.43)$$

The constant α' in equation (2.40) normally lies in the range 0.2–0.5, and times t_1 and t_2 (years) are determined relative to the completion of construction and/or first impounding.

The controlling parameters in optimizing the field compaction of rockfill, which should be near saturated to minimize later settlement, are those which apply also to earthfill compaction, i.e. layer thickness, plant characteristics and number of passes. Layer thickness may range up to 1.0–2.0 m dependent upon maximum rock size and shape. Excessively heavy plant and/or overcompaction can result in undesirable crushing and degradation of softer or weathered rockfills, with the generation of excessive fines to the detriment of drainage capability. The optimum number of passes is a compromise between further incremental gain in compacted density and cost per unit volume of compacted rockfill, and is generally determined from construction of a controlled trial fill.

Rockfill embankment analysis and design are discussed further in Hirschfeld and Poulos (1973), Jansen (1988) and Leps (1988). The use of low grade rockfill is discussed in Wilson and Evans (1990). Asphaltic concrete decks and diaphragms are discussed in Creegan and Monismith (1996) with reference to USA practice.

2.10 Small embankment dams, farm dams and flood banks

2.10.1 Small embankment dams

The principles of good practice in embankment design and construction are unrelated to the size of dam. There are nevertheless a number of points of difference which may emerge in the engineering of smaller dams, defined for the purposes of this Section as dams up to 10 m in height. The differences are essentially technical in the context of design and construction standards applied, but many of these can be linked to differences in ownership pattern and other non-technical factors. As a generalization, therefore, it is realistic to anticipate that the small dam *may* be engineered to different standards. The reasons for such differences are numerous, but on the basis of UK experience they include the following:

- small dams are more likely to be in singular private ownership;
- small dams are frequently 'low-cost' designs;
- surveillance and maintenance will be a low priority;

- the owners' resources may be limited;
- small dams are commonly of simple homogeneous section;
- problems of seepage and stability may be accentuated by minimal profiles;
- the perceived hazard may be relatively low.

Measured against the standards now applied for larger dams, there may consequently be some moderation in the standards applied in the engineering of many smaller (and essentially older) dams. Examples include standards employed in specifying the design flood (see Section 4.2 and Table 4.1), or in the incorporation of adequate seepage control measures. In the case of the first example, flood estimation for a very limited catchment may of necessity be done on an empirical basis. This may in turn lead to moderation of the standards applied for freeboard and for overflow capacity. Engineered seepage control measures, the other example identified, are critical to a good and safe design, but may be difficult to incorporate to the desirable degree other than at unacceptable cost. These and similar problems are usually linked to severe constraints on the resources available to the owner, compounded in many cases by pressure for early commissioning of the project. Collectively these and other difficulties, including underinvestment of time and money in a proper site appraisal, or problems in construction control, have historically led to significantly higher incident and failure rates for small privately owned dams (Moffat, 1982).

Minimal standards for the design of satisfactory lesser dams are suggested below:

- side slopes not steeper than 2.5:1.0;
- provision of an underdrain below the downstream shoulder;
- freeboard not less than 1.0–1.5 m;
- an allowance of not less than 8% for settlement;
- vehicular crest access provided with a width of at least 3.0 m; crest reinforced with geosynthetic if no surfaced road provided;
- adequate bottom outlet capacity, with an upstream control valve;
- good vegetative cover;
- installation of a seepage measuring weir and levelling stations (see Chapter 7);
- careful selection and placing of fill material;
- careful foundation preparation.

The planning, design and construction of small dams in the UK is addressed in detail in Kennard, Hoskins and Fletcher (1996). Design of small flood storage reservoirs is addressed in Hall, Hockin and Ellis (1993).

2.10.2 Small farm and amenity dams

Large numbers of very small privately owned reservoirs are constructed for a range of purposes. These include farm dams built to supply local irrigation needs and/or to support stock-rearing and amenity dams built to impound private fishing lakes. Most are very modest in scale, impounded by small and simple embankment dams, but it is important to recognize that the same basic principles as regards satisfactory design and construction practice apply to these dams as much as to the largest.

Low cost is clearly a dominant consideration in the engineering of small private dams, and in-service supervision and maintenance is likely to be minimal unless the dam is subject to legislation regarding safety (see Chapter 7). The design must therefore be simple, conservative and robust. Translated into practice the following general observations on small dams of this nature should be noted:

- A homogeneous profile is likely to prove most economic given the ready availability of soil with a clay content of *c.* 20–30%. Significantly higher clay contents should be avoided, as they can lead to problems with shrinkage and cracking.
- A simple zoned embankment, employing clayey soil to form an upstream low-permeability 'core' with a base width of approximately twice the height is advantageous from the point of view of stability.
- Where the local available soils are of higher permeability these may be used to construct the bulk of the embankment, with a 1 m thick blanket of clayey soil placed on the upstream face.
- Alternatives for the watertight element for a small dam of this type include a PVC or polyethylene membrane, but these should be protected by at least 0.6 m of soil.
- An embankment with a central 0.6 m thick bentonite diaphragm wall forming a 'core' is a further option. Walls of this nature can be extended below the base of the dam to form a cut-off, but they are comparatively expensive to install.
- There should be a rubble toe-drain or a base drain under the downstream shoulder.
- A simple key trench cut-off is satisfactory if a suitably impermeable clay soil is present at shallow depth. On a more permeable foundation it will be necessary to go considerably deeper to form a cut-off, e.g. with a bentonite diaphragm wall extending at least 0.6 m into a suitable low-permeability horizon.
- Face slopes should be no steeper than 2.5:1.0 upstream and 2.0:1.0 downstream.
- Overflow capacity and freeboard should be adequate to accept a design flood with a return period of not less than 100–150 yrs.

- Design freeboard should be a minimum 1.0 m and crest width not less than 2.5 m.
- Outlet pipework should be surrounded in concrete and laid in trench in the natural foundation with an upstream control valve.
- Allowance for long-term settlement should be a minimum 10% of dam height.
- Simple protection to the upstream face can be provided by rubble and suitable aquatic vegetation.
- Operational drawdown rates should be kept to less than 1 m/week.
- Provision should be made for measuring seepage flows.
- Good vegetative cover should be encouraged.

The design of small dams and farm dams is described in Stephens (1991).

2.10.3 Flood control banks and dykes

Structures of this nature differ from conventional embankment dams in a number of important regards:

1. they are of modest height but of very considerable length, with consequent variation in soil type and consistency of construction standard;
2. they are not generally 'engineered' in the sense of being a form of embankment dam;
3. operation as a dam is occasional only; this may result in leakage associated with shrinkage and cracking of more cohesive soils;
4. the section is minimal, with steep slopes;
5. they have poor resistance to local overtopping at low spots;
6. they are low-profile, at least until a catastrophic flood event;
7. they are prone to damage by root growth, by burrowing animals and by animal tracking;
8. construction is invariably 'lowest cost';
9. flood banks do not normally have a core;
10. surveillance and maintenance is frequently minimal.

With the foregoing general points of difference in mind it is suggested that flood bank designs should follow the guidelines listed below.

- side slopes no steeper than 2.0:1.0; flatter if subject to wave attack or to animal tracking;
- crest width adequate for access, including light plant, and not less than 3.0 m in width;
- settlement allowance not less than 10%;
- good vegetative cover;
- careful selection and use of soil types;

- proper stripping of foundation to below depth of weathering;
- controlled compaction in construction.

For discussion of the important question of flood bank alignment and protection reference should be made to Chapter 8.

2.11 Tailings dams and storage lagoons

Tailings dams and storage lagoons are constructed for the retention of waste fines, slurries and liquids or semi-liquids produced by industry. Examples are the tailings dams storing waste fines from mining and similar extractive industries, the extensive fly-ash storage lagoons associated with many large coal-fired power stations, and the liquid storage lagoons required by process and manufacturing industries. Storage lagoons are invariably impounded behind earthfill retaining embankments; tailings dams are commonly constructed with industrial waste materials, including fines. Tailings dams are frequently designed to accelerate de-watering and 'solidification' of the stored fines by decanting surplus surface water and controlled seepage through the body of the retaining embankment.

Recent decades have seen the size and storage capacity of the largest tailings dams and lagoons increase very significantly, and at the upper end of the scale eight have a height in excess of 150 m. In terms of intended storage capacity, Syncrude Tailings Dam in Canada is the world's largest facility of this type, with an ultimate capacity of $540 \times 10^6 \, m^3$. On completion the 18 km long retaining embankments will range up to 90 m in height. A further very large facility is Mission Tailings No 2 in the USA, completed in 1973 and holding $40 \times 10^6 \, m^3$ of tailings. As with conventional dams, however, the majority are much more modest in scale and do not exceed 15–20 m in height.

The embankments constructed for tailings disposal or for storage lagoons differ from conventional embankment dams in a number of important regards:

- tailings embankments, particularly the smaller examples, are almost invariably planned as 'minimum cost' structures. Design and construction standards, particularly on older projects, may consequently fall well short of modern embankment dam practice and utilize poorly selected and compacted low quality fills;
- the retained fines or slurry can be of very high density, exceptionally three or more times that of water;
- a common construction technique is to raise the crest of the retaining embankment and retention level of the stored waste incrementally and in parallel, i.e. the embankment functions as a dam as it is itself

being raised in stages. A retention dam constructed in this fashion is satisfactory provided the incremental raising is properly engineered and controlled. Construction by hydraulic filling is also practiced, i.e. a natural silt or the waste fines if suitable are fluidized and pumped to hydro-cyclones located at intervals along the planned embankment. These distribute the fluidized fines in controlled fashion, with the coarsest fraction settling out first and leaving the finest material to form a zone or 'core' of lower permeability, but it may be noted that hydraulic fills of this nature can be prone to liquefaction under dynamic loading.

The large modern lagoon embankments associated with the operation of many process industries generally conform with good embankment dam standards and practice. Older and/or smaller lagoons, however, may be much less capably engineered.

Common deficiencies in the design of older tailings dams and lagoon embankments are poor control of the internal seepage regime and marginal freeboard, the last often combined with limited overflow capacity. Excessive deformations and steep downstream slopes with marginal stability are also common.

DoE (1991) and ICOLD Bulletin 106 (ICOLD, 1996) provide guidelines for good practice in the design and management of tailings dams and storage lagoons; DoM and E, (1999) presents a comprehensive guide to Australian construction and operating standards for tailings storage.

2.12 Geosynthetics in embankment dams

Geosynthetics (geotextiles and geomembranes) have considerable potential in dam engineering given that issues of durability in specific applications can be resolved. A range of geosynthetics have been employed in a number of different applications both in new construction and in rehabilitation projects (Section 8.7). Their use in dams, commencing about 1970, has developed relatively slowly by comparison with other geotechnical applications. This is attributable to caution based on limited information as to the long-term durability of man-made synthetics, particularly when used internally where they cannot be readily inspected and, if necessary, replaced.

Geosynthetics can be employed to fulfil several different functions in embankment dams.

1. *Impermeable membranes (upstream or internal).* Polyvinyl chloride (PVC) and high-density polyethylene (HDPE) upstream membranes have been successfully employed in dams up to 40 m in height. The membranes, typically 3–4 mm thick, are laid in 4–6 m wide strips on a

prepared sand bed and drainage layer, and anchored at crest and toe. Careful attention must be given to ensuring the integrity of the seams and welds, and the completed membrane is normally provided with a granular protective cover supporting conventional upstream face protection.

Internal membranes, less commonly employed, are normally provided with protective transition zones on either side.

2. *Filter and drainage layers (seepage control).* Relatively thick geosynthetics with high internal transmissivity are suitable for filters or drainage layers. It is important to ensure that the anticipated stresses and/or some degree of soil particle migration do not diminish the transmissive capacity to an unacceptable degree.
3. *Earth reinforcement (stability of slopes etc.).* Geosynthetic reinforcement materials, e.g. geogrids, can be used to permit construction of steeper face slopes or to help to contain lateral deformation and spread within the embankment or in a soft foundation.
4. *Control of surface erosion (precipitation or limited overtopping flows).* The use of geosynthetics in conjunction with natural vegetation can considerably enhance erosion resistance and reduce 'gullying' due to localized run-off on the downstream face. Geogrids and mats have also been employed to provide a measure of security against modest overtopping of limited duration (Section 4.4).
5. *Separation interlayers.* Geosynthetics can be employed to act as an interlayer to ensure positive separation of fill materials at an interface. Interface functions can require the geosynthetic to act as a supporting or cushion layer, or as a low- or high-friction interlayer. An example of use in this context would be separation of earthfill from adjacent rockfill of different compressibility.

The general application of geosynthetics to dams is described by Giroud (1990, 1992). A review of the use of geosynthetic membranes is given in ICOLD Bulletin 78 (ICOLD, 1991) and of geotextiles for filter-transition layers in Bulletin 55 (ICOLD, 1986).

2.13 Upgrading and rehabilitation of embankment dams

The upgrading and rehabilitation of older embankment dams is assuming great importance within the older industrialized nations where, insofar as dams are concerned, much of the national infrastructure was constructed in the 19th and 20th centuries. The median age is therefore comparatively high. A well-engineered embankment dam is a surprisingly tolerant and durable structure. The majority of problems which arise with older dams are associated with a contemporary lack of understanding with regard to:

1. the consequencies of uncontrolled internal seepage;
2. parameters controlling the stability of slopes;
3. deformation and settlement parameters;
4. design flood estimation, overflow capacity and freeboard;
5. internal interface effects.

Internal erosion, with migration of core material, plus an undesirably high phreatic surface in the downstream shoulder, are obvious outcomes of 1. Problems arising from the absence of internal drainage and protective filters are difficult and costly to remedy. The only practicable shorter-term response may be a limit on retention level and increased surveillance. Remedial options based on sophisticated geotechnical procedures to install drains or reconstruct a damaged core may then be evaluated. Major rehabilitation work on a defective rolled-clay core seriously damaged by erosion following hydraulic fracturing is described in Vaughan *et al.*, 1970.

Inadequate stability of the downstream slope is a relatively straightforward problem. The primary causes are either an excessively steep slope or a high phreatic surface. The problem can be resolved by construction of a downstream berm of free-draining fill, e.g. a rockfill.

Excessive core settlement and loss of freeboard can be indicative of ongoing consolidation processes. Raising the core to restore freeboard is a common procedure. In a more extreme case at Ladybower, a UK dam where core settlement in excess of 1.2 m was recorded over the 50 year life of the 45 m high dam and was continuing, a major re-construction and upgrading of the dam which included re-profiling the downstream shoulder was undertaken (Vaughan, Chalmers and Mackay (2002)).

The flood discharge capacity at many older dams has been shown to be inadequate. Considerable effort has therefore been expended on reconstructing and upgrading overflow and outlet works on older dams, an exercise which is costly and, on a congested site, is very difficult. An alternative option is installation of an auxiliary overflow which can take one of several forms, e.g. a separately located erodible 'fuse-plug' (Dan, 1996) or concrete flap-gates operated by hydrostatic pressure.

Interface effects are principally associated with discontinuities and strain incompatibilities, e.g. at the interface between clay core and coarser shoulder material or an outlet culvert, or at a step in the base of a deep cut-off trench. They remain a complex aspect of dam behaviour, and while amenable to resolution in the design of a new structure they are seldom so in the case of old dams; total reconstruction is then the only option.

Recent examples of major upgrading or reconstruction works are fully described in Vaughan, Chalmers and Mackay (2000), and in Banyard, Coxon and Johnston (1992). Further descriptions of such work are presented in Macdonald, Dawson and Coleshill (1993) and in Chalmers, Vaughan and Coats (1993).

Worked Example 2.1

The outline profile of an earthfill embankment is illustrated in Fig. 2.18.

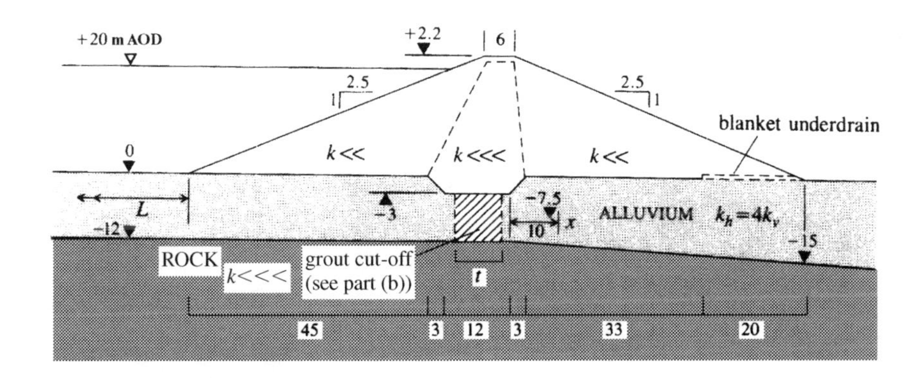

Fig. 2.18 Worked example 2.1 (Fig. 2.11)

GEOTECHNICAL DATA

The foundation alluvium has a permeability (antisotropic) of $k_h = 4k_v = 1 \times 10^{-5}\,\mathrm{m\,s^{-1}}$. The design effective permeability of the grouted cut-off zone, $k_c = 2.5 \times 10^{-7}\,\mathrm{m\,s^{-1}}$.

1. Draw the foundation flownet without a grouted cut-off in position and calculate seepage flow, Q_0, in $\mathrm{m^3\,m^{-1}\,day^{-1}}$.
2. Determine the width, t, of grouted cut-off zone required to achieve a design E_Q of 70% and compute the corresponding value of E_H.
3. Determine the corresponding porewater pressure at point X in m head.

Solution

1. Transform factor = 1/2. See flownet on Fig. 2.11: $N_F/N_d = 3/16.5 = 0.18$. k' (alluvium) $= (k_v k_h)^{1/2} = 0.5 \times 10^{-5}\,\mathrm{m\,s^{-1}}$. Therefore

$$Q_0 = (0.5 \times 10^{-5})20(0.18)(24 \times 3600) \quad (2.15)$$
$$= 1.55\,\mathrm{m^3\,m^{-1}\,day^{-1}}.$$

2. With E_Q set at 70%, $Q = 0.30\,Q_0 = 0.47\,\mathrm{m^3\,m^{-1}\,day^{-1}}$. Therefore

$$\text{'equivalent' } N_f'/N_d' = \frac{(0.18)0.47}{1.55} = 0.055.$$

Therefore

$$N'_{\mathrm{d}} = 3/0.055 = 54.5, \quad \text{say } 55.$$

The relative rate of head loss, cut-off and alluvium $= k'/k'_{\text{cut-off}}$

$$= \frac{0.5 \times 10^{-5}}{2.5 \times 10^{-7}} = 20.0/1.0.$$

Thus, n_{d} in cut-off zone, $n_{\mathrm{d}} = 20n_{\mathrm{d}} + (16.5 - n_{\mathrm{d}}) = 55$. Therefore $n_{\mathrm{d}} = 2.02$, say 2. However, the flownet mesh dimension on the transform scale $\simeq 3$ m, $\equiv 6$ m on the natural scale, and hence the cut-off thickness, $t \simeq 2 \times 6 = 12$ m.

$$E_H = h/H, \;\; H \equiv 2 \times 20H/N'_{\mathrm{d}} \equiv 40/20N'_{\mathrm{d}}.$$

Therefore

$$h = 40 \times 20/55 = 14.5\,\text{m} \quad \text{and} \quad E_H = 14.5/20 = 73\%.$$

3. $H - h = 5.5$ m head loss in the alluvium. The seepage path in alluvium $= 116 - 20 - t = 84$ m. Therefore $i = 5.5/84$ and the upstream head loss $= (48 \times 5.5)/84 = 3.1$ m; the downstream head loss $= 2.4$ m; hence the hydraulic grade line in Fig. 2.11, and porewater pressure at X $\equiv 9.3$ m head.

Worked Example 2.2

The profile of a zoned earthfill embankment founded on impervious clay is illustrated in Fig. 2.12, and steady-state piezometric levels recorded within the embankment are indicated.

DATA

Measured seepage at the V-notch, $q = 0.1\,\text{m}^3\,\text{day}^{-1}$ per m length. The core zone consists of soil A. The shoulders are of soil B downstream and coarse rockfill upstream. The drains are of fine gravel.

1. Interpret the piezometric levels and construct a flownet representative of the steady-state seepage régime.
2. From the flownet estimate the effective permeability k', of soil A and the head efficiency, E_{H} of the core.

Solutions

1. See the flownet in Fig. 2.12, noting (1) the interpretation of zone

permeabilities, and (2) the construction of the phreatic surface, with subdivision of the head, H, into increments, Δh.

2. From the flownet: $N_f = 3.6$, mean $N_d \approx 4.9$, and shape factor $N_f/N_d = 0.73$.

$$k' = \frac{0.1}{0.73 \times 29(24 \times 3600)} = 5.4 \times 10^{-8}\,\mathrm{m\,s^{-1}} \quad \text{(equation (2.15))},$$

$$E_H(\text{core}) = 14.5/29.0 = 50\%.$$

Worked Example 2.3

Figure 2.15 illustrates the section of a rolled clay embankment dam at an intermediate state of construction.

GEOTECHNICAL DATA

	Rolled clay core and shoulder
Unit weight, $\gamma(\mathrm{kN\,m^{-3}})$	21
Cohesion, c' ($\mathrm{kN\,\mu m^{-2}}$)	5
Angle of shearing resistance ϕ' (degrees)	30
Estimated pore-pressure ratio, r_u	Slices 1–3 = 0.55 Slices 4–8 = 0.45

Determine the factor of safety, F, with respect to the hypothetical failure surface indicated using the semi-rigorous Bishop analysis (equation (2.26b): slice dimensions may be scaled from the figure).

Solution

For an assumed value of $F = 1.5$, equation (2.26b) yields $F = 1.71$; iteration ($F = 1.7$) confirms $F = 1.72$.

Tabulated below are the effects on F of changing major parameters involved in the analysis; e.g. c', ϕ' and r_u:

Trial	c' ($kN\,m^{-2}$)	ϕ' (*degrees*)	r_u		F
			Slices 1–3	*Slices 4–8*	
A	5	30	0.55	0.45	1.72
B ($<c'$)	0	30	0.55	0.45	1.51
C ($<\phi'$)	5	25	0.55	0.45	1.44
D ($>r_u$)	5	30	0.65	0.55	1.41

Worked Example 2.4

The outline profile of an embankment dam and its foundations are illustrated in Fig. 2.19. The dam has a central rolled clay core flanked by shoulders of compacted rockfill.

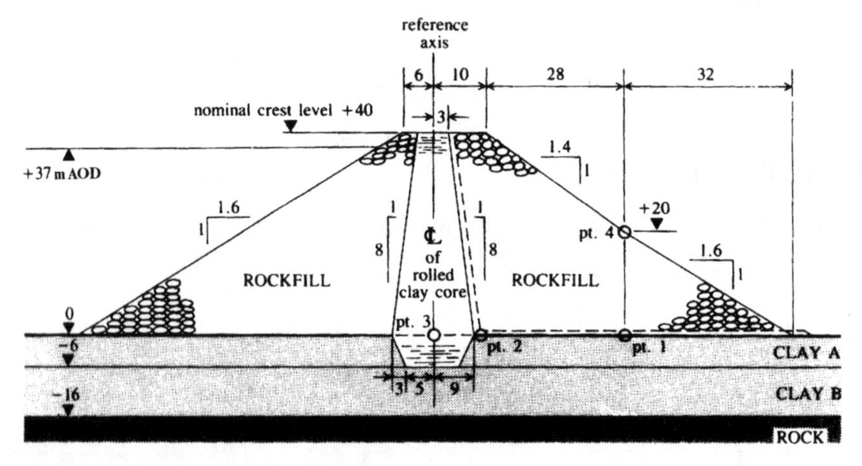

Fig. 2.19 Worked example 2.4

GEOTECHNICAL DATA

Initial groundwater level (GWL) = −2 m AOD.

1. Determine the construction crest level required to ensure that the nominal freeboard shown is not reduced by consolidation settlement.
2. Estimate the consolidation settlement which will develop at each of the points 1, 2 and 3 indicated and determine the total settlement of point 4.

Level (m AOD)	*Dam*		*Foundation*	
	Core	*Shoulder*	*Clay A*	*Clay B*
From:	*+40*	*+40*	*0*	*−6*
To:	*−6*	*0*	*−6*	*−16*
Unit weight γ(kN m^{-3})	20	21.5	20	20
Coefficient of compressibility m_v($\times 10^{-4}$ m^2 kN^{-1})	0.50	–	0.80	0.60
Equivalent modulus, E($\times 10^4$ kN m^{-2})	–	2.0	–	–
Influence factors, I, for elastic vertical stress distribution, mid-depth of strata on core.			0.98	0.95

Solutions

1. Core depth = 40 + 6 = 46 m. Mean $\Delta\sigma'_c$ for core = 20 × 46/2 = 460 kN m^{-2}. Therefore

$$\delta_{1c} = 0.5 \times 460 \times 46/10^4 = 1.06\,\text{m} \qquad \text{(equation (2.30)).}$$

Settlement under core, clay B: at mid-height,

$$\Delta\sigma'_B = 0.95 \times 20 \times 46 = 874\,\text{kN m}^{-2} \qquad \text{(equation (2.32)),}$$

$$\delta_{1B} = \frac{0.6 \times 874 \times 10}{10^4} = 0.52\,\text{m} \qquad \text{(equation (2.31));}$$

hence the nominal crest level = 41.6 m AOD.

2. *Point 1:* $\Delta\sigma' = 21.5 \times 20 = 430\,\text{kN m}^{-2}$. Therefore at mid-height in clay A,

$$\Delta\sigma'_A = 0.98 \times 430 = 421\,\text{kN m}^{-2} \qquad \text{(equation (2.32)),}$$

$$\delta_{1A} = \frac{0.8 \times 421 \times 6}{10^4} = 0.20\,\text{m} \qquad \text{(equation (2.31)),}$$

and in clay B,

$$\Delta\sigma'_B = 0.95 \times 430 = 408.5\,\text{kN m}^{-2},$$

$$\delta_{1B} = \frac{0.6 \times 408.5 \times 10}{10^9} = 0.25\,\text{m}.$$

Therefore $\delta_{\text{total}} = 0.45\,\text{m}$ at point 1.
Point 2: $\Delta\sigma' = 21.5 \times 40 = 860\,\text{kN m}^{-2}$. Therefore $\delta_{\text{total}} = 0.90\,\text{m}$ at point 2.
Point 3: $\Delta\sigma' = 20 \times 40 = 800\,\text{kN m}^{-2}$. Therefore $\Delta\sigma'$ in cut-off = $0.98 \times 800 = 784\,\text{kN m}^{-2}$,

$$\delta_c \text{ in cut-off} = \frac{0.6 \times 784.6}{10^4} = 0.24\,\text{m}$$

and $\Delta\sigma'_B = 0.95 \times 800 = 760\,\text{kN m}^{-2}$. (6 m depth cut-off balances 6 m excavated in clay A.) Therefore,

$$\delta_B = \frac{0.6 \times 760 \times 10}{10^4} = 0.46\,\text{m}.$$

Therefore $\delta_{total} = 0.68\,m$ at point 3.
Point 4: δ_{total} = rockfill settlement + $\delta_{(point\ 1)}$.

$$\sigma_{v\,mean} = \frac{21.5 \times 20}{2} = 215\,kN\,m^{-2}.$$

Therefore,

$$\epsilon_v = \sigma_v / E_r = 215/20 \times 10^4 = 1.075\% = 0.22\,m$$

and so $\delta_{total} = 0.45 + 0.22 = 0.67\,m$ at point 4.

References

Abdul Hussain, I.A., Hari Prasad, K.S. and Kashyap, P. (2004) Modelling of pore water pressure distribution in an earthen dam and evaluation of its stability, in *'Dam Engineering'*, XV (3): 197–218.

Atkinson, J.H., Charles, J.A. and Mhach, H.K. (1990) Examination of erosion resistance of clays in embankment dams. *Quarterly Journal of Engineering Geology*, 23: 103–8.

Banyard, J.K., Coxon, R.E. and Johnston, T.A. (1992) Carsington reservoir – reconstruction of the dam. *Proc. Institution of Civil Engineers; Civil Engineering*, 92 (August): 106–15.

Besley, P., Allsop , N.W.H., Ackers, J.C., Hay-Smith, D. and McKenna, J.E. (1999) Waves on reservoirs and their effects on dam protection. *Dams & Reservoirs*, 9 (3): 3–13.

Binnie, G.M. (1981) *Early Victorian Water Engineers*, Thomas Telford, London.

Bishop, A.W. (1955) The use of the slip circle in the stability analysis of slopes. *Géotechnique*, 5 (1): 7–17.

Bishop, A.W. and Bjerrum, L. (1960) The relevance of the triaxial test to the solution of stability problems, in *Proceedings of the Conference on Shear Strength of Cohesive Soil*, Boulder, CO, American Society of Civil Engineers, New York: 437–501.

Bishop, A.W. and Morgenstern, N. (1960) Stability coefficients for earth slopes. *Géotechnique*, 10 (4): 129–50.

Bridle, R.C., Vaughan, P.R. and Jones, H.N. (1985) Empingham Dam – design, construction and performance. *Proceedings of the Institution of Civil Engineers*, 78: 247–89.

BSI (1999) *Code of Practice for Site Investigation*, BS 5930, London.

Casagrande, A. (1961) Control of seepage through foundations and abutments of dams (1st Rankine Lecture). *Géotechnique*, 11 (3): 161–81.

Cedergren, H.R. (1973) Seepage control in earth dams, in *Embankment Dam Engineering – Casagrande Volume* (eds R.C. Hirschfeld and S.J. Poulos), Wiley, New York: 21–45.

—— (1977) *Seepage, Drainage and Flownets*, 2nd edn, Wiley, New York.

Chalmers, R.W., Vaughan, P.R. and Coats, D.J. (1993) Reconstructed Carsington dam: design and performance. *Proc. Institution of Civil Engineers; Water, Maritime and Energy*, 101: 1–16.

Charles, J.A. (1986) The significance of problems and remedial works at British earth dams, in *Proceedings of the Conference on Reservoirs '86*, British National Committee on Large Dams, London: 123–41.

Charles, J.A. (1998) Lives of embankment dams: construction to old age. (1998 Geoffrey Binnie Lecture). *'Dams and Reservoirs'*, 8 (3): 11–23.

Charles, J.A. and Soares, M.M. (1984) Stability of compacted rockfill slopes. *Géotechnique*, 34 (3): 61–70.

Charles, J.A. and Watts, K.S. (1980) The influence of confining pressure on the shear strength of compacted rockfill. *Géotechnique*, 30 (4): 353–67.

—— (1987) The measurement and significance of horizontal earth pressures in the puddle clay core of old earth dams. *Proceedings of the Institution of Civil Engineers. Part 1*, 82 (February): 123–52.

Charles, J.A., Abbiss, C.P., Gosschalk, E.M. and Hinks, J.L. (1991) *An Engineering Guide to Seismic Risk to Dams in the United Kingdom*, Report C1/SFB 187 (H16), Building Research Establishment, Garston.

CIRIA, CUR, CETMEF (2005) *The rock manual. The use of rock in hydraulic engineering*, 2nd edn. Construction Industry Research and Information Association, London.

Coats, D.J. and Rocke, G. (1982) The Kielder headworks. *Proceedings of the Institution of Civil Engineers*, 72: 149–76.

Craig, R.F. (2004) *Soil Mechanics*, 7th edn, Van Nostrand Reinhold, Wokingham.

Creegan, P.J. and Monismith, C.L. (1996) *Asphalt-concrete water barriers for embankment dams.* American Society of Civil Engineers, New York.

Dan, M. (1996) Fuse plug as auxiliary spillway in existing dams, in *Proceedings of Symposium on 'Repair and upgrading of dams'*, Royal Institute of Technology, Stockholm: 475–84.

Das, B.M. (1997) *Advanced Soil Mechanics* (2nd edn), 459 pp., Taylor and Francis, Washington DC.

De Mello, V.F.B. (1977) Reflections on design decisions of practical significance to embankment dams (17th Rankine Lecture). *Géotechnique*, 27 (3): 281–354.

DoE (1991) *Handbook on the design of tips and related structures.* Her Majesty's Stationery Office, London.

DoM & E (1999) *Guidelines on the safe design and operating standards for tailings storage Department of Minerals and Energy.* Government of Western Australia, Perth WA.

Dounias, G.T., Potts, D.M. and Vaughan, P.R. (1996) Analysis of progressive failure and cracking in old British dams. *Géotechnique*, 46 (4): 621–40.

Fell, R., MacGregor, P. and Stapledon, D. (1992) *Geotechnical Engineering of Embankment Dams*, Balkema, Rotterdam.

Gallacher, D. (1988) Asphaltic central core of the Megget Dam, in *Transactions of the 16th International Congress on Large Dams*, San Francisco, ICOLD, Paris, Q61, R39.

Gens, A. (1998) New trends in unsaturated soil mechanics: from fundamentals to engineering practice (BGS Touring Lecture 1998), reported in *Ground Engineering,* May: 26–8.

Gibson, R.E. and Shefford, G.C. (1968) The efficiency of horizontal drainage layers for accelerating consolidation of clay embankments. *Géotechnique*, 18 (3): 327–35.

Giroud, J.P. (1990) Functions and applications of geosynthetics in dams. *Water Power & Dam Construction*, 42 (6): 16–23.

—— (1992) Geosynthetics in dams: two decades of experience. Report. *Geotechnical Fabrics Journal*, 10 (5): 6–9; 10 (6): 22–28.

Hall, M.J., Hockin, D.L. and Ellis, J.B. (1993) *Design of flood storage reservoirs*. Construction Industry Research and Information Association, London.

Head, K.H. (1980) *Manual of Soil Laboratory Testing*, Vol. 1, Pentech, Plymouth.

Hirschfeld, R.C. and Poulos, S.J. (eds) (1973) *Embankment Dam Engineering – Casagrande Volume*, Wiley, New York.

Holton, I.R., Tedd, P. and Charles, J.A. (1999) Walshaw Dean Lower – embankment behaviour during prolonged reservoir drawdown, in *Proceedings of Conference 'The reservoir as an asset'*, British Dam Society, London: 94–104.

Hydraulics Research (1988) *Wave protection in reservoirs*. Report EX1725, Hydraulics Research, Wallingford.

ICE (1996) *Floods and reservoir safety*, 3rd edn, Institution of Civil Engineers, London.

—— (1998) *An application note to: An engineering guide to seismic risk to dams in the United Kingdom*, 40 pp., Thomas Telford, London.

ICOLD (1986) *Geotextiles as Filters and Transitions in Fill Dams*, Bulletin 55, International Commission on Large Dams, Paris.

—— (1989) *Selecting Seismic Parameters for Large Dams*, Bulletin 72, International Commission on Large Dams, Paris.

—— (1991) *Watertight Geomembranes for Dams*, Bulletin 78, International Commission on Large Dams, Paris.

—— (1993a) *Embankment Dams – Upstream Slope Protection*, Bulletin 91, International Commission on Large Dams, Paris.

—— (1993b) *Rock Materials for Rockfill Dams*, Bulletin 92, International Commission on Large Dams, Paris.

—— (1993c) *Reinforced Rockfill and Reinforced Fill for Dams*, Bulletin 89, International Commission on Large Dams, Paris.

—— (1994) *Embankment dams – granular filters and drains.* Bulletin 95, International Commission on Large Dams, Paris.

—— (1996) *A guide to tailings dams and impoundments – design, construction, use and rehabilitation.* Bulletin 106, International Commission on Large Dams, Paris.

Janbu, N. (1973) Slope stability computations, in *Embankment Dam Engineering – Casagrande Volume* (eds R.C. Hirschfeld and S.J. Poulos), Wiley, New York: 447–86.

Jansen, R.B. (ed.) (1988) *Advanced Dam Engineering*, Van Nostrand Reinhold, New York.

Jansen, R.B., Kramer, R.W., Lowe, J. and Poulos, S.J. (1988) Earthfill dam design and analysis, in *Advanced Dam Engineering* (ed. R.B. Jansen), Van Nostrand Reinhold, New York: 256–320.

Justo, J.L. (1973) The cracking of earth and rockfill dams, in *Transactions of the 11th International Congress on Large Dams*, Madrid, Vol. 4, ICOLD, Paris, Communication C.11: 921–45.

Kaptan, C. (1999) The Turkish approach to seismic design. *International Journal on Hydropower and Dams*, 4 (6): 85–93.

Kennard, M.F. (1994) 'Four decades of development of British embankment dams'. (The Geoffrey Binnie Lecture, 1994), in *Proceedings of the Conference on 'Reservoir Safety and the Environment'*, British Dam Society, Thomas Telford, London.

Kennard, M.F., Hoskins, C.G. and Fletcher, M. (1996) Small Embankment Reservoirs. Report 161, CIRIA, London.

Kennard, M.F., Lovenbury, H.T., Chartres, F.R.D. and Hoskins, C.G. (1979) Shear strength specification for clay fills, in *Proceedings of Conference on Clay Fills*, Institution of Civil Engineers, London: 143–7.

Knight, D.J. (1989) Construction techniques and control, in *Proceedings of Conference on Clay Barriers for Embankment Dams*, ICE, Thomas Telford, London: 73–86.

Kulhawy, M.F. and Gurtowski, T.M. (1976) Load transfer and hydraulic fracturing in zoned dams. *Journal of the Geotechnical Division, American Society of Civil Engineers*, 102, GT9: 963–74.

Leps, T.M. (1988) Rockfill dam design and analysis, in *Advanced Dam Engineering* (ed. R.B. Jansen), Van Nostrand Reinhold, New York, pp.368–87.

Lun, P.T.W. (1985) *A Critical Review of Hydraulic Fracturing in Dams*, Technical Report No. 138, CSIRO, Australia.

Macdonald, A., Dawson, G.M. and Coleshill, D.C. (1993) Reconstructed Carsington dam: construction. *Proc. Institution of Civil Engineers; Water, Maritime and Energy*, 101 (March): 17–30.

Maranha das Neves, E. (ed.) (1991) *Advenus in Rockfill Structures*, (NATO ASI, Series E: vol. 200) Kluwer, Dordrecht.

Marsal, R.J. (1973) Mechanical properties of rockfill, in *Embankment Dam Engineering – Casagrande Volume* (eds R.C. Hirschfeld and S.J. Poulos), Wiley, New York: 109–200.

McKenna, J.M. (1989) Properties of core materials, the downstream filter and design, in *Proceedings of Conference on Clay Barriers for Embankment Dams*, ICE, Thomas Telford, London: 63–72.

Medeiros, C.H. de A.C. and Moffat, A.I.B. (1995) A laboratory study of hydraulic fracturing using the Rowe consolidation cell, in *Proceedings of the 11th European Conference on Soil Mechanics and Foundation Engineering*, 3, Danish Geotechnical Society, Copenhagen, 3.185–3.190.

Millmore, J.P. and McNicol, R. (1983) Geotechnical aspects of Kielder Dam. *Proceedings of the Institution of Civil Engineers*, 74: 805–36.

Mitchell, R.J. (1983) *Earth Structures Engineering*, Allen & Unwin, Winchester, MA.

Moffat, A.I.B. (1982) 'Dam deterioration: a British perspective'; in *Proceedings of the Conference 'Reservoirs 82'*, British National Committee on Large Dams, London, Paper 8.

—— (2002) The characteristics of UK puddle clay cores – a review, in *Proceedings of Conference on 'Reservoirs in a changing world'*, British Dam Society, London: 581–601.

Morgenstern, N.R. (1963) Stability charts for earth slopes during rapid drawdown. *Géotechnique*, 13 (2): 121–31.

Morgenstern, N.R. and Price, V.E. (1965) The analysis of the stability of general slip surfaces. *Géotechnique*, 15 (1): 79–93.

Nandi, N., Dutta, S.C. and Roychowdhury, A. (2005) Analysis of shear failure of earthen dams, in *'Dam Engineering'*, XV (4): 255–82.

O'Connor, M.J. and Mitchell, R.J. (1977) An extension of the Bishop and Morgenstern slope stability charts. *Canadian Geotechnical Journal*, 14 (1): 144–55.

Penman, A.D.M. (1986) On the embankment dam (26th Rankine Lecture). *Géotechnique*, 36 (3): 303–48.

Penman, A.D.M. and Charles, J.A. (1975) *The Quality and Suitability of Rockfill Used in Dam Construction*, Current Paper CP87/75, Building Research Establishment, Garston.

Potts, D.M., Dounias, G.T. and Vaughan, P.R. (1990) Finite element analysis of progressive failure at Carrington dam. *Géotechnique*, 40 (1): 79–101.

Ruffle, N. (1970) Derwent dam – design considerations. *Proceedings of the Institution of Civil Engineers*, 45: 381–400.

Sarma, S.K. (1975) Seismic stability of earth dams and embankments. *Géotechnique*, 25 (4): 743–61.

—— (1980) A simplified method for the earthquake resistant design of earth dams, in *Proceedings of the Conference on Dams and Earthquakes*, ICE, Thomas Telford, London: 155–60.

Seed, H.B. (1981) Earthquake-resistant design of earth dams, in *Proceedings of Symposium on Geotechnical Problems and Practice of Dam Engineering*, Bangkok, Balkema, Rotterdam: 41–60.

Sherard, J.L. (1973) Embankment dam cracking, in *Embankment Dam Engineering – Casagrande Volume* (eds R.C. Hirschfeld and S.J. Poulos), Wiley, New York: 271–353.

—— (1985) Hydraulic fracturing in embankment dams, in *Seepage and Leakage from Dams and Impoundments* (eds R.L. Volpe and W.E. Kelly), American Society of Civil Engineers, New York: 115–41.

Sherard, J.L. and Dunnigan, L.P. (1985) Filters and leakage control in embankment dams, in *Seepage and Leakage from Dams and Impoundments* (eds R.L. Volpe and W.E. Kelly), American Society of Civil Engineers, New York: 1–30.

Stephens, T. (1991) *Handbook on small earth dams and weirs – a guide to siting, design and construction*, Cranfield Press, Cranfield.

Taylor, C.A., Dumanoglu, A.A., Trevelyan, J., Maguire, J.R., Severn, R.T. and Allen, A.C. (1985) Seismic analysis of the Upper and Lower Glendevon dams, in *Proceedings of the Conference on Earthquake Engineering in Britain*, Univ. of East Anglia, Thomas Telford, London: 351–64.

Telling, R.M., Menzies, B.K. and Simons, N.E. (1978) Cut-off efficiency, performance and design. *Ground Engineering*, 11 (1): 30–43.

Thomas, H.H. (1976) *The Engineering of Large Dams*, 2 Vols, Wiley, Chichester.

Thompson, D.M. and Shuttler, R.M. (1976) *Design of Riprap Slope Protection Against Wind Waves*, Report 61, CIRIA, London.

USBR (1987) *Design of Small Dams*, 3rd edn, US Government Printing Office, Denver, CO.

Vaughan, P.R., Chalmers, R.W. and Mackay, M. (2000) Ladybower dam: analysis and prediction of settlement due to long term operation, in *Proceedings of Conference on 'Dams 2000'*, British Dam Society, London: 360–77.

Vaughan, P.R., Dounias, G.T. and Potts, D.M. (1989) Advances in analytical techniques and the influence of core geometry on behaviour, in *Proceedings of the Conference on 'Clay Barriers for Embankment Dams'*, ICE, Thomas Telford, London: 87–108.

Vaughan, P.R., Kluth, D.J., Leonard, M.W. and Pradoura, H.H.M. (1970) Cracking and erosion of the rolled clay core of Balderhead Dam and the remedial works adopted for its repair, in *Transactions of the 10th International Congress on Large Dams*, Montreal, Vol. 1, ICOLD, Paris, Q36, R5.

Vaughan, P.R., Kovacevic, N. and Ridley, A.M. (2002) The influence of climate and climate change on the stability of embankment dam slopes, in *Proceedings of Conference on 'Reservoirs in a changing world'*, British Dam Society, London: 353–66.

Vaughan, P.R. and Soares, H.F. (1982) Design of filters for clay cores of dams. *Transactions of the American Society of Civil Engineers*, 108 (6T, 1): 17–321.

Volpe, R.L. and Kelly, W.E. (eds) (1985) *Seepage and Leakage from Dams and Impoundments.* Proceedings of Symposium, Denver, CO, American Society of Civil Engineers, New York.

Wilson, A.C. and Evans, J.D. (1990) The use of low grade rockfill at Roadford dam, in *Proceedings of Conference on the Embankment Dam*, Nottingham, British Dam Society, London.

Wilson, S.D. and Marsal, R.J. (1979) *Current Trends in Design and Construction of Embankment Dams*, American Society of Civil Engineers, New York.

Yarde, A.J., Banyard, L.S. and Allsop, N.W.H. (1996) *Reservoir dams: wave conditions, wave overtopping and slab protection.* Report SR459, HR Wallingford, Wallingford.

Chapter 3

Concrete dam engineering

3.1 Loading: concepts and criteria

3.1.1 Loads

The principal loads operative upon dams were identified in Section 1.7.2 and illustrated schematically in Fig. 1.8. The purpose of this section is to define the magnitude and mode of application of major loads in relation to the analysis of concrete dams. Where appropriate, the character of certain loads is discussed in greater depth.

This chapter assumes an understanding of fundamental principles of hydrostatics and structural theory. The equations quantifying loads and defining structural response are introduced without detailed derivation or proof; reference should be made to appropriate undergraduate texts, e.g. Featherstone and Nalluri (1995) and Case, Chilver and Ross (1993), where derivations are desired.

The transverse section of a concrete gravity dam is illustrated in Fig. 3.1; base plane X–X lies at a depth z_1 below reservoir water level. The expressions detailed in subsequent paragraphs define the pressures and resultant forces applicable to the section above X–X and all similar horizontal planes through the dam. For convenience in analysis loads are expressed per metre length of dam, i.e. they are determined for a two-dimensional transverse section with unit width parallel to the dam axis. It is similarly convenient to account for some loads in terms of resolved horizontal and vertical components, identified by the use of appropriate subscripts, P_h and P_v respectively.

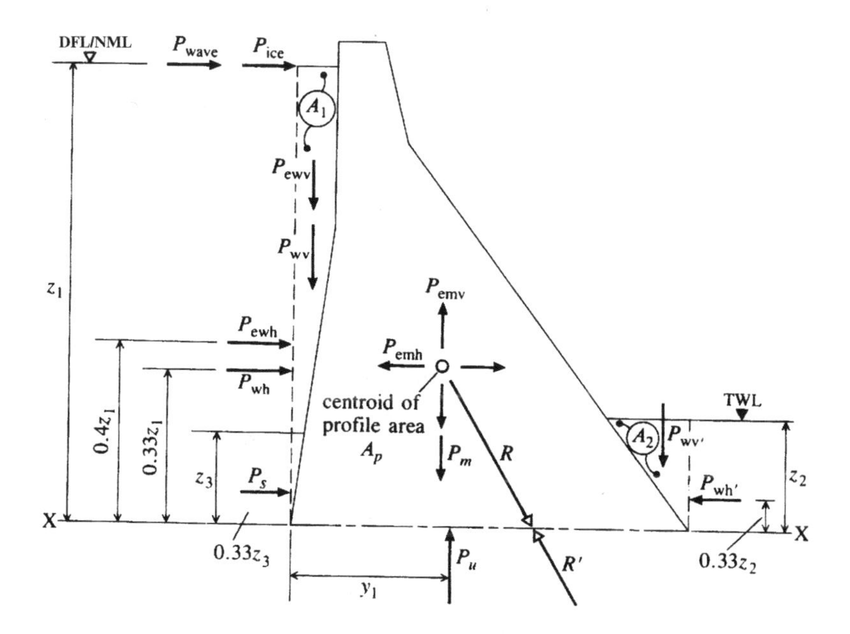

Fig. 3.1 Gravity dam loading diagram. DFL = design flood level; NML = normal maximum level, i.e. maximum retention level of spillweir; TWL = tailwater level

(a) Primary loads

WATER LOAD

The external hydrostatic pressure, p_w, at depth z_1 is expressed as

$$p_w = \gamma_w z_1 \tag{3.1}$$

where γ_w is the unit weight of water, $9.81\,\mathrm{kN\,m^{-3}}$ ($\gamma = \rho g$; symbol γ is adopted here for unit weight to be consistent with the accepted soil mechanics usage employed in Chapter 2, while ρg is used in subsequent chapters in conformity with fluid mechanics usage).

The resultant horizontal force, P_{wh}, is determined as

$$P_{wh} = \gamma_w z_1^2/2\ (\mathrm{kN\,m^{-1}}) \tag{3.2}$$

acting at height $z_1/3$ above plane X–X.

A resultant vertical force P_{wv} must also be accounted for if the upstream face has a batter or flare, as with the profile of Fig. 3.1,

$$P_{wv} = \gamma_w(\text{area } A_1)\ (\mathrm{kN\,m^{-1}}) \tag{3.3}$$

and acts through the centroid of A_1.

The pressure of any permanent tailwater above the plane considered will similarly give rise to the corresponding resultant forces P'_{wh} and P'_{wv} operative above the toe, as illustrated on the figure, and

$$P'_{wh} = \gamma_w z_2^2/2 \tag{3.2a}$$

with

$$P'_{wv} = \gamma_w(\text{area } A_2). \tag{3.3a}$$

Unit weight $\gamma_w = 10\,\text{kN m}^{-3}$ is frequently adopted for convenience in analysis. In a small number of instances, e.g. for flood diversion dams, exceptional suspended sediment concentrations following extreme flood events have been used to justify a further marginal increase in γ_w.

SELF-WEIGHT LOAD

Structure self-weight is accounted for in terms of its resultant, P_m, which is considered to act through the centroid of the cross-sectional area A_p of the dam profile:

$$P_m = \gamma_c A_p\ (\text{kN m}^{-1}). \tag{3.4}$$

γ_c is the unit weight of concrete, assumed as $23.5\,\text{kN m}^{-3}$ in the absence of specific data from laboratory trials or from core samples. Where crest gates and other ancillary structures or equipment of significant weight are present they must also be accounted for in determining P_m and the position of its line of action.

SEEPAGE AND UPLIFT LOAD

Interstitial water pressures, u_w, develop within a concrete dam and its foundation as a result of preferential water penetration along discontinuities, e.g. joint planes, cracks and fine fissures, and also by seepage within the pore structure of rock and concrete. The pressures are directly analogous to porewater pressures in soil mechanics (Chapter 2), and symbol u_w is employed in this chapter for consistency.

The theoretical long-term equilibrium pressure distribution within an 'ideal', i.e. homogeneous and intact, dam and foundation can be established from a flownet such as in the example sketched in Fig. 3.2(a). A curvilinear distribution of pressure exists across a horizontal plane, as indicated in Fig. 3.2(b). In practice, pressure distributions are governed by the nature and frequency of the random discontinuities present, e.g. joints, microfissures etc., and are thus locally indeterminate. It is therefore

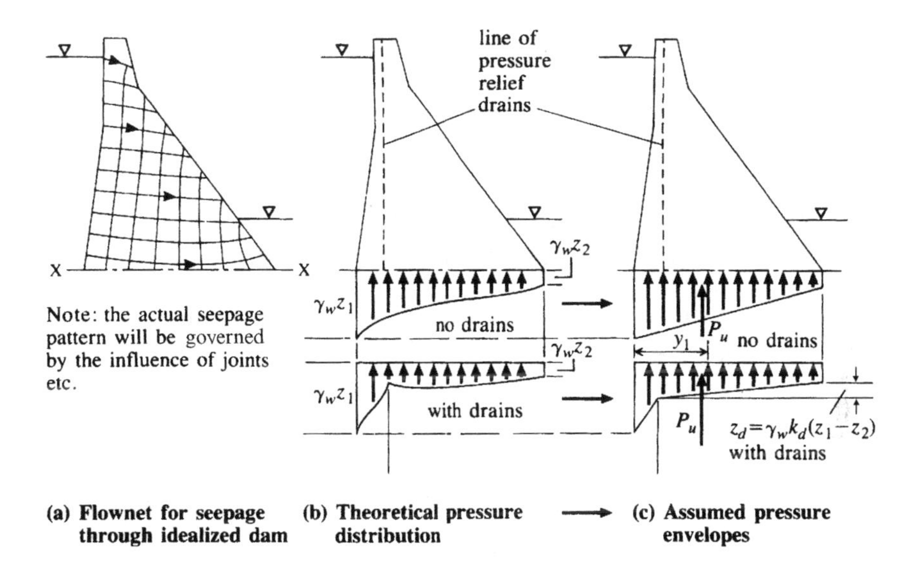

Fig. 3.2 Internal uplift and pressure envelopes

customary to assume a linear or, if pressure relief drains are installed, a bilinear pressure distribution envelope. Pressures are assumed to diminish from the external hydrostatic value $\gamma_w z_1$ at the upstream face to tailwater pressure or to zero, as appropriate, downstream. The assumed linear pressure envelopes are shown schematically in Fig. 3.2(c).

Uplift load, P_u, is represented by the resultant effective vertical components of interstitial water pressure u_w. It is referred to as internal uplift if determined with respect to a horizontal plane through the dam. Determined for an external plane it may be identified as base uplift if at the dam–rock interface, or as foundation uplift if within the underlying rock horizons. P_u is a function of the mean pressure ($u_{w\,avg}$) across a plane and of plane effective area. The latter is defined as the relative proportion of horizontal plane area, A_h, over which interstitial pressure can operate, allowing for the 'solid' mineral skeleton, i.e. effective area $A'_h = \eta A_h$, where η is an area reduction coefficient; thus

$$P_u = \eta A_h (u_{w\,avg})\ (\mathrm{kN\,m^{-1}}). \tag{3.5}$$

If no pressure relief drains are provided or if they cease to function owing to leaching and blockage, then

$$P_u = \eta A_h \gamma_w \frac{z_1 + z_2}{2}. \tag{3.6}$$

Laboratory studies using lean concretes representative of those employed in dam construction suggest that η is a complex and variable parameter, and that its value can approach unity (Leliavsky, 1958; Butler, 1981). Having regard to this and to the control over interstitial pressures exerted by local factors which are quite indeterminate, it is prudent to specify that $\eta = 1.00$ for all analytical purposes. Note that P_u is thus comparable in order of magnitude and importance with water load, P_{wh}.

Referring to the two-dimensional profile of Fig. 3.1 and plane X–X, the nominal plane area is defined by the section thickness, T, i.e. $A_h = T$. Assuming $\eta = 1.00$ and no pressure relief from drains, equation (3.6) then becomes

$$P_u = T\gamma_w \frac{z_1 + z_2}{2} \qquad (3.7)$$

with P_u acting through the centroid of the pressure distribution diagram at distance y_1 from the heel, and

$$y_1 = \frac{T}{3}\frac{2z_2 + z_1}{z_2 + z_1} \text{ (m).} \qquad (3.8)$$

In modern dams internal uplift is controlled by the provision of vertical relief drains close behind the upstream face. Formed drains rise the full height of the dam from an inspection gallery located as low as practicable in relation to the tailwater level.

The mean effective head at the line of drains, z_d, can be expressed as

$$z_d = z_2 + k_d(z_1 - z_2) \text{ (m).} \qquad (3.9)$$

The empirical coefficient k_d is a function of relief drain geometry, i.e. diameter, spacing and location relative to the upstream face. Given an efficient drainage system analysis is commonly based on assuming that $k_d = 0.33$ (Moffat, 1976; USBR, 1976). Relief drain geometry and efficiency are considered further in Section 3.5.3. Modern drains are typically of *c.* 200–250 mm diameter at 3.0–4.0 m centres.

Base and foundation uplift are controlled by a similar system of relief drains drilled from the inspection gallery to depth within the underlying rock. The standard provision of a deep grout curtain upstream of the line of relief drains and below the upstream face (Section 3.5.2), intended to limit seepage, also serves to inhibit pressures within the foundation. Its effectiveness in the latter capacity is much less certain than that of an efficient drain system, and its influence on pressure is therefore normally disregarded.

(b) Secondary loads

SEDIMENT LOAD

The gradual accumulation of significant deposits of fine sediment, notably silts, against the face of the dam generates a resultant horizontal force, P_s. The magnitude of P_s, which is additional to water load P_{wh}, is a function of the sediment depth, z_3, the submerged unit weight γ'_s and the active lateral pressure coefficient, K_a, i.e.

$$P_s = K_a \gamma'_s z_3^2/2 \ (\mathrm{kN\,m^{-1}}) \tag{3.10}$$

and is active at $z_3/3$ above plane X–X. ($\gamma'_s = \gamma_s - \gamma_w$, where γ_s is the sediment saturated unit weight, and

$$K_a \simeq \frac{1 - \sin\phi_s}{1 + \sin\phi_s}$$

where ϕ_s is the angle of shearing resistance of the sediment.)

Values of $\gamma_s = 18–20\,\mathrm{kN\,m^{-3}}$ and $\phi_s = 30°$ are representative, yielding an equivalent fluid unit weight, i.e. $K_a \gamma'_s$, of approximately $3.0\,\mathrm{kN\,m^{-3}}$. Accumulated depth z_3 is a complex time-dependent function of suspended sediment concentration, reservoir characteristics, river hydrograph and other factors (Section 4.5). Accurate prediction is inhibited by major uncertainties, but sediment load is seldom critical in design other than for smaller flood control dams and its introduction is not universal.

HYDRODYNAMIC WAVE LOAD

The transient hydrodynamic thrust generated by wave action against the face of the dam, P_{wave}, is considered only in exceptional cases. It is of relatively small magnitude and, by its nature, random and local in its influence. An empirical allowance for wave load may be made by adjusting the static reservoir level used in determining P_{wh}. Where a specific value for P_{wave} is necessary a conservative estimate of additional hydrostatic load at the reservoir surface is provided by

$$P_{wave} = 2\gamma_w H_s^2 \tag{3.11}$$

(H_s is the significant wave height, i.e. the mean height of the highest third of waves in a sample, and is reflected at double amplitude on striking a vertical face. Wave generation on reservoirs is discussed in Section 4.4.)

ICE LOAD

Ice load can be introduced in circumstances where ice sheets form to appreciable thicknesses and persist for lengthy periods. In such situations ice pressures may generate a considerable horizontal thrust near crest level. The pressures exerted on the dam are a complex function of ice thickness, scale and rate of temperature rise resulting in expansion, and the degree of restraint existing at the perimeter of the ice sheet. They may be increased by wind drag effects.

An acceptable initial provision for ice load, P_{ice}, where considered necessary, is given by $P_{ice} = 145\,kN\,m^{-2}$ for ice thicknesses in excess of 0.6m (USBR, 1976). Where ice thicknesses are unlikely to exceed 0.4m and/or will be subject to little restraint, as on a sloping face, ice load may be neglected.

In the infrequent circumstances where ice load is deemed critical, expected pressures can be estimated by reference to the charts presented in USBR (1976, 1987).

THERMAL AND DAM–FOUNDATION INTERACTION EFFECTS

Cooling of large pours of mass concrete following the exothermic hydration of cement and the subsequent variations in ambient and water temperatures combine to produce complex and time-dependent temperature gradients within a dam. Equally complex interactions develop as a result of foundation deformation or by load transfer between adjacent blocks of the dam. The prediction of such forms of interactive load response lies beyond the scope of this text. Secondary loads in very large dams can be comparable to the primary loads in order of magnitude. Their influence upon deformation and stress distribution in such cases is significant, and is discussed comprehensively in USBR (1976).

(c) *Exceptional loads*

SEISMICITY AND SEISMIC LOAD

A general introduction to seismicity and to concepts in seismic analysis is contained in Section 2.7.3, to which reference should be made.

Concrete dams are quasi-elastic structures and are intended to remain so at their design level of seismic acceleration. They should also be designed to withstand an appropriate maximum earthquake, e.g. CME (controlling maximum earthquake) or SEE (safety evaluation earthquake) (Charles *et al.*, 1991) without rupture. The possibility of structural resonance must also be investigated for higher dams, although the risk of serious resonance is considerably reduced in practice by damping effects. Seismic ground motions are in any event irregular in their magnitude, periodicity and direction. They are therefore unlikely to sustain resonance for durations much exceeding a few seconds.

The natural frequency of vibration, f_n, for a triangular gravity profile of height H(m) and base thickness T(m) constructed in concrete with an effective, or field, modulus of elasticity $E_{eff} \simeq 14\,GN\,m^{-2}$ can be approximated as

$$f_n \simeq 600\,T/H^2 \text{ (Hz)}. \quad (3.12)$$

An alternative approximate relationship is

$$f_n \simeq E_{eff}^{1/2}/0.012H \text{ (Hz)}. \quad (3.13)$$

Investigation of equations (3.12) and (3.13) indicates that resonance is unlikely to occur except in large dams. As examples, the natural frequency of vibration of monolithic gravity profiles with nominal heights of 20 m and 50 m are shown to be of the order of 15–25 Hz and 6–9 Hz respectively compared to major seismic shock frequencies of the order of 1–10 Hz.

While resonance of an entire dam is unlikely, it should be noted that vulnerable portions of a dam may be at risk due to inertia effects. High local stresses may be generated at sudden discontinuities in the profile, e.g. at a change of downstream slope to accommodate the width of a crest roadway. Careful detailing is required to minimize the risk of local overstress and cracking, as illustrated in Fig. 3.3. Similar care is necessary in the design of potentially vulnerable crest structures, e.g. gate or valve towers.

Seismic loads can be approximated using the simplistic approach of pseudostatic or seismic coefficient analysis. Inertia forces are calculated in terms of the acceleration maxima selected for design and considered as equivalent to additional static loads. This approach, sometimes referred to as the equivalent static load method, is generally conservative. It is there-

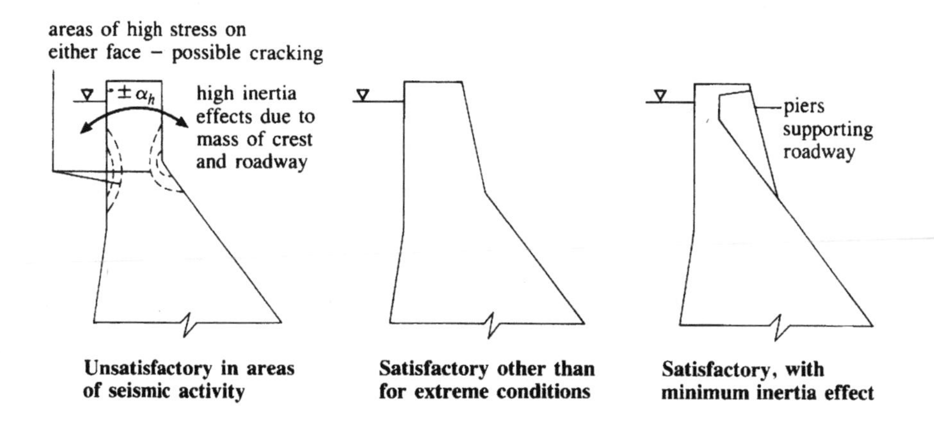

Fig. 3.3 Seismic effects and crest profiles

fore now applied only to smaller and less vulnerable concrete dams, or for purposes of preliminary analysis. For high dams, or dams in situations where seismicity is considered critical, more sophisticated procedures are necessary. In such circumstances seismological appraisal of the dam site should be followed by a comprehensive dynamic response analysis, as discussed in Chopra (1988).

PSEUDOSTATIC ANALYSIS

Pseudostatic inertia and hydrodynamic loads are determined from seismic coefficients α_h and α_v as detailed below.

INERTIA FORCES: MASS OF DAM

Horizontal force is

$$P_{emh} = \pm\alpha_h P_m \tag{3.14a}$$

and vertical force is

$$P_{emv} = \pm\alpha_v P_m. \tag{3.14b}$$

As with self-weight load, P_m, inertia forces are considered to operate through the centroid of the dam section. The reversible direction of the forces will be noted; positive is used here to denote inertia forces operative in an upstream and/or a downward sense, as indicated in Figs 3.1 and 3.4.

HYDRODYNAMIC INERTIA FORCES: WATER REACTION

An initial estimate of these forces can be obtained using a parabolic approximation to the theoretical pressure distribution as analysed in Westergaard (1933).

Relative to any elevation at depth z_1 below the water surface, hydrodynamic pressure p_{ewh} is determined by

$$p_{ewh} = C_e \alpha_h \gamma_w z_{max} \ (\mathrm{kN\,m^{-2}}). \tag{3.15}$$

In this expression z_{max} is the maximum depth of water at the section of dam considered. C_e is a dimensionless pressure factor, and is a function of z_1/z_{max} and ϕ_u, the angle of inclination of the upstream face to the vertical. Indicative values of C_e are given in Table 3.1.

The resultant hydrodynamic load is given by

$$P_{ewh} = 0.66 C_e \alpha_h z_1 \gamma_w (z_1/z_{max})^{1/2} \tag{3.16}$$

Table 3.1 Seismic pressure factors, C_e

Ratio z_1/z_{max}	*Pressure factor,* C_e	
	$\phi_u = 0°$	$\phi_u = 15°$
0.2	0.35	0.29
0.4	0.53	0.45
0.6	0.64	0.55
0.8	0.71	0.61
1.0	0.73	0.63

ϕ_u is the angle of the upstream slope to the vertical.

and acts at elevation $0.40z_1$ above X–X. As an initial coarse approximation, hydrodynamic load P_{ewh} is sometimes equated to a 50% increase in the inertia load, P_{emh}.

The resultant vertical hydrodynamic load, P_{ewv}, effective above an upstream face batter or flare may be accounted for by application of the appropriate seismic coefficient to vertical water load, P_{wv}. It is considered to act through the centroid of area A_1 thus:

$$P_{ewv} = \pm\alpha_v P_{wv}. \qquad (3.17)$$

Uplift load is normally assumed to be unaltered by seismic shock in view of the latter's transient and oscillatory nature.

3.1.2 Load combinations

A concrete dam should be designed with regard to the most rigorous adverse groupings or combinations of loads which have a reasonable probability of simultaneous occurrence. Combinations which include transitory loads of remote probability, and therefore have a negligible likelihood of occurrence in service, are not considered a valid basis for design. Such combinations may be investigated when verifying the design of the most important dams, but are generally discounted in the analysis of lesser structures.

The loads discussed in the preceding section have differing but individually distinctive operating envelopes in terms of probability of occurrence, intensity and duration. Individual load maxima which can reasonably be anticipated to act in concert under service conditions can be grouped into a structured sequence of defined load combinations for design purposes. Within such a sequence the probability of occurrence associated with the nominated load combinations diminishes as their severity is progressively raised.

Three nominated load combinations are sufficient for almost all circumstances. In ascending order of severity they may be designated as *normal* (sometimes *usual*), *unusual* and *extreme* load combinations, here denoted as NLC, ULC and ELC respectively, or by similar terms (USBR, 1976, 1987; Kennard, Owens and Reader, 1996). A tabular summary of nominated load combinations derived from representative UK (and US) practice is presented in Table 3.2.

In studying Table 3.2 it will be observed that a necessary element of flexibility is ensured by the provision of note 3(b). The nominated load combinations as defined in the table are not universally applicable. An

Table 3.2 Nominated load combinations (after Kennard, Owens and Reader, 1996)

Load source	*Qualification*[a]	*Load combination*		
		Normal (or Usual) NLC	*Unusual ULC*	*Extreme ELC*
Primary				
Water	At DFL		✓	
	At NML	✓		✓
Tailwater	At TWL	✓		✓
	Minimum		✓	
Self-weight	–	✓	✓	✓
Uplift	Drains functioning	✓	✓	
	Drains inoperative	✓*	✓*	✓*
Secondary (if applicable)				
Silt	–	✓	✓	✓
Ice	Discretionary	✓	✓[b]	✓
Exceptional				
Seismic	SEE (discretionary)			✓

a DFL = design flood level; NML = normal maximum level, i.e. maximum retention level of spillweir (or gates, if fitted); TWL = maximum tailwater level; SEE = safety evaluation earthquake (see Section 2.7.3).

b ice load should normally include both thermal expansion and wind drag but for the unusual case with a reservoir at flood level, only wind drag need be considered.

* the possibility of blocked uplift relief drains and the degree of blockage is a matter of judgement as to whether it is 'usual', 'unusual' or 'extreme' in likelihood. If deemed 'usual' or 'unusual', in that it could happen more frequently than 'extreme' events, the effect should be studied but an appropriately lower shear friction safety factor (in the range 1.25–2.00) is often accepted.

Note: US practice is in broad accord with the above, with controlling maximum earthquake (CME) substituted for SEE and the additional consideration of concrete temperature under secondary loads (USBR, 1976, 1987).

obligation remains with the designer to exercise discretion in defining load combinations which properly reflect the circumstances of the dam under consideration, e.g. anticipated flood characteristics, temperature régimes, operating rules, etc.

3.1.3 Forces, moments and structural equilibrium

The reactive forces developed in the foundation and/or abutments of the dam in response to applied loads must also be accounted for to satisfy the conditions for static equilibrium. Combination of the applied vertical and horizontal static loads equates to the inclined resultant force, R (Fig. 3.1). This is balanced by an equivalent and opposite reactive resultant R', derived from vertical reactions and the reactive horizontal resistance of the foundation. The conditions essential to structural equilibrium and so to stability can therefore be summarized as

$$\Sigma H = \Sigma V = 0 \tag{3.18a}$$

and

$$\Sigma M = 0. \tag{3.18b}$$

In equations (3.18a) and (3.18b), ΣH and ΣV respectively denote the summation of all active *and* reactive horizontal and vertical forces, and ΣM represents the summation of the moments of those forces with respect to any point.

The condition represented by $\Sigma H = \Sigma V = 0$ determines that no translational movement is possible. The further condition that $\Sigma M = 0$ proscribes any rotational movement, e.g. overturning. With respect to the latter condition some qualification is necessary in relation to the stress distribution as determined by the applied moments.

3.2 Gravity dam analysis

3.2.1 Criteria and principles

The essential criteria governing the structural competence of a gravity dam follow from consideration of equations (3.18a) and (3.18b). Assessed in relation to all probable conditions of loading, including the reservoir empty condition, the profile must demonstrate an acceptable margin of safety with regard to

1. rotation and overturning,
2. translation and sliding and
3. overstress and material failure.

Criteria 1 and 2 control overall structural stability. Both must be satisfied with respect to the profile above all horizontal planes within the dam and the foundation. The overstress criterion, 3, must be satisfied for the dam concrete and for the rock foundation.

The sliding stability criterion, 2, is generally the most critical of the three, notably when applied to the natural rock foundation. The reasons for this are associated with the influence of geological factors and are discussed in Section 3.2.3.

Assumptions inherent in preliminary analyses using the gravity method (Section 3.2.4; USBR (1976, 1987)) are as follows.

1. The concrete (or masonry) is homogeneous, isotropic and uniformly elastic.
2. All loads are carried by gravity action of vertical parallel-sided cantilevers with no mutual support between adjacent cantilevers (monoliths).
3. No differential movements affecting the dam or foundation occur as a result of the water load from the reservoir.

Stability and stress analyses are customarily conducted on the assumption that conditions of plane strain apply. Gravity analysis is therefore carried out on a two-dimensional basis, considering a transverse section of the structure having unit width parallel to the longitudinal axis of the dam, i.e. a vertical cantilever of unit width. Internal stresses are generally determined by the application of standard elastic theories. (It will be appreciated that two-dimensional analysis is inherently conservative, particularly in narrower steep-sided valleys, where the three-dimensional effects afford an additional but not readily determinate margin of safety.)

More sophisticated techniques, including finite element analysis (FEA), are applied to stress determination for larger or more complex structures, or to the investigation of specific problems (see Section 3.2.8).

Attention must be drawn to the importance of maintaining a consistent sign convention for all forces and moments considered in analysis. The convention employed throughout this chapter is illustrated in Fig. 3.4. It may be summarized as specifying all forces, loads and moments which operate in the sense of maintaining equilibrium, e.g. self-weight and self-weight moment, to be positive.

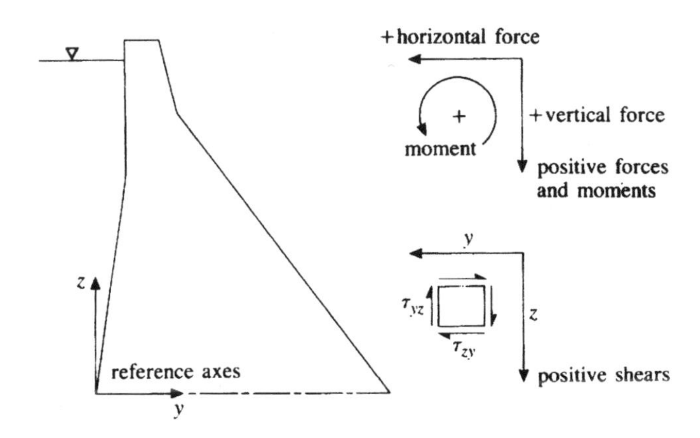

Fig. 3.4 Sign convention diagram: forces, moments and shears

3.2.2 Overturning stability

A simplistic factor of safety with respect to overturning, F_O, can be expressed in terms of the moments operating about the downstream toe of any horizontal plane. F_O is then defined as the ratio of the summation of all restoring (i.e. positive) moments, ΣM_{+ve}, to the summation of all overturning moments, ΣM_{-ve}: thus

$$F_O = \Sigma M_{+ve} / \Sigma M_{-ve}. \tag{3.19}$$

It may be noted that ΣM_{-ve} is inclusive of the moment generated by uplift load. Seismic loads are excluded from overturning calculations on account of their transient and oscillatory nature. Values of F_O in excess of 1.25 may generally be regarded as acceptable, but $F_O \geq 1.5$ is desirable (Kennard, Owens and Reader, 1996).

Overturning of a complete and intact gravity profile of significant size is, strictly, an unrealistic instability mode. The overturning moments, e.g. from water load etc., control internal stress levels, and the overturning mode is therefore closely linked to possible overstress, i.e. to criterion 3. As the nominal margin of safety against overturning is reduced the compressive stresses generated at the downstream toe will rapidly escalate and tensile stress at the upstream heel will initiate local cracking which may propagate (Section 3.2.5), leading to a reduction in sliding resistance.

Determination of a value for F_O is not universal practice owing to the link between overturning moments and stress level. Adequate stability in relation to overturning may be assured by specifying limiting stress levels for the concrete, e.g. 'no tension'.

3.2.3 Sliding stability

Sliding stability is a function of loading pattern and of the resistance to translational displacement which can be mobilized on any plane. It is conventionally expressed in terms of a factor of safety or stability factor against sliding, F_S, estimated using one or other of three definitions:

1. sliding factor, F_{SS};
2. shear friction factor, F_{SF};
3. limit equilibrium factor, F_{LE}.

Irrespective of the definition employed, the resistance to sliding on any plane within a dam will be a function of the shear resistance mobilized in the mass concrete. The horizontal construction joints (Section 3.5.4) will generally be the critical internal planes. At the base, concrete–rock bond and the resulting interface shear strength are the critical factors. Below the base interface the geological structure and shear strength parameters within the rock mass are interdependent, and collectively govern sliding stability.

Sliding resistance in the rock is a function of the surface or path investigated. It is controlled by geological discontinuities such as faulting, joints or surfaces with reduced shear resistance, etc. The geological structure of the rock foundations must therefore be thoroughly investigated, and the presence, nature, frequency and orientation of all significant discontinuities including critical intersections established. Extensive *in situ* and laboratory testing is necessary to confirm design parameters.

(a) Sliding resistance: parameters and stability factors

The resistance to sliding or shearing which can be mobilized across a plane is expressed through the twin parameters c and $\tan\phi$.

Cohesion, c, represents the unit shearing strength of concrete or rock under conditions of zero normal stress. The coefficient $\tan\phi$ represents frictional resistance to shearing, where ϕ is the angle of shearing resistance or of sliding friction, as appropriate (cf. shear strength parameters c and ϕ in soil mechanics usage; Section 2.3.2).

Envelope values for c and $\tan\phi$ recorded within mass concrete, rock, and at a concrete–rock interface are presented in Table 3.3. The great range of values encountered within foundation rocks will be noted.

Resistance to shearing within the foundation zone of a dam is determined by geological structures, rock type and rock integrity. The shearing or sliding parameters for examples of sound and inferior foundation conditions are set out in Table 3.4. The table also shows values for very low strength and potentially dangerous geological features, e.g. clay layers or

Table 3.3 Range of shearing resistance parameters

Location of plane of shearing–sliding		*Cohesion, c (MN m^{-2})*	*Friction, tan φ*
Mass concrete:	Intact	1.5–3.5	1.0–1.5
	Horizontal construction joint	0.8–2.5	1.0–1.5
Concrete–rock interface		1.0–3.0	0.8–1.8
Rock mass:	Sound	1.0–3.0	1.0–1.8
	Inferior	<1.0	<1.0

inclusions, faults, etc., which may prove critical to foundation stability. The data in Table 3.4 is illustrative; it must not be considered universally applicable to the rock types and conditions quoted. A comprehensive summary of recorded shear strength characteristics is given in Link (1969).

Considerable variation in shear strength can occur for one specific rock type within the confines of a site in consequence of local weathering or alteration. Shear strength may also be diminished by saturation in the case of some vulnerable rocks, e.g. certain shales. Illustrative examples of shear strength degradation in such circumstances are shown in Table 3.5.

Table 3.4 Foundation rock shear strength characteristics

Foundation description		*Cohesion, c (MN m^{-2})*	*Friction, tan φ*
Sound conditions			
Generally competent parent rock; few significant discontinuities in mass; no significant degree of alteration or weathering		>1.0	>1.0
Examples (see text):	gneiss	1.3	1.7
	granite	1.5	1.9
	micaschist	3.0	1.3
	sandstone	1.0	1.7
Inferior conditions			
Examples (see text):	gneiss, unaltered	0.6	1.0
	granite, weathered	0.3	1.3
	greywacke	<0.1	0.6
	limestone, open jointed	0.3	0.7
	micaschist	0.4	0.7
	sandstone	0.1	0.6
Critical foundation features			
Examples:	fault or crush zone material	<0.2	<0.3
	clay seam or clayey joint infill	<0.1	<0.2

Table 3.5 Examples of shear strength degradation

Rock type and condition		*Cohesion, $c\,(MN m^{-2})$*	*Friction, $\tan\phi$*
Gneiss 'A':	sound	1.0	1.7
	jointed–decomposed	0.4	0.5
Granite 'B':	sound	1.0	1.8
	weathered–disintegrated	0.1	0.8
Shale 'C':	dry	0.2	0.4
	saturated	0	<0.2

SLIDING FACTOR, F_{SS}

F_{SS} is expressed as a function of the resistance to simple sliding over the plane considered. It is assumed that resistance is purely frictional, and no shear strength or cohesion can be mobilized. F_{SS} can then be defined as the ratio of the summation of all horizontal load components, ΣH, to the summation of all vertical loads, ΣV, on the plane considered, i.e. for a horizontal plane:

$$F_{SS} = \Sigma H / \Sigma V. \tag{3.20}$$

If the plane is inclined at a small angle α, the foregoing expression is modified to

$$F_{SS} = \frac{\Sigma H / \Sigma V - \tan\alpha}{1 + (\Sigma H / \Sigma V)\tan\alpha}. \tag{3.21}$$

Angle α is defined as positive if sliding operates in an uphill sense. The foundation interface is frequently excavated to give a small positive overall inclination α and so raise F_{SS}.

In assessing F_{SS}, ΣH and ΣV are the respective maximum and minimum values appropriate to the loading condition under review, i.e. ΣV is determined allowing for the effect of uplift.

Applied to well-constructed mass concrete, F_{SS} on a horizontal plane should not be permitted to exceed 0.75 for the specified normal load combination. F_{SS} may be permitted to rise to 0.9 under the extreme load combination. Similar maxima for F_{SS} are applied to possible sliding at the base interface on a sound, clean and irregular rock surface, or to sliding on planes within a competent foundation. Planes of low shear resistance will require a significant reduction in the permissible maxima, e.g. F_{SS} may be limited to 0.50 or less on some limestones, schists, laminated shales and similar low-strength foundations.

SHEAR FRICTION FACTOR, F_{SF}

F_{SF} is defined as the ratio of the total resistance to shear and sliding which can be mobilized on a plane to the total horizontal load. With this approach both cohesion and the frictional components of shear strength are accounted for and

$$F_{SF} = S/\Sigma H. \tag{3.22}$$

In the above expression S is the maximum shear resistance which can be mobilized. Referring to Fig. 3.5 it may be defined as

$$S = \frac{cA_h}{\cos\alpha(1-\tan\phi\tan\alpha)} + \Sigma V\tan(\phi+\alpha)\ \ (\mathrm{kN\,m^{-1}}) \tag{3.23}$$

where A_h is the area of plane of contact or sliding (A_h is the thickness, T, for a two-dimensional section).

For the case of a horizontal plane ($\alpha = 0$), equation (3.23) simplifies to

$$S = cA_h + \Sigma V\tan\phi. \tag{3.24}$$

Substitution in equation (3.22) then gives the standard expression for the shear friction factor, i.e.

$$F_{SF} = \frac{cA_h + \Sigma V\tan\phi}{\Sigma H}. \tag{3.25}$$

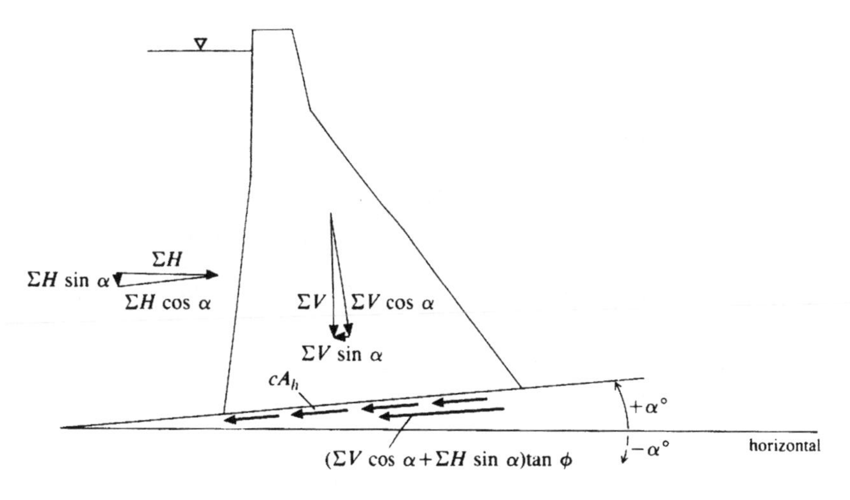

Fig. 3.5 Sliding and shearing resistance: shear friction factor

In some circumstances it may be appropriate to include downstream passive wedge resistance, P_p, as a further component of the total resistance to sliding which can be mobilized. This is illustrated in Fig. 3.6, and may be effected by modifying equation (3.22) accordingly:

$$F_{SF} = (S + P_p)/\Sigma H \tag{3.26}$$

where

$$P_p = \frac{cA_{AB}}{\cos\alpha(1 - \tan\phi\tan\alpha)} + W_w \tan(\phi + \alpha) \tag{3.27}$$

and W_W is the weight of the passive wedge, as shown.

In the presence of a horizon with low shear resistance, e.g. a thin clay horizon or clay infill in a discontinuity, as illustrated in Fig. 3.6, it may be advisable to make the assumption $S = 0$ in equation (3.26).

With the normal load combination applicable, the shear friction factor required in the foundation zone is generally $F_{SF} \geqslant 4.0$. On planes within the dam and at the base interface, $F_{SF} \geqslant 3.0$ is representative. Values

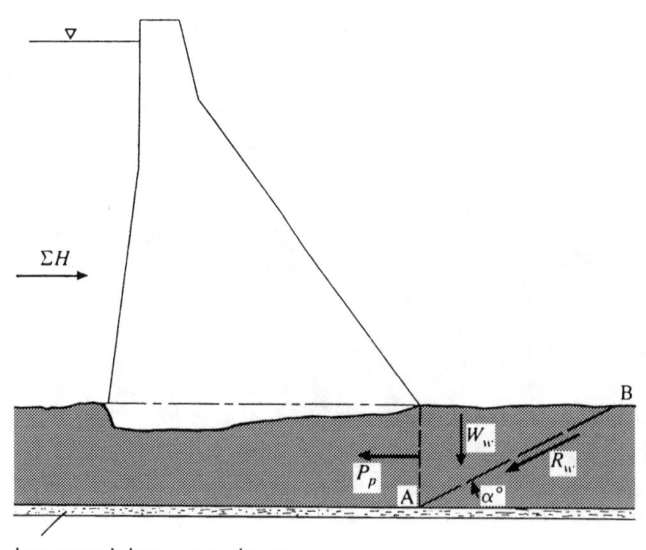

R_w = sliding resistance on inclined plane AB
$= cA_{AB} + (W_w \cos\alpha + \Sigma H \sin\alpha)\tan\phi$
where A_{AB} is the area of plane AB

Fig. 3.6 Sliding: weak seams and passive wedge resistance

Table 3.6 Recommended shear friction factors, F_{SF} (USBR, 1987)

Location of sliding plane	*Load combination*		
	Normal	*Unusual*	*Extreme*
Dam concrete/base interface	3.0	2.0	>1.0
Foundation rock	4.0	2.7	1.3

of F_{SF} required under alternative load combinations are summarized in Table 3.6 (USBR, 1987).

Acceptance of the marginal stability permissible under extreme load combination is a question of engineering judgement, and should only be contemplated for smaller structures and after the most rigorous investigation. For major dams it is recommended practice to relax the values of F_{SF} required for normal load combination by 33% for any load combination which includes seismic effects.

LIMIT EQUILIBRIUM FACTOR, F_{LE}

The limit equilibrium approach to sliding stability follows conventional soil mechanics logic in defining the limit equilibrium factor, F_{LE}, as the ratio of shear strength to mean applied shear stress across a plane, i.e.

$$F_{LE} = \tau_f / \tau \tag{3.28}$$

where τ_f is the shear strength available, and τ is the shear stress generated under the applied loading.

τ_f is expressed by the Mohr–Coulomb failure criterion (Section 2.3.2), and equation (3.28) may be rewritten accordingly:

$$F_{LE} = \frac{c + \sigma_n \tan\phi}{\tau}. \tag{3.29}$$

In the foregoing expression, σ_n is the stress acting normal to the plane of sliding.

Referring to Fig. 3.5, which illustrates a single-plane sliding mode, application of equation (3.29) with appropriate substitutions yields

$$F_{LE} = \frac{cA_h + [\Sigma V \cos\alpha + \Sigma H \sin\alpha]\tan\phi}{\Sigma H \cos\alpha - \Sigma V \sin\alpha}. \tag{3.30}$$

Note that for the case of a horizontal sliding plane ($\alpha = 0$), equation (3.30) simplifies to the expression given in equation (3.25), i.e. $F_{LE} = F_{SF}(\alpha = 0)$.

Equation (3.30) can be developed further for application to multiple-plane sliding surfaces within a complex foundation (Corns, Schrader and Tarbox, 1988). Recommended minima for limit equilibrium factors of safety against sliding are $F_{LE} = 2.0$ in normal operation, i.e. with static load maxima applied, and $F_{LE} = 1.3$ under transient load conditions embracing seismic activity.

(b) Comparative review of sliding stability factors

The expressions defining factors F_{SS}, F_{SF} and F_{LE} differ in their concept of sliding stability. They also differ in their relative rigour and sensitivity to the shearing resistance parameters c and $\tan\phi$. The apparent margin of safety against sliding failure demonstrated by a dam is therefore dependent upon which stability expression is applied. Identification of the most suitable expression requires considered assessment of the limitations of each with respect to the prevailing conditions, notably with regard to foundation complexity and integrity. Confidence in the choice of sliding expression is contingent upon the quality of the foundation investigation programme.

The shear friction stability factor, F_{SF}, is very sensitive to the input values of c and $\tan\phi$ employed. Confidence in the latter is in turn related to the quality and quantity of test data, and hence to the adequacy of the foundation investigations.

The limit equilibrium factor, F_{LE}, is a concept of relatively recent origin (USACE, 1981). In its logic it conforms to wider definitions of stability from soil mechanics practice. F_{LE} shares the input sensitivity of the shear friction factor, and is considered to be more applicable to dams resting on less competent foundations. The method is discussed fully in Nicholson (1983).

It must be stressed that values for F_{SS}, F_{SF} and F_{LE} cannot be directly correlated. The stability factor and sliding criteria most appropriate to a specific dam are determined by the designer's understanding of the conditions. An element of uncertainty must always persist with respect to conditions below the base interface, irrespective of the extent of the investigations carried out. Comprehensive studies are necessary to minimize uncertainty regarding the presence of low-strength layers or inclusions, or discontinuities containing undesirable infill material with low shear resistance. Table 3.7 provides an illustrative comparison of calculated sliding stability factors for a hypothetical simple triangular gravity profile.

Table 3.7 Comparative sliding stability factors; triangular gravity profile

Inclination of plane, α (degrees)		*c = 0; φ = 30°*		*Σc = ΣH; φ = 0°*	
	F_{SS}	F_{SF}	F_{LE}	F_{SF}	F_{LE}
−5	0.71	0.68	0.74	0.86	0.89
0	0.66	0.87	0.87	1.00	1.00
+5	0.55	1.07	1.10	1.13	1.18
+10	0.44	1.33	1.39	1.29	1.41
+15	0.34	1.52	1.79	1.44	1.75

3.2.4 Stress analysis: gravity method

Straight gravity dams are generally analysed by the gravity method of stress analysis. The approach is particularly suited to dams where adjacent monoliths or blocks are not linked by shear keys or by grouted transverse contraction joints (Section 3.5.4). The gravity method is, however, also suitable for preliminary analysis of dams where such continuity is provided. More sophisticated analytical methods are briefly referred to in Section 3.2.8.

Gravity stress analysis derives from elastic theory, and is applied to two-dimensional vertical cantilever sections on the basis of the assumptions listed in Section 3.2.1. The stress analysis makes two further assumptions, namely

4. Vertical stresses on horizontal planes vary uniformly between upstream and downstream face (the 'trapezoidal law') and
5. Variation in horizontal shear stress across horizontal planes is parabolic.

Rigorous analytical techniques reveal that assumptions 4 and 5 are less appropriate for horizontal planes near base level (compare Figs 3.7(a) and 3.7(f)). Stress concentrations develop near heel and toe, and modest tensile stresses may be generated at the heel. Gravity stress analysis is, however, adequate for the initial design of all but extremely large or geometrically complex gravity dams. In the latter instances, the need to account for influences such as, *inter alia*, dam and foundation deformation and interaction between adjacent monoliths will require the use of advanced analytical methods which are the province of specialist texts (e.g. USBR, 1976; Jansen, 1988).

The primary stresses determined in a comprehensive analysis by the gravity method are as follows:

1. vertical normal stresses, σ_z, on horizontal planes;
2. horizontal and vertical shear stresses, τ_{zy} and τ_{yz};
3. horizontal normal stress, σ_y, on vertical planes;
4. major and minor principal stresses, σ_1 and σ_3 (direction and magnitude).

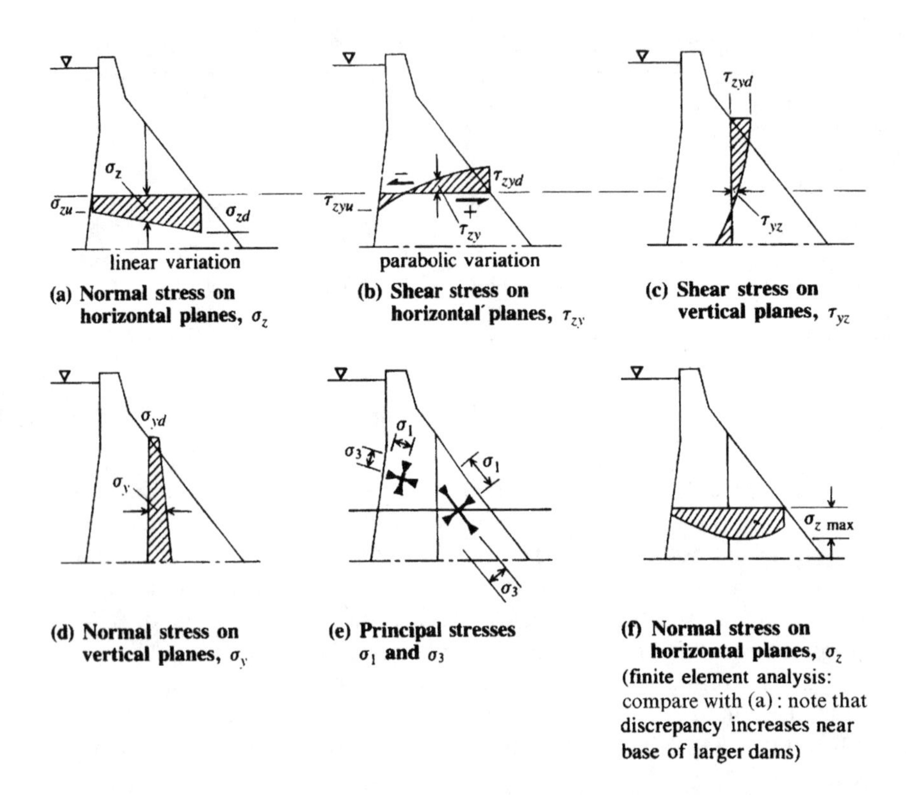

Fig. 3.7 Internal stress distribution: gravity method analysis (after USBR, 1976)

A schematic illustration of the internal variation of these stresses on the basis of gravity method analysis is provided in Figs 3.7(a)–3.7(e). The comparison between Figs 3.7(a) and 3.7(f) will be noted.

Uplift load is excluded from the equations for stress determination set out in this section. It can, if necessary, be accounted for by superposition of local uplift pressures onto calculated stresses (Zienkiewicz, 1963; Moffat, 1976). In practice the influence of internal uplift upon stress patterns is generally disregarded other than in relation to possible horizontal cracking (Section 3.2.5).

(a) Vertical normal stresses

Vertical normal stresses on any horizontal plane are determined by application of the equation for cantilever action under combined axial and bending load with suitable modifications, i.e.

$$\sigma_z = \frac{\Sigma V}{A_h} \pm \frac{\Sigma M^* y'}{I} \qquad (3.31)$$

where ΣV is the resultant vertical load above the plane considered, exclusive of uplift, ΣM^* is the summation of moments determined with respect to the *centroid* of the plane, y' is the distance from the neutral axis of the plane to the point where σ_z is being determined and I is the second moment of area of the plane with respect to its centroid.

Applied to a regular two-dimensional plane section of unit width parallel to the dam axis, and with thickness T normal to the axis, equation (3.31) may be rewritten as

$$\sigma_z = \frac{\Sigma V}{T} \pm \frac{12\Sigma Vey'}{T^3} \tag{3.32}$$

and, at $y' = T/2$,

$$\sigma_z = \frac{\Sigma V}{T}\left(1 \pm \frac{6e}{T}\right) \tag{3.33a}$$

i.e. for the reservoir full load state, at the upstream face,

$$\sigma_{zu} = \frac{\Sigma V}{T}\left(1 - \frac{6e}{T}\right) \tag{3.33b}$$

and, at the downstream face,

$$\sigma_{zd} = \frac{\Sigma V}{T}\left(1 + \frac{6e}{T}\right) \tag{3.33c}$$

where e is the eccentricity of the resultant load, R, which must intersect the plane downstream of its centroid for the reservoir full condition. (The signs in equations (3.33b) and (3.33c) interchange for the reservoir empty condition of loading.)

The eccentricity is determined by evaluating the moments, M^*, given by

$$e = \Sigma M^* / \Sigma V$$

where ΣV excludes uplift.

It is evident from equation (3.33b) that, for $e > T/6$, upstream face stress, σ_{zu}, will be negative, i.e. tensile. This is not permissible in view of the limited and unpredictable tensile strain capacity of concrete (the classic 'middle-third' rule). Total vertical stresses at either face are obtained by the addition of external hydrostatic pressures.

(b) Horizontal shear stresses

Numerically equal and complementary horizontal (τ_{zy}) and (τ_{yz}) shear stresses are generated at any point as a result of the variation in vertical normal stress over a horizontal plane.

It is normally sufficient to establish the boundary, i.e. upstream and downstream, τ values. If the angles between the face slopes and the vertical are respectively ϕ_u upstream and ϕ_d downstream, and if an external hydrostatic pressure, p_w, is assumed to operate at the upstream face, then

$$\tau_u = (p_w - \sigma_{zu}) \tan \phi_u \qquad (3.34a)$$

and

$$\tau_d = \sigma_{zd} \tan \phi_d. \qquad (3.34b)$$

Between the boundary values given by equations (3.34a) and (3.34b) the variation in shear stress is dependent upon the rate of change in vertical normal stress. A graphical solution may be used if it is considered necessary to determine the parabolic distribution generally assumed to apply.

(c) Horizontal normal stresses

The horizontal stresses on vertical planes, σ_y, can be determined by consideration of the equilibrium of the horizontal shear forces operating above and below a hypothetical element within the dam. The difference in shear forces is balanced by the normal stresses on vertical planes. Boundary values for σ_y at either face are given by the following: for the upstream face,

$$\sigma_{yu} = p_w + (\sigma_{zu} - p_w) \tan^2 \phi_u; \qquad (3.35a)$$

for the downstream face,

$$\sigma_{yd} = \sigma_{zd} \tan^2 \phi_d. \qquad (3.35b)$$

(d) Principal stresses

Principal stresses σ_1 and σ_3 may be determined from knowledge of σ_z and σ_y and construction of a Mohr's circle diagram to represent the stress conditions at a point, or by application of the equations given below: for the major principal stress,

$$\sigma_1 = \frac{\sigma_z + \sigma_y}{2} + \tau_{max}, \qquad (3.36a)$$

and for the minor principal stress,

$$\sigma_3 = \frac{\sigma_z + \sigma_y}{2} - \tau_{max}, \tag{3.36b}$$

where

$$\tau_{max} = \left(\left(\frac{\sigma_z - \sigma_y}{2}\right)^2 + \tau^2\right)^{1/2}. \tag{3.36c}$$

The upstream and downstream faces are each planes of zero shear, and therefore planes of principal stress. The boundary values for σ_1 and σ_3 are then determined as follows: for the upstream face,

$$\sigma_{1u} = \sigma_{zu}(1 + \tan^2 \phi_u) - p_w \tan^2 \phi_u, \tag{3.37a}$$

$$\sigma_{3u} = p_w; \tag{3.37b}$$

for the downstream face, assuming no tailwater,

$$\sigma_{1d} = \sigma_{zd}(1 + \tan^2 \phi_d), \tag{3.37c}$$

$$\sigma_{3d} = 0. \tag{3.37d}$$

3.2.5 Permissible stresses and cracking

The compressive stresses generated in a gravity dam by primary loads are very low, seldom exceeding 2.0–3.0 MN m^{-2} except in the largest structures. A factor of safety, F_c, with respect to the specified minimum compressive strength for the concrete, σ_c, is nevertheless prescribed; $F_c \geqslant 3.0$ is a common but seldom critical criterion. Some authorities (USBR, 1976) relate the values prescribed for F_c to the load combination applied, but qualify them in terms of absolute maxima, as shown in Table 3.8. The table also summarizes corresponding factors of safety, F_r, specified with regard to the compressive strength of the rock foundation, σ_r.

Table 3.8 Permissible compressive stresses (after USBR, 1976)

Load combination	*Minimum factor of safety on compressive strength*	
	F_c (concrete)	*F_r (rock)*
Normal	3.0 ($\sigma_{max} \ngtr 10$ MN m^{-2})[a]	4.0
Unusual	2.0 ($\sigma_{max} \ngtr 15$ MN m^{-2})[a]	2.7
Extreme	1.0	1.3

a σ_{max} is the maximum allowable compressive stress.

Horizontal cracking is sometimes assumed to occur at the upstream face if σ_{zu} (computed without uplift) falls below a predetermined minimum value:

$$\sigma_{zu\,min} = \frac{k'_d \gamma_w z - \sigma'_t}{F'_t} \tag{3.38}$$

where k'_d is a drainage factor ($k'_d = 0.4$ if drains are effective; $k'_d = 1.0$ if drains are not present, or blocked), σ'_t is the tensile strength of the concrete across a horizontal joint surface and F'_t is a factor of safety, scaled to load combination ($F'_t = 3.0$ for NLC, 2.0 for ULC and 1.0 for ELC).

Cracking is normally permissible only under the extreme load combination, except in the case of old dams. The crack is assumed to propagate to the point where $\sigma_z = p_w$. Stability and stresses are then reassessed for the uncracked thickness of the section, and the dam is considered safe if the resulting stresses remain within specified maxima and sliding stability remains adequate (USBR, 1976, 1987).

3.2.6 Upstream face flare

The upstream face of a gravity profile is frequently modified by the introduction of a significant flare, as shown in Fig. 3.8. A flare is advantageous in that it serves to lengthen the base contact, and so considerably enhances the contact area available at foundation level to distribute stress and resist sliding. The influence of a generous flare can also prove significant in relation to providing an adequate margin of safety against seismic load. For profiles having a considerable crest width in relation to their height, the

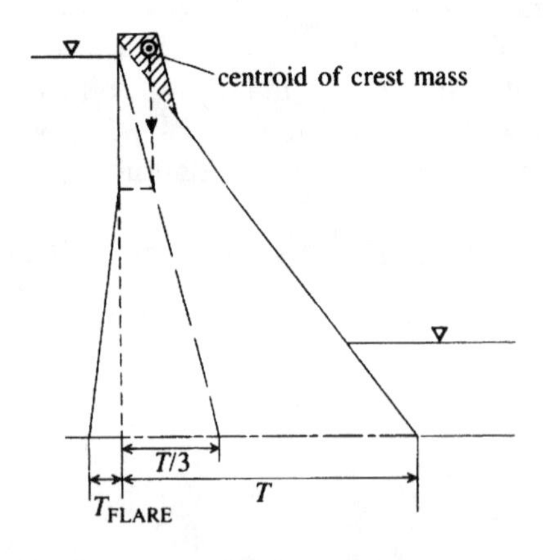

Fig. 3.8 Upstream flare diagram

flare also serves to reduce any tensile stress which may generate at the downstream face under the reservoir empty condition.

The optimum height for the flare can be set by the elevation where the line of action of the crest mass, shaded on Fig. 3.8, intersects the upstream middle third line, as shown. The flare width required can be calculated by equating the summation of the self-weight moments about the revised upstream third point to zero.

3.2.7 Gravity profile selection

The primary load régime for a gravity dam of given height is essentially fixed. Little scope exists to modify the 'standard' triangular profile significantly, and a vertical upstream face is invariably associated with a mean downstream slope equating to a ratio for base thickness to height to spillweir level of the order of 0.75:1.00. Design of a small dam may therefore be based on adopting such a geometry, checking its adequacy, and effecting any minor modification necessary.

In the case of larger dams a unique profile should be determined to match the specific conditions applicable. Two approaches are possible: the multistage and the single stage.

The multistage approach defines a profile where the face slopes are altered at suitable intervals, as in the example of Fig. 3.9. Design commences from crest level, and descends through profile stages corresponding to predetermined elevations. Each stage is proportioned so as to maintain stress levels within acceptable limits, e.g. no tension under any condition of loading. The resulting profile allows marginal economies in concrete, but may be more expensive to construct than the single-stage equivalent. Multistage profiles are now seldom employed, even on large dams.

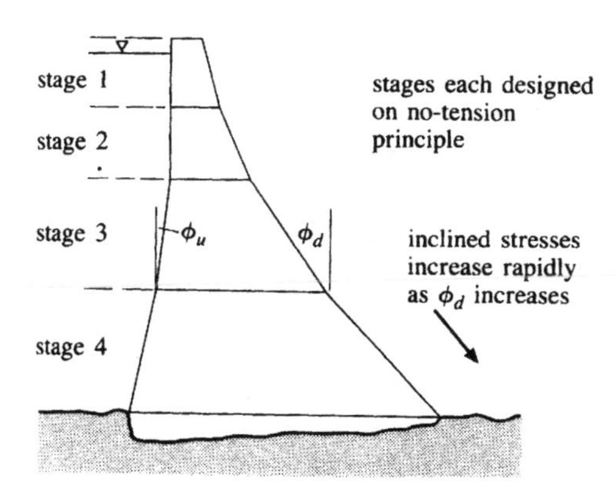

Fig. 3.9 Multistage gravity dam profile

Single-stage design is based upon definition of a suitable and uniform downstream slope. The apex of the theoretical triangular profile is set at or just above maximum retention level (or DFL) and an initial required base thickness, T, determined for each loading combination in terms of a satisfactory factor of safety against overturning, F_O. The critical value of T is then checked for sliding stability and modified if necessary before checking heel and toe stresses at base level.

An approximate definition of the downstream slope in terms of its angle to the vertical, ϕ_d, required for no tension to occur at a vertical upstream face is given by

$$\tan \phi_d = \left[1 \bigg/ \left(\frac{\gamma_c}{\gamma_w} - \eta \right) \right]^{1/2}. \qquad (3.39)$$

3.2.8 Advanced analytical methods

The simplifying assumptions on which gravity method analysis is based become progressively less acceptable as the height of the dam increases, particularly in narrower steep-sided valleys. Cantilever height changes rapidly along the axis of the dam, and interaction between adjacent monoliths results in load transfer and a correspondingly more complex structural response. This interaction is further compounded by the influence of differential foundation deformations. Where it is necessary to take account of such complexities, two alternative rigorous analytical approaches are appropriate.

(a) Trial load twist analysis

This approach is suited to situations in which, as a consequence of valley shape, significant twisting action can develop in the vertical cantilevers or monoliths. Torsional moments are generated as a result of cantilever interaction, and some water load is transferred to the steep abutments, resulting in stress redistribution. The applied loads are thus carried by a combination of cantilever action, horizontal beam action and twisting or torsion.

Trial load analysis (TLA) is conducted by subdividing the dam into a series of vertical cantilever and horizontal beam elements, each of unit thickness and intersecting at defined node points. A trial distribution of the loading is then made, with a proportion of the nodal load assigned to each mode of structural behaviour, i.e. cantilever, horizontal beam, torsion, etc. The relevant node point deflections for each response mode are then determined. An iterative solution of the resulting complex array of equations is necessary in order to make the mode deflections, δ, all match, i.e., for a correct load distribution,

$$\delta_{\text{cantilever}} = \delta_{\text{beam}} = \delta_{\text{torsion}}$$

at all node points. The approach is detailed in USBR (1976), but has now largely been displaced by more powerful and flexible finite element techniques.

(b) Finite element analysis (FEA)

The finite element approach considers the dam and that portion of the supporting foundation lying within the structural zone of influence of the dam as an assemblage of distinct elements, interconnected at node points at their vertices. A mathematical model is constructed based on determining the nodal displacements which, in turn, define the state of strain, and therefore of stress, within each element. An acceptable mathematical representation of the load response of the concrete and the foundation strata is therefore important, accuracy of representation being balanced against computational workload. The quality of the output is a function of element type, mesh configuration and mesh size. Accuracy is enhanced by a finer mesh, but the computational effort is greatly increased and may be unacceptably high. A compromise between mesh fineness and the accuracy of the solution is therefore necessary.

Analysis of problems which can be considered in terms of plane stress or plane strain, as with the gravity dam, can be conducted using a two-dimensional mesh of quadrilateral elements. Where a simplification to plane stress or plane strain conditions is not justifiable, a three-dimensional analysis is necessary, e.g. for an arch or cupola dam. Three-dimensional analysis is frequently based on the use of isoparametric hexahedronal elements having eight node points, and the computing power required is considerable. The finite element approach is an elegant, flexible and extremely powerful analytical tool. Secondary loads, e.g. temperature effects, deformation, etc., can be introduced and the method is, in principle, well suited to parametric studies. As examples, the effect of rock stiffness on structural response, or the influence of a major crack or opening in the dam can be studied. The method is particularly suited to analysing the complex shape of arch or cupola dams.

A comprehensive introduction to finite element methods is given in Zienkiewicz (1977). The application of the technique to the analysis of dams is reviewed in Clough and Zienkiewicz (1978). The effort and specific computational skills necessary for setting up an FEA application are such that many analyses are conducted using commercially available software suites. Preparation of the mathematical model using such software is nevertheless a task requiring particular expertize and considerable care in the selection of input parameters and data.

3.2.9 Stabilizing and heightening

A number of older gravity dams are underdesigned by present-day standards, having undesirably low stability. The underdesign is a reflection of past limitations on the understanding of loads, materials and dam behaviour. In some cases stability margins have been eroded by revised loading criteria, e.g. uprated flood maxima. In others, distress has resulted from progressive deterioration. Remedial action to improve stability can be taken by

1. vertical prestressing of the dam or
2. construction of a downstream supporting shoulder or berm.

(a) Prestressing

Prestressing provides an additional vertical load with a resultant line of action close to the upstream face. It requires the introduction of highly stressed steel-strand cables or tendons into the dam, anchored at crest level and at depth within the foundation. A necessary condition for prestressing is therefore the presence of rock competent to accept the high anchorage loads imposed. The principle of prestressing is illustrated in Fig. 3.10.

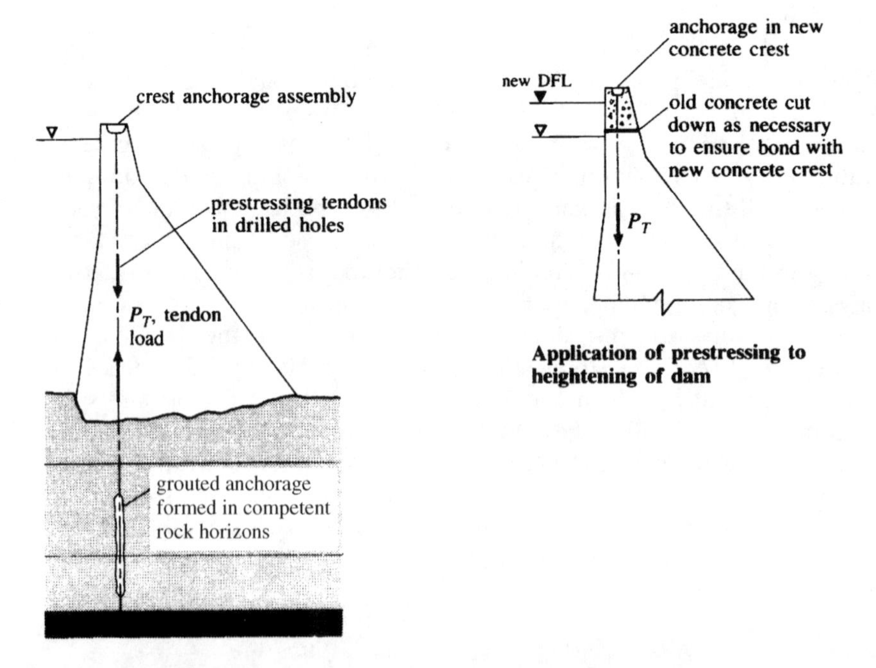

Fig. 3.10 Stabilizing and heightening: vertical prestressing

At suitable intervals along the crest holes are cored through the dam and into the foundation. Prestressing cable or tendon assemblies are inserted and a grouted lower anchorage formed as shown. The strands are carried through an upper anchorage assembly at crest level, where they are stressed to the desired load before being finally secured. The prestress load is thus distributed into the dam from the upper anchorage, and careful detailing with local reinforcement will be necessary to avoid overstress of the adjacent concrete.

The resultant prestress load operates as an adjunct to P_{m}, and so improves sliding stability. Its positive moment with respect to the toe acts to increase overturning stability. The compressive stresses generated internally must be superimposed on the stresses generated by other loads. Note that for the reservoir full condition the effect of superposition is to eliminate any upstream tensile stresses, but that under the reservoir empty condition tensile stresses may develop at the downstream face due to the prestress.

The design prestress, P_{ps} (kN m^{-1}), should satisfy the more demanding of the stability modes, i.e. overturning or sliding, having regard to the safety factor desired for each. Sliding stability generally proves the more demanding criterion. P_{ps} is determined by the inclusion of a separate prestress term alongside F_{O} and F_{SF} in the stability expressions of sections 3.2.2 and 3.2.3 and solving for P_{ps} in terms of the specified stability factors of safety. For overturning (F_O, from equation (3.19)),

$$P_{\mathrm{ps}} = \frac{F_{\mathrm{O}}(\Sigma M_{-\mathrm{ve}}) - (\Sigma M_{+\mathrm{ve}})}{y_2} \ (\mathrm{kN\,m^{-1}}) \tag{3.40}$$

where y_2 is the moment arm of P_{ps} with respect to the toe. For sliding (F_{SF}, from equation (3.25)),

$$P_{\mathrm{ps}} = \frac{(F_{\mathrm{SF}}\Sigma H - cA_{\mathrm{h}})}{\tan\phi} - \Sigma V. \tag{3.41}$$

Equivalent expressions may be derived giving P_{ps} for sliding in terms of F_{SS} or F_{LE}.

The prestressing tendons are typically located at 3–7 m centres along the crest. The prestress load required for each, P_{T} (kN), is the appropriate multiple of P_{ps}. A recent UK instance of prestressing, at Mullardoch dam, is described by Hinks *et al.* (1990).

Prestressing also provides a structurally efficient and economical approach to heightening existing dams. In its absence additional concrete must be placed to thicken as well as raise the profile in order to ensure adequate stability. Prestressing is technically preferable, avoiding questions about compatibility and structural unity of old and new concrete in a thickened profile.

(b) Downstream supporting shoulder

A supporting shoulder can be constructed, as shown in Fig. 3.11, employing compacted earthfill or rockfill. Support is provided from the positive contribution to stability made by the weight of fill above the downstream face, W_F. A further contribution is made by the horizontal 'at rest' pressure, P_{ds}, generated on plane AB.

$$W_F = \gamma(\text{area } A)\,(\text{kN m}^{-1}) \tag{3.42}$$

and may be considered to act through the centroid of the fill profile of area A above the toe, and

$$P_{ds} = K_0 \gamma z_{AB} Z\,(\text{kN m}^{-1}) \tag{3.43}$$

where K_0 is the 'at rest', or zero lateral strain, pressure coefficient and z_{AB} and Z are defined as in Fig. 3.11. P_{ds} is considered to act at height $z_{AB}/3$ above the base plane. Illustrative values for K_0 are shown in Table 3.9.

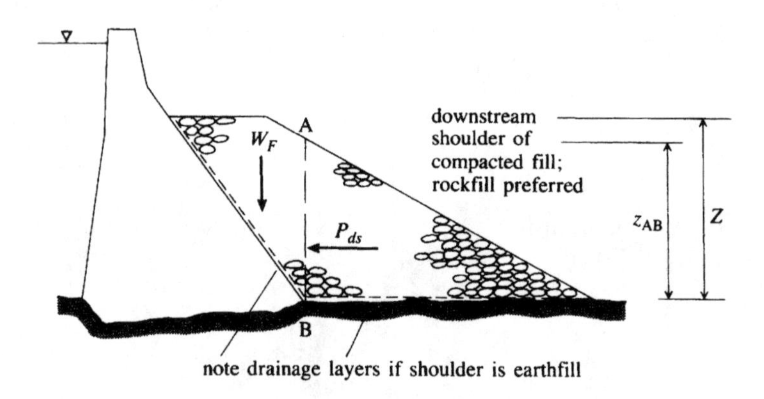

Fig. 3.11 Stabilizing: construction of downstream supporting shoulder

Table 3.9 Illustrative values for coefficient, K_0

Shoulder fill	*Coefficient, K_0*
Compacted rockfill	0.30–0.60
Compacted sand	0.45–0.60
Compacted clay	1.00–1.50
Heavily compacted clay	>1.50

The selection of compacted rockfill for the supporting shoulder offers the advantage, as opposed to earthfill, of good drainage and steeper side slopes. A further material for shoulder construction is rolled dry lean concrete (RDLC), introduced in Section 3.7.2. The construction of a downstream rockfill shoulder to improve stability of the 45 m high Upper Glendevon dam in the UK is described in Macdonald, Kerr and Coats (1994).

3.3 Buttress dam analysis

3.3.1 General

Buttress dams fall into two distinct groups, as identified in Section 1.4. Massive diamond or round-headed buttress dams representative of modern practice were illustrated schematically in Figs 1.4(b) and 1.4(c). The earlier but now largely obsolete flat slab or Ambursen buttress dam was also shown, in Fig. 1.5(c). The latter type is not considered further. Relative to the gravity dam the principal advantages of the massive buttress dam lie in its obvious economy of material and in a major reduction in uplift load. The buttress dam also offers greater ability to accommodate foundation deformation without damage. The advantages listed are offset by considerably higher finished unit costs (overall cost of the completed dam per m^3 of concrete) for the reduced quantities of concrete employed. This is attributable largely to the cost of the more extensive and frequently 'non-repetitive' formwork required. Significantly increased stresses are also transmitted into the foundation of each buttress. The criteria for foundation competence are therefore appreciably more rigorous than those applicable to the gravity dam.

3.3.2 Buttress analysis and profile design

Buttress dam analysis parallels gravity dam practice in being conducted in two phases, stability investigations preceding the determination of stresses within the profile. The structural form of the buttress dam, detailed in the example of Fig. 3.12, has two important consequences with respect to primary loads. First, uplift pressures are effectively confined to the buttress head, resulting in the modified uplift distribution shown in Fig. 3.12. Pressure relief drains are therefore only necessary in exceptional cases. As a further consequence of the form the vertical component of the water load P_{wv}, on the sloping upstream face is very much enhanced relative to any gravity profile. Stability against overturning is therefore a less meaningful design criterion.

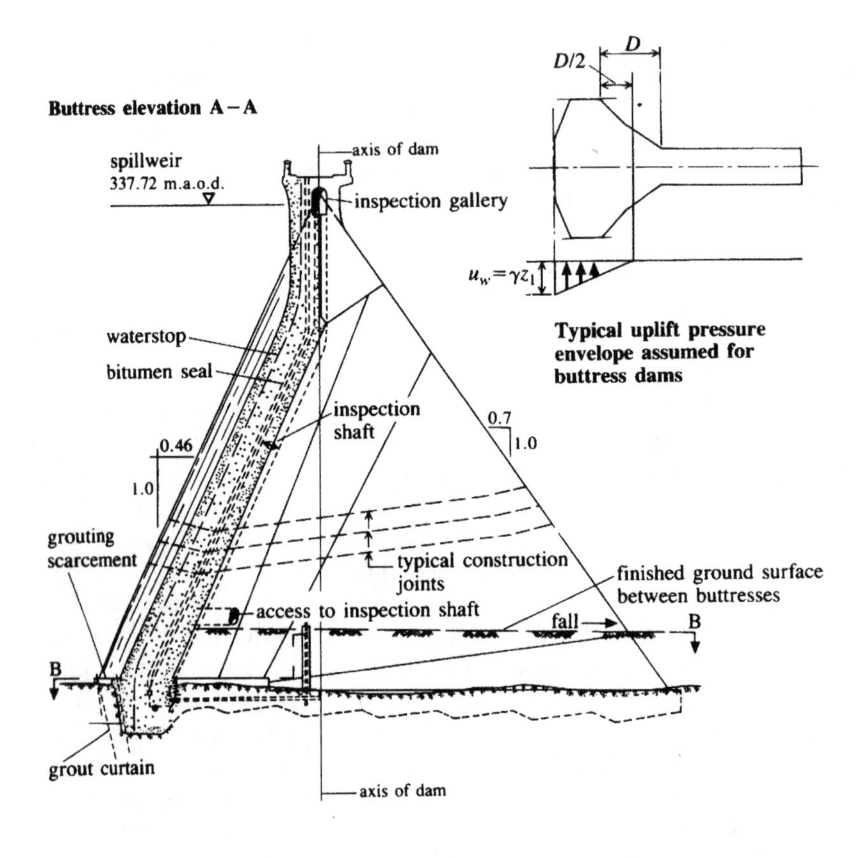

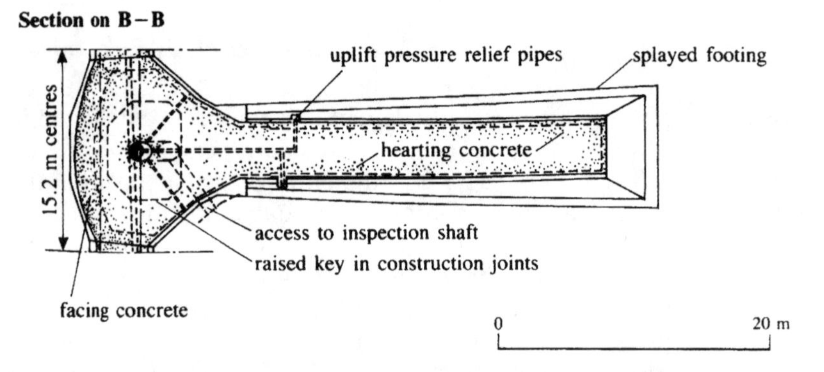

Fig. 3.12 Shira buttress dam, UK

In structural terms the massive buttress dam is constructed from a series of independent 'units', each composed of one buttress head and a supporting buttress, or web. Each unit has a length along the axis of the dam of about 12–15 m. Structural analysis is therefore conducted with respect to the buttress unit as a whole. The sliding stability of one complete unit is investigated in terms of F_{SS}, sliding factor or, more usually, F_{SF}, shear friction factor, in accordance with the principles of these approaches as outlined in Section 3.2.3. The design minimum values for F_{SS} and F_{SF} are normally comparable with those required of a gravity profile.

Stress analysis of a buttress 'unit' is complex and difficult. Modern practice is to employ finite element analyses to assist in determining the optimum shape for the buttress head to avoid undesirable stress concentrations at its junction with the web. An approximate analysis of the downstream portion of the buttress web, using modified gravity method analysis, is possible if the web is parallel sided. The sides of the buttress web are, however, generally flared towards base level, to increase contact area and hence sliding resistance and to moderate the contact stress on the foundation.

Profile design for a buttress dam is not subject to the simplifications outlined in Section 3.2.7. A trial profile is established on the basis of previous experience, the selection of a round head or a diamond head being largely at the discretion of the designer. The profile details are then modified and refined as suggested by initial stress analyses.

3.4 Arch dam analysis

3.4.1 General

The single-curvature arch dam and its natural derivative, the double-curvature arch or cupola with vertical and horizontal curvature, were introduced in Section 1.4 and in Figs 1.4(d) and 1.4(e) respectively. The valley shape and rock conditions which may favour consideration of an arched dam in preference to gravity or rockfill alternatives were outlined in Section 1.6.6 and Table 1.7.

Arch and cupola dams transfer the greater proportion of the water load to the valley sides rather than to the floor, functioning structurally by a combination of arch action and vertical cantilever action. Abutment integrity and stability are therefore critical, and the importance of this point cannot be overstated. Progressive abutment deformation or yielding in response to arch thrust results in load transfer and stress redistribution within the dam shell and in the abutment itself. In more extreme situations of significant abutment yielding or instability local overstress of the dam wall will ensue, destroying arch action and resulting in catastrophic collapse, as occurred with the 61 m high Malpasset cupola dam (France) in 1959.

The concepts of overturning and sliding stability applicable to gravity or buttress analysis have little relevance to the arch or cupola. An arch represents a stable structural form and, given that the integrity of the supporting abutments is assured, failure can occur only as a result of overstress. Arch dam design is therefore centred largely upon stress analysis and the definition of an arch geometry which avoids local tensile stress concentrations and/or excessive compressive stress. In achieving this objective it is frequently necessary to adopt varying curvatures and thicknesses between arch crown and abutment and also from crest level to base.

The arch and cupola dam offer great economies in volume of concrete. In the case of a slender cupola the saving in volume may exceed 80% of that necessary for an equivalent gravity profile. Associated savings may also be realized in foundation excavation and preparation. As in the case of buttress dams (Section 3.3), the sophisticated form of the cupola leads to very much increased finished unit costs. In financial terms, therefore, the potential overall economies may be significantly diminished. In the case of a geologically complex and difficult site they may be completely negated by stabilization costs associated with the paramount requirement of ensuring abutment integrity under all conditions.

The structural interaction between a loaded arch or cupola shell and its supporting abutments is extremely complex. This section is therefore restricted to treating the preliminary elastic analysis of single-curvature arch shells employing classical ring theory. The advanced mathematical modelling techniques required for rigorous arch analysis are identified, but their treatment lies outside the scope of this text and reference should be made to USBR (1977) and Boggs, Jansen and Tarbox (1988). The application of physical modelling methods to arch analysis is referred to in Section 16.2.

3.4.2 Arch geometry and profile

The horizontal component of arch thrust must be transferred into the abutment at a safe angle, i.e. one which will not promote abutment yielding or instability. At any elevation the arch thrust may be considered to enter the abutment as shown in Fig. 3.13. Horizontal thrust is then assumed to distribute into the rock with an included angle of 60° as indicated. In distributing through the abutment the thrust must not be aligned too closely with the valley sound rock contours or with any major discontinuity which may contribute to abutment instability. In general terms this suggests an abutment entry angle, β (Fig. 3.13), of between 45° and 70°. It is apparent that the horizontal arch radius and therefore the arch stresses

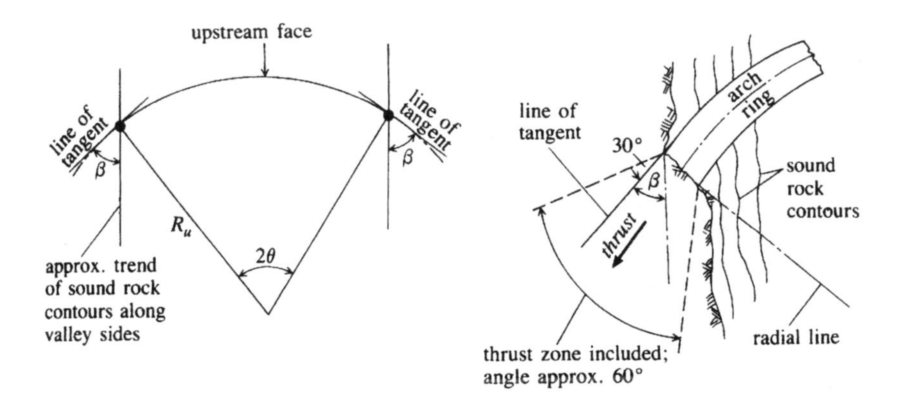

Fig. 3.13 Abutment entry angle geometry for arch dams

and the volume of the arch will be functions of the entry angle selected. The optimum value of β is determined from careful assessment of the geological structure and associated design parameters. In a perfectly symmetrical valley the minimum dam volume would theoretically occur with $2\theta = 133°$ at all elevations. Considerations of abutment entry angle preclude this, and in practice the central angle at crest level is generally limited to $2\theta \simeq 70°–110°$.

Arch and cupola profiles are based on a number of geometrical forms, the more important of which are introduced below.

(a) Constant-radius profile

The constant-radius profile has the simplest geometry, combining a vertical upstream face of constant radius with a uniform radial downstream slope. The downstream face radius therefore varies with elevation. The profile is shown schematically in Fig. 3.14(a), and it is apparent that the central angle, 2θ, reaches a maximum at crest level. A constant-radius profile is not the most economical in volume, but has the merit of analytical and constructional simplicity. The profile is suited to relatively symmetrical U-shaped valleys.

(b) Constant-angle profile

The concept of the constant-angle profile follows logically as a development of the constant-radius profile of minimum volume. Constant-angle geometry is more complex, however, since, as demonstrated by Fig. 3.14(b), it leads to a considerable upstream overhang as the abutments are approached. Excessive overhang is undesirable, as the resulting local cross-section can prove unstable under construction or for the reservoir

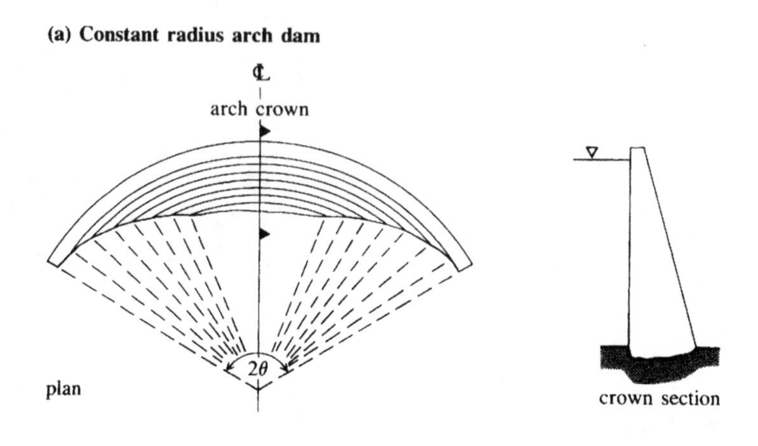

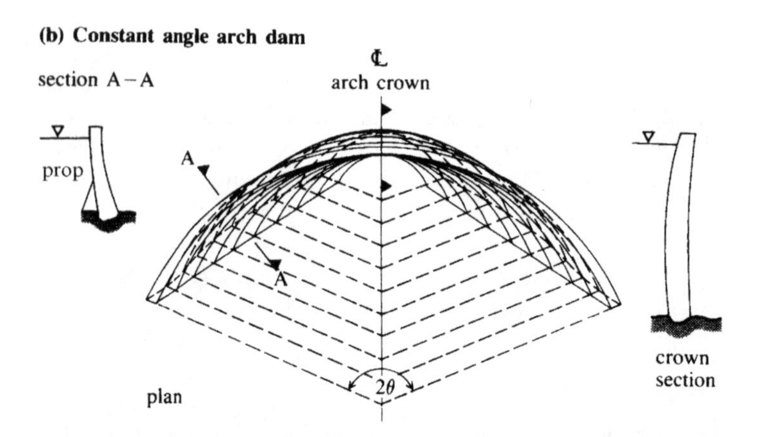

Fig. 3.14 Arch dam geometries and profiles

empty condition. To alleviate this it may be necessary to introduce an upstream prop, as indicated on the figure, or to modify the central angle 2θ. The profile is best suited to narrow and relatively symmetrical steep-sided V-shaped valleys.

(c) Cupola profile

In its developed form the double-curvature cupola has a particularly complex geometry and profile, with constantly varying horizontal and vertical radii to either face. An example of the type demonstrating the geometrical complexity is shown in Fig. 3.15. A trial geometry suitable for preliminary design purposes can be selected using the nomograms

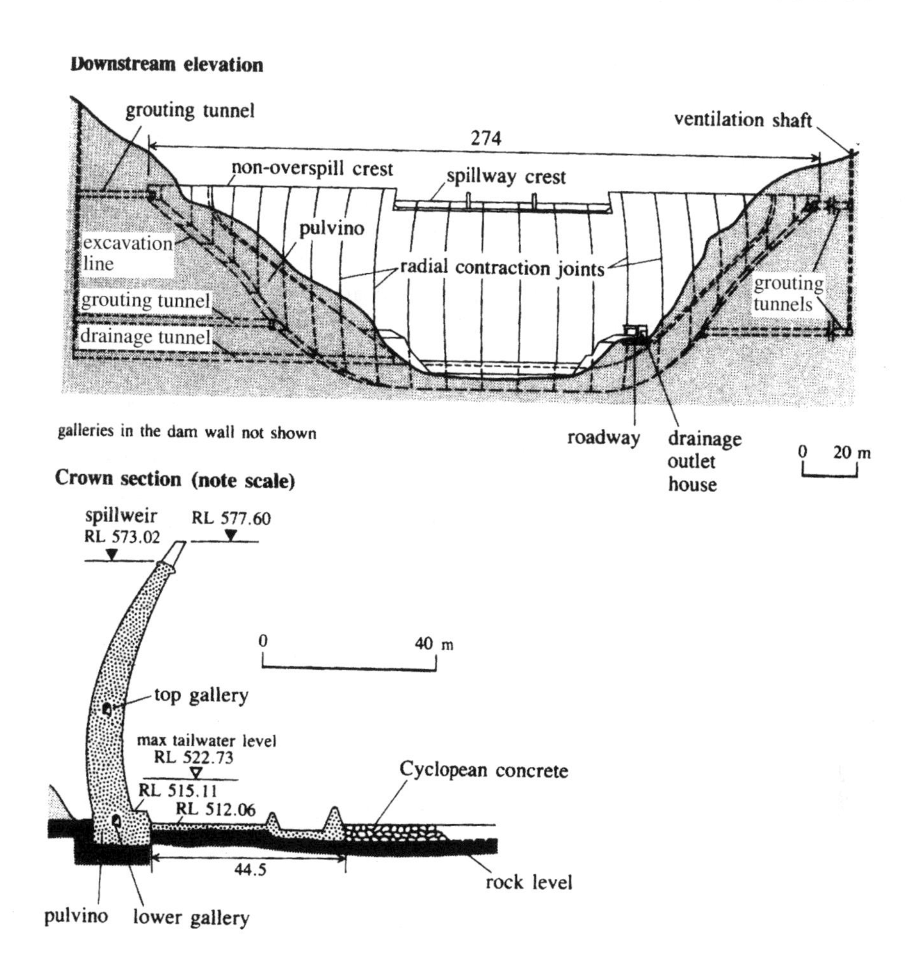

Fig. 3.15 Roode Elsberg cupola dam, South Africa

presented in Boggs (1975), and refined as necessary by mathematical or physical models.

3.4.3 Arch stress analysis: elastic ring theory

Elastic analysis based on the application of classical ring theory is adequate for the initial study of simple single-curvature arch dams of modest height. The approach is therefore appropriate to the preliminary analysis of constant-radius profiles. The more complex geometry of the constant-angle profile makes it much less suitable for analysis by this method. Ring theory is not applicable to cupola profiles.

The ring theories consider water load only, self-weight stresses being determined separately and superimposed if significant to the analysis. Uplift load is not regarded as significant except for a thick arch, and therefore may normally be neglected. For analytical purposes the dam is considered to be subdivided into discrete horizontal arch elements, each of unit height, and the important element of vertical cantilever action is thus neglected. The individual rings are then analysed on the basis of the thick ring or thin ring theories as considered most appropriate, and the horizontal tangential arch stresses determined.

(a) Thick ring stress analysis

The discrete horizontal arch elements are each assumed to form part of a complete ring subjected to uniform external radial pressure, p_w, from the water load. The compressive horizontal ring stress, σ_h, for radius R is then given by

$$\sigma_h = \frac{p_w(R_u^2 + R_u^2 R_d^2/R^2)}{R_u^2 - R_d^2} \ (\mathrm{MN\,m^{-2}}) \tag{3.44}$$

where R_u and R_d are respectively the upstream and downstream face radii of the arch element considered.

Ring stress σ_h has a maximum at the downstream face. Ring thickness T_r, equal to $R_u - R_d$, is assumed uniform at any elevation. For $R = R_d$, equation (3.44) may consequently be rewritten in terms of $\sigma_{h\,max}$, with $p_w = \gamma_w z_1$, thus

$$\sigma_{h\,max} = \frac{2\gamma_w z_1 R_u^2}{T_r(R_u + R_d)} \tag{3.45}$$

(b) Thin ring stress analysis

If mean radius R_m is very large in comparison with T_r it may be assumed that $R_m = R_u = R_d$ and, consequently, that stress σ_h through the ring element is uniform. Equation (3.45) then simplifies to the classical thin ring expression:

$$\sigma_h = \gamma_w z_1 R_u / T_r. \tag{3.46}$$

In the upper reaches of a dam equations (3.44) and (3.46) agree closely, the difference diminishing to under 2% when $R_u/T_r \geqslant 25$.

Both variants of ring theory are inexact and of limited validity for two principal reasons. First, the simplifying assumption of discrete, independent horizontal rings which are free of any mutual interaction is clearly untenable. Secondly, the assumption of uniform radial deformation implicit in elastic ring theory is similarly untenable because of restraint at the abutment, and also the fact that arch span will marginally increase owing to elastic deformation of the abutment. Arch deflection will, in turn, reduce σ_h near the crown and progressively increase σ_h as the abutments are approached.

In theory, arch thickness should therefore diminish towards the crown and increase close to the abutments. In practice it is usual to maintain a uniform thickness at any elevation for a single-curvature dam, and the maximum tangential stresses will therefore be those generated at either abutment.

Assuming that no abutment yielding occurs, an approximation of the maximum stress at the abutment can be made by application of a correction factor, K_R. In terms of thin ring theory, therefore, at the abutment

$$\sigma_h = K_R \gamma_w z_1 R_u / T_r. \tag{3.47}$$

Factor K_R is a function of 2θ and of ratio R_u/T_r, and curves for K_R are presented in Fig. 3.16. It will be noted that $K_R \rightarrow 1.0$ for high values of 2θ, i.e. the solution tends to that of pure thin ring theory.

3.4.4 Advanced arch analysis

Certain of the deficiencies of elastic ring analysis have been alluded to. Early recognition of the importance of arch–cantilever and arch–abutment interactions led to the development of trial load analysis, TLA. The principles

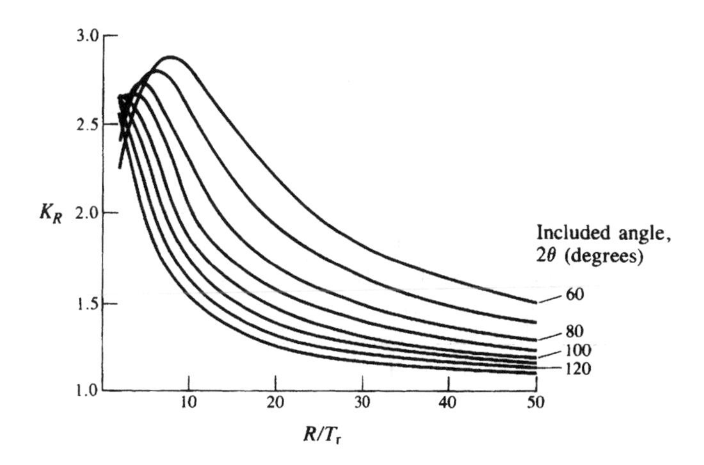

Fig. 3.16 Arch dam abutment stress correction factor K_R

underlying TLA in the context of an arch or cupola dam, i.e. arch–cantilever load distribution and node deflection matching, are essentially as outlined in Section 3.2.8, in relation to the gravity dam in a steeply sided valley.

Finite element analysis, FEA, also identified in Section 3.2.8, is extensively applied in arch dam design. Considerable effort has gone into the development of suitable software, and in its most refined three-dimensional form FEA is now the most powerful, reliable and well-proven approach to design. In this application it is, however, a highly specialist analytical method demanding experience in its application. It is also highly sensitive to the quality of the input information, notably as regards proper characterization of the load response of both concrete and rock foundation. The solutions it provides, while mathematically correct, must therefore be interpreted with some caution. Reference should be made to Clough and Zienkiewicz (1978) or USCOLD (1985) for a comprehensive introduction to FEA and other numerical/computational methods in the specific context of dam analysis and design.

3.5 Design features and construction

3.5.1 Introduction

All analyses are founded to a greater or lesser degree on assumptions with respect to load régime, material response, structural mechanisms, etc. Application of the analytical methods introduced in preceding sections represents only the initial phase of the design process. The second phase is to ensure that by good detail design the assumptions made in analysis are fulfilled to the greatest extent commensurate with rapid and economic construction.

Certain key design features reflect the engineer's approach to problems of loading, e.g. uplift relief drain systems. Other features are dictated by the characteristics of mass concrete, discussed in Section 3.6, or by pressure to rationalize and simplify in the interests of cost containment. Good detailing is not necessarily a matter of rational and formalized design. In many instances it is the application of empirical principles based on precedent and satisfactory experience. Thomas (1976) addresses design features and detailing in considerable depth, aided by an extensive selection of illustrative examples taken from international practice.

Design features divide into three major categories: those which relate to the control of external or internal seepage; those which accommodate deformation or relative movement; features which contribute to structural continuity, i.e. load transfer devices. A possible fourth category could be formed from those which simplify or facilitate construction. The most important features are introduced in succeeding sections.

3.5.2 Cut-offs and foundation grouting

The cut-offs under modern concrete dams are invariably formed by grouting. The relatively shallow key trench constructed under the heel of some dams contributes little to seepage control. Grouting below concrete dams falls into two categories, as shown in Fig. 3.17.

1. *Curtain grouting.* The purpose of curtain grouting is to form a partial cut-off to limit seepage and, in theory, to modify the downstream pressure régime. The primary grout screen or curtain is formed by drilling a regular series of holes, typically at intervals of 2–3 m, from a narrow scarcement or platform extending upstream from the heel and injecting a grout, generally cement based, under pressure. The curtain depth is frequently comparable to the height of the dam and extends beyond either abutment as required. Additional grouting may be done through intermediate or additional secondary holes introduced as necessary.
2. *Consolidation grouting.* The primary aim of consolidation grouting is to 'stiffen' and consolidate the rock in the critical contact zone immediately under the dam. It also assists in reducing seepage in the contact zone, where the rock may be more fissured or weathered than at greater depths. Care is required over the grout injection pressures employed to avoid disruption, fracturing and the opening up of horizontal fissures.

Examples of grouting practice are discussed in Thomas (1976) and Bruce and George (1982).

3.5.3 Uplift relief drains

Foundation uplift relief is effected by a line of drainage holes close downstream of the grout curtain (Fig. 3.17). The holes are generally about 75–100 mm diameter spaced at 3–5 m centres, and are drilled from the inspection gallery (see below).

Uplift within the dam is relieved by formed holes running the full height of the structure and located close to the upstream face. They should be at least 200 mm in diameter to inhibit blocking by leached out calcareous deposits, spaced at about 3 m centres, and relieve into an outlet drainage channel on the gallery invert. It is important to design the relief drain system such that drains may be reamed out or re-drilled in the event of blocking.

Relief drain efficiency is a function of drain geometry, i.e. diameter, spacing and distance to the upstream face. A comparative estimate of efficiency for different drain configurations may be obtained from Fig. 3.18 (Moffat, 1984).

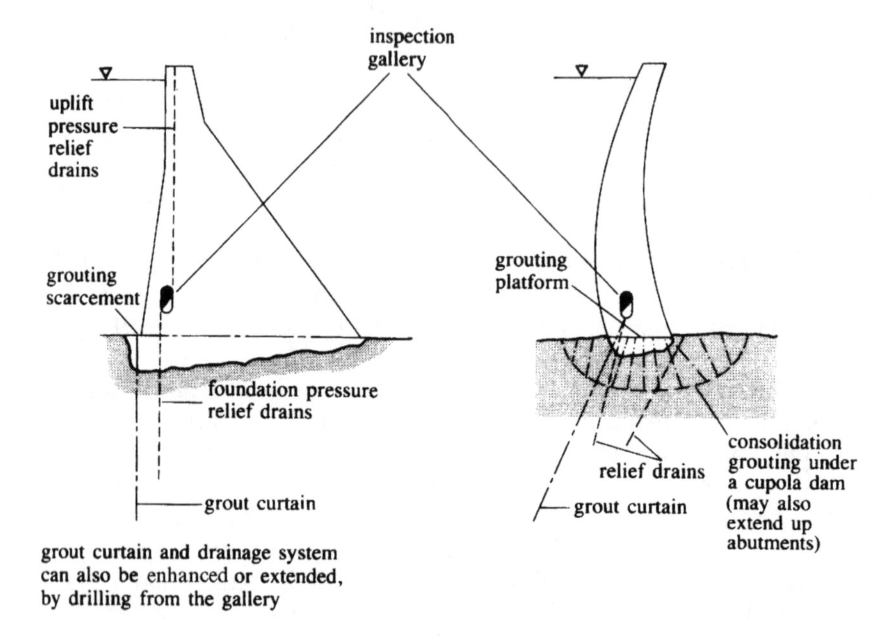

Fig. 3.17 Grouting and pressure relief drain systems

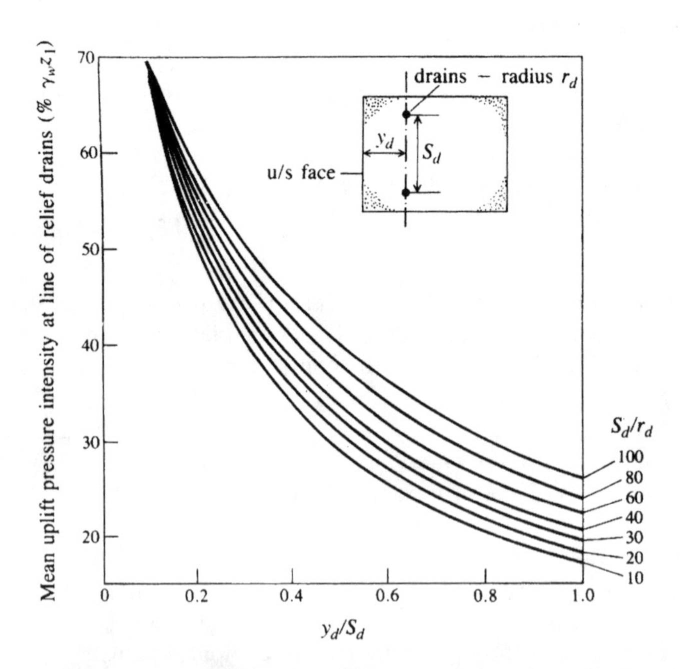

Fig. 3.18 Influence of relief drain configuration on mean internal uplift pressure intensity (after Moffat, 1984)

3.5.4 Internal design features

1. *Inspection galleries.* A low-level inspection gallery is necessary to collect seepage inflow from the uplift relief drains. The gallery also serves to give access to instruments (Section 7.2) and to internal discharge valves and pipework. Galleries should be not less than 2.0 m × 1.2 m in section, and adequate provision must be made for access, ventilation and lighting. Larger dams may also have one or more galleries at higher levels, interconnected by vertical shafts.
2. *Transverse contraction joints (interblock joints).* Vertical contraction joints are formed at regular intervals of 12–15 m along the axis of the dam. The joints are made necessary by the shrinkage and thermal characteristics of mass concrete (Section 3.6). They permit minor differential movements between adjacent blocks, and in their absence major transverse cracks will develop. To control seepage along the plane of the joint a water barrier detail similar to the example of Fig. 3.19 is formed close behind the upstream face.
3. *Construction joints (interlift joints).* Individual concrete pours within each monolith must be limited in volume and in height to reduce post-construction shrinkage and cracking. Concrete pours are therefore restricted by the regular formation of near-horizontal construction or 'lift' joints. Lift height is generally limited to 1.5–2.0 m. The lift surface is generally constructed with a stepped or uniform fall of 5–10% towards the upstream face to improve the notional resistance to sliding on that potentially weaker plane.

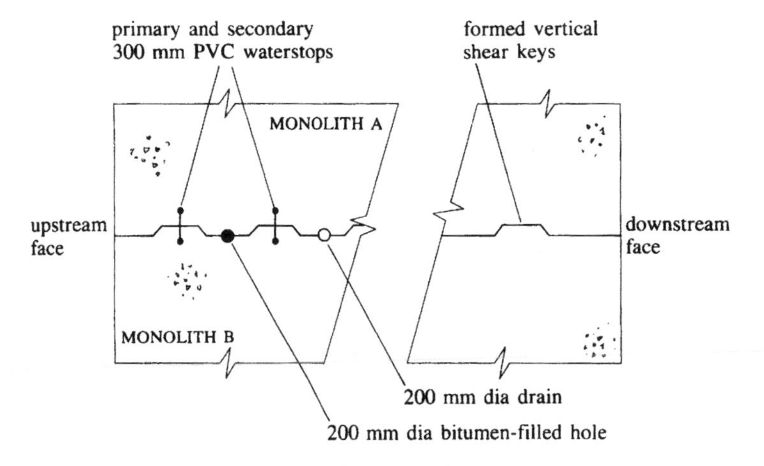

Fig. 3.19 Typical transverse contraction joint and shear key details

4. *Load transfer and continuity.* Although gravity dams are traditionally designed on the basis of independent free-standing vertical cantilevers it is normal to provide for a degree of interaction and load transfer between adjacent blocks. This is effected by forming interlocking vertical shear keys on the construction joint face (Fig. 3.19). In the case of arch and cupola dams it is essential to provide horizontal continuity to develop arch action. The construction joints of such dams are therefore grouted after the structure has cooled to its lowest mean temperature to ensure effective load transfer and monolithic arch action. Joint grouting has also been carried out on a number of gravity dams.
5. *Pulvino.* A thick perimetral concrete 'pulvino', or pad, is frequently constructed between the shell of a cupola dam and the supporting rock (Fig. 3.15). It assists in distributing load into the abutments and foundation. The shell may be separated from the pulvino by a perimetral joint to avoid fixity moments and hence induced tensile stress at the upstream face.
6. *Concrete zoning.* A zone of richer facing concrete some 1–1.5 m thick is frequently poured at either face. The richer concrete mix, which is generally also air-entrained, is provided to improve durability and, on the spillway face, resistance to cavitation damage and erosion. On the largest gravity dams the interior or hearting concrete may also be zoned for reasons of economy. Cement content, and therefore concrete strength, is then reduced in stages with increasing elevation (Fig. 3.20).

3.5.5 Construction planning and execution

Efficient and economic construction requires thorough planning of each of the principal phases, e.g. foundation preparation, concreting, etc. It is also necessary to pre-plan for the concurrent operations contributing to each

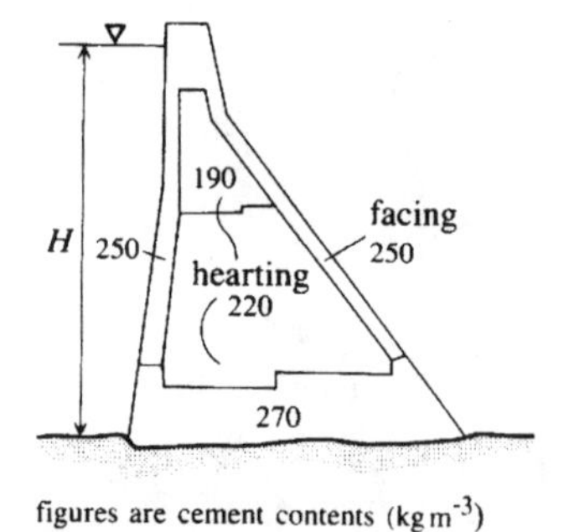

Fig. 3.20 Zoning of concretes in high dams ($H > 100$ m)

phase, e.g. rock excavation and foundation grouting etc. Detailed preplanning of all construction activities must therefore commence well in advance of site preparation, with the objective of ensuring optimum availability and utilization of all resources, i.e. finance, materials, plant and labour, throughout the construction period. It will be noted that several major activities are subject to non-concurrent peaks and troughs as construction progresses. A comprehensive account of the construction of a 40 m high curved gravity dam is presented in Tripp, Davie and Sheffield (1994).

The initial phase, site preparation, embraces the provision of a site infrastructure. This includes access roads, offices and workshops, and fixed plant. Accommodation, commissariat and other facilities for the labour force are also necessary on large or remote projects.

Preparations for river diversion, the second phase, may involve driving a bypass tunnel through the flank of the dam. As an alternative, the river may be diverted through a temporary gap formed by leaving a low-level tunnel through a block or by the omission of a complete monolith until the final closure stage is approached. Temporary cofferdams are normally required to allow the construction of diversion and other works on the river bed.

Foundation excavation and preparation is the third phase. Rock excavation should be the minimum consistent with attaining a safe foundation, the use of explosives being carefully controlled to avoid shattering and fissuring otherwise sound rock. The finished rock surface should be irregular, sound and clean. It is normal to provide a rise towards the toe, the nominal slope of about 4–15% being determined by the rock characteristics. Fissures are infilled with 'dental' concrete, and the grout curtain to control underseepage installed. Additional shallow grouting to consolidate the contact zone is carried out as necessary.

The construction operations of the fourth phase require particularly careful planning and control. Individual monoliths are raised on either the 'alternate block' or the 'shrinkage slot' principle as illustrated in Fig. 3.21. In either method the objective is to maximize shrinkage before pouring abutting lifts of concrete in adjacent blocks.

The sequence of events within each concrete pour is formwork erection, surface preparation and placing concrete, with compaction by vibrator. This is followed by an interval before formwork can be struck, and a further appreciable interval for initial shrinkage to occur before the cycle is repeated. Features which complicate or delay the concreting cycle unnecessarily are expensive, and must be kept to the minimum. Curing of the completed pours is effected by moist curing and/or by membrane curing compounds.

The final phase of project execution involves the completion of any ancillary structures, and installation and testing of valves, gates etc. It is concluded with a controlled and carefully monitored initial impounding (Chapter 7 should also be referred to).

(a) Alternate block construction: adjacent pours phased to accommodate shrinkage – lag time approx. 30–60 days

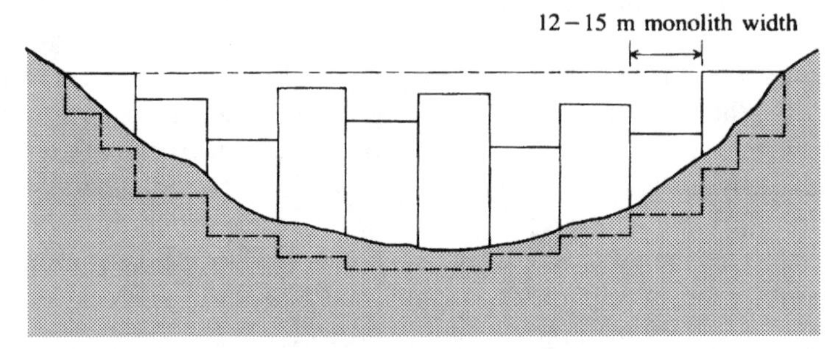

(b) Construction with contraction gaps or shrinkage slots: gaps concreted approx. 30–60 days after adjacent lifts completed

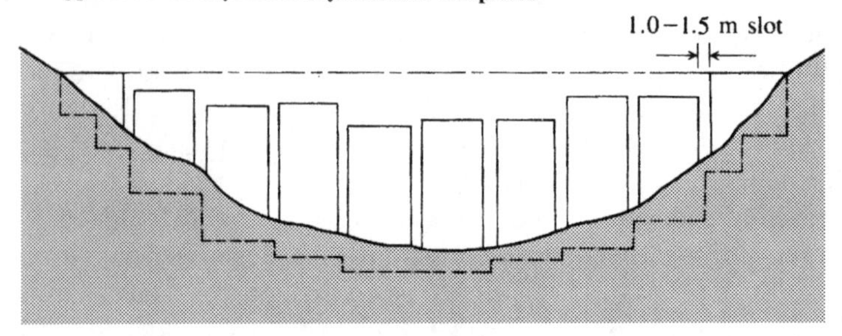

Fig. 3.21 Alternative monolith construction procedures

3.6 Concrete for dams

3.6.1 General

Concrete is a superficially inert but chemically and physically complex construction material. Its merits lie in adaptability, its use of readily available mineral resources and, above all, its low bulk cost.

Mass concrete in dams is not subjected to compressive stresses comparable with those developed in most other major structures. The volume of concrete within a dam is relatively great, however, requiring large pours and high placing rates. Several other properties therefore rank equally with strength as indices of quality and fitness for purpose. The desirable characteristics for a mass concrete for use in dams can consequently be summarized as follows: (a) satisfactory density and strength; (b) durability; (c) low thermal volume change; (d) resistance to cracking; (e) low permeability; (f) economy. The listed characteristics are all interrelated to a greater or lesser degree. As an example, strength, impermeability and durability are intimately linked for a dense, well-proportioned concrete mix. Less immediate relationships can similarly be established in other instances, e.g. between volume stability and resistance to cracking.

Concrete technology is a wide-ranging discipline with an extensive literature. For a general but comprehensive introduction to the subject, reference should be made to Neville and Brooks (1987). The coverage given in this section is necessarily brief, and is confined to outlining elements of concrete technology immediately relevant to dam construction. It may be supplemented by reference to the specialist treatment of mass concrete provided in ACI (1970) and to the review given in Jansen (1988).

The development of roller-compacted concretes (RCCs) and the prospects they offer for continuous and economic construction are introduced separately in Section 3.7.

3.6.2 Constituent materials

The primary constituents of concrete are cement, mineral aggregate and water. Secondary constituents employed in mass concrete for dams include pozzolans and other selected admixtures.

1. *Cements.* The hydration of unmodified Portland cements (ASTM type I equivalent) is strongly exothermic. The resulting temperature rise and heat gain in large pours is unacceptable in relation to consequent problems of shrinkage, heat dissipation and cracking. It is therefore preferable to employ a low heat (ASTM type IV) or modified Portland cement (ASTM type II) if available. Thermal problems can also be alleviated by the use of pozzolan-blended Portland cements (ASTM type 1P). In the absence of special cements, partial replacement with pulverized fuel ash (PFA) and/or cooling are also effective in containing heat build-up. (PFA for concrete is a quality-assured and processed waste product from coal-fired power stations.)
2. *Aggregates.* The function of the coarse and fine aggregate is to act as a cheap, inert bulk filler in the concrete mix. A maximum size aggregate (MSA) of 75–100 mm is considered the optimum, with rounded or irregular natural gravels generally preferable to crushed rock aggregates. In the fine aggregate range, i.e. <4.67 mm size, natural sands are similarly preferable to crushed fines. Aggregates should be clean and free from surface weathering or impurities. The petrographical, thermal and moisture characteristics should be compatible with hydrated cement paste. A smooth, well-graded particle size distribution curve for the combined aggregates will ensure maximum packing density for the compacted concrete mix.

3. *Water.* Water for use in concrete should be free of undesirable chemical contamination, including organic contaminants. A general standard is that the water should be fit for human consumption.
4. *Pozzolans.* Pozzolans are siliceous–aluminous substances which react chemically with calcium hydroxide from the cement to form additional cementitious compounds. PFA, an artificial pozzolan, is now almost universally employed, if available, in partial replacement (25–50%) of cement. The introduction of PFA reduces total heat of hydration and delays the rate of strength gain for the concrete up to about age 90–180 days. Long-term strength is generally slightly enhanced and certain aspects of durability may be improved, but strict quality control of the PFA is necessary. The use of PFA may also assist in marginally reducing overall concrete costs.
5. *Admixtures.* The most commonly used admixtures are air-entraining agents (AEAs). They are employed to generate some 2–6% by volume of minute air bubbles, significantly improving the long-term freeze–thaw durability of the concrete. They also reduce the water requirement of the fresh concrete and improve its handling qualities. Water reducing admixtures (WRAs) are sometimes employed to cut the water requirement, typically by 7–9%. They are also effective in delaying setting time under conditions of high ambient temperatures.

3.6.3 Concrete mix parameters

The parameters which are principally responsible for controlling the properties of concretes manufactured with specific cement and aggregates are as follows: cement content, C (kg m^{-3}); water content, W (kg m^{-3}); water:cement ratio (by weight). Some further influence can be exerted through the addition of PFA and/or the use of other admixtures such as AEA and WRA.

The desirable primary characteristics of density, adequate strength, durability and impermeability are favourably influenced by increasing cement content and/or reducing the water:cement ratio. Thermal characteristics and volume stability, on the other hand, are improved by restricting the cement content of the mix. Economy is similarly dependent upon minimizing the cement content. A balance has therefore to be sought between upper and lower bounds for each parameter, limits which are set by contradictory requirements.

The dominant considerations in selecting a suitable concrete mix are controlling thermal characteristics and attaining the minimum cost consistent with adequate strength and durability.

The *in situ* properties of the mature concrete are dependent upon attaining maximum density through effective compaction. The ability to

Table 3.10 Characteristics of mass concrete for dams

Characteristics	*Concrete mix*	
	Hearting	*Facing*
Cement (C) + PFA (F) (kg m^{-3})	150–230	250–320
$F/(C+F)$ (%)	20–35	0–25
Water:$(C+F)$ ratio	0.50–0.70	0.45–0.65
90-day compressive strength, σ_c (MN m^{-2})	18–30	25–40
$\dfrac{\text{Tensile strength}}{\text{Compressive strength}}$ (σ_t/σ_c)	0.10–0.15	0.07–0.10
Unit weight, γ_c (kN m^{-3})	23–25	
Modulus of elasticity, E (GN m^{-2})	30–45	
Poisson ratio	0.15–0.22	
Shrinkage (% at 1 year)	0.02–0.05	
Coefficient of thermal expansion ($\times 10^{-6}$ °C^{-1})	9–12	

achieve this is largely controlled by the physical characteristics of the fresh concrete, notably its cohesiveness and workability. It is therefore again related to the mix proportions, principally in terms of the water, cement and fines contents.

Satisfactory mix proportions are dependent upon balancing the several conflicting demands. The range within which each of the principal parameters may be varied, is, in practice, subject to severe constraints.

Indicative mix proportions and properties are summarized in Table 3.10. It may be noted that it is the characteristics of the mature mass concrete at ages in excess of 90–180 days which are significant in dam construction.

3.6.4 Concrete production and placing

High production rates are necessary on large projects and require a carefully planned central materials handling facility and concrete batching plant. Efficient utilization of the plant is made difficult by inevitable variations in demand caused by the need to programme pours in accordance with the optimum monolith and lift construction schedules.

Transportation of concrete may be by travelling overhead cableway and skips on a compact site. Trucks may be preferable where the site area is extensive, and increasing interest is being shown in the use of conveyor systems on large projects.

Concrete lifts are normally formed in at least two layers, and compacted by poker vibrators. The cost efficiency and effectiveness of the compaction process may be improved on larger dams by the use of immersion vibrators mounted on suitable tracked plant running on the surface of the concrete pour.

Planning of concreting operations is a vital but difficult task. Details of planning methods and production, placing and compacting techniques are given in Thomas (1976) and Jansen (1988).

It should be noted that the primary objective in concrete production and placing over the period of construction is to ensure *uniformity* and *consistency of quality* in the mature finished product.

3.7 The roller-compacted concrete gravity dam

3.7.1 General

The concrete gravity dam shares with the embankment the central attributes of simplicity of concept and adaptability, but conventional mass concrete construction rates, unlike those for embankment construction, remain essentially as they were in the 1950s. The volume instability of mass concrete due to thermal effects imposes severe limitations on the size and rate of concrete pour, causing delay and disruption through the need to provide contraction joints and similar design features. Progressive reductions in cement content and partial replacement of cement with PFA have served only to contain the problem. Mass concrete construction remains a semi-continuous and labour-intensive operation of low overall productivity and efficiency.

In some circumstances the technical merits of the gravity dam and the embankment may be evenly balanced, selection resting on estimated construction cost. Economic advantage will almost invariably favour the embankment, particularly if constructed in compacted rockfill. In some instances, however, factors such as locating a spillway of sufficient capacity etc. may indicate the concrete gravity dam as being a preferable design solution, provided that the cost differential lies within acceptable limits. Despite advances in embankment dam engineering, therefore, there remains a strong incentive to develop a cheaper concrete gravity dam.

The problem of optimizing concrete dam construction and reducing costs can be approached in several ways. In the absence of progress towards an ideal cement and a dimensionally stable concrete the most promising lines of approach may be classified as follows:

1. a reappraisal of design criteria, particularly with regard to accepting modest tensile stresses;
2. the development of improved mass concretes through the use of admixtures to enhance tensile strength and to modify stress–strain response, and/or the use of modified cements with reduced thermal activity;

3. the development of rapid continuous construction techniques based on the use of special concretes.

Neither of the first two approaches is capable of offering other than a token reduction in cost. The third option offers the greatest potential through financial benefits associated with a shortening of construction period by up to 35% combined with a lower-cost variant of concrete.

The concept of dam construction using roller-compacted concrete (RCC), first developed in the 1970s, is based primarily on approach 3. Several variants of RCC have now been developed and offer the prospect of significantly faster and cheaper construction, particularly for large gravity dams.

3.7.2. Alternatives in roller-compacted concrete

Research has resulted in the emergence of three principal approaches to developing RCCs. In the first, RCC is conceived as a low-cost fill material, offering the maximum possible economy consistent with satisfactory strength and durability and suitability to continuous construction techniques, i.e. compaction by roller. An example of this is provided by rolled dry lean concrete (RDLC) (Moffat, 1973; Moffat and Price, 1978). The second approach, developed in Japan, is closer to conventional lean hearting concretes. It is confusingly identified as the rolled-concrete dam, or RCD, concept of RCC. The third approach to RCC is the concept of a dense, high-paste content material, and is exemplified by high PFA content concrete (Dunstan, 1981). Variants of RCC are the subject of continuing development, particularly in the USA, Japan, China and South Africa.

Two distinct approaches to the design of an RCC dam have developed. In the traditional or concrete technology approach, RCCs of the high-paste and RCD types are employed to build what is in all important respects a conventional gravity dam. The more radical 'geotechnical' approach is based upon optimizing the design profile to construction in a lean RCC such as RDLC. In this approach the concrete is handled as an earthfill, and compacted at or near its optimum moisture content in thin layers. The logical extension of this approach is to draw upon embankment dam design principles in developing an optimum profile. The dam following this approach is therefore optimized for construction using a lower strength and relatively permeable RCC gravity profile in conjunction with a horizontally slipformed high-quality upstream concrete membrane. A profile for an RDLC dam constructed on this principle is illustrated in Fig. 3.22(a). An equivalent high-paste RCC profile is shown in Fig. 3.22(b).

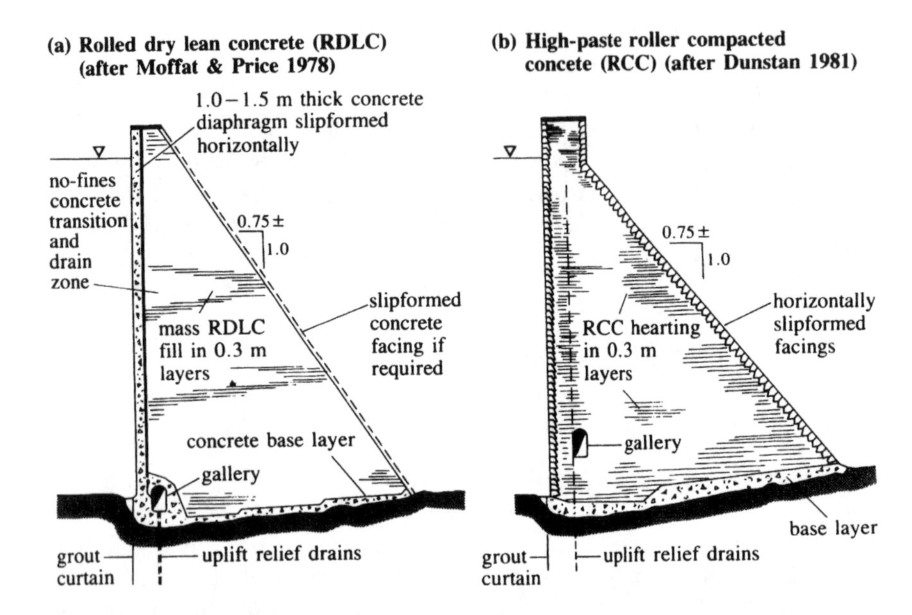

Fig. 3.22 Illustrative first-generation roller-compacted concrete (RCC) gravity dam profiles

Construction in RDLC and some other RCCs permits an intensively mechanized construction process, with concrete delivered by conveyor belt and handled by standard earthmoving and compaction plant. Construction joints, if considered necessary, may be sawn through each successive layer of concrete after placing. Experience in the design of RCC dams, including the construction procedures employed, has been published in Hansen (1985), and an overview of early US developments is given in Dolen (1988).

The RCC approach is best suited to wide valleys, giving scope for unobstructed 'end-to-end' continuous placing. The construction savings realized are at a maximum for high-volume dams and arise from a 25–35% reduction in construction time as well as from reduced unit costs for the RCC. In its low-cost 'geotechnical' format (e.g. RDLC) RCC is particularly suited to more remote sites where importation of cement and/or PFA is difficult or expensive.

The characteristics of the major variants of RCC are compared with those of a conventional hearting concrete in Table 3.11.

Table 3.11 Characteristics of RCCs for dams

Characteristics	*RCC type*			*Conventional lean hearting concrete*
	Lean RCC (RDLC)	*RCD*	*High-paste RCC*	
Cement (C) + PFA (F) (kg m^{-3})	100–125	120–130	>150	150–230
$F/(C+F)$ (%)	0–30	20–35	70–80	20–35
Water:($C+F$) ratio	1.0–1.1	0.8–0.9	0.5–0.6	0.5–0.7
90-day compressive strength, σ_c (MN m^{-2})	8–12	12–16	20–40	18–40
Unit weight, γ_c (kN m^{-3})	←	23–25	→	22–25
Layer thickness (m)	0.3	(lifts) 0.7–1.0	0.3	(lifts) 1.5–2.5
Contraction joints	Sawn	Sawn	Sawn or formed	Formed

3.7.3 Developments in roller-compacted concrete dam construction

The RCC dam has developed rapidly since construction of the earliest examples in the early 1980s, and in excess of 200 large dams had been completed in RCC by 2000. The majority of RCC dams have been gravity structures, but the RCC technique has been extended to a number of arch-gravity and thick arch dams. As confidence has grown RCC has been used for progressively larger dams, and has been employed for the major part of the 7.6×10^6 m^3 volume of the 217 m high Longtan gravity dam in China. In a number of recent instances the RCC gravity dam option has been selected in preference to initial proposals for the construction of a rockfill embankment.

The early RCC dams were noted for problems associated with relatively high seepage and leakage through the more permeable RCC, and for a degree of uncontrolled cracking (Hollingworth and Geringer, 1992). A relatively low interlayer bond strength also prompted some concern, particularly in the context of seismic loading. The philosophy of RCC dam design has in consequence evolved, with emphasis being placed on optimizing design and detailing to construction in RCC rather than using RCC to construct a conventional gravity dam. This trend has led to the common provision of an 'impermeable' upstream element or barrier, e.g. by a slip-formed facing (Fig. 3.22 and also New Victoria dam, Australia (Ward and Mann, 1992)).

An alternative is the use of a PVC or similar synthetic membrane placed against or just downstream of a high-quality concrete upstream face. In the case of the 68 m high Concepcíon gravity dam, Honduras, a 3.2 mm PVC geomembrane backed by a supporting geotextile drainage layer was applied to the upstream face of the RCC (Giovagnoli, Schrader and Ercoli, 1992). Recent practice has also moved towards control of cracking by sawn transverse joints, or by the cutting of a regular series of slots to act as crack inducers.

The very considerable cost savings attaching to RCC construction are dependent upon plant and RCC mix optimization, and hence continuity of the RCC placing operation. This in turn requires that design features which interfere with continuous unobstructed end-to-end placing of the RCC, e.g. galleries, internal pipework, etc., must be kept to the minimum and simplified. Experiments with retrospectively excavating galleries by trenching and by driving a heading in the placed RCC fill at Riou, France, have proved successful (Goubet and Guérinet, 1992).

Vertical rates of raising of 2.0–2.5 m $week^{-1}$ are attainable for RDLC and high-paste RCCs compared with 1.0–1.5 m $week^{-1}$ for RCD construction. As one example, the Concepcíon dam, Honduras, referred to earlier was raised in seven months. A lean RCC mix (cement content 80–95 $kg\,m^{-3}$) was employed for the $290 \times 10^3\,m^3$ of RCC fill, and a continuous mixing plant was used in conjunction with a high-speed belt conveyor system. Placing rates of up to 4000 m^3 day^{-1} were ultimately attained (Giovagnoli, Schrader and Ercoli, 1992).

The employment of RCC fill has also been extended to the upgrading of existing dams, e.g. by placing a downstream shoulder where stability is deficient (Section 3.2.9). RCC has also been applied to general remedial works and to raising or rebuilding older dams. The benefits of RCC construction have also been appropriate, in special circumstances, to the construction of smaller dams, e.g. Holbeam Wood and New Mills in the UK (Iffla, Millmore and Dunstan, 1992).

ICOLD Bulletin 126 (ICOLD, 2004) provides a comprehensive overview of the use of RCC for dam construction. US developments are discussed in Hansen (1994). Design options with respect to upstream face construction have been reviewed in some detail by Schrader (1993).

Construction in RCC is recognized as providing the way forward in concrete dam engineering. Extensive reviews of current issues in RCC dam design and construction are presented in Li (1998), and in Berga *et al.* (2003). Major issues discussed include the need, or otherwise, for a conventional concrete upstream face, and the question of resistance to high seismic loading, where dynamic tensile strength of the interlayer bond between successive layers of RCC will be critical.

The recently completed 95 m high RCC gravity dam, at Platanovryssi, Greece, located in a seismic zone, is described in Stefanakos and Dunstan (1999). The design peak ground acceleration corresponding to the MCE at Platanovryssi was determined as 0.385 g, equating to a maximum dynamic tensile stress of *c.* 2 MN/m^{-2}, requiring a cylinder compressive strength for the RCC of 28 MN/m^2. The high-paste RCC employed contained 50 kg/m^3 of PFA. Platanovryssi does not have an upstream face of conventional concrete, and vertical contraction 'joints' were induced at 25 m intervals by slots sawn in the upstream face in conjunction with steel crack inducers vibrated into the RCC. The 'joints' were subsequently sealed by a 600 mm wide external waterstop bonded to the face. Seepage through the dam body diminished to a satisfactory 10–12 l/s over the first 12 months' operational service.

The first use of RCC in Turkey, for the 124 m high by 290 m long Cine gravity dam (originally planned as a rockfill embankment with a clay core) is presented in Ozdogan (1999). The low-paste RCC used for Cine has a cement content of 70 kg/m^3, with 90 kg/m^3 of PFA and 88 l/m^3 of water. Target 180 day compressive strength was specified as 24 MN/m^2.

The concept of the RCC dam has attained maturity, most of the initial concerns with regard to the performance of the earliest RCC dams such as the higher permeability leading to greater seepage flows and reduced durability, and the issue of providing a durable and aesthetically satisfactory facing, having now been successfully addressed. RCC mix selection and design is now a well-understood sector of concrete technology. Gravity and even arch dam profiles have been rationalized for rapid and economic construction in RCC. Recent developments in RCC dam practice have therefore centred upon refining the overall construction process, optimizing concrete production and placing operations to raise output and enhance consistency of the product at a reduced gross finished unit cost/m^3 for the RCC.

Reviews of contemporary RCC dam technology and practice, together with descriptions of completed projects and others as yet in the planning stage, are presented in Berga *et al.* (2003) and ICOLD (2004). Understanding of the RCC mix is discussed in Schrader (2005), and the same author considers the construction of RCC dams in relation to more difficult foundation conditions in a later paper (Schrader, 2006).

Construction of the Miel 1 dam, Colombia, the tallest RCC dam in the world at the time of it's completion in 2002, is described in Santana and Castell (2004). The body of the dam was divided into zones according to design stress level, with four different RCC mixes employed with cement contents ranging from 85 to 150 kg/m^3. The incidence of cracking due to thermal stress was controlled by cutting transverse joints in the freshly placed RCC at intervals of 18.5 m, a similar longitudinal joint being cut in the lower levels of the dam. Average RCC mix parameters for Miel 1 are quoted as being:

compacted density :	24.7 kN/m^3
Poisson ratio:	0.23
angle of shearing resistance:	45 deg
strain capacity:	80 microstrain
thermal diffusivity:	0.00335 m^2/h
thermal expansion:	7×10^{-6} m/deg C.

3.8 Upgrading of masonry and concrete dams

The upgrading of older masonry or concrete dams is in most instances the response to a recognized inadequacy in design and/or construction standards when reviewed in context with present-day good practice. The most common deficiencies relate to overturning stability under more extreme loading conditions, with the possibility of unacceptable levels of tensile stress being generated at the heel. Concerns over sliding are rather less common, but almost all concerns over stability can be identified with slenderness of the dam profile, i.e. excessive stresses attributable to underdesign. A common feature in underdesigned profiles is inadequate provision for the effects of internal seepage and uplift pressure, the incorporation of internal relief drains not becoming near-universal practice until *c*. 1950–55.

Upgrading the stability of a gravity dam by direct methods such as prestressing or constructing a downstream supporting shoulder was discussed in Section 3.2.9. In the case of dams which do not have adequate provision for uplift relief, but have an internal inspection gallery stability can be enhanced by drilling a screen of relief drains from crest to gallery and, if required, from gallery into the underlying rock foundation.

Gallacher and Mann (2002) describe and discuss major upgrading work undertaken to stabilize the 49 m high masonry-faced Tai Tam Tuk dam, Hong Kong. Stability of the dam was deemed acceptable under normal loading conditions but inadequate for seismic loading, and stabilization was accomplished by drilling internal relief drains. The paper discusses the influence of the installed drainage screen on seepage and uplift and the consequent effects on stability under extreme loading.

An alternative approach to the alleviation of seepage-related problems lies in the application of a protective upstream membrane. Reservations have been expressed with regard to the long-term durability and integrity of such membranes, but the availability of durable high-performance synthetic membranes has led to a reappraisal of the technique and to increasing use of upstream membranes, particularly in Italy. Scuero and Vaschetti (1998) describe a proprietary system and it's successful application to dams of different type. A later development has been the underwater application of a protective membrane of this type to an arch dam in California, described in Scuero, Vaschetti and Wilkes (2000). The latter also give a useful summary of the alternative approaches to controlling seepage and deterioration.

Cases of in-service deterioration of the exposed faces of concrete dams are not uncommon. This is particularly the case with dams built before 1955–60, in a time of limited understanding as to the nature and characteristics of mass concrete and of what constituted good concreting practice. In most instances the deterioration is associated with internal seepage, particularly at badly executed construction joints, resulting in the leaching out of hydrated cementitious compounds and some regression in strength and impermeability/durability. The damage is generally localized and quite superficial in relation to the overall integrity of a gravity or massive buttress dam. The longer-term effects are considerably more serious in the case of Ambursen-type buttress dams and thin multiple-arch structures (see Figs 1.5c) and d)), where the nominal hydraulic gradient through the slender concrete sections is very much greater, giving rise to high rates of leaching and attrition. It is for this reason, *inter alia*, that dams of these types are no longer favoured.

Superficial physical deterioration of mass concrete can also be caused by severe freeze-thaw action or the action of aggressive reservoir water, particularly on concrete mixes which had a high initial water/cement ratio or were poorly compacted.

The in-service deterioration of masonry in gravity dams is generally confined to degradation of the bedding mortar in the joints. It is therefore relatively superficial and of little immediate significance to the overall integrity of a massive dam.

Masonry sourced from a dense, competent parent rock is almost invariably highly durable. Dressed masonry was generally limited to upstream and downstream faces, providing a highly durable and aesthetically attractive external finish. Cost considerations normally dictated that the interior or hearting be of cyclopean masonry, i.e. very large semi-dressed or undressed rocks bedded in concrete and/or with the intervening void spaces packed with smaller stones and mortar.

Sims (1994) discusses the ageing of masonry dams with particular reference to examples from Indian experience. Bettzieche and Heitefuss (2002) describe in-service rehabilitation of older masonry dams in Germany with regard to the retrospective drilling of relief drainage and driving of inspection or drainage galleries. They describe the successful employment of drill-and-blast techniques and, in the case of the 51 m high Ennepe dam a tunnel-boring machine, for driving internal galleries. The successful application of underwater working techniques to the in-service rehabilitation of bottom outlet works at the Ennepe and Moehne dams is discussed in Heitefuss and Kny (2002).

Recent years have seen considerable advances made in regard to rehabilitation options. Modern rehabilitation techniques are assuming ever-greater importance in relation to cost-effective upgrading of ageing masonry and concrete dams to conform to the standards of today.

Worked Example 3.1

Note that the unit weight of water, γ_w, is taken as $10\,kN\,m^{-3}$ throughout these worked examples.

The profile of an old concrete gravity dam is shown in Fig. 3.23. The structural competence of the dam is to be reviewed in relation to planned remedial work.

Design criteria. Normal load combination (NLC): water load (to design flood level (DFL)) + self-weight + uplift + silt. Uplift: no provision in original design. Sediment: friction angle, $\phi_s = 30°$; submerged unit weight: $\gamma'_s = 15\,kN\,m^{-3}$.

Concrete characteristics: core samples. Unit weight, $\gamma_c = 24\,kN\,m^{-3}$; unit shear resistance, $c = 600\,kN\,m^{-2}$; angle of shearing resistance, $\phi_c = 35°$.

1. Analyse the stability of the profile with respect to plane X–X for the NLC, using shear friction factor, F_{SF}, for sliding stability.
2. Determine the vertical normal stresses and major principal stresses at either face.

Solutions

Full uplift applies as no drains are operative. The sediment load, $K_a = (1 - \sin\phi_s)/(1 + \sin\phi_s) = 0.33$.

The profile is subdivided into the elements A and B, identified in Fig. 3.23, for convenience.

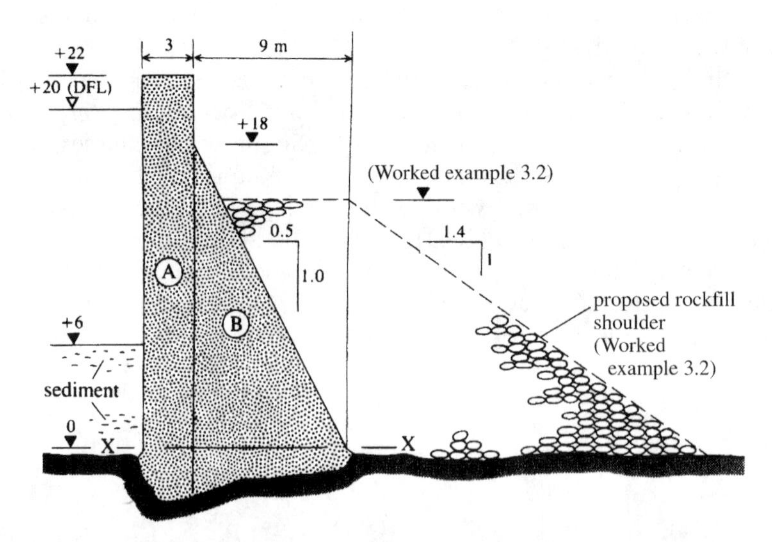

Fig. 3.23 Worked examples 3.1 and 3.2

1. Load–moment table (all moments are relative to toe):

Load	*Moment arm (m)* ←	↑	*Vertical (kN)* −↑	+↓	*Horizontal (kN)* →	*Moments (kNm)* (−)	(+)
Water	–	6.66	–	–	2000	13 330	–
Uplift	8.00	–	1200	–	–	9 600	–
Sediment	–	2.00	–	–	90	180	–
Self-weight A	10.50	–	–	1584	–	–	16 632
Self-weight B	6.00	–	–	1944	–	–	11 664
		Σ	1200	3528	2090	23 110	28 296

Equation (3.19):

$$F_{\mathrm{O}} = \frac{\Sigma \overset{\frown}{M} +}{\Sigma \overset{\frown}{M} -} = \frac{28\,296}{23\,110} = 1.22 \text{ (low, but acceptable)}$$

Equation (3.25):

$$F_{\mathrm{SF}} = \frac{cT + \Sigma V \tan \phi_{\mathrm{c}}}{\Sigma H} = \frac{600 \times 12 + 2328 \times 0.70}{2090} = 4.22 \text{ (acceptable)}$$

2. Moment table 2 (moments* relative to centroid of X–X):

Load element	*Moment arm (m)* ←	↑	*Moments** (−)	(+)
Water	–	6.66	13 300	–
Sediment	–	2.00	180	–
Self-weight A	4.50	–	–	7128
Self-weight B	0	–	–	0
		Σ	13 510	7128

$$e = \frac{\Sigma M^*}{\Sigma V} = \frac{-6382}{3528} = -1.81\,\mathrm{m}$$

i.e. e lies downstream of the centroid. Hence, in equation (3.33),

$$\sigma_z = \frac{3528}{12}\left(1 \pm \frac{6 \times 1.81}{12}\right)$$

which gives (equation (3.33b)) $\sigma_{zu} = 0.03\,\mathrm{MN\,m^{-2}}$ and (equation (3.33c)) $\sigma_{zd} = 0.56\,\mathrm{MN\,m^{-2}}$ (Section 3.2.5 should be referred to regarding this very low value for σ_{zu}). Major principal stresses (equations (3.37a) and (3.37c)) are

$$\sigma_{1u} = 0.03\,\mathrm{MN\,m^{-3}}$$

and

$$\sigma_{1d} = 0.56(1 + 0.25) = 0.70\,\mathrm{MN\,m^{-2}}.$$

Worked Example 3.2

The overturning stability factor of the dam in worked example 3.1 is to be revised to $F_O = 1.6$ to improve the stress distribution and obviate cracking. The foundation is unable to take prestressing anchorage loads, and a downstream supporting shoulder of compacted rockfill is to be added as suggested in Fig. 3.23. Determine the height of rockfill shoulder required.

Compacted rockfill. Unit weight, $\gamma_r = 19\,\mathrm{kN\,m^{-3}}$; coefficient, K_0 (estimated) = 0.60.

Solution

$$F_O = 1.60 = \frac{\Sigma \overset{\frown}{M} +}{\Sigma \overset{\frown}{M} -} \quad \text{(equation (3.19))}.$$

Hence, $\Sigma \overset{\frown}{M}+$ required from shoulder = $(1.6 \times 23110) - 28296 = 8680\,\mathrm{kN\,m^{-1}}$. Therefore,

$$\frac{K_0 \gamma_r h^2}{2}\frac{h}{3} + \frac{\gamma_r h}{2}\frac{(0.5h)^2}{3} = 8680$$

and

$$0.141h^3 = 8680/19$$

giving $h = 14.8\,\mathrm{m}$, say 15 m.

Worked Example 3.3

The profile of the major monolith of a buttress dam is illustrated in Fig. 3.24. The stability of the dam is to be reviewed in relation to updated design criteria.

Normal load combination (NLC). Water load (to design flood level (DFL)) + self-weight + uplift (no pressure relief drains).

Static stability. Overturning, $F_O > 1.5$; sliding (shear friction factor), $F_{SF} > 2.4$.

Concrete characteristics: core samples. Unit weight, $\gamma_c = 23\,\mathrm{kN\,m^{-3}}$; unit shear resistance, $c = 500\,\mathrm{kN\,m^{-2}}$; angle of shearing resistance (internal friction), $\phi_c = 35°$.

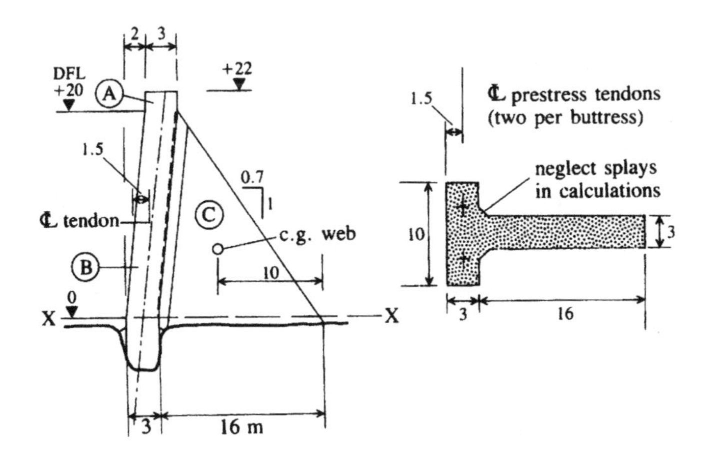

Fig. 3.24 Worked example 3.3

1. Analyse the static stability of the buttress unit with respect to plane X–X under NLC and in relation to the defined criteria for F_O and F_{SF}.
2. Concern is felt with regard to stability under possible seismic loading. Dynamic stability criteria are specified as $F_O = 2.0$; $F_{SF} = 3.2$, and will be met by prestressing as shown. Determine the prestress load required in each inclined tendon.

Solutions

1. All calculations relating to stability refer to the monolith as a complete unit. Uplift is considered to act only under the buttress head, and the profile is subdivided into the elements A, B and C, identified in Fig. 3.24, for convenience. The load–moment table (all moments are relative to toe) is as follows:

Load	*Moment arm (m)*		*Vertical (kN)*		*Horizontal (kN)*	*Moments (kN m)*	
	←	↑	−↑	+↓	$\overrightarrow{-}$	↶ (−)	↷ (+)
Water:							
horizontal	–	6.66	–	–	20 000	133 330	–
vertical	18.33	–	–	2 000	–	–	36 660
Uplift	18.00	–	3000	–	–	54 000	–
Self-weight A	15.50	–	–	1 380	–	–	21 390
Self-weight B	16.50	–	–	13 800	–	–	227 700
Self-weight C	10.00	–	–	11 040	–	–	110 400
		Σ	3000	28 220	20 000	187 330	396 150

Equation (3.19):

$$F_{\mathrm{O}} = \frac{\Sigma \overset{\frown}{M} +}{\Sigma \overset{\frown}{M} -} = \frac{396\,150}{187\,330} = 2.11 \text{ (satisfactory).}$$

Equation (3.25):

$$F_{\mathrm{SF}} = \frac{cA_{\mathrm{h}} + \Sigma V \tan\phi_{\mathrm{c}}}{\Sigma H} = \frac{500 \times 78 + 25\,220 \times 0.70}{20\,000} = 2.83 \text{ (satisfactory).}$$

2. Seismic criteria, $F_{\mathrm{SF}} = 2.83$ unsatisfactory (<3.2); prestress load/buttress $= P_{\mathrm{ps}}$; inclination to vertical, $\tan\phi_{\mathrm{u}} = 0.10$; $\phi_{\mathrm{u}} = 5.7°$. From equation (3.41),

$$F_{\mathrm{SF}} = 3.20 = \frac{cA_{\mathrm{h}} + (\Sigma V + P_{\mathrm{ps}} \cos\phi_{\mathrm{u}}) \tan\phi}{\Sigma H - P_{\mathrm{ps}} \sin\phi_{\mathrm{u}}}$$

$$= \frac{500 \times 78 + (25\,220 + P_{\mathrm{ps}} \times 0.99) \times 0.70}{20\,000 - P_{\mathrm{ps}} \times 0.10}$$

gives $P_{\mathrm{ps}} = 7230$ kN per buttress $= 3615$ kN per tendon.

Worked Example 3.4

A constant-radius cylindrical arch dam with a vertical upstream face is to be constructed in a symmetrical valley. The idealized valley profile consists of a trapezoid with a base 50 m wide and sides at 45°. The base is at 100 m AOD, the spillway crest at 140 m AOD. Design flood level (DFL) will be 1.0 m above the full length crest spillway, which has a structural thickness of 1.5 m. The maximum permitted horizontal arch stress, assumed uniform through the arch thickness, is 2.5 MN m^{-2}.

1. Select a suitable geometry in plan and determine the crest level arch stress under DFL conditions.

2. Determine a profile for the dam using thin-ring theory and, assuming that DFL loading applies, calculate the thickness required at vertical intervals of 10 m.
3. Confirm the upstream and downstream face stresses at mid-height in terms of thick-ring theory and estimate the abutment stress.

Solutions

1. Select central subtended angle, $2\theta = 90°$ (range 70°–110°). Hence $R_u = 65/[\sin(90° - \theta)] = 92$ m. From equation (3.46),

$$\sigma_h = \frac{p_w R_u}{T_T} = \frac{10 \times 1 \times 92}{1.5} = 0.61\,\text{MN m}^{-2}.$$

2. Using equation (3.46), and taking $\sigma_h = 2.5\,\text{MN m}^{-2}$,

Level (m AOD)	*T (m)*
140	1.50 (given)
130	4.05
120	7.73
110	11.41
100	15.09

3. At level 120, from equations (3.44) and (3.45), with $R_u = 92$ m, $R_D = 92 - 7.73 = 84.27$ m, and

$$p_w = \frac{10 \times 21}{10^3} = 0.21\,\text{MN m}^{-2},$$

giving (equation (3.44))

$$\sigma_{hu} = 0.21\left[(92)^2 + \frac{(92)^2 \times (84.27)^2}{88.13^2}\right] \Big/ [(92)^2 - (84.27)^2]$$

$$= 0.21\,\frac{(15\,565)}{1363} = 2.50\,\text{MN m}^{-2}$$

and (equation (3.45))

$$\sigma_{hd} = \frac{0.42 \times (92)^2}{(7.73)(92 + 84.27)} = 2.61\,\text{MN m}^{-2}$$

(note the variation in stress through the arch ring). To correct for maximum abutment stress, $R_u/T = 11.4$, and hence K_r (Fig. 3.17) $= 1.9$, giving $\sigma_{hd\,max}$ at abutment $= 2.61 \times 1.9 = 4.96\,\text{MN m}^{-2}$.

References

ACI (1970) *Mass Concrete for Dams and Other Massive Structures*, American Concrete Institute, New York.

Berga, L., Buil, J.M., Jofre, C. and Chonggang, S. (eds.) (2003) *RCC dams – roller compacted concrete dams.* Balkema, Lisse.

Bettzieche, V. and Heitefuss, C. (2002) Rehabilitation of old masonry dams at full reservoir level – a comparison of successful rehabilitation projects, in *Proceedings of Conference on 'Reservoirs in a changing world'*, British Dam Society, London: 155–66.

Boggs, H.L. (1975) *Guide for Preliminary Design of Arch Dams*, Engineering Monograph No. 36, USBR, Denver.

Boggs, H.L., Jansen R.B. and Tarbox, G.S. (1988) Arch dam design and analysis, in *Advanced Dam Engineering* (ed. R.B. Jansen), Van Nostrand Reinhold, New York.

Bruce, D.A. and George, C.R.F. (1982) Rock grouting at Wimbleball Dam. *Géotechnique*, 32 (4): 323–48.

Butler, J.E. (1981) The influence of pore pressure upon concrete. *Magazine of Concrete Research*, 33 (114): 3–17.

Case, J., Chilver, A.H. and Ross, C.T.F. (1993) *Strength of Materials and Structures*, 3rd edn, Edward Arnold, London.

Charles, J.A., Abbiss, C.P., Gosschalk, E.M. and Hinks, J.L. (1991) *An Engineering Guide to Seismic Risk to Dams in the United Kingdom*, BRE Report C1/SFB 187 (H16), Building Research Establishment, Garston.

Chopra, A.K. (1988) Earthquake response analysis of concrete dams, in *Advanced Dam Engineering* (ed. R.B. Jansen), Van Nostrand Reinhold, New York.

Clough, R.W. and Zienkiewicz, O.C. (1978) *Finite Element Methods of Analysis and Design of Dams*, Bulletin 30, International Commission on Large Dams, Paris.

Corns, C.F., Schrader, E.K. and Tarbox, G.S. (1988) Gravity dam design and analysis, in *Advanced Dam Engineering* (ed. R.B. Jansen), Van Nostrand Reinhold, New York.

Dolen, T.P. (1988) Materials and mixture proportioning concepts for roller compacted concrete dams, in *Advanced Dam Engineering* (ed. R.B. Jansen), Van Nostrand Reinhold, New York.

Dunstan, M.R.H. (1981) *Rolled Concrete for Dams – a Laboratory Study of the Properties of High Flyash Content Concrete*, Technical Note 105, CIRIA, London.

Featherstone, R.E. and Nalluri, C. (1995) *Civil Engineering Hydraulics*, 3rd edn, Blackwell Scientific, Oxford.

Gallacher, D. and Mann, R.J. (1998) Stabilisation of Tai Tam Tuk dam, Hong Kong, in *Proceedings of Conference on 'The prospect for reservoirs in the 21st century'*, British Dam Society, London: 373–87.

Giovagnoli, M., Schrader, E.K. and Ercoli, F. (1992) Concepcíon dam: a practical solution to RCC problems. *Water Power & Dam Construction*, 44 (2): 48–51.

Goubet, A. and Guérinet, M. (1992) Experience with the construction of the Riou dam. *Water Power & Dam Construction*, 44 (2): 14–18.

Hansen, K.D. (1994) Built in the USA-RCC dams of the 1990s. *Water Power & Dam Construction*, 46 (4): 24–32.

Heitefuss, C. and Kny, H.J. (2002) Underwater work as a means for the rehabilitation of large hydraulic structures under full operation and unrestricted water supply, in *Proceedings of Conference on 'Reservoirs in a changing world'*, British Dam Society, London: 167–78.

Hinks, J.L., Burton, I.W., Peacock, A.R. and Gosschalk, E.M. (1990) Post-tensioning Mullardoch dam in Scotland. *Water Power & Dam Construction*, 42 (11): 12–15.

Hollingworth, F. and Geringer, J.J. (1992) Cracking and leakage in RCC dams. *Water Power & Dam Construction*, 44 (2): 34–6.

ICOLD (2004) *Roller-compacted concrete dams – state of the art and case histories.* Bulletin 126, International Commission on Large Dams, Paris.

Iffla, J.A., Millmore, J.P. and Dunstan, M.R.H. (1992) The use of RCC for small flood alleviation dams in the UK. *Water Power & Dam Construction*, 44 (2): 40–4.

Jansen, R.B. (1988) Conventional concrete for dams, in *Advanced Dam Engineering* (ed. R.B. Jansen), Van Nostrand Reinhold, New York.

Kennard, M.F., Owens, C.L. and Reader, R.A. (1996) *Engineering Guide to the Safety of Concrete and Masonry Dam Structures in the UK*, Report 148, CIRIA, London.

Leliavsky, S. (1958) *Uplift in Gravity Dams*, Constable, London.

Li, E. (ed.) (1999) *Proceedings of International Symposium on RCC dams*, Chengdu. Chinese Society for Hydro Electric Engineering, Beijing, 2 vols.

Link, H. (1969) The sliding stability of dams. *Water Power*, 21 (5): 172–9.

Macdonald, A., Kerr, J.W. and Coats, D.J. (1994) Remedial works to upper Glendevon dam, Scotland, in *Transactions of the 18th International Congress on Large Dams*, Durban, ICOLD, Paris, Q68 R69.

Moffat, A.I.B. (1973) A study of dry lean concrete applied to the construction of gravity dams, in *Transactions of the 11th International Congress on Large Dams*, Madrid, ICOLD, Paris, Q43, R21.

—— (1976) *A Review of Pore Pressure and Internal Uplift in Massive Concrete Dams*, Technical Note 63, CIRIA, London.

—— (1984) *Uplift: The problem and its significance*, Supplementary Paper, BNCOLD Conference, Cardiff, BNCOLD, London.

Moffat, A.I.B. and Price, A.C. (1978) The rolled dry lean concrete gravity dam. *Water Power & Dam Construction*, 30 (7): 35–42.

Neville, A.M. and Brooks, J.J. (1987) *Concrete Technology*, Longman, Harlow.

Nicholson, G.A. (1983) *Design of Gravity Dams on Rock Foundations: Stability Assessment by Limit Equilibrium and Selection of Shear Strength Parameters*, Technical Report GL-83-13, US Army Waterways Experiment Station, Vicksburg.

Ozdogan, Y. (1999) Cine: Turkey's first RCC dam. *International Journal on Hydropower and Dams*, 6 (4): 62–64.

Santana, H. and Castell, E. (2004) RCC record breaker, in *International Water Power and Dam Construction,* 57 (2): 28–33.

Schrader, E.K. (1993) Design and facing options for RCC on various foundations. *Water Power & Dam Construction*, 45 (2): 33–8.

—— (2005) Roller-compacted-concrete: understanding the mix, in *HRW,* 12 (6): 26–9.

—— (2006) Building roller-compacted-concrete dams on unique foundations, in *HRW*, 14 (1): 28–33.

Scuero, A.M. and Vaschetti, G.L. (1998) A drained synthetic geomembrane system for rehabilitation and construction of dams, in *Proceedings of Conference on 'The prospect for reservoirs in the 21st century'*, British Dam Society, London: 359–72.

Scuero, A.M., Vaschetti, G.L. and Wilkes, J.A. (2000) New technologies to optimise remedial work in dams: the underwater installation of waterproofing revetments, in *Proceedings of Conference on 'Dams 2000'*, British Dam Society, London: 298–310.

Sims, G.P. (1994) Ageing of masonry dams. *Proc. Institution of Civil Engineers; Water, Maritime and Energy*, 106 (Mar.): 61–70.

Stefanakos, J. and Dunstan, M.R.H. (1999) Performance of Platanovryssi dam on first filling. *International Journal on Hydropower and Dams*, 6 (4): 139–142.

Thomas, H.H. (1976) *The Engineering of Large Dams*, 2 vols, Wiley, Chichester.

Tripp, J.F., Davie, J. and Sheffield, P. (1994) Construction of Maentwrog new dam. *Proc. Institution of Civil Engineers; Water, Maritime and Energy*, 106 (Dec.): 299–310.

USACE (1981) *Sliding Stability for Concrete Structures*, Technical Letter 1110-2-256, US Army Waterways Experiment Station, Vicksburg, MISS.

USBR (1976) *Design of Gravity Dams*, US Government Printing Office, Denver, CO.

—— (1977) *Design of Arch Dams*, US Government Printing Office, Denver, CO.

—— (1987) *Design of Small Dams*, 3rd edn, US Government Printing Office, Denver, CO.

USCOLD (1985) *Current United States Practice for Numerical Analysis of Dams*, Report of USCOLD Committee, USCOLD, New York.

Ward, R.J. and Mann, G.B. (1992) Design and construction aspects of New Victoria dam. *Water Power & Dam Construction*, 44 (2): 24–9.

Westergaard, H.M. (1993) Water pressure on dams during earthquakes. *Transactions of the American Society of Civil Engineers*, 119: 126.

Zienkiewicz, O.C. (1963) Stress analysis of hydraulic structures including pore pressure effects. *Water Power*, 15 (3): 104–8.

—— (1977) *The Finite Element Method*, McGraw-Hill, London.

Chapter 4

Dam outlet works

4.1 Introduction

Dam outlet works consist generally of spillways and bottom (high-head) outlets. Spillways are basically dam appurtenances ensuring a safe passage of floods from the reservoir into the downstream river reach. The spillway design depends primarily on the design flood, dam type and location, and reservoir size and operation. The design of bottom outlet works depends primarily on the purpose of the reservoir and the sediment inflow and deposition in the reservoir.

Spillways may be classified in several ways: according to function as (main) service, emergency and auxiliary spillways; according to mode of control as free (uncontrolled) or gated (controlled) spillways; according to hydraulic criteria, i.e. type, as overfall, side channel, chute, shaft, siphon and tunnel spillways. In the following text this last type classification will be used.

Apart from economics, the main factors governing the choice of spillway for a given project are the reliability and accuracy of flood prediction, the duration and amount of spillage, seismicity of project site, topography and geology, and the dam type.

In the case of gated spillways the gates may be operated manually, by remote control, or automatically, depending on the level of water in the reservoir. Rigid control regulations are required for non-automatic operation in order to prevent an artificial major flood downstream and/or not to lose valuable water from the reservoir. When controlling the reservoir outflow by spillway gates prior to or during a flood period, reliable flood forecasting methods have to be developed. Gates are of decreasing usefulness in lowering the reservoir level as the dam height increases unless they are submerged as, for example, in tunnel spillways. If the local conditions (e.g. seismic activity, lack of confidence in maintenance and operating skills, or difficulties of access) mean that there are doubts about the dependability of the gates, it might be better to opt for an ungated spillway. Regardless of how reliable gate operation may be, it is often stipulated that the spillway must be adequate to prevent the overtopping of the dam should one or more gates fail to open.

Although the emphasis in this chapter is on the hydraulics of outlet works, some of the more general aspects, such as the selection of design flood, flood routing, freeboard and reservoir sedimentation, have to be dealt with as well. Throughout this and the next two chapters the terms cavitation, aeration, and energy dissipation will appear; although the last two can be treated in relation to specific designs and spillway types, cavitation and its prevention requires a more general discussion before being dealt with in individual outlet works designs.

Energy dissipators and gates and valves are the subject of the next two chapters, while models of outlet works are discussed in Chapter 16. For the treatment of some aspects of the effects of flow over spillways on water quality see Section 9.1.7.

4.2 The design flood

The selection of the design flood (reservoir inflow) hydrograph is one of the most important tasks in dam design; it depends on the dam location and the type of dam and the procedure for its determination. Only the basic principles for its estimation and selection can be touched upon here; a survey of methods for the selection of the design flood can be found, e.g. in ICOLD (1992).

The methods used for the calculation of floods developed from historical records of maximum observed floods, empirical and regional formulae, flood envelope curves and flood frequency analysis to modern methods based on rainfall analysis and conversion to runoff. In the UK the five volume *Flood Estimation Handbook* published by the Institute of Hydrology (1999) (now Centre for Ecology and Hydrology) largely superseded the previous Flood Studies Report (NERC 1993) and placed the emphasis from generalized regression equations to techniques for transferring hydrological data from gauged to ungauged catchments. In the USA the US Army Corps of Engineers Hydrologic Engineering Centre produced a series of software packages (USACE, 2002) providing methods of modelling catchments and including reservoir operation simulation.

Methods used for the selection of design floods in different countries vary (see, e.g., Minor, 1998) and are contained either in recommendations and guidelines (issued by professional bodies) or even in legislation. In many – but not all – cases the highest standard uses the PMF – probable maximum flood – i.e. the flood hydrograph resulting from the probable maximum precipitation (PMP) and (where applicable) snowmelt, coupled with the worst flood producing catchment conditions that can be realistically expected in the prevailing meteorological conditions. A discussion of the rainfall–runoff method and unit hydrograph procedure is beyond the scope of this text but for further particulars see, e.g., Linsley, Kohler and Paulhus (1988), Wilson (1990), Shaw (1994), Institute of Hydrology (1999), Gosschalk (2002) and US Army Corps of Engineers (2002).

The PMF may also approximately be determined as a multiple of a flood of a certain return period (e.g. 150 years, Q_{150} ($m^3 s^{-1}$)) which, in turn, is given by flood frequency analysis or by an equation correlating the flood with the catchment area, the index of catchment permeability, the stream frequency (number of junctions on a 1:25 000 scale map divided by catchment area), the net one-day rainfall of a certain return period (e.g. 5 years), and incorporating a regional constant (NERC, 1993). In UK conditions the PMF is then about 5 times the Q_{150} value; similarly the PMF corresponds to about 3 times the 1000 year flood and twice the 10 000 year flood.

The definition of PMF implies that it is not a fixed value, as its determination (apart from location) depends on the reliability of information, the advance of technical knowledge, and the accuracy of analysis; thus PMF can – and should – be periodically reviewed (see also Reed and Field (1992)). Its probability cannot be determined, as it represents events that are so rare that no observed data are available to establish it, and thus it can only be treated deterministically. (For estimating probabilities of extreme floods, see also National Research Council (1988).)

Table 4.1 is an example of the current recommendations of the Institution of Civil Engineers (UK) for reservoir flood standards (ICE, 1996).

In the case of category b and c dams (Table 4.1) an alternative economic optimization study could be used (see also ICOLD, 1997) in which the chosen flood minimizes (on a probability basis) the sum of the spillway and damage costs; however, this reservoir inflow flood may not drop below the specified minimum.

In some cases of dams in category d (Table 4.1) a cost–benefit analysis and economic evaluation may also be used as a complementary measure.

For small embankments designed for a limited life of L (years) (e.g. cofferdams for the diversion of rivers during dam construction) the risk r of exceeding the design flood of a return period T (years) can be computed as

$$r = 1 - (1 - 1/T)^L. \tag{4.1}$$

Fahlbusch (1999) points out that the safety levels adopted for major dam construction cofferdams are often far too low (usually 30 years for embankments and 20 years for concrete dams), whereas to be consistent with the safety levels for the finished dams, diversion works for construction periods 2–4 years should be designed for floods 200–800 years return

Table 4.1 Flood, wind and wave standards by dam category (ICE, 1996)

Dam category	*Potential effect of a dam breach*	*Initial reservoir condition standard*	*Reservoir design flood inflow*		*Concurrent wind speed and minimum wave surcharge allowance*
			General	*Minimum standard if overtopping is tolerable*	
a	Where a breach could endanger lives in a community	Spilling long-term average inflow	Probable maximum flood (PMF)	10 000-year flood	Mean annual maximum hourly wind speed
b	Where a breach (i) could endanger lives not in a community or (ii) could result in extensive damage	Just full (i.e. no spill)	10 000-year flood	1000-year flood	Wave surcharge allowance not less than 0.6 m
c	Where a breach would pose negligible risk to life and cause limited damage	Just full (i.e. no spill)	1000-year flood	150-year flood	Mean annual maximum hourly wind speed Wave surcharge allowance not less than 0.4 m
d	Special cases where no loss of life can be foreseen as a result of a breach and very limited additional flood damage would be caused	Spilling long-term average inflow	150-year flood	Not applicable	Mean annual maximum hourly wind speed Wave surcharge allowance not less than 0.3 m

period as during the advanced stages of construction a breach of the unfinished dam could be as destructive as that of the finished structure.

An example of a slightly different approach can be found in the report of the Committee on Spillway Design Flood Selection (of the Committee on Surface Hydrology of the Hydraulics Division of ASCE (ASCE, 1988), which recommends three categories of dams depending on the failure consequences and the level of effort required to select a design flood.

Category 1 includes dams 'where the identification of failure consequences indicates potential for loss of life or other social and economic losses that unarguably warrant the use of PMF as the safety design flood'.

Category 2 includes dams 'where the social and economic consequences of failure are not large enough to categorically require the use of PMF as the safety design flood and thus require a detailed analysis to determine the safety design flood'.

Category 3 includes smaller dams 'where the cost of construction is relatively small and the failure damage is low and confined to the owner'.

When comparing the UK guidelines with those used in other European countries Law (1992) concludes that the UK may be oversafe, but that slackening of guidelines would require more certainty about flood estimation precision and that community concern and panel engineer responsibility are unlikely to run risks in a world with non-stationary climate. Cassidy (1994) further to basic research of hydrological processes, which could reduce inherent uncertainties, recommends also surveys of social attitudes to acceptable risks and in economic analysis to take into consideration the dam owner's ability to pay for damages resulting from failure as well as for the reconstruction of the dam.

Thus the difficult selection of a 'safe' reservoir inflow design flood involves complex issues (including moral values) and requires identifying and evaluating a mix of economic, social and environmental impacts.

4.3 Flood routing

To determine the spillway design discharge we must convert the inflow hydrograph of the design flood into the outflow by flood routing which, in turn, is a function of the spillway type, size, and its operation, and of the reservoir area. This is, therefore, a typical iterative design procedure in which the outflow at the dam, determining the size (and type) of spillway, will depend on the inflow and on the spillway type and size.

Reservoir flood routing (which is a special case of general open channel flood routing – Section 8.6) can be carried out using any of the established methods (iteration, Puls, Goodrich, initial outflow value) depending on the size of the reservoir, the time step Δt chosen, and the accuracy required.

All methods are based on the continuity equation, in the form:

$$I - O = \mathrm{d}V/\mathrm{d}t \tag{4.2}$$

where I is the inflow, O the outflow, and V the reservoir storage. In a finite difference form, equation (4.2) can be written as

$$\bar{I} - \bar{O} = \Delta V/\Delta t \tag{4.3a}$$

or

$$\frac{V_2 - V_1}{\Delta t} = \frac{I_1 + I_2}{2} - \frac{O_1 + O_2}{2}. \tag{4.3b}$$

In equation (4.3a), O refers to the spillway outflow. If other regulated outflows O_R (bottom outlets, irrigation outlets, hydroelectric power, etc.) are present then these should be included in the form $\bar{O}_R$.

The solution of equation (4.3), which contains two unknowns, V_2 and O_2 (Δt is chosen), is possible only because in reservoir routing there is a unique relationship between water level and storage (this follows from the assumption of a horizontal water level in the reservoir) as well as water level and outflow; therefore there is also a unique relationship between outflow O and storage volume V.

Denoting by h the head above the spillway crest, and by A the reservoir area at level h,

$$A = f_1(h) \tag{4.4}$$

$$V = f_2(h) \tag{4.5a}$$

or

$$\Delta V = A\,\Delta h \tag{4.5b}$$

$$O = f_3(h) \tag{4.6}$$

$$O = f_4(V). \tag{4.7}$$

Equations (4.3a) and (4.7) together yield the solution in a numerical, graphical, or semigraphical procedure. For example, by rewriting equation (4.3b) as

$$\frac{2V_2}{\Delta t} + O_2 = I_1 + I_2 + \frac{2V_1}{\Delta t} - O_1 \tag{4.8}$$

we have on the right-hand side of the equation only known quantities enabling us to establish $O = f(t)$ from the relationship $O \times (2V/\Delta t + O)$, which we can derive from equation (4.7) for a chosen Δt (Worked example 4.1).

Equation (4.7) applies to a free-flowing spillway without gates (for a water level in the reservoir above the crest). In the event of the flow being controlled by one or more gates it will apply to one position of the gate(s) only. A series of similar relationships is therefore required for each successive position of the gate(s), making flood routing with a gated spillway much more complicated than for a free (ungated) spillway. Therefore, it is often assumed in this case that, for a major flood, the gates have been raised and the reservoir emptied to crest level prior to the flood, the routing of which is then treated in the conventional manner applicable for a free-flowing spillway.

In design we thus first estimate the maximum outflow from the reservoir (for a given inflow hydrograph), and then choose the spillway size and type and test our assumption by routing the inflow with the chosen spillway. In this procedure it is assumed that the crest and maximum permissible water levels are fixed by other considerations (reservoir use, flooding upstream, economic height of the dam, etc.). In flood routing we assume the initial reservoir level to be as high as can be expected at the start of a major flood (usually at spillway crest level).

In general, narrow gated spillways require higher dams and can, therefore, be highly effective in flood routing. Wide free or gated spillways save on dam height, but are usually not very effective for regulating floods. Thus the spillway type influences the benefits to be gained from flood control. The required size of the spillway – and hence its cost – decreases with increase of the dam height, which in turn increases the dam cost; combining the two costs (dam and spillway), the level of the crest for a minimum total cost can be ascertained, although usually the crest level is determined by other considerations such as reservoir operation.

4.4 Freeboard

Freeboard is the vertical distance between the top of the dam and the full supply level in the reservoir; the top of the dam is the highest watertight level of the structure and could thus be the top of a watertight parapet. The freeboard has several components:

1. rise in reservoir level due to flood routing (flood surcharge; Section 4.3);
2. seiche effects;
3. wind set-up of the water surface;
4. wave action and run-up of waves on the dam.

The last three components are often considered as the freeboard proper, or wave freeboard. Sometimes (in gated spillways) an additional component is introduced to cover malfunction of the gates. A further component may be required to account for the effect of impulse waves generated by shore instabilities (rock falls, landslides, snow avalanches etc.); for the hydraulics of these waves and their effect on the dam see, e.g., Huber (1997), Vischer and Hager (1998) and ICOLD (2002). In embankment dams the total freeboard must also include an allowance for the settlement of the dam and foundations (Chapter 2). Thus freeboard determination involves engineering judgement, statistical analysis, and consideration of the damage that would result from the overtopping of a dam.

Seiches, periodic undulations of the reservoir, are usually ignored, particularly in medium-sized reservoirs, and their effect is included in a safety margin added to the other freeboard components. Thomas (1976) mentions seiches up to 0.5 m high in some of the very large reservoirs.

Wind set-up (or wind tide (Roberson, Cassidy and Chaudry, 1998)), s (m) results from the shear induced by continuous wind (or regular gusts in one direction). Its value will depend on the reservoir depth, d_r (m), wind fetch, F (km) (the maximum free distance which wind can travel over the reservoir), the angle of the wind to the fetch, α, and the wind speed U (km h^{-1}) measured at a height of 10 m. The Zuider Zee equation can be used as a guide:

$$s = U^2 F \cos\alpha/(kd_r) \tag{4.9}$$

where k is a constant – about 62 000; using m units throughout and U (ms^{-1}) equation (4.9) can be written as

$$s = k_1 \frac{U^2}{gd_r} F \cos\alpha \tag{4.9a}$$

where $k_1 \cong 2 \times 10^{-6}$.

To produce this set-up the wind must blow for a certain time; the shorter the fetch and higher the wind speed the shorter the time. Typical values are 1 h for a fetch of 3 km, 3 h for $F = 20$ km, 8 h for $F = 80$ km and $U = 40$ km h^{-1}, or 4 h for $U = 80$ km h^{-1}. It must also be remembered that the wind speed at 10 m above the surface of a new reservoir will be higher than that recorded over the original topography. The ratio of wind speed over water to speed over land will be about 1.1–1.3 for effective fetches of 1–12 km.

Allowances for wave height and the run-up of wind-generated waves are the most significant components of freeboard.

As a basis for a *wave height* computation, H (m) (crest to trough) can be estimated from:

$$H = 0.34F^{1/2} + 0.76 - 0.26F^{1/4} \tag{4.10}$$

(for large values of fetch ($F > 20$ km) the last two terms may be neglected).

A modification of equation (4.10) includes a provision for wind speed:

$$H = 0.032\,(UF)^{1/2} + 0.76 - 0.24(F)^{1/4}. \tag{4.11}$$

In the case of medium-sized reservoirs and preliminary design stages of large reservoirs, wave freeboard, f_w, is usually taken as $0.75H + c^2/2g$, where c is the wave propagation velocity (m s^{-1}) which, in turn, is approximately given by $c = 1.5 + 2H$; thus

$$f_w = 0.75H + (1.5 + 2H)^2/2g. \tag{4.12}$$

(Equation (4.12) assumes that the height of the wave crest above the reservoir level is approximately $0.75H$.) For a more detailed treatment of the subject see, for example, Saville, Clendon and Cochran (1962), US Corps of Engineers (1962), Falvey (1974), Zipparo and Hasen (1993) and Sentürk (1994).

Using the concept of significant wave height, H_s (the mean height of the highest third of the waves in a train with about 14% of waves higher than H_s (see also Sections 14.9 and 14.10 for further details), e.g. ICE (1996) recommends the use for the design wave height, H_d, of a multiple of H_s ranging from $0.75H_s$ for concrete dams to $1.3H_s$ for earth dams with a grassed crest and downstream slope and $1.67H_s$ for dams with no water carryover permitted. H_s (m) can be determined quickly from Fig. 4.1 as a function of the wind velocity (m s^{-1}) and fetch (m) based on the simplified Donelan/JONSWAP equation $H_s = UF^{0.5}/1760$.

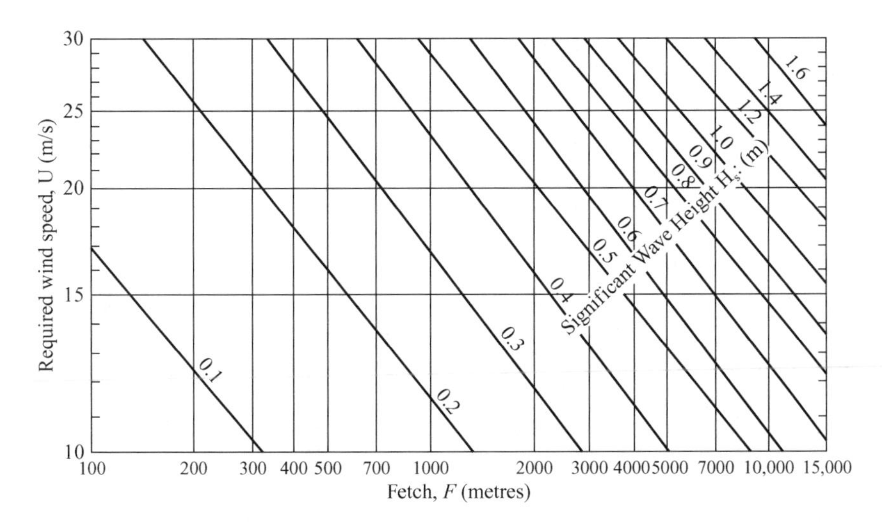

Fig. 4.1 Relationship between fetch, windspeed and significant wave height (ICE, 1996)

The *run-up* of waves on the upstream dam face, i.e. the maximum vertical height attained by a wave running up a dam face, referred to the steady water level without wind action, depends primarily on the wave height, the depth in front of the dam, the geometry and material of the upstream face of the dam, and the approach conditions in the reservoir.

The wave run-up for a typical vertical face in deep water is equal to H_d, but can attain values over $2H_d$ for a smooth slope of 1 in 2 (ICE, 1996).

It may be appropriate to test several alternative wind/fetch directions to determine the maximum value of significant design wave height. The wind speed used in Fig. 4.1 should include adjustments for open water, duration, direction, altitude and return period (see ICE, 1996) (the combined effect of all adjustments is often close to 1). When determining the final value of the freeboard all the surcharge components should be summed; cross-reference can also be made to the last column of Table 4.1.

The probability of an extreme wind coinciding with the maximum flood level in the reservoir is usually very small, and thus it may be deemed unnecessary to superimpose all the extreme conditions for freeboard determination, particularly in the case of concrete dams. Equally, it is a matter of engineering judgement of the possible damage from some water passing over the dam and the cost of a large freeboard for a wave run-up which will ultimately determine the adopted freeboard value. It is generally accepted that well chosen grass on an embankment can withstand water velocities of up to $2\,m\,s^{-1}$ for prolonged periods and up to $5\,m\,s^{-1}$ for brief periods (less than 1 h); reinforced grass waterways perform substantially better (CIRIA, 1987).

4.5 Sedimentation in reservoirs

The assessment of economic viability, safety and cultural considerations as well as the environmental and social impact assessment should form an integral part of any large dam project; this assessment has many facets, one of the most important being the estimation of the sediment deposition in a reservoir and its 'life'.

The loss of storage is only one deleterious effect of sedimentation in reservoirs; others are increased flood levels upstream of the reservoir, retrogression of the river bed and water levels downstream of the dam, the elimination of nutrients carried by the fine sediments, the effect of sedimentation on the reservoir water quality, etc. At present, many reservoirs have a life expectancy of only 100 years. A useful life of a reservoir less than, say, 200 years should certainly be a matter of concern, and one has to consider whether the drastic environmental effects are outweighed by the economic advantages during a relatively short effective life (see also Section 1.2.4).

A comprehensive guide to sedimentation in lakes and reservoirs has been published by Unesco (Bruk, 1985), Morris and Fan (1998) and Batuca and Jordaan (2000).

Sediment run-off in many rivers is continuously increasing – mainly as a result of human influence. Sediment concentration in rivers fluctuates greatly and is a function of sediment supply and discharge (Sections 8.2.3 and 8.4.5). In some rivers it can be extremely high, with hyperconcentrations over $200\,g\,l^{-1}$. In the Yellow River basin mudflows containing up to $1600\,g\,l^{-1}$ have been recorded (Bruk, 1985), while concentrations of $5000\,mg\,l^{-1}$ (ppm) are certainly not unusual on many Asian, but also some other rivers (concentrations expressed in $mg\,l^{-1}$ or ppm are effectively the same up to about $7000\,mg\,l^{-1}$; here 'concentration' is the ratio of the mass of dry sediment to the total mass of the suspension). In many Indian reservoirs the annual loss of storage due to sedimentation is between 0.5% and 1%. At Tarbela on the Indus in Pakistan, the loss has been about 1.5% per year. At the Three Gorges Project on the Yangtze River in China the sedimentation problem is one of the key issues and subject of long-term and on-going studies of various design controls; according to a report by the China Yangtze Three Gorges Project Development Corporation (CTGPC) the annual average sediment load entering the reservoir is at present 526×10^6 tons. However, the morphology of the reservoir is favourable for sediment flushing (see below). It is estimated that worldwide over $30\,km^3$/year of storage are lost by sedimentation; the total storage loss up to the year 2000 is of the order of $570\,km^3$, i.e. about 10% of the gross stored volume.

The relative bulk density of the deposited sediment varies enormously as consolidation takes place over the years, and can reach values of up to 2.0 (usually about 1.2–1.6).

The detailed computation of the amount of sediment deposited in a reservoir requires not only knowledge of the quantity and composition of the incoming sediment but also of the reservoir operation and cross-sections along the reservoir.

The ratio of sediment load, W, left in suspension at the end of a reach of length L, depth of flow y, and velocity V to the initial load W_0 can be expressed as

$$W/W_0 = e^{-KL/yV} \tag{4.13}$$

where K is a constant which is a function of the sediment particle fall velocity w_s (Section 8.2.3). A further treatment of equation (4.13), which is best carried out by mathematical modelling, is beyond the scope of this text.

For preliminary studies the use of trap efficiency curves is sufficient. The most frequently used version is the graph constructed by Brune (1953), plotting the percentage of trapped sediment versus the ratio of the reservoir capacity (m^3) and the annual inflow (m^3) (Fig. 4.2). The graph has to be used in time steps (1–10 years depending on the accuracy required), as at the end of each period the reservoir volume will be decreased by the amount of the settled sediment, and thus the trap efficiency in the next period will be decreased.

It has to be appreciated that the curve in Fig. 4.2, when applied to the reservoir as a whole, does not take into account the shape of the reservoir (width to depth and depth to length ratios) or the graded sediment transport; these parameters can be included only in more refined mathematical modelling (Reeve, 1992).

A useful concept for comparing different projects, as far as the deposition of sediment in the reservoir is concerned, is the so-called half-life of a reservoir, i.e. the time required to lose half of the storage volume (Worked example 4.2).

The results obtained by the application of the trap efficiency curve have to be treated with caution as they can differ considerably from the results of more detailed mathematical modelling which usually gives less favourable results.

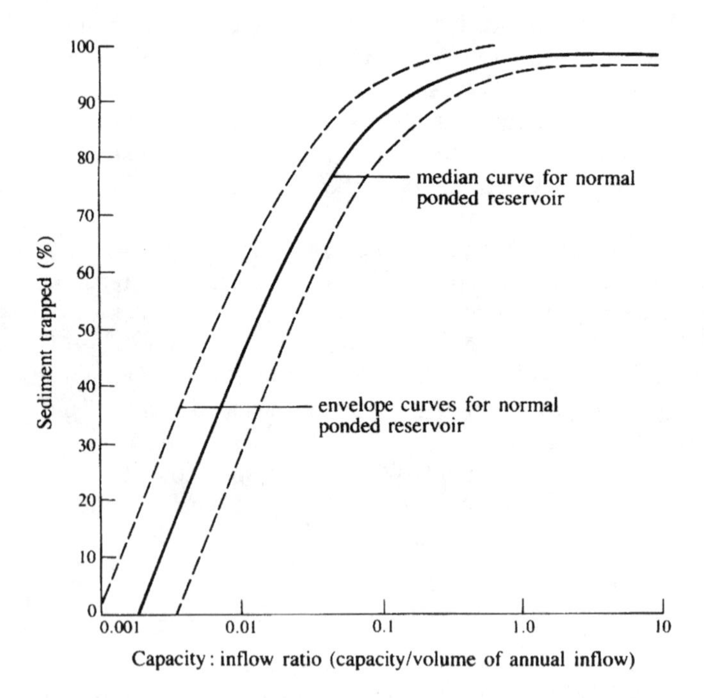

Fig. 4.2 Reservoir trap efficiency (Brune, 1953). The capacity and the volume of the annual inflow must be in the same units of measurement. The upper enveloping curve is to be used when inflowing sediment is highly flocculated or coarse. The lower enveloping curve is to be used when inflowing sediment is colloidal or fine

The various deposition stages in a reservoir are shown schematically in Fig. 4.3.

Reservoir capacity can be preserved by (a) minimizing the sediment input into the reservoir, (b) maximizing the sediment throughflow, or (c) the recovery of storage.

Minimizing sediment input is by far the most effective measure, and can be achieved by optimal choice of the location of the reservoir, the prevention of erosion in the catchment by soil conservation methods (afforestation, terracing, vegetation cover, etc.), the trapping of sediment in traps or by vegetative screens on the tributaries upstream of the reservoir, or by bypassing heavily sediment-laden flows during floods from an upstream diversion structure to downstream of the dam; (in a diversion tunnel a compromise has to be achieved between sedimentation and abrasion; a velocity of about $10\,m\,s^{-1}$ is usually acceptable).

Maximizing sediment throughflow requires flow regulation during floods (sluicing) and/or flushing during a reservoir drawdown. Under certain conditions the sediment-laden inflow does not mix with the water in the reservoir but moves along the old river bed as a density current towards the dam, where it can be drawn off by suitably located and operated outlets (density current venting). In principle, the development of density currents requires a significant difference between the density of the incoming flow and the water in the reservoir, a large reservoir depth, and favourable morphological conditions (steep, straight old river bed).

The techniques used for assessing the effect of reservoir operation on sedimentation and its numerical modelling are discussed, e.g., by Atkinson (1998); for further information on the flushing of sediment from reservoirs see White (2001).

The recovery of storage can be achieved by flushing deposited sediment, a technique which is effective only when combined with a substantial reservoir drawdown, by siphoning or dredging; in the latter case either

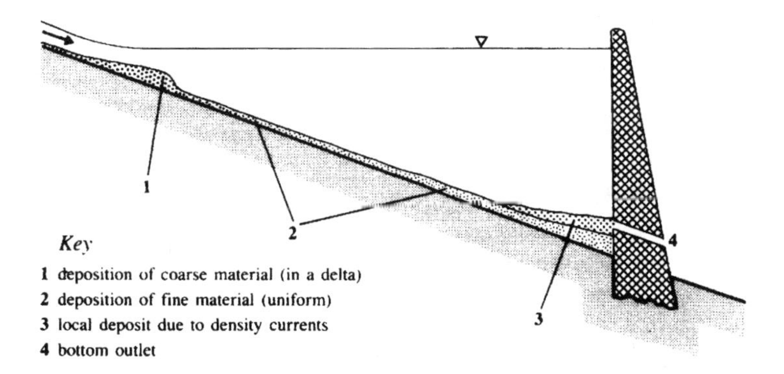

Fig. 4.3 Deposition stages in a reservoir

conventional methods, particularly a suction dredger with a bucket wheel, or special techniques (e.g. pneumatic or jet pumps) may be used. The environmental impact of all techniques for maximizing sediment throughflow or/and the recovery of storage require very careful consideration.

For further discussion of methods of dealing with sedimentation in reservoirs see ICOLD (1999) and Batuca and Jordaan (2000).

4.6 Cavitation

Cavitation occurs whenever the pressure in the flow of water drops to the value of the pressure of the saturated water vapour, p_v (at the prevailing temperature); cavities filled by vapour, and partly by gases excluded from the water as a result of the low pressure, are formed. When these 'bubbles' are carried by the flow into regions of higher pressure, the vapour quickly condenses and the bubbles implode, the cavities being filled suddenly by the surrounding water. Not only is this process noisy, with disruption in the flow pattern, but – more importantly – if the cavity implodes against a surface, the violent impact of the water particles acting in quick succession at very high pressures (of the order of 1000 atm), if sustained over a period of time, causes substantial damage to the (concrete or steel) surface, which can lead to a complete failure of the structure. Thus cavitation corrosion (pitting) and the often accompanying vibration is a phenomenon that has to be taken into account in the design of hydraulic structures, and prevented whenever possible (Knapp, Daily and Hammit, 1970; Galperin *et al.*, 1977; Arndt, 1981).

Low pressures – well below atmospheric pressure – will occur at points of separation of water flowing alongside fixed boundaries, particularly if the flow velocity is high. Thus there are two factors, pressure p and velocity u, which influence the onset of cavitation. They are combined with density ρ in the cavitation number, σ, which is a form of the Euler number:

$$\sigma = 2(p - p_v)/\rho u^2. \tag{4.14}$$

Cavitation occurs if the cavitation number falls below a critical value σ_c which is a function of the geometry and can vary widely. As an example, the incipient cavitation number for sloping offsets and triangular protrusions, as determined from data by Wang and Chou (Cassidy and Elder, 1984), is shown in Fig. 4.4.

According to Ball and Johnson (Cassidy and Elder, 1984), a 3 mm perpendicular offset into the flow can cause cavitation at velocities as low as 11 m s^{-1}; for an equally high recess from the flow the critical velocity is about 32 m s^{-1}. In spillway design we certainly should be very wary of cavitation problems at velocities exceeding 35 m s^{-1}, even if the spillway

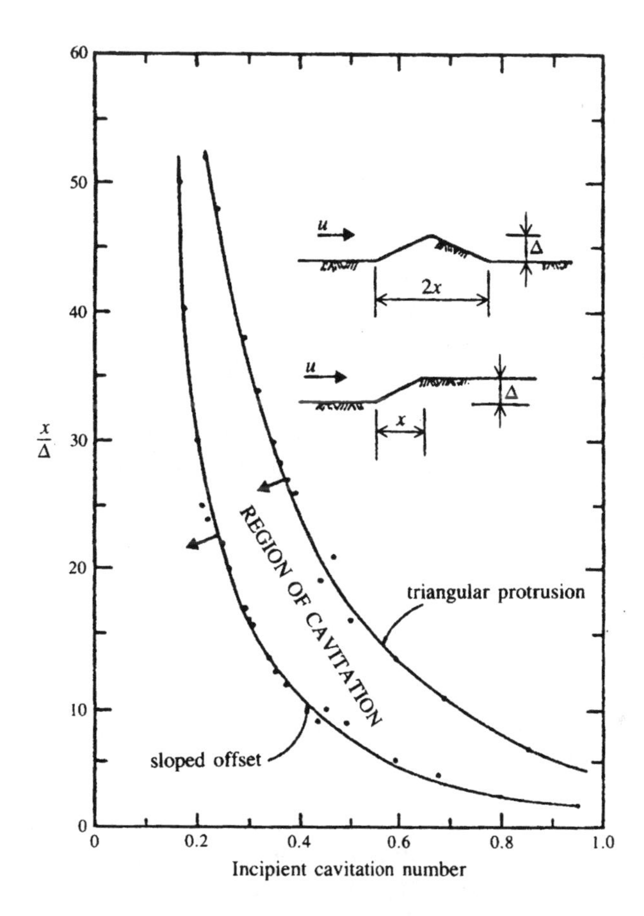

Fig. 4.4 Incipient cavitation number for sloped protrusions (Cassidy and Elder, 1984)

surface is 'smooth' and well constructed. (Elder reported values of absolute roughness of about 1 mm for the 5.5 m diameter Appalachia tunnel, which is probably the smoothest concrete surface obtainable without steel trowelling of the finished surface.)

A value of σ_c of around 0.2 is sometimes considered (Falvey, 1990) when assessing the critical velocity on 'smooth' concrete surfaces; another possibility is to express σ_c as a function of the Darcy–Weisbach friction factor λ (see eq. 8.4) of the surface (e.g. $\sigma_c = 4\lambda$ (Cassidy and Elder, 1984)).

The value of p_v in equation (4.14) is a function of atmospheric pressure and temperature ($p_v \simeq 10$ m $H_2O = p_0$ for 100 °C; $p_v = 6.5$ m for 90 °C and 0.5 m for 30 °C). Although it is generally assumed that the onset of cavitation occurs when $p = p_v$ ($\simeq 0$ for normal water temperatures, i.e. at 10 m below $p_0/\rho g$), the presence of dissolved gases and/or particles in suspension can cause cavitation at higher pressures; thus it is advisable to avoid pressures below about 7 m vacuum (3 m absolute) in hydraulic engineering design. It has to be stressed, however, that in turbulent flows the mean pressure may be well above the danger limit but cavitation can still occur owing to fluctuating instantaneous pressures that fall below the limit. To assess the danger of cavitation in this case it is necessary to analyse the turbulent pressure fluctuations, e.g. under the hydraulic jump in a stilling basin (Chapter 5).

To ascertain the danger of cavitation in any particular situation, model (Chapter 16) or prototype measurements are necessary, or the design has to follow well-established principles. Should there be a clear danger of cavitation damage then either the design or the mode of operation of a particular structure has to be changed, or some other safeguard has to be applied. The most frequent of these is the introduction of air at the endangered parts, i.e. artificial aeration, preventing the occurrence of extremely low pressures. The use of special epoxy mortars can also substantially postpone the onset of cavitation damage on concrete surfaces, and is a useful measure in cases where the cavitation is not frequent or prolonged.

4.7 Spillways

4.7.1 Overfall spillways

The basic shape of the overfall (ogee) spillway is derived from the lower envelope of the overall nappe flowing over a high vertical rectangular notch with an approach velocity $V_0 \simeq 0$ and a fully aerated space beneath the nappe ($p = p_0$), as shown in Fig. 4.5 (e.g. Creager, Justin and Hinds, 1945; US Bureau of Reclamation, 1987).

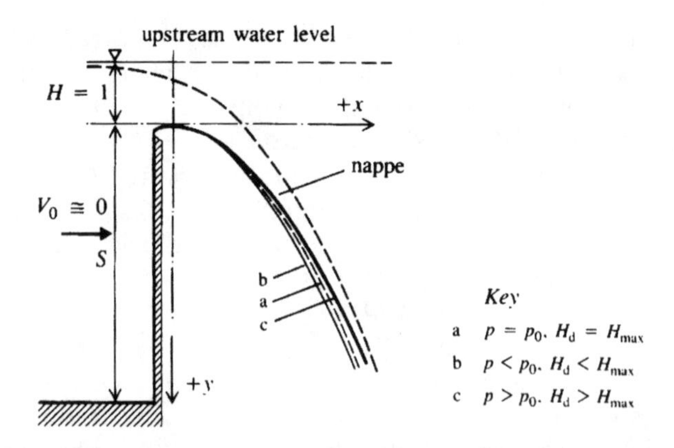

Fig. 4.5 Overfall spillway shape derivation

For a notch of width b, head h, and coefficient of discharge C'_d, the discharge equation is

$$Q = \frac{2}{3}\sqrt{2}g^{1/2}bC'_d\left[\left(h + \frac{\alpha V_0^2}{2g}\right)^{3/2} - \left(\frac{\alpha V_0^2}{2g}\right)^{3/2}\right] \tag{4.15}$$

which, for $V_0 \simeq 0$, reduces to

$$Q = \frac{2}{3}\sqrt{2}g^{1/2}bC'_d h^{3/2}. \tag{4.16}$$

C'_d is about 0.62 (Section 8.4).

Scimeni (1937) expressed the shape of the nappe in coordinates x and y, measured from an origin at the highest point, for a unit value of H (Fig. 4.6) as

$$y = Kx^n \tag{4.17}$$

with $K = 0.5$ and $n = 1.85$.

As the nappes for other values of H are similar in shape, equation (4.17) can be rewritten as

$$y/H = K(x/H)^n \tag{4.18a}$$

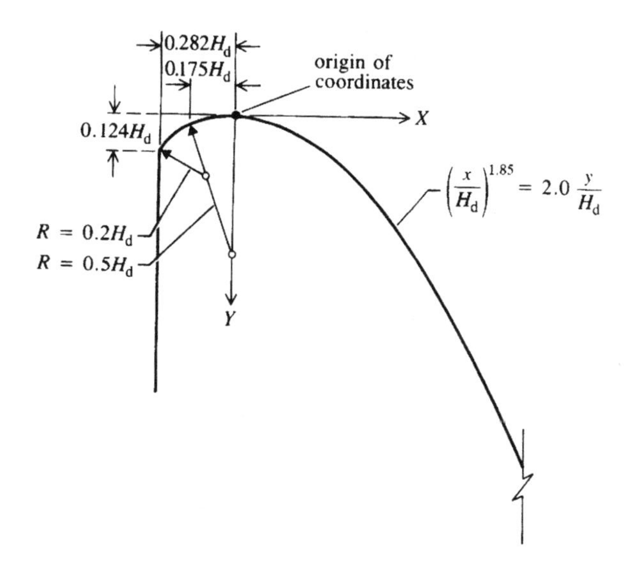

Fig. 4.6 Standard spillway crest (after US Army Waterways Experimental Station, 1959)

or

$$y = Kx^n H^{1-n} = 0.5x^{1.85} H^{-0.85} \tag{4.18b}$$

(curve a in Fig. 4.5 (Novak and Čábelka, 1981)). Note that the head above the new crest (origin) H is smaller than the head h above the crest of the sharp-edged notch from which the shape of the overall spillway (equation (4.18a)) was derived. As, for $K=0.5$, the pressures acting on the surface defined by equation (4.18) are atmospheric (p_0), for $K>0.5$ (curve b) the pressures acting on the spillway will be negative ($p<p_0$), and for $K<0.5$ (curve c) positive ($p>p_0$).

For an overflow spillway we can thus rewrite equation (4.16) as

$$Q = \frac{2}{3}\sqrt{2}g^{1/2} b C_d H^{3/2} \tag{4.19a}$$

or

$$Q = g^{1/2} b C_1 H^{3/2} \tag{4.19b}$$

or

$$Q = C_2 b H^{3/2} \text{ (C_2 has dimensions L}^{1/2}\text{T}^{-1}\text{).} \tag{4.19c}$$

(As $H<h$ in equation (4.19), $C_d>0.62$ for atmospheric pressures.)

There are three possibilities for the choice of the relationship between the design head H_d used for the derivation of the spillway shape and the maximum actual head H_{max}:

$$H_d \gtreqless H_{max}. \tag{4.20}$$

For $H_d=H_{max}$ the pressure is atmospheric and $C_d=0.745$. For $H_d>H_{max}$ the pressure on the spillway is greater than atmospheric and the coefficient of discharge will be $0.578<C_d<0.745$. (The lower limit applies for broad-crested weirs with $C_d=1/\sqrt{3}$, and is attained at very small values of H_{max}/H_d (say, 0.05). For $H_d<H_{max}$ negative pressures result, reaching cavitation level for $H=2H_d$ with $C_d=0.825$. For safety it is recommended not to exceed the value $H_{max}\simeq 1.65H_d$ with $C_d\simeq 0.81$, in which case the intrusion of air on the spillway surface must be avoided, as otherwise the overfall jet may start to vibrate.

Some further details of the standard overfall spillway with $H_d=H_{max}$ are shown in Fig. 4.6 (US Army Waterways Experimental Station, 1959; US Bureau of Reclamation, 1987).

For gated spillways, the placing of the gate sills by $0.2H$ downstream from the crest substantially reduces the tendency towards negative pressures for outflow under partially raised gates. The discharge through partially raised gates may be computed (Fig. 4.7(a)) from

$$Q = \frac{2}{3}\sqrt{2}g^{1/2}bC_{d_1}(H^{3/2} - H_1^{3/2}) \tag{4.21a}$$

with $C_{d_1} = 0.6$ or, better, from

$$Q = C_{d_2}ba(2gH_e)^{1/2} \tag{4.21b}$$

where a is the distance of the gate lip from the spillway surface, and H_e the effective head on the gated spillway ($\simeq H$) ($0.55 < C_{d_2} < 0.7$).

For slender dam sections, e.g. in arch dams, it may be necessary to offset the upstream spillway face into the reservoir in order to gain sufficient width to develop the spillway shape (Fig. 4.7(b)); the effect on the coefficient of discharge is negligible. For details of discharge coefficients for irregular overfall spillways see, for example, Bradley (1952).

In equations (4.19) and (4.21a), b refers to the spillway length. In the case of piers on the crest (e.g. gated spillways) this length has to be reduced to

$$b_c = b - knH \tag{4.22}$$

where n is the number of contractions and k is a coefficient which is a function of H and pier shape; ($0 < k < 0.09$ with the upper limit for semicircular-nosed piers and $H/H_d = 0.2$). For further details see, e.g., Lencastre (1987), Hager (1988), ASCE (1995) and Vischer and Hager (1998).

In the case of concrete gravity dams, the spillway shown in Fig. 4.6 is continued by a plane surface (forming a tangent to it and coinciding with the downstream dam face) to the foot of the dam and into the stilling basin.

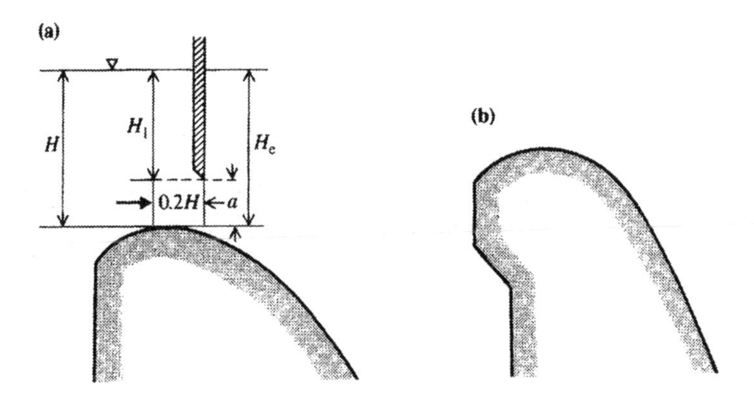

Fig. 4.7 Overfall spillway with (a) gates and (b) offset

However, the spillway may terminate in a vertical face with a free-falling jet (e.g. in arch dam spillways) or in a ski jump shape followed by a more or less dispersed jet. As these arrangements (apart from the case where necessitated by the dam shape as, for example, in arch dams) are primarily used to enhance energy dissipation, they will be discussed further in Chapter 5.

An important design feature is the point at which self-aeration of the overfall nappe (in contact with the spillway) starts. The mechanics of the phenomenon of self-aeration will be dealt with in Section 4.7.3; the distance L_i (m) of the inception point from the crest can be estimated using Hickox's equation:

$$L_i = 14.7q^{0.53} \simeq 15q^{1/2} \tag{4.23}$$

where q is the specific discharge (discharge for unit length, $m^2 s^{-1}$).

In 1980 ICOLD (International Commission on Large Dams) conducted a survey of spillways (of all types) covering 123 structures, with 71 operating over 100 days (Cassidy and Elder, 1984). Most of those where erosion of the surface was reported were operating with maximum velocities of flow over $30 m s^{-1}$ and specific discharges over $50 m^2 s^{-1}$. The survey covered cases with $20 m s^{-1} < V_{max} < 40 m s^{-1}$ and $5 m^2 s^{-1} < q < 200 m^2 s^{-1}$. Not all damage was necessarily due to cavitation, though most of it was. As shown in Section 4.5 even 'smooth' concrete surfaces require great caution with velocities over $30–35 m s^{-1}$, corresponding to a head of over 50 m, if no losses are taken into account; actually, substantially higher dams may have velocities well below these values if energy losses are included (Chapter 5). The role of large values of q is apparent in conjunction with equation (4.23), as the onset of self-aeration, and thus even more so the beginning of air contact with the spillway surface (affording cavitation protection), is delayed in proportion to $q^{1/2}$. For further discussion of the importance of q in energy dissipation and downstream erosion see Chapter 5.

4.7.2 Side-channel spillways

Side-channel spillways are mainly used when it is not possible or advisable to use a direct overfall spillway as, e.g., at earth and rockfill dams. They are placed on the side of the dam and have a spillway proper, the flume (channel) downstream of the spillway, followed by a chute or tunnel. Sometimes a spillway that is curved in plan is used, but most frequently it is straight and more or less perpendicular to the dam axis (Fig. 4.8); the latter is certainly the case in a gated spillway.

The spillway proper is usually designed as a normal overfall spillway (Section 4.7.1). The depth, width, and bed slope of the flume must be designed in such a way that even the maximum flood discharge passes with

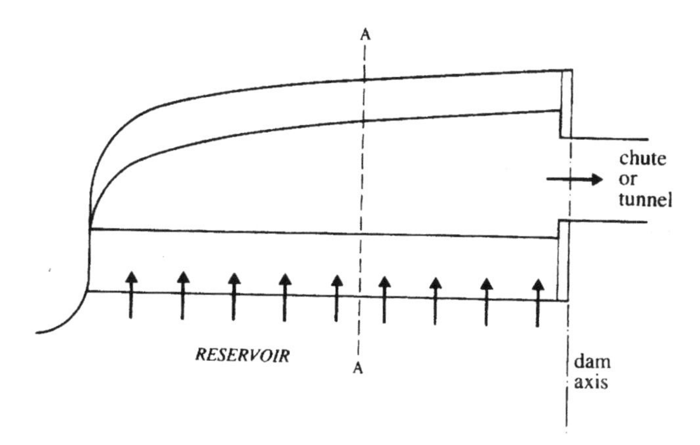

Fig. 4.8 Side-channel spillway: plan

a free overfall over the entire horizontal spillway crest, so that the reservoir level is not influenced by the flow in the channel. The width of the flume may therefore increase in the direction of the flow (Fig. 4.8). From the energy dissipation point of view, the deeper the channel and the steeper the side facing the spillway, the better; on the other hand, this shape is in most cases more expensive to construct than a shallow wide channel with a gently sloping side. The result is usually a compromise, as shown in Fig. 4.9.

The flow in a side-channel spillway is an example of a spatially varied non-uniform flow that is best solved by the application of the momentum principle, assuming that the lateral inflow into the channel (Fig. 4.9) has no momentum in the direction of flow, but that there is substantial energy dissipation in the channel. Taking the slope of the channel, S_0, and the resistance (friction slope S_f) into account results, for a cross-section of the channel, A, wetted perimeter, P, depth y^+ of the centre of gravity of section A, and length (in the direction of flow) Δx, in

$$\frac{\mathrm{d}M}{\mathrm{d}x} = \frac{\mathrm{d}}{\mathrm{d}x}\left(\frac{Q^2}{gA} + Ay^+\right) = A(S_0 - S_f). \tag{4.24}$$

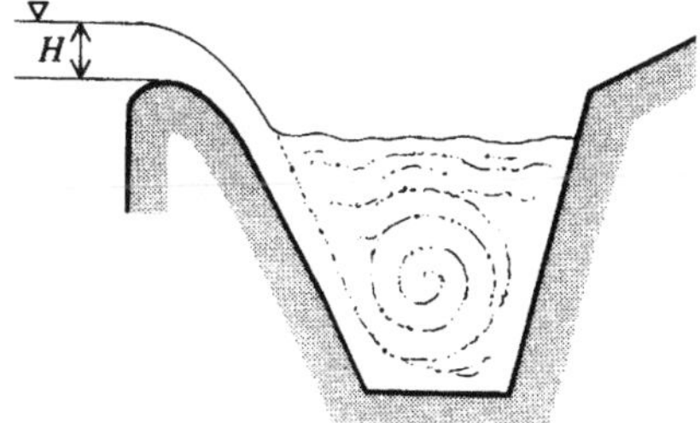

Fig. 4.9 Side-channel spillway: section A–A

Equation (4.24) can be solved by numerical methods, and also results in the differential equation of flow:

$$\frac{\mathrm{d}y}{\mathrm{d}x} = \left(S_0 - S_\mathrm{f} - \frac{2Q}{gA^2}\frac{\mathrm{d}Q}{\mathrm{d}x}\right)/(1 - Fr^2). \tag{4.25}$$

The flow in the channel will, certainly in the upstream part, be subcritical, but at the downstream end may change into supercritical. Critical flow occurs at $Fr = 1$, but as $\mathrm{d}y/\mathrm{d}x$ has a finite value for critical flow

$$S_0 - S_\mathrm{f} - \frac{2Q}{gA^2}\frac{\mathrm{d}Q}{\mathrm{d}x} = 0.$$

For $Q = 2/3 C_\mathrm{d}\sqrt{2}g^{1/2}H^{3/2}x = qx$, $\mathrm{d}Q/\mathrm{d}x = q$ (with $q = 2/3\ C_\mathrm{d}\sqrt{2}g^{1/2}H^{3/2}$ = constant).

For $Fr^2 = Q^2B/gA^3 = 1$ (B is the water surface width in the channel), the above condition leads to

$$S_0 = \frac{gP}{C^2B} + \frac{2}{B}\left(\frac{q^2B}{gx}\right)^{1/3} \tag{4.26}$$

when C is the Chézy coefficient.

For $P = P_\mathrm{c}$ and $B = B_\mathrm{c}$ (critical section) we can calculate from equation (4.26) the critical slope $S_{0\mathrm{c}}$ with control at the outflow from the channel at $x = L$. For a critical section to occur inside the channel, either the slope S_0 must be larger than the value given by equation (4.26) for $x = L$, $P = P_\mathrm{c}$, and $B = B_\mathrm{c}$, or the length L of the spillway (channel) must be larger than

$$x = 8q^2\left[gB_\mathrm{c}^2\left(S_0 - \frac{gP_\mathrm{c}}{C^2B_\mathrm{c}}\right)^3\right]^{-1}. \tag{4.27}$$

Equation (4.25) can be integrated for a rectangular channel section and for $S_0 \simeq S_\mathrm{f} = 0$, resulting in

$$x/L = \frac{y}{y_L}\left(1 + \frac{1}{2Fr_L^2}\right) - \left(\frac{y}{y_L}\right)^3 \frac{1}{2Fr_L} \tag{4.28}$$

giving the relationship between x and the depth of flow in the channel y; index L refers to the outflow (end) section ($x = L$). For a critical depth y_c at $x = L$, equation (4.28) yields

$$x/L = \frac{y}{y_\mathrm{c}}\left[1.5 - 0.5\left(\frac{y}{y_\mathrm{c}}\right)^2\right]. \tag{4.29}$$

Returning to equation (4.24), after substituting for

$$dM/dx = A\frac{dy}{dx} + \frac{d}{dx}\left(\frac{VQ}{g}\right)$$

the scheme for numerical integration could, for two consecutive sections, be expressed as follows (Chow, 1983):

$$\Delta y = -\frac{\alpha Q_1(V_1 + V_2)}{g(Q_1 + Q_2)}\left(\Delta V + \frac{V_2 \Delta Q}{Q_1}\right) + (S_0 - S_f)\,\Delta x. \tag{4.30}$$

Equation (4.30) can be solved by trial and error for Δy, with Q_1, V_1, S_0, Δx, Q_2 and channel shape known, and an assumed value of y_2 (and therefore V_2) which must agree with $\Delta y = y_2 - y_1$ given by equation (4.30). The solution starts from the control section and proceeds upstream; the control section is either the outflow from the channel or the critical depth section inside the channel, the position of which is determined from equation (4.27). Should the computation show that there would be a substantial length of non-modular outflow in the upstream part of the side-channel spillway (or, alternatively, that the channel is unnecessarily large) the design parameters have to be altered.

4.7.3 Chute spillways

A chute spillway is a steep channel conveying the discharge from a low-overfall, side-channel, or special shape (e.g. labyrinth) spillway over the valley side into the river downstream. The design of chute spillways requires the handling of three problems associated with supercritical flow: waves of interference, translatory waves, and (self-)aeration.

Interference waves (cross-waves, standing waves) are shock waves which occur whenever the supercritical flow is interfered with, at inlets (Fig. 4.10), changes of section, direction, or slope, bridge piers, etc. They are stationary waves, the position of which depends on discharge; their main significance is that they require an increased freeboard and higher chute side walls, as water tends to ‘pile up’ at points where the waves meet the side walls (e.g. points B and D in Fig. 4.10). The waves can also create additional difficulties in energy dissipators if they persist so far downstream (which is rarely the case because once the flow becomes aerated they disappear). They can be minimized by carefully shaping any changes necessary in cross-section, and making transitions in direction and slope as gradual as possible. If the chute is relatively long a very gradual reduction in width commensurate with the flow acceleration can produce some savings in cost; near its outlet the chute may also be widened gradually to reduce the flow per unit width and the depth (and cavitation risk), and to improve energy dissipation. All other forms of interference with the flow are best avoided.

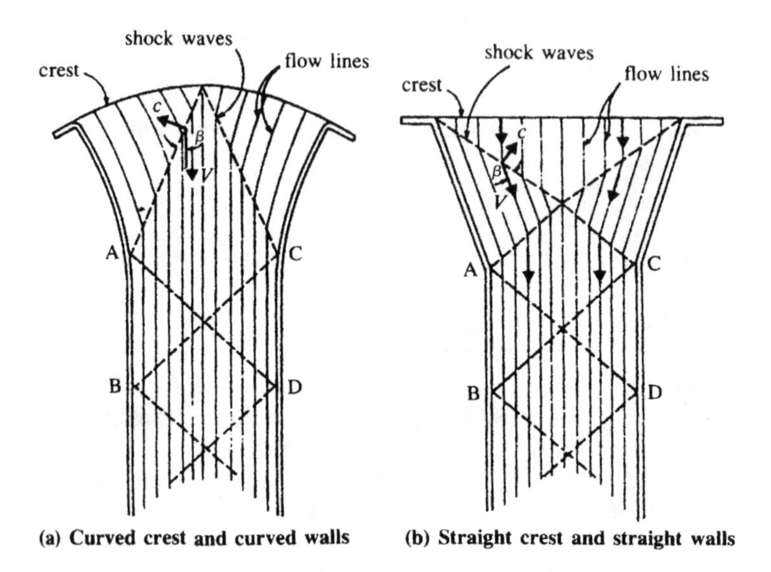

Fig. 4.10 Interference waves at a chute spillway inlet (Cassidy and Elder, 1984)

The basic treatment of the hydraulics of shock waves can be found for example in Henderson (1966) and of design considerations in Chow (1983) (following the classical work of Ippen and Knapp). Briefly, for an inclination β of the shock wave to the direction of flow and the ratio of the downstream and upstream depths y_2 and y_1 close to 1, $Fr_1 \sin\beta = 1$. For values of y_2/y_1 appreciably bigger than 1,

$$Fr_1 \sin\beta = \left[\frac{1}{2}\frac{y_2}{y_1}\left(\frac{y_2}{y_1}+1\right)\right]^{1/2}. \tag{4.31}$$

Substituting for $y_2/y_1 = \tan\beta/\tan(\beta-\theta)$ (from continuity and momentum equations) with θ as the wall deflection angle results in

$$\frac{\tan\beta}{\tan(\beta-\theta)} = \frac{1}{2}[(1+8Fr_1^2\sin^2\beta)^{1/2}-1]. \tag{4.32}$$

The above equations can be simplified to

$$y_2/y_1 = \sqrt{2}Fr_1\sin\beta - 1/2, \tag{4.33}$$

$$\beta = \theta + 3/(2\sqrt{2}Fr_1). \tag{4.34}$$

For small values of β and $y_2/y_1 > 2$ this may be written as

$$y_2/y_1 = 1 + \sqrt{2}Fr_1\theta. \tag{4.35}$$

In this case the maximum depth at the wall y_w (important for the freeboard)(occurs at a distance about $1.75y_1Fr_1$ from the point of wall deflection and is given by

$$y_w/y_1 = 1 + \sqrt{2}Fr_1\theta(1 + Fr_1\theta/4). \tag{4.36}$$

The reduction of shock waves can be achieved by one (or a combination) of three methods: *reduction in the shock number* $Fr_1\theta$ (certainly the most effective method), *wave interference* (e.g. in channel contractions – applicable to one approach flow only) or by bottom and/or sides *reduction elements* (shock diffractors). Modification of chute geometry including 'banking' and curved inverts is often successful, but not necessarily economical. Bottom elements include vanes or oblique steps (Ellis, 1989). For further details see also ICOLD (1992b) and for a comprehensive treatment of shock waves and flow through chute expansions, contractions and bends Vischer and Hager (1998).

Gate piers on the crest of spillways also induce shock waves; formerly, (not very successful) attempts were made to reduce the resulting 'rooster tails' by specially shaped downstream pier ends, but today rectangular pier ends are often used, accepting the resulting shock waves (Vischer, 1988).

Translatory waves (waves of translation, roll waves) originate under certain conditions from the structure of supercritical flow and, as their name implies, move with the flow right into the stilling basin. They again require a higher freeboard, and by imparting unsteady flow impulses to the stilling basin may even cause its failure (Arsenishvili, 1965). They occur at slopes with $0.02 < S < 0.35$ (which is a very wide range covering most chute spillways) and with long chutes, but even then can be avoided if the ratio of the depth to the wetted perimeter is greater than 0.1. Another criterion for the appearance of translatory waves is the Vedernikov number – Ved; they form, if $\text{Ved} = k\varphi Fr > 1$, where k is a constant ($k = 2/3$) and φ the channel shape factor, $\varphi = 1 - R\ \text{dP/dA}$, and if the chute length L is greater than $L > -9.2\ V_0^2/(gS_0)\ ((1 + 2/3\varphi/\text{Ved})/(1 - \text{Ved}))$ (Jain 2000). Roll waves can also be avoided by introducing artificial roughness to the chute surface which is, of course, contrary to cavitation control. The best method is to design the spillway with a depth: perimeter ratio larger than 0.1 for the maximum discharge, and to accept roll waves at low flows when they are not dangerous to the stilling basin and do not require additional freeboard.

(Self-)aeration is by far the most important feature of supercritical flow. Although beneficial for energy dissipation and cavitation protection, it causes an increase of the depth of flow (bulking) and thus requires an increase of the chute side walls.

The transition from critical flow at the crest, through supercritical non-uniform non-aerated flow to non-uniform partially and fully aerated flow to, finally, uniform aerated flow is shown schematically in Fig. 4.11; the flow depth, y_a, can be compared with that of uniform non-aerated flow,

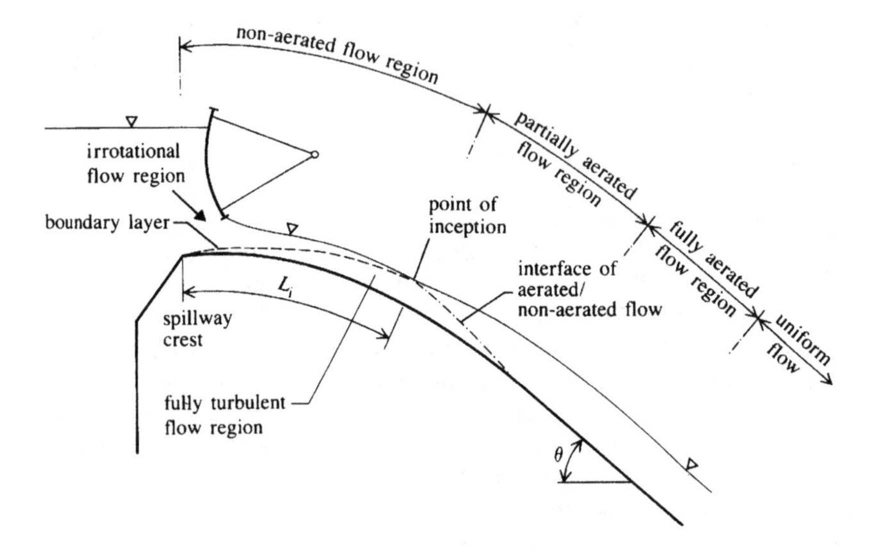

Fig. 4.11 Development of flow on a chute spillway (Ackers and Priestley, 1985)

y_0 to obtain the increase, $\Delta y = y_a - y_0$, for the aerated flow. The inception of aeration occurs at the point where the turbulent boundary layer penetrates the full depth of the flow. This position can be calculated by combining the equations of non-uniform (non-aerated) flow with the equation of the turbulent boundary layer growth or (sufficient for preliminary design purposes) from equation (4.23). For aeration to begin the energy of the turbulent surface eddies must exceed the energy of the surface tension. Thus in model experiments often no air is actually entrained at all (or its inception is delayed), because of the disproportionally high surface tension, although the 'inception point' can be clearly distinguished by the beginning of a rough water surface. Although closely connected with drop ejection and impact the whole mechanism of self-aeration of supercritical flow is more complex and subject of further research (Rein, 1998, Ferrando and Rico, 2002).

Ackers and Priestley (1985) and Wood (1991) quote a simple equation for the growth of the boundary layer δ with distance L:

$$\delta/L = 0.0212(L/H_s)^{0.11}(L/k)^{-0.10} \qquad (4.37)$$

where H_s is the potential flow velocity head and k is the equivalent roughness value. Equation (4.37) was derived primarily from field data for concrete gravity dam spillways.

Water levels in the non-uniform non-aerated flow can be determined by standard non-uniform flow calculations. The computation of the partially and fully aerated non-uniform flow is beyond the scope of this text but various approaches can be found in, for example Haindl (1984), Ackers and Priestley (1985), Naudascher (1987), Falvey (1990), ICOLD (1992b) and Wilhelms and Gulliver (2005). As a very rough estimate, it is possible to take the distance from the inception of aeration to the point where air reaches the spillway (and thus may provide some cavitation protection) to be given also by equation (4.23).

The important depth of the uniform aerated flow, y_a, can be estimated in several ways. Writing the average air concentration as $C = Q_a/(Q_a + Q)$, (Q_a is the discharge of air), the ratio of the water discharge Q to the total discharge of the mixture of air and water as $\rho_1 = Q/(Q_a + Q)$ and the ratio of air to water discharge as $\beta = Q_a/Q$,

$$C = 1 - \rho_1 = \beta/(\beta + 1). \tag{4.38}$$

For a rectangular chute:

$$\rho_1 = Q/(Q_a + Q) = y_0/y_a \tag{4.39}$$

where y_0 is the depth of the non-aerated (uniform) flow.

Wood (1991) defines C by using the depth, y', where the air concentration is 90%, as

$$C = 1 - y_0/y'. \tag{4.40}$$

For a quick assessment of y_a we may use the approximate equations

$$y_a = c_1 y_c = c_1 (q^2/g)^{1/3} \tag{4.41}$$

with the coefficient c_1 in the range $0.32 < c_1 < 0.37$, or

$$(y_a - y_0)/y_0 = 0.1(0.2Fr^2 - 1)^{1/2}. \tag{4.42}$$

Experiments by Straub and Anderson (1958) have shown that the ratio of the aerated spillway friction factor, λ_a, to the non-aerated one, λ, decreases with an increase of air concentration approximately according to (Ackers and Priestley, 1985)

$$\lambda_a/\lambda = 1 - 1.9C^2 \tag{4.43}$$

for $C < 0.65$. For $C > 0.65$, λ_a/λ remains constant at 0.2.

Anderson's data (Anderson, 1965) for rough and smooth spillways after curve fitting and recalculation to SI units give the following equations for air concentration. For rough spillways (equivalent roughness 1.2 mm),

$$C = 0.7226 + 0.743 \log S/q^{1/5} \tag{4.44}$$

for $0.16 < S/q^{1/5} < 1.4$. For smooth spillways,

$$C = 0.5027(S/q^{2/3})^{0.385} \tag{4.45}$$

for $0.23 < S/q^{2/3} < 2.3$ (q in equations (4.44) and (4.45) is in $m^2 s^{-1}$). When applying any one of these equations it is important not to exceed the (experimental) limits given above; even so, the equations tend to underestimate C at the lower end of the range.

From equation (4.38) it is evident that y_0 should really be determined from the water discharge component of the water and air mixture (assuming both parts to have equal velocities). As seen from equations (4.43) and (4.48) the aerated flow friction coefficient is smaller than the non-aerated one; this means that the water component depth $y_0' = \int_0^\infty (1 - C)\,dy$ is smaller than y_0 for $C > 0$. According to Straub and Anderson for $0 < C < 0.7$, $1 > y_0'/y_0 > 0.75$, with significant departures of y_0'/y_0 from 1 only for $C > 0.4$ ($C = 0.4$, $y_0'/y_0 = 0.95$).

Falvey (1980) developed an equation for $0 < C < 0.6$ and smooth spillways (slope θ) which includes the effect of surface tension in the form of Weber number $We(V/(\sigma/\rho y_0)^{1/2})$;

$$C = 0.05Fr - (\sin\theta)^{1/2} We/(63Fr). \tag{4.46}$$

For relatively narrow spillways, Hall (1942) suggested the use of Manning's equation for aerated flow:

$$V = \frac{R_a^{2/3}}{n} S^{1/2} \tag{4.47}$$

and

$$V = \frac{R^{2/3}}{n_a} S^{1/2} \tag{4.48}$$

where $n_a < n$ (equation (4.43)) and

$$\frac{1 - \rho_1}{\rho_1} = c_1 \frac{V^2}{gR} + c_2 \tag{4.49}$$

where for smooth concrete surfaces $c_1 \simeq 0.006$ and $c_2 \simeq 0$.

Writing $R = by/(b + 2y) = q/(V + K)$, where $K = 2q/b$, equations (4.47)–(4.49) result in

$$V = \frac{S^{1/2}}{n} \left(\frac{q}{V + K}\right)^{2/3} \left[1 + c_1 \frac{V^2}{gq}(V + K)\right]^{2/3} \Big/ \left[1 + \frac{2c_1 V^2}{gb}\right]^{2/3}. \tag{4.50}$$

For small values of K, equation (4.50) results in

$$V \simeq \frac{S^{1/2} q^{2/3}}{n V^{2/3}} \left(1 + c_1 \frac{V^3}{gq}\right)^{2/3}. \quad (4.51)$$

Equations (4.50) and (4.51) are implicit equations for the velocity which, when solved, give ρ_1 (equation (4.49)) and thus the depth of the uniform aerated flow y_a (Worked example 4.4).

Wood (1991) recalculated Straub's data and plotted the average concentration (%) as a function of the chute slope θ. The result can be approximated as follows:

$$\text{for } 0° < \theta < 40°, \quad C = (3/2)\theta;$$

$$\text{for } 40° < \theta < 70°, \quad C = 45 + 0.36\theta. \quad (4.52)$$

Together with equation (4.43) this permits the computation of y_a (see worked example).

An even simpler way of finding C (ICOLD, 1992b and Vischer and Hager, 1998) is given by

$$C = 0.75(\sin\theta)^{0.75} \quad (4.53)$$

with the concentration at the chute surface given by

$$C_0 = 1.25\,\theta^3 \quad (4.54)$$

with θ in radians for $0 < \theta° < 40$ and

$$C_0 = 0.65 \sin\theta \quad (4.54a)$$

for $\theta > 40°$.

It must be appreciated that uniform aerated flow is reached only at a very considerable distance from the spillway crest, and in many chute spillways may never be reached at all, particularly if the specific discharge q is large ($q > 50\,\text{m}^2\,\text{s}^{-1}$). From the point of view of cavitation protection also a minimum air concentration is required in contact with the spillway (about 7%); equations (4.41)–(4.53) yield only the average concentration, which should probably exceed 35% to provide the minimum necessary concentration of air at the spillway surface (see, for example, Wood, 1991).

In order to provide cavitation protection in cases where there is no air in contact with the spillway, or the air concentration is insufficient and the velocities high enough to make cavitation damage a real possibility, 'artificial' aeration by aerators has been developed (Pinto, 1991; Volkart and Rutschmann, 1991). These aerators have the form of deflectors (ramps), offsets, or grooves, or a combination of two or all three of them (Fig. 4.12). The most frequent shape used on spillways is the combination of a deflector with an offset. Air is supplied to the spillway surface automatically through air ducts as the flow separation causes the pressure downstream of the aerator to drop below atmospheric; the air duct(s) must be designed to control this pressure drop within acceptable limits. As air bubbles will not stay in contact with the spillway but rise to the water surface, a series of aerators is sometimes considered necessary to achieve the minimum required air concentration at the spillway surface. The higher the flow velocity, the larger can be the distance between the aerators; roughly, the distance in metres should be equal to about 2 times the mean flow velocity (in $\mathrm{m\,s^{-1}}$). However, recent research (Kramer 2004) seems to indicate that for certain flow conditions the average minimum air concentration never falls below a critical value, suggesting that one aerator may be sufficient to protect a chute from cavitation damage.

For the computation of the air discharge through an aerator Pinto (1991) recommends (from prototype measurements)

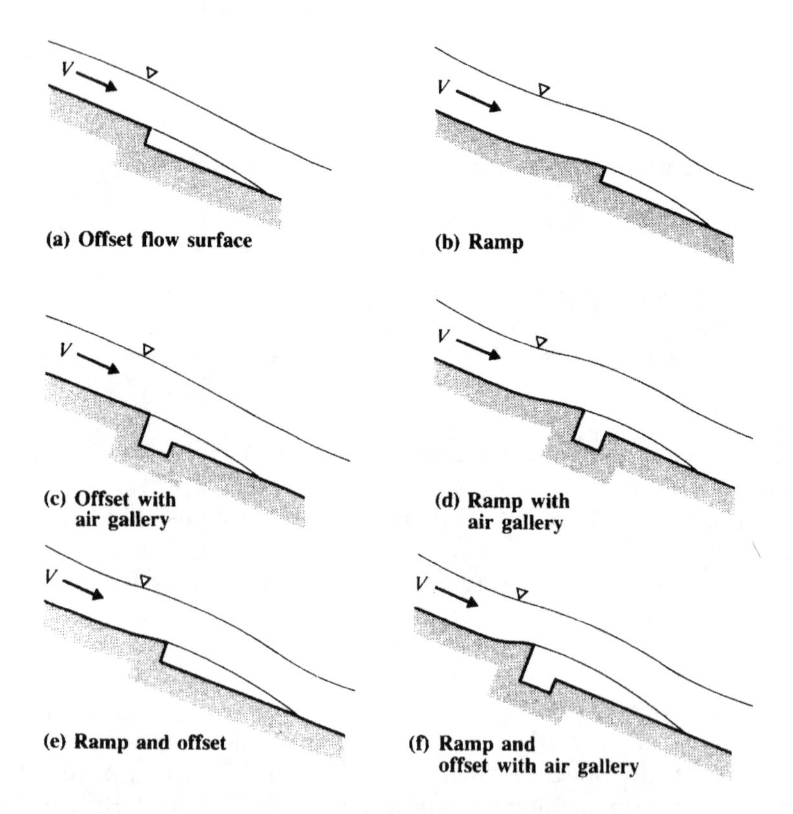

Fig. 4.12 Types of aeration facilities (after Cassidy and Elder, 1984)

$$\beta = q_a/q = 0.29(Fr - 1)^{0.62}(D/y)^{0.59} \tag{4.55}$$

where Fr is the Froude number just upstream of the aerator, and $D = cA/b$, with A the aerator control orifice area, c the discharge coefficient and b the chute width. The total air flow for an average negative pressure under the nappe at the aerator Δp is then

$$Q_a = q_a b = cA(2\Delta p/\rho_a)^{1/2}. \tag{4.56}$$

For further discussion and details of aerators see ICOLD (1992b).

Self-aerated flow in steep partially filled pipes and tunnels resembles flow in steep rectangular channels (chutes) but the effect of the geometry of the conveyance and the enclosed space results in differences in air entrainment and concentration.

Thus Volkart (1982) on the basis of laboratory and field measurements proposed for the air concentration C:

$$C = 1 - 1/(0.02(Fr - 6)^{1.5} + 1) \tag{4.57}$$

or for $C < 0.3$

$$C = 0.044Fr - 0.3 \tag{4.58}$$

(where $Fr = V/(gR)^{1/2}$ with both the velocity and hydraulic radius applying to the water component only).

Equation (4.57) defines also $Fr = 6$ as a limiting value for the beginning of air entrainment; this in turn can be translated into a limiting slope S (using the Manning–Strickler equation)

$$S = (36.4/D)^{1/3} gn^2. \tag{4.59}$$

For $Fr < 17$ and $S < 1$ the air concentration for flow in partially filled pipes is smaller than for open rectangular channels (for further details see Volkart (1982)).

The air supply to the flow must also take into account the air demand for the air flow above the air–water 'interface'; (see also Sections 4.7.4, 4.8 and 6.6).

4.7.4 Shaft spillways

A shaft ('morning glory') spillway consists of a funnel-shaped spillway, usually circular in plan, a vertical (sometimes sloping) shaft, a bend, and a tunnel terminating in an outflow. Shaft spillways can also be combined with a draw-off tower; the tunnel may also be used as part of the bottom outlets or even a turbine tailrace. The main components and rating curve of the spillway are shown schematically in Fig. 4.13.

The shape of the shaft spillway is derived in a similar manner to the overfall spillway from the shape of the nappe flowing over a sharp-edged circular weir (Wagner, 1956; Fig. 4.14). Clearly, in this case the shape for atmospheric pressure on the spillway is a function of H_s/D_s, where H_s is the head above the notch crest of diameter D_s. For ratios $H_s/D_s<0.225$ the spillway is free-flowing (i.e. with crest control) and for $H_s/D_s>0.5$ the overfall is completely drowned. For the free overfall the discharge is given by

$$Q=\frac{2}{3}C_d\pi D_c\sqrt{2}g^{1/2}H^{3/2} \tag{4.60}$$

(curve 1 in Fig. 4.13, and for the drowned (submerged) régime (with orifice control) by

$$Q=\frac{1}{4}C_{d_1}\pi D^2[2g(H+Z)]^{1/2} \tag{4.61}$$

(curve 3 in Fig. 4.13), where D is the shaft diameter, D_c is the crest diameter ($D_c<D_s$), H is the head of the reservoir level above the crest ($H<H_s$), z is the height of crest above the outflow from the shaft bend, and C_d and C_{d_1} are discharge coefficients (Fig. 4.13).

Figure 4.15 is a plot of the discharge coefficient (in the form $m=2/3C_d$ irrespective of the type of flow at the shaft inlet) for two values of s/D_s as a function of $H_s/D_s \simeq H/D_c$ (defined in Fig. 4.15). In equation (4.60) in the range $0.1<\dfrac{H}{D_c}<0.25$ the coefficient of discharge C_d may be

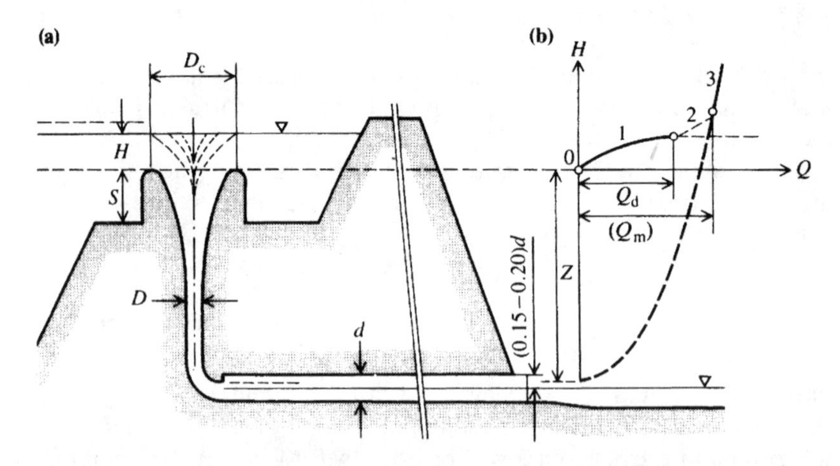

Fig. 4.13 Shaft spillway

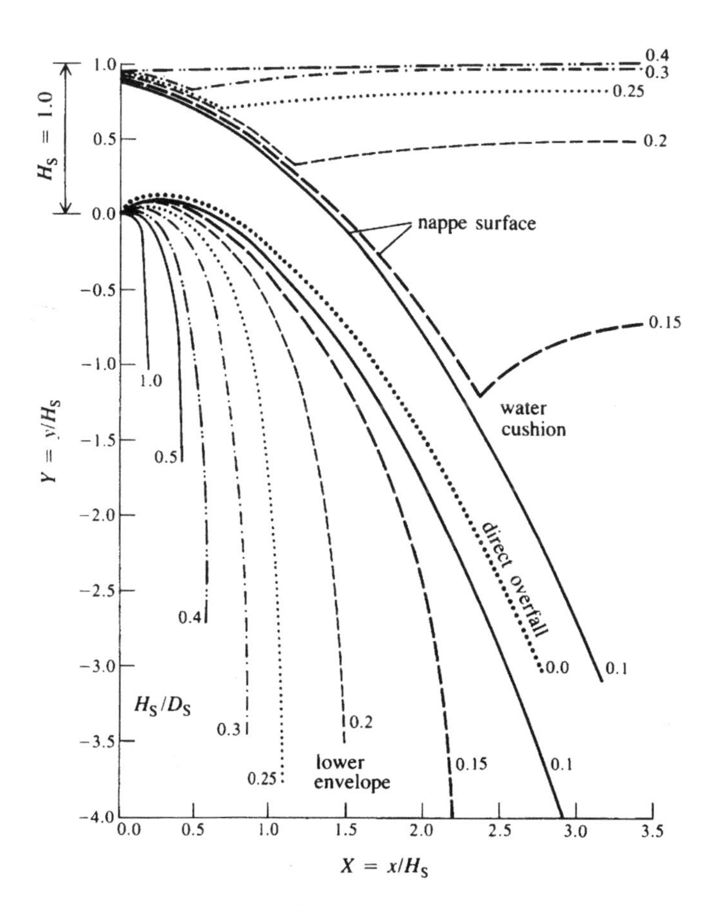

Fig. 4.14 Shapes of shaft spillways (after Wagner, 1956)

approximated by $C_d < 0.773\left[1 - 0.4\dfrac{H}{D_c}\right]$; for $\dfrac{H}{D_c} < 0.1$ the coefficient has practically the same value as for a straight-crested overfall (0.745).

C_{d_1} in equation (4.61) has to take into account losses in the shaft and bend (and possibly also in the tunnel for 'pipe control', i.e. when the tunnel is flowing full – Fig. 4.17(a)).

In the case of a drowned spillway (or a spillway in the transition region) a vortex, which reduces the spillway capacity, is formed; to prevent it, piers or other anti-vortex devices are sometimes used, as is the case for free-flowing spillways if these are placed close to the reservoir bank or even in a cut in the bank; in this situation the anti-vortex device actually plays a role in erosion prevention (Fig. 4.16). The coefficient $m = 2/3C_d$ is, in this case, reduced to $m' = m\sigma_1\sigma_s$, where $\sigma_1 = f(l/D_c,\ H/D_c)$ and $\sigma_s = \phi(S/D_c,\ H/D_c)$ with $0.5 < \sigma_1 < 1$ and $0.8 < \sigma_s < 1.0$. For optimum performance $a \simeq 1.75D_c$, $l > 6D_c$, and $0 < n' < 0.5$ (see Fig. 4.16).

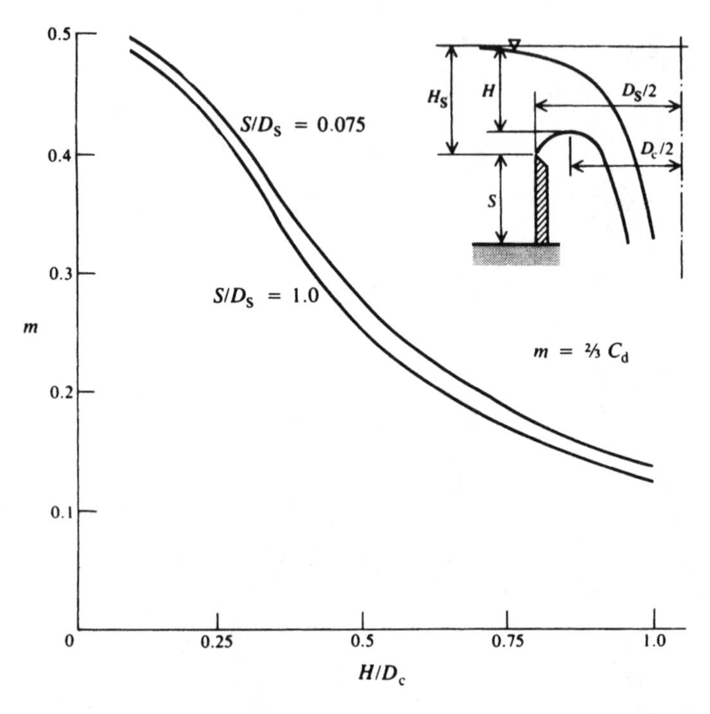

Fig. 4.15 Discharge coefficients of shaft spillways (after US Bureau of Reclamation, 1987)

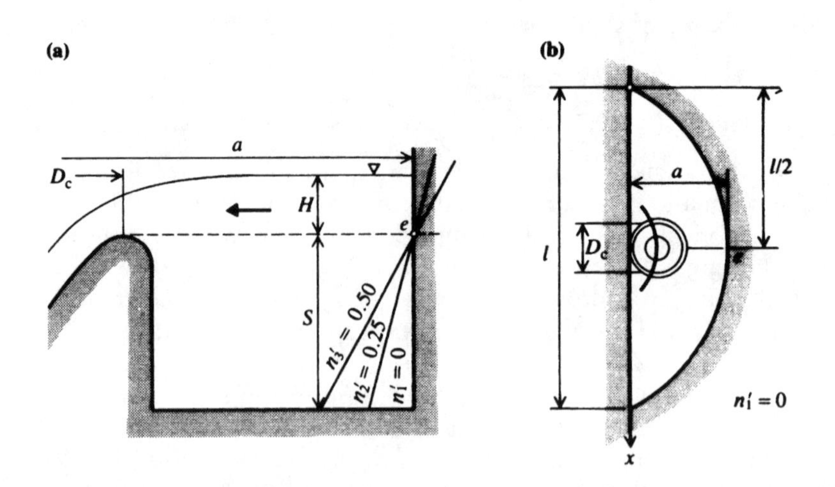

Fig. 4.16 Shaft spillway placed in a cut in the bank (after Novak and Čábelka, 1981)

Although, in some of the older shaft spillways placed well away from the banks and operating under free-flow conditions, anti-vortex devices were also provided, it is in fact advantageous in these cases to induce a vortex at the inflow by placing low curved vanes below the spillway crest. These vanes do not measurably reduce the discharge, but by inducing spiral flow along the shaft walls they substantially reduce vibration and pressure fluctuations induced by an otherwise free-falling jet, particularly at low (and thus frequent) discharges.

As for free flow, $Q \propto H^{3/2}$, whereas for submerged flow even a small increase in Q results in a substantial increase of the required head H (Fig. 4.13), for safety shaft spillways are usually designed to operate only as free-flowing, i.e. with $H_s/D_s \simeq H/D_c < 0.225$. For values of $H/D_c < 0.1$ and close to a bank, a wide-crested spillway may be advantageous.

The tunnel below shaft spillways can be designed for four different flow régimes shown in Fig. 4.17, depending on the relative position of the tunnel soffit and the downstream water level. The configuration in Fig. 4.17(d) is the most favourable one from the point of view of stability of flow in the tunnel and prevention of vibrations. The aerated transition from the shaft bend into the tunnel is an important feature of the design. Its purpose is to establish a control section for the shaft and bend and, at the same time, to stabilize the free or pressure flow (with a stable jump) in the tunnel. For some further design features and details of shaft spillways, see Novak and Čábelka (1981), Haindl (1984), US Bureau of Reclamation (1987), Vischer and Hager (1998) and Ervine (1998).

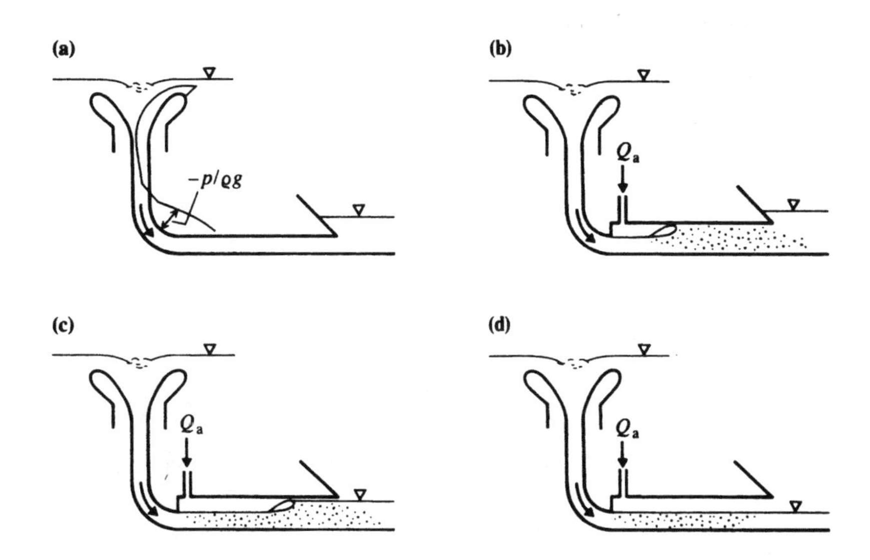

Fig. 4.17 Shaft spillway and tunnel configurations (after Novak and Čábelka, 1981)

4.7.5 Siphon spillways

Siphon spillways are closed conduits in the form of an inverted U with an inlet, short upper leg, throat (control section), lower leg, and outlet.

For very low flows a siphon spillway operates as a weir; as the flow increases, the upstream water level rises, the velocity in the siphon increases, and the flow in the lower leg begins to exhaust air from the top of the siphon until this primes and begins to flow full as a pipe, with the discharge given by

$$Q = C_d A (2gH)^{1/2} \tag{4.62}$$

where A is the (throat) cross-section of the siphon, H is the difference between the upstream water level and siphon outlet or downstream water level if the outlet is submerged and

$$C_d = \frac{1}{(k_1 + k_2 + k_3 + k_4)^{1/2}}$$

where k_1, k_2, k_3 and k_4 are head loss coefficients for the entry, bend, exit, and friction losses in the siphon.

Once the siphon is primed the upstream water level falls but the siphon continues to operate, even though this level may have fallen below the crest of the siphon; a further increase in the upstream water level produces only a small increase in the discharge. Priming normally occurs when the upstream level has risen to about one-third of the throat height. Well-designed siphons can thus control the upstream water level within fairly close limits.

When the discharge decreases this process is reversed, with a hysteresis effect which may cause instabilities; these are reduced by using multiple siphons with differential crest heights or air-regulated siphons, of which the design in Fig. 4.18 is an example. This siphon spillway at Spelga dam (Poskitt and Elsawy, 1976) was built to augment the flow over an older overfall spillway; there are four batteries of three siphons, each with a special 'duck's bill' inlet for air regulation.

Another type of air-regulated siphon spillway developed from a model study has a downward projecting hood with an upstream baffle and a series of holes which can be selectively sealed to compensate, if necessary, for any scale effects and to ensure the required head-discharge relationship (Hardwick and Grant, 1997).

At the apex of the siphon the pressure falls below atmospheric, with the lowest pressure (and greatest danger of cavitation) on the inside of the bend.

Apart from negative pressures, and stability and regulation difficulties, siphons have to be protected against blocking by floating debris and freezing; their main advantage over other types of spillways is the

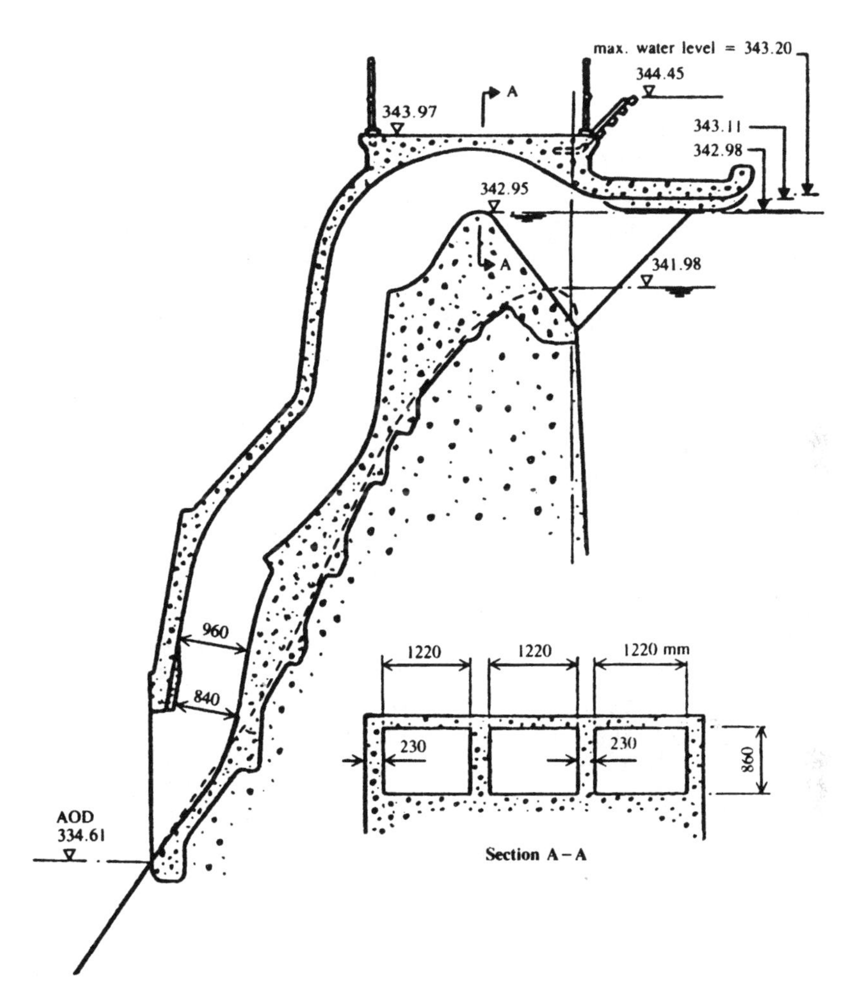

Fig. 4.18 Siphon spillway, Spelga Dam, UK (Poskitt and Elsawy, 1976)

increase in specific discharge and the automatic control of water levels within a small range. They are also a very useful alternative when considering the augmentation of the capacity of existing spillways; on the other hand, once installed, there is very little scope of increasing the design discharged as $Q \propto H^{1/2}$ (equation (4.62)).

For further information on siphon spillways, see BHRA (1976), Ervine and Oliver (1980), Ervine (1998) and Ackers (2000).

4.7.6 Stepped spillways

Stepped spillways (ordinary overfall spillways followed by steps on the spillway face) have been used for a very long time – the earliest was built around 700 BC (Chanson, 1995). In modern times their use was rather limited but during recent decades they have been attracting increased attention mainly because of new material technologies (RCC dams and prefabricated blocks); they are used mainly for auxiliary but also for main spillways, where their enhanced energy dissipation contributes to the economy of overall design. For examples of stepped spillways see Chanson (1995).

For stepped spillways the crucial problems are the flow régime (nappe or skimming flow) and the associated questions of air entrainment (and thus cavitation protection) and energy dissipation. In the *skimming flow* régime water flows over the stepped spillway as a stream skimming the steps and cushioned by recirculating fluid trapped between them, whereas the *nappe flow* with fully developed hydraulic jumps and a succession of free fall jets is essentially a flow in a series of drop structures – see Section 10.3.

The early systematic investigation of hydraulic conditions on cascades with a slope <1:1 was carried out by Essery and Horner (1978) who produced useful design curves for nappe flow.

Denoting the step height h and the length l Chanson (1995) showed that full onset of the skimming flow régime is characterized by a value of critical depth y_c $(=(q^2/g)^{1/3})$ given by

$$y_c/h = 1.057 - 0.465h/l. \tag{4.63}$$

Chamani and Rajaratnam (1999) suggest the equation

$$h/l = 0.405(y_c/h)^{-0.62} \tag{4.64}$$

for the upper boundary of nappe flow and

$$h/l = \sqrt{0.89\left[\left(\frac{y_c}{h}\right)^{-1} - \left(\frac{y_c}{h}\right)^{-0.34} + 1.5\right] - 1} \tag{4.65}$$

as the lower boundary for the skimming flow. Equation (4.65) gives results agreeing with experiments for $h/l > 1$; for $h/l < 0.8$, y_c/h was found to be almost constant at 0.8 Boes and Hager (2003b) suggest for the onset of skimming flow:

$$y_c/h = 0.91 - 0.14 \tan\theta \tag{4.66}$$

There is a transition zone between the two flow regimes and any differences between the experimentally derived equations 4.63–4.66 are mainly due to the difficulties in the definition of the onset of skimming flow as well as to possible scale effects.

The same as for any rough spillway surface, the point of inception of self-aeration on stepped spillways is closer to the crest than in the case of smooth spillways due to the faster growth of the turbulent boundary layer thickness (equation (4.37)). According to Boes and Hager (2003a) the blackwater distance L_i from the crest to the air entrainment inception point is

$$L_i = 5.9\, y_c^{6/5}/((\sin\theta)^{7/5}\, h^{1/5}) \tag{4.67}$$

(this can be compared with equation (4.23) for smooth spillways).

Equation 4.67 demonstrates again the great importance of the unit discharge q and the relatively small influence of the step height h.

The uniform equivalent clear water depth y_w can be computed from (Boes and Hager 2003b):

$$y_w/h = 0.215\,(\sin\theta)^{-1/3} \tag{4.68}$$

i.e. y_w is independent of the step height.

The uniform flow depth for the mixture of air and water y_{90} (important for spillway side wall height) as given by the authors is:

$$y_{90}/h = 0.5\, Fr*^{(0.1\tan\theta + 0.5)} \tag{4.69}$$

where $Fr* = q_w/(g \sin\theta\, h^3)^{1/2}$

Although the enhanced aeration and earlier air inception point on stepped spillway surfaces act as cavitation protection, there still is a risk of cavitation (due to negative pressures on the vertical step faces) in the region of blackwater flow upstream of the inception point and particularly downstream of it before sufficient entrained air reaches the spillway surface. According to the discussion of the Boes and Hager (2003a) paper the velocity at the inception point should not exceed about 15 m/s to avoid risk of cavitation.

For further details of the design of stepped spillways and their flow structure see Chanson (1995, 2001), Chanson and Toombes (2002), Boes and Hager (2003a, 2003b) and Chanson (2004). For energy dissipation on stepped spillways see Section 5.2.

Although superficially similar in terms of type of flow, spillways using prefabricated blocks have lower steps, smaller slopes and present additional design problems to stepped spillways on large (RCC) dams.

Gabion stepped spillways may have a rather limited life, but earth dam spillways protected by prefabricated interlocking concrete blocks have been used quite extensively. Russian engineers pioneered the design of concrete wedge blocks and Pravdivets and Bramley (1989) described in some detail several configurations; Baker (1994) included in his investigations flat blocks and reinforced grass and reported stability for downward sloping wedge-shaped blocks for velocities up to about $7\,\mathrm{m\,s^{-1}}$. In spite of their proven stability attention must be paid to the detailing of the block edges, the drainage slots connecting the low pressure zone on top of the blocks to the drainage layer below them and to the subsoil or underlayer consisting of granular material and possibly geotextiles (Hewlett *et al.*, 1997).

4.7.7 Other spillways

The previous six sections have dealt with the main and most frequent types of spillways. Their combinations, as well as other types of spillways, are also sometimes used. If of unusual shape, their design should be developed with the aid of model studies (Chapter 16), particularly if they are intended to convey major discharges.

Unusual spillway shapes (in plan) are often associated with the desire to increase the effective spillway length (even if the specific discharge, q, may be decreased owing to interference to flow); examples are a spillway with a *'duck bill'* crest, *labyrinth* or *special shape* (shaft) spillways (Fig. 4.19 – note also the aeration vent for the shaft bend). The discharge coefficients of these spillways are usually (but not always) somewhat smaller than those of conventional spillways described earlier. The main objective of model studies is to ascertain these coefficients and the modular limits of the spillway.

Fuse plugs are used as auxiliary spillways. They are basically broad-crested weirs with a crest higher than the main spillway crest but below the maximum water level, and an earth embankment on top of the spillway, designed to fail at a predetermined reservoir level. The sudden flow after the fuse plug failure must be taken into account when choosing the site of the auxiliary spillway, which usually discharges into a (side) valley other than the main spillway. The downstream face of the weir must be suitably protected, e.g. by concrete plain or wedge-shape blocks (Section 4.7.6) or reinforced grass (CIRIA, 1987).

Tunnelled spillways, free flowing or under pressure, usually convey flow from side channel or shaft spillways (Sections 4.7.2, 4.7.3 and 4.7.4). Large-capacity outlets placed below the dam crest and controlled by gates are usually called *orifice* or *submerged* spillways (Fig. 4.20). Apart from the gates (Chapter 6) their important design features are the inlet arrangement and associated head losses and the prevention of vortices (Section 4.8), and the control and effect of the outflow jet (Chapter 5). All of these aspects are again best studied on suitable scale models.

The elimination of a conventional spillway is a unique feature of some rockfill dams; this is achieved by the inclusion of an impervious wall,

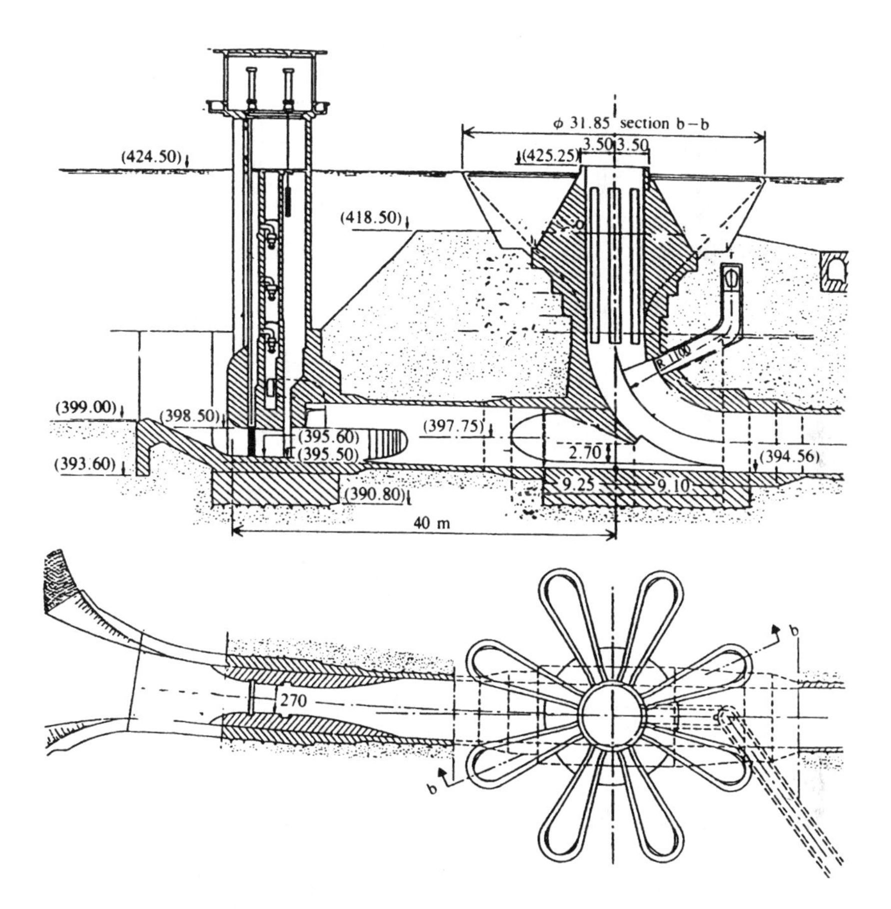

Fig. 4.19 Oued Sarno dam, Algeria; general arrangement of spillway and outlets (after ICOLD, 1987)

lower than the upstream water level, inside the dam, with flow passing through the main body of the rockfill (Lawson, 1987).

4.8 Bottom outlets

Bottom outlets are openings in the dam used to draw down the reservoir level. According to the type of control gates (valves) (see Section 6.3) and the position of the outflow in relation to the tailwater, they operate either under pressure or free flowing over part of their length. The flow from the bottom outlets can be used as compensating flow for a river stretch downstream of the dam where the flow would otherwise fall below acceptable limits: outlets can also serve to pass density (sediment-laden) currents through the reservoir (Section 4.5).

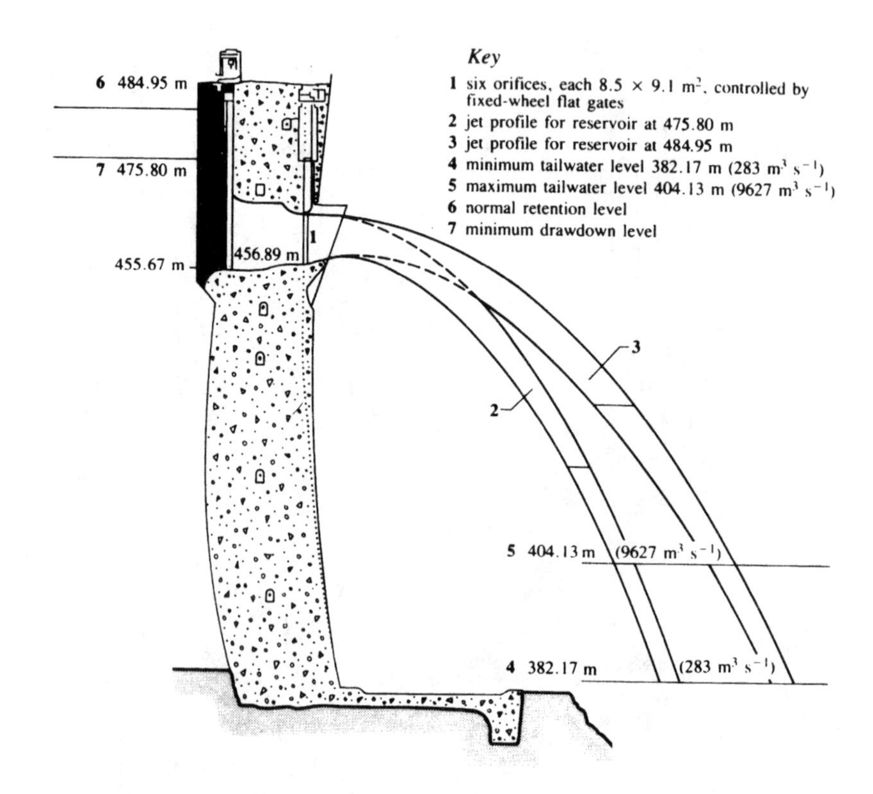

Fig. 4.20 Kariba dam spillway, Zimbabwe (after ICOLD, 1987)

Large bottom openings serve as submerged spillways (Section 4.7.7) and their capacity can be used for flushing sediment from the reservoir (Section 4.5) and during the dam construction (if only one opening is available its blockage must be prevented). A typical arrangement for a large bottom spillway is shown in Fig. 4.21, and for a bottom outlet of a smaller capacity in Fig. 4.22. Note the inlet construction and shape designed to reduce the head loss, the flared outflow section (to aid energy dissipation downstream of the outlet) and the air vent to protect the junction of the spillway and outlet against possible cavitation damage (in case of their joint operation). The outflow area has in this case been reduced to 85% of the bottom outlet area to provide cavitation protection (at the cost of a 10% reduction in outlet capacity). Should the outlet terminate in a regulating valve, then a similar area reduction should be provided (for the full valve opening see Section 6.3).

The head loss at an inlet is expressed as $h = \xi V^2/2g$; for an inlet with grooves and a screen, $0.15 < \xi < 0.34$. Although the lower values of the coefficient ξ are usually associated with inlets with curved walls (giving a gradually decreasing cross-section), sometimes the same (or an even better) result can be achieved by a transition formed by several plane sur-

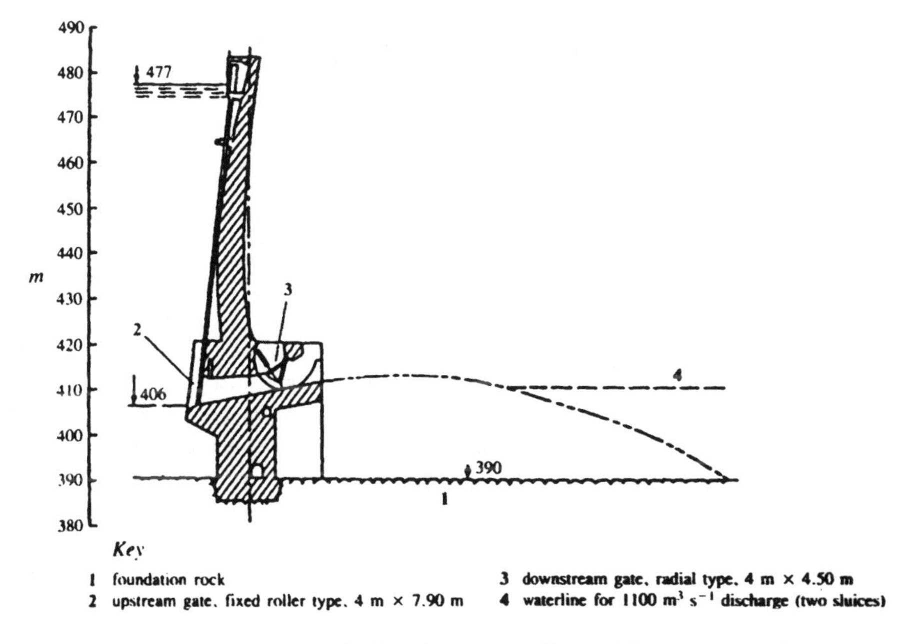

Fig. 4.21 Sainte-Croix arch dam bottom spillway (Thomas, 1976)

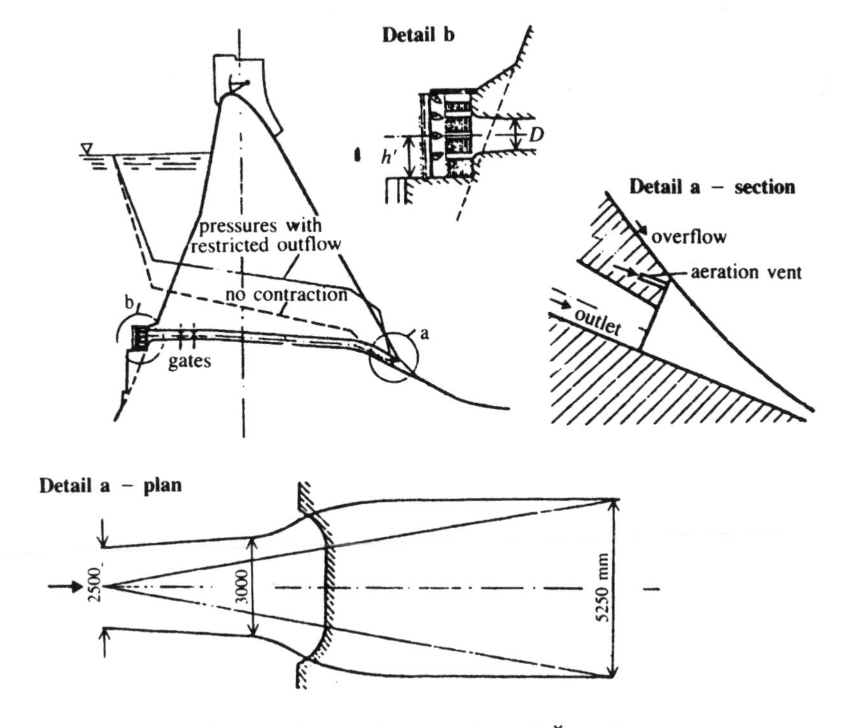

Fig. 4.22 Bicaz dam outlet works (Novak and Čábelka, 1981)

faces, at the expense of allowing small negative pressures to develop on them. To prevent the formation of vortices in the water level upstream of the inlet, its axis should be submerged sufficiently and/or vortex suppression devices (e.g. floating rafts) should be used (Knauss, 1987). It may be advantageous to combine the outflow part of bottom outlets with the spillway energy dissipator; in many cases, however, a separate outlet energy dissipator is provided (Section 5.5). For further details of bottom outlet model testing and design, see Novak and Čábelka (1981); for details of the protection of inlets against ice and their operation in winter conditions, see Ashton (1988).

The flow in the outlet downstream of a gate can be free with or without a hydraulic jump (see also Fig. 4.17). In their study of the jump in circular conveyancies, diameter D, with a free downstream surface Stahl and Hager (1999) defined the supercritical Froude number as $Fr_1 = Q/(gDy_1^4)^{1/2}$ and obtained experimentally a simple relationship for the sequent depths ratio for $1.5 < Fr_1 < 6.5$ and $0.2 < y/D < 0.7$:

$$y_2/y_1 = 1.00\ Fr_1^{0.9} \tag{4.70}$$

The important choking discharge can be approximated from equation 4.70 by putting $y_2 = D$ and $D/y_1 = Fr_1$.

The air demand of a hydraulic jump in a conduit where the flow downstream of the jump is under pressure has been well researched and the original Kalinske and Robertson (1943) equation has been shown to be valid for a range of conditions ($\beta = Q_a/Q$):

$$\beta = 0.0066\ (Fr_1 - 1)^{1.4} \tag{4.71}$$

However, Haindl (1984) points to an increase of the entrainment capacity of a jump with increasing conduit size and quotes a coefficient 0.015 in equation (4.71) for the envelope of field data.

For further details of flow at tunnel bends see Gisonni and Hager (1999).

For a comprehensive discussion of air demand and entrainment in bottom outlet tunnels, see Speerli (1999) and Hager (1999) (see also Section 6.6).

Worked Example 4.1

The inflow hydrograph into a reservoir is given by the first two columns in the lower table on this page. The relationship between the storage volume in the reservoir (taking storage at crest level as zero) and the head above the crest is given by the first two columns of the upper table (this was obtained from equation (4.5)). A simple overfall spillway is to be designed which would limit the maximum head on the spillway crest to 3.00 m.

Solution

Initially, assume a spillway length of 200 m and a constant discharge coefficient of 0.75. Thus, the spillway capacity is given by (equation (4.19))

$$Q = \frac{2}{3} C_d \sqrt{2} g^{1/2} b H^{3/2} = \frac{2}{3} 0.75 \times 4.43 \times 200 H^{3/2} = 443 H^{3/2} \ (\text{m}^3\,\text{s}^{-1}).$$

Note that the maximum inflow of 3300 $\text{m}^3\,\text{s}^{-1}$ has to be reduced by flood routing to $443 \times 3^{3/2} = 2300\ \text{m}^3\,\text{s}^{-1}$ or less.

Taking the time step in the computation (routing period) as 10 h = 36 000 s, compute the following:

H (m)	V ($\times 10^6$ m^3)	O (m^3 s^{-1})	$\frac{2V}{\Delta t} + O$ (m^3 s^{-1})
0.5	45	156	2656
1.0	90	443	5443
1.5	138	814	8480
2.0	188	1253	11 697
2.5	243	1751	15 251
3.0	300	2302	18 969

Using the above values and the given inflow hydrograph, now compute the following table (interpolating values from above as necessary) – equation (4.8):

T (h)	I (m^3 s^{-1})	$I_1 + I_2$	O (m^3 s^{-1})	$\frac{2V}{\Delta t} - O$ (m^3 s^{-1})	$\frac{2V}{\Delta t} + O$ (m^3 s^{-1})
0	200		0	0	0
10	960	1160	68	1 024	1 160
20	1720	2680	263	3 176	3 704
30	2480	4200	679	6 018	7 376
40	3240	5720	1259	9 220	11 738
50	2860	6100	1761	11 798	15 320
60	2480	5340	2031	13 076	17 138
70	2100	4580	2107	13 441	17 656
80	1720	3820	2049	13 163	17 261
90	1340	3060	1895	12 433	16 223
100	960	2300	1678		14 733

The maximum outflow is 2107 $\text{m}^3\,\text{s}^{-2}$ (<2300 $\text{m}^3\,\text{s}^{-1}$) and the maximum head on the spillway is $H = (2107/443)^{2/3} = 2.83$ m. For a preliminary design this is a satisfactory result. Note that the maximum outflow is equal to the inflow at that time. Should a more accurate result be required, a slightly shorter spillway could be considered, as well as the variation of the coefficient of discharge (see worked example 4.3).

Worked Example 4.2

The inflow to a reservoir has an average sediment concentration of 800 ppm. If the volume of the reservoir is $100 \times 10^6 \,\mathrm{m}^3$ and the annual flow of the river is $900 \times 10^6 \,\mathrm{m}^3$, determine the approximate 'half-life' of the reservoir. Assume that the average porosity of the settled sediment over this period is 0.4.

Solution

For the empty reservoir the ratio of capacity and annual inflow is $100/900 = 0.111$; for the half-full reservoir this ratio is $50/900 = 0.055$.

From Fig. 4.2 the trap efficiency of the full reservoir is about 89%, and for the half-full reservoir 71%. The average trap efficiency is thus about 80%.

The annual sediment discharge is

$$Q_s = \frac{900 \times 10^6 \times 800}{10^6 \rho_s} \rho \;(\mathrm{m}^3\,\mathrm{year}^{-1}).$$

The annual bulk discharge, assuming $\rho = 10^3$ and $\rho_s = 2650\,\mathrm{kg\,m}^{-3}$, is

$$Q_b = \frac{900 \times 10^6 \times 800}{10^3 \times 2650(1-0.4)} = 452\,830\,\mathrm{m}^3\,\mathrm{year}^{-1}.$$

The half-life of the reservoir is

$$\frac{50 \times 10^6}{452\,830 \times 0.8} = 138 \text{ years}.$$

Worked Example 4.3

The overfall spillway in example 4.1 has been designed for a head of 2.80 m. What will the discharge be for heads of 0.20 m and 1.50 m, and what is the maximum discharge that can be passed over this spillway (assuming the dam freeboard to be high enough and the spillway to be well constructed) without cavitation?

Solution

At the design head

$$Q = 2/3 \times 0.75\sqrt{2}g^{1/2} \times 200 \times 2.8^{3/2} = 590.66 \times 0.75 \times 2.8^{3/2} = 2075\,\mathrm{m}^3\,\mathrm{s}^{-1}.$$

For $H = 0.20\,\mathrm{m}$, $H/H_d = 0.20/2.80 = 0.071$; thus $C_d \simeq 0.58$ and

$$Q = 590.66 \times 0.58 \times 0.2^{3/2} = 31\,\mathrm{m}^3\,\mathrm{s}^{-1}.$$

For $H = 1.50\,\text{m}$, $H/H_d = 1.50/2.80 = 0.536$; by interpolation between 0.578 ($H/H_d = 0.05$) and 0.75 ($H/H_d = 1$), $C_d = 0.666$ and

$$Q = 590.66 \times 0.666 \times 1.5^{3/2} = 723\,\text{m}^3\,\text{s}^{-1}.$$

The maximum head for no cavitation. $H_{max} = 1.65 H_d = 1.65 \times 2.8 = 4.62\,\text{m}$. For this condition, $C_d = 0.81$, and therefore

$$Q_{max} = 590.66 \times 0.81 \times 4.62^{3/2} = 4757\,\text{m}^3\,\text{s}^{-1}.$$

Worked Example 4.4

A long chute 2 m wide, of slope 0.25, and with $n = 0.012$ carries a discharge of $7.5\,\text{m}^3\,\text{s}^{-1}$. Using various methods, estimate the average air concentration and the depth of the fully aerated uniform flow.

Solution

The discharge per unit width, $q = 7.5/2 = 3.75\,\text{m}^3\,\text{s}^{-1}\,\text{m}^{-1}$.

The depth of the uniform non-aerated flow, y_0 from q is

$$q = \left(\frac{2y_0}{2 + 2y_0}\right)^{2/3} y_0 \frac{S^{1/2}}{n}, \quad 3.75 = \left(\frac{2y_0}{2 + 2y_0}\right)^{2/3} y_0 \frac{(0.25)^{1/2}}{0.012}.$$

By trial and error, $y_0 = 0.259\,\text{m}$, $V = q/y_0 = 14.48\,\text{m}\,\text{s}^{-1}$ and $R = 0.206\,\text{m}$.

$$Fr = V/(gR)^{1/2} = 10.19.$$

1. From equation (4.41),

$$y_a = c_1(q^2/g)^{1/3}, \quad 0.361 < y_a(\text{m}) < 0.418$$

for

$$C = 1 - \rho_1 = 1 - y_0/y_a, \quad 28 < C(\%) < 38.$$

2. From equation (4.42),

$$\frac{y_a - y_0}{y_0} = 0.1(0.2Fr^2 - 1)^{1/2} = 0.1\left(0.2\frac{V^2}{gR} - 1\right)^{1/2}$$

$$= 0.1(0.2 \times 10.19^2 - 1)^{1/2} = 0.445,$$

$$\frac{y_a - 0.259}{0.259} = 0.445, \quad y_a = 0.374\,\text{m}, \quad C = 31\%.$$

3. From equation (4.44) for $S/q^{1/5} = 0.1919$ (>0.16),

$C = 0.7226 + 0.743 \log 0.1919 = 0.189 = 19\%$, $y_a = 0.320$ m.

From equation (4.45) for $S/q^{2/3} = 0.10 < 0.23$ (outside experimental limit),

$C = 0.5027(0.10)^{0.385} = 0.207 = 21\%$, $y_a = 0.328$ m.

The equivalent roughness $k = 2a$ is obtained from (Section 8.2.2)

$$\frac{R^{1/6}}{n} = 18 \log \frac{6R}{a + \delta/7}$$

with

$$\delta = \frac{11.6\nu}{U_*} + \frac{11.6 \times 10^{-6}}{(gRS)^{1/2}} = 0.16 \times 10^{-4}\,\text{m},$$

$$a = 3.423 \times 10^{-4} - \frac{0.16 \times 10^{-4}}{7}.$$

Thus $a \gg \delta/7$, i.e. the flow regime is hydraulically rough with $k = 0.7$ mm. (Note that k is smaller than the 1.2 mm applicable to Anderson's experiments; also, equation (4.44) underestimates C by about 0.1 for very low values of $S/q^{1/5}$ when compared with actual experimental results.)

As explained in Section 4.7.3 in the above computations (1–3) the uniform flow depth for the water component in the water–air mixture y_0' should have been used instead of y_0. On the other hand for C about 0.3 the difference between these two values is negligible (from Straub and Anderson $y_0'/y_0 = 0.98$).

4. From equation (4.52) with $\theta = 14°$ ($S = \tan\theta = 0.25$) $C = 21\%$. As $\lambda = n^2 8g/R^{1/3}$, for non-aerated flow $\lambda = 0.0191$ and from equation (4.43)

$$\lambda_a = 0.0191\ (1 - 1.19 \times 0.21^2) = 0.0175.$$

From the Darcy–Weisbach equation $y_0'^2 R_0' = Q^2\lambda_a/b^2 8gS = 0.0125$ and $y_0' = 0.25$ m (giving $y_0'/y_0 = 0.965$ and $y_a' = 0.25/(1 - 0.21) = 0.316$ m. The aerated flow velocity is $V = 3.75/0.25 = 15.0\,\text{m s}^{-1}$ (>14.48).

Using equation (4.53) $C = 0.75(\sin 14)^{0.75} = 26\%$.

5. From equation (4.50), for $K = 2q/b = 3.75\,\text{m s}^{-1}$. By trial and error, $V = 19.45\,\text{m s}^{-1}$ (velocity of aerated flow).

$$\frac{1 - \rho_1}{\rho_1} = c_1 \frac{V^2}{gR} = 0.006 \frac{19.45^2}{9.81 \times 0.206} = 1.124 = \frac{y_a}{y_0} - 1.$$

Thus $y_a = 2.124 y_0 = 0.550$ m; $C = 1 - (0.259/0.550) = 53\%$.

Note that Hall's equation is really applicable for spillways with a ratio $b/y_0 < 5$, where the wall-generated turbulence tends to produce more intensive aeration than in wide(r) channels (in this case $b/y_0 = 7.7$). In summary: not all of equations (4.41)–(4.54) are strictly applicable in this case; the range of y_a and C computed for $y_0 = 0.259$ m is

$$0.32 < y_a(\text{m}) < 0.55, \quad 19 < C(\%) < 53.$$

The most likely value is $y_a \simeq 0.40$ m, $C \simeq 35\%$. Adding freeboard for safety, the side walls of the chute should be about 0.7 m high.

References

Ackers, J. (2000) Early siphon spillways, in *Dams 2000, Proc. British Dam Society, Bath* (ed. P. Tedd), Thomas Telford, London.

Ackers, P. and Priestley, S.J. (1985) Self-aerated flow down a chute spillway, in *Proceedings of the 2nd International Conference on the Hydraulics of Floods and Flood Control*, Cambridge, British Hydromechanics Research Association, Cranfield, Paper A1.

Anderson, A.G. (1965) Influence of channel roughness on the aeration of high-velocity, open channel flow, in *Proceedings of the 11th International Association for Hydraulic Research Congress*, Leningrad, Vol. 1, Paper 1.37.

Arndt, R.E.A. (1981) Recent advances in cavitation research, in *Advances in Hydroscience*, Vol. 12 (ed. V.T. Chou), Academic Press, New York.

Arsenishvili, K.I. (1965) *Effect of Wave Formation on Hydroengineering Structures*, Israel Program for Scientific Translations, Jerusalem.

ASCE (1988) *Evaluation Procedures for Hydrologic Safety of Dams*, Report of the Task Committee on Spillway Design Flood Selection, ASCE, New York.

—— (1995) *Hydraulic Design of Spillways*, Technical engineering and design guides adapted from US Army Corps of Engineers, No. 12, ASCE, New York.

Ashton, G.D. (1988) Intake design for ice conditions, in *Developments in Hydraulic Engineering*, Vol. 5 (ed. P. Novak), Elsevier Applied Science, London.

Atkinson, E. (1998) Reservoir operation to control sedimentation: techniques for assessment, in *Proc. 10th Conf. BDS*, Bangor (ed. P. Tedd), Thomas Telford, London.

Baker, R. (1994) Using pre-cast concrete blocks to prevent erosion of dams and weirs. *Hydro Review Worldwide*, 2 (2): 40–8.

Batuca, D.J. and Jordaam, J.M. (2000) *Silting and Desilting of Reservoirs*, A.A. Balkema, Rotterdam.

BHRA (1976) *Proceedings of the Symposium on Design and Operation of Siphons and Siphon Spillways*, BHRA Fluid Engineering, Cranfield.

Boes, R.M. and Hager, W.H. (2003a) Two-phase flow characteristics of stepped spillways, *Journal of Hydraulic Engineering*, ASCE, 129, No. 9: 651–70.

Boes, R.M. and Hager, W.H. (2003b) Hydraulic design of slipped spillways, *Journal of Hydraulic Engineering*, ASCE, 129, No. 9: 671–9.

Bradley, J.N. (1952) *Discharge Coefficients for Irregular Overfall Spillways*, Engineering Monographs No. 9, Bureau of Reclamation, CO.

Bruk, S. (ed.) (1985) *Methods of Computing Sedimentation in Lakes and Reservoirs*, UNESCO, Paris.

Brune, G.M. (1953) Trap efficiency of reservoirs. *Transactions of the American Geophysical Union*, 34 (3).

Cassidy, J.J. (1994) Choice and computation of design floods and their influence on dam safety. *Hydropower & Dams*, 1 (January): 57–67.

Cassidy, J.J. and Elder, R.A. (1984) Spillways of high dams, in *Developments in Hydraulic Engineering*, Vol. 2 (ed. P. Novak), Elsevier Applied Science, London.

Chamani, M.R. and Rajaratnam, N. (1999) Onset of skimming flow on stepped spillways, *Journal of Hydraulic Engineering*, ASCE, 125, No. 9: 969–71.

Chanson, H. (1995) *Hydraulic Design of Stepped Cascades, Channels, Weirs and Spillways*, Pergamon, Oxford.

—— (2001) *The Hydraulics of Stepped Chutes and Spillways*, Balkema, Lisse.

—— (2004) *The Hydraulics of Open Channel Flows: An Introduction*, 2nd edn., Butterworth-Heinemann, Oxford.

Chanson, H. and Toombes, L. (2002) Air-water flows down stepped chutes, *Int. Journal of Multiphase Flow*, 28, No. 11: 1737–61.

Chow, V.T. (1983) *Open Channel Hydraulics*, McGraw-Hill, New York.

CIRIA (1987) *Design of Reinforced Grass Waterways*, Report 116, Construction Industry Research and Information Association, London.

Creager, W.P., Justin, J.D. and Hinds, J. (1945) *Engineering for Dams*, Wiley, New York.

Ellis, J. (1989) *Guide to Analysis of Open-channel Spillway Flow*, Technical Note 134, Construction Industry Research and Information Association, London.

Ervine, A. (1998) Air entrainment in hydraulic structures: a preview, *Water, Maritime and Energy, Proceeding of the Institution of Civil Engineers* 130 (September): 142–53.

Ervine, D.A. and Oliver, G.S.C. (1980) The full scale behaviour of air regulated siphon spillways. *Proceedings of the Institution of Civil Engineers, Part 2*, 69: 687–706.

Essery, I.T.S. and Horner, M.W. (1978) *The Hydraulic Design of Stepped Spillways*, Report No. 33, 2nd edn, CIRIA, London.

Fahlbush, F.E. (1999) Spillway design floods and dam safety, *Hydropower and dams*, No. 4: 120–7.

Falvey, H.T. (1974) Prediction of wind wave heights. *Journal of Waterways, Harbours and Coastal Division*, Paper 3524; *Proceedings of the American Society of Civil Engineers*, 89 (WW2): 1–12.

—— (1980) *Air Water Flow in Hydraulic Structures*, US Bureau of Reclamation, Engineering Monograph 41, US Department of Interior, Washington, DC.

—— (1990) Cavitation in chutes and spillways, Engineering Monograph 42, *Water Resources Technical Publication*, Bureau of Reclamation, Denver.

Ferrando, A.M. and Rico, J.R. (2002) On the incipient aerated flow in chutes and spillways, *Journal of Hydraulic Research*, IAHR, 40, No. 1: 95–7.

Galperin, R.S., Sokolov, A.G., Semenkov, V.M. and Tserdov, G.N. (1977) *Cavitation of Hydraulic Structures*, Energiya, Moscow.

Gisonni, C. and Hager, W.H. (1999) Studying flow at tunnel bends, *Hydropower and Dams*, No. 2: 76–9.

Gosschalk, E.M. (2002) *Reservoir Engineering: Guidelines for Practice*, Thomas Telford, London.

Hager, W.H. (1988) Discharge characteristics of gated standard spillways. *Water Power & Dam Construction*, 40 (1): 15–26.

—— (1999) *Wastewater Hydraulics: Theory and Practice*, Springer-Verlag, Berlin.

Haindl, K. (1984) Aeration at hydraulic structures, in *Developments in Hydraulic Engineering*, Vol. 2 (ed. P. Novak), Elsevier Applied Science, London.

Hall, L.S. (1942) Open channel flow at high velocities. *Proceedings of the American Society of Civil Engineers*, 68: 1100–40.

Hardwick, J.D. and Grant, D.J. (1997) An adjustable air regulated siphon spillway, *Proc Instn Civ Engrs Water-Maritime-Energy*, 124, June: 95–103.

Henderson, F.M. (1966) *Open Channel Flow*, Macmillan, London.

Hewlett, H.W.M., Baker, R., May, R.W.P. and Pravdivets, Y.P. (1997) *Design of Stepped-block Spillways*, CIRIA Special Publication 142.

Huber, A. (1997) Quantifying impulse wave effects in reservoirs, *Proc. 19th Congress of Large Dams, Florence*, ICOLD, Paris.

ICE (1996) *Floods and Reservoir Safety*, 3rd edn, Institution of Civil Engineers, London.

ICOLD (1987) *Spillways for Dams*, Bulletin 58, International Commission on Large Dams, Paris.

—— (1992a) *Spillways, Shockwaves and Air Entrainment*, Bulletin 81, International Commission on Large Dams, Paris.

—— (1992b) *Selection of Design Flood-current Methods*, Bulletin 82, International Commission on Large Dams, Paris.

—— (1997) *Cost of Flood Control in Dams – Review and Recommendations*, Bulletin 108, International Commission on Large Dams, Paris.

—— (1999) *Dealing with Reservoir Sedimentation*, Bulletin 115, International Commission on Large Dams, Paris.

—— (2002) *Reservoir Landslides: Investigation and Management*, Bulletin 124, International Commission on Large Dams, Paris.

Institute of Hydrology (1999) *Flood Estimation Handbook*, CEH Institute of Hydrology, Wallingford.

Jain, S.C. (2001) *Open Channel Flow*, Wiley and Sons, New York.

Kalinske, A.A. and Robertson, J.M. (1943) Closed conduit flow, *Trans. ASCE*, 108, 1435–47, 1513-16.

Knapp, R.T., Daily, J.W. and Hammit, F.G. (1970) *Cavitation*, McGraw-Hill, New York.

Knauss, J. (ed.) (1987) *Swirling Flow Problems at Intakes, IAHR Hydraulic Structures Design Manual*, Vol. 1. Balkema, Rotterdam.

Kramer, R. (2004) *Development of Aerated Chute Flow*, Mitteilungen No. 183, VAW-ETH, Zürich.

Law, F.M. (1992) A review of spillway flood design standards in European countries including freeboard margins and prior reservoir level, in *Water Resources and Reservoir Engineering* (eds N.M. Parr, J.A. Charles and S. Walker), *Proceedings of the 7th Conference of the British Dam Society*, Thomas Telford, London.

Lawson, J.D. (1987) Protection of rockfill dams and cofferdams against overflow and throughflow – the Australian experience. *Transactions of the Institution of Engineers of Australia, Civil Engineering*, 79 (3), Paper C1659, 10 pp.

Lencastre, A. (1987) *Handbook of Hydraulic Engineering*, Ellis Horwood, Chichester.

Linsley, R.K., Kohler, M.A. and Paulhus, J.L.H. (1988) *Hydrology for Engineers*, 3rd edn, McGraw-Hill, New York.

Minor, H.E. (1998) Report of the European R & D Working Group 'Floods' in Dam Safety, *Proc of the Int Symp on New Trends and Guidelines on Dam Safety*, Barcelona (ed. L. Berga), Balkema, Rotterdam.

Morris, G.L. and Fan, J. (1998) *Reservoir Sedimentation Handbook*, McGraw-Hill, New York.

National Research Council (1988) *Estimating Probabilities of Extreme Floods*, National Academy Press, Washington, DC.

Naudascher, E. (1987) *Hydraulik der Gerinne und Gerinnebauwerke*, Springer, Vienna.

NERC (1993) *Flood Studies Report* (extended reprint of the 1975 edition), Natural Environment Research Council, London.

Novak, P. and Čábelka, J. (1981) *Models in Hydraulic Engineering – Physical Principles and Design Applications*, Pitman, London.

Pinto de S., N.L. (1991) Prototype aerator measurement, in *Air Entrainment in Free Surface Flows* (ed. I.R. Wood), IAHR Hydraulic Structures Design Manual, Vol. 4, Balkema, Rotterdam.

Poskitt, F.F. and Elsawy, E.M. (1976) Air regulated siphon spillways at Spelga dam. *Journal of the Institution of Water Engineers and Scientists*, 30 (4): 177–90.

Pravdivets, Y. and Bramley, M. (1989) Stepped protection blocks for dam spillways. *Water Power & Dam Construction*, 41 (July): 49–56.

Reed, D.W. and Field, E.K. (1992) *Reservoir Flood Estimation: Another Look*, IH Report 114, Institute of Hydrology, Wallingford.

Reeve, C.E. (1992) Trapping efficiency of reservoirs, in *Water Resources and Reservoir Engineering* (eds N.M. Parr, J.A. Charles and S. Walker), *Proceedings of the 7th Conference of the British Dam Society*, Thomas Telford, London.

Rein, M. (1998) Turbulent open channel flows: drop generation and selfaeration. *Journal of Hydraulic Engineering*, ASCE, 124, No. 1: 98–102.

Roberson, J.A., Cassidy J.J. and Chaudry, M.H. (1998) *Hydraulic Engineering*, 2nd edn, John Wiley & Sons, New York.

Saville, T.A., Clendon, E.W. and Cochran, A.L. (1962) Freeboard allowance for waves on inland reservoirs. *Journal of Waterways Harbours, Coastal Division, American Society of Civil Engineers*, 88 (WW2): 93–124.

Scimeni, E. (1937) Il profilo delle dighe sfiranti. *L'Energia Elettrica.*

Sentürk, F. (1994) *Hydraulics of Dams and Reservoirs*, Water Resources Publications, Colorado.

Shaw, E.M. (1994) *Hydrology in Practice*, 3rd edn, Chapman & Hall, London.

Speerli, J. (1999) *Strömungsprozesse in Grundablassstollen*, Mitteilungen No. 163, VAW-ETH, Zürich.

Stahl, H. and Hager, W.H. (1999) Hydraulic jump in circular pipes, *Canadian J. of Civil Engineering*, 26: 368–73.

Straub, L.G. and Anderson, A.G. (1958) Experiments of self-aerated flows in open channels. *Proceedings of the American Society of Civil Engineers, Journal of the Hydraulics Division*, 84 (Hy 7), 1890/1–1890/35.

Thomas, H.H. (1976) *The Engineering of Large Dams*, Wiley, Chichester.

US Army Corps of Engineers (2002) HEC-RAS *River Analysis Systems*, version 3.1, Hydrologic Engineering Center, Davis, CA.

US Army Waterways Experimental Station (1959) *Hydraulic Design Criteria*, USA Department of the Army, Washington, DC.

US Bureau of Reclamation (1987) *Design of Small Dams*, 3rd edn, US Department of the Interior, Washington, DC.

US Corps of Engineers (1962) *Waves in Inland Reservoirs*, Technical Memorandum 132, Washington, DC.

Vischer, D. (1988) Recent developments in spillway design. *Water Power & Dam Construction*, 40 (1): 10–15.

Vischer, D.L. and Hager, W.H. (1998) *Dam Hydraulics*, John Wiley & Sons, Chichester.

Volkart, P.U. (1982) Selfaerated flow in steep, partially filled pipes. *Journal of the Hydraulic Division*, ASCE, 108, HY 9: 1029–46.

Volkart, P. and Rutschmann, P. (1991) Aerators on spillways, in *Air Entrainment in Free Surface Flows* (ed. I.R. Wood), IAHR Hydraulic Structures Design Manual, Vol. 4, Balkema, Rotterdam.

Wagner, W.E. (1956) Morning-glory shaft spillways: determination of pressure controlled profiles. *Transactions of the American Society of Civil Engineers*, 121: 345–83.

White, W.R. (2001) *Evacuation of Sediments from Reservoirs*, Thomas Telford, London.

Wilhelms, S.C. and Gulliver, J.S. (2005) Bubbles and waves description of self-aerated spillway flow, *J. of Hydraulic Research*, 43, No. 5: 522–31.

Wilson, E.M. (1990) *Engineering Hydrology*, 4th edn, Macmillan, London.

Wood, I.R. (1991) Free surface air entrainment on spillways, in *Air Entrainment in Free Surface Flows* (ed. I.R. Wood), IAHR Hydraulic Structures Design Manual, Vol. 4, Balkema, Rotterdam.

Zipparo, V.J. and Hasen, H. (1993) Davis' *Handbook of Applied Hydraulics*, 4th edn, McGraw-Hill, New York.

Chapter 5

Energy dissipation

5.1 General

Energy dissipation at dams and weirs is closely associated with spillway design, particularly with the chosen specific discharge q, the difference between the upstream and downstream water levels (H_*) and the downstream conditions. Chapter 4 dealt mainly with the actual spillway inlet works and certain standard types of conduits conveying the flow from the spillway inlet, i.e. chutes, tunnels, etc. In this chapter the main concern is the concept of energy dissipation during the whole passage of the flow from the reservoir to the tailwater and, in particular, the stilling basin (energy dissipator) design.

The magnitude of energy that must be dissipated at high dams with large spillway discharges is enormous. For example, the maximum energy to be dissipated at the Tarbela dam service and auxiliary spillways could be 40 000 MW, which is about 20 times the planned generating capacity at the site (Locher and Hsu, 1984).

In the design of energy dissipation, environmental factors have to be considered; some of the most important ones are the effect of dissolved gases supersaturation on fish in deep plunge pools, and of spray from flip bucket jets which can result in landslides and freezing fog.

The passage of water from a reservoir into the downstream reach involves a whole number of hydraulic phenomena such as the transition into supercritical flow, supercritical non-aerated and aerated flow on the spillway, possibly flow through a free-falling jet, entry into the stilling basin with a transition from supercritical to subcritical flow, and echoes of macroturbulence after the transition into the stream beyond the basin or plunge pool. It is, therefore, best to consider the energy dissipation process in five separate stages, some of which may be combined or absent (Novak and Čábelka, 1981) (Fig. 5.1):

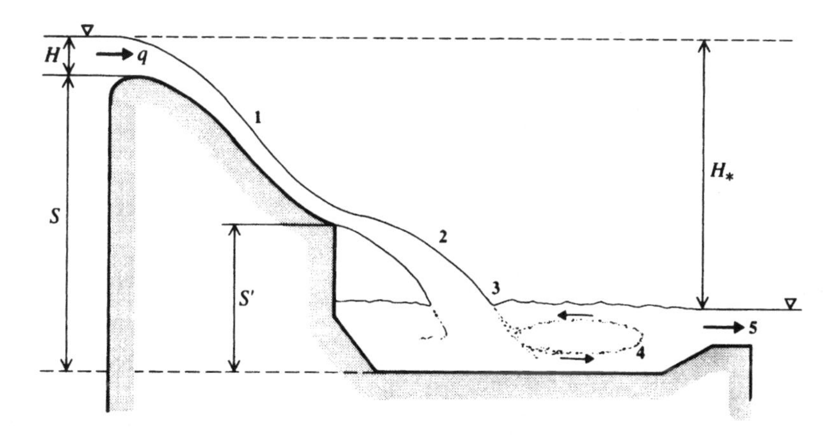

Fig. 5.1 Definition sketch for the five phases of energy dissipation

1. on the spillway surface;
2. in a free-falling jet;
3. at impact into the downstream pool;
4. in the stilling basin;
5. at the outflow into the river.

Stages 1–3 are dealt with in Section 5.2 and 4 and 5 in Section 5.3.

5.2 Energy dissipation on spillways

5.2.1 Energy dissipation on the spillway surface

The energy head loss on the spillway surface can be expressed as

$$e = \xi \alpha V'^2/2g \tag{5.1}$$

where V' is the (supercritical) velocity at the end of the spillway, α is the Coriolis coefficient and ξ is the head loss coefficient, related to the velocity coefficient φ (the ratio of actual to theoretical velocity) by

$$1/\varphi^2 = 1 + \xi. \tag{5.2}$$

The ratio of the energy loss, e, to the total energy E (i.e. the relative energy loss) is

$$\frac{e}{E} = \frac{\xi V'^2}{2g} \bigg/ \left(\frac{V^2}{2g} + \frac{\xi V'^2}{2g} \right) = \frac{\xi}{1+\xi} = 1 - \varphi^2. \tag{5.3}$$

For a ratio of the height S of the spillway crest above its toe (or in case of a spillway with a free-falling jet as in Fig. 5.1 (S–S') above the take-off point) and the overfall head H, with $S/H < 30$ (or $\frac{S-S'}{H} < 30$), and for smooth spillways (Novak and Čábelka, 1981),

$$\varphi_1 \simeq 1 - 0.0155 S/H. \tag{5.4}$$

For a given S, φ_1 increases as H increases, i.e. if for a given discharge Q the spillway width b decreases and thus q increases (equation (4.19)). Thus, for $S/H = 5$, $\varphi_1 = 0.92$ and the relative head loss is 15%, whereas for $S/H = 25$, $\varphi_1 = 0.61$ and the loss is 62%.

The value of ξ could be increased (and φ decreased) by using a rough spillway or by placing baffles on the spillway surface. However, unless aeration is provided at these protrusions, the increased energy dissipation may be achieved only by providing an opportunity for cavitation damage (Section 4.6).

Stepped spillways *may* provide an opportunity for additional energy dissipation (when compared with smooth spillways) pending on the value of the unit discharge (q) (see also Section 4.7.6). E.g. Rice and Kadavy (1994) compared (using models) the energy loss on a smooth and stepped spillway for a 17 m high dam; the result agreed broadly with equation (5.4) for the smooth spillway and showed a 2–3 times higher energy loss for the stepped alternative (steps 0.61 m high and 1.52 m deep). Stephenson (1991) confirms the importance of the unit discharge on the efficiency of energy dissipation on cascade spillways and Boes and Hager (2003) give a detailed analysis of the friction factor for the skimming flow regime demonstrating the effect of the chute slope (friction factor decreases with the slope) which is much larger than that of the relative roughness (see also equation (4.68)).

5.2.2 Ski-jump spillways

In many modern spillway designs, increased energy dissipation is achieved by using free-falling jets, either at the end of a 'ski-jump' or downstream of a flip bucket (Figs 5.2 and 5.3).

The ski-jump spillway was first used by Coyne (1951), and was later further developed by detailed model studies. Its use brings substantial

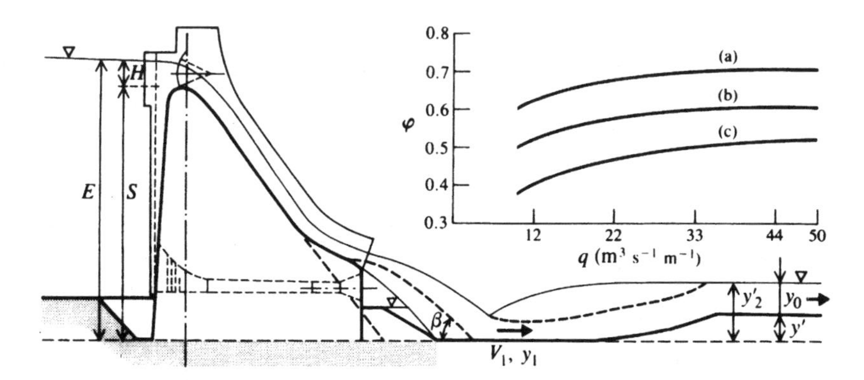

Fig. 5.2 Comparison of (a) normal spillway, (b) ski-jump spillway and (c) ski-jump spillway with jet splitters (Novak and Čábelka, 1981)

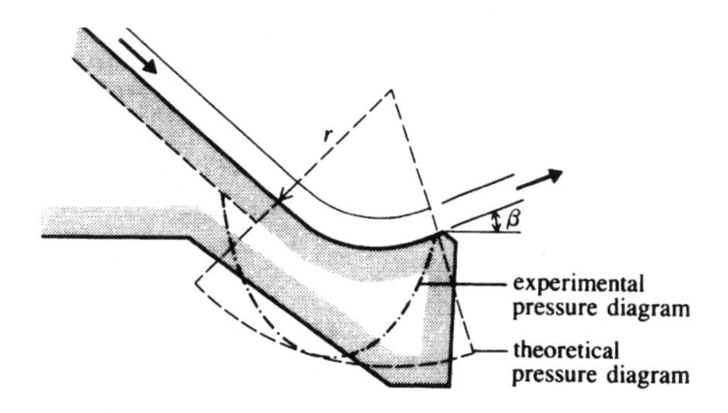

Fig. 5.3 Flip bucket

economies where geological and morphological conditions are favourable, and particularly where the spillway can be placed over the power station or at least over the bottom outlet works (Novak and Čábelka, 1981).

The head loss in the jet itself, whether solid or (more frequently) disintegrated, is not very substantial – only about up to 12% (Hoření, 1956). The energy loss on a ski-jump spillway can be substantially enlarged by splitting the overfall jet into several streams (see Fig. 5.2) or by using two spillways with colliding jets.

A substantial benefit for energy dissipation from jet spillways is in the third phase at impact into the downstream pool. Here most of the energy losses in the first three phases occur through the collision of masses of water, and through the compression of air bubbles, both those contained in the overfall jet as well as those drawn into the downstream pool at the point of impact. The decrease in energy in this phase may be, therefore, enhanced by having a dispersed and intensively aerated jet before impact.

The combined energy loss in the first three phases of energy dissipation can be expressed from a velocity coefficient $\varphi_{1\text{–}3}$, which can be determined in model tests from the theoretical supercritical flow depth conjunctive to the subcritical depth needed to form a stable jump downstream of a (ski-jump) spillway. Even if this value of φ may be subject to scale effects (the prototype φ is likely to be smaller because of increased aeration), the model studies give a very good indication of the relative merits of various designs.

Generally (Fig. 5.1),

$$\varphi_{1\text{–}3} = f(S'/S, q, \text{geometry}) \qquad (5.5)$$

where S' is the height of the 'take-off' point above the reference datum.

A comparison of $\varphi_{1\text{–}3}$ for three designs is shown in Fig. 5.2: curve a, a normal spillway ending in a stilling basin; curve b, a ski-jump spillway without baffles at the take-off edge; curve c, a spillway with baffles which split the jet so that the air entrainment and energy dissipation is enhanced and the pressures on the stilling basin floor are reduced. The increase of φ (and decrease of relative energy loss) with q is again demonstrated. By using a suitable design, the values of φ have been reduced throughout by a factor about 0.7; this results in substantial energy losses, e.g. for $\varphi_{1\text{–}3} = 0.5$, $e/E = 75\%$ (equation (5.3)). The optimum (lowest) value of $\varphi_{1\text{–}3}$ is attained for $S'/S \simeq 0.6$; at this value the overall nappe gains a sufficient velocity and degree of turbulence while flowing over the upper part of the overall spillway to disperse effectively on the baffles at the take-off edge, and has a sufficiently long free fall through air to aerate intensively and to break up the jet core (Novak and Čábelka, 1981).

The disintegration of a falling circular jet of diameter D was studied by Ervine and Falvey (1987), who showed that a complete decay of the solid inner core occurs after a length of fall L, with L/D in the range 50–100. For flat jets, which are more relevant to spillway design, Hoření (1956) established experimentally that the length of fall from the crest required for total jet disintegration (for q in $m^2 s^{-1}$) is

$$L = 5.89q^{0.319} \simeq 6(q)^{1/3} \text{ (m)}. \qquad (5.6)$$

5.2.3 Flip bucket

The flip bucket (Fig. 5.3) is a version of a ski-jump spillway that is usually used as an end to a chute or tunnel spillway whenever the geological and topographical conditions are suitable. Flip buckets (just as ski-jump spillways) are usually tailor-made for a given project, and the designs are developed with the aid of scale models.

The key parameters for the flip bucket design are the approach flow velocity and depth, the radius r of the bucket, and the lip angle β. For a two-dimensional circular bucket the pressure head can be computed for irrotational flow; experimental data confirm these values for the maximum pressure head but (in contrast to the theory) show a non-uniform pressure distribution (Fig. 5.3). At low flow the bucket acts like a stilling basin with water flowing over the lip and the downstream face; the foundation of the flip bucket has, therefore, to be protected against erosion. As the flow increases a 'sweep-out' discharge is attained at which point the flip bucket starts to operate properly with a jet.

The jet trajectory is hardly affected by air resistance for velocities below $20\,\mathrm{m\,s^{-1}}$, but for velocities of $40\,\mathrm{m\,s^{-1}}$ the throw distance can be reduced by as much as 30% from the theoretical value, given by $(v^2/g)\sin 2\beta$.

The designer's main concern is usually to have the impact zone as far as possible from the bucket to protect the structure against retrogressive erosion. Many designs with skew jets and various three-dimensional forms of flip buckets have been developed. Heller *et al.* (2005) give an analysis of ski-jump hydraulics and Locher and Hsu (1984) discuss further the flip bucket design.

5.3 Stilling basins

5.3.1 Hydraulic jump stilling basin

The stilling basin is the most common form of energy dissipator converting the supercritical flow from the spillway into subcritical flow compatible with the downstream river régime. The straightforward – and often best – method of achieving this transition is through a simple submerged jump formed in a rectangular cross-section stilling basin. Vischer and Hager (1995) give an overview of the hydraulics of various energy dissipators.

Hydraulic jumps have been investigated by many researchers, more recently by Rajaratnam (1967) and Hager, Bremen and Kawagoshi (1990), who also extended this investigation to a jump with a control sill (Hager and Li, 1992). The implications of the hydraulics of the jump for the submerged jump stilling basin have been studied by Novak (1955).

Referring to the notation in Fig. 5.4 and to equations (5.1) and (5.2) we can write

$$E = y_1 + \frac{\alpha q^2}{2g\varphi^2 y_1^2}. \tag{5.7}$$

$$y_2 = \frac{y_1}{2}\left[-1 + \left(1 + 8\frac{q^2}{g y_1^3}\right)^{1/2}\right]. \tag{5.8}$$

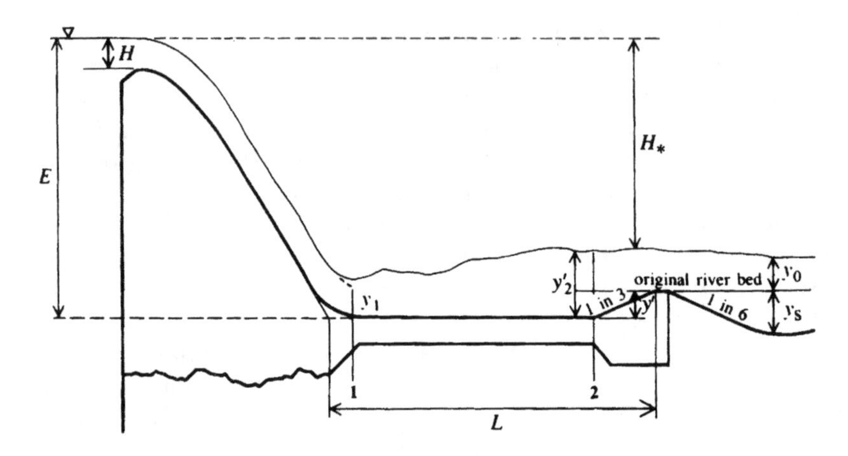

Fig. 5.4 Definition sketch for hydraulic jump stilling basin

The stilling basin depth is then given by

$$y' = y_2' - y_0 = \sigma' y_2 - y_0 \tag{5.9}$$

and the length by

$$L = K(y_2 - y_1) \tag{5.10}$$

where σ' and K are coefficients (derived from laboratory and field experiments).

When applying equations (5.7)–(5.10) we start with a known discharge q and the corresponding downstream depth y_0. For a suitably chosen φ (Section 5.2) and a value of E corresponding to the total energy available above the stilling basin floor, y_1 can be computed from equation (5.7), y_2 from equation (5.8) and y' from equation (5.9) (from a chosen value of safety coefficient). E is, of course, initially not known and thus it is best to apply the above procedure by iteration, initially assuming $y' = 0$, i.e. taking the energy datum at the downstream river-bed level. This computation, carried out for several discharges, can produce five alternatives:

1. $y_2 > y_0$ throughout the range of q;
2. $y_2 = y_0$ throughout the range of q;
3. $y_2 < y_0$ throughout the range of q;
4. $y_2 > y_0$ only at high discharges;
5. $y_2 > y_0$ only at low discharges.

Case 1 is the most frequent one, and shows that a stilling basin is required for all discharges in order to produce a submerged jump. For safety the same is required in case 2 (which is really only a theoretical possibility).

For $y_2 < y_0$ no stilling basin is necessary, and a horizontal apron protecting the river bed downstream of the dam is sufficient, as a submerged jump will result naturally. The stilling basin design for case 4 has to be based on the maximum discharge (the same as for case 1) and, for case 5, on the discharge giving a maximum difference between y_2 and y_0 (with $Q_d < Q_{max}$); this can result in a small stilling basin at the toe of the dam followed by a horizontal apron (or vice versa), or in a sloping apron design.

Where the result of the first computation shows that a stilling basin is required, the procedure is repeated for a new value of E (in equation (5.7)) which takes into account the lowering of the energy datum by a sufficient amount (see Worked example 5.1).

The values of the coefficients σ' and K in equations (5.9) and (5.10) can be taken (Novak and Čábelka, 1981) as $1.1 < \sigma' < 1.25$ and $4.5 < K < 5.5$, where the lower value of K applies for $Fr_1 > 10$ and the higher for $Fr_1 \leqslant 3$. Because at low supercritical Froude numbers the jump is not well developed and can be unstable, it is rather difficult to design an economically dimensioned basin in these cases without model studies.

Equations (5.8) and (5.10), and thus the design under discussion, apply to basins with a horizontal floor only. In sloping channels the value of y_2/y_1 increases with the slope; for a slope $S_0 = 0.2$, y_2/y_1 is twice the value of a horizontal channel, with the same Froude number.

The quoted values of K, and particularly of σ', are fairly low (as dictated by economy) and dependent on a good assessment of the coefficient φ, and particularly of the downstream depth y_0 which, in turn, depends usually on an assumed value of Manning's n for the river. If conservative values of n and φ are assumed (i.e. low n and high φ) then a small value of σ' (say, 1.1) is sufficient, otherwise a higher value may have to be selected. It is also very important to assess the possible long-term river-bed degradation downstream of the dam, which could result in a lowering of the downstream water levels and of y_0.

A simple end sill with a 1 in 3 slope is usually as good as more complicated sills (Section 5.3.3).

It is evident from equations (5.7)–(5.10) that the lower the value of φ (the higher the value of ξ) the smaller will be the required stilling basin; φ in equation (5.7) refers to the total losses between the spillway crest and entry into the stilling basin, i.e. to $\varphi_{1\text{-}3}$ (equation (5.5)).

The energy loss in the fourth and fifth phases of energy dissipation (Section 5.1) can be expressed as

$$e_{4,5} = (y_2 - y_1)^3 / 4y_2y_1. \tag{5.11}$$

Downstream of the jump at the outflow from the basin there is still a substantial proportion of excess energy left, mainly due to the high turbulence of flow which can be expressed (Novak and Čábelka, 1981) as

$$e_5 = (\alpha' - \alpha)V_0^2/2g \tag{5.12}$$

where α' is the increased value of the Coriolis coefficient reflecting the high degree of turbulence and uneven velocity distribution; $2<\alpha'<5$ for $3<Fr_1<10$, while $\alpha \simeq 1$.

From equations (5.11) and (5.12) we obtain

$$\frac{e_4}{e_{4,5}} = 1 - \frac{e_5}{e_{4,5}} \simeq 1 - 4(\alpha'-1)\frac{1+(1+8Fr_1^2)^{1/2}}{[-3+(1+8Fr_1^2)^{1/2}]^3}. \tag{5.13}$$

Equation (5.13) shows that the efficiency of energy dissipation in the jump itself within the stilling basin decreases with the Froude number, leaving up to 50% of the energy to be dissipated downstream of the basin at low Froude numbers (Section 5.3.3).

The hydraulic jump entrains a substantial amount of air additional to any incoming aerated flow. A constant air concentration throughout the jump ($\bar{C}_1 = \bar{C}_2$) results in a lower height of the jump than for the case without air, while for $\bar{C}_1 > 0$ and $\bar{C}_2 \simeq 0$ (which is a more realistic assumption) a slightly higher y_2 is needed than for no air (Naudascher, 1987). Thus the main significance of the presence of air in the jump region is the requirement of higher stilling basin side walls due to the higher depth of flow (equation (4.33)). The effect of air entrainment by hydraulic jumps on oxygen concentration in the flow is briefly discussed in Section 9.1.7.

The highly turbulent nature of the flow in the hydraulic jump induces large pressure fluctuations on the side walls and particularly on the floor of the basin which, in turn, could lead to cavitation. Using a cavitation number, σ, in the form $(\overline{p'^2})^{1/2}/(1/2\rho V_1^2)$ (equation (4.14)), where p' is the deviation of the instantaneous pressure p from the time-averaged pressure $\bar{p}$ ($p = f(t)$ can be obtained from pressure transducer records), the relationship between σ and x/y_1 (where x is the distance from the toe of the jump) for a free and submerged jump at $Fr_1 \simeq 5$ is shown in Fig. 5.5 (Narayanan, 1980; Locher and Hsu, 1984). Assuming the length of the jump to be approximately $6(y_2 - y_1)$, the hydrostatic pressure at the point of maximum pressure fluctuations, i.e. in a free jump at $x/y_1 = 12$, will be $\rho g \bar{y}$, with

$$\bar{y} = y_1 + \frac{y_2 - y_1}{6(y_2 - y_1)} 12 y_1 = 3 y_1.$$

For $\sigma = 0.05$, cavitation will occur if

$$p_0 + \rho g \bar{y} - 0.05 k \rho \frac{V_1^2}{2} = p_v \simeq 0 \tag{5.14}$$

where $k = p'/(\overline{p'^2})^{1/2} > 1 \; (1<k<5)$.

The value of k can be computed from equation (5.14) and assuming, for example, a normal distribution of pressure fluctuations, the intermittency factor, i.e. the proportion of time for which k is exceeded (the probability of occurrence of cavitation) can be computed (worked example

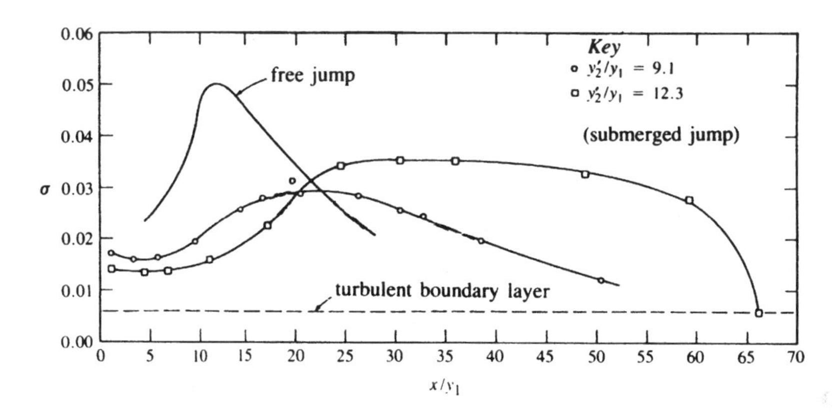

Fig. 5.5 Comparison of streamwise variation of pressure fluctuations in free and submerged jumps for *Fr* = 5 (after Locher and Hsu, 1984)

5.1). For $k > 5$ there is practically no cavitation danger; for $k = 3.5$ there is a 2% probability of the occurrence of cavitation during the time of basin operation (for well constructed basins). In reality, this probability of cavitation could be appreciably higher because the inception of cavitation will occur at pressures higher than p_v, irregularities in the basin floor will cause local pressure reduction, and the incoming flow upstream of the jump may already have a substantial degree of turbulence (Narayanan, 1980).

Physical models provide, with certain safeguards, a suitable tool for evaluating the amplitude and frequency characteristics of macroturbulent pressure fluctuations and for assessing the tendency towards cavitation with intermittent cavitation in prototype (Lopardo, 1988) (Chapter 16).

Potential cavitational damage is not the only danger in hydraulic jump stilling basins (as well as in other types of basins). Probably the most serious structural problem is the effect of uplift pressures due to the dam drainage system or the tailwater level or the water table in the basin bank. This pressure excess over the hydrostatic pressure in the basin is aggravated by the macroturbulent pressure fluctuations underneath (and on the side of) the jump. Although the pressure fluctuations have an unequal spatial distribution (which alleviates this part of the problem) it is only sensible to design the floor slab for the more severe of the two situations: either the full downstream uplift pressure applied over the whole area of the floor with the basin empty or the uplift pressure head equal to the root mean square (r.m.s.) value of pressure fluctuations of the order of $0.12V_1^2/2g$ (V_1 is the inlet supercritical velocity) applied under the whole full basin (this is half of the maximum point pressure fluctuation resulting from σ index max $= 0.05$ and $k_{max} = 5$). Furthermore, all contraction joints should be sealed, no drain openings should be provided, and floor slabs should be as large as possible and connected by dowels and reinforcement (ICOLD, 1987).

Farhoudi and Narayanan (1991) investigated the forces acting on slabs of different lengths and widths beneath a hydraulic jump giving further details of effects of slab size, position and width–length ratio. Pinheiro, Quintela and Ramos (1994) give a summary and comparative analysis of methodologies for computing the hydrodynamic forces acting on hydraulic jump stilling basin slabs.

The prevention of vibration of basin elements (due to the turbulence of the flow) also requires massive slabs, pinned to the foundation when possible (see also Fiorotto and Salandin, 2000).

Abrasion of concrete in the basin could take place if this is also used for bottom outlets carrying abrasive sediment (although this is unlikely to happen for velocities below $10\,\mathrm{m\,s^{-1}}$, or from sediment drawn into the basin from downstream either by bad design or operation. The basin should be self-cleaning to flush out any trapped sediment.

Uplift, abrasion, and cavitation are, of course, closely connected, and provision for maintenance and repairs should be considered in the basin design.

The discharge used in stilling basin design is in most cases the spillway design (maximum) discharge (in cases 1, 2 and 4 above). This, however, need not always be the case. Sometimes it may be more economical to take a calculated risk and design the basin for a smaller and more frequently occurring discharge (say Q_{1000} or smaller instead of PMF – Chapter 4) and carry out repairs should this chosen Q be exceeded. Great care and experience is necessary when opting for this alternative.

5.3.2 Other types of stilling basins

Although the stilling basin based purely on a simple hydraulic jump works well and relatively efficiently, in certain conditions other types of basins may produce savings in construction costs. Standard basins were developed with *baffles, chute blocks and special end sills* by the USBR (Bradley and Peterka, 1957; Peterka, 1963; US Bureau of Reclamation, 1987). An example of a basin with chute and baffle blocks – USBR Type III – which can be used for velocities $V < 18.2\,\mathrm{m\,s^{-1}}$ and $q < 18.6\,\mathrm{m^2\,s^{-1}}$ is shown in Fig. 5.6. As this basin is shorter than others, the temptation is to use it outside these limits; however, the danger of cavitation damage in these cases is substantial and great care must be exercised in the design and positioning of the blocks. Basco (1969) and Nothaft (2003) carried out a detailed investigation of the trend in design of baffled basins and of drag forces, pressure fluctuations, and optimum geometry; the whole area of baffled basins is also reviewed by Locher and Hsu (1984).

The *plain and slotted roller bucket* dissipators developed mainly in the USA (Peterka, 1963) (Fig. 5.7) require substantially higher tailwater

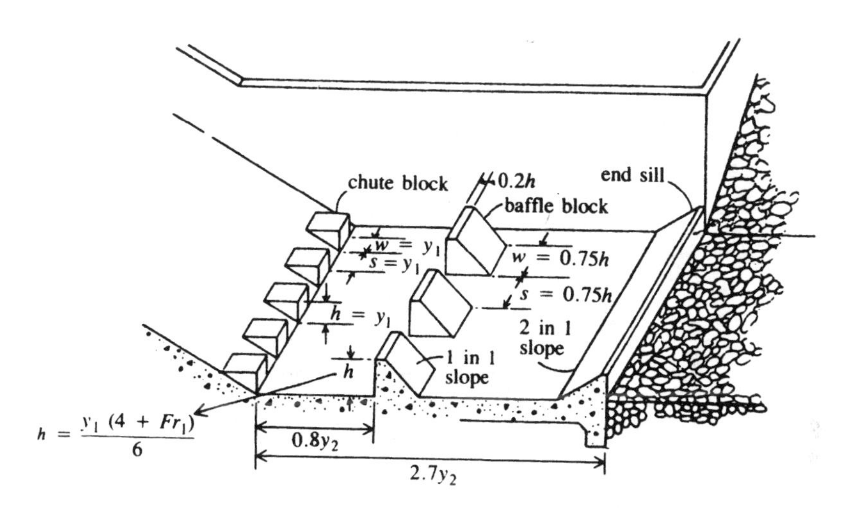

Fig. 5.6 Stilling basin with chute blocks and baffles, USBR Type III (after US Bureau of Reclamation, 1987)

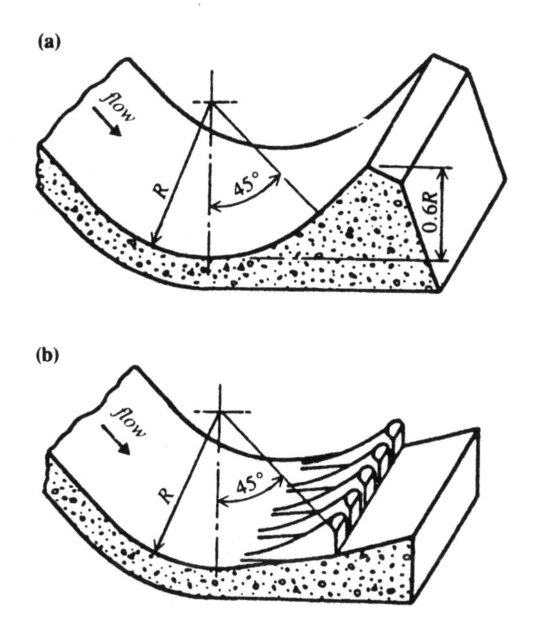

Fig. 5.7 Plain and slotted roller buckets (US Bureau of Reclamation, 1987)

levels than conventional hydraulic jump basins and, in the case of gated spillways, symmetrical gates operation (to prevent side rollers which could bring sediment into the bucket which, in turn, could damage the dissipator).

The stilling basin with a *surface regime hydraulic jump* uses a submerged small shallow flip bucket (Fig. 5.8(b)); the theory and its application

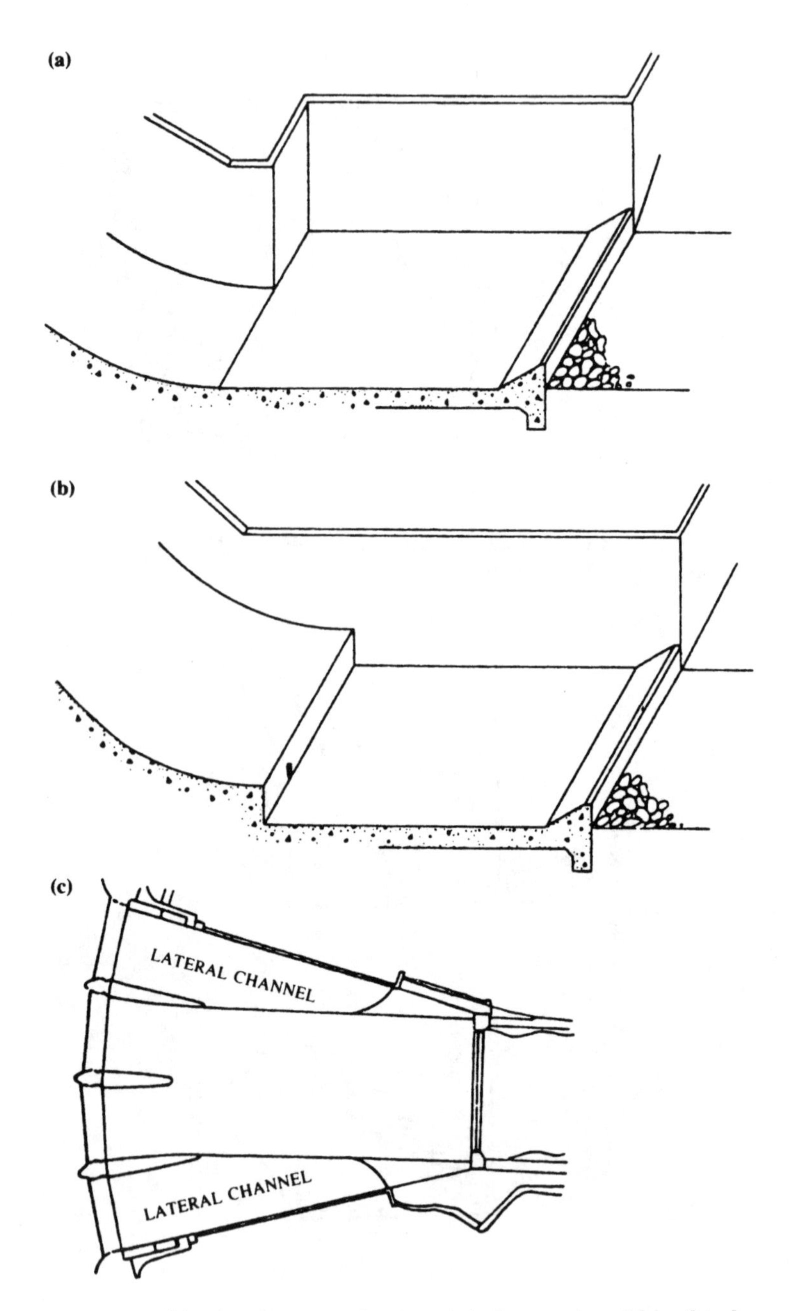

Fig. 5.8 Spatial hydraulic jump basins: (a) change in width; (b) change in depth; (c) flow from lateral channels (after Locher and Hsu, 1984)

to low dams was developed particularly in the USSR (e.g. Skladnev, 1956) and is reviewed by Novak and Čábelka (1981). This type of basin is really only one example of *spatial hydraulic jump* basins. Others involve a sudden change in width (Fig. 5.8(a)) or a jump combined with side inflows form chutes (Fig. 5.8(c)).

Yasuda and Hager (1995) investigated the formation of a hydraulic jump in a *linearly contracting channel* with the conclusion that although there was substantial agreement with the classical jump (equation (5.8)) there were significant differences in its structure. Bremen and Hager (1993) investigated jumps in *abruptly expanding channels* with the conclusion that jumps with the toe just in the upstream channel are more efficient than classical jumps, but in stilling basin construction this advantage has to be offset against the fact that they may become asymmetric for expansion ratios larger than 1.4 and – being longer – require more excavation volume than basins based on the classical jump.

For gated barrages various types of stilling basins other than those already discussed have been developed; one of the most common is a *sloping apron (glacis)* followed by a horizontal sill with or without appurtenances. The shape of the basin is usually dependent on the morphology of the river bed and the amount of excavation needed for its construction and its function has to be considered together with various modes of gate operation. For further details see Section 9.1.

All these and other types of stilling basins and energy dissipators are best developed with the aid of scale model studies (Novak and Čábelka, 1981).

Mason (1982) carried out a survey of 370 dissipators constructed since 1950; the survey included rock basins (unprotected bed with submerged roller bucket), simple jump basins, baffle basins and various types of dissipators involving free trajectory jets. The results of the survey are summarized in Fig. 5.9, and confirm that to operate successfully baffle basins need a certain minimum inflow velocity (head), but their range is limited by cavitation problems for $H > 30$ m. Simple hydraulic jumps have been used for heads that are larger than indicated on the figure ($H > 50$ m) but great care is needed in design and construction. Various types of jet dissipators need a minimum head ($H > 10$ m) to work properly.

5.3.3 Erosion downstream of stilling basins

It has been demonstrated in Section 5.3.1 that at the outflow from the basin there remains a certain proportion of energy to be dissipated. Because of this and the uneven velocity distribution, there will always be some local erosion downstream of the basin. To eliminate this is almost impossible and, above all, uneconomical. The main purpose of the basin is

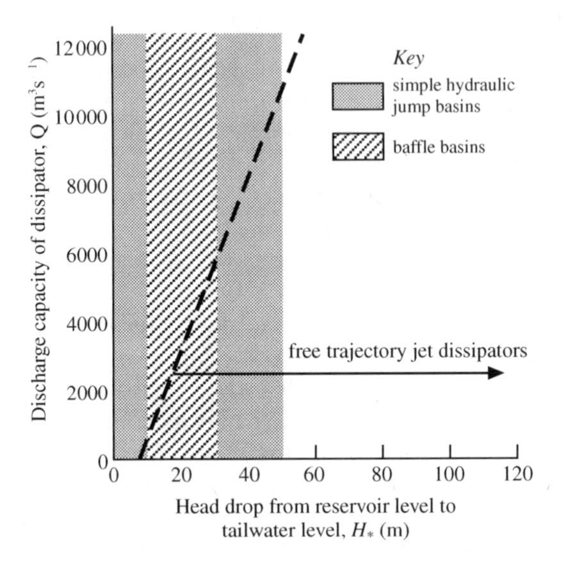

Fig. 5.9 Preferred ranges of use for the main types of dissipators (after Mason, 1982)

to reduce and localize the scour (in a position where it can be controlled and is not harmful to the dam), not to eliminate it. The end sill of the basin has to be protected against retrogressive erosion and/or designed in such a way as to encourage sediment accumulation against the sill rather than erosion (Fig. 5.4).

The extent and depth of the local scour depend on hydraulic parameters, geological structure (erodibility index (Annandale, 1995)) and basin geometry. Several methods are in use for its computation including model studies, but perhaps the simplest is to estimate the scour depth as a percentage of the depth which would occur at the foot of a free overfall without a basin; this in turn can be computed from several equations (Schoklitsch, Veronese, Jaeger, etc.). Using Jaeger's form Novak (1955) expresses the scour depth (downstream of hydraulic jump basins as determined from model experiments of relatively low head structures – say $S < 10\,\mathrm{m}$ – with coarse sand and limited field observations) as

$$y_s = 0.55[6H_*^{0.25}q^{0.5}(y_0/d_{90})^{1/3} - y_0] \tag{5.15}$$

where y_s is the scour depth below the river bed (m), H_* is the difference between upstream and downstream water levels (m), y_0 is the tailwater depth (m), q is the specific discharge ($\mathrm{m^2s^{-1}}$), and d_{90} is the 90% grain size of sediment forming the river bed (mm). Equation (5.15) thus indicates that the stilling basin reduces the potential scour by 45–50%.

For erosion downstream of an apron see e.g. Dey and Sarkar (2006).

5.4 Plunge pools

Downstream of free-falling jets (jet, ski-jump spillways, flip buckets) energy dissipation takes place in stilling basins, or more frequently, in plunge pools, usually excavated fully or partially in the stream bed during dam construction, but sometimes only scoured by the action of the jet itself. The scour and protection works have been reviewed by, for example, Hartung and Häusler (1973), and the equations of the maximum scour y'_s (m) by Locher and Hsu (1984), Breusers and Raudkivi (1991), Annandale (2006) and Liu (2005) (who considered both the energy dissipation mechanism and submerged jet diffusion properties in the scour development).

Bollaert and Schleiss (2005) using investigations of transient water pressures in rock fissures developed a physically based model evaluating the time evolution of scour in jointed rocks. Pagliara *et al.* (2006) examined the effect of the densimetric Froude number, jet impact angle and air content, tailwater elevation, sediment nonuniformity and upstream flow Froude number on the plane plunge pool scour with the result that the densimetric Froude number (involving average jet velocity and the particle determining size) had the main influence on the scour.

The general form of the simplified equation for the maximum scour depth, y'_s, measured from the tailwater surface ($y'_s = y_s + y_0$) is

$$y'_s = Cq^x H_*^y \beta^w / d^z \tag{5.16a}$$

where C is a coefficient, β is the angle of the flip bucket with the horizontal, d is the particle size (mm) and q and H_* are the same as in equation (5.15); the range of the coefficient C and exponents x, y, z is

$$0.65 < C < 4.7, \quad 0.5 < x < 0.67, \quad 0.1 < y < 0.5,$$

$$0 < z < 0.3, \quad 0 < w < 0.1.$$

Thus the range of y'_s is wide, as is only to be expected because equation (5.16a) covers a wide range of structures with different designs, degrees of air entrainment, and geological conditions.

Accepting the possibility of substantial 'errors', a simplified equation by Martins (1975) with $x = 0.6$, $y = 0.1$, $w = z = 0$ and $C = 1.5$ can be used:

$$y'_s = 1.5q^{0.6}H_*^{0.1} \text{ (m)} \tag{5.16b}$$

(in equation (5.16b) H_* is the drop from the reservoir level to the flip bucket lip).

To reduce the variations in the powers in equation (5.16a) Mason (1989) investigated the effect of air content in the plunge pool and proposed an equation (applicable to model and prototype) where air entrainment replaces the effect of H_*:

$$y_s = 3.39\,(q^{0.6}\,(1+\beta)^{0.3}\,y_0^{0.16})/(g^{0.3}\,d^{0.06}) \tag{5.16c}$$

where $\beta = q_a/q = 0{,}13\,(1 - v_c/v)\,(H_*/t)^{0.446}$; t is the jet thickness and v_c the minimum velocity required for air entrainment (1.1 m/s).

Equation (5.16c) can be approximated by an earlier author's equation

$$y_s = 3.27\,(q^{0.60} H_*^{0.05} y_0^{0.15})/(g^{0.3} d^{0.10}). \tag{5.16d}$$

Lopardo (1997) proposed the equation

$$y_s/H_* = 2.5\,(q/(g\,H_*^3)^{0.5})^{0.5}(\text{m}) \tag{5.16e}$$

which translates to :

$$y_s = 1.41\,q^{0.5}\,H_*^{0.25}(\text{m}) \tag{5.16f}$$

Equation (5.16e) is based on over 60 laboratory and 17 prototype results and with a safety factor 1.3 covers all recent field data, e.g. from Tarbela and several dams in China, Africa and South America; it can give higher results than (5.16b), (5.16c) or (5.16d) (particularly when the safety factor 1.3 is applied). Equations (5.16c–e) are dimensionally correct and all equations (5.16) confirm again the decisive influence of the unit discharge in dam design (see also Sections 4.7.1 and 5.2.1).

For flip bucket spillways Tarajmovich proposed the equation (1978):

$$y_s = 6 y_{cr} \tan \beta_1 \tag{5.17}$$

where $y_s = y_s' - y_0$ (as in equation (5.15)), y_{cr} is the critical depth, and β_1 is the upstream angle of the scour hole, which is a function of the flip bucket exit angle β but does not vary widely ($14° < \beta_1 < 24°$ for $10° < \beta < 40°$).

The pressure fluctuations on the floor of a plunge pool underneath a plunging jet can be very considerable; it will be a function of plunge length, pool depth, jet size and shape and can reach a substantial percentage of the head H_* (Ervine, Falvey and Withers, 1997). If the plunge pool is not deep enough to absorp the energy of the jet it may have to be provided with slabs forming a stilling basin floor (see Section 5.3.1). In this case slab joints should be located outside the zone affected by the jet deflection to avoid strong uplift forces in case of a joint waterstop fracture or if using open joints (which may be advantageous with effective foundation drainage). For further details of forces acting on plunge pool slabs and their design see Melo *et al.* (2006).

5.5 Energy dissipation at bottom outlets

The flow from outlets occurs most frequently in a concentrated stream of high velocity. The outlet may terminate below or above the downstream water level, with or without an outlet regulating valve at its end (Sections 4.8 and 6.3). These variations in design are also reflected in the methods of energy dissipation.

The two main design trends are either to disperse artificially and to aerate the outflow jets (outflow above tailwater with or without control gate at its end) or to reduce the specific discharge at entry into the stilling basin. This basin may be a common one with the spillway – the best solution when feasible (e.g. Fig. 5.2) – or a separate one. The reduction in specific discharge for a high-velocity stream can be achieved either by depressing the soffit of the outlet simultaneously with its widening, or by using blocks and sills or guide walls just downstream of the outlet and before the entry into the stilling basin, or in the basin itself, or by a combination of the various methods. The first method has the advantage of avoiding cavitation and/or abrasion and is particularly effective when used in conjunction with a spillway stilling basin because the stream from the outlet can be suitably directed into the basin (Fig. 4.22).

When *submerged deflectors* are used as, for example, in outflows from tunnels, care must be taken in shaping the end of the deflector to avoid cavitation. For further design details, see US Bureau of Reclamation (1987) and Novak and Čábelka (1981).

For the design of a *gradually widening transition* for the free supercritical outflow from an outlet which terminates in a separate stilling basin, Smith (1978), on the basis of work by Rouse (1961), recommends for an initial width B_0 (at the outlet), final width B_1 (at the entry into a hydraulic jump basin), and straight side walls diverging at an angle θ from outflow axis, the equations

$$B_1 = 1.1Q^{1/2}, \tag{5.18}$$

$$\tan\theta = \left(\frac{B_1}{B_0} - 1\right)^{1/3} \Big/ (4.5 + 2Fr_0) \tag{5.19}$$

where $Fr_0 = V_0/(gB_0)^{1/2}$ for a square conduit or $Fr_0 = V_0(gD)^{1/2}$ for a circular conduit section. This rate of flare produces a reasonably uniform and steady distribution of flow at the start of the jump and also permits a continued expansion of the flow at the same rate through the stilling basin; for further details of the jump computation and basin design, see Smith (1978).

For small-capacity outlets (5–10 $m^3 s^{-1}$) under high heads, *vertical stilling wells* provide a compact means of energy dissipation.

Sudden expansion energy dissipators (Locher and Hsu, 1984) which utilize the principle of energy loss at a sudden enlargement are a fairly recent development. Although almost inevitably associated with cavitation, this occurs away from the boundaries without undue danger to the structure. As an example of this type of structure, the section of one of the three circular cross-section expansion chambers of the New Don Pedro dam, designed to pass a total of $200\,m^3\,s^{-1}$ at a gross head of 170 m, is shown in Fig. 5.10. The chambers dissipate about 45 m of head and the remainder is dissipated in pipe resistance and the 9.14 m diameter tunnel downstream of the gates.

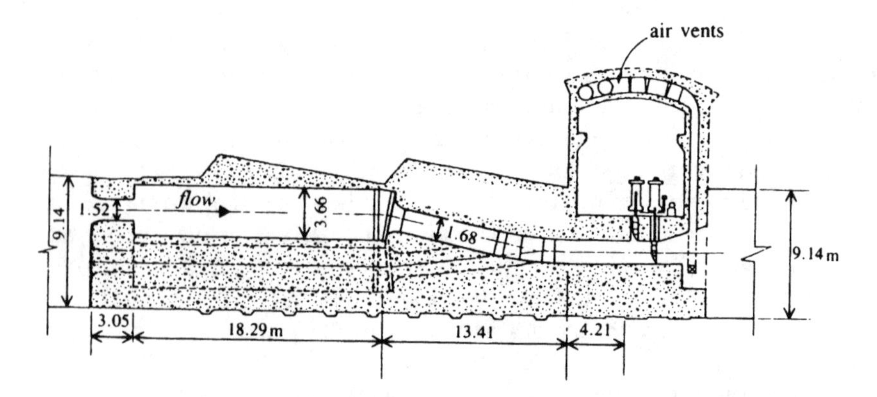

Fig. 5.10 Sudden expansion dissipator, New Don Pedro dam; dimensions in metres (Locher and Hsu, 1984)

Worked Example 5.1

Design a hydraulic jump stilling basin for the maximum discharge of $25\,m^3\,s^{-1}\,m^{-1}$ flowing from an overfall spillway, with the spillway crest 50 m above the downstream gravel river bed with a slope $S_0 = 0.001$ and $n = 0.028$. Check the possibility of cavitation in the basin floor and estimate the scour downstream of the basin; compare this with the depth of a plunge pool downstream of a flip bucket.

Solution

For the design head $C_d \simeq 0.75$ (Section 4.7.1).

From $q = 2/3 C_d \sqrt{2} g^{1/2} H^{3/2}$,

$$H = \left(\frac{25 \times 3}{2 \times 0.75 (19.62)^{1/2}} \right)^{2/3} = 5.032\,\text{m}.$$

From equation (5.4), for $S = 50\,\text{m}$,

$$\varphi = 1 - 0.0155\frac{S}{H} = 1 - 0.0155\frac{50}{5.032} = 0.846.$$

For $n = 0.028$, $d^{1/6} = 0.028/0.04$; thus $d = 0.118\,\text{m}$ (Section 8.2).

For the design discharge, the uniform flow depth y_0 from $q = y_0^{5/3}S_0^{1/2}/n$,

$$y_0 = \left(\frac{0.028 \times 25}{(0.001)^{1/2}}\right)^{3/5} = 6.416\,\text{m},$$

$$\frac{RS_0}{d(2.65-1)} = \frac{6.416 \times 0.001}{0.118 \times 1.65} = 0.033 < 0.05 \text{ (Shields)}$$

(Section 8.2); the river bed is therefore stable.

From equation (5.7) with the datum at river bed level,

$$50 + 5.032 = y_1 + \frac{25^2}{19.6 \times 0.846^2 y_1^2};$$

by trial and error $y_1 = 0.907\,\text{m}$.

From equation (5.8) for $Fr_1^2 = q^2/gy_1^3 = 25^2/(9.81 \times 0.907^3) = 85.38$,

$$y_2 = \frac{0.907}{2}[-1 + (1 + 8 \times 85.38)^{1/2}] = 11.40\,\text{m}.$$

As $y_2 \gg y_0$ a stilling basin is required.

For $\sigma = 1.2$, from equation (5.9),

$$y' = 1.2 \times 11.40 - 6.416 = 7.264\,\text{m}.$$

Assume $y' = 7.50\,\text{m}$ (σ' will be reduced by lowering the datum) and repeat the computation:

$$E = 50 + 5.032 + 7.50 = 62.532, \quad \varphi = 1 - (0.0155 \times 57.5)/5.032 = 0.823.$$

From $62.532 = y_1 + 25^2/(19.62 \times 0.823^2 y_1^2)$, $y_1 = 0.873\,\text{m}$, $Fr_1^2 = 95.75$, and $y_2 = 11.65\,\text{m}$, $\sigma' = 1.19$ (satisfactory).

Check the design for a smaller discharge, say $10\,\text{m}^3\,\text{s}^{-1}$:

$$y_0 = \left(\frac{0.028 \times 10}{(0.001)^{1/2}}\right)^{3/5} = 3.7\,\text{m}.$$

Assume $C_d \simeq 0.65$; thus $H = 10 \times 3/(2 \times 0.65\sqrt{19.62}) = 2.69$ m, and $\varphi = 1 - (0.0155 \times 57.5)/2.69 = 0.67$. $57.5 + 2.69 = 60.19 = y_1 + 10^2/(19.6 \times 0.67^2 y_1^2)$; $y_1 = 0.436$ m. $Fr_1^2 = 123$; $y_2 = 6.62$ m; $\sigma' = (7.5 + 3.7)/6.62 = 1.69 \gg 1.19$. The basin designed for the maximum discharge is more than adequate for lower discharges.

From equation (5.10) the stilling basin length $L = K(11.65 - 0.873)$; for $Fr_1 = \sqrt{95.75} = 9.78$ and $K \simeq 4.5$; $L = 4.5 \times 10.777 = 48.49\,\text{m} \simeq 50$ m.

From equation (5.14) for $V_1 = 25/0.873 = 28.63\,\text{m s}^{-1}$ and $y_1 = 0.873$ m,

$$k = \left(\frac{p_0}{\rho g} + \bar{y}\right) \bigg/ \left(0.05\,\frac{V_1^2}{2g}\right) = \frac{(10 + 3y_1)19.62}{0.05 \times 28.63^2} = 6.04 > 5.$$

Thus, theoretically, there is no cavitation danger, even for the maximum discharge.

From equation (5.15) for $d_{90} = 118$ mm, $y_0 = 6.416$ m, and $H_* = 55.032 - 6.416 = 48.616$ m, $y_s = 0.55[6 \times 48.616^{0.25} \times 25^{0.5}(6.416/118)^{1/3} - 6.416] \simeq 13$ m. As $S = 50 \gg 10$ m, this equation is not really applicable and the computed scour depth is far too high. A more realistic result is about 50% of this value, i.e. about 6.5 m. This maximum scour is likely to occur at a distance $L + 6 \times 6.5 \simeq 90$ m from the toe of the dam and is thus harmless for the dam, but requires suitable bank protection.

If a flip bucket were used the probable required depth of the plunge pool would, from equation (5.16b), be

$$y_s' \simeq 1.5 \times 25^{0.6} \times 48.616^{0.1} = 15.25\,\text{m}.$$

Thus $y_s = 15.25 - 6.416 \simeq 9$ m below river bed level. (This erosion would, however, very probably be closer to the toe of the dam unless a chute, diverging the flow from the dam, were used.)

From equation (5.17) for $14° < \beta_1 < 24°$, $y_s = 6(25^2/9.8)^{1/3} \tan \beta_1$, i.e. $6.0 < y_s < 10.6$ m; this agrees reasonably well with the previous result of $y_s \simeq 9$ m. However, equations (5.16d) and (5.16e) give bigger results (16 m and 12 m).

References

Annandale, G.W. (1995) Erodibility. *Journal of Hydraulic Research*, IAHR, 33, No. 4: 471–94.

—— (2006) *Scour Technology: Mechanics and Engineering Practice*, McGraw-Hill, New York.

Basco, D.R. (1969) *Trends in Baffled Hydraulic Jump Stilling Basin Design of the Corps of Engineers since 1947*, Miscellaneous Paper H-69-1, US Army Engineers Waterways Experimental Station, Vicksburg.

Boes, R.M. and Hager, W.H. (2003) Hydraulic design of stepped spillways, *Journal of Hydraulic Engineering ASCE*, 129, No. 9: 671–9.

Bollaert, E.F.R. and Schleiss, A.J. (2005) Physically based model for evaluation of rock scour due to high-velocity jet impact, *Journal of Hydraulic Engineering ASCE*, 131, No. 3: 153–65.

Bradley, J. and Peterka, A.J. (1957) The hydraulic design of stilling basins. *Journal of the Hydraulics Division, American Society of Civil Engineers*, 83 (Hy 5), Papers 1401–6, 130 pp.

Bremen, R. and Hager, W.H. (1993) T-jump in abruptly expanding channel. *Journal of Hydraulic Research*, 31 (1): 61–78.

Breusers, H.N.C. and Raudkivi, A.J. (1991) *Scouring*, IAHR Hydraulic Structures Design Manual, Vol. 2, Balkema, Rotterdam.

Coyne, A. (1951) Observation sur les déversoirs en saut de ski, in *Transactions of the 4th Congress of ICOLD*, New Delhi, Vol. 2, Report 89: 737–56.

Dey, S. and Sakar, A. (2006) Scour downstream of an apron due to submerged horizontal jets, *Journal of Hydraulic Engineering ASCE*, 132, No. 3: 246–57.

Ervine, D.A. and Falvey, H.T. (1987) Behaviour of turbulent water jets in the atmosphere and in plunge pools. *Proceedings of the Institution of Civil Engineers, Part 2*, 83: 295–314.

Ervine, D.A., Falvey, H.T. and Withers, W. (1997) Pressure fluctuations on plunge pool floors. *Journal of Hydraulic Research*, IAHR, 35, No. 2: 257–79.

Farhoudi, J. and Narayanan, R. (1991) Force on slab beneath hydraulic jump. *Journal of Hydraulic Engineering, American Society of Civil Engineers*, 117 (1): 64–82.

Fiorotto, V. and Salandin, P. (2000) Design of anchored slabs in spillway stilling basins. *Journal of Hydraulic Engineering*, ASCE, 126, No. 7: 502–12.

Hager, W.H. and Li, D. (1992) Sill-controlled energy dissipator. *Journal of Hydraulic Research*, 30 (2): 165–81.

Hager, W.H., Bremen, R. and Kawagoshi, N. (1990) Classical hydraulic jump: length of roller. *Journal of Hydraulic Research*, 28 (5): 592–608.

Hartung, F. and Häusler, E. (1973) Scour, stilling basins and downstream protection under free overfall jets at dams, in *Proceedings of the 11th Congress on Large Dams*, Madrid: 39–56.

Heller, V., Hager, W.H. and Minor, H.-E. (2005) Ski-jump hydraulics, *Journal of Hydraulic Engineering ASCE*, 131, No. 5: 347–55.

Hoření, P. (1956) Disintegration of a free jet in air, in *Práce a Studie*, No. 93, VUV, Prague (in Czech).

ICOLD (1987) *Spillways for Dams*, Bulletin 58, International Commission on Large Dams, Paris.

Liu, P. (2005) A new method for calculating depth of scour pit caused by overflow water jets, *Journal of Hydraulic Research*, IAHR, 43, No. 6: 695–701.

Locher, F.A. and Hsu, S.T. (1984) Energy dissipation at high dams, in *Developments in Hydraulic Engineering*, Vol. 2 (ed. P. Novak), Elsevier Applied Science, London.

Lopardo, R.A. (1988) Stilling basin pressure fluctuations (invited lecture), in *Proceedings of the International Symposium on Model–Prototype Correlation of Hydraulic Structures* (ed. H. Burgi), American Society of Civil Engineers, New York: 56–73.

—— (1997) Hydrodynamic actions on large dams under flood conditions, *3rd Int. Conf. on River Flood Hydraulics*, Stellenbosch, HR Wallingford.

Martins, R.B.F. (1975) Scouring of rocky river beds by free jet spillways. *Water Power & Dam Construction*, 27 (4): 152–3.

Mason, P.J. (1982) The choice of hydraulic energy dissipators for dam outlet works based on a survey of prototype usage. *Proceedings of the Institution of Civil Engineers, Part 1*, 72: 209–19.

—— (1989) Effects of air entrainment on plunge pool scour, *Journal of Hydraulic Engineering ASCE*, 115, No. 3: 385–99.

Melo, J.F., Pinheiro, A.N. and Ramos, C.M. (2006) Forces on plunge pool slabs: influence of joint location and width, *Journal of Hydraulic Engineering ASCE*, 132, No. 1: 49–60.

Narayanan, R. (1980) Cavitation induced by turbulence in stilling basins. *Journal of the Hydraulics Division, American Society of Civil Engineers*, 106 (Hy 4), 616–19.

Naudascher, E. (1987) *Hydraulik der Gerinne und Gerinnebauwerke*, Springer, Vienna.

Nothaft, S. (2003) Die hydrodynamische Belastung von Störkörpern, *Wasserbau und Wasserwirtschaft*, No. 95, Technische Universität, München.

Novak, P. (1955) Study of stilling basins with special regard to their end sills, in *Proceedings of the 6th Congress of the International Association of Hydraulics Research*, The Hague, Paper C15.

Novak, P. and Čábelka, J. (1981) *Models in Hydraulic Engineering – Physical Principles and Design Applications*, Pitman, London.

Pagliara, S., Hager, W.H. and Minor, H.-E. (2006) Hydraulics of plane plunge pool scour, *Journal of Hydraulic Engineering ASCE*, 132, No. 5: 450–61.

Peterka, A.J. (1963) *Hydraulic Design of Stilling Basins and Energy Dissipators*, Engineering Monograph 25, US Bureau of Reclamation, CO.

Pinheiro, A.N., Quintela, A.C. and Ramos, C.M. (1994) Hydrodynamic forces in hydraulic jump stilling basins, in *Proceedings of the Symposium on Fundamentals and Advancements in Hydraulic Measurements and Experimentation*, Buffalo, NY, August (ed. C.A. Pugh), American Society of Civil Engineers, New York: 321–30.

Rajaratnam, N. (1967) Hydraulic jumps, in *Advances in Hydroscience*, Vol. 4 (ed. V.T. Chow), Academic Press, New York.

Rice, C.E. and Kadavy, K.C. (1994) Energy dissipation for RCC stepped spillway, in *Hydraulic Engineering 1994, Proceedings of the 1994 Buffalo Conference*, Vol. 1 (eds G.V. Cotroneo and R.R. Rumer), ASCE, New York.

Rouse, H. (1961) *Fluid Mechanics for Hydraulic Engineers*, Dover Publications, New York.

Skladnev, M.F. (1956) Limits of bottom and surface hydraulic jumps, *Izvestiya Vsesoyuznogo Nauchno-Issledovatel'skogo Instituta Gidrotekhniki*, 53 (in Russian).

Smith, C.D. (1978) *Hydraulic Structures*, University of Saskatchewan.

Stephenson, D. (1991) Energy dissipation down stepped spillways, *Int. Water Power and Dam Construction*, 43, No. 9: 27–30.

Tarajmovich, I.I. (1978) Deformations of channels below high-head spillways on rock foundations. *Gidrotekhnicheskoe Stroitel'stvo*, (9): 38–42 (in Russian).

US Bureau of Reclamation (1987) *Design of Small Dams*, 3rd edn, US Department of the Interior, Washington, DC.

Vischer, D.L. and Hager, W.H. (1995) *Energy Dissipators*, IAHR Hydraulic Structures Design Manual, Vol. 9, Balkema, Rotterdam.

Yasuda, Y. and Hager, W.H. (1995) Hydraulic jump in channel contraction. *Canadian Journal Civil Engineering* 22: 925–33.

Chapter 6

Gates and valves

6.1 General

Although this chapter deals mainly with gates and valves used in dam engineering some parts of its text have a wider or special application, e.g. gates used in river engineering at weirs and at tidal barrages and surge protection works.

The main operational requirements for gates and valves are the control of floods, watertightness, minimum hoist capacity, convenience of installation and maintenance and above all failure free performance and avoidance of safety hazards to the operating staff and the public. Despite robust design and precautions, faults can occur and the works must be capable of tolerating these faults without unacceptable consequences.

Gates may be classified as follows:

1. position in the dam – crest gates and high-head (submerged) gates and valves;
2. function – service, bulkhead (maintenance) and emergency gates;
3. material – gates made of steel, aluminium alloys, reinforced concrete, wood, rubber, nylon and other synthetic materials;
4. pressure transmission – to piers or abutments, to the gate sill, to the sill and piers;
5. mode of operation – regulating and non-regulating gates and valves;
6. type of motion – translatory, rotary, rolling, floating gates, gates moving along or across the flow;
7. moving mechanisms – gates powered electrically, mechanically, hydraulically, automatically by water pressure or by hand.

Because of the variety of criteria many gate and valve designs are in use; only the salient features of some of the more important types can be dealt with here.

In the following sections some broadly valid ranges of heads and spans for some of the most frequently used types of gates are given. There are, however, several outstanding examples of large span or high-head gates considerably exceeding these parameters particularly in modern flood control schemes and surge barriers where special types of gates are also often used (see Section 6.4).

Because of the multitude of their functions and sizes there is great scope for innovation in gate and valve design both in details (e.g. seals, trunnions), as well as in conceptual design (automatic level control by hinged flap gates, gates and valves used in water distribution systems, 'hydrostatic' gates etc.).

Both crest gates and gates in submerged outlets require provision for emergency operation and maintenance. Emergency closure gates, maintenance gates, stop-logs, grappling beams, rail mounted gantry cranes, screens and their cleaning devices (in case of submerged gates) are some of the necessary appurtenances. For further design and construction details of various types of gates and valves see Leliavsky (1981), Zipparo and Hasen (1993), Erbisti (2004) and particularly Lewin (2001).

6.2 Crest gates

6.2.1 Pressure transmission

The basic feature of the structural design of crest gates is the method used for transferring the pressure acting on them.

1. Pressure transmission to piers and abutments is used by plain vertical lift gates and stop-logs, radial gates and roller gates; the gates may be designed for flow over or under them or for a combined flow condition.
2. Pressure transmission to the gate sill is used, for example, by sector (drum) gates (with upstream or downstream hinge), roof (bear-trap) gates, pivot-leaf (flap) gates, roll-out gates, and inflatable gates.
3. Pressure transmission both to piers and sill, e.g. some types of flap gates and floating (pontoon) gates.

6.2.2 Plain gates

Plain (vertical lift) gates, designed as a lattice, box girder, a grid of horizontal and vertical beams and stiffeners, or a single slab steel plate, may consist of single or double section (or even more parts can be involved in the closure of very high openings); in the case of flow over the top of the gate it may be provided with an additional flap gate. The gates can have slide or wheeled support (e.g. Fig. 4.20). In the latter case fixed wheels (most frequent type), caterpillar or a roller train (Stoney gate) may be used; for fixed wheels their spacing is reduced near the bottom. The gate seals are of specially formed rubber (Fig. 6.1).

The gate weight, G, is related to its span, B(m), and the load, P(kN), by

$$G = k(PB)^n. \tag{6.1}$$

For slide gates with $PB > 200\,\text{kN m}$, $k = 0.12$ and $n = 0.71$; for wheeled gates and $PB > 270\,\text{kN m}$, $k = 0.09$ and $n = 0.73$. The usual range of heads for single plain gates is $1 < H(\text{m}) < 15$, and of spans $4 < B(\text{m}) < 45$.

The product of span B and head at the bottom of the gate H is usually kept below $200\,\text{m}^2$ (Smith, 1978); in special cases sliding gates with larger values have been used (e.g. Scheldt, The Netherlands, $250\,\text{m}^2$) (see also Section 6.4).

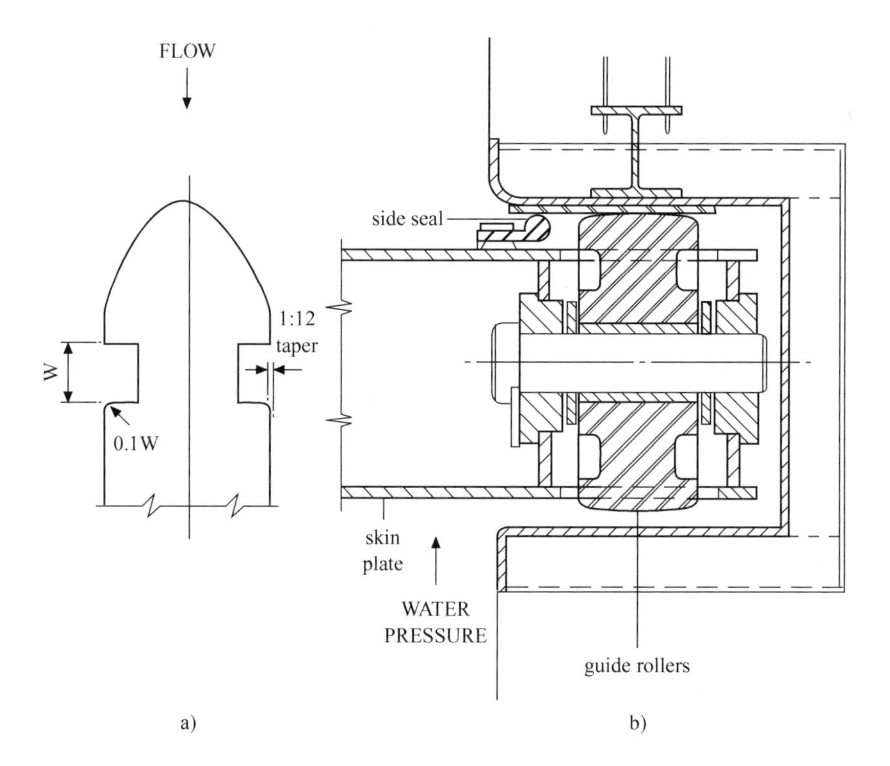

Fig. 6.1 Vertical lift gate (Lewin, 2006)

The pier face downstream of the gate slots may be protected against the effect of flow separation and possible cavitation (see Section 4.6) either by recessing it against the upstream pier face (by about 0.1 w, where w is the width of the slots) and rounding off the downstream corner by a radious 0.1 w or by a bevel of a slope about 1:12 (Fig. 6.1); for further details see, e.g. Ball (1959).

6.2.3 Radial gates

Radial (Tainter) gates are usually constructed as portals with cross-bars and arms (straight, radial, or inclined), but could also be cantilevered over the arms. Their support hinges are usually downstream but (for low heads) could also be upstream, resulting in shorter piers.

In equation (6.1), $0.11 < k < 0.15$ and $n \simeq 0.07$ for $PB > 150$ kN m. The usual range of heads and spans for radial gates is $2 < H(\mathrm{m}) < 20$ and $3 < B(\mathrm{m}) < 55$, with $(BH)_{max} \simeq 550\,\mathrm{m}^2$. Radial gates may be designed for more than 20 MN per bearing.

The advantages of radial over vertical lift gates are smaller hoist, higher stiffness, lower (but longer) piers, absence of gate slots, easier automation and better winter performance.

The gate is usually hoisted by cables fixed to each end to prevent it from twisting and jamming (Fig. 6.2). If the cables are connected to the bottom of the gate its top can be raised above the level of the hoist itself, if the layout of the machinery allows it.

6.2.4 Drum and sector gates

Drum and sector gates are circular sectors in cross-section. Drum gates are hinged upstream (Fig. 6.3) and sector gates downstream; in the latter case the hinge is usually below the spillway crest by about $0.1H$ to $0.2H$. Gates on dam crests are usually of the upstream hinge type, with the hinge about $0.25H$ above the downstream gate sill and a radius of curvature of $r \simeq H$. The heads can be as high as 10 m and the spans 65 m. Drum gates float on the lower face of the drum, whereas sector gates are usually enclosed only on the upstream and downstream surfaces.

Both types of gates are difficult to install, and require careful maintenance and heating in winter conditions; their main advantages are ease of automation and absence of lifting gear, fast movement, accuracy of regulation, ease of passing of ice and debris, and low piers.

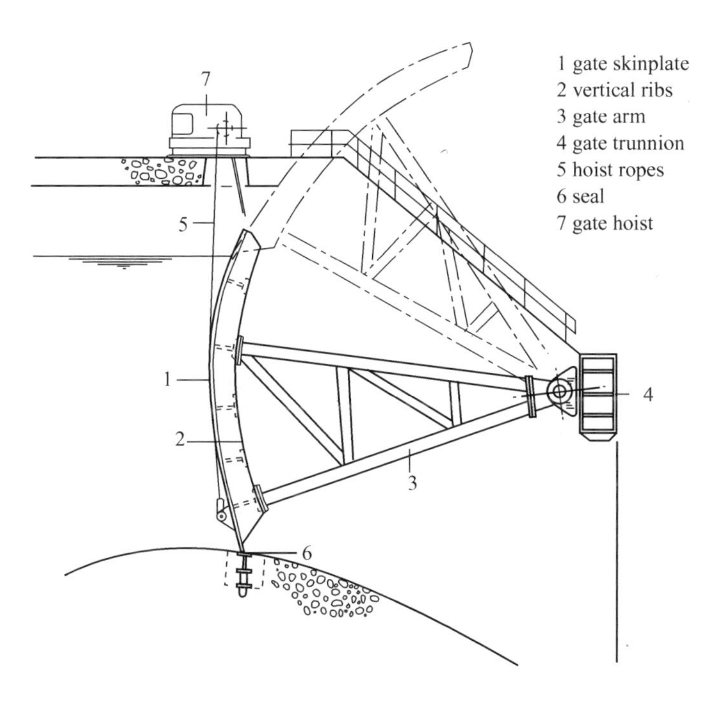

Fig. 6.2 Radial gate (Lewin, 2006)

6.2.5 Flap gates

Flap gates (bottom-hinged gates) (Fig. 6.4) are one of the simplest and most frequently used types of regulating gates used mainly on weirs and barrages (rarely on dam crests), either on their own or in conjunction with plain lift gates. They were developed as a replacement for wooden flash-boards, originally as a steel-edged girder flap, which was later replaced by a torsion-rigid pipe; further development was achieved by placing the pipe along the axis of the flap bearing, with the skin plate transmitting the water pressure to cantilevered ribs fixed to the pipe. Next in use were torsion-rigid gate bodies with curved downstream sides (fish-belly gates), with torsion-rigid structures using prism-shaped sections being the latest development in flap gates.

The heads for which flap gates (on their own) are being used may be as high as 6 m and the span up to about 30 m; for larger spans several flaps connected to each other, but with each actuated by its own hydraulic hoist, may be used (Brouwer, 1988).

Flap gates provide fine level regulation, easy flushing of debris and ice, and are cost effective and often environmentally more acceptable

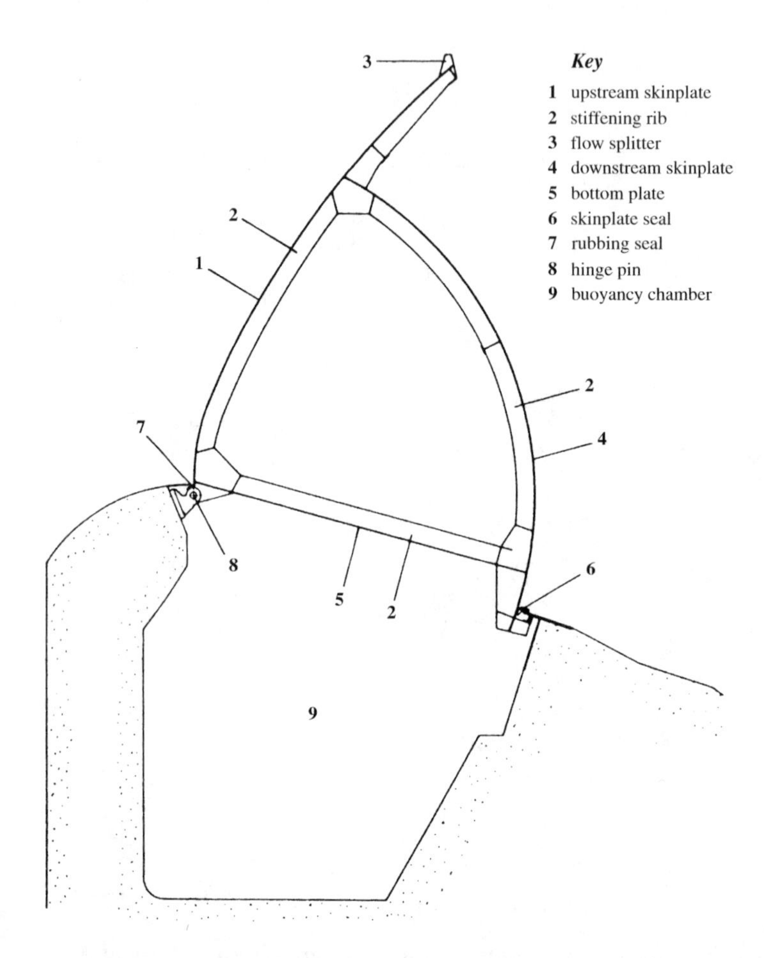

Fig. 6.3 Drum gate (after Thomas, 1976)

than other types of gates; they require protection against freezing and are particularly sensitive to aeration demand and vibration, which may be prevented, however, by the use of splitters at the edge of the flap (Fig. 6.3).

6.2.6 Roller gates

A roller gate consists of a hollow steel cylinder, usually of a diameter somewhat smaller than the damming height; the difference is covered by a steel attachment, most frequently located at the bottom of the cylinder (in the closed position). The gate is operated by rolling it on an inclined track (Fig. 6.5). Because of the great stiffness of the gate, large spans (of up to 50m) may be used, but roller gates require substantial piers with

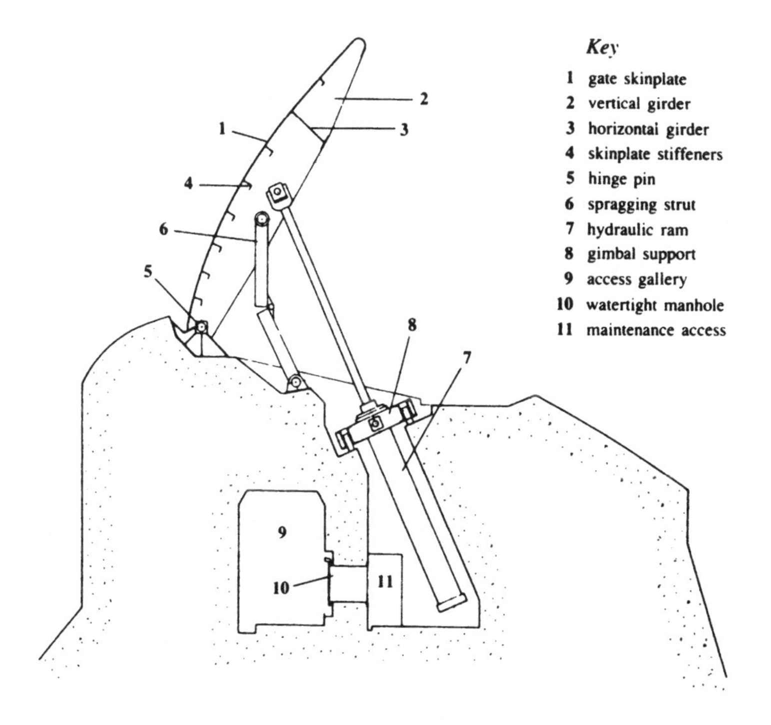

Fig. 6.4 Flap gate (after Thomas, 1976)

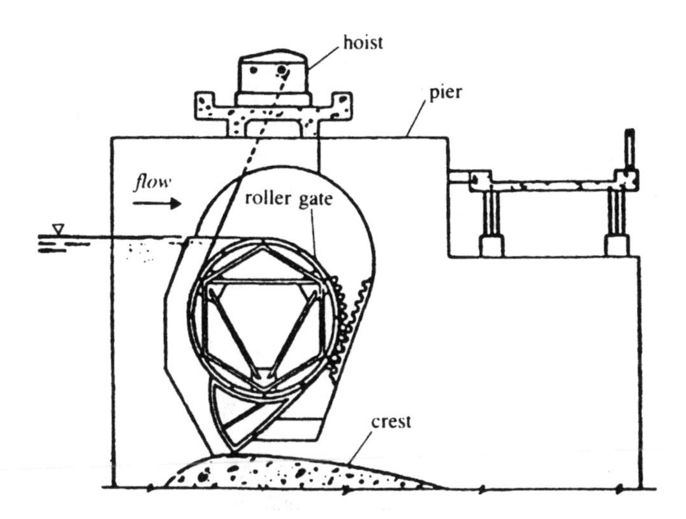

Fig. 6.5 Roller gate (after US Army Waterways Experimental Station, 1959)

large recesses. New roller gate installations are not being used nowadays partly because large units of this type are very vulnerable to single point failure.

6.2.7 Fabric gates

Inflatable rubber or fabric gates can be pressurized by air, water or both. They usually have an inner shell and an outer casing, and can be used to close very large spans (over 100 m) with heads up to 6 m. The strength of fabric used exceeds 100 kN m^{-1}.

The main advantages of fabric gates are low cost, low weight, absence of lifting mechanism, little need for maintenance, acceptance of side slope (at river banks), and ease of installation; the disadvantage is that they can easily be damaged and may have a limited life.

6.2.8 Overspill fusegates

Fusegates for installation on previously ungated spillways use the principle of fuse plugs (Section 4.7.7) without the disadvantages particularly of the sudden large increase in discharge and the loss of the plug.

Designed and used first in France under the name Hydroplus fusegate (in 1991 Lussas irrigation dam with 10 gates 2.15 m high and 3.5 m wide) (Lemperière, 1992) their height can be up to 75% of the free spillway head (saving of storage) and they can pass moderate floods as a spill over their labyrinth-shaped top. For higher floods the independent free-standing units turn about a downstream sill, if a bottom chamber filled through an inlet well generates an uplift pressure beneath the unit. After the flood has receded the overturned units can be reinstalled on the crest or, if damaged, replaced by new ones; they remain on the spillway chute below the crest only for velocities lower than 3 m s^{-1}, otherwise being washed away. Should the need for their recovery or replacement occur frequently, the fusegates may become uneconomical.

Since their first use much bigger fusegates have been installed. E.g. during the reconstruction of the Terminus dam in California (to pass a revised PMF) six concrete units each 11.7 m wide, 6.5 m high with 13 m long crest have been used. The opening (overturning) of the units is staggered with the first tipping at water levels 7.8 m and the last at 9.3 m above the dam crest (Kocahan 2004).

In Russia fusegates 1.75 m high and 3.15 m wide as well as a prototype of a new folding fusegate type gate were successfully tested next to a labyrinth spillway in very harsh winter conditions (Rodionov *et al.*, 2006).

6.3 High-head gates and valves

6.3.1 General

High-head (submerged) gates and valves transmit the load to the surrounding structure either directly through their support, e.g. plain (vertical lift), radial, or ring follower gates, or through the shell encasing the valve. The most common valves of the latter type are *non-regulating* disc (butterfly) valves (turning about a horizontal or vertical axis), cylindrical or sphere (rotary) valves. The main advantage of the latter is the clear water passage when fully open and hence a very low head loss coefficient, but their cost is higher than that of butterfly valves. The most frequently used *flow regulating* valves in closed pipe systems are sphere valves or pressure reducing valves. For terminal discharge regulation a frequent choice are needle, tube, hollow-jet and particularly Howell–Bunger (fixed-cone dispersion) valves.

6.3.2 High-pressure gates

Plain (vertical lift) gates (Fig. 4.19) are sliding, wheeled or moving on rollers or caterpillars. The load on them may reach 3000–4000 kN m^{-1} with heads up to 200 m and gate areas up to 100 m^2 (usually 30–50 m^2). For optimum conditions it is best to contract the pressure conduit upstream of the gate and to provide deflectors downstream to aid aeration as an anti-cavitation measure (see also Section 4.7.3). The conduit face downstream of the gate slots should be protected against cavitation in the same way as for crest gates (see Section 6.2.2). Even better protection is provided for non-regulating lift gates by ring followers, which close the gate grooves and ensure a smooth passage for flow through the fully opened gate. Other developments of this type of ring follower gate are paradox and ring seal gates (Zipparro and Hasen, 1993).

Radial gates are normally hinged downstream (Fig. 4.21), but are sometimes used in the reversed position with arms inside or, more frequently, outside the conduit at the end of which the regulating gate is installed.

A special feature of some high-head Tainter gates is the use of eccentric trunnions (Buzzel, 1957) which permit a gap to be formed between the seals and the seal seats before opening the gate so that only moments caused by the gate weight and hinge friction have to be overcome. This type was used, for example, at the Dongjiang dam for gates with span 6.4 m and height 7.5 m operating under a head of 120 m (Erbisti, 1994).

Cylinder gates with all hydraulic forces counterbalanced are frequently used in tower type intakes; gate diameters up to 7.00 m with maximum heads about 70 m have been used.

6.3.3 High-head valves

The *cone dispersion* (Howell–Bunger) valve (Fig. 6.6) is probably the most frequently used type of regulating valve installed at the end of outlets discharging into the atmosphere. It consists of a fixed 90° cone disperser, upstream of which is the opening covered by a sliding cylindrical sleeve. It has been used for heads of up to 250 m, and when fully open its discharge coefficient is 0.85–0.9. The fully opened valve area is about 0.8 of the conduit area.

The Howell–Bunger valve is cheaper, less robust but more reliable and with a better discharge coefficient than the needle or hollow-jet valves, and should really be used in dam design only when discharging into the atmosphere (although in some special cases it has been used in conduits followed by a ring hydraulic jump (Haindl, 1984)). The fine spray associated with the operation of the valve may be undesirable, particularly in cold weather; sometimes, therefore, a fixed large hollow cylinder is placed at the end of the valve downstream of the cone, resulting in a ring jet valve. The discharge coefficient is reduced in this case to about 0.75–0.80.

The *needle* valve, (and its variation the *tube* valve), has a bulb-shaped fixed steel jacket, with the valve closing against the casing in the

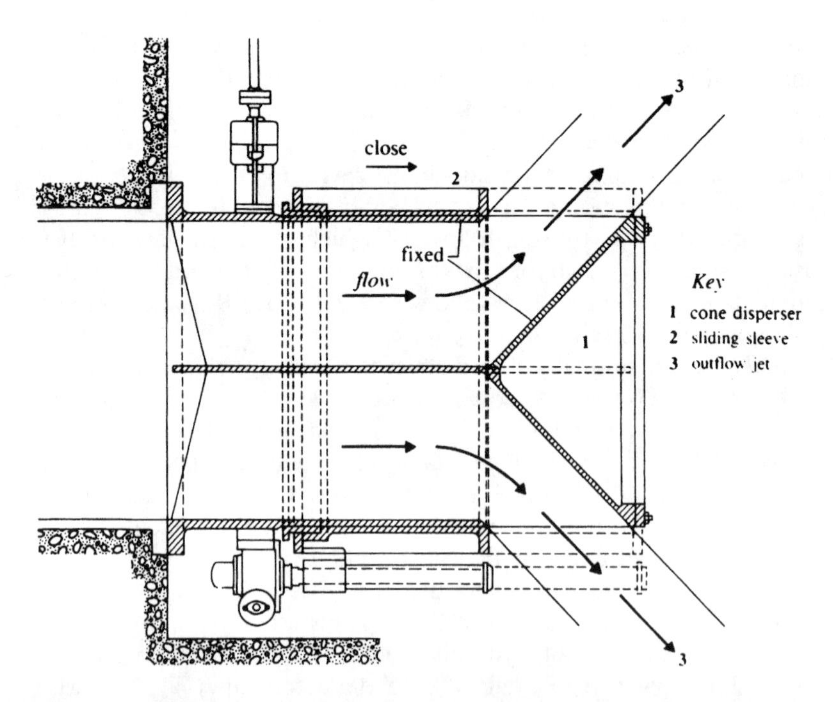

Fig. 6.6 Cone dispersion valve (after Smith, 1978)

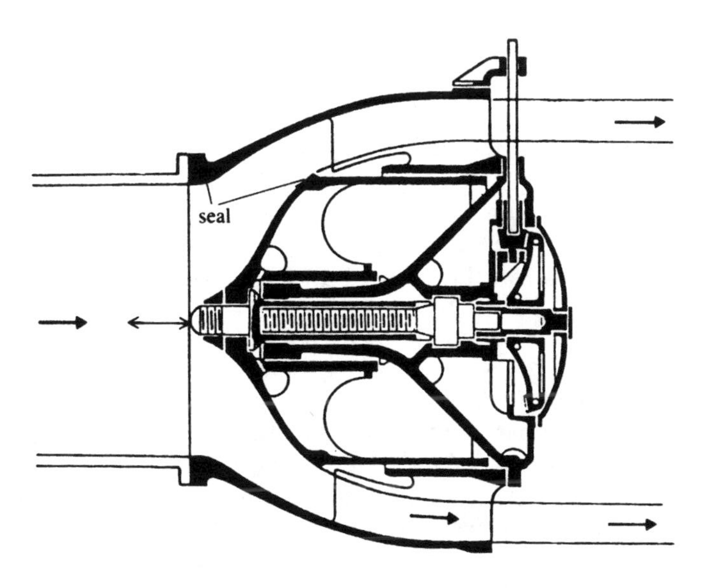

Fig. 6.7 Hollow-jet valve (after Smith, 1978)

downstream direction. When open, the valves produce solid circular jets and can also be used in submerged conditions. The valves may suffer from cavitation damage and produce unstable jets at small openings, and are expensive as they have to withstand full reservoir pressures.

Most of those disadvantages are overcome in the *hollow-jet* valve (Fig. 6.7), which closes in the upstream direction (when closed the valve body is at atmospheric pressure); because of this the valve is, of course, not suitable for use in submerged conditions. The discharge coefficient of the fully open valve is about 0.7.

6.4 Tidal barrage and surge protection gates

Due to the large spans and heads to be controlled the design of tidal barrages often built for flood control and surge protection purposes provides a special challenge in the design, installation and operation of gates. The following text gives just some parameters of gates used in a few of these works.

The main storm-surge protection barrage of the *Delta works on the Western Scheldt in the Netherlands* completed in 1986 has 62 lift gates of heights 5.9–11.9 m and 42 m span. The storm surge *Maeslant barrier in the navigation way (Nieuwe Waterweg) to Rotterdam*, completed in 1997 and forming part of the same overall scheme has two floating horizontal radial/sector gates capable of closing the 360 m wide channel. Each gate is 21.5 m high, 8 m wide and 210 m long; the two horizontal arms of each gate are 220 m long. The gates have a unique ball (pivot) joint 10 m in diameter. Under normal conditions the gates are 'parked' in two dry docks (one on each bank). To operate the gates the docks are flooded, the gates are floated and swivelled into position in the waterway. When contact between the gates is established, they are flooded and sink to the bed closing the channel. After the flood danger has passed the whole process is reversed and the gates are towed back to their 'parking' position.

Large vertical lift gates need a high superstructure required to lift the gate clear and to allow sufficient room for navigation. This high superstructure is not only expensive but also environmentally intrusive; furthermore the gate in its raised position may be subjected to high wind loads. A solution to these problems adopted, e.g. in the single gate *tidal barrier on the River Hull* (UK), is to rotate the gate through 90° to a horizontal plane when 'parked' in the raised position. Another possibility of course is the use of different types of gates as, e.g., on the *tidal barrage on the River Tees* (UK) completed in 1994 where 4 buoyant fish-belly gates 13.5 m long, 8.1 m high and 2 m thick (Norgrove 1996) have been used.

The *Thames barrier* protecting London against flooding from tidal surges completed in 1982 uses a novel concept of a rising sector gate (see Fig. 6.8). The 20 m high gates are attached at both ends to discs (about

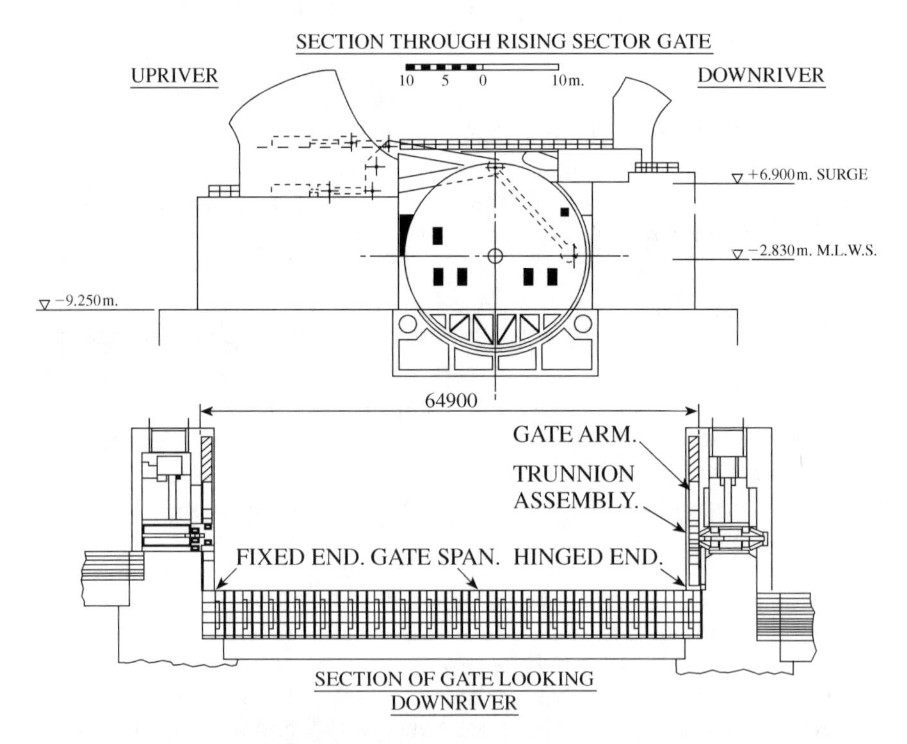

Fig. 6.8 Thames barrier gates

24 m diameter) the rotation of which raise or lower the gates, which in the open position are level with the river bed and in the flood control mode are vertical (Lewin, 2001). The barrage has four 61 m wide rising sector gates (protecting the navigation passages), a further two 31.5 m wide rising sector gates and four 31.5 m falling radial gates.

The protection of *Venice* against flooding is a great engineering challenge. To avoid piers in the navigation channels connecting the Venice Lagoon and the Adriatic it is proposed to use in the three channels four barriers each 400 m wide, with bottom hinged buoyant 20 m wide, 18–28 m long and 3.6–5 m deep flap gates recessed in closed position in reinforced concrete caissons with a hydraulic sediment ejector system (i.e. altogether 79 gates). In their operating position the gates are at an angle approximately 45°; they are not linked and move independently under wave action with leakage between the gates (and the hinges). An experimental full-scale module was built and operated 1988–92 (Lewin and Scotti, 1990, Bandarin, 1994); in 2003 agreement was reached to proceed with the project. For further details see Lewin and Eprim (2004).

6.5 Hydrodynamic forces acting on gates

The main force acting on gates is usually due to hydrodynamic pressures caused by nonuniform turbulent flow with subsidiary forces caused by waves, ice, impact of floating bodies etc. The hydrodynamic forces are usually considered in two parts: time averaged mean component (or steady flow part) and fluctuating components induced by various excitation mechanisms (see Section 6.6).

6.5.1 Forces on low-head gates

In their closed position gates and valves are subjected to hydrostatic forces determined by standard procedures for forces acting on plane and curved surfaces. For radial or sector gates the vector of the resultant force must pass through the gate pivot and thus its moment will be zero; equally it can be converted into one tending to open or close the gate by placing the pivot above or below the centre of curvature of the skin plate.

The computation of hydrodynamic forces acting on partially opened gates is far more complicated as it is closely related to flow conditions. Although it is tempting to express these using various coefficients of discharge and contraction, this approach is acceptable only in the simplest of cases (see Worked example 6.1), but hardly sufficient in the final design, where a full analysis of the flow conditions and factors influencing them is really necessary.

The most important factors on which flow over and under the gates will depend are the geometry of the upstream and downstream water passage, the geometry and position of the gate and its accessories such as seals, supports, etc., the upstream head, whether the flow through the gate is free with atmospheric pressure downstream of it or submerged, the flow Froude and Reynolds – and possibly even Weber – numbers, the degree of turbulence of the incoming flow and aeration of the space downstream of the gates. A detailed treatment of the subject of the prediction of flow conditions and related hydrodynamic forces acting on gates is outside the scope of this book, but can be found, for example, in Naudascher (1987, 1991). For an example of conditions in case of wicket gates with a partial opening, see de Bejar and Hall (1998).

Generally the gate design is a difficult area partly because of the complexity of hydraulic conditions indicated above and partly because the design has to satisfy conflicting demands: vibration damping may conflict with keeping the forces for the gate operation to a minimum; the need to avoid vibrations may conflict with the optimum shape and strength required by the flow conditions and loading; the optimum shape of the edges and seals may conflict with water tightness of the closed gate, etc.

It is often useful to carry out a potential flow analysis and use this for the determination of the hydrodynamic force (Rouse, 1950), which (in some simpler cases) can also be obtained by using the momentum equation (see Worked example 6.1).

For gates with *overflow* the discharge coefficient can be expressed by the same equation as for an overfall spillway, i.e.

$$C_d = Q \Big/ \frac{2}{3}(2g)^{1/2} b h^{3/2} \tag{6.2}$$

where h is the head of the upstream water level above the crest (or top edge) of the gate. The total head, H, can also be used in equation (6.2) with an appropriate change of C_d (for dam crest gates $H \simeq h$). The coefficient will depend largely on the geometry of the overflow and the amount of air which has access to ventilate the nappe. Very broadly, the range of C_d is similar to that for overfall spillways, i.e. $0.55 < C_d < 0.8$; for further details see, for example, Hager (1988) and Naudascher (1987, 1991).

For gates with *underflow* the discharge and flow field will depend primarily whether the outflow is *free* (modular) or *drowned*. For free flow the form of equation (4.21b) may be used:

$$Q = C_d b a (2gH)^{1/2} \tag{6.3}$$

(a is the gate opening and H the upstream head), where the coefficients of discharge C_d and contraction C_c are related by

$$C_d = C_c \Big/ \left(1 + C_c \frac{a}{H}\right)^{1/2}. \qquad (6.4)$$

For a vertical gate the coefficient of contraction C_c varies slightly between 0.6 and 0.61 (Henderson, 1966) with free flow conditions generally for $y_3/y_1 < 0.5$ (see Figs 6.9 and 6.10). For non-modular flow the conditions indicated in Fig. 6.9 apply; equation (6.3) may still be used but the

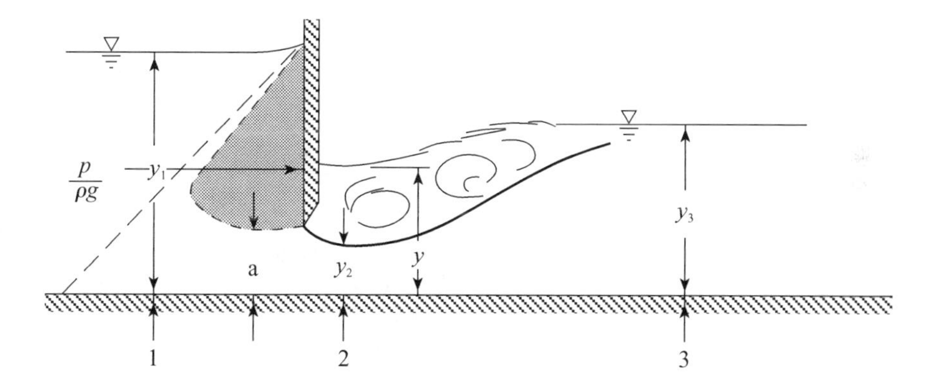

Fig. 6.9 Flow under a vertical gate

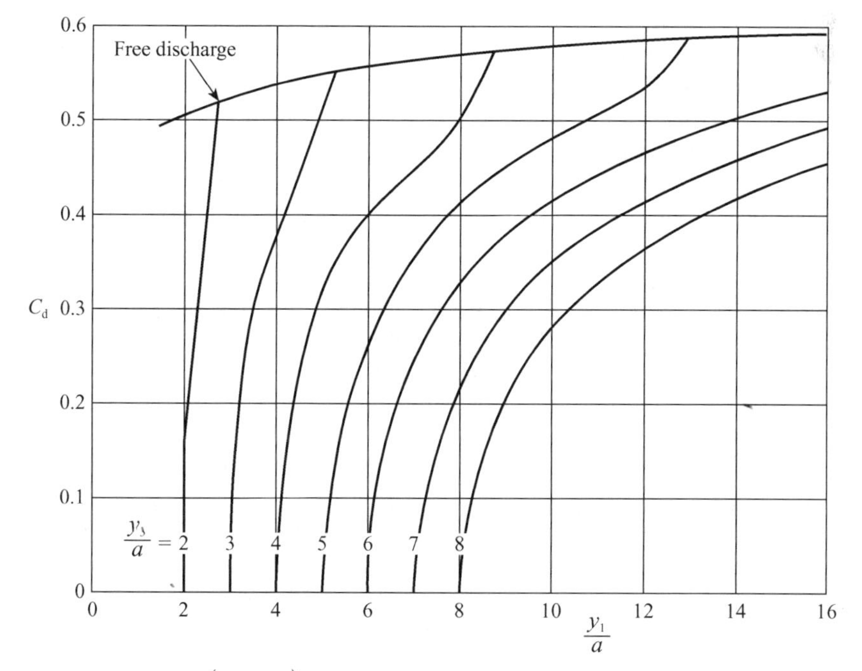

Fig. 6.10 $C_d = \phi\left(\frac{y_1}{a}, \frac{y_3}{a}\right)$

discharge coefficient will depend on the gate opening and upstream and downstream water depths – for guidance see Fig. 6.10 (e.g. Henderson, 1966, see also Worked example 6.1). For other types of gates and further particulars, see Hager (1988) and Naudascher (1987, 1991).

For gates with planar and cylindrically shaped skin plates (Tainter gates) with free flow and a downstream horizontal apron and the outflow edge of the gate inclined at θ to the flow ($\theta < 90°$ and varies according to the gate position) Toch (1955) suggested for C_c the equation

$$C_c = 1 - 0.75\frac{\theta}{90} + 0.36\left(\frac{\theta}{90}\right)^2. \tag{6.5}$$

6.5.2 Forces on high-head gates

In closed positions hydrostatic forces determined by standard procedures will again apply. For vertical lift gates, the hoist lifting force has to be dimensioned to overcome the gate weight, frictional resistance and, most importantly, the *downpull forces* resulting from the fact that during the gate operation the pressure along the bottom edge of the gate is reduced (to atmospheric or even smaller pressures), whereas the pressure acting on the top of the gate is practically the same as under static conditions (i.e. full reservoir pressure). This condition applies both to gates located within a conduit or at the upstream face of the dam or an intake (it should be noted that the seals for the gates at the intakes have to be on the downstream side, and for gates in conduits are usually in the same position). The downpull (or uplift) forces acting on a gate with a given housing and seal geometry and the gate vibrations have to be analysed at various operating conditions using theoretical considerations and experience from field trials, as well as model experiments, if necessary.

The lifting force of a gate will also be influenced by the geometry of the bottom edge and the seal. In order to avoid negative pressures and downpull there should be no separation of the flow until the downstream edge of the gate is reached. By investigating the shape of a jet flowing under a sharp-edged plate opening by an amount a under a head h, Smetana (1953) determined the shapes of outlet jet surface curves for various h/a ratios in the same manner as shapes for overfall spillway crests were determined (Section 4.7). For $h/a > 3.3$ this curve had a constant shape, while for ratios $h/a < 3.3$ the curves were flatter and of varying shapes. By forming the bottom edge of the gate according to these curves it is possible to avoid separation of flow, increased downpull, vibration, and resulting instabilities (for further details, see Novak and Čábelka (1981)).

Forces acting on an emergency gate closing into the flow (with the downstream gate/valve open and failing to close) can be particularly complicated with the shape and size of the passage between the two gates playing an important part.

6.6 Cavitation, aeration, vibration of gates

Cavitation, air demand and vibration of gates are closely connected and influence each other.

For a general discussion of *cavitation* see Section 4.6; some instances of cavitation at gates and valves also have been mentioned in previous sections. In principle cavitation danger exists in any situation where there is flow separation without sufficient air supply, e.g. in case of flow under a gate in a high-head conduit or downstream of gate slots (see Section 6.2.2).

The determination of the *air demand* and the associated design of *air vents* is important both for crest gates with an overfalling nappe and for high-head gates and their downstream conduits.

For crest gates Novak and Čábelka (1981) quote for the air demand the equation proposed by Hickox, giving an air vent diameter D:

$$D = C\frac{H^{1.82}b^{0.5}}{p^{0.82}} \tag{6.6}$$

where for a flow over the gate only $C = 0.004$ for all parameters in metres (p is the negative head underneath the overfall jet), or the procedure of Wisner and Mitric, which takes into account the length of contact of the overfall jet with air and which, for $p/H > 0.1$, gives about the same results as equation (6.6). The air demand both for flow in a conduit with a hydraulic jump and for flow under pressure downstream of the jump has been discussed in Section 4.8.

For air vents of conduit gates, Smith (1978) quotes the equation of Campbell and Guyton for the ratio of air and water discharge β:

$$\beta = 0.04(Fr - 1)^{0.85} \tag{6.7}$$

where Fr refers to the flow just downstream of the gate. In using equation (6.7) it is assumed that the maximum air demand occurs with the conduit flowing half full. When dimensioning the air vent the air velocity should not exceed $45\,\mathrm{m\,s^{-1}}$ and the pressure drop should be limited to 2m (of water).

Using the coefficient 0.015 in equation (4.71) gives identical results to equation (6.7) for Froude numbers about 6.

Aydin (2002) after investigating on a model the air demand required to keep the pressure drop downstream of a high-head gate during emergency closure to relatively insignificant values concludes that a ratio of the effective air shaft length (i.e. accounting also for head losses) to the conduit diameter should be less than 40.

Vibration of gates, if it occurs, can be dangerous – seal leakage, intermittent flow attachment, inadequate venting being the main causes. Some mention of vibration and flow instabilities already has been made in previous sections.

In principle the fluctuating components of hydrodynamic forces can be induced by different excitation mechanisms and are closely related to the structural vibrations which they in turn may induce. The main categories of these excitation mechanisms are *extraneously induced excitation* (EIE) (e.g. pressure fluctuations due to the turbulent flow downstream of a control gate and impinging on the gate), *instability induced excitation* (IIE) (e.g. flow instabilities due to vortex shedding), *movement induced excitation* (MIE) (e.g. self-excitation due to the fluctuation of a gate seal) and *excitation due to (resonating) fluid oscillator* (EFO) (e.g. surging water mass in a shaft downstream of a tunnel gate). Flow induced vibrations can be forced by turbulence or flow structure – see EIE and IIE above; they can also be amplified by body vibrations synchronizing with random turbulence excitation or, the flow excitation is purely induced by the body vibration itself – *self-excitation* or *negative damping* – see MIE and EFO above. In the design of gates it is desirable for their excitation frequencies to be remote from resonance frequencies (unless there is high damping) and negative damping (self-excitation) should be avoided.

Further treatment of the subject involving the computation of flow induced gate vibrations including the bulk modulus of water, Young's modulus of the gate material, the excitation and natural frequencies as well as all relevant hydrodynamic parameters is beyond the scope of this text; for this and further details see Kolkman (1984), Naudascher and Rockwell (1994), ICOLD (1996) and Lewin (2001). For a brief mention of physical modelling of hydraulic structures vibrations see Chapter 16.

6.7 Automation, control and reliability

There are two aspects of *automation* in connection with dam crest gates: gates that move automatically according to reservoir level and installations with automatic control of gate movement. In the former category are, e.g. drum and sector gates, some types of fabric and flap gates and crest gates using fluid dynamics systems (Townshend, 2000) (gates with buoyancy tanks connected by ducted radial arms to an upstream axle). Fusegates (see Section 6.2.8) can also be classed in this category. Automatic gates are advantageous in remote locations but are not restricted to them.

In *control* systems electromechanical controls have largely been replaced by electronic closed-loop systems. Programmable logic control (PLC) is now general with duplicated controllers for the control of critical functions. Di Stefano *et al.* (1976) deal comprehensively with feedback and control systems and Lewin (2001) illustrates the application of such systems to automatic methods of gate control. The requirements for *remote control* are that two independent means of communication are available to transmit key information determining the operation of spillway gates and for monitoring the condition of essential equipment.

The *reliability* requirement of gates must be consistent with their safety role and the overall dam safety. Various aspects of design flood selection, freeboard and flood routing dealt with in Chapter 4 and dam safety instrumentation and surveillance dealt with in Chapter 7 are here of relevance. Some of the general reliability principles for the gates are that the equipment must be able to perform its duty throughout its operational life, no single random failure of operating equipment should prevent the gate operation and without adding to the system complexity failure of any kind is revealed to the site staff. Redundancy of equipment should be incorporated into the flood discharge systems to achieve a defence in depth against potential equipment failure.

Worked Example 6.1

A vertical lift gate discharges into a canal; the upstream depth is 3.0m and the downstream depth is (i) 0.5m and (ii) 2.10m. Estimate the discharge and the force acting on the gate in case (i) and the force in case (ii) assuming the discharge remains constant (as in case (i)).

(i) As $y_3/y_1 \ll 0.5$ free flow conditions will apply; assuming the coefficient of contraction $C_c = 0.605$ the gate opening is $a = 0.5/0.605 = 0.826$ (see Fig. 6.9).

$$C_d = \frac{C_c}{\left(1 + C_c \dfrac{a}{H}\right)^{1/2}} = \frac{0.605}{\left(1 + 0.605 \dfrac{0.826}{3}\right)^{1/2}} = 0.56 \text{ (see equation (6.4)}$$

(neglecting the upstream velocity head, i.e. $H = y_1$)

$$q = C_d\, a\, (2gy_1)^{1/2} = 0.56 \times 0.826\, (19.6 \times 3)^{1/2} = 3.547\,\text{m}^2\,\text{s}^{-1}.$$

From the momentum equation (assuming the Boussinesq coefficient $\beta = 1$)

$$\rho g\left(\frac{1}{2}y_1^2 - \frac{1}{2}y_3^2\right) - P = \rho q(V_3 - V_1)$$

$$9.8 \times 10^3(3^2/2 - 0.5^2/2) - P = 3.547(3.547/0.5 - 3.547/3) \times 10^3 (= 20.97 \times 10^3)$$

$$P = (42.92 - 20.97)10^3 = 21.95\,\text{kN}\,\text{m}^{-1}.$$

For comparison the hydrostatic force on the gate is

$$P = (3 - 0.826)^2 \rho g/2 = 23.15\,\mathrm{kN\,m^{-1}}.$$

(ii) *First trial*
Referring to Fig. 6.9 assume gate opening $a = 1.0\,\mathrm{m}$

$$y_2 = 0.605 \times 1.0 = 0.605\,\mathrm{m}$$

$$V_2 = 3.547/0.605 = 5.86\,\mathrm{m\,s^{-1}};\ Fr_2 = V_2/(gy_2)^{1/2} = 2.40$$

conjugate depth to y_2

$$y_3^1 = \frac{y_2}{2}\left(-1 + \sqrt{1 + 8Fr_2^2}\right) = \frac{0.605}{2}\left(-1 + \sqrt{1 + 8 \times 2.4^2}\right) = 1.77\,\mathrm{m}.$$

As $y_3 = 2.1 > y_3^1$ $(= 1.77)$ submerged flow conditions exist at the gate.
Assuming no energy less between sections 1 and 2 from the energy equation:

$$y_1 + V_1^2/(2g) = y + V_2^2/(2g)$$

$$3 + 3.547^2/(19.6 \times 3^2) = y + 5.86^2/19.6 \quad y = 1.32\,\mathrm{m}.$$

From the momentum equation between sections 2 and 3:

$$y^2/2 + qV_2/g = y_3^2/2 + qV_3/g$$

$$1.32^2/2 + 3.547 \times 5.86/9.81 = 2.1^2/2 + 3.547^2/9.81 \times 2.1$$

$$2.99 \neq 2.816$$

Second trial
Assume gate opening $a = 0.92\,\mathrm{m}$

$$y_2 = 0.605 \times 0.92 = 0.557\,\mathrm{m};\ V_2 = 6.37\,\mathrm{m\,s};\ Fr_2 = 2.73;\ y_3^1 = 1.89\,\mathrm{m}\ (<2.1).$$

The same procedure as in the first trial results in $y = 1.0\,\mathrm{m}$ and

$$y^2/2 + qV_2/g = 2.803 \cong 2.816.$$

Therefore the gate opening is $a = 0.925\,\mathrm{m}$.
Computing C_d from q for this gate opening:

$$C_d = \frac{q}{a\sqrt{2gy_1}} = \frac{3.547}{0.925(19.6 \times 3)^{1/2}} = 0.5.$$

Referring to Fig. 6.9 for $y_1/a = 3/0.925 = 3.24$ and $y_3/a = 2.1/0.925 = 2.27$. C_d can be estimated also as about 0.5; (this procedure could be used for computing q from given values of y_1, y_3 and a).

For estimating the force acting on the gate we use again the momentum equation between sections 1 and 2 (Fig. 6.9).

$$\rho g\left(\frac{1}{2}y_1^2 - \frac{1}{2}y^2\right) - P = \rho q(V_2 - V_1)$$

$$9.8 \times 10^3\left(\frac{3^2}{2} - \frac{1}{2}\right) - P = 10^3 \times 3.547(6.37 - 1.18) = 18.41 \times 10^3$$

$$P = 20.83\,\text{kN}\,\text{m}^{-1}.$$

The above is, of course, only a rough estimate as the hydrodynamic pressure on the gate can fluctuate considerably because of the pulsating pressure on its downstream side (due to the submerged hydraulic jump).

References

Aydin, I. (2002) Air demand behind high-head gates during emergency closure, *J. of Hydraulic Research*, 40, No. 1: 83–93.

Ball, J.W. (1959) Hydraulic characteristics of gate slots. *Proceedings of the American Society of Civil Engineers, Journal of the Hydraulics Division*, 85 (HY 10): 81–140.

Bandarin, F. (1994) The Venice project: a challenge for modern engineering, *Proceedings of Institution of Civil Engineers, Civil Engineering*, 102, Nov.: 163–74.

Béjar, L.A. de and Hall, R.L. (1998) Forces on edge-hinged panels in gradually varied flow. *Journal of Hydraulic Engineering*, ASCE, 124, No. 8: 813–21.

Brouwer, H. (1988) Replacing wooden flashboards with hydraulically operated steel flap gates, *Hydro Review*, June: 42–50.

Buzzel, D.A. (1957) Trends in hydraulic gate design. *Transactions of the American Society of Civil Engineers*, Paper 2908.

Di Stefano, J.J., Stubbard, A.R. and Williams, I.J. (1976), *Feedback and Control Systems*, McGraw-Hill, New York.

Erbisti, P.C.F. (2001) *Design of Hydraulic Gates*, A.A. Balkema, Rotterdam.

Hager, W.H. (1988) Discharge characteristics of gated standard spillways. *Water Power & Dam Construction*, 40 (1): 15–26.

Haindl, K. (1984) Aeration at hydraulic structures, in *Developments in Hydraulic Engineering*, Vol. 2 (ed. P. Novak), Elsevier Applied Science, London.

Henderson, F.M. (1966) *Open Channel Flow*, Macmillan, New York.

ICOLD (1996) *Vibrations of Hydraulic Equipment for Dams*, Bulletin 102, International Commission on Large Dams, Paris.

Kocahan, H. (2004) They might be giants, *Int. Water Power and Dam Construction*, March, 32–5.
Kolkman, P. (1984) Gate vibrations, in *Developments in Hydraulic Engineering*, Vol. 2 (ed. P. Novak), Elsevier Applied Science, London.
Leliavsky, S. (1981) *Weirs*, Design Handbooks in Civil Engineering, Vol. 5, Chapman & Hall, London.
Lemperière, F. (1992) Overspill fusegates. *Water Power & Dam Construction*, 44 (July): 47–8.
Lewin, J. (2001) *Hydraulic Gates and Valves in Free Surface Flow and Submerged Outlets*, 2nd ed, Thomas Telford, London.
—— (2006) Personal communication.
Lewin, J. and Scotti, A. (1990) The flood prevention scheme of Venice, experimental module, *J. Int. Water and Environmental Management*, 4, No. 1: 70–6.
Lewin, J. and Eprim, Y. (2004) *Venice Flood Barriers*, paper to the British Dam Society, ICE, London.
Naudascher, E. (1987) *Hydraulik der Gerinne und Gerinnebauwerke*, Springer, Vienna.
—— (1991) *Hydrodynamic Forces*, IAHR Hydraulic Structures Design Manual, Vol. 3, Balkema, Rotterdam.
Naudascher, E. and Rockwell, D. (1994) *Flow Induced Vibrations*, IAHR Hydraulic Structures Design Manual, Vol. 7, Balkema, Rotterdam.
Novak, P. and Čábelka, J. (1981) *Models in Hydraulic Engineering: Physical Principles and Design Applications*, Pitman, London.
Norgrove, W.B. (1995) River Tees barrage – a catalyst for urban regeneration, *Proceedings of Institution of Civil Engineers, Civil Engineering*, 108, August: 98–110.
Rodionov, V.B., Lunatsi, M.E. and Rayssiguier, J. (2006) Small Hydro in Russia: Finding a spillway gate to withstand harsh winters, *HRW*, 14, No. 1: 34–9.
Rouse, H. (1950) *Engineering Hydraulics*, Wiley, New York.
Smetana, J. (1953) *Outflow Jet from Under a Gate*, ČSAV, Prague (in Czech).
Smith, C.D. (1978) *Hydraulic Structures*, University of Saskatchewan.
Thomas, H.H. (1976) *The Engineering of Large Dams*, Wiley, London.
Toch, A. (1955) Discharge characteristics of Tainter gates. *Transactions of the American Society of Civil Engineers*, 120: 290.
Townshend, P.D. (2000) Towards total acceptance of full automated gates, in Tedd, P. (ed.) *Dams 2000*, Thomas Telford, London.
US Army Waterways Experimental Station (1959) *Hydraulic Design Criteria*, US Department of the Army, Washington, DC.
Zipparro, V.J. and Hasen, H. (1993) *Davis' Handbook of Applied Hydraulics*, 4th edn, McGraw-Hill, New York.

Chapter 7

Dam safety: instrumentation and surveillance

7.1 Introduction

7.1.1 A perspective on dam safety

Reservoirs constitute a potential hazard to downstream life and property. The floodplain at risk in the event of catastrophic breaching may be extensive, densely populated and of considerable economic importance. In such instances dam failure can result in unacceptable fatalities and economic damage.

Catastrophic failures involving large modern dams are rare, but major disasters at Malpasset (France 1959), Vaiont (Italy 1963), Teton (USA 1976) and Macchu II (India 1979) in particular had a seminal influence on all matters relating to dam safety. The four named disasters are briefly put into perspective in Table 7.1. Collectively with the numerous lesser failures of old or smaller dams occurring during the same period, those four disasters focused international attention on dam safety and surveillance to significant effect.

Surveys and statistical analyses of failures and other serious incidents have been published by the International Commission on Large Dams (ICOLD, 1995). Jansen (1980) provides a more detailed review of a number of major dam disasters.

Catastrophic failure of a dam, other than as the direct result of an extreme flood event, is invariably preceded by a period of progressively increasing 'structural' distress within the dam and/or its foundation. Dam surveillance programmes and instrumentation are intended to detect symptoms of distress and, where possible, to relate those symptoms to specific problems at the earliest possible stage.

Instruments strategically placed within or on a dam are not of themselves a guarantee against serious incident or failure. Their prime function

Table 7.1 Selected major dam disasters 1959–1993

Dam (year of event) (Country)	*Type, height (m) (year of completion)*	*Event*
Malpasset (1959) *(France)*	Arch, 61 m (1954)	Foundation failure and abutment yield; total collapse (421 killed)
Vaiont reservoir (1963) *(Italy)*	Arch, 262 m (1960)	Rockslide ($200 \times 10^6 m^3$) into reservoir; splash-wave 110 m over dam crest; dam intact (*c.* 2000 killed)
Teton (1976) *(USA)*	Embankment, 93 m (completing)	Internal erosion from poor cut-off trench design; total destruction (11 killed, damage $\$500 \times 10^6$)
Macchu II (1979) *(India)*	Embankment and gravity, 26 m (1972)	Catastrophic flood; gate malfunction; overtopping and embankment washout (estimated 2000 killed)
Tous (1982) *(Spain)*	Embankment, 127 m (1979)	Failed by overtopping during extreme flood event (16 killed, *c.* $\$500 \times 10^6$ damage)
Gouhou (1993) *(China)*	Embankment, 70 m (1985)	Failure of upstream deck leading to local instability and erosion, with rapid breaching (*c.* 400 killed)

is to reveal abnormalities in behaviour, and so to provide early warning of possible distress which may have the potential to develop into a serious incident or failure. Numbers of instruments installed are of less importance than the selection of appropriate equipment, its proper installation at critical locations, and intelligent interpretation of the resulting data within an overall surveillance programme. The effectiveness of the latter is determined by many factors, including the legislative and administrative framework within which procedures and responsibilities have been established.

The objective of this chapter is to identify parameters of primary significance to the integrity of dams, and to outline the instrumentation and techniques employed in surveillance. The chapter includes a review of alternative philosophies in reservoir safety legislation and of organizational provisions for surveillance and inspection of dams. Recent publications by the International Commission on Large Dams (ICOLD, 1987, 1988, 1992) provide guidelines on dam safety and monitoring. Issues regarding dam safety are regularly addressed at ICOLD Congresses and in principal journals.

7.1.2 Tailings dams and lagoons

A particular hazard is represented by tailings dams and storage lagoons. International concern regarding the safety of such facilities has been heightened as a result of a number of serious failures, e.g. Stava (Italy (1985)), Aznalćollar (Spain (1998)) and Baia Mare (Romania (2000)).

The Stava failure released $0.19 \times 10^6\,m^3$ of tailings slurry from a fluoride mine and led to 269 deaths (Berti *et al.*, 1988). Two embankments in series, the upper one 29 m high and the lower 19 m, collapsed in cascade, the slurry travelling up to 4 km. The initiating mechanism was rotational instability of the upper dam following raising of the phreatic surface as a result of problems with the decanting culvert.

Aznalćollar, with a height of 28 m at time of failure, released some $7 \times 10^6\,m^3$ of acidic tailings which inundated 2.6×10^3 ha of farmland, causing very considerable environmental and economic damage (Olalla and Cuellar, 2001). Baia Mare, newly constructed with tailings from an old impoundment employed as fill, had a surface area of *c.* 80 ha. The retaining embankment failed by local overtopping, and the cyanide plume following the release of *c.* $1 \times 10^5\,m^3$ of contaminated liquid was detectable four weeks later some 1500 km away at the mouth of the Danube (Penman, 2002).

The potential hazard associated with tailings dams and lagoons is raised very significantly by several factors which distinguish them from conventional embankment dams and reservoirs (see also Section 2.11):

- the risk of tailings dams breaching may be much enhanced by inadequate levels of supervision and surveillance;
- the environmental and economic consequences of serious breaching are frequently heightened by the toxic nature of the outflow;
- a tailings dam or lagoon may be discontinued and/or abandoned, and left unsupervised, following the cessation of active operation.

The management of tailings dams in the UK and the legislation applicable to their construction and operation are discussed in DoE (1991a), and an international perspective on the safety of tailings storage dams has been published in ICOLD (2001).

7.2 Instrumentation

7.2.1 Application and objectives

The provision of monitoring instruments is accepted good practice for all new dams of any magnitude. In parallel with this, a basic level of instrumentation is now frequently installed retrospectively to monitor existing dams.

In the context of new dams, instrumentation data are interpreted in a dual rôle: to provide some confirmation of the validity of design assumptions and also to establish an initial datum pattern of performance against which subsequent observations can be assessed. With existing dams, particularly elderly or less adequate structures, suites of instruments may be installed to provide a measure of reassurance. In such instances they serve to detect significant and abnormal deviations in the long-term behaviour of the dam. In other applications to existing dams instruments may be required to record specific parameters of behaviour in response to an acknowledged or suspected design deficiency or behavioural problem.

The scope and degree of sophistication of individual suites of instruments varies greatly. Careful attention to specification, design, and correct installation of all but the simplest instrument arrays is critical to their satisfactory performance. Responsibility for the planning and commissioning of monitoring installations is therefore best retained at an experienced and relatively senior level within an appropriate organization, e.g. the design agency, owner, or state authority.

Suites of instruments may, for convenience, be classified according to the primary function of the installation.

1. *Construction control:* verification of critical design parameters with immediate looped feedback to design and construction.
2. *Post-construction performance:* validation of design; determination of initial or datum behavioural pattern.
3. *Service performance/surveillance:* reassurance as to structural adequacy; detection of regressive change in established behavioural pattern; investigation of identified or suspected problems.
4. *Research/development:* academic research; equipment proving and development.

Areas of possible overlap between certain of the functional classifications are self-evident.

For construction control or research purposes absolute values of specific parameters and the observed trend in those parameters may be of equal importance. This is less the case when the primary function of the instrument array is to monitor long-term performance. Absolute values may then be considered of secondary importance to the early detection of changes and deviant trends in behavioural pattern which are not attributable with absolute certainty to an observed change in loading régime, e.g. in retained water level or other identifiable influence. Detection and analysis of change in a previously established 'normal' cycle or pattern of behaviour is the essential justification for all suites of instruments installed primarily for surveillance purposes.

7.2.2 Parameters in monitoring dam behaviour

The most significant parameters in monitoring dam behaviour are as follows:

1. seepage and leakage (quantity, nature (e.g. turbidity), location and source);
2. settlement and loss of freeboard in embankments (magnitude, rate);
3. external and internal deformation (magnitude, rate, location);
4. porewater pressures and uplift (magnitude, variation);

Certain key parameters are of primary concern regardless of the type of dam considered, e.g. seepage and external movement or deflection; others are relevant essentially to a specific type of dam, e.g. porewater pressures in relation to earthfill embankments. The relative significance of individual parameters may also reflect the nature of a problem under investigation, e.g. the settlement of an old embankment, where progressive deformation is suspected.

The desirable minimum provisions for monitoring and surveillance on all dams, to be installed retrospectively if necessary, should account for the measurement of seepage flows and crest deformations. The latter provision is of particular importance in relation to detecting settlement on embankment crests as an indicator of possible internal distress and of local loss of freeboard.

The supreme importance of seepage flow as a parameter cannot be overstressed. Regular monitoring should be standard practice for all but the smallest of dams. Serious problems are invariably preceded by a detectable change in the seepage regime through or under the dam which is unrelated to changes in the retained water level or to percolation of precipitation falling on the downstream slope. Direct observation of the seepage quantity and turbidity is relatively simple, with internal drain systems conducted to calibrated V-notch weirs. Ideally, a number of weirs are each positioned to collect the flow from specific lengths of the dam, permitting identification of the approximate location of any change in the seepage régime.

The principal parameters are set out in greater detail in Table 7.2, with identification of the instruments employed. The table also lists indicative examples of defects or problems which may be associated with abnormally high values or significant trends in each parameter.

7.2.3 Instruments: design principles

Monitoring instruments are required to function satisfactorily under very harsh environmental conditions and for essentially indeterminate periods

Table 7.2 Primary monitoring parameters and their relationship to possible defects

Parameter	*Instruments*	*Measurement*	*Illustrative defect*	*Dam type*
Seepage	Drains–underdrains to V-notch weirs (ideally several, 'isolating' sections of dam–foundation)	Seepage flow quantity, and nature of seepage water, e.g. clear or turbid	Could indicate initiation of cracking and/or internal erosion	E–C[a]
Porewater pressure (E); uplift (C)	Piezometers	Internal water pressure in earthfill	Leaking core, or incipient instability	E
		Internal water pressure in concrete or rock foundation	Instability, sliding	C
Collimation	Precise survey (optical or electronic)	Alignment	Movement or deformation	E–C
Settlement	Precise survey (surface)	Crest settlement	Tilting (C) or loss of freeboard (E), e.g. core subsidence, or foundation deformation	E–C
	Settlement gauges (internal)	Internal or relative settlement		
External deformation	Precise survey (surface) Photogrammetry (E), pendula or jointmeters (C)	Surface deflection	Local movement, instability	E–C
Internal deformation or strain: vertical horizontal	Inclinometers, strain gauges or duct tubes	Internal relative displacements	Incipient instability	E
Stress or pressure	Pressure cells	Total stress	Hydraulic fracture and internal erosion	E

a E = embankment dams, C = concrete dams.

of time, possibly several decades. As guidelines underwriting sound design it is therefore desirable that instruments be

1. as simple in concept as is consistent with their function,
2. robust and reliable,
3. durable under adverse environmental and operating conditions and
4. acceptable in terms of 'through-life' costs (i.e. the sum of purchase, installation, support and monitoring costs).

A sound principle is to retain the sophisticated and vulnerable sensing elements, e.g. electronic components and transducers, above ground level wherever possible. In such instances it may also be advantageous to make the above-ground elements readily transportable, e.g. by use of compact portable transducer units to monitor porewater pressures from piezometers. Additional advantages associated with the use of a transportable sensing element lie in greater physical security and avoidance of the need to construct large and costly instrument houses for fixed measuring equipment.

Instrument capability has developed significantly over recent years, and reliable and robust equipment is now readily available. More recent developments have tended to concentrate upon provision for automatic or semi-continuous interrogation of instruments, with a facility for storage and/or automatic transmission of data to a central location. The enhanced capability and complexity are reflected in high costs and a greater risk of component, and thus system, malfunction or failure. Such sophistication is therefore generally justifiable only in more exceptional circumstances. For most dams instrumentation at a relatively unsophisticated and basic level will prove adequate for routine monitoring and surveillance.

7.2.4 Instruments: types and operating principles

A brief review of the more common instruments is provided below to demonstrate important principles of operation and measurement. Reference should be made to Dunnicliff (1988) and Penman, Saxena and Sharma (1999), or manufacturers' literature, for comprehensive details of instruments, their operating principles and characteristics.

(a) Collimation, settlement and external deformation

Precise survey techniques, using optical or electronic distance measuring (EDM) equipment or lasers, are employed to determine the relative vertical and horizontal movement of securely established surface stations. Collimation and levelling to check crest alignment on concrete dams may be supplemented by tilt data from pendula installed in internal shafts, and from inverted floating pendula anchored at depth within boreholes in the foundation rock. The relative movement and tilt of adjacent monoliths may also be determined from simple mechanical or optical joint-meters. In the case of embankments, the longitudinal crest settlement profile is of prime importance. Embankment deformation data may also be derived from the use of sensitive borehole inclinometers to determine profiles of internal horizontal displacement incrementally through the height of a dam in both planes.

(b) Porewater pressures

Hydraulic piezometers are of two types: the simple Casagrande standpipe and the 'closed-circuit' twin-tube type. The Casagrande piezometer, Fig. 7.1, uses an electrical dipmeter to record the phreatic level; integrity of the

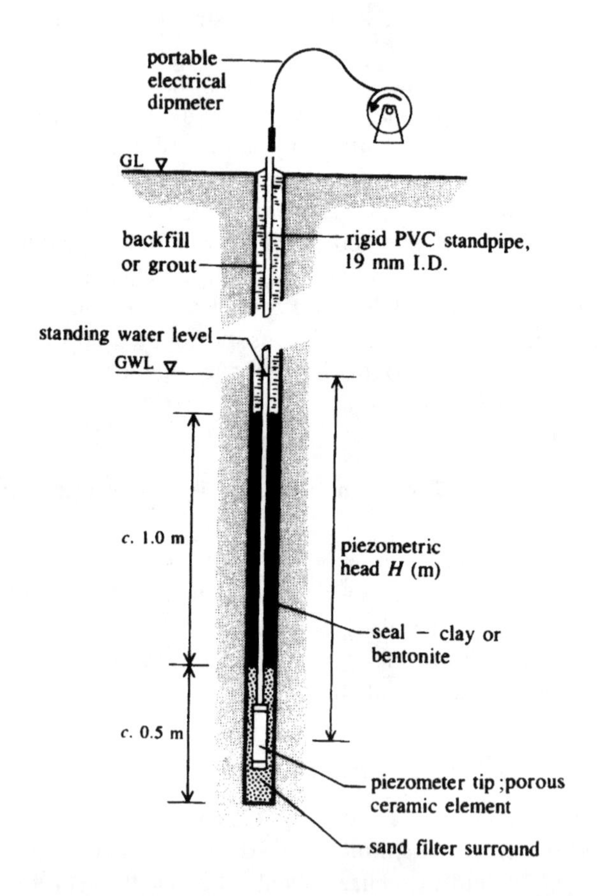

Fig. 7.1 Casagrande-type standpipe piezometer in borehole

clay seal is critical. The type is readily installed in existing dams and is suitable where the phreatic surface is sensibly static; its merits lie in simplicity of operation, reliability and low cost.

The closed-circuit hydraulic piezometer, best exemplified by the Bishop type instrument illustrated in Fig. 7.2, offers more rapid response to porewater pressure change. It is suited to lower-permeability soils and to non-saturated soils, and therefore to determining negative as well as positive porewater pressures in compacted earthfill. The twin 3 mm bore hydraulic leads are permanently filled with de-aired de-ionized water and can be laid for considerable distances (>200 m) to a suitably located instrument house, where measurement is effected by a pressure transducer or, in older installations, a mercury manometer. De-airing is required at intervals to flush out occlusions of air or water vapour entering the hydraulic leads from the fill or from the pore fluid. The use of a fine-pored (1μ nominal pore size) 'high air-entry value' ceramic element (Fig. 7.2) considerably reduces the frequency with which de-airing is required. Consideration must be given to the maximum elevation of the hydraulic leads relative to piezometer tip, readout point and anticipated porewater pressure range to avoid problems associated with negative pressures in the leads. The Bishop piezometer has proved effective, durable, and reliable, and has been widely installed in embankments during construction. It may also be employed to estimate *in-situ* permeability.

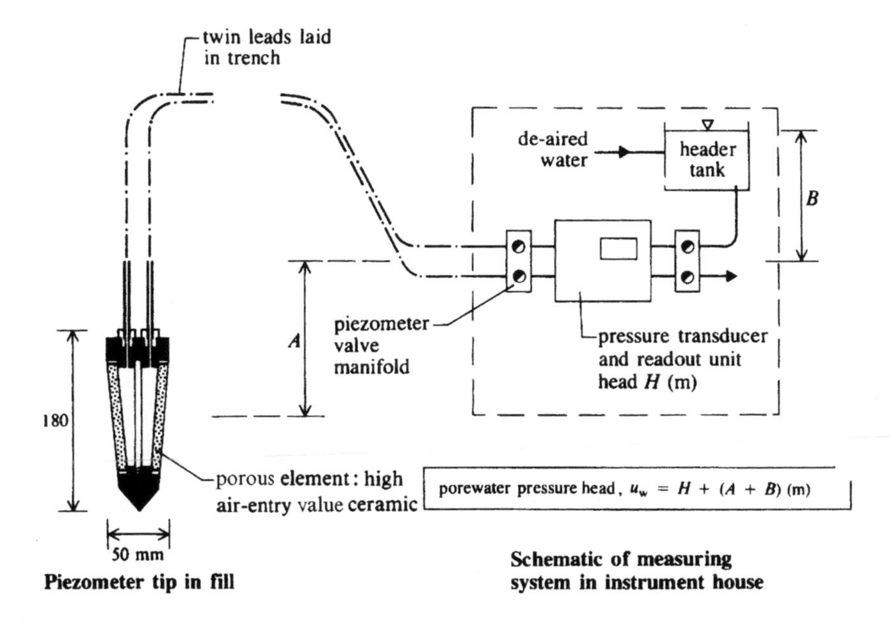

Fig. 7.2 Bishop-type twin-tube hydraulic piezometer

The electrical piezometer and the pneumatic piezometer offer rapid response to change in phreatic level, but may prove to be relatively expensive and less flexible in use than the hydraulic alternatives. They offer the advantage of requiring only a small terminal measuring chamber, rather than the costly instrument house generally required for twin-tube hydraulic instruments. Both are therefore suited to isolated installations of limited numbers of piezometers. The principle of a pneumatic piezometer tip is illustrated in Fig. 7.3. It functions by application of a known and controlled gas back-pressure to balance the porewater pressure operating on the diaphragm. The pressure balance is indicated by deflection of the diaphragm, allowing the gas to vent to a flow indicator and is confirmed at closure of the diaphragm when gas pressure is slowly reduced. The electrical piezometer, which is less common and is not illustrated, senses pressure via resistance strain gauges bonded to a steel diaphragm or by use of a vibrating wire strain gauge.

(c) Internal settlement and deformation

Vertical tube extensometer gauges with external annular plate magnet measuring stations located at vertical intervals of about 3 m can be installed during embankment fill construction. Most are based on the principles illustrated in Fig. 7.4. A probe on a calibrated tape is lowered down the tube to detect the plate magnet stations, allowing the determination of relative levels and increments of internal settlement to ±2 mm. The prin-

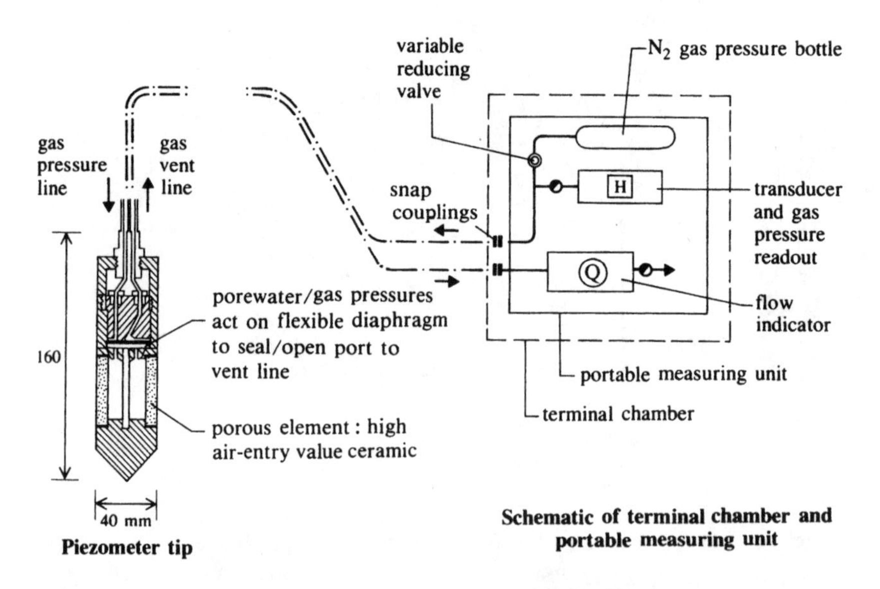

Fig. 7.3 Pneumatic piezometer

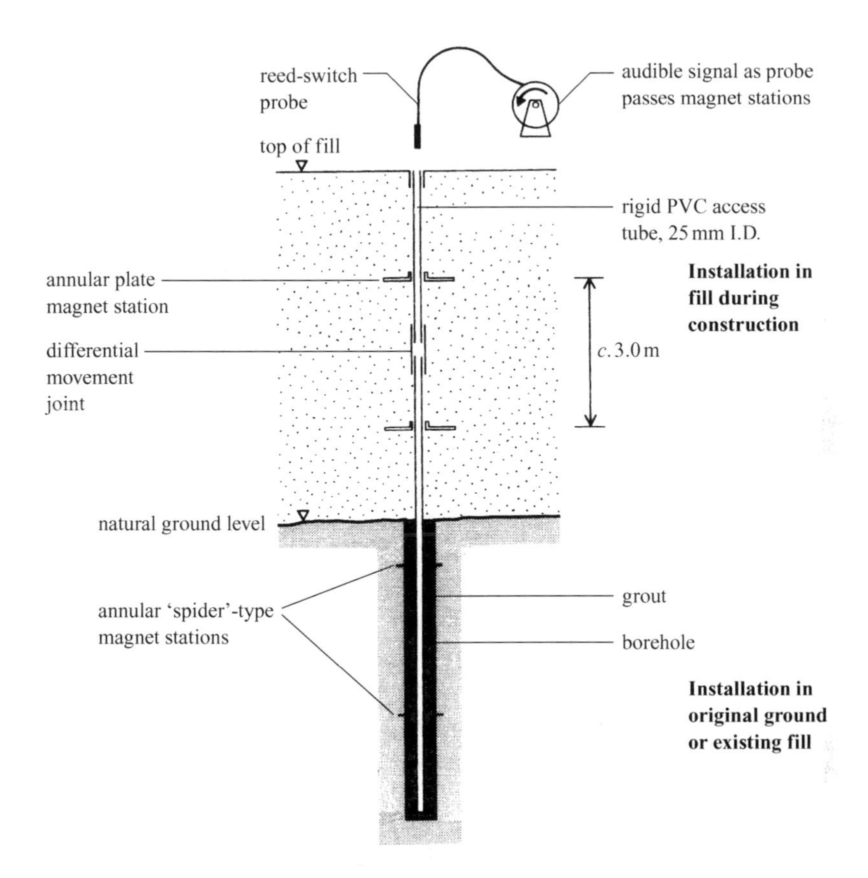

Fig. 7.4 Vertical settlement: magnet extensometer

ciple can be extended to existing fills and foundations as shown, using special 'snap-out' spider-type magnet stations designed for installation at suitable horizons within boreholes. Individual point-settlement gauges, generally operated on the manometer principle and illustrated in Fig. 7.5, are sometimes preferable for the more plastic fill materials, e.g. in cores.

Horizontal components of internal deformation may be determined by the borehole inclinometer referred to previously or by adaptation of the principle of the vertical extensometer gauge. In the latter instance, the tubes are laid in trenches within the shoulder fill, with the magnet stations or collars installed at suitable horizontal intervals. A special duct-motor device can then be employed to 'walk' the probe through the tube and back, determining the location of each station as it passes.

It is customary to provide fixed-length rock extensometers (not illustrated), to monitor the abutment response of arch dams. Similar extensometers may also be appropriate for other types of concrete dams if foundation deformability is a critical parameter.

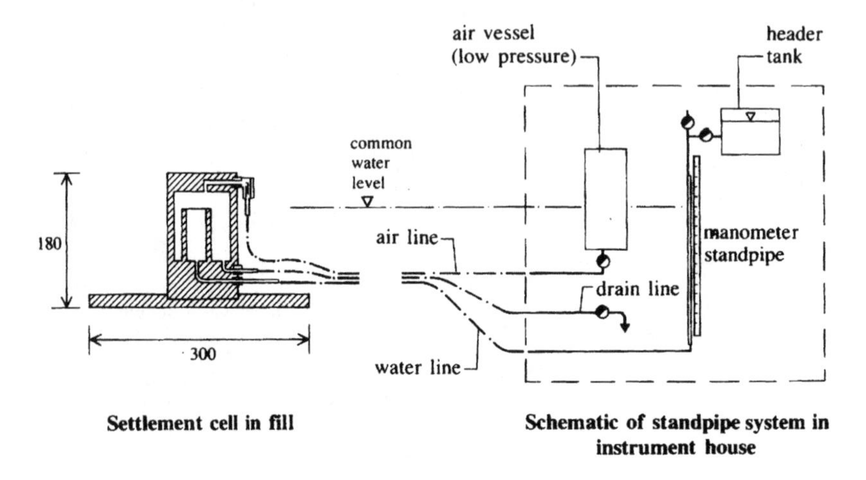

Fig. 7.5 Hydraulic point-settlement cell

(d) Internal stress

The direct determination of stress level in a continuum such as rock, concrete, or soil presents the most serious difficulties. Theory dictates the use of a disc-shaped sensing element of high aspect ratio. Material constraints render impossible the ideal of a sensor which exactly matches, at all stress levels, the stiffness of the continuum in which it is emplaced. The greatest circumspection is thus required in interpreting pressures or stresses recorded in the field. Commercially available devices for use in concrete and rock generally consist of an oil-filled metal disc linked to a pressure transducer which responds to fluid pressure as a function of external stress on the face of the disc. Similar devices are sometimes employed in embankment fills, an alternative being the hollow disc earth pressure cell, incorporating electrical strain gauges mounted on the internal faces of the cell.

Indirect determination of stress from measured strains, e.g. in mass concrete, is similarly fraught with serious difficulties associated with non-linear and time-dependent stress–strain response. Uncertainties associated with stress determination severely degrade the value of the recovered data, and the installation of field instrumentation for this purpose is seldom advisable. (Instrumentation of this nature is generally associated with research projects.)

7.2.5 Indirect investigation of leakage and seepage

It is frequently necessary to investigate and locate suspected leakage and seepage paths within the body of an embankment or through it's foundation. In circumstances where the situation is not deemed critical, and time permits a systematically planned investigation, piezometers may be installed to determine the local phreatic profile within a suspect location. Where time is considered to be critical, and/or there is no clear evidence as to the approximate location of a suspected seepage path, the use of fluorescein dyes or radioactive tracers may provide a quick method of determining entrance and exit points and hence notional seepage path. More sophisticated techniques which have proved successful in specific instances include monitoring temperature and/or resistivity profiles within an embankment (Johansson, 1996), and the use of fibre optics to monitor temperature gradients and leakage (Aufleger *et al.*, 2005). In the latter instance anomalies in the temperature regime can relate to the presence of internal seepage flows and may also highlight other negative aspects of behaviour.

7.2.6 Instrumentation planning

The planning and specification of a comprehensive suite of instruments involves a logical sequence of decisions:

1. definition of the primary purpose and objectives;
2. definition of observations appropriate to the dam considered;
3. determination of the locations and numbers of measuring points for the desired observations;
4. consideration of the time period to be spanned, i.e. long- or short-term monitoring;
5. consideration of the optimum sensing mode in relation to the desired rapidity of response, required accuracy, etc.;
6. selection of hardware appropriate to the task as defined under 1–5.

Step 3 is one of particular importance and sensitivity. Instruments must cover known critical features of the dam, but for purposes of comparison some should also be placed at locations where 'normal' behaviour may be anticipated. In the case of a new dam at least two sections should be instrumented, including the major section. It is good practice to draft an ideal instrumentation plan in the first instance, and then to progressively eliminate the less essential provisions until an adequate, balanced, and affordable plan is determined.

At the installation and setting-to-work stage success is dependent upon attention to detail. Points to be considered and resolved in advance include procedures for the commissioning and proving of the instruments, for the determination of 'datum' values and for the special training of monitoring personnel. Detailed consideration must also be given at this stage to data-handling procedures (Sections 7.2.6 and 7.3). It is advisable to consider instrumentation programmes in terms of the overall 'system' required, i.e. instruments, installation, commissioning, monitoring, and data management and interpretation.

An illustrative instrumentation profile for the major section of a new earthfill embankment is illustrated in Fig. 7.6. Comprehensive instrumentation programmes of this type are described in Millmore and McNicol (1983) and in Evans and Wilson (1992). A valuable review of embankment instrumentation experience in the UK is given by Penman and Kennard (1981); the rôle of instrumentation is discussed in Penman (1989), in Charles, Tedd and Watts (1992) and in Charles *et al.* (1996).

For new-construction dams presenting little danger to life in the event of breaching a modest level of instrumentation is generally adequate. As a minimum, provision should be made for monitoring seepage flow through the internal drainage system and, in the case of embankment dams, secure levelling stations should be established at intervals of 20 m or 25 m along the crest for monitoring settlement. Instrumentation at this level may also be appropriate on lesser dams, including small farm or amenity dams, if their location suggests a possible hazard to life in the event of breaching.

The retrospective instrumentation of existing embankment dams, in particular those considered to represent a significant potential hazard, is now common practice. The limitations on the type of instrument which can be installed retrospectively are self-evident, and the datum values against which subsequent change may be assessed now become those

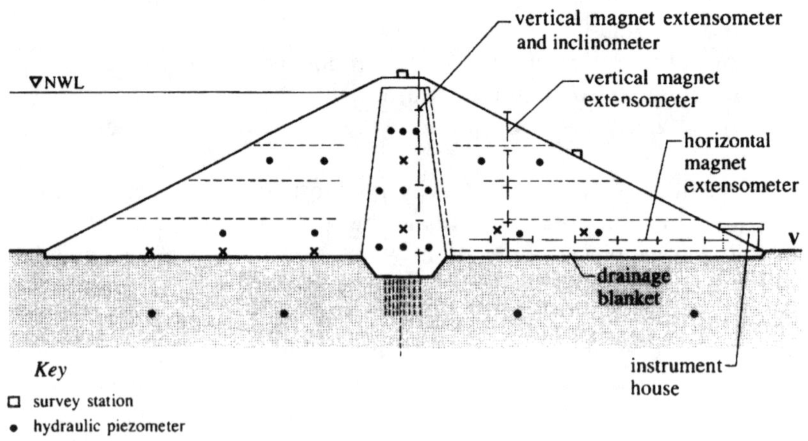

Fig. 7.6 Schematic instrument layout: major section of new embankment dam

existing at time of installation, possibly many years after completion of the dam. This renders interpretation of the significance of an observed change in behavioural pattern more difficult than in the case of a dam which has been routinely monitored since construction and first impounding. The desirable minimum installation for retrospective instrumentation is once again provision for monitoring seepage flows and, on embankments, crest settlement and deformation. It may in some instances be appropriate to make further provision for monitoring local piezometric and/or deformation profiles, e.g. where culverts and similar works run transversely through the body of the embankment.

Initial impounding and the first few years of operation represent the most critical phase for a new dam, as foundation and structure interact and progressively adjust to the imposed loadings. Failings in design or construction which impinge upon structural integrity and safety will generally become apparent at this early stage, given that an appropriate monitoring regime is in operation. First impounding should consequently take place at a controlled and modest rate, with the response of dam and foundation particularly closely monitored during filling and over the critical first few years.

The procurement and installation of all but the most basic level of instrumentation, i.e. provision for monitoring seepage and settlement, requires care in planning and execution. In the case of the more extensive instrumentation arrays common for larger dams it is always advisable to plan in consultation with the specialist manufacturers and suppliers. Considerable advantage is to be derived from entering an appropriate contractual arrangement with the selected provider to cover procurement, installation, setting-to-work and proving of the instrumentation. The contract may also be extended to include training of the technical staff who will subsequently take responsibility for in-service monitoring.

The level of instrumentation provided on embankment dams is almost invariably more comprehensive and more complex than that for concrete dams of similar size, where only seepage flows and alignment may be provided for. The instrumentation of embankment dams, from selection through installation to data processing, is discussed in the context of surveillance in Penman, Saxena and Sharma, 1999.

7.2.7 Data acquisition and management

Logical planning of data acquisition and processing is essential if the purpose of an instrumentation programme is to be fully realized. Unless observations are reliable and the information is interpreted quickly, the value of a programme will be severely diminished. Operating procedures must be carefully defined and the individual responsibilities of personnel clearly identified. Within the operating plan the frequency of monitoring should be determined on a rational basis, reflecting the objectives and the individual parameters under scrutiny. It is, in any event, subject to amendment in the context of the information retrieved.

Detailed prescription of periodicity is a question of common sense allied to engineering judgement. An excess of data will prove burdensome and may confuse important issues; too little information will raise more questions than it resolves. Excessive complexity in a system, whether in terms of equipment or the operating skills required, similarly diminishes its utility. A judicious balance is therefore always required, and care must be taken to ensure that the 'system' remains sufficiently responsive and flexible. The monitoring routine should provide for observations at the different seasons and with significant changes in retained water level. Representative monitoring frequencies for embankments may range as indicated in Table 7.3.

Routines for prompt processing of field data must be established, giving careful consideration to the optimum form of presentation. Charts and overlays are generally the most satisfactory method, with parameters plotted against retained water levels and precipitation as on Fig. 7.7. Illustrative schematic diagrams are shown in Johnston *et al.* (1999). It may sometimes prove useful to superimpose predetermined 'safe limit' envelopes for certain key parameters (e.g. porewater pressures) on such plots.

7.3 Surveillance

Dams of all types require regular surveillance if they are to be maintained in a safe and operationally efficient state. As with all structures, they are subject to a degree of long-term but progressive deterioration. Some of the latter may be superficial in relation to structural integrity, but the possibility of concealed and serious internal deterioration must be considered. Older dams will have been designed and constructed to standards which may no longer be considered adequate, e.g. in terms of their spillway discharge capacity or structural stability.

Table 7.3 Representative monitoring frequencies

Parameter	*Frequency*
Water level	Daily wherever possible
Seepage	Daily or weekly
Piezometers	Once or twice weekly (e.g. construction), extending to three to six monthly (routine)
Settlement–deformation	Daily (e.g. suspected serious slip), extending to three to six monthly (routine)

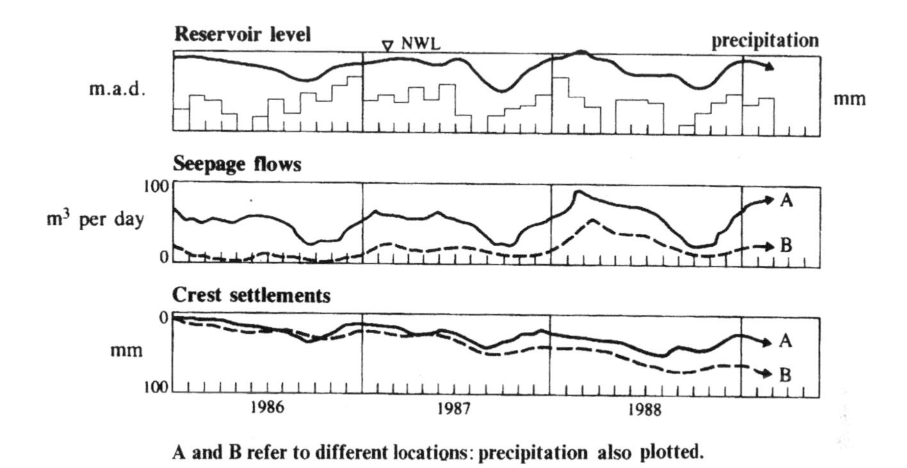

A and B refer to different locations: precipitation also plotted.

Fig. 7.7 Representative monitoring plots: embankment dam

The primary objective of a surveillance programme is to minimize the possibility of catastrophic failure of the dam by the timely detection of design inadequacies or regressive changes in behaviour. A further objective is to assist in the scheduling of routine maintenance or, when necessary, of major remedial works.

Surveillance embraces the regular and frequent observation and recording of all aspects of the service performance of a dam and its reservoir. It includes routine observation and inspection, the monitoring and assessment of seepage and instrumentation data and the recording of all other relevant information, including hydrological records. Less frequent but more rigorous statutory inspections by specialist engineers are also carried out as part of a comprehensive surveillance programme and may include a detailed investigation and complete reappraisal of the dam.

Daily external observation of a large dam may be considered to represent the ideal; in practice, problems of access or of availability of personnel may render weekly or even monthly observation a more realistic target. The surveillance procedures practised by two large UK dam-owners are discussed in Beak (1992) and Robertshaw and Dyke (1990).

Routine inspection should cover all readily accessible parts of the dam and of its associated components (e.g. spillways, gates, valves and outlet works). Visual inspection should also extend to the area downstream of the dam, including the mitres and abutments, and to any parts of the reservoir perimeter designated as requiring observation. Additional effort may be directed at particular locations or to specific signs of some possible deterioration, e.g. sources of suspected seepage or leakage and, in the case of embankments, localized crest settlement and slope deformation. The inspecting engineer must be alert to change, whether favourable or unfavourable, between successive visits.

The conduct of second-level inspections, i.e. the relatively infrequent but rigorous inspections carried out to comply with imposed statutory or other obligations, lies beyond the scope of this text. Inspection routines and procedures, including comprehensive checklists appropriate to different types of dam, are detailed in the SEED Manual (USBR, 1983), Johnston *et al.* (1999) and Kennard, Owens and Reader (1996).

7.4 Dam safety legislation

7.4.1 Legislative patterns

Legislation to cover the construction and safe long-term operation of dams has assumed greater importance as the number and size of dams at risk has steadily increased. The situation is dynamic rather than static, and most countries have introduced or reviewed national legislation in the wake of catastrophic disasters such as those identified in Table 7.1.

National legislation falls into one of two generic patterns. In the first, the legislation is precise and detailed, and is operated through some measure of direct state control. It may adopt a detailed, prescriptive approach to design standards, to construction and to inspection and surveillance. In such cases supervision and periodic inspection may be the responsibility of a state authority or inspectorate. The second pattern of legislation adopts a much less prescriptive approach. State involvement is minimal, with most responsibility devolved to nominated agencies or authorities. An example of the first pattern is provided by French legislation, while the second, but with responsibility placed in the hands of nominated individual engineers, is exemplified in the British approach outlined under Section 7.4.3.

Advantages and disadvantages are apparent in either approach. A centralized inspectorate or design bureau has the merit of ensuring uniformity of standards, but may prove to be insufficiently flexible and responsive to specific situations and to the peculiarities of individual dams. A system which minimizes state prescription and devolves responsibility is, however, open to the counter-criticism of possible variation in the standards applied. In reviewing legislation it is also important to appreciate that legislative provision does not, of itself, guarantee safety. Legislation is only as effective in operation as the associated administrative structure will allow. It must be considered in context with the national perspective regarding numbers, age, type, size and ownership of the dam population.

The applicability of any legislation must be clearly defined. This is generally prescribed in terms of a minimum reservoir storage capacity above natural ground level and/or a minimum height of dam. Legislation must also adequately prescribe responsibility for enforcement and for the proper supervision of design, construction and operation, and must detail arrangements for surveillance, periodic major inspections and abandonment.

7.4.2 Representative national approaches

Illustrative examples of legislative and administrative approaches are summarized below.

In France, the initial legislation of 1898 has been extended and revised at intervals, notably following the Malpasset disaster of 1959. A standing technical committee advises on the design and construction of major dams, and all significant dams are subject to annual inspection by a government department. The provisions are detailed and quite prescriptive, and apply to dams exceeding 20 m height.

In Switzerland, regulations originated in 1934 and have been modified at intervals subsequently. Application of the legislation is entrusted to a federal department operating in conjunction with consultants and military authorities.

The qualifying criteria for national legislation to apply in Italy are set at a height of 15 m or storage capacity of $1 \times 10^6 m^3$, with smaller dams subject to control by the regional authorities. Spanish dam safety legislation sets the same qualifying criteria at national level. Legislation across Europe is summarized and discussed in Charles and Wright (1996).

In the United States of America federal dams are subject to control through the appropriate government agency, e.g. the Corps of Engineers and the Bureau of Reclamation. Non-federal dams are the subject of state legislation, but the degree of control exercised at state level is very variable. A US national survey completed in 1983 identified 68 000 dams, of which approximately 40% were considered to present a significant potential hazard. Of 8818 dams subjected to further screening and inspection, 33% were classified as technically inadequate, with 2% categorized as 'emergency unsafe' (Duscha, 1983).

International guideline recommendations covering the safety of dams have been published (UNESCO, 1967). These provided a suggested framework for use in drafting appropriate national legislation. The recommendations favoured a national authority empowered to examine and approve all proposals to construct, enlarge or repair dams and be responsible for approving operating and surveillance schedules.

7.4.3 UK legislation: the Reservoirs Act (1975)

UK legislation was first introduced in 1930, following two small but catastrophic failures in 1925. The legislative framework underwriting dam safety is now provided by the Reservoirs Act (HMSO, 1975) and is explained in detail in an associated guide (ICE, 2000). A review of the British system cannot be divorced from an appreciation of the circumstances in which it is required to function. Population densities and reservoir numbers in the UK are relatively high, particularly in the traditional heavy industrial regions. As a compounding factor, 83% of the 2650 UK dams subject to statutory legislation are earthfill embankments, and are therefore inherently more susceptible to certain risks, e.g. flood overtopping. The median age of the dam population is also relatively high at an estimated 100–105 years, and over 30% of dams are owned by small firms or by individuals. The composition of the UK dam population is discussed in Tedd, Skinner and Charles (2000). In terms of downstream hazard it is reported that some 75% of the reservoirs subject to statutory legislation are Category A or B (Table 4.1 and ICE, 1996), i.e. they are considered to represent a significant risk to life and/or would cause extensive damage to property in the event of breaching. Reviews of the incidence of deterioration and remedial works on British dams have been published by Moffat (1982) and Charles (1986).

The essential features of the UK Reservoirs Act can be summarized as follows.

1. The legislation applies to all 'raised reservoirs' with a storage capacity in excess of 25 000 m^3 above the natural level of any part of the adjacent land.
2. The owner is entirely responsible in law for maintaining a reservoir in safe condition.
3. The registration of reservoirs and enforcement of the Act in England and Wales are the responsibility of the Environment Agency. In Scotland, the responsibility is currently (2006) vested in the regional authorities, but will transfer to the Scottish Environmental Protection Agency. (The Reservoirs Act is not applicable to Northern Ireland).
4. The design and construction of new reservoirs, and the statutory inspection or major alteration of existing reservoirs, must be carried out under the direction of a nominated Panel Engineer, i.e. a senior engineer appointed to a statutory specialist panel.
5. Statutory inspections by a Panel Engineer are required at intervals not exceeding 10 years.
6. Owners are required to nominate a Supervising Engineer, qualified and empanelled as such under the Act, to maintain continuity of supervision and surveillance between statutory inspections.

7. The owner is required to maintain certain fundamental records, e.g. water levels, spillway flows, leakages, etc.
8. The Panel Engineer nominated under 4 carries personal legal responsibility for the statutory certificates and reports required of him.
9. The Act includes provisions to ensure the safe discontinuance or abandonment of a reservoir.
10. Statutory empanelment is an *'ad hominem'* appointment made on the basis of relevant experience at an appropriately senior level. Appointment is for five years and is renewable.

British legislation therefore provides for a two-tier system of surveillance. Immediate responsibility for ongoing surveillance rests in the hands of the Supervising Engineer, who is required to inspect at least annually, with an appointed Panel Engineer responsible for periodic statutory inspections. In the course of the latter, the Panel Engineer can make 'recommendations in the interest of safety' and may initiate such rigorous and detailed investigations as are thought necessary, e.g. a reappraisal of spillway capacity etc. The principle of investing statutory responsibility for inspections in the individual Panel Engineer is a most valuable feature of the British approach, encouraging rapidity of response and flexibility.

7.5 Reservoir hazard and risk assessment

7.5.1 Hazard and risk assessment

Classification of reservoir hazard in terms of the scale of human and economic loss to be anticipated in the event of a catastrophic dam failure is increasingly common. At its simplest, hazard level is assessed and rated on a descriptive categorization, e.g. 'low', 'significant' or 'high' (USBR, 1988). An element of discretion may then be applied in defining the surveillance régime to be applied to a specific dam. The limitations implicit in such subjective appraisal and classification are evident.

A more rigorous assessment of reservoir hazard and associated risks has the potential to significantly enhance the effectiveness of dam surveillance programmes. In particular, reservoir risk assessment can be a valuable aid in ranking priorities for the allocation of limited resources available for remedial work or for optimizing surveillance. Detailed and more explicit hazard analyses can also be employed in contingency planning, embracing emergency scenarios for selected high-hazard reservoirs. The latter may include the preparation of inundation maps (Section 7.5.2).

The nominal hazard level can be expressed adequately in terms of the estimated risk of catastrophic breaching and/or the predicted worst-event loss (or cost) as the two primary determinants. Hazard levels are readily compared or ranked in relation to either risk or loss provided that either is defined as the decisive parameter. In UK practice a simple hazard categorization is embedded in the procedure for the determination of the design flood and spillway size. (See Section 4.2 and Table 4.1).

Several options have been developed for reservoir hazard analysis, ranging from simplistic expressions involving the reservoir capacity and height of the dam to sophisticated techniques of probabilistic risk assessment (PRA). The cumbersome methodologies for PRA were derived from those developed for high-risk or sensitive activities, e.g. in the nuclear and chemical industries. Options in hazard analysis are reviewed in Moffat (1988, 1995), with proposals for a quantitative reservoir hazard rating (RHR). The RHR was conceived as a semi-rigorous but relatively simplistic expression, intended primarily for the ranking of hazard levels and priorities within a population of dams. It was designed principally for general application in first-level screening and for the identification of high-hazard reservoirs.

Alternative methodologies have been comprehensively discussed and reviewed (Binnie and Partners, 1992). The application of relatively simplistic rapid assessment techniques to hazard ranking a population of dams is discussed in Thompson and Clark (1994).

A further approach to risk assessment in the context of surveillance and reservoir management, developed for use in the UK, is presented in Hughes *et al.* (2000). This approach is based on the semi-empirical but systematic and logical concept of 'Failure Modes, Effects and Criticality Analysis' (FMECA). The procedure offers a balance between the extremes of relying solely upon qualitative and subjective engineering judgement on the one hand and the rigour of an expensive and difficult probabilistic risk analysis based on limited statistical evidence on the other.

The FMECA-based risk assessment procedure has three stages:

Stage 1: potential failure impact assessment;
Stage 2: from Stage 1, selection of one of three options, i.e. detailed FMECA, standard FMECA, or no further assessment necessary;
Stage 3: the FMECA assessment, to the degree selected in Stage 2.

Stage 1 requires application of a rapid method of determining potential flood water levels downstream of the dam, together with a semi-quantitative assessment of flood impact in relation to lives, infrastructure and property.

In Stage 2, the magnitude of the potential impact as assessed in Stage 1 is used as the basis of selection. Where the assessed impact score is considered to be minimal, no further work is necessary. For medium and high impacts, FMECA assessment is required with, in the case of high impact scores, the consideration of alternative failure scenarios and consequences.

Stage 3, the FMECA assessment, provides for review of the several elements and failure modes associated with the type of dam. This is achieved through special 'Location, Cause, Indicator' (LCI) charts which consider the components of the dam and the mechanisms of deterioration or failure. A numerical value is assigned to each of these in terms of consequence, likelihood of occurrence, and degree of confidence in these values. The outcome is the identification of specific problems together with aggregate numerical criticality and risk scores. The latter can be deployed to assist in the prioritization of resource allocation and in assessing the need for more rigorous dam-break analysis and contingency planning.

The application of risk assessment methodologies in the management of dam safety programmes has aroused considerable interest, with co-ordinated research effort extending across several countries. A Canadian perspective on the issue is presented in Hartford and Stewart (2002), and it is further explored in ICOLD Bulletin 130 (ICOLD, 2005). Methodologies for quantitative risk assessment (QRA) have aroused particular interest in a number of quarters in recent years, and QRA is addressed in some depth in Hartford and Baecher (2004).

In the UK, a QRA methodology based on the concept of an integrated approach to the assessment of overall risk to a dam from specific agencies, e.g. extreme flood events etc., has recently been developed (Brown and Gosden, 2004). This incorporates features of the FMECA approach to risk assessment proposed in Hughes *et al.* (2000), which it was intended to supersede.

Brown and Gosden (2004) defines a procedure for determination of a screening level assessment of the absolute risk represented by an individual dam and reservoir. Absolute risk is then compared with the standards for public safety deemed applicable in other contexts such as high-hazard industries and acceptability determined. The estimated absolute risk can also be ranked relative to that determined for other reservoirs within a population. The procedure requires estimation of the annual probability of failure associated with:

1. an extreme rainfall event;
2. presence of an upstream reservoir;
3. loss of internal stability (embankment and appurtenant works (e.g. outlet works));
4. a matrix of other significant threats (e.g. seismicity etc.).

(The extreme rainfall event of 1 above should not be regarded as in conflict with Section 4.2 and Table 4.1).

The overall annual probability of failure is estimated, the individual probabilities subsequently being reviewed to confirm that they in turn represent appropriate estimates.

The consequence of breaching and failure is estimated from

consideration of the predicted dam failure hydrograph embracing a sequence of impact assessments embracing likely loss of life and third-party damage. Overall probability of failure and consequence are then considered in combination to determine the 'tolerability of risk', with the latter assessed in terms of the predicted likely annual loss of life, consequence, and probability of failure. The outcome is then subject to comparative review to determine whether a more detailed and exhaustive risk assessment is to be recommended.

Viewed in the context of dam surveillance QRA methodologies have been criticized on a number of grounds, notably their dependence upon the estimated probabilities of failure associated with specific threats to the integrity of a dam of particular type. A feature of quantitative methodologies is therefore their dependence upon access to historical data identifying incidents, where the integrity of a dam has been compromised, and lesser occurrences have threatened fitness for purpose. Historical data must, by it's nature, be less than complete and may be of questionable reliability and accuracy. National data will, in addition, reflect the national perspective as regards the composition and nature of the dam population and the prevailing legislative provision for dam safety. In the interests of developing a verifiable UK database of incidents and lesser events impinging upon fitness for purpose it has been suggested, for the UK, that a national incident and occurrence reporting system should be introduced (Moffat, 2001).

The complexity of QRA and the superficial precision implicit in the output probabilities have also been the subject of criticism. More fundamentally, in the specific context of dam risk assessment and management, where each dam within a population is quite unique, reservations have been expressed with regard to the operational utility of the outcome, whether quantitative (Brown and Gosden, 2004) or semi-quantitative (Hughes *et al.*, 2000). The role of risk assessment in safety management is discussed in Charles, Tedd and Skinner (1998).

7.5.2 Dam-break analysis, flood wave propagation and inundation mapping

The extension of hazard analysis to include dam-break contingency planning is a sensitive issue. It is, however, a logical and prudent step when applied selectively to reservoirs commanding a populous and vulnerable floodplain. Rigorous dam-break analysis involves a balanced consideration of hydrological, hydraulic, environmental, and geotechnical and structural parameters applicable to the dam and floodplain being studied. An empirical first approximation for dam-breach formation characteristics for embankments has been suggested in MacDonald and Langridge-Monopolis (1984); an alternative initial approximation is presented in Dewey and Gillette (1993). Dam breaching and the hydraulics of dam-break floodwaves are discussed in some depth in Vischer and Hager (1998).

Time-dependent floodwave propagation downstream of a breached dam is extremely complex. It is a function of site-specific parameters including reservoir capacity and breaching characteristics for the dam under review, e.g. the progressive erosion of an embankment, as at Teton or, at the other extreme, the sudden monolithic collapse of the Malpasset arch dam. The advance of the floodwave across the floodplain will in turn be governed by a further range of determinants, many difficult to replicate in a mathematical model, e.g. differences in terrain and surface, including the influence of urban development and/or vegetative cover. Certain of the difficulties inherent in modelling the floodplain adequately have been eased with the advent of digital mapping techniques, but considerable specialist expertize and computational effort is required to produce a high-quality mathematical representation of the breach development process for an embankment and of the subsequent floodwave as it spreads across the floodplain. Sophisticated commercial software is now available for dam-break analysis and the preparation of downstream inundation maps, much of the software deriving from the early 'DAMBRK' and 'DAMBRK UK' programs (Fread, 1984 and DoE, 1991a, 1991b). Later dam-break modelling programs include FLD-WAV (Fread and Lewis, 1998) and the recently developed HEC-RAS. An overall review of dam-break flood analysis methodologies is published in ICOLD (1998); simplified predictive equations for estimation of the dam-break flood are presented in Zhou and Donnelly (2005).

The European CADAM (Concerted Action on Dam-break Modelling) project reviewed dam-break modelling practice in depth and identified questions regarding performance (Morris, 2000). Particular attention was paid to the two critical issues of breach formation and floodwave propagation, with results from physical and computer modelling of the latter extrapolated to real valleys and calibrated against records from the 1959 Malpasset arch dam failure. Guidelines and recommendations for modelling derived from the CADAM project are published in Morris and Galland (2000). Following the conclusion of CADAM a successor European project, IMPACT (Investigation of Extreme Flood Processes and Uncertainty), included within its scope further investigation of the central issues of breach formation mechanism and floodwave propagation. The programme included a study of breach formation through a series of 22 small-scale (0.6 m high) laboratory models and, in the field, a series of five controlled breachings on embankments of 4.5 m to 6 m height. Alternative dam sections, some homogeneous section some not, some employing cohesive and others non-cohesive soils, were studied in the two series of tests. The project outcomes are presented and discussed in IMPACT (2005).

Dam-break inundation mapping exercises have now been conducted for high-profile reservoirs in a number of countries. Examples of inundation mapping experience in the UK, where circumstances are such that quite modest dams may represent a significant *potential* hazard by virtue of their location relative to centres of population, are described by Tarrant, Hopkins and Bartlett (1994) and Claydon, Walker and Bulmer (1994).

7.5.3 Contingency and emergency planning

The logical corollary to inundation mapping for reservoirs in high-hazard locations, e.g. sited in close proximity to sizeable communities, schools, major industrial units etc., is the preparation of a contingency or emergency response plan. Two principal elements are required within the overall plan:

- the organisational action plan, for implementation by the owner and/or operator of the dam in the event of a major emergency;
- the response plan, prepared in consultation with the civil authorities, for action in conjunction with the emergency services.

The organizational or internal action plan is designed to document operating procedures etc., appropriate to all foreseeable emergency scenarios for the site under review. For dams in vulnerable locations it may be appropriate to give consideration to identifiable 'external' threats to safety such as accidental aircraft collision, or the action of subversive organizations, in generating project-specific 'Standard Response Procedures' (Moffat, 1998).

Operating staff at all levels must be made familiar with the response plan, which should be subject to periodic reappraisal. The scope of the plan should range from defining levels of incident and internal responsibilities to procedures for alerting the emergency services, civil authorities and the population at risk. The plan should state the appropriate on-site response to likely incidents, e.g. the activation of bottom outlets etc., and should provide documented instruction and guidance for staff in key positions.

The second element, the emergency response plan, must be prepared in close collaboration with the appropriate civil authorities including, in the UK, the county Emergency Planning Officer. The response plan should embrace the 'worst case' scenario, i.e. rapid and catastrophic breaching of the dam. The content of the plan is determined by, *inter alia*, the inundation map, the nature of the flood plain, and the numbers, locations and accessibility of persons under threat. The response plan is necessarily comprehensive and quite detailed, but it must retain an essential element of flexibility.

All stages of the response must be considered and documented, from alerting those at risk and the timely mobilization and deployment of emergency services through to an evacuation strategy, the latter including provision of support and shelter. The response must also have regard to the capabilities and availability of key emergency services, e.g. police and, if available, civil aid organizations. For incidents requiring very large-scale evacuation of people from heavily populated urban areas it may be necessary to plan for military aid in terms of vehicles and personnel. The principles of emergency planning for major incidents are fully described in Alexander (2002).

Emergency response plans are subject to periodic review, most notably if extensive new developments take place on the downstream floodplain. Live rehearsal of an emergency response plan is generally quite impracticable, and proving of the plan must therefore effectively be confined to desk exercises. The latter can be made quite testing if external events or operational constraints, e.g. weather and incidental flooding obstructing evacuation routes, or non-availability of personnel, etc., are injected into the exercise narrative.

It will be appreciated that emergency planning in the context of dam safety can be a highly sensitive and emotive issue. It is consequently essential that discretion is observed with regard to public disclosure, particularly with respect to the identification of specific dams or locations.

References

Alexander, D.E. (2002) *Principles of emergency planning and management.* Terra Publishing, Harpenden.

Aufleger, M., Conrad, M., Perzlmeier, S. and Porras, P. (2005) Improving a fiber optics tool for monitoring leakage, in *HRW* (Sept.): 18–23.

Beak, D.C. (1992) Implementation of the Reservoirs Act 1975 and monitoring of dams, in *Proceedings of the Conference on Water Resources and Reservoir Engineering*, British Dam Society, London, pp. 287–300.

Berti, G., Villa, F., Dovera, D., Genevois, R. and Brauns, J. (1988) The disaster of Stava/Northern Italy, in *Proceedings of Specialty Conference on Hydraulic Fill Structures*, American Society of Civil Engineers, New York: 492–510.

Binnie & Partners (1992) *Review of Methods and Applications of Reservoir Hazard Assessment*, Report, Contract PECD 7/7/309, Department of the Environment, London.

Brown, A.J. and Gosden, J.D. (2004) *Interim guide to quantitative risk assessment for UK reservoirs.* Thos. Telford Ltd., London.

Charles, J.A. (1986) The significance of problems and remedial works at British earth dams, in *Proceedings of the Conference 'Reservoirs 86'*, British National Committee on Large Dams, London: 123–41.

Charles, J.A., Tedd, P., Hughes, A.K. and Lovenbury, H.T. (1996) *Investigating Embankment Dams; a Guide to the Identification and Repair of Defects.* Building Research Establishment Report CI/SFB 187, 81 pp. Construction Research Communications, Watford.

Charles, J.A., Tedd, P. and Skinner, H.D. (1998) The role of risk analysis in the safety management of embankment dams, in *Proceedings of Conference 'The prospects for reservoirs in the 21st century'*, British Dam Society, London: 1–12.

Charles, J.A., Tedd, P. and Watts, K.S. (1992) The role of instrumentation and monitoring in safety procedures for embankment dams, in *Proceedings of the Conference on Water Resources and Reservoir Engineering*, British Dam Society, London.

Charles, J.A. and Wright, C.E. (1999) European dam safety regulations from a British perspective, in *Proceedings of Conference 'The reservoir as an asset',* British Dam Society, London: 180–91.

Claydon, J.R., Walker, R.A. and Bulmer, A.J. (1994) Contingency planning for dam failure, in *Proceedings of the Conference 'Reservoir Safety and the Environment'*, British Dam Society, London: 224–35.

Dewey, R.L. and Gillette, D.R. (1993) Prediction of embankment dam breaching for hazard assessment, in *Proceedings of the Conference on Geotechnical Practice in Dam Rehabilitation*, Geotechnical Special Publication 35, American Society of Civil Engineers, New York.

DoE (1991a) *Handbook on the design of tips and related structures.* Her Majesty's Stationery Office, London.

—— (1991b) Dambreak Flood Simulation Program: DAMBRK UK. Binnie & Partners for Department of the Environment, London.

Dunnicliff, J. (1988) *Geotechnical Instrumentation for Monitoring Field Performance*, Wiley, New York.

Duscha, L.A. (1983) National program of inspection of non-federal dams, in *Proceedings of the Conference on New Perspectives on the Safety of Dams*, Massachusetts Institute of Technology, Cambridge, MA, 14 pp.

Evans, J.D. and Wilson, A.C. (1992) The instrumentation, monitoring and performance of Roadford Dam during construction, in *Proceedings of the Conference on Water Resources and Reservoir Engineering*, British Dam Society, London.

Fread, D.L. (1984) *DAMBRK: the NWS Dambreak Flood Forecasting Model*, US National Weather Service, Washington, DC.

Fread, D.L. and Lewis, J.M. (1998) *The NWS FLDWAV model: theoretical background/user document.* National Weather Service, Silver Springs.

Hartford, D.N.D. and Baecher, G.B. (2004) *Risk and uncertainty in dam safety.* Thos. Telford Ltd, London.

Hartford, D.N.D. and Stewart, R.A. (2002) Risk assessment and the safety case in dam safety evaluation, in *Proceedings of Conference 'Reservoirs in a changing world',* British Dam Society, London: 510–19 .

HMSO (1975) *The Reservoirs Act* (and associated Statutory Instruments), Her Majesty's Stationery Office, London.

Hughes, A.K., Hewlett, H.W.M., Morris, M.W., Sayers, P., Moffat, A.I.B., Harding, A. and Tedd, P. (2000) *Risk Management for UK Reservoirs.* Construction Industry Research and Information Association, Report C452, CIRIA, London.

ICE (1996) *Floods and reservoir safety,* 3rd edn. Institution of Civil Engineers, London.

—— (2000) *A Guide to the Reservoirs Act 1975.* Institution of Civil Engineers, London.

ICOLD (1987) *Dam Safety Guidelines*, Bulletin 59, International Commission on Large Dams, Paris.

—— (1988) *Dam Monitoring: General Considerations*, Bulletin 60, International Commission on Large Dams, Paris.

—— (1992) *Improvement of Existing Dam Monitoring*, Bulletin 87, International Commission on Large Dams, Paris.

—— (1995) *Dam Failures – Statistical Analysis*, Bulletin 99, International Commission on Large Dams, Paris.

—— (1998) *Dam-break flood analysis – Review and Recommendations*, Bulletin 111, International Commission on Large Dams, Paris.

—— (2001) *Tailings dams: risk of dangerous occurrences: Lessons learnt from practical experience.* Bulletin 121, International Commission on Large Dams, Paris.

—— (2005) *Risk assessment in dam safety management – a reconnaissance of benefits, methods and current applications.* Bulletin 130, International Commission on Large Dams, Paris.

IMPACT (2005) *IMPACT – Final Technical Report.* HR, Wallingford.

Jansen, R.B. (1980) *Dams and Public Safety*, US Water and Power Resources Service, Denver, CO.

Johansson, S. (1996) Seepage monitoring in embankment dams by temperature and resistivity measurements, in *Proceedings of Symposium on 'Repair and upgrading of dams'*, Royal Institute of Technology, Stockholm: 288–97.

Johnston, T.A., Millmore, J.P., Charles, J.A. and Tedd, P. (1999) *An Engineering Guide to the Safety of Embankment Dams in the United Kingdom* (2nd edn), Construction Research Communications, Watford.

Kennard, M.F., Owens, C.L. and Reader, R.A. (1996) *Engineering Guide to the Safety of Concrete and Masonry Dam Structures in the UK*, Report 148, Construction Industry Research and Information Association, London.

MacDonald, T.C. and Langridge-Monopolis, J. (1984) Breaching characteristics of dam failures. *Journal of Hydraulic Engineering, American Society of Civil Engineers*, 110 (5): 576–86.

Millmore, J.P. and McNicol, R. (1983) Geotechnical aspects of the Kielder dam. *Proceedings of the Institution of Civil Engineers, Part 1*, 74: 805–36.

Moffat, A.I.B. (1982) Dam deterioration: a British perspective, in *Proceedings of the Conference 'Reservoirs 82'*, British National Committee on Large Dams, London, Paper 8.

—— (1988) Embankment dams and concepts of reservoir hazard analysis, in *Proceedings of the Conference 'Reservoirs 88'*, British National Committee on Large Dams, London, Paper 6.4.

—— (1995) Hazard analysis as an aid to effective dam surveillance, in *Proceedings of International Conference on Dam Engineering*, Malaysian Water Association, Kuala Lumpur: 549–58.

—— (1998) Contribution on 'Developments in legislation and practice', in discussion volume, *Conference 'The prospect for reservoirs in the 21st century'*, British Dam Society, London: 40–1 .

—— (2001) 'DIOR – Dam Integrity: Occurrence Reporting', *unpublished discussion, 'Supervising engineers forum 2001'*, British Dam Society, London .

Morris, M.W. (2000) CADAM: A European Concerted Action Project on Dambreak Modelling, in *Proceedings of Conference 'Dams 2000'*, British Dam Society, Thomas Telford, London: 42–53.

Morris, M.W. and Galland, J.C. (eds) (2000) *Dambreak modelling guidelines and best practice.* HR, Wallingford.

Olalla, C, and Cuellar, V. (2001) Failure mechanism of the Aznaćollar Dam, Seville, Spain. *Geotechnique,* 51 (3): 399–406.

Penman, A.D.M. (1989) The design and use of instrumentation, in *Proceedings of the Conference on Clay Barriers for Embankment Dams*, Thomas Telford, London: 131–48.

—— (2002) Tailings dam incidents and new methods, in *Proceedings of Conference 'Reservoirs in a changing world'*, British Dam Society, London: 471–83.

Penman, A.D.M. and Kennard, M.F. (1981) Long-term monitoring of embankment dams in Britain, in *Proceedings of the ASCE Conference 'Recent Developments in Geotechnical Engineering for Hydro Projects'*, American Society of Civil Engineers, New York: 46–67.

Penman, A.D.M., Saxena, K.R. and Sharma, V.M. (1999) *Instrumentation, monitoring and surveillance: embankment dams.* Balkema, Rotterdam.

Robertshaw, A.C. and Dyke, T.N. (1990) The routine monitoring of embankment dam behaviour, in *Proceedings of the Conference on The Embankment Dam*, British Dam Society, London, pp. 177–84.

Tarrant, F.R., Hopkins, L.A. and Bartlett, J.M. (1994) Inundation mapping for dam failure – lessons from UK experience, in *Proceedings of the Conference 'Reservoir Safety and the Environment'*, British Dam Society, London: 282–91.

Tedd, P., Skinner, H., and Charles, J.A. (2000) Developments in the British national dams database, in *Proceedings of Conference 'Dams 2000'*, British Dam Society, London: 181–9.

Thompson, G. and Clark, P.B. (1994) Rapid hazard ranking for large dams, in *Proceedings of the Conference 'Reservoir Safety and the Environment'*, British Dam Society, London: 306–15.

UNESCO (1967) *Recommendations Concerning Reservoirs*, UN Educational, Scientific, and Cultural Organization, Paris.

USBR (1983) *Safety Evaluation of Existing Dams*, US Bureau of Reclamation, Denver, CO.

—— (1988) *Downstream Hazard Classification Guidelines*, ACER Technical Memorandum 11, US Bureau of Reclamation, Denver, CO.

Vischer, D.L. and Hager, W.H. (1998) *Dam Hydraulics*, Wiley, Chichester.

Zhou, R.D. and Donnelly, R.C. (2005) Dam break flood estimation by simplified predicting equations, in *'Dam Engineering'*, XVI (1): 81–90.

Part Two

Other hydraulic structures

Chapter 8

River engineering

8.1 Introduction

The subject of river engineering is a very wide-ranging one, and only some of the more important aspects can be covered in this book. Other chapters deal with structures which are usually regarded as an integral part of river engineering, but which (in the authors' view) have merited separate chapters: diversion works, including intakes (Chapter 9); cross drainage and drop structures (Chapter 10); inland navigation (Chapter 11). Thus only the methods of hydrometry, i.e. the measurement of water and bed levels, discharges and water quality, with their associated structures, as well as river improvement structures, i.e. structures mainly designed to create favourable geometric and kinematic conditions and/or to protect river banks, or to aid navigation and flood control, are briefly discussed in this chapter.

As this is also the introductory chapter to the whole subject of river engineering (as outlined above), some basic principles of open-channel flow and a brief discussion of river morphology and régime and of flood routing are also included in the next sections.

The main reason for the complexity of river engineering is that river flow in alluvium has no really fixed boundaries and geometry compared with, say, pipe flow or open-channel flow in rigid canals. Adding to this the complexity introduced by the changing boundary roughness in sediment-transporting streams, and the dependence of the flow on water and sediment discharge and sediment availability, the difficulties of a rigorous treatment of the hydraulics of flow in alluvium are increased manifold. It is therefore hardly surprising that, despite a very substantial and sustained research effort, our knowledge in the field of sediment transport and river morphology is still somewhat sketchy. Nevertheless, the growing data bank together with the application of physical principles and computer-aided analysis are steadily advancing our ability to handle design problems in river engineering (Kennedy, 1983), which at present also relies (and always will rely) heavily on experience.

8.2 Some basic principles of open-channel flow

8.2.1 Definitions

Open-channel flow may be laminar or turbulent, depending on the value of the Reynolds number $Re = VR/v$, where V is the mean flow velocity, R is the hydraulic radius $R = A/P$ (A is the cross-sectional area of the flow, and P the wetted perimeter) and v is the coefficient of kinematic viscosity. In rivers and canals there is invariably turbulent flow. Furthermore, open-channel flow may be steady if the discharge, Q, is a function of distance only (or constant), and unsteady if Q is also a function of time t. Additionally, steady flow may be uniform (with Q and depth y and hence velocity $V = Q/A$ constant) or non-uniform. Non-uniform flow may be rapidly or gradually varying (Q = constant, V and y varying with position x) or spatially varying ($Q = f(x)$). In natural rivers flow is normally unsteady, whereas in canalized rivers and canals flow is predominantly steady non-uniform or uniform. The flow can be either supercritical ($Fr > 1$) or subcritical ($Fr < 1$), where Fr is the Froude number defined by $Fr^2 = \alpha Q^2 B/(gA^3)$, with B the water surface width and α the Coriolis coefficient derived from the velocity distribution in the section. In rivers and canals the flow is in most cases subcritical. Canals are usually prismatic open channels, while rivers are generally non-prismatic. The boundaries of open channels formed by the bed and sides may be fixed, e.g. in artificial concrete-lined channels, or movable as in rivers or unlined canals in alluvium.

8.2.2 Some basic equations

Only a few basic concepts can be touched upon here. For further and more detailed treatment of open-channel flow the reader is referred to books on the subject, e.g. Henderson (1966), Chow (1983), French (1986), Graf (1998) and Chadwick , Morfett and Borthwick (2004).

For uniform flow, the bedslope, S_0, energy gradient, S_e, and friction slope, S_f, are all equal.

Denoting by τ_0 the mean shear stress on the channel perimeter P and the ratio $A/P = R$ (the hydraulic radius), from the balance between gravity and frictional resistance we obtain

$$\tau_0 = \rho g R S_0 = \rho U_*^2 \tag{8.1}$$

where U_* is the shear velocity.

Since in fully turbulent flow $\tau_0 \propto V^2$, equation (8.1) leads to the well-known Chézy equation for uniform flow:

$$V = C(RS_0)^{1/2} \tag{8.2}$$

(the dimensions of C are $\mathrm{L}^{1/2}\mathrm{T}^{-1}$).

The 'coefficient' C can be expressed as

$$C = (8g/\lambda)^{1/2} \tag{8.3}$$

where λ is the friction coefficient in the Darcy–Weisbach equation:

$$h_\mathrm{f} = (\lambda L/D)V^2/2g = \lambda LV^2/8gR. \tag{8.4}$$

λ can be expressed from boundary layer theory as

$$1/\lambda^{1/2} = 2\log\left[6R\Big/\left(\frac{k}{2} + \delta'/7\right)\right] \tag{8.5}$$

where k is the roughness 'size' and δ' $(=11.6\nu/U_*)$ is the thickness of the laminar sublayer.

Another frequently used expression is the Manning equation, using a constant n which is a function of roughness:

$$V = (1/n)R^{2/3}S_0^{1/2} \tag{8.6}$$

(i.e. $C = R^{1/6}/n$; the dimensions of n are $\mathrm{TL}^{-1/3}$). According to Strickler, $n \approx 0.04d^{1/6}$, where d is the roughness (sediment) size (in metres) (Worked example 8.1).

From equations (8.1), (8.2) and (8.3) it follows that

$$U_* = V(\lambda/8)^{1/2}. \tag{8.7}$$

From Bernoulli's equation it follows that for a general non-prismatic channel and non-uniform flow

$$-S_0 + \mathrm{d}y/\mathrm{d}x - (\alpha Q^2/gA^3)[B\ \mathrm{d}y/\mathrm{d}x + (\partial A/\partial b)(\mathrm{d}b/\mathrm{d}x)] + S_\mathrm{f} = 0$$

and thus

$$\mathrm{d}y/\mathrm{d}x = [S_0 - S_\mathrm{f} + (\alpha Q^2/gA^3)(\partial A/\partial b)(\mathrm{d}b/\mathrm{d}x)]/(1 - Fr^2). \tag{8.8}$$

For a prismatic channel, $\mathrm{d}b/\mathrm{d}x = 0$ and equation (8.8) reduces to

$$\mathrm{d}y/\mathrm{d}x = (S_0 - S_\mathrm{f})/(1 - Fr^2). \tag{8.9}$$

Introducing the channel conveyance, $K = CAR^{1/2}$ (i.e. the discharge for slope = 1), equation (8.9) becomes

$$dy/dx = S_0\{[1 - (K_0/K)^2]/(1 - Fr^2)\} \quad (8.10)$$

(equation (8.10) implies that $\tau_0 = \rho g R S_f$ – equation (8.1) – as well as $\tau_0 = \rho g R_0 S_0$). Equation (8.10) can be conveniently used to analyse and compute various surface profiles in non-uniform flow as, generally, $K^2 \propto y^N$, where the exponent N is called the hydraulic exponent. Numerical methods may also be used to solve equation (8.9) or (8.10) for prismatic channels, and must be used for non-prismatic channels.

For unsteady flow the treatment of continuity and Bernoulli's equation yields (Saint Venant equation)

$$S_f = S_0 - \partial y/\partial x - (V/g)(\partial V/\partial x) - (1/g)(\partial V/\partial t). \quad (8.11)$$

The first term on the right-hand side of equation (8.11) signifies uniform flow, and the first three terms signify non-uniform flow. From continuity it follows that a change of discharge in Δx must be accompanied by a change in depth in Δt, i.e.

$$\partial Q/\partial x + B\,\partial y/\partial t = 0 \quad (8.12)$$

(for no lateral discharge in Δx); thus

$$A\,\partial V/\partial x + V\,\partial A/\partial x + \partial y/\partial t = 0. \quad (8.13)$$

The first term on the left-hand side of equation (8.13) represents the prism storage and the second the wedge storage (Section 8.6).

For a rectangular channel, equation (8.13) becomes

$$y\,\partial V/\partial x + V\,\partial y/\partial x + \partial y/\partial t = 0. \quad (8.14)$$

The solution of the above equations can be achieved only by numerical techniques applied, for example, to their finite difference form (Cunge, Holly and Verwey, 1980).

In the case of rapidly varying unsteady flow, a surge is formed which has a steep front with substantial energy dissipation (analogous to a moving hydraulic jump). From the momentum and continuity equations (y_1 and V_1 refer to the section ahead of the surge height Δy moving with celerity c),

$$c = -V_1 \pm \{g[(A_1 + \Delta A)\Delta y/\Delta A + (A_1 + \Delta A)y_1/A_1]\}^{1/2}. \quad (8.15)$$

For a rectangular section, equation (8.15) converts into

$$c = -V_1 \pm [g(y_1 + 3\Delta y/2 + \Delta y^2/2y_1)]^{1/2} \quad (8.16)$$

which for small surges (e.g. in navigation canals) results in

$$c \approx -V_1 \pm [g(y_1 + 3\Delta y/2)]^{1/2}. \quad (8.17)$$

For the flow velocity $V_1 = 0$ and small Δy, equation (8.17) reduces to $c \simeq (gy_1)^{1/2}$ (see also Chapter 14).

8.2.3 Sediment transport

A full discussion of sediment transport in open channels is clearly outside the scope of this brief text, but a few fundamental aspects have to be included here.

From the point of view of source sediment, transport can be divided into washload comprising very fine material moving in rivers and canals in suspension and bed material load moving as bedload and suspended load depending on sediment size and velocity. For river engineering and navigation canals bedload is the important element of sediment transport, as it determines the morphological erosion and sedimentation aspects; suspended load may be important in river engineering only in reservoir sedimentation and, exceptionally, in sedimentation at canal intakes.

The important properties of sediment and sediment transport are the sediment size (d), shape, density ρ_s (usually $2650\,\text{kg}\,\text{m}^{-3}$), fall velocity ($w_s$), bulk density and porosity, and sediment concentration (C) (volumetric, ppm, or $\text{mg}\,\text{l}^{-1}$). According to size, we usually distinguish clay ($0.5\,\mu\text{m} < d < 5\,\mu\text{m}$), silt ($5\,\mu\text{m} < d < 60\,\mu\text{m}$), sand ($0.06\,\text{mm} < d < 2\,\text{mm}$) and gravel ($2\,\text{mm} < d < 60\,\text{mm}$).

The fall velocity can be approximately expressed by the equation

$$w_s = (4gd\Delta/3C_D)^{1/2} \quad (8.18)$$

where $\Delta = (\rho_s - \rho)/\rho$ and C_D is a drag coefficient dependent on the Reynolds number $Re = w_s d/\nu$. For values of $Re < 1$ (very fine sediment) $C_D = 24/Re$ which leads to Stokes law, while for larger sizes with $Re > 10^3$ C_D becomes constant and is a function of grain shape only (usually $C_D \approx 1.3$ for sand particles). The fall velocity varies therefore with $d^{1/2}$ to d^2.

The threshold of sediment motion (incipient motion) is given by a critical value of the shear stress which, for a plane sediment bed, is given by the Shields criterion:

$$\tau_c = c(\rho_s - \rho)gd \quad (8.19)$$

where, according to various authors, c varies between 0.04 and 0.06. The condition of validity of equation (8.19) is that $Re\ (= wd/\nu) > 10^3$. As, from equation (8.1), $\tau_0/\rho = U_*^2$, equation (8.19) can also be written as

$$Fr_d^2 = U_*^2/gd\Delta = c. \tag{8.20}$$

For a sediment particle on a slope (e.g. the side slope of a canal) inclined at an angle β to the horizontal, the critical shear stress is reduced by a factor of $\{1 - (\sin^2\beta/\sin^2\varphi)\}^{1/2}$, where φ is the natural angle of stability of the non-cohesive material. (For stability, naturally $\beta < \varphi$.) The average value of φ is about 35°.

On the other hand, the maximum shear stress induced by the flow on a side slope of the canal is usually only about $0.75\rho g y S$ (instead of $\rho g R S$ applicable for the bed of wide canals – equation (8.1) and Worked example 8.2). Thus, in designing a stable canal in alluvium it is necessary to ascertain whether the bed or side slope stability is the critical one for channel stability. In a channel which is not straight the critical shear stresses are further reduced by a factor between 0.6 and 0.9 (0.6 applies to very sinuous channels).

Investigations into bedload transport have been going on for decades without a really satisfactory all-embracing equation being available to connect the fluid and sediment properties. This is due mainly to the complexity of the problem including the effect of different bed forms on the mode and magnitude of bedload transport, the stochastic nature of the problem and the difficulty of verifying laboratory investigation in prototype. Nevertheless, substantial advances have been made. Most of the approaches used can be reduced to a correlation between the sediment transport parameter, $\phi = q_s/d^{3/2}(g\Delta)^{1/2}$, where q_s is the sediment transport (in $m^3 s^{-1} m^{-1}$) and $Fr_d^2 = 1/\psi = U_*^2/\Delta g d$, where ψ is called the flow parameter (ψ can also contain an additional parameter – the ripple factor – to account for the effect of bed form (Graf, 1984)). The power of Fr_d^2 in many correlations varies between 2 and 3, i.e. q_s varies as V^n with $4 < n < 6$, demonstrating the importance of a good knowledge of the velocity field in the modelling and computation of bedload transport, particularly when using two- or three-dimensional models.

Examples of simplified ϕ and ψ correlations are the Meyer–Peter and Muller equation

$$\phi = (4/\psi - 0.188)^{3/2}, \tag{8.21}$$

the Einstein–Brown equation

$$\phi = 40\psi^{-3} \tag{8.22}$$

and the Engelund–Hansen equation

$$\phi = 0.4\psi^{-5/2}/\lambda. \tag{8.23}$$

It must be emphasized that the full application of the above and other more sophisticated equations requires further reading (e.g. Ackers, 1983; Garde and Ranga Raju, 1985; Graf, 1984; Vanoni, 1975), and the equations are quoted here only to demonstrate the correlation and trend.

Thus, although the relationship between the transport of bed sediment and the flow and even the relationship between the hydraulic resistance and channel sedimentary features, particularly the bed configuration (which in turn is a function of sediment characteristics and discharge), are broadly known, the third required equation for alluvial channel computation, relating the flow parameters and the erosive resistance of the banks, still by and large eludes a physically based formulation. Nevertheless, the minimum stream power concept or other optimization methods show most promise in this area, where we otherwise fall back on régime equations (Section 8.3) which synthesize the physical functions into groups of formulae describing the channel geometry (Ackers, 1983).

8.3 River morphology and régime

River morphology is concerned with channel configuration and geometry, and with longitudinal profile; it is time dependent and varies particularly with discharge, sediment input and characteristics, and with bank material. River morphology can be substantially influenced by engineering works, although this influence is not necessarily beneficial. Natural river channels are straight (usually only very short reaches), meandering, i.e. consisting of a series of bends of alternate curvature connected by short, straight reaches (crossings), or braided, i.e. the river divides into several channels which continuously join and separate. The various stages of a schematized river (de Vries, 1985) are shown in Fig. 8.1.

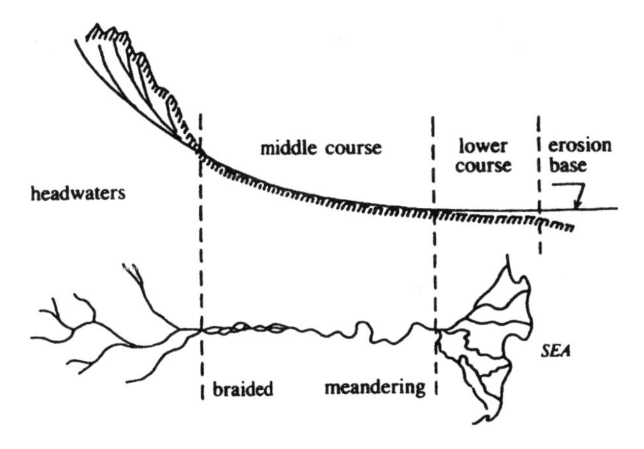

Fig. 8.1 Schematized river (after de Vries, 1985)

Bends can be divided into free (surface), limited (entrenched) and forced (deformed), with ratios of radius of curvature to width ranging from about 3 (forced) to about 7.5 (limited) (5 for free bends). In free and limited bends the depth gradually increases to a maximum downstream of the apex of the bend; bends are characterized by spiral flow and triangular sections, with the maximum depth and velocity at the concave bank, and maximum sediment transport at the convex bank and the talweg (line of maximum depth) deviating from the river centreline, as shown in Fig. 8.2.

Crossings are relatively straight reaches between alternate bends and are approximately rectangular in section (Fig. 8.3).

Meandering rivers usually have a ratio of channel to valley length greater than 1.5 (Petersen, 1986) with the meander length (the distance between vertices of alternate bends) about 10 times the stream width. The ratio of meander length to width varies between 2 and 4 (for further details see, for example, Jansen, van Bendegom and van den Berg (1979)).

In rivers the mean cross-sectional velocity varies, from about $0.5\,\mathrm{m\,s^{-1}}$ at low flows to $4.0\,\mathrm{m\,s^{-1}}$ at floods, but there are exceptions to these values. The maximum velocity in a section usually exceeds the mean by 25–30%.

River régime is concerned with the channel geometry. A river in alluvium is considered to be in régime if its channel is stable on a long-term average. Short-term changes will occur with changes of discharge and sediment transport, and the concept of 'stability' here clearly differs from the one defined by the critical limiting tractive force implying no motion of sediment on the bed and banks.

The cross-section and longitudinal slope for a régime channel will primarily be a function of discharge, with the width B, depth y and slope S

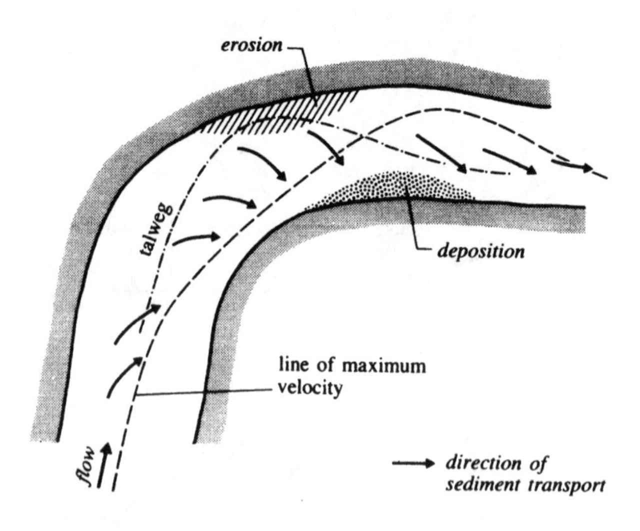

Fig. 8.2 Flow in a bend

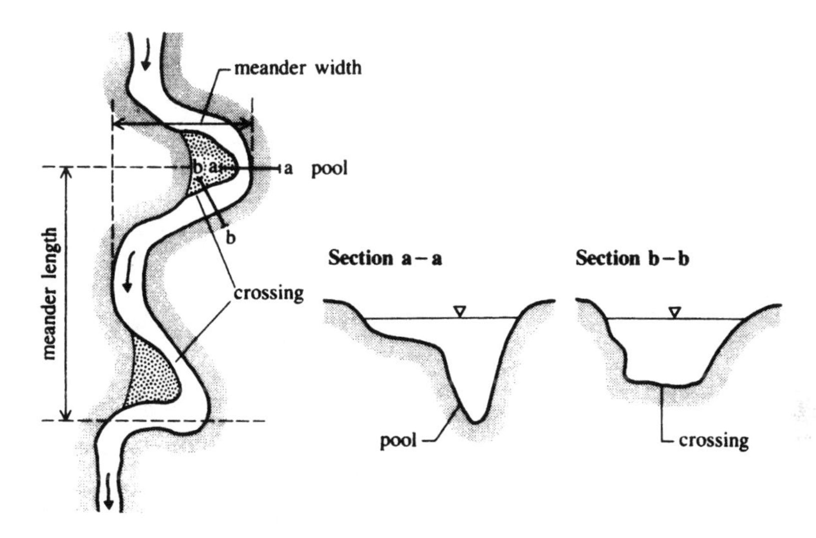

Fig. 8.3 Meandering stream

providing three degrees of freedom for self-adjustment of the channel. The relationship between these three parameters and discharge – proceeding in a river system in the downstream direction – has been based mainly on measurements carried out on the Indian subcontinent and is usually expressed as (Blench, 1969)

$$B \propto Q^{1/2}, \tag{8.24a}$$

$$y \propto Q^{1/3}, \tag{8.24b}$$

$$S \propto Q^{-1/6}. \tag{8.24c}$$

Lacey and Pemberton (Ackers, 1983) generalized the basic régime equation into

$$V = aR^{(b+1)/2}S^b \tag{8.25}$$

where a and b vary with sediment diameter. The power b is $1/4 < b < 1$ with the lower limit for $d > 2\,\text{mm}$ and the upper for $0.2\,\text{mm} > d > 0.1\,\text{mm}$. Lacey's original equation

$$V = 0.635(fR)^{1/2} \tag{8.26}$$

where $f = (2500d)^{1/2}$ (d in m, V in m s^{-1}), combined with equations (8.25) or (8.24), leads to the basic régime statement

$$R^{1/2}S \propto d. \tag{8.27}$$

In contrast, the critical tractive force theory leads to (equation (8.19))

$$B \propto Q^{0.46}, \tag{8.28a}$$

$$y \propto Q^{0.46}, \tag{8.28b}$$

$$S \propto Q^{-0.46} \tag{8.28c}$$

and

$$RS \propto d. \tag{8.29}$$

Both the régime concept and critical tractive force theory lead to a relatively weak dependence of velocity on discharge. The régime concept results in

$$V \propto Q^{1/6} \tag{8.30}$$

and the critical tractive force theory in

$$V \propto Q^{0.08}. \tag{8.31}$$

Generally, the critical tractive force approach would be more associated with coarse material (gravel) and upland river reaches, and the régime concept with fine material in lowland river reaches and canals.

The above relationships apply to changes in cross-section in the downstream direction. At any one river section different 'at a station' relationships apply. Characteristically, they are (Leopold, Wolman and Miller, 1964)

$$B \propto Q^{0.26}, \tag{8.32a}$$

$$y \propto Q^{0.4}, \tag{8.32b}$$

$$V \propto Q^{0.34}, \tag{8.32c}$$

$$S \propto Q^{0.14}. \tag{8.32d}$$

Braided river reaches are usually steeper, wider, and shallower than individual reaches with the same Q; indeed, braiding may be regarded as the incipient form of meandering (Petersen, 1986).

In a river with constantly changing values of Q, the problem arises as to what to consider as an appropriate discharge in the above equations.

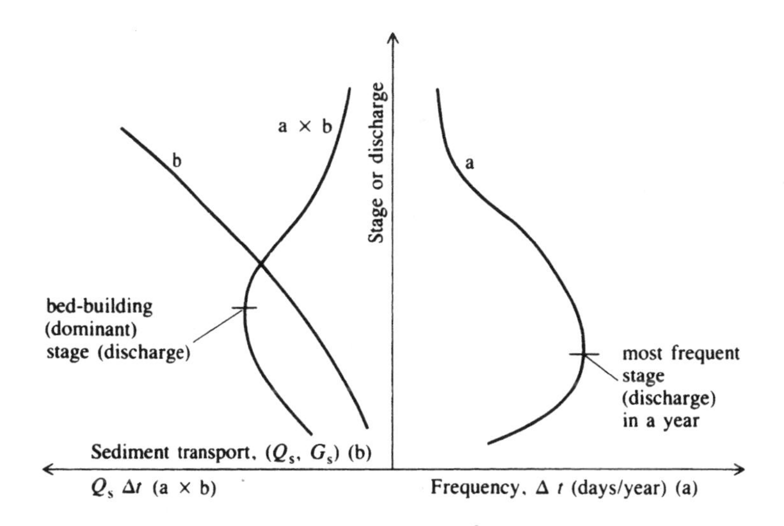

Fig. 8.4 Bed-building (dominant) stage (discharge)

The concept of a 'dominant discharge', i.e. a steady discharge giving rise to the same pattern of meanders, slope and channel geometry as the annual sequence of discharges, is frequently used. It can be defined as bankful discharge, or a discharge of a certain return period, or perhaps best as the discharge associated with the 'bed building water level' at which the largest total volume of sediment transport per year occurs, and is determined as shown in Fig. 8.4.

8.4 River surveys

8.4.1 Mapping

The river engineer needs reliable maps for river investigations and for the design and execution of engineering works. Maps compiled by the usual terrestrial surveying techniques are often incomplete and unreliable, but aerial photographic techniques and satellite images can provide accurate valuable data on river systems.

8.4.2 Water levels (stages)

Water levels or stages are obtained from gauges (non-recording or recording types) installed at gauging stations. Figure 8.5 shows an automatic

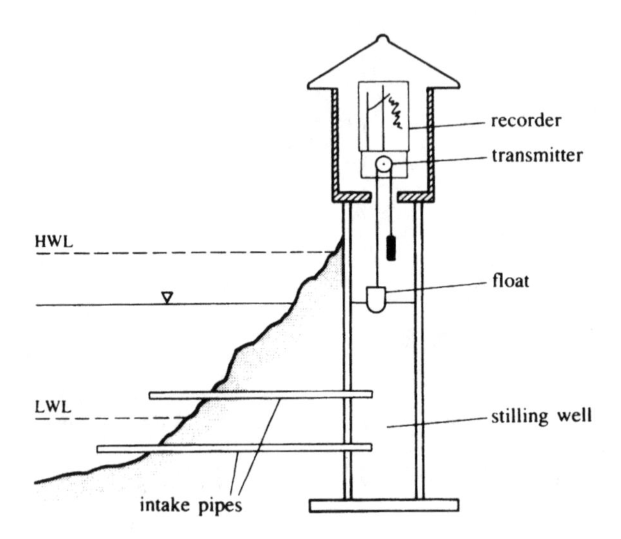

Fig. 8.5 Water level (stage) recording station

recording type gauge. Shallow streams and low-flow stages may be waded and measured with the help of a graduated staff. In alluvial streams with fast-moving flows the gauge may be suspended from a fixed reference point, e.g. a bridge parapet wall.

Gauge recordings are indispensable for design purposes and for hydrological studies. The data are equally valuable for the purposes of navigational studies and for high (flood predictions) and low (water quality, aquatic life) water management. Stage readings at a gauging station combined with discharge measurements yield stage–discharge (rating curves) relationships (Section 8.4.4).

Telemetry is used to transfer water level data from remote stations, either continuously or at predetermined intervals (Herschy, 1986).

8.4.3 Bed levels

The simplest method of establishing bed levels is by using a graduated sounding rod with a base plate fixed at the bottom end. The base plate prevents the rod from penetrating into the river bed and also helps to keep it in a vertical position.

The bed topography can also be obtained by sounding techniques from the water surface. In river engineering bed level measurements have to be made periodically. The most commonly used sounding instrument is the echo or supersonic sounder, which can be mounted aboard the survey vessel. The sounder works on a principle based on the propagation speed of sound in water, and measurement of the corresponding time lapse. Since the distance recorded is the shortest distance between the transducer and river bed, care must be taken in using the sounders in an irregular river bed. In such cases high-frequency sounders (small angle of spread from transducer) are preferable. Optical instruments such as the sextant and the range finder, and electronic instruments such as a radiolog system, are commonly used to establish the location of sounded points. Appropriate data processing (hand or computerized) finally produces the bed profile.

8.4.4 Discharges

(a) Area–velocity methods

1. *Floats.* River discharges are usually computed from velocity measurements, the simplest method being timing the movement of a float over a known distance. Several types of float are in use, providing data from which mean flow velocities can be derived (BSI, 1969a, b, 1986; Herschy, 1999). The float technique is more often used for reconnaissance estimates of flow. Surface floats are very useful to identify flow patterns, e.g. in estuarial surveys.
2. *Current meters.* The flow velocities are measured using current meters at systematically distributed points over the cross-section. The area of the cross-section is determined from soundings and the discharge obtained as:

$$Q = \Sigma A_i V_i. \tag{8.33}$$

In the case of shallow streams, a miniature type current meter attached to a graduated staff may be operated by the gauger wading across the section, measuring both the stages and velocities simultaneously. Wide rivers may be gauged from a boat held in position along a fixed alignment across the section. Sometimes, permanent installations such as a cable car (gauger travels with car and operates current meter) or a cableway (gauger operates current meter from the bank with mechanical winches) are used to carry out velocity and water level measurements.

(b) Dilution methods

Dilution gauging is an alternative solution in streams with steep gradients, and in shallow torrents where conventional current metering techniques cannot be used.

A solution of a known strength of tracer (e.g. NaCl, rodamine dye, etc.) is injected into the flow at a constant rate (or by sudden injection) over the cross-section and samples are taken at the downstream end at regular time intervals. As soon as the strength of the samples has obtained a constant value (plateau level) the flow rate, Q, in the channel can be calculated using the equation

$$Q = qN \tag{8.34}$$

where q is the rate of injection of the tracer and $N = C_1/(C_2 - C_0)$, C_1 being the concentration of the injected solution, C_2 the plateau concentration of the samples and C_0 the background concentration already present in the stream.

The gulp (or sudden) injection or integration method results in the following equation:

$$Q = VC_1 \Big/ \int^{t} (C_t - C_0)\mathrm{d}t \tag{8.35}$$

where V is the volume of solution injected, and C_t is the concentration of tracer at the sampling point during a period of time dt.

(c) Gauging structures

Discharges (modular flows) can also be measured using the standard gauging structures (artificial control) such as the broad-crested weir, the Venturi flume and the Crump weir, etc., with one single measurement of the stage upstream of the structure with respect to its sill level. Section 8.5 provides further details.

(d) Modern techniques of river gauging

The existing methods of river gauging are not always suitable, e.g. for navigable rivers with locks, and non-obtrusive techniques of gauging are preferred in such cases (Cole, 1982).

1. *Electromagnetic flow gauge.* A large coil buried across and beneath the river bed is magnetized by an external source and the water moving through the coil induces an electromotive force (e.m.f.) which is proportional to the average velocity through the cross-section. The probe system (Herschy, 1999) to sense the induced e.m.f. is very sophisticated and needs highly skilled personnel. The installation is expensive both to set up and to operate.
2. *Ultrasonic flow gauge.* Acoustic sound pulses transmitted through the water from transmitters from one side of the river are received by

sensors at the other side. By installing transmitters and receivers on each bank along a line joining them at an angle across the stream, the difference in travel times of the two pulses in the two opposite directions is measured; this is related to the mean velocity of the stream. This method again requires sophisticated electronic equipment (Herschy, 1999) to process the data from the transmitters and receivers, and can only be used in weed-free, clear water and stable bed streams. It is expensive to install and operate, although a high accuracy of flow measurement is possible.

3. *Integrating floats.* This technique is based on the principle of moving floats, the floats being air bubbles released from nozzles located at regular intervals from a compressed air line laid across the bed of the stream. The areal spread (Sargent, 1981) of the rising bubbles at the water surface (photographed with special equipment) gives the measure of the mean velocity of the stream.

(e) Natural control

The natural control of a river reach is a particular section (gauging station) where discharges and their corresponding stages are measured and a unique relationship (called the rating curve) between them is established. This relationship can then be used to estimate discharges from observed stages.

The rating curve, once established, has to be checked periodically for its validity and, if necessary, adjustments have to be made. Considerable deviations may occur as a result of morphological changes of the river bed (scouring, deposition, vegetal growth, etc.), the presence of flood waves and any changes introduced along the reach of the river in the vicinity of the control section, both upstream and downstream of it.

The rating curve generally assumes the form

$$Q = a(H \pm z)^b \tag{8.36}$$

where H is the stage, and a, b, and z (stage at zero discharge) are the parameters relating to the control section.

If three values of discharges Q_1, Q_2, and Q_3 are selected from the established rating curve such that $Q_2 = (Q_1 Q_3)^{1/2}$, the stage at zero discharge z can be obtained from (WMO, 1980)

$$z = (H_1 H_3 - H_2^2)/(H_1 + H_3 - 2H_2) \tag{8.37}$$

where H_1, H_2, and H_3 are the stages corresponding, respectively, to the three discharges.

8.4.5 Sediment transport

The sediment concentration in rivers varies enormously with time and between continents, countries, and even catchments, e.g. from an average 15 000 ppm at the mouth of the Hwang Ho River to 10 ppm in the Rhine delta (de Vries, 1985) (Section 4.5).

Sediment data are essential for the study of the morphological problems of a river. From the existing sediment transport equations and the hydraulic parameters of the river it is possible to estimate the rates of sediment transport; however, the computed results could lead to large errors and differ by several orders of magnitude. Thus, whenever possible, actual sampling is the more reliable method of measuring sediment transport rates in a river.

Suspended sediment (concentration) samples can most simply be collected by spring-loaded flap valve traps, or by samplers consisting of a collecting pipe discharging into a bottle. Continuous or intermittent pumped samplers are also being used. Point-integrating or depth-integrating sediment samplers, with nozzles oriented against and parallel to the flow and samplers shaped to achieve a true undistorted stream velocity at the intake, are used for measuring suspended sediment discharge. The methods used for measuring suspended sediment concentration and discharge are summarized in ISO 4363/1993. The US series of integrating samplers (particularly the US P-61) developed by the US Geological Survey are frequently used (e.g. Jansen, van Bendegom and van den Berg, 1979). Sediment from the bed is collected for further analysis of size, shape, etc. by various types of grabs (ISO 4364/1977).

Bedload transport measurement can, in principle, be carried out as follows:

1. using a sediment sampler, i.e. a device placed temporarily on the bed, and disturbing the bedload movement as little as possible;
2. using other methods consisting of (i) surveying sediment deposits at river mouths or, in smaller streams, collected in trenches, (ii) differential measurement between normally suspended sediment load and total load, including the bedload brought temporarily into suspension in a river section with naturally or artificially increased turbulence (turbulence flumes), (iii) dune tracking, (iv) remote sensing, (v) tracers, and (vi) acoustic dectectors.

Quantitative measurements of bedload is extremely difficult and there is probably no universally satisfactory method, although some reasonably well functioning samplers have been developed. Their efficiency (i.e. the ratio of actually measured sediment transport to that occurring without the presence of the sampler) has to be tested in the laboratory for the range of field conditions in which they are to be used (Novak, 1957; Hubbell *et al.*, 1981).

8.4.6 Water quality

The quality of water is judged in relation to its use (e.g. irrigation, drinking, etc.) and its suitability to the aquatic life. Quality is defined by several parameters such as its pH value, BOD, COD, TOC, etc., and for each criterion there can be a variety of determinants.

These parameters are usually determined by laboratory tests on samples collected from the river water, carried out either *in situ* (mobile laboratory) or in central laboratory facilities. Continuous *in situ* monitoring is preferable in order to avoid possible changes in characteristics due to transportation and the time taken between sampling and analysis. The quality standards are formulated on a statistical basis, and are more flexible for uses of abstracted water since it is possible to improve it by treatment.

Dissolved oxygen (DO) sensors, pH meters and suspended solids (SS) measuring devices, continuous monitors of temperature, ammonia, etc., are commonly used to measure water quality. The reader is referred to James (1977) for the quality requirements of water in relation to its uses and to Kiely (1998) and Tebbutt (1998) for discussion of water quality parameters and their control.

8.5 Flow-measuring structures

Flow-measuring (artificial control) structures are built across or in the streams to be gauged. The flow is diverted through the structure, creating critical flow conditions (flow upstream of the structure being subcritical), thus producing a unique relationship between the discharge, Q, and stage, h (upstream water level above the sill crest of the structure), in the form

$$Q = Kh^n \tag{8.38}$$

where K and n are dependent on the type of structure which, in turn, depends on the size of the stream and the range of flows to be gauged and on the sediment load carried. Gauging structures can be divided into two categories: (a) weirs – sharp-crested (thin plate) weirs or notches, spillways, broad-crested weirs, or Crump weirs; (b) flumes – Venturi, Parshall, or steep slope (Bos, Replogle and Clemmens, 1984).

A weir raises the upstream water level above its sill, water flowing over it under critical conditions. They are not suitable in debris- or sediment-laden streams, as deposition may take place upstream of the weir, thus changing the approach conditions of the flow. Floating debris and ice may also dent the crest of sharp-crested weirs.

Flumes are suitable for small streams carrying sediment-laden flows. The contraction introduced in the channel can be designed to create critical flow conditions at the throat (contracted section). Sometimes a raised sill (Featherstone and Nalluri, 1995) in a throatless flume may be introduced to create critical flow conditions. The water level in the upstream subcritical flow is then directly related to the discharge passing through the flume. The length of the flume may be reduced by providing a hump at the throat, which ensures critical flow conditions. The rise in upstream water levels is smaller than in the case of weirs.

Different types of weirs and flumes in use are summarized in Table 8.1.

The behaviour of the flow-measuring structures largely depends on the downstream water level (dependent on downstream channel control) which increases with discharge. The increased level may drown the structure and the unique free-flow (modular) relationship between the stage and discharge is lost. It is not always practicable to set the weir sill at higher elevations to avoid submergence because of the consequent flooding and waterlogging problems upstream of the structure. On the other hand, the drop in the sill level of the Parshall flume can considerably increase the measuring range of modular flows; and such a structure can conveniently be used as a drop- and flow-measuring structure in irrigation canals. However, equation (8.38) may also be used to gauge non-modular flows with appropriate correction factors (Water Resources Board, 1970; Bos, 1976; Ranga Raju, 1993) (Chapter 9).

The weir and flume structures are not normally suitable for measuring flood flows. In such cases open-channel methods such as the slope-area method, the constant fall method, etc. may be used. If there are no radical changes in the cross-section of a natural control with rising stages (e.g. no floodplains) extreme flood flows can also be predicted by special techniques such as extending the available stage–discharge curves.

8.6 River flood routing

Flood routing is the process of transforming an inflow passing through a river reach (or a reservoir) into the outflow. During this process the flood inflow hydrograph changes its shape: its peak is usually lowered and its base extended, i.e. the flood subsides. The usual task is to determine the peak reduction – attenuation of the flood between inflow and outflow – and the time lag between the peaks. Flood subsidence is controlled both by the frictional and by the local resistance and acceleration terms in the equation of motion and by storage.

Table 8.1 Types of weir

Type	*Shape*	*K*	*n*	*Remarks*
Sharp-crested weir	Rectangular (Fig. 8.6(a))	$2/3C_dC_v(2g)^{1/2}b$	3/2	b effective width of notch; to measure moderate to large discharges
$C_d = \alpha + \beta(h/P)$, where $\alpha = 0.602$ and $\beta = 0.075$ for $b/B = 1$; $\alpha = 0.592$ and $\beta = 0.011$ for $b/B = 1/2$				
	Triangular (Fig. 8.6(b))	$8/15C_dC_v(2g)^{1/2}\tan\theta/2$	5/2	θ included angle; to measure small flows
$C_d = f(h/P, P/B, \theta) \simeq 0.58 - 0.61$ (see BSI, 1969a, b, 1986)				
	Compound weirs (Fig. 8.6(c))			To measure wide range of flows; sensitive to approach conditions and submergence
Broad-crested weir	Rectangular	$0.544C_dC_vg^{1/2}b$	3/2	To measure large flows; less sensitive to approach conditions and submergence (Table 9.2 and Worked example 9.1)
$C_d = f(h$, crest length L, h/b, roughness of the crest$) \simeq 0.85 - 0.99$ (for recent studies, see Ranga Raju (1993))				
Spillways: K and n values are the same as those of sharp-crested rectangular weirs but C_d may vary (Chapter 4)				
Crump weir (Water Resources Board, 1970)	Sharp crested with 1:2 upstream and 1:5 downstream slopes (Fig. 8.21)	$C_dC_vg^{1/2}b$	3/2	Fairly constant value of C_d; to measure moderate flows; less sensitive to approach conditions; good prediction of submerged (non-modular) flows (Worked example 8.3)
Flumes	Venturi	$0.544C_dC_vg^{1/2}b$	3/2	b throat width; to measure wide range of flows; copes with sediment and debris-laden flows; increased non-modular flow range with reasonable estimates (see Bos, 1976)
$C_d = f(L/b, h/L) \simeq 0.95$–$0.99$ (see BSI, 1969a, b, 1986)				
	Parshall (Fig. 8.7)	K and n vary with size of flume; design tables available (Bos, 1976)		
	Throatless flume: raised bed (hump) in stream	K and n values are the same as those of broad-crested weirs; cheap		To measure moderate flows (see Featherstone and Nalluri, 1995)
	Steep slope stream flume: supercritical approach flow: special flume (Harrison and Owen, 1967)			

$C_vh^n = H^n$; $H = h + V_a^2/2g$; $V_a = Q/B(h+P)$, where H is the energy head and V_a is the approach velocity; $C_v(>1.0)$ is a function of the discharge coefficient C_d, b/B and $h/(h+P)$, where B is the channel width and P is the height of the sill. Solutions for C_v (graphical or analytical) are available (e.g. see BSI, 1969a, b, 1986; Ackers *et al.*, 1978)

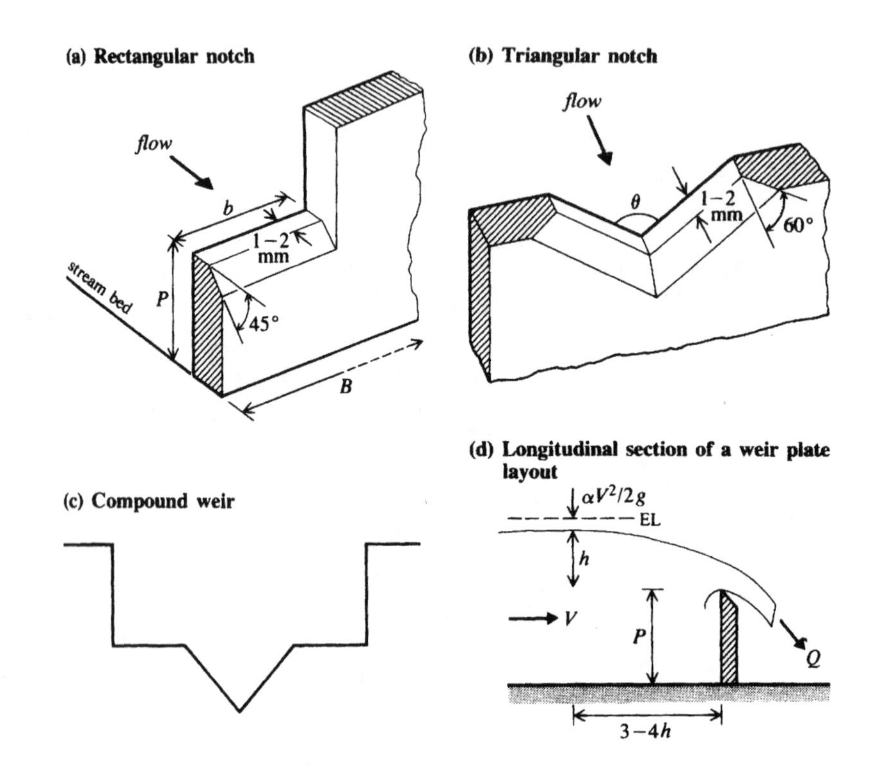

Fig. 8.6 Thin-crested weirs

In *reservoir flood routing* only the storage effects are considered, as the influence of impulsive motion of the inflow on the momentum of the outflow is neglected and with the assumption of a horizontal water surface the storage effects are quickly transmitted (Section 4.3); in river (channel) flood routing the combined effect of storage, resistance and acceleration is manifest.

There are three possible approaches to *channel flood routing.*

1. The *hydraulic method (exact dynamic solution)* is based on a numerical solution of the finite difference form of the Saint Venant equation(s) (equations (8.11)–(8.14)); its further discussion is outside the scope of this text, but see for example Abbott and Minns (1998) and Cunge, Holly and Verwey (1980). For this method data on channel shape and friction coefficients are required.
2. *Diffusion analogy (approximate dynamic solution)* is based on the analogy between the flood transformation and the dispersion–diffusion process. Depending on the approximation of the full dynamic equation various methods have been developed (e.g. the variable parameter diffusion method). The determination of various parameters requires knowledge of at least one flood event and of the

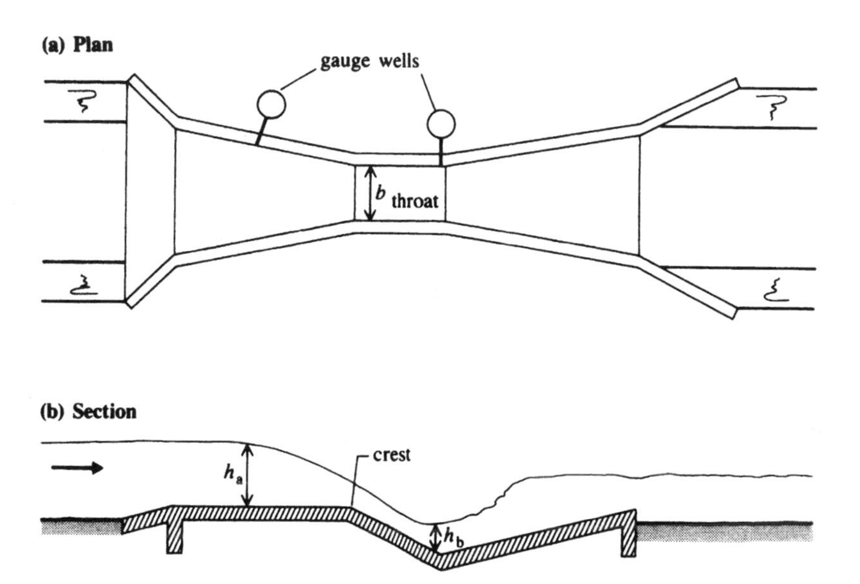

Fig. 8.7 Parshall flume

channel geometry. For further details see for example Price (1985) and Raudkivi (1979).

3. The *hydrologic method (kinematic solution)* is the simplest and most frequently used method, particularly after the connection between its parameters K and x – described below – (Muskingum method) and the channel physical characteristics had been established by Cunge (Muskingum–Cunge method) thus in effect bridging the gap between the last two methods. Again its application requires at least the data from one previous flood (e.g. Shaw, 1994).

The Muskingum method is basically an extension of the reservoir routing methods taking into account that the storage (V) in a channel is a function of both the inflow (I) and the outflow (O), i.e. that in addition to the prism storage there is a (positive or negative) wedge storage (equation (8.13)). Thus

$$V = K[O + x(I - O)] \tag{8.39}$$

where K is a storage constant (units of time) and x a weighting factor. It is evident that in the case of a reservoir $x = 0$ (equation (4.7)). Combining equation (8.39) with continuity ($I - O = \mathrm{d}V/\mathrm{d}t$; equation (4.2) or (4.3)) and eliminating the storage terms leads to the routing equation (Δt is the routing period):

$$O_2 = c_0 I_2 + c_1 I_1 + c_2 O_1 \tag{8.40}$$

with

$$c_0 = -\frac{Kx - 0.5\Delta t}{K - Kx + 0.5\Delta t}, \quad c_1 = \frac{Kx + 0.5\Delta t}{K - Kx + 0.5\Delta t}, \quad c_2 = \frac{K - Kx - 0.5\Delta t}{K - Kx + 0.5\Delta t}$$

$$(c_0 + c_1 + c_2 = 1).$$

In the original Muskingum method K and x (which are assumed to be constant for a given channel) are determined empirically from a plot of V against $xI + (1-x)O$ (equation (8.39)) selecting the plot which for a chosen value of x gives the best approximation to a straight line (V is determined from the previously known I and O hydrographs). However, Cunge (1969) has shown that

$$K = \Delta L / w \tag{8.41a}$$

and

$$x = 0.5 - \frac{\bar{Q}_p}{2S\bar{B}w\Delta L} \tag{8.41b}$$

(w is the average speed of the flood peak Q_p, ΔL the length of the reach, S the average bed slope and $\bar{B}$ the average channel width), giving a better physical interpretation of the routing parameters.

8.7 River improvement

8.7.1 General

The objectives of river improvement works are to aid navigation, to prevent flooding, to reclaim or protect land or to provide water supply for irrigation, hydropower development or domestic and industrial use.

The design of river improvements works in general builds upon the principles discussed in Sections 8.2 and 8.3 and should be based on fluvial geomorphology and wider river engineering aims (Thorne, Hey and Newson, 1997) and river mechanics (Hey, 1986; Yalin, 1992; Knighton, 1998; Yalin, Ferreira and da Silva, 2001; Julian, 2002).

It is extremely important in river training to adopt a holistic approach and to incorporate environmental impact assessment and socio-economic considerations in any design.

Flood-protection works include high-water river training (mainly by dykes), diversion and flood-relief channels with or without control structures, and flood-control reservoirs. Flood protection schemes require a careful cost–benefit analysis to determine a suitable design discharge which depends on the type of land, structures and property to be protected and the processes involved. The return period of this discharge may range from 1 to 100 years and in very special cases (large settlements, ancient historic monuments, nuclear installations, etc.) may be substantially higher (Jaeggi and Zarn, 1990).

The design of flood protection schemes nearly always involves the computation of stage in two- or multistage channels. A considerable body of basic research conducted in laboratory flumes (some of them up to 10m wide) and prototype channels has been carried out worldwide particularly during the last 25 years. It has provided fairly comprehensive information on stage–discharge curves, shear stress distribution and turbulence characteristics in the main channel and flow over the flood plain(s) and the rôle of the apparent shear plane between the main channel and flood plain(s) (e.g. Keller and Rodi, 1988; Knight and Samuels, 1990; Ackers, 1993). Based on this research, guidelines for the design of straight and meandering channels using zonal calculations and the coherence concept (ratio of the single section conveyance – Section 8.2.2 – to the conveyance obtained by summing the conveyances of the separate flow zones) have been developed (Wark, James and Ackers, 1994).

The interaction of in- and overbank flow on sediment transport has been investigated e.g. by Wormleaton, Sellin and Bryant (2004). A more detailed discussion of the whole topic of hydraulics of multistage channels is beyond the scope of this text; for further information see, e.g., Knight *et al.* (1999) and Chadwick, Morfett and Borthwick (2004).

For *navigation purposes* the main river improvement works are those which provide sufficient depth and/or stabilize the river channel in a suitable form, and provide bank protection against wave action, particularly on constricted waterways (Chapter 11).

The principal methods used to *improve river channels* are river regulation and dredging; on navigable rivers canalization, construction of lateral canals and flow improvement by reservoir construction and operation have also to be considered. In planning river improvement works, both the upstream situation and historical factors have to be taken into account, as the river is an evolving system; a good design has to try to assess this evolution even if the purpose of the improvement works is to stabilize the situation at least over the life of the design. A typical example of an upstream man-made influence is the effect of a reservoir construction on the downstream river morphology, which has to be taken into account in the planning and execution of river training works. The methods used for estimating this effect are spatial interpolation, use of régime relationships (Section 8.3), use of one- or two-dimensional mathematical models or a combination of all these methods (Brierley and Novak, 1983).

In river *regulation or training* the river may be encouraged to pursue its natural course or it may be straightened; the latter requires great sensitivity and should be used only with caution and due regard to environmental constraints. In the upstream reaches the main problem is the short-term and seasonal variation of flow, high velocity, channel instability and shoal formation. In the middle and lower reaches it is often necessary to raise river banks and carry out works reducing the channel width, e.g. groynes, longitudinal training walls etc. Dredging, using mechanical or suction dredgers, is the most effective means of estuarine river regulation, but its impact is often only temporary.

An efficient river training system will try to maintain and improve the natural sequence of bends of a meandering river, while preserving sufficient depth (e.g. for navigation) at low flows and suppressing unduly sharp bends and excessive velocities. This is mainly achieved by groynes, jetties, longitudinal dykes, and embankments and ground sills (see Section 8.7.2).

Even today, the guidelines for river regulation follow the 'laws' formulated by Fargue from his experience with the River Garonne in the latter part of the 19th century (laws of deviation, greatest depth, trace, angle, continuity and slope), which in principle advocate the avoidance of sharp discontinuities in river planform and longitudinal section, and the following of natural river forms and meanders appropriate to the river concerned (Section 8.3).

The normalization of the River Rhine upstream of Mannheim, carried out in the 19th century, is shown in Fig. 8.8: this was only partially successful because of the use of curves of too large a radius, and excessive shortening of the river (Jansen, van Bendegom and van den Berg, 1979). One must realize that the construction of cut-offs, which drastically shorten the river meander, increases the channel slope and can result in upstream erosion and downstream sediment deposition unless accompanied by carefully designed local training works, to preserve the tractive

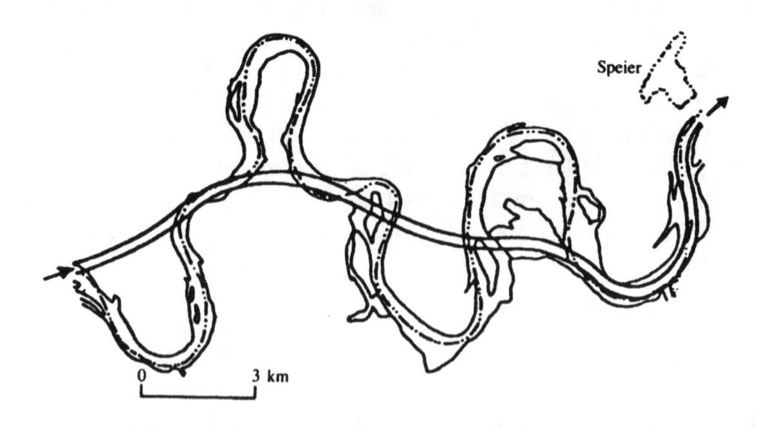

Fig. 8.8 Normalization of the Rhine (Jansen, van Bendegom and van den Berg, 1979)

force of the stream. The change of slope may require narrowing (for increasing slope) or widening of the river which, if left to self-adjustment, may take decades to be achieved.

The combination of the stage duration curve with a sediment rating curve (established by direct measurement (Section 8.4) or by computation (Section 8.3) can be used as shown in Fig. 8.9, to estimate the annual sediment run-off; the comparison of sediment duration curves corresponding to different stage–discharge curves in turn then gives an estimate of the increase or decrease of sediment transport due to a change in width of a river channel (Fig. 8.10) (UNECAFE, 1953).

After the selection of alignment and cross-section river training should proceed from upstream taking into account the requirements of '*high*', '*mean*' and '*low*' water training.

A suitable arrangement for this type of water river training is shown in Fig. 8.11, and the use of the above guidelines for the improvement of a confluence of two rivers is illustrated in Fig. 8.12 (Schaffernak, 1950).

A special case of river training is the junction or crossing of a river and navigation canal (or harbour entrance from a navigable river). Because the canal usually requires a fairly wide entry for navigation purposes, support for the river flow is lost and unwelcome sediment deposition may occur in the canal mouth. A solution to this problem is provided by specially formed circular enlargement of the canal ('Thijsse egg' developed in the Netherlands) with a circulation that prevents or at least reduces the sedimentation in the canal mouth (Fig. 8.13); for further particulars see Jansen *et al.* (1979).

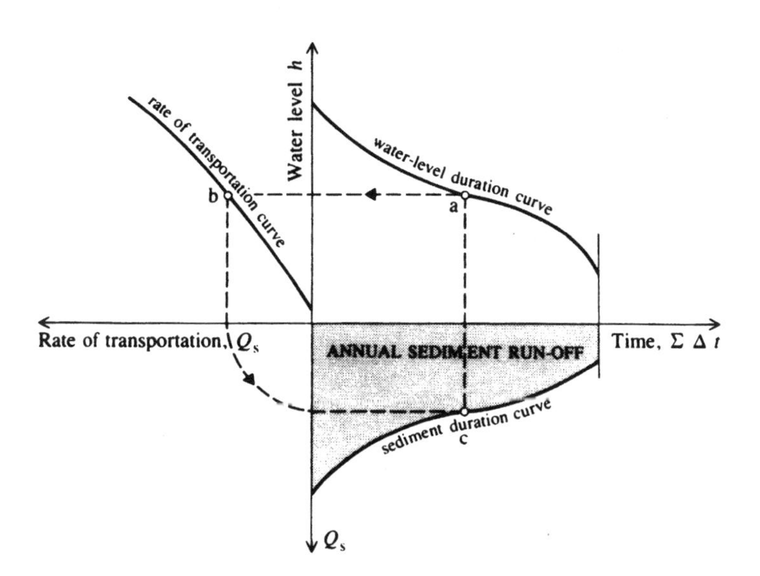

Fig. 8.9 Determination of annual sediment run-off

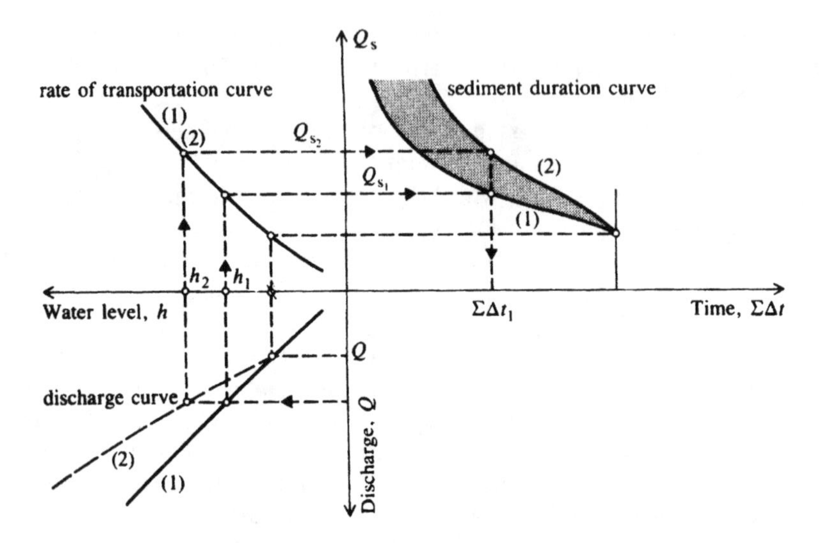

Fig. 8.10 Increase of sediment run-off due to channel contraction

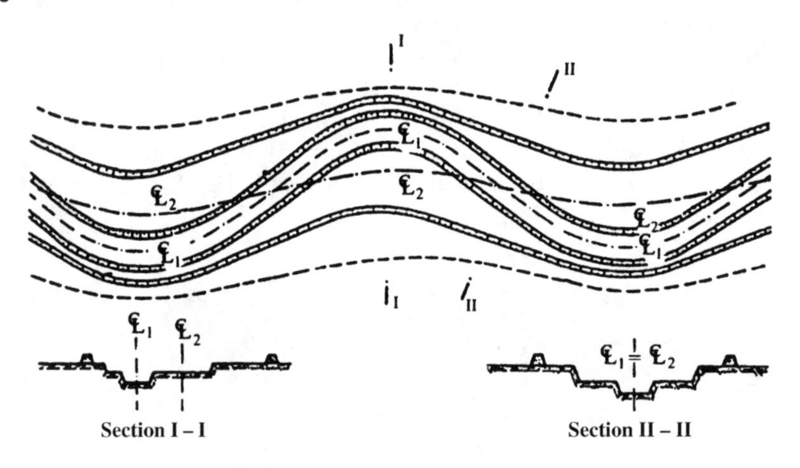

Fig. 8.11 High, mean and low water training

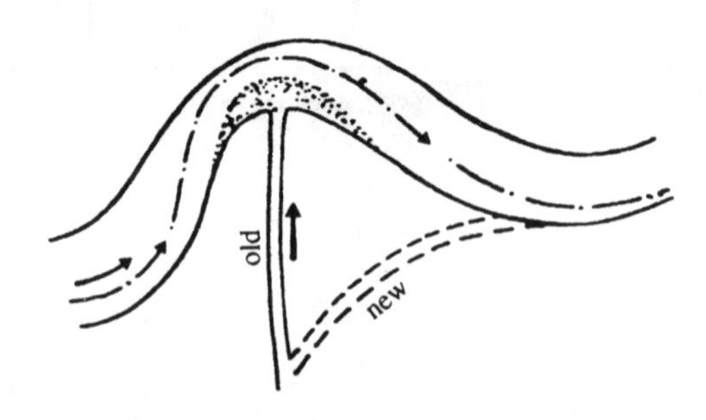

Fig. 8.12 River junction improvement

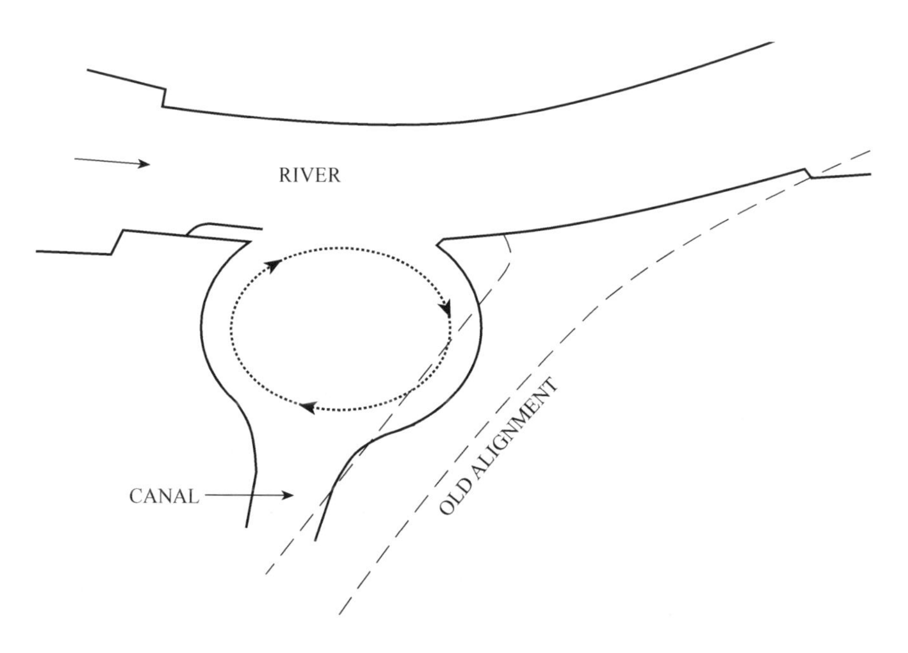

Fig. 8.13 Connection of canal and river (after Jansen *et al.*, 1979)

8.7.2 Groynes, dykes, vanes, and sills

Groynes are small jetties, solid or permeable, constructed of timber, sheet piling, vegetation, and stone rubble, etc. They usually project into the stream perpendicularly to the bank, but sometimes are inclined in the upstream or downstream direction. The main purpose of groynes is to reduce channel width and to remove the danger of scour from the banks; their ends in the stream are liable to scour, with sediment accumulation between them. As their effect is mainly local, the spacing between groynes should not exceed about five groyne lengths, but usually is appreciably smaller; a spacing of about two lengths results in a well-defined channel for navigation; the larger the ratio of groyne spacing to river width, the stronger the local acceleration and retardation, and thus the greater the hindrance to shipping. Details of groyne design are strongly influenced by economic factors, and a cost–benefit analysis for the determination of their height, spacing, length, and material is usually necessary; in more important cases this may have to be backed up by model studies (Chapter 16). Permeable groynes usually constructed from vegetation (e.g. tree tops) exert a less severe effect on the flow than solid ones, and by slowing down the current aid deposition of sediment in the space between them.

Examples of the use of groynes, with some construction details of a solid groyne, are shown in Figs 8.14 and 8.15. Note the combination of

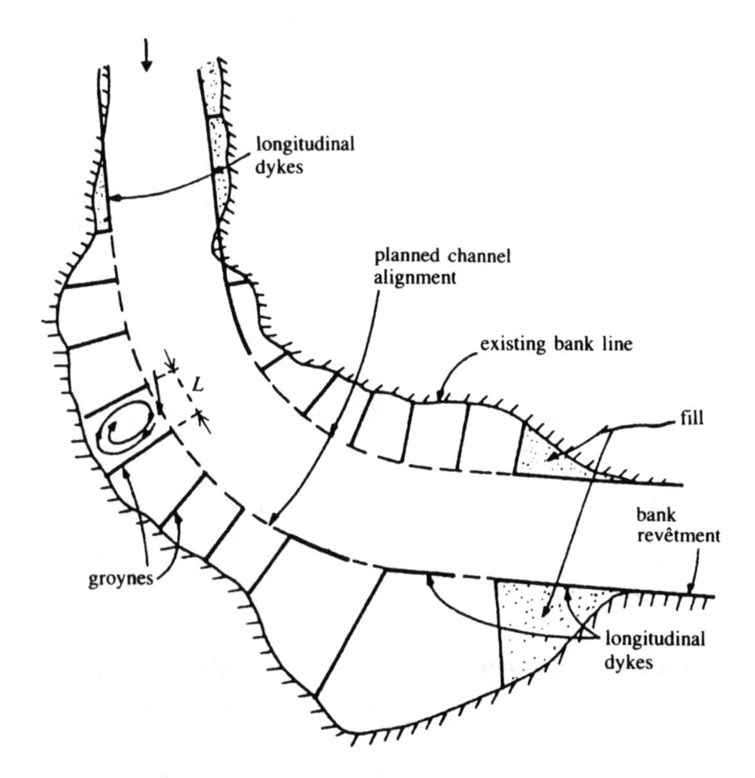

Fig. 8.14 River training by groynes (after Kinori and Mevorach, 1984)

some groynes with short longitudinal dykes to protect the head against erosion and aid sediment deposition between the groynes; in mountain streams it is advisable to offset the small dyke slightly into the space between the groynes.

Longitudinal *dykes* (training walls) are usually more economical than groynes and – if properly positioned – equally or even more effective. The material used is again rubble, stone, or fascine work (on soft river beds). Training walls may be single – on one side of the channel – or double. An example of the combined use of groynes and dykes in river training is shown in Fig. 8.16; note the open end of the training wall and its connection with the old river bank. High water dykes are used in flood protection (Fig. 8.11) and their design and construction has been discussed in Chapter 2; due consideration must be given to their protection against erosion by high-velocity flows and undermining by scour from a meandering river.

In some instances a series of stream *deflectors (vanes)* constructed of wood panels or metal (e.g. floating drums with sheet metal vanes), placed at a suitable angle (often almost parallel to the bank) and depth, can be used to either divert an eroding flow from the river bank or, on the other hand, to induce bed erosion and local deepening of the flow. Details of

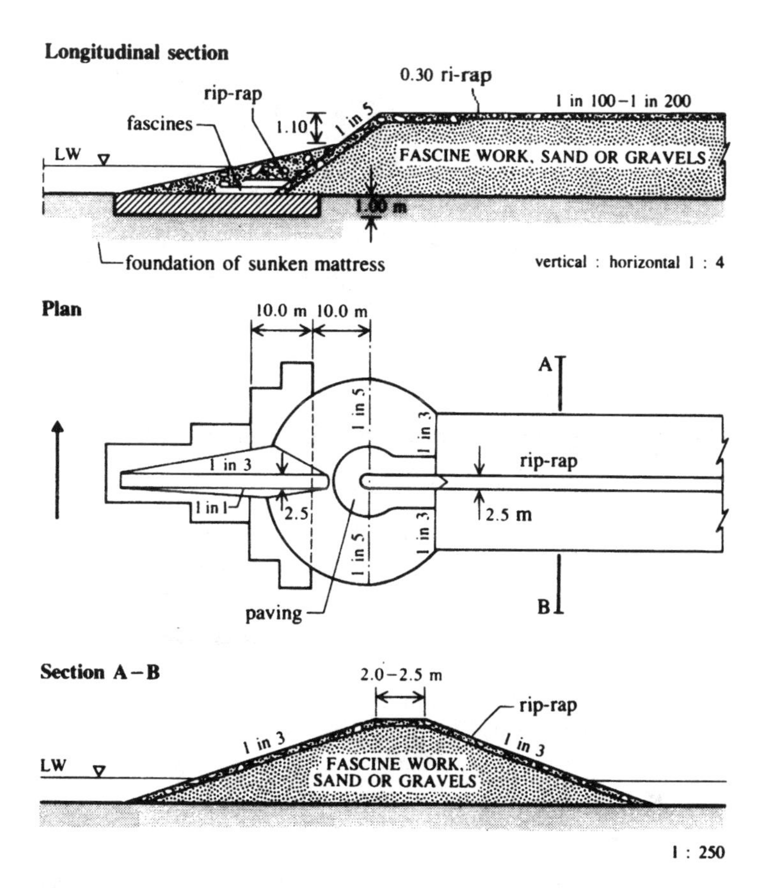

Fig. 8.15 Groyne construction

their location are best determined by model studies or experiments *in situ*. Deflectors rely in their action on the induced spiral flow and secondary currents as well as on the rectified main direction of flow. An example of the use of a series of floating vanes to rectify the flow and protect a bank in a large river is shown in Fig. 8.17.

Channel beds liable to substantial erosion can be stabilized by *ground sills* or – more extensively (and expensively) – by a series of drop structures. Ground sills usually span the whole width of the river channel, with the greatest height at each bank and a gentle slope to the stream centre. Rubble mounds, cribs filled with rubble, and concrete are some of the materials more frequently used for ground sills.

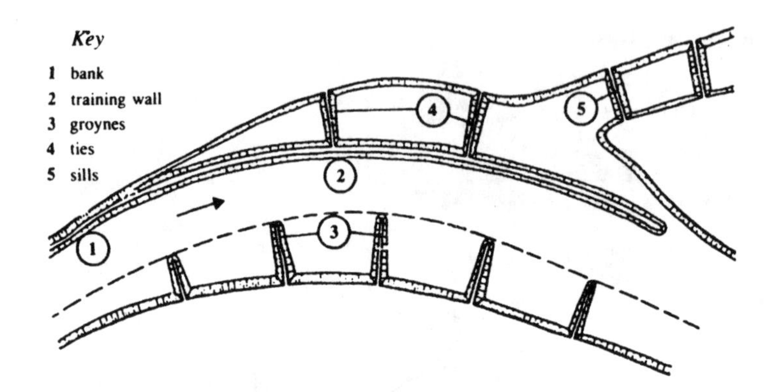

Fig. 8.16 River training with groynes and training wall

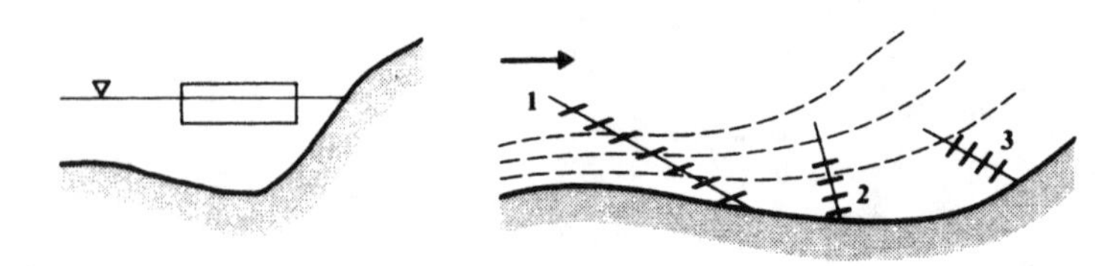

Fig. 8.17 Use of deflectors for bank protection

8.7.3 Bank protection

Bank protection is carried out by planting, faggotting (faggots or fascines are bundles of branches, usually willow), thatching, wattling, mattresses, rubble, stone pitching, gabions, bagged concrete, concrete slabs, asphalt slabs, prefabricated concrete interlocking units with or without vegetation, articulated concrete mattresses, soil–cement blocks, asphalt and asphaltic concrete, geotextiles (woven and non-woven fabrics, meshes, grids, strips, sheets and composites of different shapes and constituents), used tyres, etc., all used with or without membrane linings (e.g. nylon, rubber, polythene, etc.). The choice of material is influenced by the extent of the area to be protected, hydraulic conditions, material availability, material and labour costs, access to the site, available mechanization, soil conditions, design life, required impermeability, robustness, flexibility, roughness, durability, environmental requirements, etc.

Geotextiles and related materials have found increasing use in civil engineering projects since about the late 1950s (mainly in coastal engineering) and have been widely used in bank protection since about 1970. Non-wovens were initially used in Europe while the use of woven monofilaments originated in the USA; both are now used worldwide. When used as turf reinforcement in bank erosion control, velocities up to about 4 m s^{-1}

are withstood without damage and geotextiles placed between the soil and riprap are cheaper and give greater control during construction than a graded aggregate filter, particularly in underwater applications (see, e.g., PIANC, 1987 and Pilarczyk, 1999). The use of geotextiles as a concrete form has been developed for heavy duty (IFAI, 1992).

Unanchored inflatable tubes or *geotubes* are sleeves of high strength geotextile fabric (nylon, polyester, polypropylene, polyethylene), thickness from 0.8 to 4 mm, filled usually with sand (or with slurry mixture or even concrete). Since the 1980s geotubes have been marketed to replace sandbags in flood protection and have found increasing use in coastal and river engineering. Their height ranges from 0.7 to 2.5 m (in an elliptical cross-section) and length from 15 to 30 m (exceptionally appreciably longer tubes have been used in some coastal engineering projects).

There are many commercially available products used in bank protection, and the manufacturers should be able to provide information about the best conditions for their use, their durability, and about equivalent roughness sizes or coefficients of friction and maximum permissible velocities. For natural materials these velocities would typically range from $0.5\,m\,s^{-1}$ for fine sand to $1.5\,m\,s^{-1}$ for shingle, $1.85\,m\,s^{-1}$ for hard clay and $0.8–2.1\,m\,s^{-1}$ for various types of grass (and soil conditions). Stone-filled wire-mesh mattresses and gabions can withstand velocities over $5\,m\,s^{-1}$ if thicker than 0.30 m. Manning's coefficient for this type of mattress can vary from 0.016 (corresponding to a roughness size of about 3.5 mm) for a channel lined with mattresses, grouted, and sealed with sand asphalt mastic, with a smooth finish, to 0.027 ($k = 125$ mm) for a channel lined with gabions and filled with unselected quarry stones.

It is important to appreciate that any protective facing of banks must be continued to the river bed and be provided with a good footing (foundation). A good filter adapted to suit the subsoil is essential, as is drainage of sufficient capacity underneath more or less impermeable revêtment (see also Section 2.6.3 and Worked example 9.1). Where permeable revêtment is used (the majority of cases in river training) it must provide sufficient drainage from the slope without air being trapped. An example of the use of fascines with a foundation of heavy brush rollers, used for the protection of a bank, is shown in Fig. 8.18. The use of gabions and gabion mattresses for bank strengthening is illustrated in Fig. 8.19, and bank protection with a flexible mattress of prefabricated concrete elements (Armorflex) on a geotextile layer is shown in Fig. 8.20; this could also be covered by vegetation, making the bank protection almost invisible. For further details of bank protection design, materials and construction, see for example Brandon (1987, 1989), Hemphill and Bramley (1989) and Escarameia (1998).

For further information on resistance of vegetation to flow and its effect on bank stability see, e.g., Thorne (1990), Copin and Richards (1990), Oplatka (1998) (which includes also an extensive bibliography on plant biology and mechanics) and Armanini *et al.* (2005).

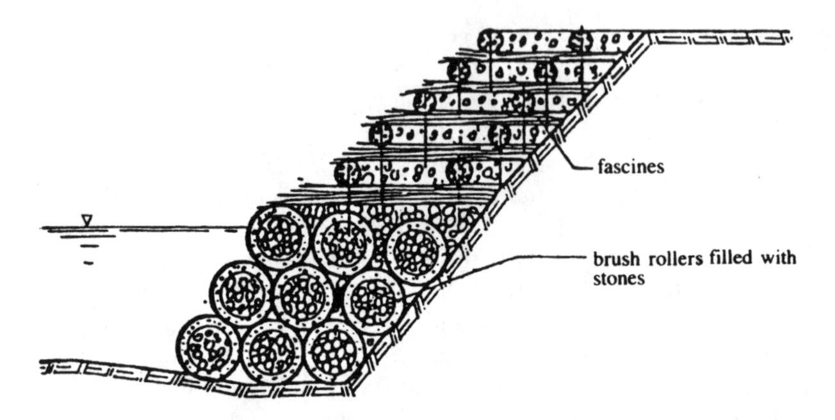

Fig. 8.18 Bank protection with fascines and brush rollers filled with stones

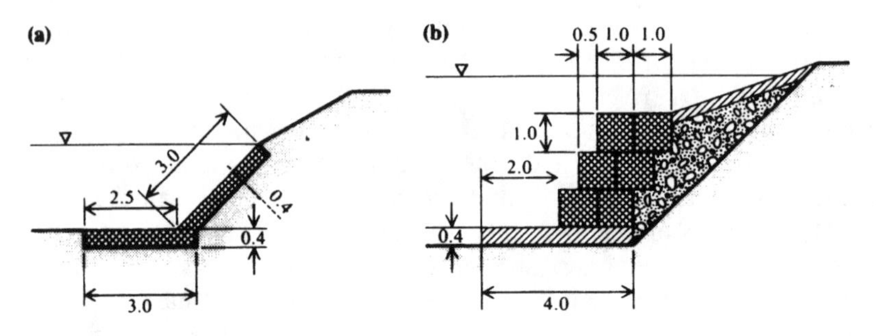

Fig. 8.19 Bank protection with gabions; dimensions in metres

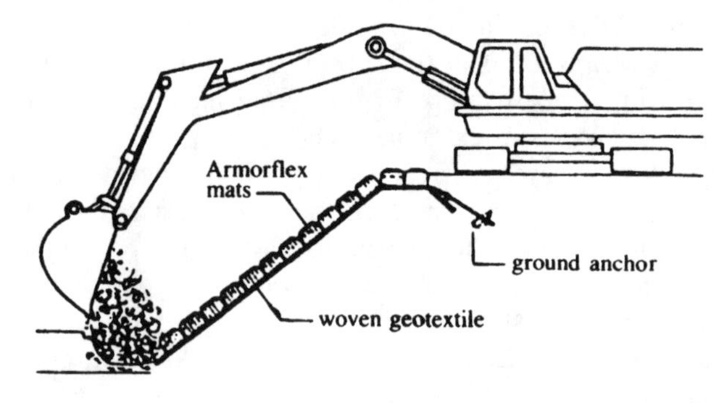

Fig. 8.20 Bank protection with flexible mattress of concrete elements

Worked Example 8.1

A wide open channel has a depth of flow 1.7 m and mean velocity 2.5 m s^{-1}. Taking $\rho_s = 2650$ kg m^{-3}, find the minimum size of bed material needed to obtain a stable bed.

Solution

Taking c in equation (8.19) as 0.056 (Shields) and using the Manning–Strickler equation (8.6),

$$d = RS/(\Delta \times 0.056) = 11RS = 11(Vn)^2/R^{1/3} = (0.04V)^2 \times 11 \times (d/R)^{1/3}.$$

Therefore

$$d^{1/3} = 0.133V/R^{1/6} = 0.133 \times 2.5/1.7^{1/6} = 0.303, \quad \text{or}$$
$$d = 0.028\,\text{m} = 28\,\text{mm}.$$

Worked Example 8.2

A channel 2.0 m deep, 15 m bed width, with 1:2 ($V{:}H$) side slopes is excavated in gravel of $d = 50$ mm. What is the maximum permissible channel slope, and what discharge can the channel carry without disturbing its stability? Take $\varphi = 37°$ and the bed critical shear stress as $0.97\rho g y S$.

Solution

For a particle on the bed, $\tau_{bc} = \rho g \Delta d \times 0.056 = 10^4 \times 1.65 \times 0.05 \times 0.056 = 46.2$ N m^{-2}. For a particle on the side, $\tau_{sc} = \tau_{bc}\ (1 - \sin^2\theta/\sin^2\varphi)^{1/2} = 30.9$ N m^{-2} (note that $\tan\theta = 1/2$). Thus

$$0.75\rho g y S_1 = 30.9\,\text{N m}^{-2}\ \text{(side)},$$

$$0.97\rho g y S_2 = 46.2\,\text{N m}^{-2}\ \text{(bed)}.$$

Therefore $S_1 = 0.865S_2$ and as $S_1 < S_2$ the side stability is decisive. The permissible channel slope is $S_1 = 30.9/(10^4 \times 0.75 \times 2) = 2.06 \times 10^{-3}$. The permissible discharge is

$$Q = AR^{2/3}S^{1/2}/n = [38/(0.04 \times 0.05^{1/6})][38/(15 + 4\sqrt{5})]^{2/3}(2.06 \times 10^{-3})^{1/2}$$
$$= 96\,\text{m}^3\,\text{s}^{-1}.$$

Worked Example 8.3

Design a Crump weir to be installed to measure the flow from a small catchment. The cross-section of the river where the structure is to be built is shown in Fig. 8.21, and existing current meter gaugings of the river flow are tabulated below.

Staff gauge reading, h′ (m)	*Measured discharge, Q ($m^3 s^{-1}$)*
0.12	0.75
0.15	1.10
0.19	1.21
0.26	1.80
0.32	2.30
0.42	4.36
1.04	15.40

The staff gauge reading is 0.00 when the flow depth in channel is 0.37 m (=z). Additional data are as follows:

1. The dry weather flow range is 0.5–1.0 $m^3 s^{-1}$.
2. The 20-year return flood is 25 $m^3 s^{-1}$ and the 100-year return flood is 60 $m^3 s^{-1}$.
3. The river channel upstream of the proposed site lies in a flood plain (used for cattle grazing).

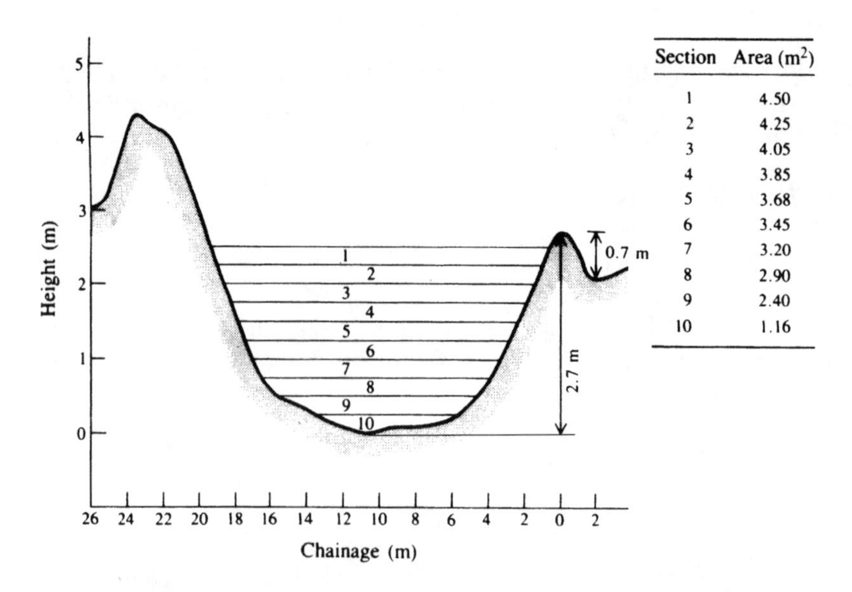

Section	Area (m^2)
1	4.50
2	4.25
3	4.05
4	3.85
5	3.68
6	3.45
7	3.20
8	2.90
9	2.40
10	1.16

Fig. 8.21 River cross-section

Solution

Figure 8.22 shows the details of a Crump weir profile. Figure 8.23 shows a typical layout of the design flow chart based on which appropriate software could be developed to solve the problem. The detailed solution and design calculations are described below.

Select the width, b, of the weir approximately as the width of the bed B, i.e. $b \simeq B = 10\,\mathrm{m}$ (Fig. 8.21). The minimum head over the crest to avoid surface tension effects is approximately 0.06 m. Therefore, using the Crump weir equation for discharge (modular),

$$Q = C_d g^{1/2} b H_1^{3/2}. \tag{i}$$

With $C_d = 0.626$ (Water Resources Board, 1970), H_1 for the minimum flow of $0.5\,\mathrm{m^3\,s^{-1}}$ is $0.0864\,\mathrm{m} > 0.06\,\mathrm{m}$, and is satisfactory. A log–log plot of the stage–discharge curve (with stage $= h' + z$ in cm) is plotted in Fig. 8.24.

From the given cross-sectional details of the stream, the area A (Fig. 8.21) is calculated for different stages, and the velocity computed. Hence the energy head H_2 downstream of the weir is calculated from

$$h' + z + V^2/2g.$$

The tailwater rating curve (hence H_2 versus Q) is plotted in Fig. 8.25. Using equation (i) the upstream rating curve (the weir formula gives H_1 versus Q) is also plotted on Fig. 8.25.

(a) Cross section of Crump weir

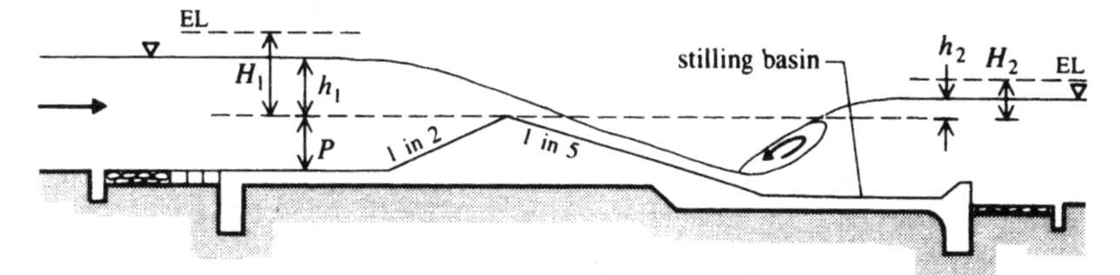

(b) Plan

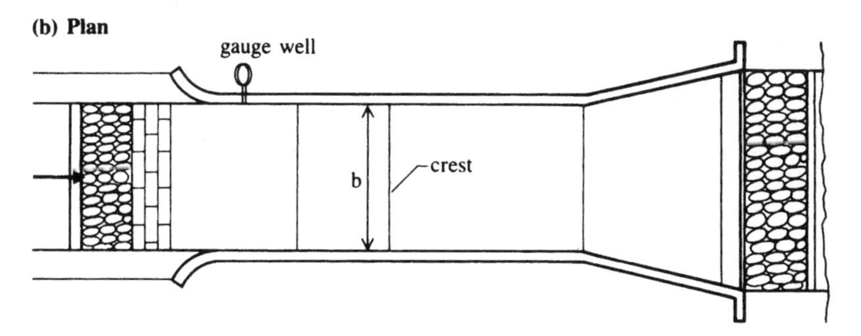

Fig. 8.22 Crump weir layout

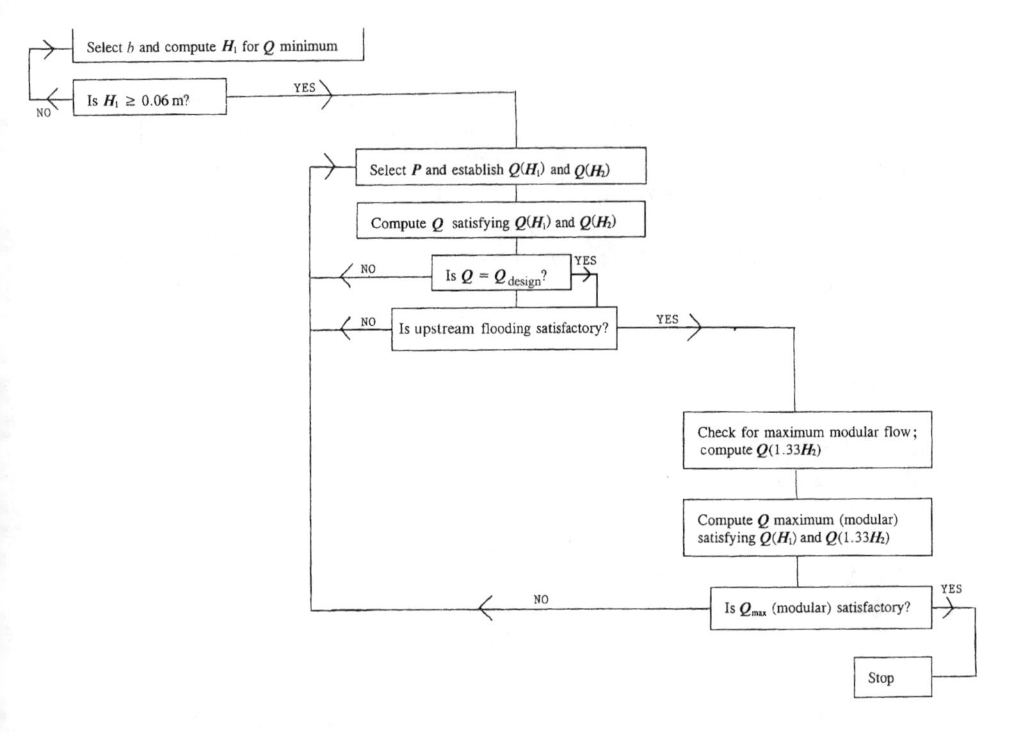

Fig. 8.23 Flow chart for the design of 'Crump weir'

SELECTION OF CREST HEIGHT, P

Large heights are uneconomical and create upstream flooding, whereas small heights create frequent non-modular (drowned) flow conditions. The limiting modular ratio, H_2/H_1 for the Crump weir is 0.75 (Water Resources Board, 1970) and a design modular flood of around 1/2 to 1 year return periods is considered as the norm. From the given information on the flood flows in the river, the 1-year return flood is around $5\,\mathrm{m^3\,s^{-1}}$ (log–log plot between Q and T). The limiting modular ratio suggests that $H_1 = 1.33H_2$ and, choosing a crest height of 0.7 m (as first trial), the curve Q versus $1.33H_2$ is plotted on Fig. 8.25, which intersects the Q versus H_1 plot at $Q = 9.4\,\mathrm{m^3\,s^{-1}}$. This corresponds to a 3.3-year return period flood, which is far in excess of the design norm. Hence select a lower value of P.

A crest height of 0.6 m produces a modular limit flood of $4.6\,\mathrm{m^3\,s^{-1}}$ with a return period of 1.1 years. This is satisfactory and hence a crest height of 0.6 m is adopted for the design.

CHECK FOR SPECIFICATIONS

1. Non-modularity: the modular limit discharge of $4.6\,\mathrm{m^3\,m^{-1}}$ corresponds to $H_2 = 0.29$ m and $H_1 = 0.39$ m (Fig. 8.25). Thus $H_2/H_1 = 0.74$ and is satisfactory.

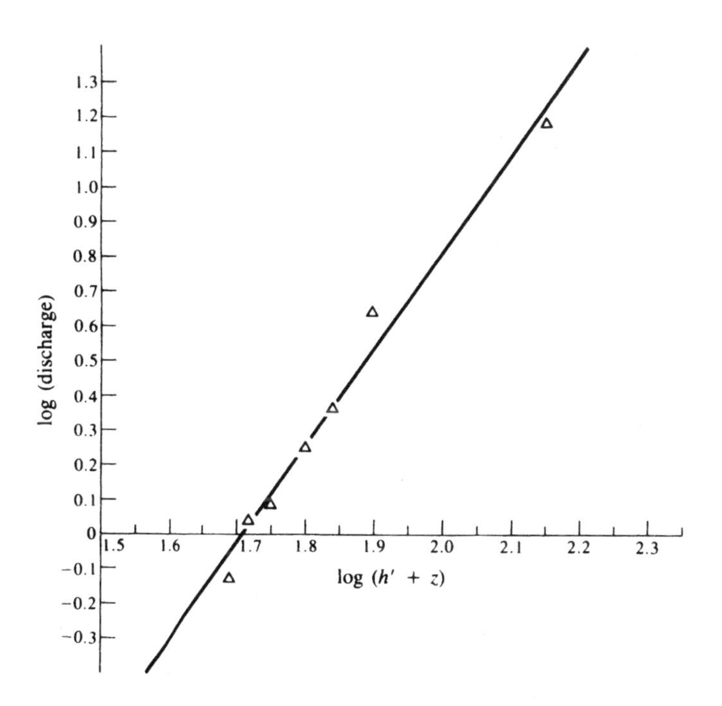

Fig. 8.24 log(h' + z) versus log Q

2. Figure 8.21 suggests that the maximum stage before overtopping of banks occurs is 2.7 m (from the bed level). Allowing a reasonable groundwater table of 0.3 m below ground level, the maximum allowable stage preventing overtopping and waterlogging is 1.7 m. Therefore the maximum water level upstream of the weir corresponding to this stage, $h_1 = 1.7 - 0.6 = 1.1$ m. Hence the discharge, Q, that can spill over the weir without flooding is 26.75 $m^3 s^{-1}$ (from Fig. 8.24). Note the following:

$$Q = C_d C_v g^{1/2} b h_1^{3/2} \text{ (weir formula)} \quad \text{(ii)}$$

$$C_v h_1^{3/2} = H_1^{3/2} \quad \text{(iii)}$$

where

$$H_1 = h_1 + V_1^2/2g \quad \text{(iv)}$$

and

$$V_1 = Q/[B(h_1 + P)] \quad \text{(v)}$$

B being the mean width of the river in height $(h_1 + P)$.

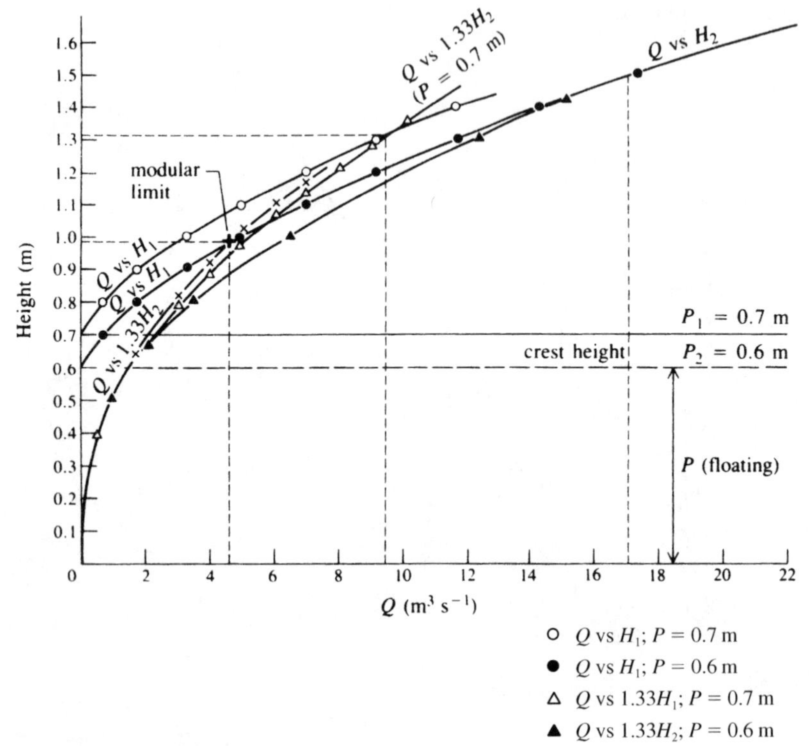

Fig. 8.25 Design curves

Combining the above gives

$$\frac{1}{2}C_{\rm d}^{2}[(b/B)(h_1/(h_1+P))]^2C_{\rm v}^{2}-C_{\rm v}^{2/3}+1=0. \qquad \text{(vi)}$$

The solution of this equation – tabular or graphical (Water Resources Board, 1970) – gives

$$C_{\rm v}=1.175.$$

The discharge so computed has a return period of about 25 years, which is quite acceptable for the stated land use upstream of the structure.

Similar calculations indicate that a flood flow of around 73 m^3 s^{-1} with a return period of 175 years clearly floods the upstream land (the water level above the sill is around 2 m). Hence the presence of the proposed structure will not cause any major flooding problems.

The modular flow equation can be used to construct the stage–discharge curve. The following table lists the data points:

$h_1 (m)$	C_v	Q $(m^3 s^{-1})$
0.10	1.003	0.62
0.20	1.009	1.78
0.30	1.018	3.30
0.38	← by iteration ←	4.60

NON-MODULAR FLOW RATING CURVE

The following procedure is used. Assume the non-modular flow, Q, to be, say, $6\,m^3 s^{-1}$. Then $H_2 = 0.375$ m (Fig. 8.25). Estimating H_1 at 0.450 m, $H_2/H_1 = 0.833$. The drowned flow (non-modular) correction factor, f, can be obtained from the plot in Fig. 8.26 (Water Resources Board, 1970) between f and H_2/H_1. Thus $f = 0.965$, and hence

$$Q = fQ_{\text{modular}}$$
$$= 0.965 \times 5.95 \; (Q_{\text{modular}} = C_d g^{1/2} b H_1^{3/2})$$
$$= 5.74\,m^3 s^{-1} < 6.0\,m^3 s^{-1}.$$

Repeating with different estimates of H_1 gives $Q \simeq 6\,m^3 s^{-1}$ for a value of $H_1 = 0.46$ m. The iterative solution of the energy equation (equation (iv)) gives $h_1 = 0.44$ m. The following table lists the data for the non-modular flow rating curve:

Q $(m^3 s^{-1})$	h_1 (m)
6.0	0.44
8.0	0.55
10.0	0.64
12.0	0.72
14.0	0.80
17.0	0.91

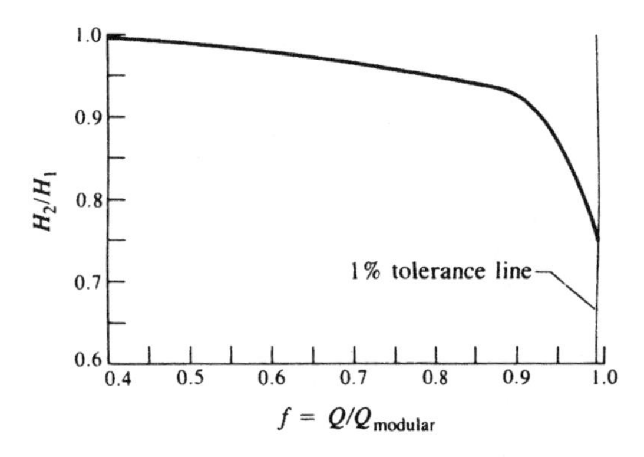

Fig. 8.26 Correction factor f for non-modular flows

Note that the proposed weir can only be used as a gauging structure for up to $17\,m^3\,s^{-1}$ (Fig. 8.25) with a return period of 10 years. If this is not acceptable either the crest height has to be increased or its length decreased. Both alternatives create undesirable conditions upstream of the structure (frequent flooding).

References

Abbott, M.B. and Minns, A.W. (1998) *Computational Hydraulics*, 2nd edn, Aldershot, Ashgate.

Ackers, P. (1983) Sediment transport problems in irrigation systems design, in *Developments in Hydraulic Engineering*, Vol. 1 (ed. P. Novak), Elsevier Applied Science, London.

—— (1993) Flow formulae for straight two-stage channels. *Journal of Hydraulic Research*, 31 (4): 509–31.

Ackers, P., White, W.R., Perkins, J.A. and Harrison, A.J.M. (1978) *Weirs and Flumes for Flow Measurement*, Wiley, New York.

Armanini, A., Righetti, M. and Grisenti, P. (2005) Direct measurement of vegetation resistance in prototype scale, *J. of Hydraulic Research*, IAHR 43, No. 5: 481–7.

Blench, T. (1969) *Mobile Bed Fluviology*, Alberta University Press, Calgary.

Bos, M.G. (ed.) (1976) *Discharge Measurement Structures*, Report 4, Laboratory of Hydraulics and Catchment Hydrology, Wageningen.

Bos, M.G. Replogle, J.A. and Clemmens, A.J. (1984) *Flow Measuring Flumes for Open Channel Systems*, Wiley, New York.

Brandon, T.W. (ed.) (1987, 1989) *River Engineering*, Parts I and II, Institute of Water and Environmental Management, London.

Brierley, S.E. and Novak, P. (1983) The effect of a reservoir on the morphology of a gravel-bed river with tributaries, in *Proceedings of the 20th Congress of the International Association for Hydraulic Research*, Moscow, Vol. VI: 49–57.

BSI (1969a) *Measurement of Liquid Flow in Open Channels: Velocity-Area Methods*, BS 3680, Part 3, British Standards Institution, London.

—— (1969b) *Measurement of Liquid Flow in Open Channels: Weirs and Flumes*, BS 3680, Part 4, British Standards Institution, London.

—— (1986) *Measurement of Liquid Flow in Open Channels: Dilution Methods*, BS 3680, Part 2, British Standards Institution, London.

Chadwick, A., Morfett, J. and Borthwick, M. (2004) *Hydraulics in Civil and Environmental Engineering*, 4th edn, E & FN Spon, London.

Chow, V.T. (1983) *Open Channel Hydraulics*, McGraw-Hill, New York.

Cole, J.A. (ed.) (1982) *Advances in Hydrometry*, Proceedings of the 1st International Symposium, Publication 134, International Association of Hydrological Science, Exeter.

Copin, N.J. and Richards, I.G. (1990) *Use of Vegetation in Civil Engineering*, CIRIA, Butterworth, London.

Cunge, J.A. (1969) On the subject of flood propagation computation method (Muskingum method). *Journal of Hydraulic Research*, 7 (2): 205–30.

Cunge, J., Holly, F.M.Jr. and Verwey, A. (1980) *Practical Aspects of Computational River Hydraulics*, Pitman, London.

Escarameia, M. (1998) *River and Channel Revêtments*, Thomas Telford, London.

Featherstone, R.E. and Nalluri, C. (1995) *Civil Engineering Hydraulics*, 3rd edn, Blackwell Scientific, Oxford.

French, R.H. (1986) *Open Channel Hydraulics*, McGraw-Hill, New York.

Garde, R.J. and Ranga Raju, K.G. (1985) *Mechanics of Sediment Transportation and Alluvial Stream Problems*, 2nd edn, Wiley Eastern, New Delhi.

Graf, W.H. (1984) *Hydraulics of Sediment Transport*, Water Resources Publications, Littleton, CO.

—— (1998) *Fluvial Hydraulics: Flow and Transport Processes in Channels of Simple Geometry*, John Wiley & Sons, Chichester.

Harrison, A.J.M. and Owen, M.W. (1967) A new type of structure for flow measurement in steep streams. *Proceedings of the Institution of Civil Engineers, London*, 36: 273–96.

Hemphill, R.W. and Bramley, M.E. (1989) *Protection of River and Canal Banks*, CIRIA and Butterworth, London.

Henderson, F.M. (1966) *Open Channel Flow*, Macmillan, New York.

Herschy, R.W. (ed.) (1986) *New Technology in Hydrometry*, Adam Hilger, Bristol.

—— (ed.) (1999) *Hydrometry – Principles and Practices*, 2nd edn, Wiley, Chichester.

Hey, R.D. (1986) River mechanics. *Journal of the Institution of Water Engineers and Scientists*, 40 (2): 139–58.

Hubbell, D.W., Stevens, H.H., Skinner, J.V. and Beverage, J.P. (1981) Recent refinements in calibrating bedload samplers, in *Proceedings of Special Conference, Water Forum 81*, American Society of Civil Engineers, San Francisco, CA.

IFAI (1992) *A Design Primer: Geotextiles and Related Materials*, Industrial Fabrics Association International, St Paul, MN.

Jaeggi, M.N.R. and Zarn, B. (1990) A new policy in designing flood protection schemes as a consequence of the 1987 floods in the Swiss Alps, in *Proceedings of the International Conference on River Flood Hydraulics* (ed. W.R. White), Wiley, Chichester, Paper C2: 75–84.

James, A. (1977) Water quality, in *Facets of Hydrology* (ed. J. Rodda), Wiley, New York, Chapter 7.

Jansen, P. Ph., van Bendegom, L., van den Berg, J., de Vries, M. and Zanen, A. (1979) *Principles of River Engineering: the Non-tidal Alluvial River*, Pitman, London.

Julien, P.Y. (2002) *River Mechanics*, Cambridge University Press.

Keller, R.J. and Rodi, W. (1988) Prediction of flow characteristics in main channel/flood plain flows. *Journal of Hydraulic Research*, 26 (4): 425–41.

Kennedy, J.F. (1983) Reflections on rivers, research and Rouse. *Journal of the Hydraulic Engineering Division, American Society of Civil Engineers*, 109 (10): 1253–71.

Kiely, G. (1998) *Environmental Engineering*, McGraw-Hill, New York.

Kinori, B.Z. and Mevorach, J. (1984) *Manual of Surface Drainage Engineering*, Elsevier, Amsterdam.

Knight, D.W. and Samuels, P.G. (1990) River flow simulation: research and

development. *Journal of the Institution of Water and Environmental Management*, 4: 163–75.
Knight, D.W. *et al.* (1999) The response of straight mobile bed channels to inbank and overbank flows, in *Proceedings of Institution of Civil Engineers, Water, Maritime and Energy*, 136, No. 4: 211–24.
Knighton, D. (1998) *Fluvial Forms and Processes: A New Perspective*, Arnold
Leopold, L.B., Wolman, M.G. and Miller, J.P. (1964) *Fluvial Processes in Geomorphology*, W.H. Freeman, San Francisco, CA.
Novak, P. (1957) Bedload meters – development of a new type and determination of their efficiency with the aid of scale models, in *Proceedings of the International Association for Hydraulic Research 7th Congress*, Lisbon, Paper A9.
Novak, P. and Čábelka, J. (1981) *Models in Hydraulic Engineering – Physical Principles and Design Applications*, Pitman, London.
Oplatka, M. (1998) *Stabilität von Weiden-verbauungen an Flussufern*, Mitteilungen 156, Vesuchsanstalt fur Wasserbau, Hydrologie und Glaziologie, ETH, Zürich.
Petersen, M.S. (1986) *River Engineering*, Prentice-Hall, Englewood Cliffs, NJ.
PIANC (1987) *Guidelines for the Design and Construction of Flexible Revêtments Incorporating Geotextiles for Inland Waterways*, Report of Working Group 4 of the Permanent Technical Committee 1, Permanent International Association of Navigation Congresses, Brussels.
Pilarczyk, K.W. (1999) *Geosynthetics and Geosystems in Hydraulics and Coastal Engineering*, Balkema, Rotterdam.
Price, R.K. (1985) Flood routing, in *Developments in Hydraulic Engineeering*, Vol. 3 (ed. P. Novak), Elsevier Applied Science, London.
Ranga Raju, K.G. (1993) *Flow Through Open Channels*, 3rd edn, Tata McGraw-Hill, New Delhi.
Raudkivi, A.J. (1979) *Hydrology: An Advanced Introduction to Hydrological Processes and Modelling*, Pergamon, Oxford.
Sargent, D.M. (1981) The development of a viable method of streamflow measurement using the integrating float technique. *Proceedings of the Institution of Civil Engineers*, 71 (2): 1–15.
Schaffernak, F. (1950) *Flussmorphologie und Flussbau*, Springer, Vienna.
Shaw, E.M. (1994) *Hydrology in Practice*, 3rd edn, Chapman & Hall, London.
Tebbutt, T.H.Y. (1998) *Principles of Water Quality Control*, 2nd edn, Elsevier.
Thorne, C. (1990) Effects of vegetation on riverbank erosion stability, in *Vegetation and Erosion*, J. Wiley & Sons, Chichester.
Thorne, C.R., Hey, R.D. and Newson, M.D. (eds) (1997) *Applied Fluvial Geomorphology for River Engineering and Management*, J. Wylie & Sons, Chichester.
UNECAFE (1953) *River Training and Bank Protection*, Flood Control Series No. 4, UN Economics Commission for Asia and the Far East, Bangkok.
Vanoni, V.A. (ed.) (1975) *Sedimentation Engineering*, American Society of Civil Engineers, New York, 745 pp.
Vries, M. de (1985) *Engineering Potamology*, International Institute for Hydraulic and Environmental Engineering, Delft.
Wark, J.B., James, C.S. and Ackers, P. (1994) *Design of Straight and Meandering Channels; Interim Guidelines on Hand Calculation Methodology*, R&D Report 13, National Rivers Authority, Bristol, 86 pp.

Water Resources Board (1970) *Crump Weir Design*, Technical Note TN8, Water Resources Board, Reading.

WMO (1980) *Manual on Streamflow Gauging*, Vols 1 and 2, WMO No. 519, World Meteorological Organization.

Wormleaton, P.R., Sellin, R.H.Y. and Bryant, T. (2004) Conveyance in a two-stage meandering channel with a mobile bed, *J. of Hydraulic Research*, 42, No. 5: 492–505.

Yalin, M.S. (1992) *River Mechanics*, Pergamon, Oxford.

Yalin, M.S., and Ferreira da Silva, A.M. (2001) *Fluvial Processes*, IAHR Monograph, IAHR, Delft.

Chapter 9

Diversion works

9.1 Weirs and barrages

9.1.1 General

Weirs and barrages are relatively low-level dams constructed across a river to raise the river level sufficiently or to divert the flow in full, or in part, into a supply canal or conduit for the purposes of irrigation, power generation, navigation, flood control, domestic and industrial uses, etc. These diversion structures usually provide a small storage capacity. In general, weirs (with or without gates) are bulkier than barrages, whereas barrages are always gate controlled. Barrages generally include canal regulators, low-level sluices to maintain a proper approach flow to the regulators, silt-excluder tunnels to control silt entry into the canal and fish ladders for migratory fish movements.

Weirs are also used to divert flash floods to the irrigated areas or for ground water recharging purposes. They are also sometimes used as flow-measuring structures. Figure 9.10 gives a detailed description of the various parts of a typical barrage constructed on rivers flowing over permeable beds (see also Baban, 1995).

The site selection of a barrage depends mainly on the location and elevation of the off-take canal, and a site must be selected where the river bed is comparatively narrow and relatively stable. The pondage requirement and interference with the existing structures such as bridges, urban development, valuable farmland, etc., must be considered, as well as available options to divert the flow during construction.

9.1.2 Cofferdams

Cofferdams are temporary structures used to divert water from an area where a permanent structure has to be constructed. They must be as watertight as practicable, relatively cheap and, if possible, constructed of locally available materials.

Diversion facilities such as tunnels or canals, provided to divert the flow from the site area, are sometimes used as part of the permanent facilities (e.g. penstocks, spillways, sluices, conveyances to turbines, or discharge channels from turbines, etc.). If the construction work proceeds in two stages, part of the structure completed in the first stage may be used as a diversion facility (spillway or sluice) during the second stage of construction (Fig. 9.1) (Linsley and Franzini, 1979; Vischer and Hager, 1997).

The selection of the design flood for these diversion works depends on the risk that one is prepared to take (see equation (4.1)). For example, a more conservative design flood has to be considered for situations where overtopping during construction would have disastrous results. If overtopping is permitted, care must be taken to strengthen the top and the downstream slope of the cofferdam to minimize erosion. The overtopping flow must be spread over the longest possible length of the cofferdam, thus reducing the concentration of flow. The control of floating debris is another essential requirement in order to minimize the clogging of diversion tunnels, especially during the high-flow season.

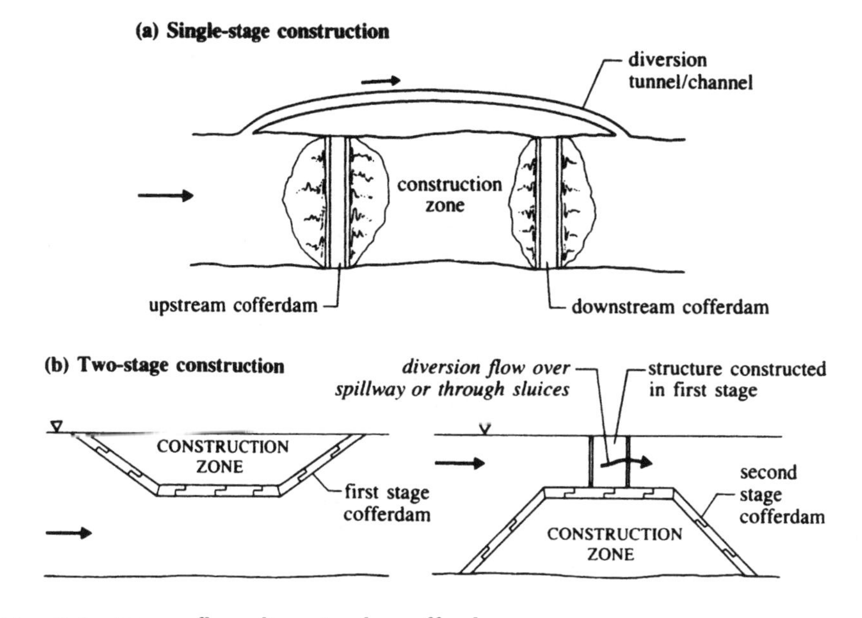

Fig. 9.1 Streamflow diversion by cofferdams

9.1.3 Barrage components

(a) Guide banks

Guide banks direct the main river flow as centrally as possible to the diversion structure. They also safeguard the barrage from erosion and may be designed so that a desirable curvature is induced to the flow for silt exclusion from the canals. The side slopes of the guide banks must be protected by stone pitching, with a sufficient 'self-launching' stone apron at the lowest feasible level (Fig. 9.2). The top levels of the guide banks will depend on the maximum increase in the flood level upstream of the barrage. The afflux (level difference between the headwater and tailwater during the passage of maximum flood flow) results in a backwater curve upstream of the barrage, and flood banks have to be provided along the upstream reach of the river to contain the flood flow.

(b) Wing walls

Wing walls flanking the barrage and supporting the abutting earth bunds are designed as retaining walls. Cut-off walls (taken below the scour levels) below the wings and abutment walls at both sides, in addition to the upstream and downstream sheetpile cut-offs across the river, form an enclosed compartment providing good weir foundation conditions.

(c) Gates

Gates used on barrages are of the same type as those used on spillway crests (Chapter 6). Vertical lift and Tainter gates are most frequently used to control the flow rate over the crest (sill) of a barrage. The discharge capacity of a gated crest depends on the conditions of free (Chapter 4) or

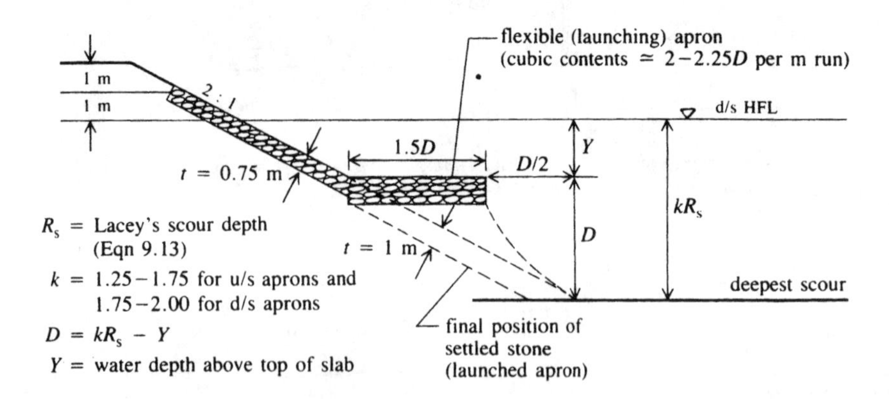

Fig. 9.2 Design criteria for launching aprons on weir and guide banks

submerged flows below the gate. For completely raised gates, Table 8.1 for free flow and Table 9.1 for submerged flow conditions should be referred to. Section 6.5 and equations (6.2)–(6.5) together with Figs 6.9 and 6.10 summarize the hydraulic conditions for flow under and over vertical lift and Tainter gates.

(d) Regulators

The structures controlling diversion into a supply canal are called regulators (Section 9.2). The design principles are the same as those used in the design of barrages, except that the regulators are a smaller version of barrages. The entry sill of a regulator must be such that it permits entry of the maximum flow at various pondage levels. Another important consideration in designing the regulator is silt exclusion from canals (Section 9.2). Silt-excluder tunnels are often provided in the barrage bays adjacent to the regulator, so that the heavier silt-laden bottom layers of water bypass through the tunnels (Fig. 9.3).

(e) Dividing wall

The dividing wall is built at right angles to the axis of the weir, separating the weir and the undersluices (Fig. 9.3). It usually extends upstream beyond the beginning of the regulator and downstream to the launching apron (talus). The sluice bay floor level is generally kept as low as possible to create pool conditions (for silt settlement and its exclusion) and the dividing wall separates the two floor levels of the weir. The downstream extension of the dividing wall provides a barrier between the stilling basin and scouring bay, in order to avoid cross-currents. A properly designed dividing wall can also induce desirable curvature to the flow for sediment exclusion from the canal-head regulator. The dividing wall may also serve as one of the side walls of the fish ladders (Section 9.3) and be used as a log chute.

(f) Weir block and stilling basin

The weir block of the barrage is designed either as a gravity structure (the entire uplift pressure due to seepage is resisted by the weight of the floor) or as a non-gravity structure (the floor, relatively thinner, resists the uplift by bending). It may be of different forms, e.g. a sloping weir with upstream and downstream glacis, a vertical drop weir, an ogee weir, or a labyrinth weir (zig-zag crest).

In some cases the barrage has a raised sill (crest) (e.g. Fig. 9.10). The advantage of this arrangement is a reduced height of the gates; however, the height of the sill must not exceed a value which could result in a raised (maximum) upstream water level. A hydraulically suitable design for a low sill is the so called 'Jambor sill' (Jambor 1959) which consists of a part

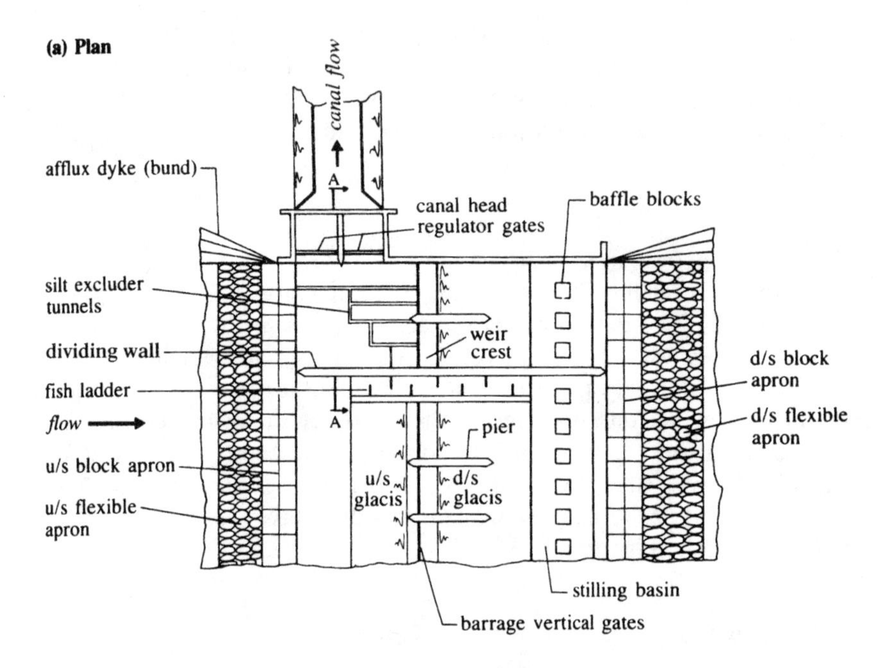

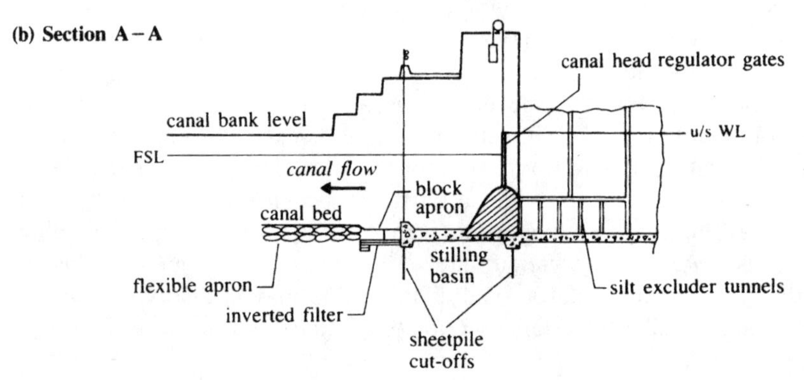

Fig. 9.3 Details of barrage and head regulator

cylindrical shape with a smooth upstream and downstream transition (typically a sloping apron with a slope 1:2 to 1:2.25). For a height s of the crest above the approach channel bed (e.g. 3.69 in Fig. 9.10) and the upstream (approach) depth y_0 (6.00 in Fig. 9.10) computation using the potential flow theory will result in a radius R of the cylindrical sill which would not influence (reduce) the discharge capacity of the structure (compared with the case without the raised sill) for a given upstream water level. Typical values for s/y_0 are $0.15 < s/y_0 < 0.3$ resulting in $2.5 < R/s < 15$. For further details of the computation, particularly for the case of non-modular flow see e.g. Doležal (1968).

The sloping surface downstream of the weir crest (glacis) and its transition into the stilling basin should be designed so that a hydraulic jump occurs on it over the full range of discharges. Details of stilling basin design are given in Chapter 5.

Protection works, such as cut-off piles, aprons and an inverted filter, are provided both upstream and downstream of the impervious floor of the weir block (see also worked examples 9.1 and 9.2). Chapters 5 and 8 also provide further details of the design of low-head weirs.

(g) Navigation lock

Special provisions must be made at the barrage site if the river is navigable. Navigation locks with appropriate approaches, etc. must be provided (Chapter 11).

9.1.4 Failures of weir foundations on permeable soils and their remedies (see also Section 2.6)

(a) Exit gradient (G_e) and piping

The exit gradient is the hydraulic gradient (Fig. 9.4) of the seepage flow under the base of the weir floor. The rate of seepage increases with the increase in exit gradient, and such an increase would cause 'boiling' of surface soil, the soil being washed away by the percolating water. The flow concentrates into the resulting depression thus removing more soil and creating progressive scour backwards (i.e. upstream). This phenomenon is called 'piping', and eventually undermines the weir foundations.

The exit gradient ($\tan\theta$; Fig. 9.4) according to the creep flow theory proposed by Bligh (Khosla, Bose and Taylor, 1954) is

$$G_e = H_s/L \tag{9.1}$$

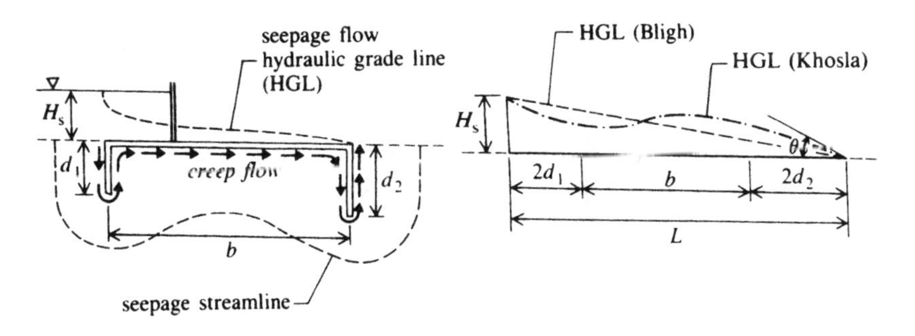

Fig. 9.4 Seepage flow hydraulic gradients

where L is the total creep length equal to $2d_1 + b + 2d_2$, d_1 and d_2 being the depths of the upstream and downstream cut-off piles respectively and b the horizontal floor length between the two piles; H_s, the seepage head, is the difference in the water levels upstream and downstream of the weir.

The piping phenomenon can be minimized by reducing the exit gradient, i.e. by increasing the creep length. The creep length can be increased by increasing the impervious floor length and by providing upstream and downstream cut-off piles (Fig. 9.4).

(b) Uplift pressures

The base of the impervious floor is subjected to uplift pressures as the water seeps through below it. The uplift upstream of the weir is balanced by the weight of water standing above the floor in the pond (Fig. 9.5), whereas on the downstream side there may not be any such balancing water weight. The design consideration must assume the worst possible loading conditions, i.e. when the gates are closed and the downstream side is practically dry.

The impervious base floor may crack or rupture if its weight is not sufficient to resist the uplift pressure. Any rupture thus developed in turn reduces the effective length of the impervious floor (i.e. reduction in creep length), which increases the exit gradient.

The provision of increased creep lengths and sufficient floor thickness prevents this kind of failure. Excessively thick foundations are costly to construct below the river bed under water. Hence, piers can sometimes be extended up to the end of the downstream apron and thin reinforced-concrete floors provided between the piers to resist failure by bending.

The two criteria for the design of the impervious floor are as follows.

1. *Safety against piping.* The creep length is given by

$$L = cH_s \tag{9.2}$$

where c is the coefficient of creep ($= 1/G_e$).

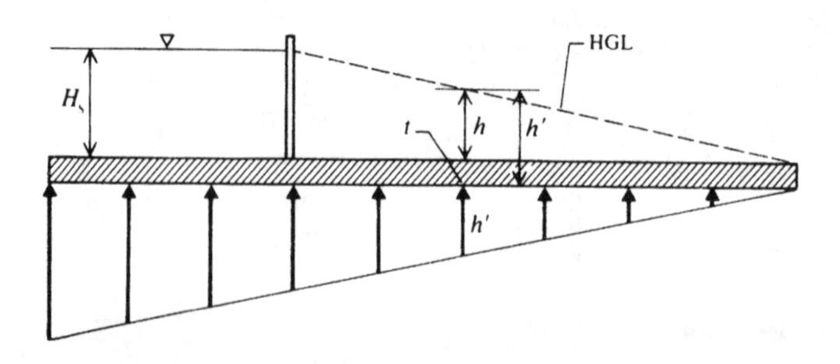

Fig. 9.5 Uplift pressure under impervious floor

2. *Safety against uplift pressure (Fig. 9.5).* If h' is the uplift pressure head at a point under the floor, the pressure intensity is

$$p = \rho g h' \text{ (N m}^{-2}). \tag{9.3}$$

This is to be resisted by the weight of the floor, the thickness of which is t and the density ρ_m (for concrete, $\rho_m = 2240\,\text{kg m}^{-3}$). Therefore,

$$\rho_m g t = \rho g h'$$

giving

$$h' = S_m t$$

where S_m is the relative density of the floor material. Thus we can write

$$h' - t = S_m t - t$$

which gives

$$t = (h' - t)/(S_m - 1) = h/(S_m - 1) \tag{9.4}$$

where h is the pressure head (ordinate of hydraulic gradient) measured above the top of the floor. A safety factor of around 1.5 is usually adopted, thus giving the design thickness of the concrete floor as

$$t \simeq 1.2h. \tag{9.5}$$

The design will be economical if the greater part of the creep length (i.e. of the impervious floor) is provided upstream of the weir where nominal floor thickness would be sufficient.

The stilling basin area of the weir is subjected to low pressures (owing to high velocities) which, when combined with excessive uplift pressures, may rupture the floor if it is of insufficient thickness. Usually, the floor is constructed in mass concrete without any joints, and with a hard top surface to resist the scouring velocities over it.

(c) Approach slab

The provision of a concrete slab upstream of the weir section (sloping down gradually from the 2 horizontal to 1 vertical slope of the crest section – Fig. 9.10) increases the seepage length, with a corresponding reduction in the exit gradient. It will also provide a smooth erosion-resistant transition for the accelerating flows approaching the weir. The upstream end of the approach slab is firmly secured to the upstream sheet-piling or to the vertical concrete cut-off. It is usually monolithic with the weir section in order to provide additional resistance to sliding.

9.1.5 Pressure distribution under the foundation floor of a weir/barrage

Applying the theory of complex variables (of potential theory) involved with the seepage flow under a flat floor (see Section 2.6) a Laplace differential equation can be formulated which on integration with appropriate boundary conditions suggests that the pressure head (P) at any point beneath the floor is a fraction, ϕ, of the total head, H_s (see Fig. 9.4). Thus a solution to Laplace equation (Khosla *et al.*, 1954; Leliavsky, 1965) can be written as

$$\phi = P/H_s = (1/\pi)\cos^{-1}(2x/b) \tag{9.6}$$

for the underside of the floor where b is the total floor length and x is the distance from the centre of the floor to the point where the uplift pressure head is P.

Equation (9.6), based on the potential theory of the seepage flow, suggests entirely different distribution to that of Bligh's creep theory (i.e. linear distribution – see Fig. 9.5).

In reality the weir foundations are composite in construction consisting of floor slabs (horizontal or sloping), pilings or cut-off structures and a direct solution of the Laplace equation is not feasible. Khosla *et al.* (1954), in dealing with this problem, introduced a method of independent variables splitting the composite weir/barrage section into a number of simple forms of known analytical solutions and by applying some corrections in transferring the results to the composite section.

The simple standard forms of a composite section are:

i) a straight horizontal floor of negligible thickness with a sheet pile at either end;
ii) a straight horizontal floor of finite thickness (depressed floor) with no cut-off piles;
iii) a straight horizontal floor of negligible thickness with an intermediate sheet pile.

Khosla *et al.* produced pressure charts (see Figs 9.6 and 9.7) from which the pressures at key points (junction points of the floor and the pile and the bottom corners of the depressed floor) of these elementary forms can

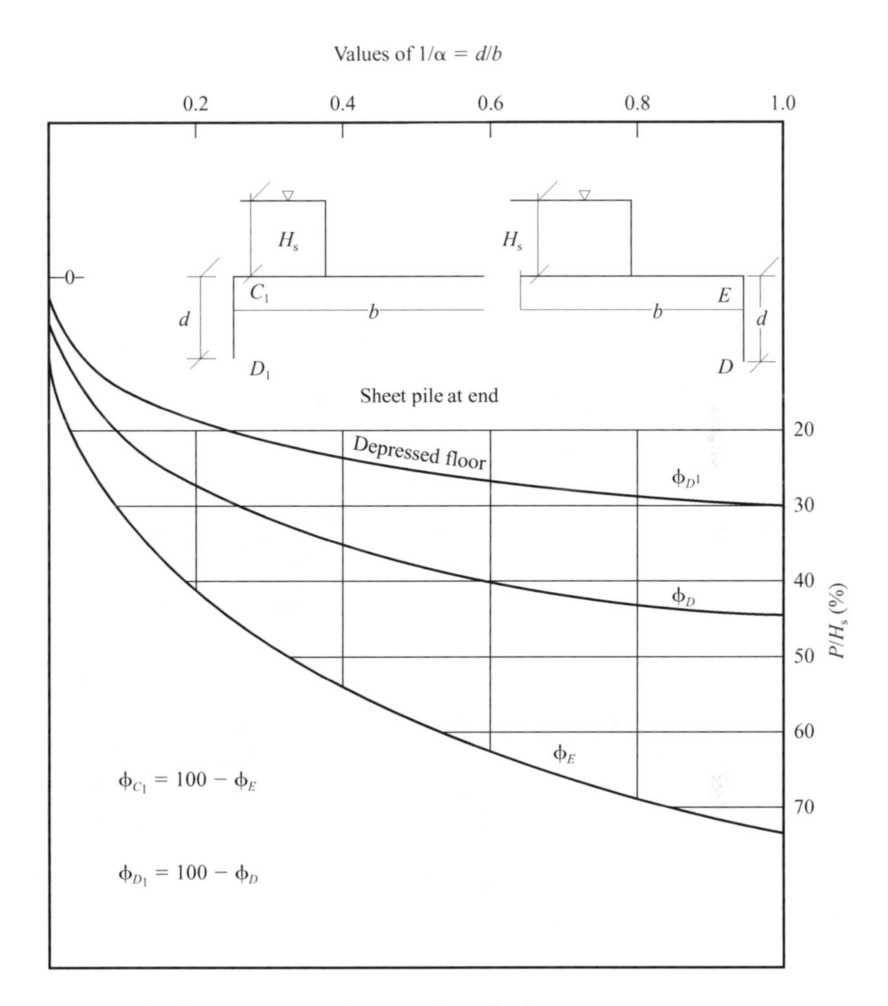

Fig. 9.6 Khosla's pressure chart with end pile

be read off. These are then corrected for the mutual interference of piles, the floor thickness and the slope of the floor, if any.

It must be realized that this procedure assumes throughout a uniform permeability coefficient (k) (see Section 2.3.4).

(a) Pressure distribution corrections

(i) CORRECTION FOR MUTUAL INTERFERENCE OF PILES (C%) (SEE FIG. 9.8)

$$C = 19\sqrt{D/b_1}(d + D)/b \qquad (9.7)$$

where D = the depth of the pile line whose influence has to be determined on the neighbouring pile of depth d, d = the depth of the pile on which the

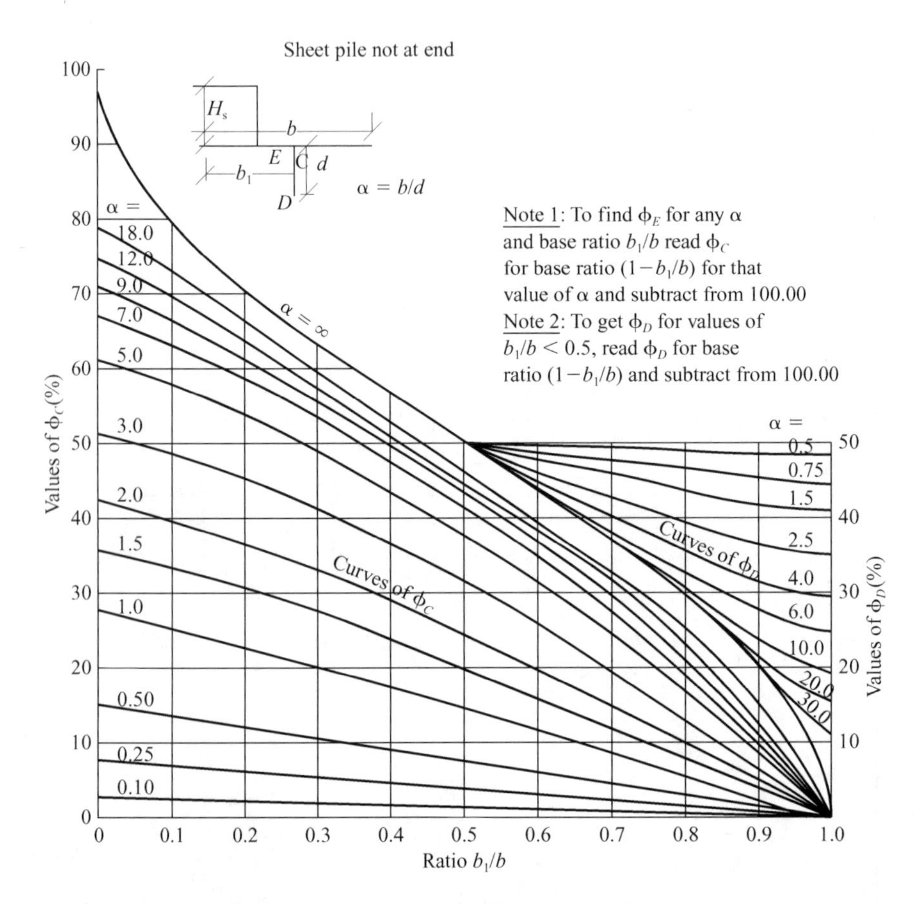

Fig. 9.7 Khosla's pressure chart with intermediate pile

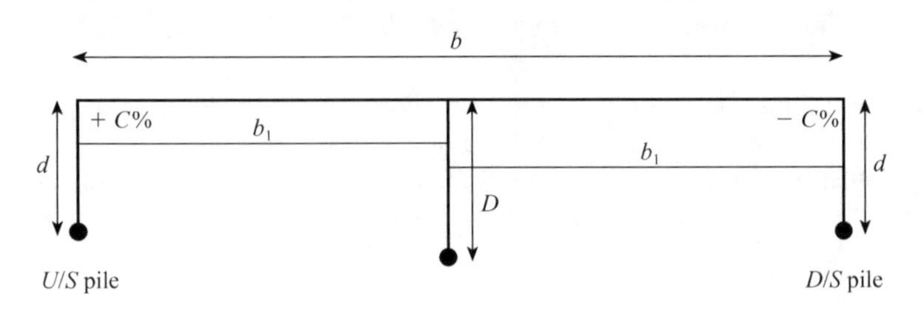

Fig. 9.8 Mutual interference of piles

effect of the pile depth D is required to be found, b = the total length of the floor, and b_1 = the distance between the two mutually interfering piles.

The correction computed by equation (9.7) is additive or subtractive depending upon whether the pile d is located upstream or downstream of pile D.

In the case of the intermediate piling shallower than the end pile ($D < d$) and $b_1 \geq 2d$ the mutual interference is negligible.

(ii) CORRECTION FOR FLOOR THICKNESS

The correction for floor thickness is interpolated linearly over the length of the piling once the pressures at the key points are computed using Figs 9.6 and 9.7. In equation (9.7) when computing the *corrections to the mutual interference with floor thickness* only the net depths of the pilings (i.e. $D - t$, $d - t$, where t is the floor thickness) instead of D and d are used.

(iii) CORRECTION FOR THE SLOPING FLOOR

The depths of the pilings to be used in the correction equation (equation (9.7)) are always measured from the top level of the piling, for which the correction is computed – see worked example 9.2.

The slope correction is only applicable to the key points of the pile line fixed at the beginning or the end of the slope. Correction charts following this approach were constructed by Khosla *et al.* (1954).

The (theoretical) percentage correction for sloping floors is negligible for slopes less than 1:7; it is $2.3 < (\%) < 4.5$ for slopes 1:7 to 1:3, 6.5% for slope 1:2 and 11.2% for 1:1; the actual correction is obtained by multiplying the appropriate value by b_2/b_1, b_2 being the horizontal distance of the sloping floor. This correction is additive for down slopes (following the direction of seepage flow) and subtractive for rising slopes.

(b) Actual exit gradient (G_e)

On the basis of the potential theory of flow Khosla *et al.* (1954) suggested the actual exit gradient given by (with one downstream end cut-off pile of depth d with the total base length of b)

$$G_e = H_s / \pi d \sqrt{\lambda} \tag{9.8}$$

where $\lambda = \frac{1}{2}(1 + (1 + \alpha^2)^{1/2})$, α being b/d.

For alluvial soils the critical (safe) exit gradient is around 1 in 1, and with a safety factor of around 5.5 the permissible exit gradients are about 1:6 to 1:7 for fine sand, 1:5 to 1:6 for coarse sand and 1:4 to 1:5 for shingle.

9.1.6 Barrage width and scour depth

(a) Barrage width

The barrage width must be sufficient to pass the design flood safely. The present trend is to design the barrage for a 100–150 year frequency flood and provide a breaching section along the main earth bund, located at a safe distance from the barrage itself. The breach section acts as a fuse plug, and its provision is a more economical solution than providing a large spillway capacity in the barrage.

The minimum stable width of an alluvial channel is given by the régime equation (Chapter 8)

$$B = 4.75Q^{1/2} \tag{9.9}$$

where B is the waterway width (measured along the water surface and at right angles to the banks) in metres, and Q is the maximum flood discharge in $m^3 s^{-1}$.

(b) Regime scour depths

The river bed is scoured during flood flows and large scour holes (not to be confused with local scour, Section 5.3.3) may develop progressively adjacent to the concrete aprons which may cause undermining of the weir structure. Such flood scour depth below HFL corresponding to a regime width (equation (9.9)) is called régime scour depth (or more precisely régime hydraulic radius), R_s, estimated by the following (Lacey's) formula

$$R_s = 0.475(Q/f)^{1/3} \tag{9.10}$$

if the actual waterway provided is greater or equal to the régime width (equation (9.9)) and

$$R_s = 1.35(q^2/f)^{1/3} \tag{9.11}$$

if the waterway provided is less than the régime width, where R_s is measured from the high flood level (HFL) and f is Lacey's silt factor (Singh, 1975):

$$f = 1.75d^{1/2} \tag{9.12}$$

where d is the mean diameter of the bed material (in mm) and q is the discharge per unit width of channel (see also Section 8.3).

Weir failure due to scour can be prevented by extending the sheet-pile cut-offs to a level sufficiently below the régime scour depth across the full width of the river (Fig. 9.9).

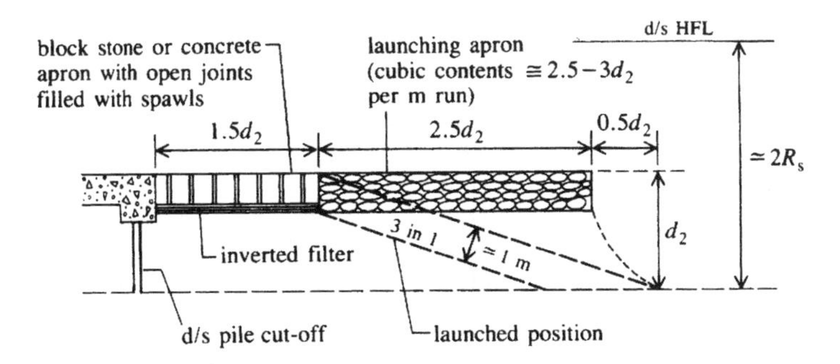

Fig. 9.9 Inverted filter and flexible (launching) apron

(c) Concrete aprons and inverted filter

The aprons are of plain concrete blocks of about 1 m × 1.5 m × 0.75 m deep, cast *in situ*. The downstream apron is laid with 70–100 mm open joints filled with spawls (broken stones), so that the uplift pressure is relieved. An inverted filter of well-graded gravel and sand is placed under the concrete apron (Fig. 9.9) in order to prevent the loss of soil through the joints. The upstream apron is laid watertight so that the uplift pressure and downward flow is reduced (due to the increase in creep length). Aprons of boulder or stone are laid downstream and upstream of the concrete aprons (Fig. 9.10).

9.1.7 Effect of barrages on river water quality

The flow of water over and/or under barrage gates as well as over spillways and weirs results in aeration with a beneficial effect on oxygen concentration levels in the downstream river reach.

The most frequent case is aeration at overfalls. On the basis of laboratory experiments (Avery and Novak, 1978) with results corroborated by extensive field measurements (Novak and Gabriel, 1997) the oxygen deficit ratio r (ratio of upstream to downstream oxygen deficit) at 15 °C is given by

$$r_{15} - 1 = kFr_{\mathrm{J}}^{1.78} Re_{\mathrm{J}}^{0.53} \tag{9.13}$$

where $k = 0.627 \times 10^{-4}$ (for water without salinity), $Fr_{\mathrm{J}} = (gh^3/2q_{\mathrm{J}}^2)^{0.25}$ and $Re_{\mathrm{J}} = q_{\mathrm{J}}/v$, h is the difference between the upstream and downstream water level and q_{J} is the specific discharge ($\mathrm{m^2\,s^{-1}}$) at impact into the downstream pool; this equals the specific discharge at a solid weir crest ($q = q_{\mathrm{J}}$) but with air access below the nappe (e.g. at flow over gates) $q = 2q_{\mathrm{J}}$. Equation (9.13) (which is dimensionless) can be transformed for the given value of (k) (h (m), q ($m^2\,s^{-1}$)) into:

$$r - 1 = 0.18\, h^{1.34}\, q^{-0.36} \tag{9.13a}$$

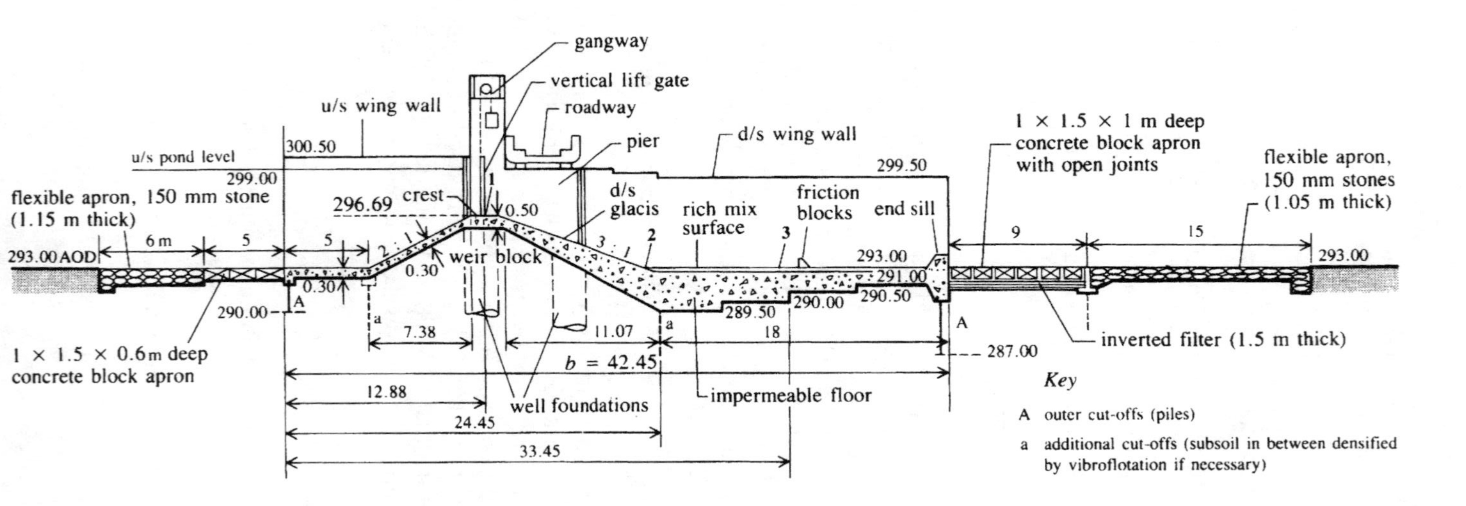

Fig. 9.10 Cross-section through a barrage showing details of the foundations; all dimensions in metres

The limits of applicability of equation (9.13) are $h \leq 6q^{1/3}$ (m) (equation (5.6) should be referred to, i.e. the overfalling jet should not disintegrate) and the downstream pool depth d should be bigger than $0.0041\ Re_J^{0.39} Fr_J^{0.24}$ (m) (or $d > 0.909\ h^{0.18}\ q^{0.27}$ (m, m^2s^{-1})) (as the major part of oxygen uptake occurs in the downstream pool).

For the outflow under a gate with a hydraulic jump the deficit ratio is

$$r_{15} - 1 = Fr_1^{2.1}\, Re^{0.75} \tag{9.14}$$

(Fr_1 is the jump supercritical Froude number and $Re = q/\nu$).

The temperature correction from r_T to r_{15} is given by

$$(r_T - 1)/(r_{15} - 1) = (1 + 0.046T)/1.69. \tag{9.15}$$

Gulliver *et al.* (1998) reviewed various prediction equations for oxygen transfer at hydraulic structures and concluded that for flow over sharp crested weirs equation (9.13) in the form ($Fr = 8gh^3/q^2)^{0.25}$, $Re = q/\nu$)

$$E_{20} = 1 - (1/(1 + 0.24 \times 10^{-4} Fr^{1.78} Re^{0.53}))^{1.115} \tag{9.16}$$

gives the best results when tested against field measurements (E_{20} is the transfer efficiency indexed at 20 °C $E = 1 - \frac{1}{\pi}$).

For ogee spillway crests they recommend

$$E_{20} = 1 - \exp(-0.263h/(1 + 0.215q) - 0.203d) \tag{9.17}$$

and for gated sills

$$E_{20} = -1 - \exp[-0.0086(hq/s) - 0.118] \tag{9.18}$$

where s is the submergence of the gate lip.

For oxygen transfer at cascades see Chanson (1994) and for further treatment of the whole subject see Novak (1994).

Aeration at hydraulic structures is usually, but not always, beneficial. If the upstream water is fully or nearly saturated with oxygen then further oxygen enrichment can lead to oxygen supersaturation that may have detrimental effects as it can cause gas bubble disease in fish. This situation is more likely to occur at high head structures with high velocities of flow than at barrages. The problem can be alleviated by some structural measures and in any case is mostly very localized and does not propagate far downstream of the dam.

Worked Example 9.1

This example considers the design of a glacis-type weir. The following data are used: maximum flood discharge $= 1800\,\text{m}^3\,\text{s}^{-1}$; HFL before construction $= 300.00$ m AOD; river bed level $= 293.00$ m AOD; normal upstream pond level $= 299.00$ m AOD; allowable afflux $= 1$ m; permissible exit gradient $= 1$ in 6; silt factor $f = 1$; crest level of canal regulator $= 297.50$ m AOD; FSL downstream of canal regulator $= 296.00$ m AOD; canal bed level downstream of regulator $= 293.50$ m AOD.

Design the various elements of the weir foundations using Bligh's theory. Also determine the waterway required for the canal head regulator in order to draw a flow of $100\,\text{m}^3\,\text{s}^{-1}$.

Solution

The régime width (equation (9.9)) of the upstream waterway, $B = 4.75\sqrt{1800} \simeq 200$ m. Adopting the gross length of the weir as 200 m and assuming 20 spans of 10 m each centre-to-centre of piers, and a pier thickness of 1.5 m,

$$\text{clear waterway} = 200 - 19 \times 1.5 = 171.5\,\text{m}.$$

Neglecting the pier and abutment contractions and assuming the weir to be broad crested ($Q = 1.7bH^{3/2}$ – modular flow; Chapters 4 and 8),

$$\text{total head over the crest,}\quad H = (1800/(1.7 \times 171.5))^{2/3} = 3.36\,\text{m},$$

$$\text{velocity of approach,}\quad V = 1800/(200 \times 7) = 1.3\,\text{m}\,\text{s}^{-1},$$

$$\text{approach velocity head} \simeq 0.08\,\text{m},$$

therefore

$$\text{upstream total energy level (TEL)} = 300.00 + 0.08 = 300.08\,\text{m AOD}$$

$$b_{\text{effective}} = b_{\text{clear}} - 2(nk_{\text{P}} + k_{\text{a}})H$$

where n is the number of piers (=19), k_{P} is the pier coefficient (=0.01 for semicircular piers), and k_{a} is the abutment coefficient (=0.1 for 45° wing walls). Therefore,

$$b_{\text{e}} = 171.5 - 2(19 \times 0.01 + 0.1) \times 3.36 = 169.55\,\text{m}$$

and the actual energy head,

$$H = (1800/1.7 \times 169.55)^{2/3} = 3.39\,\text{m}.$$

Hence the RL of weir crest = 300.08 − 3.39 = 296.69 m AOD. Since the allowable afflux is 1 m,

$$\text{downstream HFL} = 300.00 - 1.00 = 299.00\,\text{m AOD}.$$

Check whether the flow is free (modular) or submerged (non-modular). For the flow to be modular, i.e. not affected by submergence, the ratio H_2/H_1, where H_1 and H_2 are the upstream and downstream heads above the weir crest, is less than 0.75 (BSI, 1969; Bos, 1976):

$$\text{upstream head,} \quad H_1 = 300.00 - 296.69 = 3.31\,\text{m}$$

and

$$\text{downstream head,} \quad H_2 = 299.00 - 296.69 = 2.31\,\text{m}.$$

Therefore, the submergence ratio,

$$H_2/H_1 = 2.31/3.39 = 0.7\ (<0.75).$$

Hence the flow is modular, and the weir discharges the design flow with the desired upstream and downstream water levels.

Note that if a structure is submerged its discharge capacity is reduced. The submerged flow, Q_s, may be estimated by using the modular flow (Q_m) equation with a correction factor, f (i.e. $Q_s = fQ_m$). The correction factor depends on the type of structure, the submergence ratio H_2/H_1, and the ratio P_2/H_1, where P_2 is the crest height above the downstream channel bed (Table 9.1).

The régime scour depth (equation (9.10)),

$$R_s = 0.475(1800/1)^{1/3} = 5.78\,\text{m}.$$

Provide cut-offs for (1) upstream scour depth = 1.75 × 5.78 = 10.11 m and (2) downstream scour depth = 2.00 × 5.78 = 11.96 m. The RL of the bottom of the upstream cut-off = 300.00 − 10.11 = 289.89 m AOD and the RL of the bottom of the downstream cut-off = 299.00 − 11.96 = 287.04 m AOD. Therefore the depth of upstream cut-off pile, $d_1 = 293.00 - 289.89 \simeq 3.0\,\text{m}$, and the depth of the downstream cut-off pile, $d_2 = 293.00 - 287.00 \simeq 6.0\,\text{m}$. The pool level, i.e. the upstream storage level or canal FSL upstream of the head regulator = 299.0 m AOD. Therefore, the maximum seepage head (assuming tailwater depth = 0),

$$H_s = 299.00 - 293.00 = 6.00\,\text{m}.$$

Table 9.1 Correction factors for submerged (non-modular) flows

Type of structure	H_2/H_1	f	*Remarks*
Broad-crested weir (Ranga Raju, 1993)	≤0.75	1.0	Upstream and downstream faces vertical or sloping
	0.80	0.95	vertical faces
	0.85	0.88	
	0.90	0.75	
	0.95	0.57	
	0.80	≃1	Upstream face 1:5, downstream face 1:2
	0.85	0.95	
	0.90	0.82	
	0.95	0.62	
	0.80	≃1	Upstream face 1:1, downstream face 1:2
	0.85	0.98	
	0.90	0.90	
	0.95	0.73	
WES Spillway (USA) (Chapter 4)	≤0.3	≃1	$P_2/H_1 \geq 0.75$
	0.6	0.985	$P_2/H_1 = 0.75$
		0.982	1.50
		0.963	2.5
	0.8	0.92	$P_2/H_1 = 0.75$
		0.91	2.0
		0.88	3.0
	0.95	0.6	$P_2/H_1 = 0.75$
		0.55	2.0
		0.45	3.0
Sharp-crested weirs:			
rectangular	$[1-(H_2/H_1)^{3/2}]^{0.385}$		
triangular	$[1-(H_2/H_1)^{5/2}]^{0.385}$		
Crump weir: Fig. 8.27			

The exit gradient, $G_e = 1/6 = H_s/\pi d_2 \lambda^{1/2}$ (equation (9.8)), which gives

$$\lambda = (6 \times 6/(\pi \times 6))^2 = 3.65$$

and, from $\lambda = \frac{1}{2}[1 + (1 + \alpha^2)^{1/2}]$,

$$\alpha(=b/d_2) = 6.22;$$

therefore $b = 6.22 \times 6 = 37.32\,\text{m}$.

Providing an upstream glacis of 2 in 1, 5 m of horizontal apron (safety factor for additional creep length), a downstream glacis of 3 in 1, and an 18 m long stilling basin (USBR type III basin – Chapter 5, Fig. 5.6 – length $\simeq$ 3 times sequent depth), the total floor length,

$$b = 42.45\,\text{m} > 37.32\,\text{m}$$

(Fig. 9.10). The total creep length (Bligh) $= 2 \times 3 + 42.45 + 2 \times 6 = 60.45$ m. Therefore the rate of head loss (gradient) $= H/L = 6/60.45 = 0.0992$. The head loss in the upstream cut-off $= 0.0992 \times 6 = 0.595$ m, and in the downstream cut-off $= 0.0992 \times 12 = 1.19$ m. The total head loss to the gate (point 1) $= 0.595 + 0.0992 \times 12.88 = 1.87$ m. The total head loss up to end of the downstream glacis (point 2) $= 0.595 + 0.0992 \times 24.45 = 3.02$ m. The total head loss up to the mid-point of the stilling basin (point 3) $= 0.595 + 0.0992 \times 33.45 = 3.91$ m. The RL of the HGL at

$$\text{point 1} = 299.00 - 1.87 = 297.13\,\text{m AOD},$$

$$\text{point 2} = 299.00 - 3.02 = 295.98\,\text{m AOD},$$

$$\text{point 3} = 299.00 - 3.91 = 295.01\,\text{m AOD}.$$

The uplift pressure head above the top surface of the structure at

$$\text{point 1}, h = 297.13 - 296.69 = 0.44\,\text{m},$$

$$\text{point 2}, h = 295.98 - 293.00 = 2.98\,\text{m},$$

$$\text{point 3}, h = 295.01 - 293.00 = 2.01\,\text{m}.$$

The thickness of the concrete at

$$\text{point 1 (crest)} \simeq 1.2h \simeq 0.5\,\text{m},$$

$$\text{point 2 (upstream of stilling basin floor)} \simeq 1.2h \simeq 3.5\,\text{m},$$

$$\text{point 3 (centre of stilling basin floor)} \simeq 1.2h \simeq 2.5\,\text{m}.$$

Adopt a stepped floor thickness downstream of the downstream glacis as shown in Fig. 9.10 and a nominal thickness of 0.3 m for the upstream glacis and the horizontal floor slab (uplift pressures are less than the weight of water above).

DESIGN OF APRONS

For the downstream concrete block apron with open joints filled with spawls,

$$\text{length} = 1.5d_2 = 9\,\text{m}.$$

Use cast *in situ* concrete blocks of 1 m × 1.5 m × 1 m deep, laid open jointed over a graded inverted filter construction (design details later).

For the downstream flexible apron of broken stone,

$$\text{length} = 2.5d_2 = 15\,\text{m}.$$

Volume contents per metre length $= 2.63d_2 = 2.63 \times 6 = 15.78\,\text{m}^3\,\text{m}^{-1}$. Therefore the thickness of the apron = 15.78/15 = 1.05 m. The minimum stone size is given by (US Army Waterways Experimental Station, 1959)

$$d \simeq kV^2/2g\Delta,$$

where V is the mean velocity, $\Delta = \dfrac{\rho_s - \rho}{\rho} = (1.65)$ and $k = 1$ for calm flow and 1.4 for highly turbulent flow, or (Peterka – see Bos, 1976)

$$d = 0.032\Delta^{1/2}V^{9/4}.$$

The mean velocity, $V = 1800/(200 \times 6) = 1.5\,\text{m}\,\text{s}^{-1}$, and therefore the stone size, $d = 0.1 - 0.15\,\text{m}$.

Select stones that are almost cubical in shape with a size of around 150 mm or more (not flat slabs) for the downstream flexible apron.

For the upstream impervious concrete block apron,

$$\text{length} = 1.5d_1 \simeq 5\,\text{m}.$$

Use cast *in situ* concrete blocks 1 m × 1.5 m × 0.6 m deep.

UPSTREAM FLEXIBLE APRON

For this apron,

$$\text{length} = 2d_1 = 6\,\text{m}.$$

The volume contents per metre length $= 2.25d_1 = 2.25 \times 3 = 6.75\,\text{m}^3\,\text{m}^{-1}$. Therefore, the thickness of the apron $= 6.75/6 \simeq 1.15\,\text{m}$. Adopt the same graded stone as recommended for the downstream apron.

For the inverted filter design, the layers in a filter should be such that the filter is as follows (see also Section 2.6.3):

1. It should be permeable (to relieve uplift pressures), i.e.

$$\Delta H_n/D_n < \Delta H_{n-1}/D_{n-1} < \Delta H_{n-2}/D_{n-2} < \ldots$$

where ΔH_n is the head loss over the nth layer of thickness D_n (Fig. 9.11). This is satisfied if

   (a) d_{15} filter/d_{15} base $= 5 - 40$, where the filter and the base are two adjacent layers, the filter being the top layer, and
   (b) d_5 of any layer > 0.75 mm (to avoid clogging of filter layers) if possible.

2. It should be soil tight (to prevent the loss of fine material) – two conditions must be satisfied:

   (a) d_{15} filter/d_{85} base ≤ 5, and
   (b) d_{50} filter/d_{50} base $= 5 - 60$.

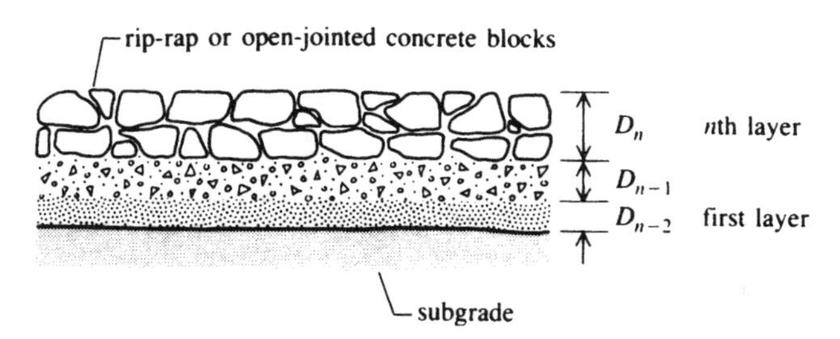

Fig. 9.11 Filter construction

The following ranges of the shape and gradation of the grains are recommended:

Shape of grains	*d_{15} (filter)/d_{15} (base)*	*d_{50} (filter)/d_{50} (base)*
Homogeneous round grains (gravel)	5–10	5–10
Homogeneous angular grains (broken gravel, rubble)	6–20	10–30
Well-graded grains	12–40	12–60

Provided that the sieve curves of the top layer and subgrade are known, the sieve curves for the intermediate layers can be plotted so that their bottom tails (small-diameter grains) run about parallel to that of the subgrade; however, it is more economical to use local materials that have a reasonably suitable grain-size distribution rather than composing a special mixture. A typical example of plotting the sieve curves of a filter construction is shown in Fig. 9.12.

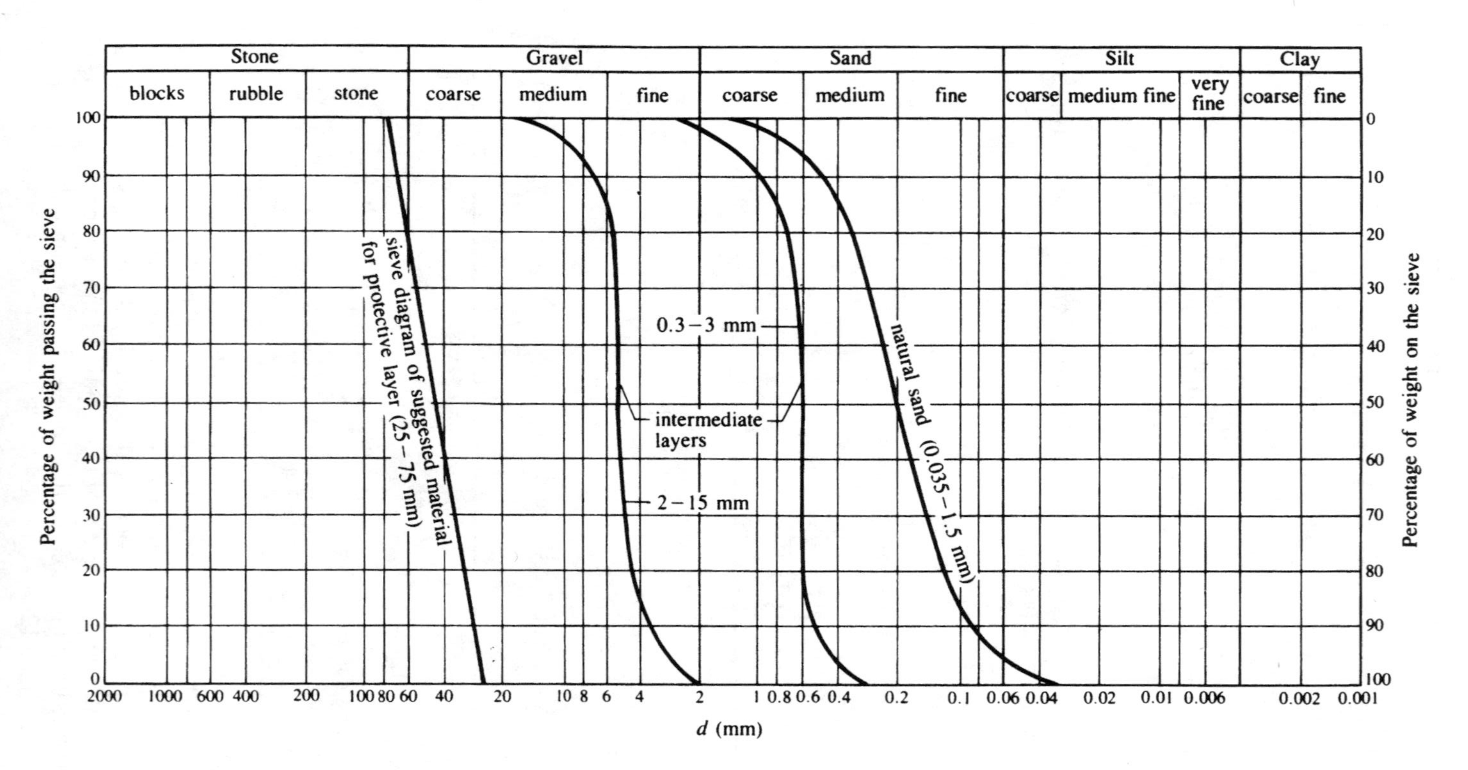

Fig. 9.12 Example of sieve curves for intermediate layers

The following filter layer thickness must be regarded as minimum requirements (for fair grain-size distribution throughout the layer) for a filter construction in dry conditions:

sand, fine gravel	0.05–0.10 m;
gravel	0.10–0.20 m;
stones	1.50–2 times the largest diameter.

ALTERNATIVE FILTER CONSTRUCTIONS

These include (a) single rip-rap or concrete blocks with open joints filled with spawls (broken stones) on a nylon filter, (b) nylon–sand mattresses and (c) gabions on fine gravel, etc.

The filter is subject to damage at either end as the subgrade may be washed out at these end joints (upstream end with the concrete structure and downstream end with unprotected channel) if no special measures are taken. The following examples (Figs 9.13 and 9.14) show the recommendations that the thickness of the filter construction be increased at these joints.

At the head regulator waterway,

$Q_{\text{canal}} = 100\,\text{m}^3\,\text{s}^{-1}$.

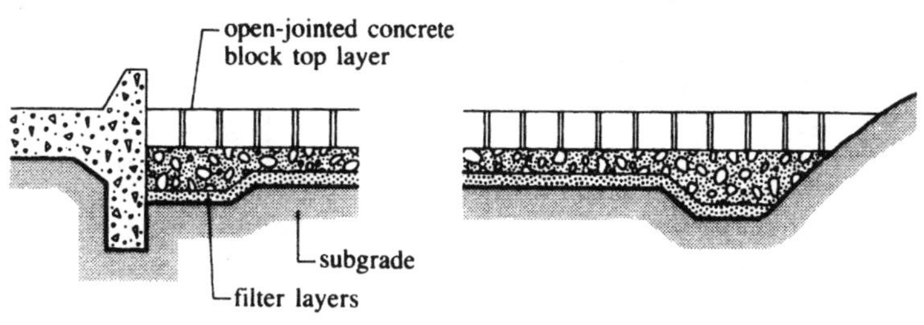

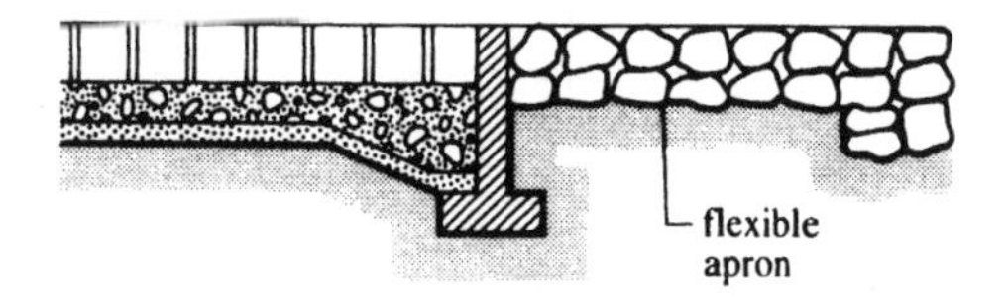

Fig. 9.13 Filter joints

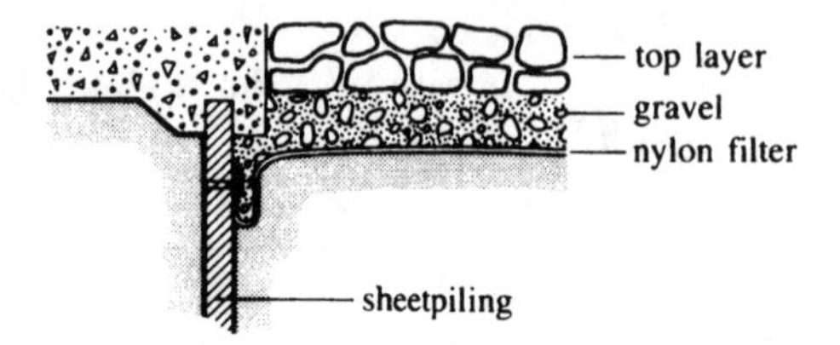

Fig. 9.14 Filter construction with nylon fabric

The pond level = 299.00 m AOD and the regulator's crest level = 297.50 m AOD. Therefore the head causing flow, $H = 1.50$ m (neglecting velocity of approach).

Check whether the flow is modular or non-modular:

$$\text{upstream head,} \quad H_1 = 299.00 - 297.50 = 1.50\,\text{m};$$

$$\text{downstream head,} \quad H_2 = 296.00 - 297.50 = -1.50\,\text{m}.$$

The crest height above the downstream bed level, $P_2 = 297.50 - 293.50 = 4$ m. Therefore the submergence ratio $H_2/H_1 = -1$ and $P_2/H_1 = 2.67$, suggesting free-flow conditions over the weir. Assuming a broad-crested weir ($Q = 1.7 b_e H^{3/2}$), the minimum effective width, $b_e = 100/(1.7 \times 1.5^{3/2}) \simeq 32$ m.

Providing four equal spans of 10 m with piers 1.5 m thick,

$$\text{clear waterway provided} = 40 - 3 \times 1.5 = 35.50\,\text{m},$$

$$\text{end contractions} = 2(3 \times 0.01 + 0.1) \times 1.5 = 0.39\,\text{m}.$$

Therefore the b_e provided = 35.50 − 0.39 = 35.11 m, which is satisfactory.

Worked Example 9.2

The figure below (Fig. 9.15) is a proposed weir floor with three vertical pilings. Examine the uplift pressure distribution under the floor of the weir and compare the result with Bligh's creep flow theory.
Pressure distributions at key points:

(A) UPSTREAM PILING 1

Figure 9.6 gives the pressures (from ϕ_D and ϕ_E curves) at D and E of the *extreme downstream pile* – see the definition sketch on this figure – as a first estimate which are then corrected for the floor thickness.

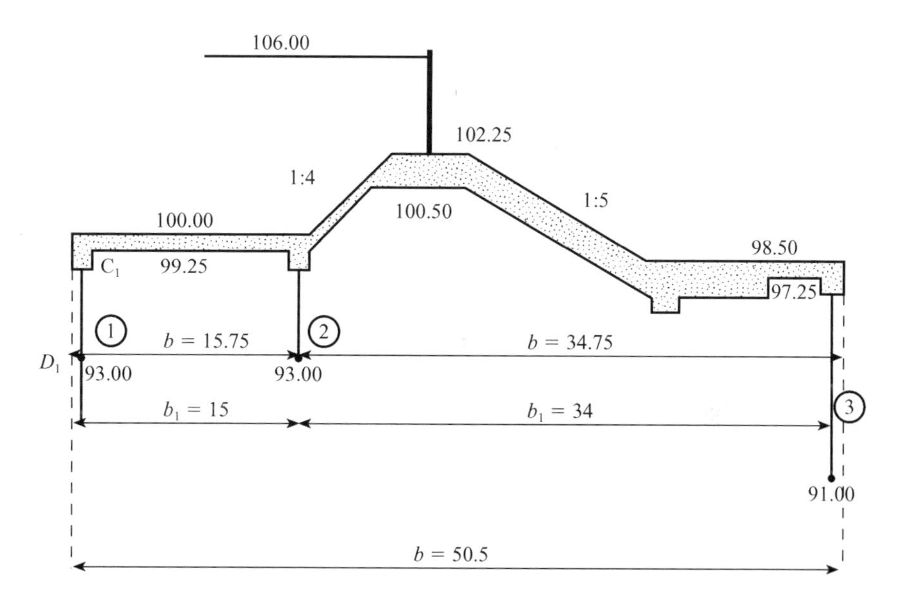

Fig. 9.15 Weir foundation design (worked example 9.2) (all dimensions in m)

Theoretical d (from the floor level) $= 100.00 - 93.00 = 7\,\text{m}$
Total floor length, $b = 50.5\,\text{m}$
$\therefore 1/\alpha = d/b = 7/50.5 = 0.139.$

From Fig. 9.6 $P_D/H = 23\%$, if the piling had been located at the downstream end of the floor. Since the piling 1 under question is located at the upstream end, the law of reversibility gives

$$P_{D1}/H = 100 - 23 = 77\% \text{ (see Fig. 9.15)}$$

Similarly from the chart we can get P_E/H (at $1/\alpha = 0.139$) $= 33\%$ and hence (point E of the downstream pile corresponds to C_1 of the pile at upstream end) by the law of reversibility,

$$P_{C1}/H = 100 - 33 = 67\%.$$

Correction for floor thickness

Total pressure drop on the d/s face of the piling $= 77 - 67 = 10\%$ over a height of 7 m; since the floor is 0.75 m thick $(100 - 99.25)$ the corresponding correction

$$= (10/7)0.75 = 1.07\% \ (+).$$

CORRECTION FOR THE MUTUAL INTERFERENCE OF PILES (SEE FIG. 9.15)

Interference of pile 2 on pile 1

True depths of the piles: $D = d = 99.25 - 93.00 = 6.25\,\text{m}$ and the distance between them, $b_1 = 15\,\text{m}$

$$\therefore \text{the correction, } C \text{ (equation (9.10))} = 19(6.25/15)^{1/2}(6.25 + 6.25)/50.5 = 3.03\% \ (+).$$

Therefore the corrected pressure behind (downstream side) the first piling

$$P_{C1}/H = 67 + 1.07 + 3.03 = 71.10 \approx 71\% - d/s \text{ of pile 1.}$$

(B) INTERMEDIATE PILING – PILE 2 (SEE FIG. 9.15)

Theoretical depth of piling, $d = 7\,\text{m}$ (pile 1) and hence $\alpha = 50.5/7 = 7.21$ and the distance b' from the piling to the upstream end of the floor = 15.75 m.

$$\therefore b'/b = 15.75/50.50 = 0.312; \text{ thus } 1 - b'/b = 0.688.$$

From the pressure chart (left side diagram of Fig. 9.7) the pressure at C – *see the notes on the chart for computations* – corresponding to $1 - b'/b = 0.688$ on x-axis, $P_C/H = 31\%$ which gives $P_E/H = 100 - 31 = 69\%$.

Also, P_D/H (from the right-hand diagram of Fig. 9.7) = 36% which gives the actual

$$P_D/H = 100 - 36 = 64\% \text{ since } b'/b\ (=0.312) < 0.5.$$

For P_C/H direct use of left-hand diagram of Fig. 9.7 gives (for $\alpha = 7.21$ and $b'/b = 0.312$)

$$P_C/H = 53\%.$$

Correction for floor thickness

For $P_E/H = (69 - 64)0.75/7 = -0.54\%$
For $P_C/H = (64 - 53)0.75/7 = +1.18\%$.

Correction for mutual interference

INFLUENCE OF PILE 1 ON PILE 2 – ON P_E/H

As $D = d$, i.e. the piles 1 and 2 are of the same length, this effect is the same as the effect of pile 2 on pile 1. However, this correction is subtractive and $= -3.03\%$.

INFLUENCE OF PILE 3 ON P_C/H (SEE FIG. 9.15)

Now D (measured from the top level of pile 2) $= 99.25 - 91.00 = 8.25$ m.
Also, $d = 99.25 - 93.00 = 6.25$ m.

$$\therefore \text{The correction, } C = 19\sqrt{8.25/34}(8.25 + 6.25)/50.5 = +2.69\%.$$

Correction for sloping floor for P_C/H

The slope being $\frac{1}{4}$ the theoretical correction is around 3.3% (see Section 9.1.5 a (iii)) and the corrected slope with the horizontal length of the sloping floor, b_2 (100.50 − 99.25) being 5 m

$$= 3.3\ (5/34) = -0.48\%.$$

∴ The corrected pressure behind (E) the pile 2 is

$$P_E/H = 69.00 - 0.54 - 3.03 = 65.43 \approx 65\%$$

and ahead (C) of the pile,

$$P_C/H = 53.00 + 1.18 + 2.69 - 0.48 = 56.39 \approx 56\%.$$

(C) DOWNSTREAM END PILING – PILE 3 (SEE FIG. 9.15)

Theoretical pile depth, $d = 98.50 - 91.00 = 7.50$ m

$$1/\alpha = d/b = 7.5/50.50 = 0.148 \text{ and}$$

P_E/H (directly from pressure chart – see Fig. 9.5) $= 36\%$.

Similarly, $P_D/H = 26\%$.

CORRECTION FOR FLOOR THICKNESS, t (98.50 − 97.25) = 1.25 m

$= (36 - 26)1.25/7.5 = -1.67\%$.

Correction for interference of pile 2 on pile 3 (see Fig. 9.15)

Depth of pile 2, $D = 97.25 - 93.00 = 4.25$ m
Depth of pile 3, $d = 97.25 - 91.00 = 6.25$ m.

$$\therefore \text{The correction, } C = 19\sqrt{4.25/34}\ (4.25 + 6.25)/50.50 = -0.073\%.$$

Note: *If the intermediate pile (pile 2, D) is shallower than the end pile (pile 3, d), i.e.* $D < d$ *and* $b_1 \geq 2d$ *(=12.50 m) the mutual interference correction is negligible.* The corrected pressures behind (E) of the pile 3,

$$P_E/H = 36.00 - 1.67 - 0.073 = 34.26 \approx 34\% \text{ and } P_D/H = 26\%.$$

The actual pressure distributions are summarized in the following table:

Pressures	*Behind pile 1*	*Middle pile 2*	*Middle pile 2*	*Ahead of pile 3*
		Ahead	*Behind*	
$\phi = P/H$	$\phi_C = 71\%$ $\phi_D = 77\%$	$\phi_E = 65\%$ $\phi_D = 64\%$	$\phi_C = 56\%$ $\phi_D = 64\%$	$\phi_E = 34\%$ $\phi_D = 26\%$

Bligh's theory suggests linear pressure distribution with a safe exit gradient (see equation (9.1)),

$$G_e \text{ (Bligh)} = 6/(50.50 + 2(7 + 7 + 7.5)) = 1 \text{ in } 15.6.$$

The actual exit gradient (after Khosla *et al.* – see equation (9.8)) = 1 in 7.75.

9.2 Intakes

9.2.1 Introduction

The intake structure (or head regulator) is a hydraulic device constructed at the head of an irrigation or power canal, or a tunnel conduit through which the flow is diverted from the original source such as a reservoir or a river. The main purposes of the intake structure are (a) to admit and regulate water from the source, and possibly to meter the flow rate, (b) to minimize the silting of the canal, i.e. to control the sediment entry into the canal at its intake, and (c) to prevent the clogging of the entrance with floating debris.

In high-head structures the intake can be either an integral part of a dam or separate; for example, in the form of a tower with entry ports at various levels which may aid flow regulation when there is a wide range of fluctuations of reservoir water level. Such a provision of multilevel entry also permits the withdrawal of water of a desired quality.

The layout of a typical intake structure on a river carrying a heavy bedload is shown in Fig. 9.16. The following are its major appurtenances:

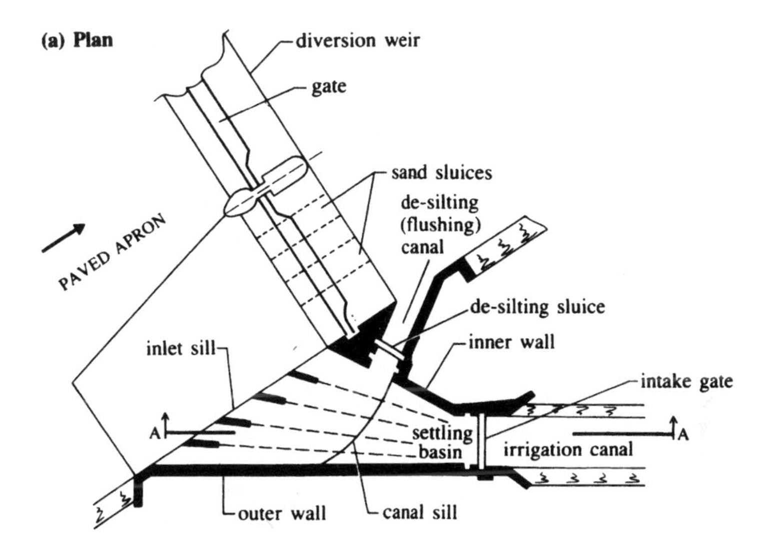

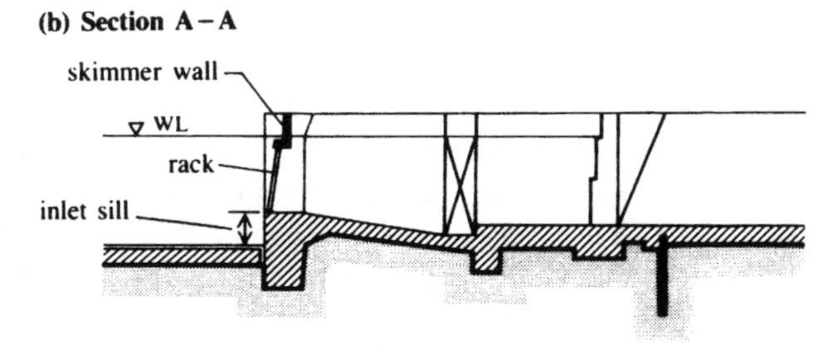

Fig. 9.16 Canal intake on a river carrying heavy bedload (after Mosonyi, 1987)

1. the raised inlet sill to prevent entry of the bedload of the river;
2. the skimmer wall (with splitter pier) at the inlet to trap floating ice and debris;
3. the coarse rack (trash rack) to trap subsurface trash, equipped with either manual or automatic power-driven rack cleaning devices;
4. the settling basin (sand trap) followed by a secondary sill (entrance sill) diverting the bottom (sediment-laden) layers towards the desilting canal;
5. the flushing (desilting) sluice to flush the deposited silt;
6. the intake (head regulator) gates to control the flow rate into the canal;
7. the scouring (tunnel) sluices in the diversion weir to flush the bedload upstream of the inlet sill.

The desilting canal with its flushing sluices may be omitted if the sediment load which would settle in the settling basin is negligible; however, smaller grain size sediment (silt) is always likely to enter the canal, and maintenance of minimum velocities in the canal is essential to avoid its silting up.

9.2.2 Location and alignment of an intake

The river reach upstream of the intake should be well established with stable banks. As the bottom layers of the flow around a bend are swept towards its inside (convex) bank (see Section 8.3), it is obvious that the best location for an intake (to avoid bedload entry) is the outer (concave) bank, with the intake located towards the downstream end of the bend. This choice of location from the sediment exclusion point of view is not always possible, and other considerations such as the pond (command) levels and their variations, navigation hazards, and location of the diversion structure, pump/power house, and outfalls must be considered.

An offtake at 90° to the main flow is the least desirable one. The structure should be aligned to produce a suitable curvature of flow into the intake, and a diversion angle of around 30°–45° is usually recommended to produce this effect; in addition, an artificial bend (Fig. 9.17), a groyne island (Fig. 9.18) or guide vanes (Fig. 9.19) may be designed to cause the required curvature of flow (see Avery, 1989). Model tests are desirable in deciding on the location and alignment of any major intake structure (Novak and Čábelka, 1981).

The entrance losses at an intake depend upon the change in direction of the flow (entering the intake), the extent of contraction and the type of trash rack provided at the inlet. They are expressed in terms of the velocity head as $KV^2/2g$.

The entrance loss due to a change in direction of flow (intake at an angle α with the main stream) is given by

$$\Delta h_\alpha = V^2/2g - \epsilon V_0^2/2g \qquad (9.19)$$

where V_0 is the velocity of the main stream at the inlet, and ϵ is around 0.4 for $\alpha = 90°$ and 0.8 for $\alpha = 30°$.

In the case of the inlet having a sill constructed with curved abutments and piers, the head loss, Δh_c, is given by

$$\Delta h_c \simeq 0.3V^2/2g. \qquad (9.20)$$

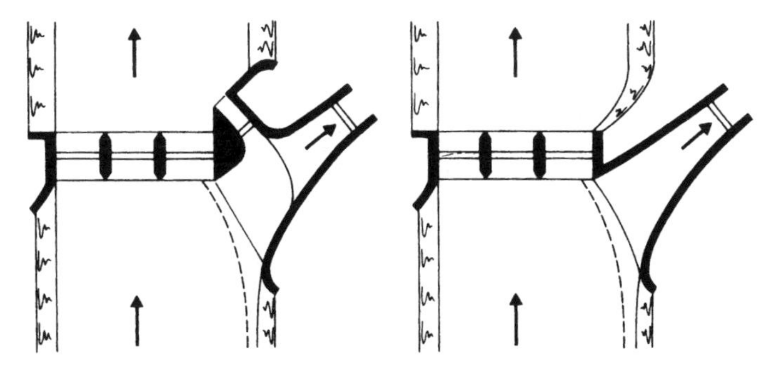

Intakes for straight reaches with bed contraction

(a) With de-silting canal (b) Without de-silting canal

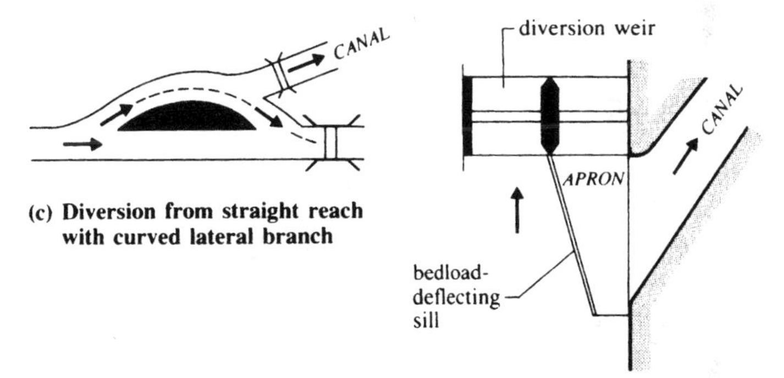

(c) Diversion from straight reach with curved lateral branch

(d) Intake with bedload-diverting sill

Fig. 9.17 Intake layouts with induced curvature to flow (after Mosonyi, 1987)

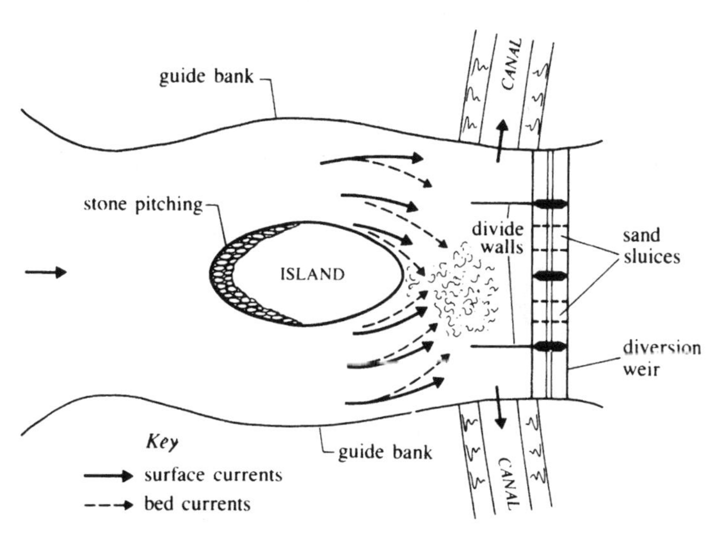

Fig. 9.18 Use of artificial groyne (e.g. island) to induce desired curvature to flow at intakes

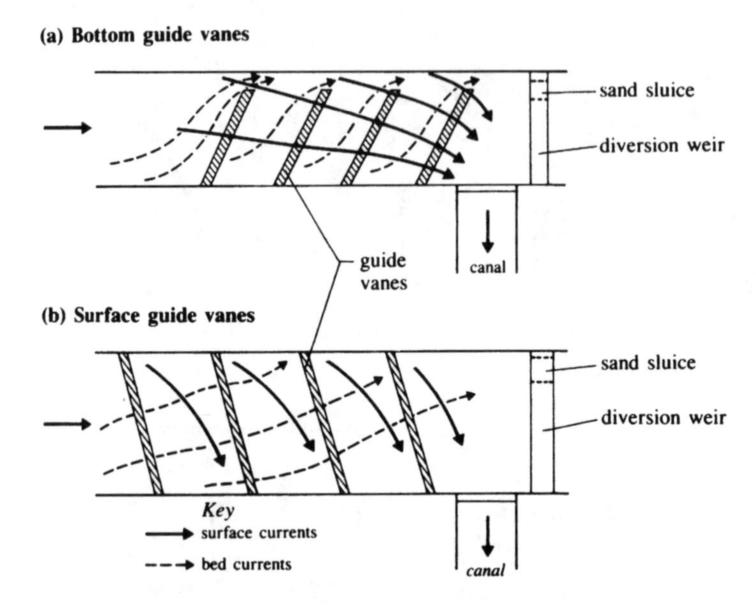

Fig. 9.19 Guide vanes layouts upstream of intake for sediment exclusion

Equations (9.19) and (9.20) suggest the maximum entrance loss at the inlet:

$$\Delta h_e = 1.3V^2/2g - \epsilon V_0^2/2g. \tag{9.21}$$

The rack losses, Δh_r, can be expressed by (Fig. 9.20)

$$\Delta h_r = \beta(s/b)^{4/3} \sin \delta V^2/2g \text{ (Kirschmer's formula)} \tag{9.22}$$

(with flow parallel to rack bars), where β is a coefficient which depends on the type of rack bar (Table 9.2).

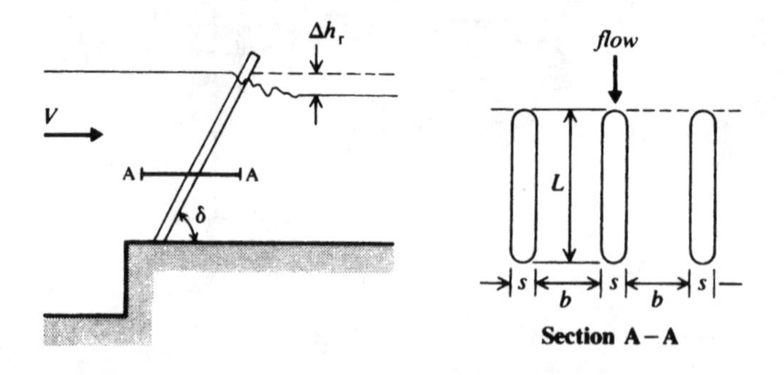

Fig. 9.20 Rack loss

Table 9.2 Values of β for parallel flow

Type of rack bar	β
Square nose and tail, $L/s = 5$	2.42
Square nose and semicircular tail, $L/s = 5$	1.83
Semicircular nose and tail, $L/s = 5$	1.67
Round	1.79
Airfoil	0.76

Meusburger (2002) provides a comprehensive discussion of inlet rack losses including an extensive bibliography.

9.2.3 Silt control at headworks

(a) Silt excluder

The silt excluder is a device constructed in the river bed just upstream of the regulator to exclude silt from the water (source) entering the canal. It is so designed that the top and bottom layers of flow are separated with the least possible disturbance, the top sediment-free water being led towards the canal while the bottom sediment-laden water is discharged downstream of the diversion structure through undersluices. The device basically consists of a number of tunnels (Fig. 9.21) in the floor of the deep pocket of the river, isolated by a dividing wall. The sill level of the regulator is kept the same as that of the top level of the roof slab of the tunnels.

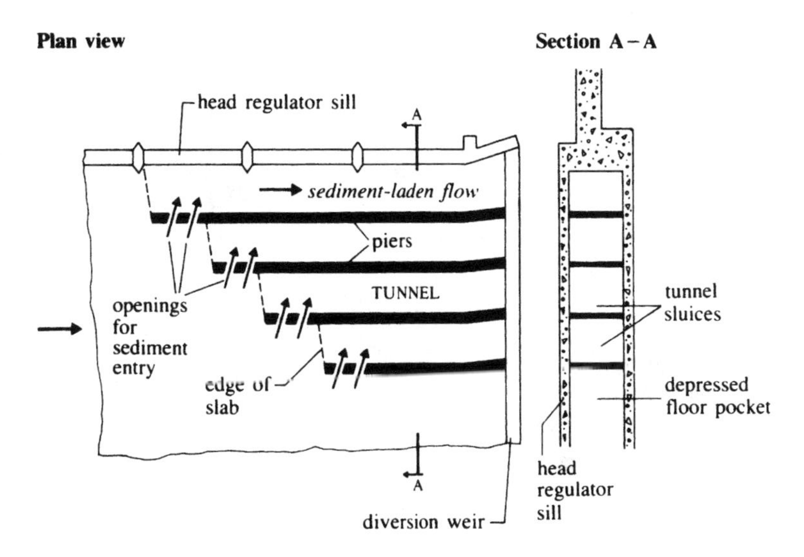

Fig. 9.21 Silt excluder (undertunnel type)

The capacity of the tunnel(s) is usually kept at about 20% of the canal discharge, and they are designed to maintain a minimum velocity of 2–3 $m\,s^{-1}$ (to avoid deposition in tunnels).

(b) Silt ejector or extractor

The silt ejector is a device constructed on the canal downstream of the head regulator but upstream of the settling basin (if any), by which the silt, after it has entered the canal, is extracted.

1. *Vane type ejector.* The layout of a vane type ejector is shown in Fig. 9.22. A diaphragm at the canal bed separates the top layers from the

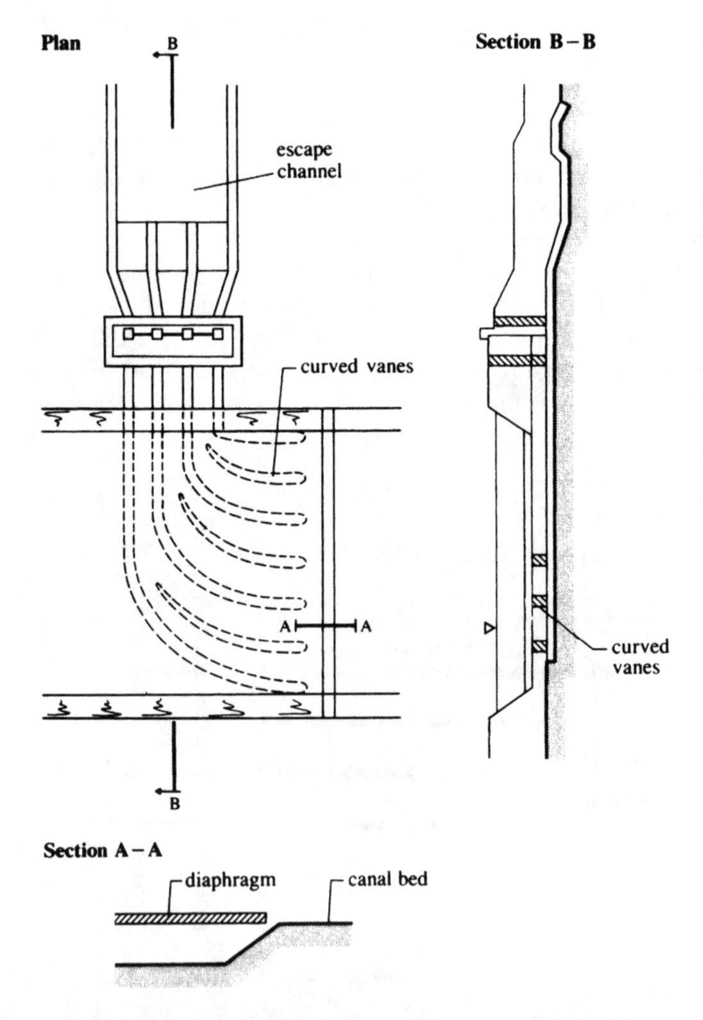

Fig. 9.22 Silt ejector (vane type)

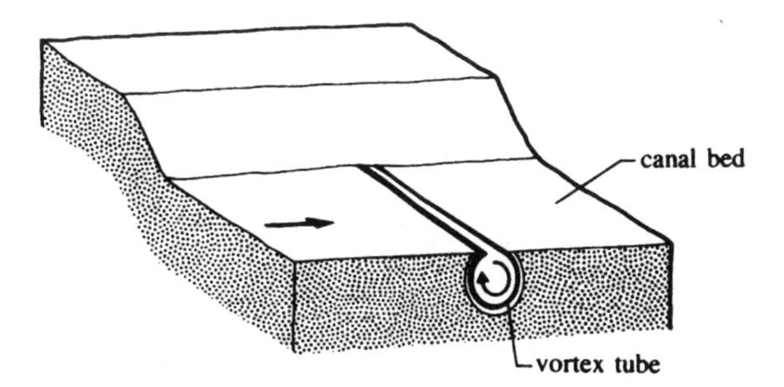

Fig. 9.23 Vortex tube type silt ejector

bottom ones. On entering the depressed area of the canal bed, the bottom sediment-laden layers are diverted by the curved vanes towards the escape chamber. The design should be such that the entry disturbances are minimal; the streamlined vane passages accelerate the flow through them, thus avoiding deposition.

2. *Vortex tube type ejector.* The vortex tube ejector (Fig. 9.23) consists of a pipe with a slit along its top, placed across the bottom of the canal at an angle of around 30°–90° to the direction of flow. The vortex motion within the tube draws the sediment into it, and he wall velocities along the tube eventually eject the sediment at its discharge end. A properly designed vortex tube ejector can be more efficient than any other conventional ejector, with less water loss.

9.2.4 Settling basin

The settling basin is a device placed on the canal downstream of its head regulator for the removal of sediment load which cannot be trapped by the conventional excluders or ejectors. It consists of an enlarged section of the channel where the flow velocity is sufficiently low so that the fine sediment settles on the bed (Fig. 9.24). The settled sediment is removed by sluicing, flushing or dredging.

The following equation may be used to design a settling basin:

$$W = W_0 e^{-w_s x/q} \tag{4.13}$$

where W is the weight of sediment leaving the basin, W_0 is the weight of sediment entering the basin, w_s is the fall velocity of a sediment particle (equation (8.18)), q is the discharge per metre width of the basin and x is the length of the settling basin. Alternatively

$$x = cD_s V/w_s \tag{9.23}$$

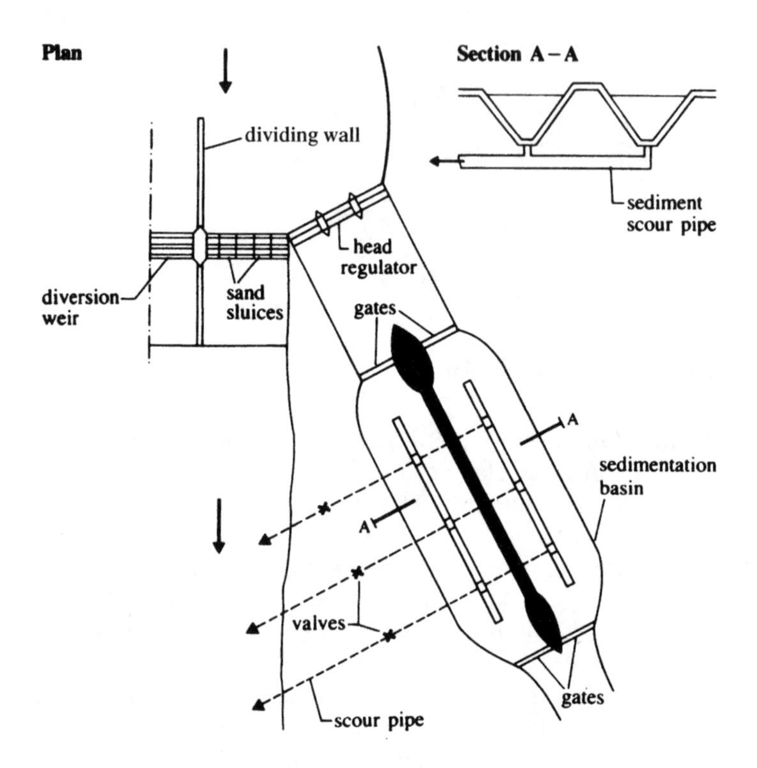

Fig. 9.24 Arrangement of settling basin

where D_s is the depth of the settling basin, V is the mean velocity in the basin, and c is the safety factor (1.5–2).

On the basis of the steady-state two-dimensional dispersion equation, the fraction of removal, f, at a distance x in a sedimentation tank of depth y, can be obtained (assuming isotropic dispersion) from

$$x/y = [-12V\log(1-f)]/(10w_s - U_*) \quad (9.24)$$

where U_* is the shear velocity in the basin, given by equation (8.7).

Equation (9.24) gives reasonably satisfactory results when computing each fraction of removed sediment independently, if the concentration is small.

For a detailed discussion of sediment exclusion at inlets see Ortmanns (2006).

9.2.5 Canal outlets

The canal outlet is an intake structure for small canals, usually designed for proportional distribution of irrigation water supplies from the parent canal. These structures do not require any control works in the parent canal and function automatically with no manual control. Except for routine inspection they require very little maintenance (Mazumdar, 1983).

The flexibility of an outlet (F) is the ratio of the fractional change of discharge of the outlet ($\mathrm{d}q/q$) to the fractional change of discharge of the parent (distribution) canal ($\mathrm{d}Q/Q$). Thus,

$$F = (\mathrm{d}q/q)/(\mathrm{d}Q/Q) \tag{9.25}$$

where q and Q are the outlet and parent canal discharges respectively, given by

$$q = kH^m, \tag{9.26a}$$

$$Q = cD^n, \tag{9.26b}$$

H is the head acting on the outlet and D is the depth of water in the distribution canal (Fig. 9.25).

Combining equations (9.25) and (9.26) gives

$$F = (m/n)(D/H)\mathrm{d}H/\mathrm{d}D. \tag{9.27a}$$

Since $\mathrm{d}H = \mathrm{d}D$ (change in D results in an equal change in H), equation (9.27a) becomes

$$F = (m/n)(D/H). \tag{9.27b}$$

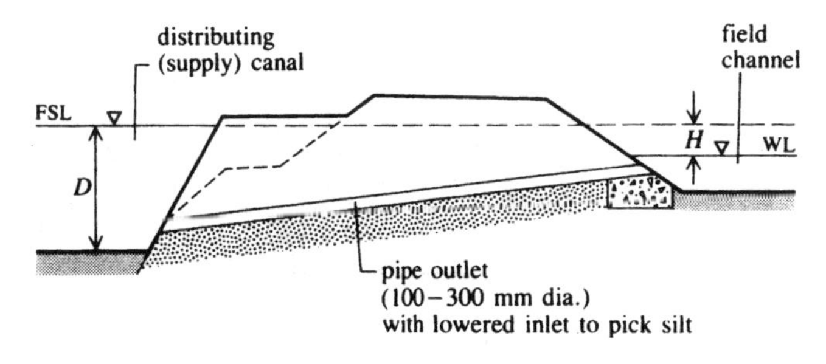

Fig. 9.25 A submerged pipe outlet (non-modular type)

The design of a proportional outlet ($F=1$) is thus governed by the condition

$$H/D = m/n. \tag{9.28}$$

The ratio H/D is called the setting ratio of the outlet.

For an orifice or pipe-type proportional outlet ($m=1/2$) from a trapezoidal channel ($n \simeq 5/3$), the setting ratio H/D is 0.30, whereas for an open flume (Crump type, $m=3/2$) proportional outlet it is 0.90.

(a) Classification of outlets

1. *Non-modular type.* This is an outlet in which the discharge depends upon the difference in level (H) between the water levels in the distributing (supply) channel and the water course (field channel). A submerged pipe outlet (Fig. 9.25), or an orifice, is a common example of this type. Referring to the pipe outlet layout in Fig. 9.25, any conventional pipe flow solution will give the outlet discharge, e.g. an iterative procedure combining Darcy–Weisbach and Colebrook–White equations.
2. *Semi-modular or flexible type.* The discharge through this type of outlet (semimodule) is only affected by the change in the water level of the distributing canal. A pipe outlet discharging freely, an open flume outlet and an adjustable orifice outlet are the common types of flexible outlet. As the discharge through this type of outlet is independent of the water level in the field channel, interference by farmers is minimal. The pipe outlet is usually constructed with a flexibility $F<1$, so that the outlet discharge changes by a smaller percentage than the change in the distributing channel discharge. An open flume type outlet is shown in Fig. 9.26. The entrance is so designed that the flume takes a fair share of the silt. The discharge through this type of outlet (e.g. Crump) is given by

$$Q = CbH^{3/2}\ (\mathrm{m^3\,s^{-1}}) \tag{9.29}$$

where b is the throat width (in m), H is the head over the crest (in m), and C equals 1.71 (theoretically $0.544g^{1/2}$; actual range 1.60–1.66). An adjustable orifice semimodule (AOSM) type outlet, consisting of an orifice followed by a gradually expanding downstream flume, is shown in Fig. 9.27. This type of outlet is commonly used in the Indian subcontinent, and is considered to be one of the best forms of outlet.
3. *Modular type.* This type of outlet (rigid or invariant module) delivers a constant discharge within set limits, irrespective of the water level fluctuations in the distributing channel and/or field channel. Gibb's

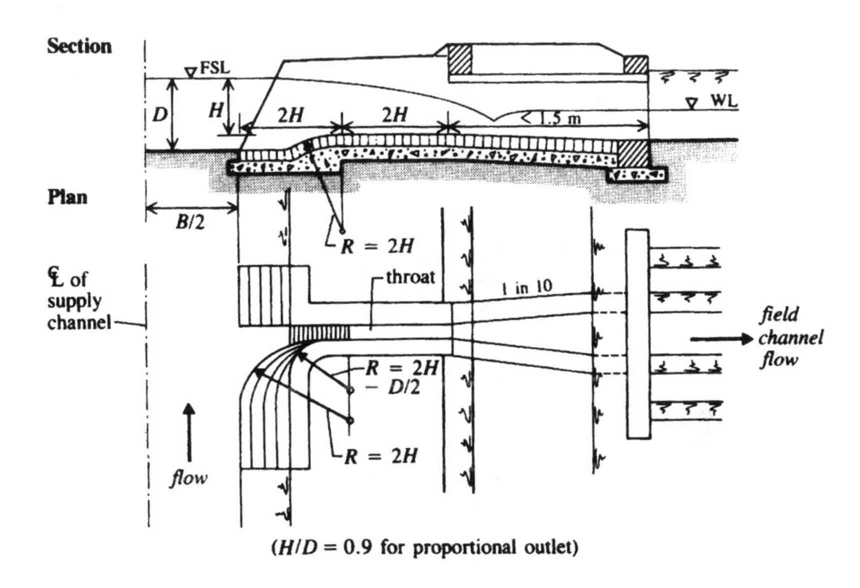

Fig. 9.26 Open flume type farm outlet (India and Pakistan)

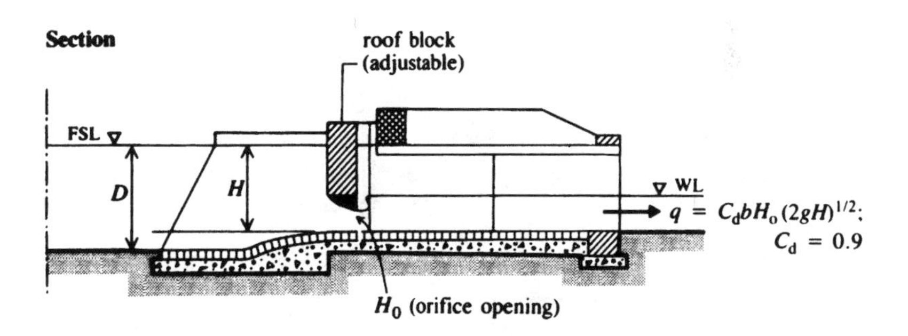

Fig. 9.27 Adjustable orifice semi-module (AOSM). Plan layout to conform to the open flume type (Fig. 9.26). $H/D = 0.3$ for proportional outlet

module (Fig. 9.28) is one of the several types of rigid module outlet. It has a semicircular eddy chamber in plan, connected by a rising inlet pipe. A free vortex flow is developed within the inlet pipe, thus creating a rise in the water level at the outer circumference of the eddy chamber (since $Vr =$ constant for free vortex flow). The excessive energy of the entering flow (due to the increase in the head causing the flow) is dissipated by the baffle walls supported from the roof, thereby keeping the discharge constant over a wide range of

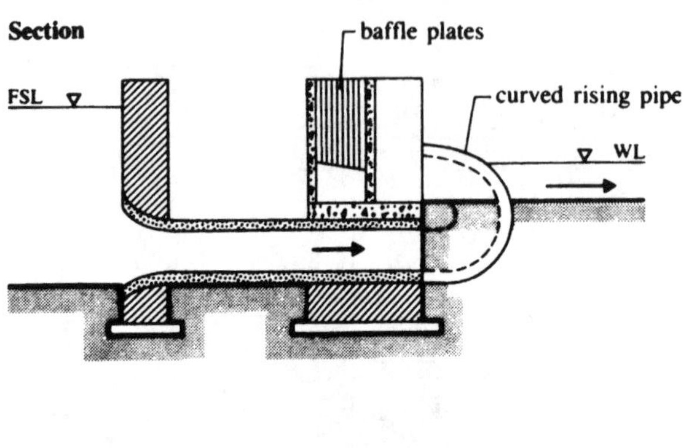

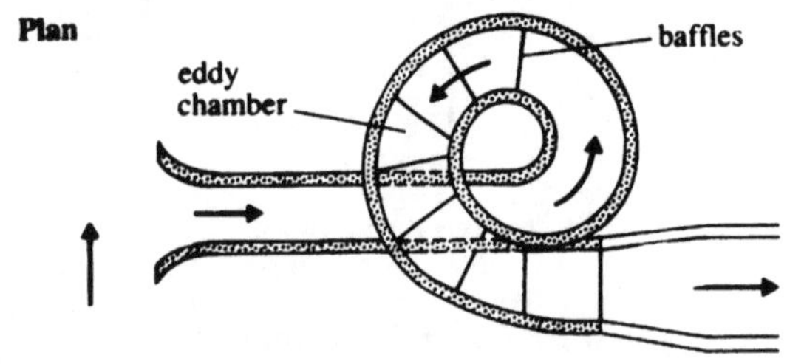

Fig. 9.28 Gibb's type rigid module

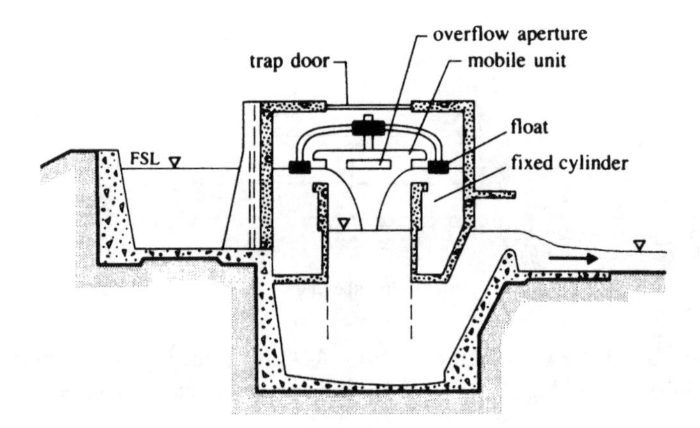

Fig. 9.29 Autoregulator (Italian-type rigid module)

variations in the head. An Italian design rigid module outlet is shown in Fig. 9.29. It consists of a cylindrical sleeve with circumferential apertures (which act as weirs), floating in a fixed outer cylinder. This ensures a constant head, causing a flow through the apertures irrespective of the water level in the distribution canal (Water and Water Engineering, 1956). The Neyrpic orifice type of rigid module outlet (French design) is shown in Fig. 9.30; this facilitates the

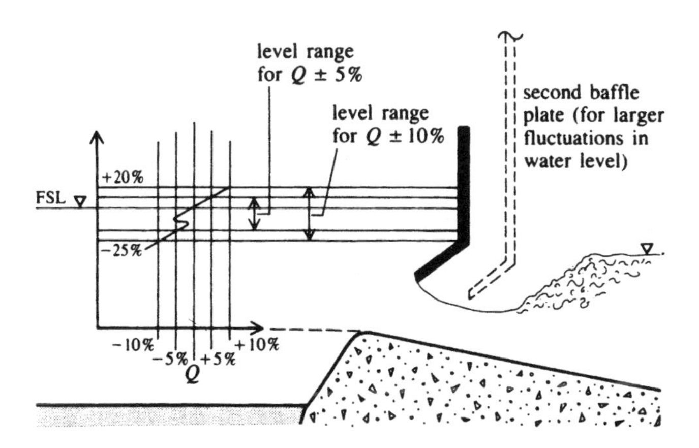

Fig. 9.30 Orifice-type rigid module (single-baffle layout)

drawing of an almost constant discharge over a wide range of water-level fluctuations in the distribution channel. With the increase in head the contraction of the jet increases, thus offsetting any increase in the corresponding discharge. In order to cope with a much wider range of water-level fluctuations, a double baffle plate orificed module may be used. If the water-level fluctuations are beyond tolerable limits for constant flow in the off-taking canal, auxiliary equipment such as downstream and upstream level gates (Kraatz and Mahajan, 1975) must be installed in the distribution system.

Worked Example 9.3

Design an intake structure for a minor canal, using the following data:

	Supply canal		*Offtake*
	Upstream	*Downstream*	
Discharge ($m^3 s^{-1}$)	4.0	3.5	0.5
Bed level (m AOD)	120.00	119.94	119.86
FSL (m AOD)	120.96	120.84	120.36
Bed width (m)	8.0	7.0	2.0
Angle of offtake			60°

Solution

The upstream full supply depth, $D = 120.96 - 120.00 = 0.96$ m. The water depth in the offtake $= 120.36 - 119.86 = 0.50$ m. The working head =

120.96 − 120.36 = 0.60 m. Adopting a proportional open flume intake (outlet), $H/D = 0.9$. Therefore the head, H, over the crest = $0.9D = 0.9 \times 0.96 = 0.864$ m. The crest level = 120.96 − 0.864 = 120.096 m AOD. As the crest level is greater than the bed level of the offtake, this may be adopted. From the flume formula, the throat width, $b = Q/CH^{3/2} = 0.5/1.6 \times 0.864^{3/2} = 0.39$ m.

Note that suggested values of C are as follows:

Q ($m^3 s^{-1}$)	*Offtake angle, C values*	
	60°	*45°*
<0.55	1.60	1.61
0.55–1.5	1.61	1.63

The width of the offtake canal = 2.0 m. Therefore the flume ratio, $f_r =$ 0.39/2 = 0.195. The crest length = $2H$ = 1.728 m; say 1.75 m.

For the glacis and stilling basin (cistern), the discharge intensity (assuming 67% for splay), $q = 0.67 \times 0.50/0.39 = 0.86\,m^3 m^{-1}$. The working head, $h = 120.96 - 120.36 = 0.60$ m. The critical depth, $h_c = (q^2/g)^{1/3} =$ 0.42 m, and therefore $h/h_c = 0.60/0.42 = 1.43$. The non-dimensional energy curve, h/h_c versus E/h_c, gives the relationship

$$E/h_c = h/h_c + \frac{1}{2}(h_c/h)^2$$

from which $E = 0.706$ m. Therefore the depth of the cistern below the bed level of offtake

= E − water depth in offtake

= 0.206 m (minimum depth ≥ 0.075 m).

Therefore the floor level of the cistern = 119.86 − 0.206 = 119.654 m AOD, and hence the depth of the cistern floor below the crest, $d =$ 120.096 − 119.654 = 0.442 m. The length of the downstream glacis = $2.5d$ = 1.10 m. The length of the cistern = offtake flow depth + working head = 1.10 m.

DOWNSTREAM EXPANSION LENGTH

1. 1.5 × (offtake bed width − b) ≃ 2.4 m, or
2. length of glacis + length of cistern = 2.20 m, or
3. with a minimum splay of 1 in 10, length required = f_r/splay = 1.95 m.

Therefore adopt the largest of the three, i.e. 2.4 m. The downstream curtain wall depth below the cistern floor level is one-half of the offtake flow depth = 0.25 m. Adopt a minimum depth of 0.5 m. The upstream curtain wall depth is one-third of the FSL flow depth = 0.32 m. Adopt a minimum depth of 0.5 m.

DOWNSTREAM BED PROTECTION

Length of protection = offtake flow depth + working head = 1.10 m. Provide a gravel apron, about 1 m long and 150 mm thick, downstream of the cistern. The upstream approach length of the crest

$$\simeq \{[5H - (\text{crest level} - \text{bed level})](\text{crest level} - \text{bed level})\}^{1/2}$$

$$= 0.64\,\text{m}.$$

The radius joining the crest $= 2H \simeq 1.728$ m. The details of the layout are shown in plan and section in Fig. 9.31.

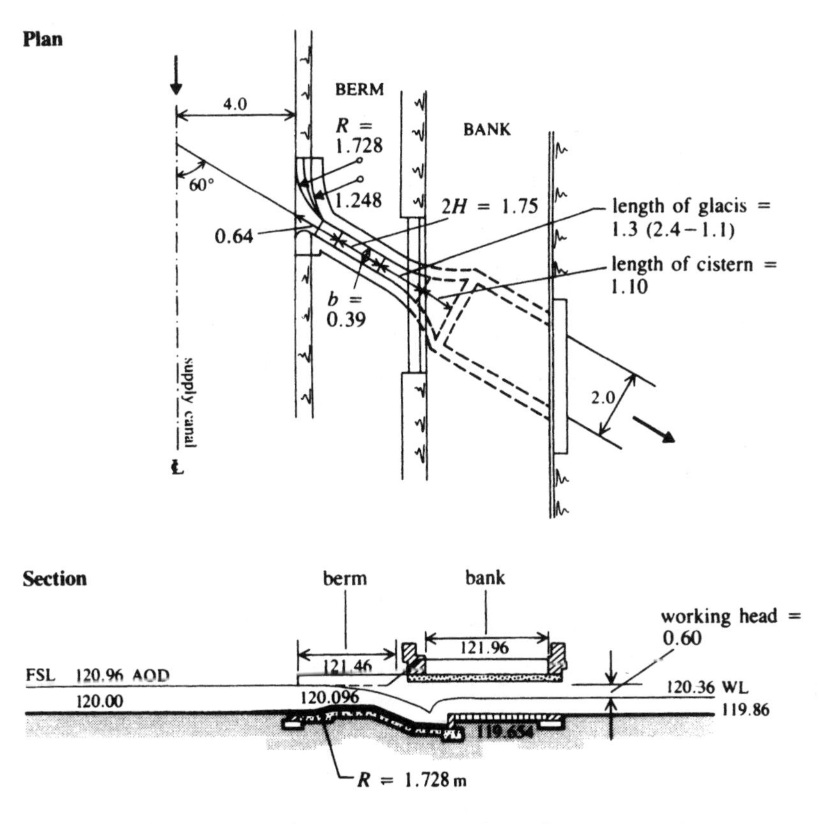

Fig. 9.31 Intake structure for a minor canal (outflow = 0.5 m³ s⁻¹; proportional setting, *H*/*D* = 0.9); all dimensions in metres

Worked Example 9.4

Design a submerged pipe outlet for a ditch to discharge $15\,\mathrm{l\,s^{-1}}$ from a supply canal, using the following data:

	Supply canal	*Field channel*
Bed level (m AOD)	100.00	100.00
FSL (m AOD)	101.00	100.90

Pipe outlet: length of pipe = 30 m. Concrete pipes in multiples of 50 mm diameters with a k value of 0.1 mm are available.

Solution

The working head, $H = 101.00 - 100.90 = 0.10\,\mathrm{m}$. The depth of flow in the supply canal, $D = 101.00 - 100.00 = 1.00\,\mathrm{m}$. Therefore the pipe setting, $H/D = 0.1/1.0 = 0.1 < 0.3$, and so the pipe outlet functions as a hyperproportional one with a flexibility of $F = 0.3/0.1 = 3$.

PIPE DESIGN

Combining Colebrook–White and Darcy–Weisbach equations (Featherstone and Nalluri, 1995),

$$Q = -2A(2gdS)^{1/2}\log[k/3.7d + 2.51\nu/d(2gdS)^{1/2}]$$

where d is the pipe diameter, $A = \pi d^2/4$, k is the pipe roughness, and S is the friction gradient. Neglecting minor losses, $S = 0.1/30 = 0.0033$, and we obtain

d(mm)	100	150	200
$Q(\mathrm{l\,s^{-1}})$	4.0	12.0	26.0

If minor losses are included (here they cannot be neglected), 200 mm pipe gives $18\,\mathrm{l\,s^{-1}}$ which is quite adequate.

Worked Example 9.5

Design a suitable outlet to discharge $50 \, l \, s^{-1}$ from a canal with a full supply depth of 1 m. The working head available is 150 mm.

Solution 1: open flume outlet

For non-submergence of the flume, i.e. modular flow conditions, the minimum working head is approximately 0.2 times the head over the crest. Therefore the maximum head over the crest, $H = 0.15/0.2 = 0.75$ m. Hence the throat width, b (from the weir formula) $\simeq 0.05$ m. Therefore, adopt the minimum value of $b = 0.06$ m; this gives the head, $H = 0.65$ m, and hence the minimum working head, $h = 0.2 \times 0.65 = 0.13$ m, which is satisfactory, as the available head is 0.15 m. This sets $H/D = 0.65/1.0 = 0.65$, and the flexibility, $F = 0.9/0.65 > 1$. The design may be acceptable but may not draw its fair share of silt from the supply canal because of the excessive sill height of 0.35 m (depth of flow − head over sill).

Solution 2: pipe and open flume outlet

An open flume type outlet is expensive, particularly if the supply canal bank is very wide, and in such cases the pipe semimodule is used. The outlet is also suitable for drawing its share of silt, with its lead pipe set at or near the bed level (Fig. 9.32). The pipe delivers the water into a tank on the downstream side, to which an open flume or an orifice semimodule is fitted.

$Q = 50 \, l \, s^{-1}$; $D = 1.0$ m; $h = 0.15$ m.

Assume a bank width of 10 m and adopt a concrete pipe with a k value of 0.1 mm. The head loss through the pipe, $h_1 = (1.5 + \lambda L/d)V^2/2g$. Assuming a pipe diameter, $d = 300$ mm, $V = 0.707 \, m \, s^{-1}$ and $V^2/2g = 0.0255$ m. Therefore the Reynolds number, $Re = 2 \times 10^5$; $k/d = 3.3 \times 10^{-4}$, and so $\lambda = 0.0175$ from the Moody chart. Hence the head loss $= 0.0531$ m, giving the available working head for the semimodule as $0.15 - 0.0531 = 0.0969$ m. Therefore the maximum head over the crest, $H = 0.0969/0.2 = 0.484$ m. The throat width, b (from weir formula) $= 0.093$ m. Therefore provide a throat width of 10 cm, which gives $H = 0.46$ m and thus $h_{minimum} = 0.092$ m, which is satisfactory as the available head is 0.0969 m. The layout of the proposed design of the pipe and open flume outlet is shown in Fig. 9.32(a). Two other alternative proposals which are in use are shown in Figs 9.32(b) and (c).

(a) Pipe and open flume type outlet

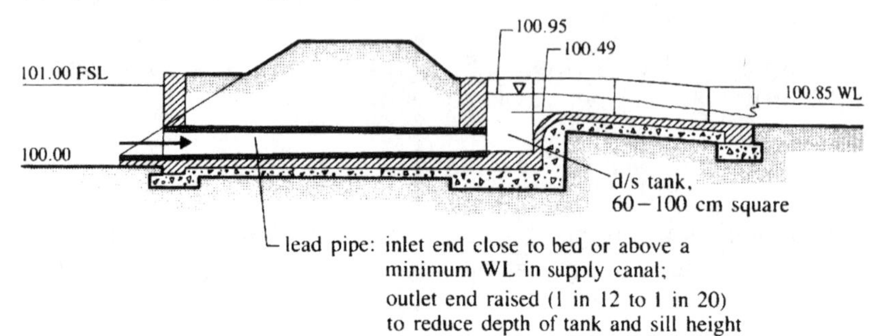

(b) Pipe and AOSM outlet

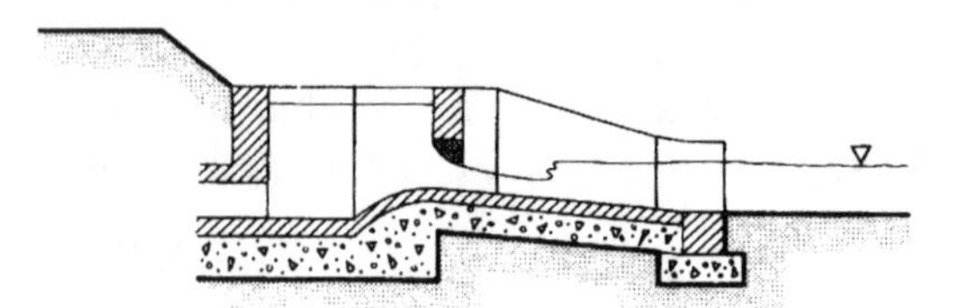

(c) Pipe and Jamrao type orifice module

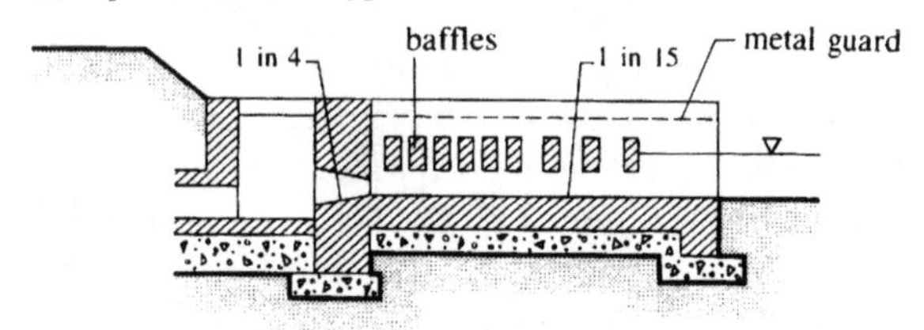

Fig. 9.32 Pipe semimodule arrangements

9.3 Fish passes

9.3.1 Introduction

Environmental and fishery protection interests require the provision of appropriate facilities as an integral part of dams and weirs. Fish protection facilities consist of fish passes for safe migration (ladders, lifts) and screens. The design of a satisfactory fish passage must be based on the study of fish behaviour and swimming performance, hydraulics and general ecological flow management. A discussion of these aspects with a comprehensive list of references can be found in Katopodis (2005).

Salmon and trout are the main migratory fish. The smolts (young salmon) travel to the ocean 1–2 years after spawning in fresh water, while the kelts (fully grown salmon) return to their spawning grounds after spending 1–3 years in the sea. The design criteria and layout of the facilities, largely depending on the type of fish and whether they are migrating downstream or upstream, are best established by monitoring existing structures and by model testing.

Fish-passing facilities may be divided into two groups:

1. facilities for upstream migrant fish, consisting of fish ladders, fish locks (lifts), tramways, and arrangements for trapping and trucking the fish;
2. special facilities for a safe passage for the migrating smolts, which consist of the arrangements to collect the migrating fish in the forebay (headwater) at fine-mesh screens (fixed or moving types) and directing them to safe bypass systems through which they are transferred downstream.

Fish passes (for both upstream and downstream migratory fish) must be designed so that the fish are able to find the entrance to the system provided for their passage and swim through without undue effort and unusual risk of injury.

9.3.2 Upstream fish-passing facilities

(a) Fish ladder (pool and traverse fish pass)

This is an artificial upstream fish passage, most commonly used for heads up to 20 m consisting of (1) a fish entrance, (2) a fish ladder proper and (3) a fish exit. Sometimes an auxiliary (additional) water supply is also provided to attract fish to the entrance.

The fish ladder proper consists of a series of traverses (cross-walls) and pools circumventing an obstruction (such as a weir or dam) for the fish to migrate to the head waters in easy stages. This is achieved by creating a series of drops of around 300 mm–450 mm between pools (Fig. 9.33) on a gradient of around 1 in 8 to 1 in 15 (for high heads). Rest pools of a larger size (normally twice the size of an ordinary pool) are also provided after every 5 6 pools.

The actual arrangement of pools and traverses is chosen according to a particular obstruction; a low level weir or dam may need a fish pass of shallow gradient (corresponding to the surrounding gradient of the land) whereas a tightly folded pass may be necessary in case of a high weir or dam.

The fish pass is designed to take a fixed proportion of the flow over the main weir or spillway. This is normally achieved by siting the invert

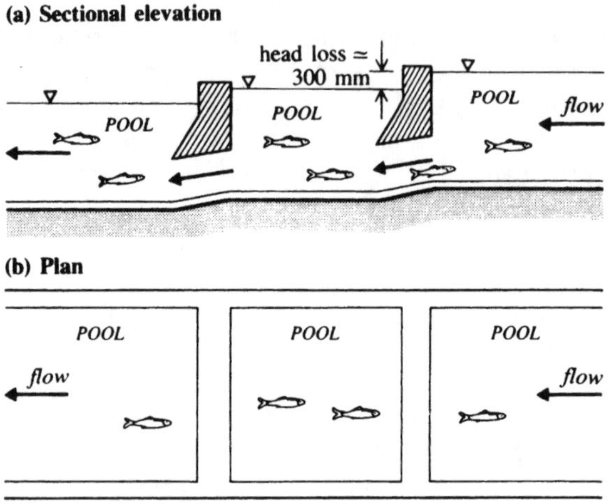

Fig. 9.33 Fish ladder

level of the uppermost notch lower than the adjoining weir or spillway crest; the discharge and head calculations may be achieved by using appropriate weir–notch or orifice formulae.

The entrance to the fish pass (ladder) must be located in the downstream white water parallel to the main flow, whereas the exit (at the upstream end) should be well within the reservoir away from the spillway structure. The pools are usually 1–2 m deep, 2–5 m long and 2–10 m wide, depending on the number of fish migrating. Cross-walls may be provided with staggered notches–orifices, with the velocities in the ladder being around $0.5\,\mathrm{m\,s^{-1}}$. Guiny *et al.* (2005) investigated the efficiency of different passages through the baffles in a fish pass (orifice, slot, weir) and found – with limited data – a strong preference of the migrating juvenile Atlantic salmon for the orifice type.

The maximum swimming speed U ($\mathrm{m\,s^{-1}}$) of a fish is a function of the fish length L (m) and the water temperature T(°C) and is predicted (Zhou, 1982) by

$$U = 0.7L/2t \tag{9.30}$$

where t is the muscle twitch contraction time (related to tail beat frequency) given by

$$t = 0.17L^{0.43} + (0.0028 - 0.0425L^{0.43})\ln T - 0.0077. \tag{9.31}$$

Equations (9.30) and (9.31) are recommended for the ranges of $L = 0.05\,\mathrm{m}$–$0.80\,\mathrm{m}$ and $T = 2\,°\mathrm{C}$–$18\,°\mathrm{C}$. The fish may be able to swim at this speed for a specified length of time (endurance time, t_m) which is also a function of fish length and water temperature and is given by

$$t_m = E/(|P_c - P_r|) \tag{9.32}$$

where E (energy store) $\simeq 19400L^3$, P_c (chemical power) $\simeq 0.97e^{-0.0052T} U^{2.8}L^{-1.15}$ and P_r (power supplied) $\simeq 48L^3$. It is suggested that the swimming speed and endurance times as proposed by equations (9.30) and (9.32) should not be exceeded in the design of a fish pass. Some other examples of fish ladder passes in use are the weir-type fish entrance at Priest Rapides, and the slotted fish entrances at Wanapum dam (Columbia River) in the United States.

Flow control structures such as flood relief channels should be so designed that the fish are not stranded; the minimum sluice opening should be around $0.3\,m \times 0.3\,m$ and the water velocity not more than $3\,m\,s^{-1}$ (a head difference of about 450 mm and a discharge rate of $0.27\,m^3\,s^{-1}$).

(b) Denil fish pass

The Denil fish pass (Denil 1936; see Beach (1984)) consists of closely spaced baffles (for energy dissipation) and set at an angle to the axis of the channel to form secondary channels for the passage of the fish. A typical Denil fish pass with a channel width (b) (of say 0.9 m) may consist of simple single-plane baffles with rectangular opening over a V-shape opening (similar to a compound notch) set at a spacing of about $2/3b$ (0.6 m) sloping upstream at an angle of 45° to the channel bed whose slope itself should not exceed 1:4; the width of the opening is approximately $0.58b$, whereas the height of the sill of the V-shape opening above the channel bed is around $0.24b$ with its top level (from bed) at about $0.47b$. Large rest pools (3 m long $\times$ 2 m wide $\times$ 1.2 m deep) are provided at vertical intervals of 2 m. An example of the Denil fish pass complex is at Ennistyman on the river Inagh in the Republic of Ireland.

(c) Fish lift (Borland type)

The fish lifts are primarily intended for high impounding structures and the arrangements of the Borland fish lift are shown in Fig. 9.34. Its operation is very similar to navigation locks; however, in the fish lock throughflow is maintained (bypass pipes) to induce the fish into and out of the chamber. The fish are attracted into the lower pool (chamber) by the downstream flow through the pool. The chamber is then closed at its downstream end, and the fish are induced to swim out into the reservoir by the flow through the exit from the lift. The operation of a fish lock is cyclic,

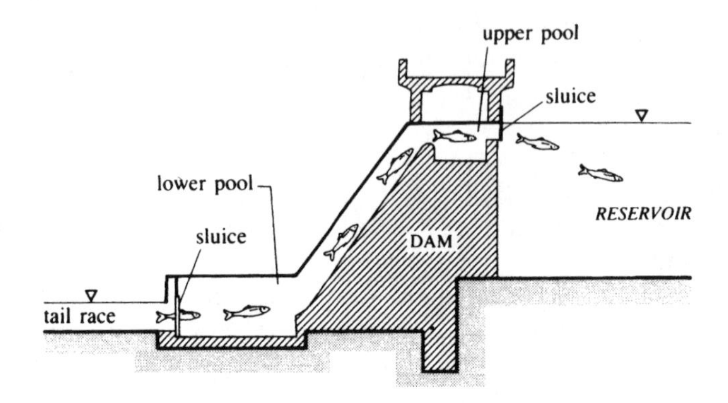

Fig. 9.34 Borland fish lift

with variable times according to a specific installation and its operating procedures. Examples of Borland lift passes (Aitken, Dickerson and Menzies, 1966) are in both Scotland and Ireland.

(d) Fish traps and transportation

At high-head structures, fish ladders are neither economical nor practical. Instead, tramways or cableways are used. The installation consists of hoppers into which fish swim; they are then transported to the reservoir by tramway or cableway. Fish may also be unloaded into a tank truck (filled with aerated and refrigerated water) which conveys them to either the reservoir or a hatchery where they are carefully unloaded.

(e) Fish-barrier dams

These are low-head weirs (with an electrical field if necessary) which stop the upstream migration of fish and induce them to swim into a fish ladder or hopper situated downstream of the barrier (with a proper entrance to attract fish) from where they are transported by tank trucks.

9.3.3 Downstream fish-passing facilities

The seaward-migrating fish are normally protected from spillways and turbine intakes by screens (Aitken, Dickerson and Menzies, 1966) which divert or deflect them (Fig. 9.35) into safer bypass arrangements. The velocity of approach to screens is generally kept at around $0.5\,\mathrm{m\,s^{-1}}$, whereas for louvre diverters it is around $1\,\mathrm{m\,s^{-1}}$ to maintain the required turbulence levels. Fish screens are usually of either mechanical or electrical type.

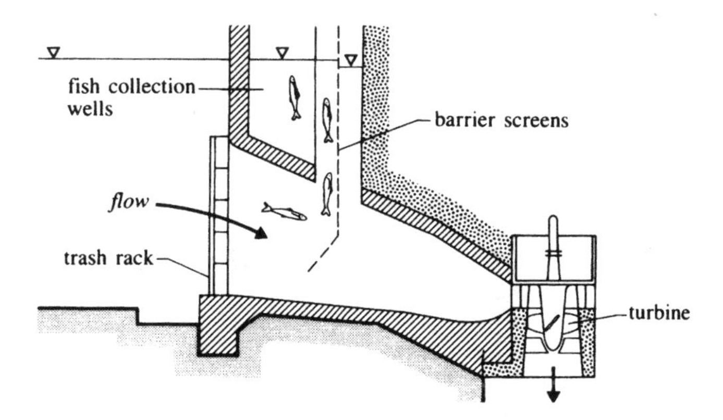

Fig. 9.35 Fish screen at hydropower intake

The mechanical screens include the rotary (drum) type or travelling screens with water jets or stationary screens (with duplicate numbers for maintenance purposes) with louvre guides. One of the recent trends (Odgaard, Cherian and Elder, 1987) in screen design in the turbine intake is to suspend a screen in the intake water passage to direct fish toward and into a gate well (Fig. 9.35), for subsequent collection and release downstream of the dam.

Electrical screens consist of live electrodes, vertically suspended along a ground conductor, which generate electrical pulses across the flow thus forming a barrier (electrical screen) to stop and divert the fish into safer passages for their onward (downstream) migration.

Recent technology suggests a 'wall' (sound barrier) that emanates from a Bio Acoustic Fish Fence (BAFF) consisting of a perforated pipe down which compressed air is pumped. Every few metres, pneumatic sound transducers vibrate to produce low frequency sound waves which become trapped in the air bubbling up, out of the pipe, creating a 'wall' of sound to the surface of water. Thompsett (2000) suggests that diverting fish with sound barrier systems (acoustic screens) ahead of the turbine installation are generally less expensive, require little maintenance and reduce loss of head in comparison to traditional steel/mechanical screen diversion systems.

Protection of downstream migrating fish, particularly eel at hydroelectric power stations could also include turbine management (Hadderingh and Bruijs, 2003) and special turbine design.

References

Aitken, P.L., Dickerson, L.H. and Menzies, W.J.M. (1966) Fish passes and screens at waterpower works. *Proceedings of the Institution of Civil Engineers*, 35: 29–57.

Avery, P. (1989) *Sediment Control at Intakes – a Design Guide*, British Hydromechanics Research Association, Cranfield.

Avery, S. and Novak, P. (1978) Oxygen transfer at hydraulic structures. *Journal of the Hydraulics Division, American Society of Civil Engineers*, 104 (HY 11): 1521–40.

Baban, R. (1995) *Design of Diversion Weirs: Small Scale Irrigation in Hot Climates*, John Wiley & Sons, Chichester.

Beach, M.H. (1984) *Fish Pass Design – Criteria for the Design and Approval of Fish Passes and Other Structures to Facilitate the Passage of Migratory Fish in Rivers*, Fisheries Research Technical Report 78, Ministry of Agriculture, Fisheries and Food, London.

Bos, M.G. (ed.) (1976) *Discharge Measurement Structures*, Laboratorium voor Hydraulica en Afvoerhydrologie, Wageningen.

Breusers, H.N.C. and Raudkivi, A.J. (1991) *Scouring*, IAHR Hydraulic Structures Design Manual, Vol. 2, Balkema, Rotterdam.

BSI (1969) *Measurement of Liquid Flow in Open Channels: Weirs and Flumes*, BS 3680, Part 4, British Standards Institution, London.

Chanson, H. (1994) *Hydraulic Design of Stepped Cascades, Channels, Weirs and Spillways*, Pergamon, Oxford.

Dolezal, L. (1968) L'influence d'un barrage fixe peu élevé sur le débit maximum, Société Hydro-technique de France, X^{mes} *Journées de L'hydraulique* Rapport 10, Question 111: 1–4.

Featherstone, R.E. and Nalluri, C. (1995) *Civil Engineering Hydraulics*, 3rd edn, Blackwell Scientific, Oxford.

Guiny, E., Ervine, A. and Armstrong, J.D. (2005) Hydraulic and biological aspects of fish passes for Atlantic salmon, *J. of Hydraulic Engineering*, ASCE, 131, No. 7: 542–53.

Gulliver, J.S., Wilhelms, S.C. and Parkhill, K.L. (1998) Predictive capabilities in oxygen transfer at hydraulic structures. *Journal of Hydraulic Engineering*, ASCE, 124, No. 7: 664–71.

Hadderingh, R.H. and Bruijs, M.C.M. (2003) Hydroelectric power stations and fish migration, *Journées d'Élude du Cebedeau, Tribune de l'Eau*, No. 619–21, 89–97.

Jambor, F. (1959) Mögliche Erhöhung und Entwicklun der gesten Wehrschwelle, sowie Gestaltung der damit verbundenen Wehrkonstruktionen, im besonderern des Sektorwehres, *Bautechnik*, 36 (6): 221–8

Katopodis, C. (2005) Developing a toolkit for fish passage, ecological flow management and fish habitat works, *J. of Hydraulic Research*, IAHR, 43, No. 5: 451–67.

Khosla, A.N., Bose, N.K. and Taylor, E.M. (1954) *Design of Weirs on Permeable Foundation*, Publication No. 12, Central Board of Irrigation and Power, New Delhi.

Kraatz, D.B. and Mahajan, I.K. (1975) *Small Hydraulic Structures – Irrigation and Drainage*, FAO, Rome, Papers 26/1 and 26/2.

Leliavsky, S. (1965) *Design of Dams for Percolation and Erosion: Design Text Books in Civil Engineering*, Vol. III, Chapman and Hall, London.

Linsley, R.K. and Franzini, J.B. (1979) *Water Resources Engineering*, 3rd edn, McGraw-Hill, New York.

Mazumdar, S.K. (1983) *Irrigation Engineering*, Tata McGraw-Hill, New Delhi.

Meusburger, H. (2002) Energieverluste an Einlaufrechen von Flusskraftwerken, *Mitteilungen*, No. 179, VAW-ETH, Zürich.

Mosonyi, E. (1987) *Water Power Developments*, 3rd edn, Vol. 1, Hungarian Academy of Sciences, Budapest.

Novak, P. (1994) Improvement of water quality in rivers by aeration at hydraulic structures, in *Water Quality and its Control* (ed. M. Hino), IAHR Hydraulic Structures Design Manual, Vol. 5, Balkema, Rotterdam.

Novak, P. and Čábelka, J. (1981) *Models in Hydraulic Engineering*, Pitman, London.

Novak, P. and Gabriel, P. (1997) Oxygen uptake at barrages of the Elbe cascade, in *Proceedings of the 27th Congress of the International Association for Hydraulic Research*, San Francisco, Vol. D, 489–94.

Odgaard, A.T., Cherian, M.P. and Elder, R.A. (1987) Fish diversion in hydropower intake. *Journal of Hydraulic Engineering, Proceedings of the American Society of Civil Engineers*, 113 (4): 505–19.

Ortmanns, C. (2006) Entsander von Wasserkraftanlagen, *Mitteilungen*, No. 193, VAW-ETH, Zürich.

Ranga Raju, K.G. (1993) *Flow Through Open Channels*, 3rd edn, Tata McGraw-Hill, New Delhi.

Singh, B. (1975) *Fundamentals of Irrigation Engineering*, 5th edn, Nemchand, Roorkee.

Thompsett, A. (2000) *Diverting Fish with a Sound Barrier*, HRW (Hydro Review Worldwide), Vol. 8, No. 3.

US Army Waterways Experimental Station (1959) *Hydraulic Design Criteria*, US Department of the Army, Washington, DC.

Vischer, D.L. and Hager, W.H. (1997) *Dam Hydraulics*, John Wiley & Sons, Chichester.

Water and Water Engineering (1956) New automatic flow regulator. *Water and Water Engineering*, 60: 250–7.

Zhou, Y. (1982) *The Swimming Behaviour of Fish in Towed Gears: a Re-examination of the Principles*, Scottish Fisheries Working Paper 4, Department of Agriculture and Fisheries, Scotland.

Chapter 10

Cross-drainage and drop structures

10.1 Aqueducts and canal inlets and outlets

10.1.1 Introduction

The alignment of a canal invariably meets a number of natural streams (drains) and other structures such as roads and railways, and may sometimes have to cross valleys. Cross-drainage works are the structures which make such crossings possible. They are generally very costly, and should be avoided if possible by changing the canal alignment and/or by diverting the drains.

10.1.2 Aqueducts

An aqueduct is a cross-drainage structure constructed where the drainage flood level is below the bed of the canal. Small drains may be taken under the canal and banks by a concrete or masonry barrel (culvert), whereas in the case of stream crossings it may be economical to flume the canal over the stream (e.g. using a concrete trough, Fig. 10.1(a)).

When both canal and drain meet more or less at the same level the drain may be passed through an inverted siphon aqueduct (Fig. 10.1(d)) underneath the canal; the flow through the aqueduct here is always under pressure. If the drainage discharge is heavily silt laden a silt ejector should be provided at the upstream end of the siphon aqueduct; a trash rack is also essential if the stream carries floating debris which may otherwise choke the entrance to the aqueduct.

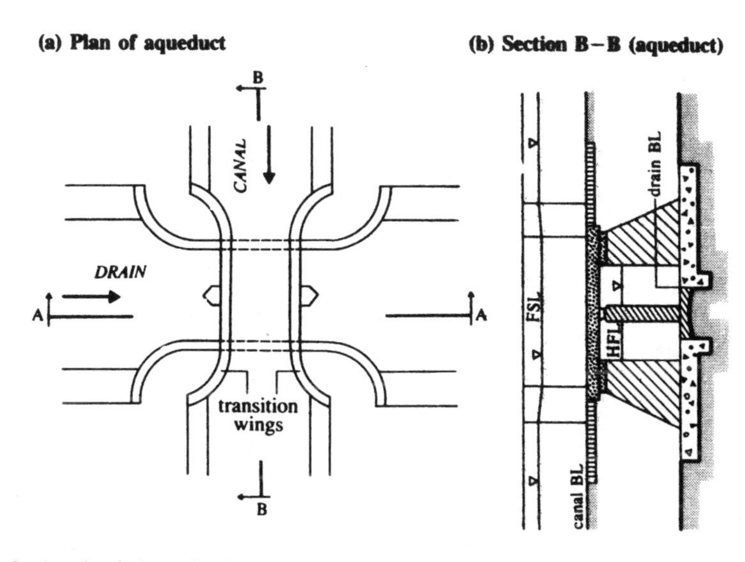

(c) Section A–A (aqueduct)

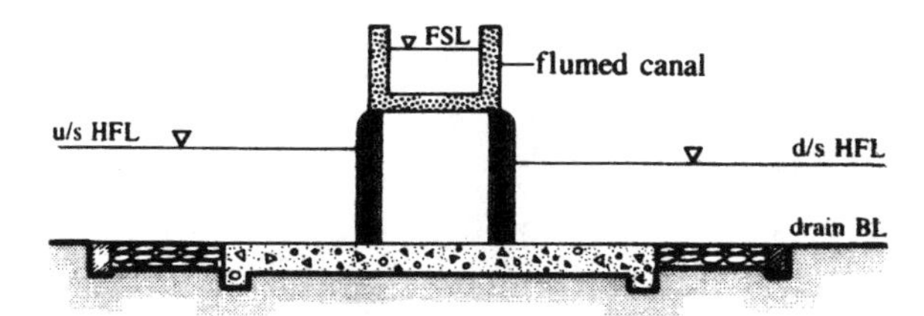

(d) Section A–A (inverted siphon aqueduct)

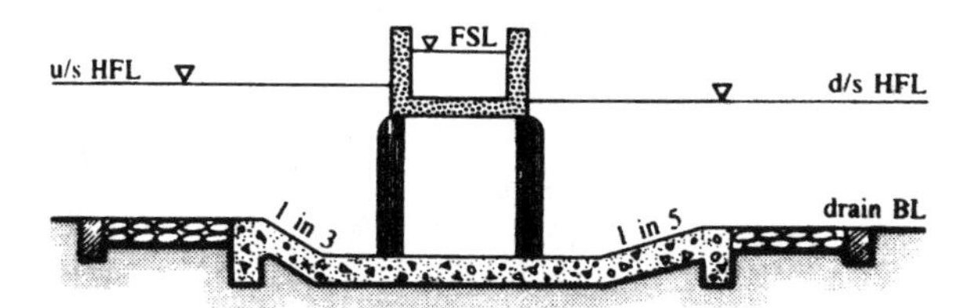

Fig. 10.1 Layout of an aqueduct

10.1.3 Superpassage

In this type of cross-drainage work, the natural drain runs above the canal, the canal under the drain always having a free surface flow. The superpassage is called a canal siphon or simply an inverted siphon if the canal bed under the drain is lowered to accommodate the canal flow, which will always be under pressure. The layouts of the superpassage and canal siphon are similar to those shown in Figs 10.1(a) and 10.1(b), with the canal and drain interchanged.

10.1.4 Level crossing

Level crossing facilities are provided when both the drain and the canal run at more or less the same level. This is more frequently used if either of the flows occurs for a short period (e.g. flash floods in the drain); in addition, the mixing of the two bodies of water must also be acceptable (quality considerations).

The plan layout of a level crossing with two sets of regulators, one across the drain and the other across the canal, is shown in Fig. 10.2. Normally, the canal regulator regulates its flow with the drain regulator kept closed. Whenever the flash floods occur, the canal gates are closed and drainage gates opened to let the flood flow pass.

10.1.5 Canal inlets and outlets

When the drainage flow is small it may be absorbed into the canal through inlets. The flow in the canal may be balanced, if necessary (in the case of small canals), by providing suitable outlets (or escapes). The inlet and outlet structures must also be provided with energy dissipators wherever necessary.

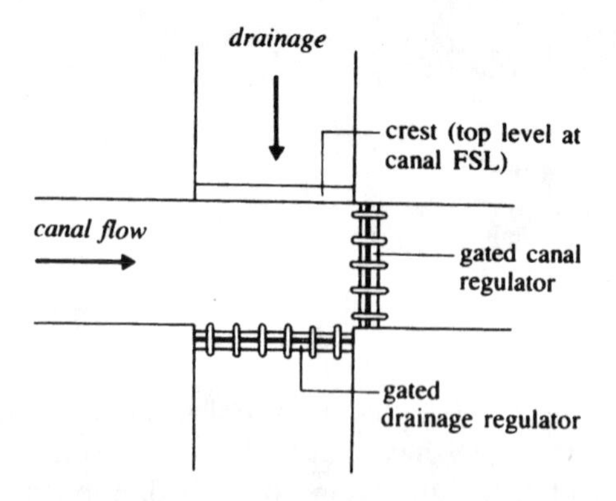

Fig. 10.2 Level crossing

The following worked example highlights the various aspects of the hydraulic design of a cross-drainage structure.

Worked Example 10.1

Design a siphon aqueduct for the following data:

	Canal	*Stream*
Discharge ($m^3 s^{-1}$)	30	500
Bed level (m AOD)	200.00	198.00
Canal FSL (m AOD)	202.00	
Bed width (m)	25.00	
Canal side slopes	1.5:1 V	
Stream HFL (m AOD)		200.50

The general terrain level is 200.00 m AOD.

Solution

DRAINAGE WATERWAY

Perimeter $P = 4.75Q^{1/2}$ (régime width, equation (9.9)) $\approx$ 106 m. Providing 12 piers of 1.25 m thickness, we have 13 spans of 7 m each. Therefore waterway provided $= 13 \times 7 + 12 \times 1.25 = 106$ m (satisfactory). Assuming a maximum velocity through the siphon barrels of $2\,m\,s^{-1}$, height of barrel $= 500/(13 \times 7 \times 2) = 2.747$ m. Provide rectangular barrels, 7 m wide and 2.75 m high (shown in Fig. 10.5).

CANAL WATERWAY

Since the drainage width is large (106 m at the crossing) it is economical to flume (concrete, $n = 0.014$) the canal. Adopt a maximum flume ratio of 0.5. Therefore the flumed width of the canal (trough) $= 0.5 \times 25 = 12.5$ m. Providing a splay of 2:1 in contraction and a splay of 3:1 in expansion (Hinds, 1928),

$$\text{length of transitions in contraction} = 12.5\,\text{m},$$

$$\text{length of transitions in expansion} = 18.75\,\text{m}.$$

$$\text{The length of the trough from abutment to abutment} = 106\,\text{m}.$$

DESIGN OF FLUMED SECTION WITH TRANSITIONS

Referring to Fig. 10.3, the following results can be obtained to maintain a constant depth of flow of 2.0 m (given). The calculations are achieved from section 44 and proceed towards section 11 as tabulated below:

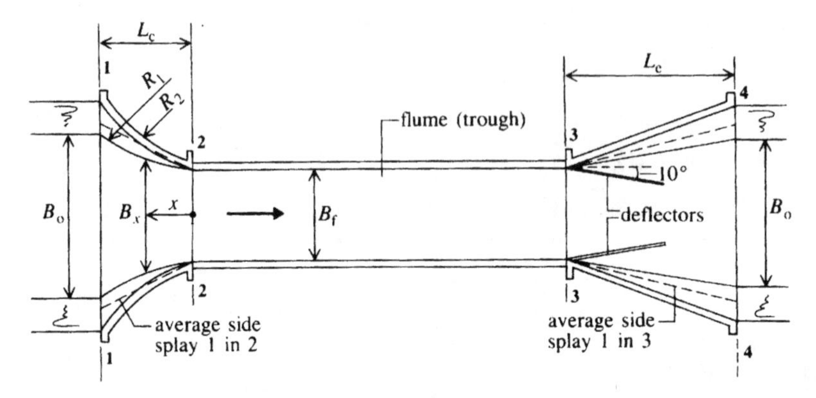

(L_c, L_e: contraction and expansion transition lengths. Length of deflector = ½ L_e; triangular wedge shaped with its height equal to u/s FSL)

Fig. 10.3 Transition with cylindrical inlet and linear outlet. L_c and L_e are contraction and expansion lengths

	Section						
	44		*33*		*22*		*11*
Width (m)	25.00		12.50		12.50		25.00
Area of flow (m^2)	56.00		25.00		25.00		56.00
Velocity ($m s^{-1}$)	0.536		1.20		1.20		0.536
Losses (m)		(expansion) 0.017		(friction) 0.017		(contraction) 0.012	
Water surface level (m AOD)	202.000		201.959		201.976		202.406
Velocity head (m)	0.015		0.073		0.073		0.015
TEL (m AOD)	202.015		202.032		202.049		202.061
Flow depth (m)	2.00		2.00		2.00		2.00
Bed level (m AOD)	200.00		199.959		199.976		200.046

Note that the contraction loss $= 0.2(V_2^2 - V_1^2)/2g$; the expansion loss $= 0.3(V_3^2 - V_4^2)/2g$; the flume friction loss $= V_f^2 n^2 L_f / R_f^{4/3}$ (the suffix f denotes the flume, and o the original canal – $V_f = V_3 = V_2$; $V_o = V_4 = V_1$).

DESIGN OF TRANSITIONS

For a constant depth of flow the transition may be designed such that the rate of change of velocity per metre length of transition is constant. This approach yields the bed width of the transition at a distance x from the flume section as

$$B_x = B_o B_f L / [L B_o - (B_o - B_f) x]$$

which, modified after experimental studies (UPIRI, 1940), gives

$$x = LB_o^{3/2}[1 - (B_f/B_x)^{3/2}]/(B_o^{3/2} - B_f^{3/2})$$

where L is length of the transition.

The following table shows the calculated geometries of the transition provided:

B_x (m)	12.5	15.0	17.5	20.0	22.5	25.0	
x (m)	0	4.64	7.69	9.73	11.31	12.5	(contraction)
x (m)	0	6.96	11.53	14.59	16.96	18.75	(expansion)

The transitions are streamlined and warped to avoid any abrupt changes in the width.

Transitions with a cylindrical inlet with an average splay of 2:1 and a linear outlet with a splay of 3:1 provided with flow deflectors (Fig. 10.3; Ranga Raju, 1993) have been found to perform better than lengthy curved expansions.

As the flow is accelerating in a contracting transition and the energy loss is minimal any gradual contraction with a smooth and continuous boundary should be satisfactory, e.g. an elliptical quadrant is an alternative to a cylindrical quadrant for inlet transitions. The bedline profile for an elliptical quadrant transition has the equation

$$\left[\frac{x}{2(B_o - B_f)}\right]^2 + \left[\frac{y}{0.5(B_o - B_f)}\right]^2 = 1$$

and the length of transition given by

$$L_c = 2(B_o - B_f).$$

At any location (x) from flume end of the transition y is computed and the bed width B_x calculated by

$$B_x = B_o - 2y.$$

The side slope (m) of the transition ($m = 0$ for flume section and $m \geq 2$ for canal side slope) and bed elevation may be varied linearly along the transition length.

The expansion experiences considerable energy loss and care must be exercised in designing a hydraulically satisfactory transition.

On the basis of theoretical and experimental investigations Vittal and Chiranjeevi (1983) proposed the following design equations for bed width and the side slopes of an expanding transition. The bed widths B_x are fixed by

$$\frac{B_x - B_f}{B_0 - B_f} = \frac{x}{L}\left[1 - \left(\frac{1-x}{L}\right)^n\right]$$

where

$$n = 0.80 - 0.26m_o^{1/2}$$

and the transition length, $L = 2.35(B_0 - B_f) + 1.65m_o y_o$, y_o being the flow depth in the canal, and m_o its side slope. The side slopes (m) along the transition are given by

$$\frac{m}{m_o} = 1 - \left(1 - \frac{x}{L}\right)^{1/2}.$$

Using the constant specific energy condition in the transition between canal and flume the depth in the flume, y_f, and depths (y_x) along the transition length can be obtained. The energy balance between adjacent sections within the transition with expansion loss as $0.3(V_i^2 - V_{i+1}^2)/2g$ gives the bed elevations to be provided at successive sections so that the specific energy remains constant throughout the transition. Worked example 10.2 provides detailed design calculations for an expanding transition based on the Vittal and Chiranjeevi method.

WATER SURFACE PROFILE IN TRANSITION

The water surface in the transition may be assumed as two smooth parabolic curves (convex and concave) meeting tangentially. Referring to Fig. 10.4, the following equations give such profiles in transitions:

inlet transition, $y = 8.96 \times 10^{-4}x^2$;

outlet transition, $y = 2.33 \times 10^{-4}x^2$.

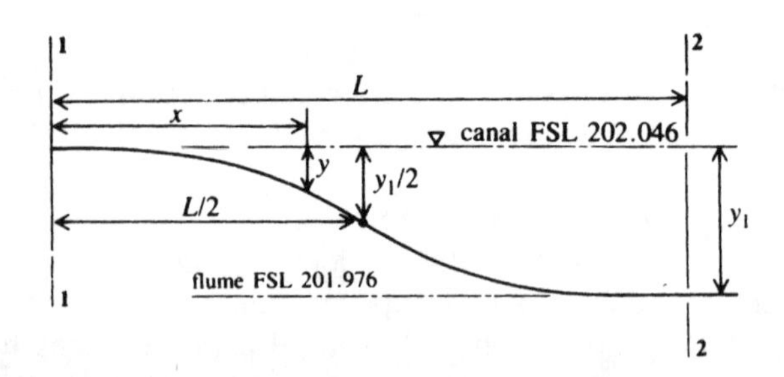

Fig. 10.4 Water surface profile in transition (inlet)

A highway 6m wide is provided alongside the canal by dividing the flume into two compartments by a 0.3m thick partition. The entire trough (flume section) can be designed as a monolithic concrete structure. Provide side walls and a bottom slab of about 0.4m (to be fixed by the usual structural design methods).

SIPHON BARRELS

Thirteen barrels, each 7m wide and 2.75m high, are provided; assume that the effective roughness, $k = 0.6$mm (concrete). The length of the barrel, $L = 12.50 + 0.30 + 2 \times 0.40 = 13.60$m. The head loss through the barrel, $h_f = (1.5 + \lambda L/4R)V^2/2g$. The velocity through the barrel, $V = 500/(13 \times 7 \times 2.75) = 1.998\,\mathrm{m\,s^{-1}}$. The hydraulic radius, $R = 7 \times 2.75/\{2(7 + 2.75)\} = 0.987$m. Therefore the Reynolds number $= 4VR/\nu = 8 \times 10^6$ and $k/4R = 1.5 \times 10^{-4}$. Hence, from Moody's chart, the friction factor $\lambda = 0.015$, giving $h_f = 0.316$m. Therefore, the upstream HFL $= 200.500 + 0.316 = 200.816$m AOD.

The uplift pressures on the roof of the barrel are as follows. The RL of the bottom of the trough $= 200.00 - 0.40 = 199.60$m AOD. The entry loss at the barrel $= 0.5V^2/2g = 0.102$m. Therefore the pressure head inside the barrel just downstream of its entry $= 200.816 - 0.102 - 199.600 = 1.114\,\mathrm{m} \simeq 11\,\mathrm{kN\,m^{-2}}$.

The most critical situation arises when the canal is empty and the siphon barrels are full. The weight of the roof slab $= 0.4 \times 2.4 \times 9.81 = 9.42\,\mathrm{kN\,m^{-2}}$ (assuming the relative density of concrete to be 2.4). Hence the roof slab needs additional reinforcement at its top to resist the unbalanced pressure forces (uplift pressures).

The total weight of the trough (when empty) needs to be checked against the total upward force and suitable anchorages to piers provided, if necessary. Equally, the trough floor slab has to be checked when it is carrying water at FSL and the level in the drainage is low, i.e. barrels running part full.

The uplift on the floor of the barrel (assuming the barrel floor thickness to be 1m initially) is as follows:

RL of the bottom of the barrel $= 199.60 - 2.75 - 1.00 = 195.85$m AOD;

RL of the drainage bed $= 198.00$m AOD.

Therefore the static uplift on the floor $= 198.00 - 195.85 = 2.15$m (the worst condition with the water table as the drain bed level). The seepage head (a maximum when the canal is at FSL and the drainage is empty) $= 202.00 - 198.00 = 4.00$m.

In spite of the three-dimensional seepage flow pattern, Bligh's creep length may be approximated as follows. Creep flow commences from the beginning of the upstream transition (downstream of this the floor is impervious) and enters the first barrel floor; from its centre the flow follows downstream of the drain and emerges at the end of the impervious concrete floor of the barrel. Therefore the total creep length can be approximated as

> inlet transition length + $\frac{1}{2}$ barrel span + $\frac{1}{2}$ length of barrel impervious floor.

Let us assume that the total length of the impervious floor of the barrel is 25 m, consisting of the following:

length of barrel	= 13.60 m
pier projections, 2×0.8	= 1.60 m
downstream ramp (1:5), 1.15×5	= 5.75 m
upstream and downstream cut-offs, 2×0.3	= 0.60 m
total floor length	= 21.55 m

Therefore provide the upstream floor (1:3) length $= 25.00 - 21.55 = 3.45$ m. The total creep length $= 12.5 + 7/2 + 25/2 = 28.5$ m. The creep length up to the centre of the barrel $= 12.5 + 7/2 = 16.0$ m. Therefore the seepage head at the centre of the barrel $= 4(1 - 16.0/28.5) = 1.75$ m. The total uplift is then $2.15 + 1.75 = 3.90\,\text{m} \simeq 38\,\text{kN}\,\text{m}^{-2}$, and the weight of the floor $= 1.00 \times 2.4 \times 9.81 = 23.54\,\text{kN}\,\text{m}^{-2}$. Hence additional reinforcement has to be designed to resist the unbalanced uplift forces.

UPSTREAM AND DOWNSTREAM PROTECTION WORKS

The scour depth, R_s (regime scour depth, equation (9.10)) $= 0.47(500/1)^{1/3} = 3.73$ m. The upstream cut-off below HFL $= 1.5R = 5.6$ m. Therefore

$$\text{RL of upstream cut-off wall} = 200.816 - 5.60 = 195.00 \text{ (say) m AOD.}$$

The downstream cut-off below HFL $= 1.75R_s = 6.53$ m. Therefore

$$\text{RL of downstream cut-off wall} = 200.50 - 6.53 = 194.00 \text{ (say) m AOD,}$$

$$\text{downstream apron length} = 2.5(198.00 - 194.00) = 10\,\text{m},$$

$$\text{upstream apron length} = 2.0(198.00 - 195.00) = 6\,\text{m}.$$

The detailed layout (longitudinal section) of the design is shown in Fig. 10.5.

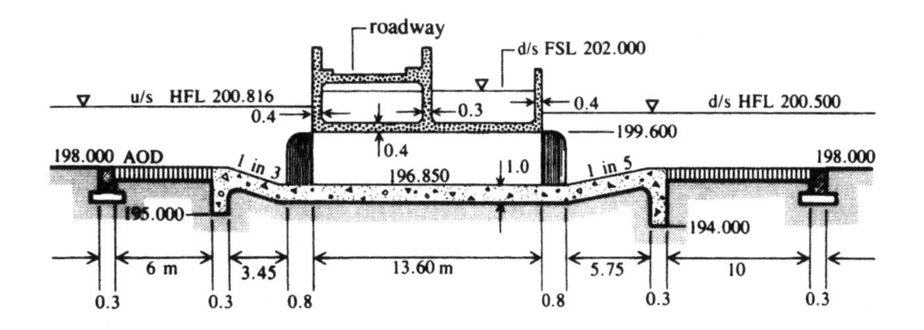

Fig. 10.5 Longitudinal section of the siphon aqueduct; all dimensions in metres

Worked Example 10.2

Design an expanding transition for the flume–canal layout of worked example 10.1 using the Vittal and Chiranjeevi (1983) method.

Solution

Design discharge $= 30\,m^3 s^{-1}$; bed width of canal, $B_o = 25$ m; bed width of flume, $B_f = 12.5$ m; side slope of canal, $m_o = 1.50$; bed level of canal = 200.00 m AOD; depth of flow in canal, $y_o = 2$ m (Worked example 10.1). Length of transition, $L = 2.35(25 - 12.5) + 1.65 \times 1.5 \times 2 \approx 36$ m. Bed width in transition, B_x, with $n = 0.8 - 0.26(1.5)^{1/2} = 0.482$,

$$B_x = 12.5 + (12.5x/36)[1 - (1 - x/36)^{0.482}].$$

Side slope in transition,

$$m = 1.5[1 - (1 - x/36)^{1/2}].$$

The complete set of calculations is presented in the table below:

Number	*Distance from upstream end of transition, x (m)*	*y_x (m)*	*m*	*B_x (m)*	*Δz (m)*	*Bed elevation (m AOD)*
1	36	2.000	1.50	25.000		200.000
2	27	1.985	0.75	17.069	0	200.000
3	18	1.965	0.44	15.737	0.001	199.999
4	9	1.950	0.20	12.545	0.002	199.998
5	0	1.935	0	12.500	0.003	199.997

10.2 Culverts, bridges and dips

10.2.1 Introduction

Highways cross natural drainage channels or canals, and provision must be made for appropriate cross-drainage works. The alignment of a highway along ridge lines (though it may be a circuitous route with less satisfactory gradients) may eliminate the cross-drainage work, thus achieving considerable savings.

Highway cross-drainage is provided by culverts, bridges and dips. Culverts are usually of shorter span (<6m), with the top not normally forming part of the road surface like in a bridge structure. They are submerged structures buried under a high-level embankment. On the other hand, if the embankment is a low-level one, appropriate armouring protection works against overtopping during high floods have to be provided. Such a low-level structure (sometimes called a 'dip') in the absence of the culvert is often economical if the possible traffic delays do not warrant a costly high-level structure such as a bridge, keeping the road surface above all flood levels. A culvert combined with a dip (lowered road surface) is an attractive solution for small perennial streams with occasional flash floods; however, appropriate traffic warning systems/signs have to be incorporated.

Bridges are high-level crossing structures which can be expensive for large rivers. It is therefore essential to protect them even from rare floods. It is often advantageous to allow overtopping of part of the approach embankment, which may act as a fuse plug, to be replaced if necessary, after the flood event. Such an alternative route for the water avoids the overtopping of the bridge deck and, in addition, reduces the scouring velocities which may otherwise undermine the foundations of the structure.

10.2.2 Culverts

The culvert consists essentially of a pipe barrel (conveyance part) under the embankment fill, with protection works at its entrance and exit. At the entrance a head wall, with or without wing walls, and a debris barrier are normally provided. If necessary, an end wall with energy-dissipating devices is provided at the exit.

The culvert acts as a constriction and creates a backwater effect to the approach flow, causing a pondage of water above the culvert entrance. The flow within the barrel itself may have a free surface with subcritical or supercritical conditions depending on the length, roughness, gradient, and upstream and downstream water levels of the culvert. If the upstream head is sufficiently large the flow within the culvert may or may not fill the barrel, and its hydraulic performance depends upon the combination of entrance and friction losses, length of barrel, and the downstream backwater effects (Fig. 10.6).

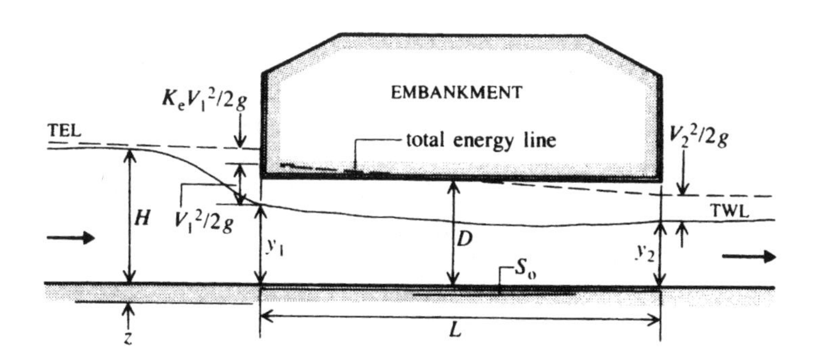

Fig. 10.6 Flow through a culvert

The various flow types that can exist in the pipe barrel of a culvert are shown in Table 10.1. The hydraulic design of the culvert is based on the characteristics of the barrel flow, and Worked examples 10.3 and 10.4 highlight calculations involving all the six types of flow listed in Table 10.1.

Table 10.1 Types of flow in the barrel of a culvert (Chow, 1983)

Type	*H/D*	*Exit depth* y_2	*Flow type*	*Length* L	*Slope* S_0	*Control*	*Remarks*
Submerged entrance conditions							
1	>1.0	$>D$	Full	Any	Any	Outlet	Pipe flow
2	>1.2	$<D$	Full	Long	Any	Outlet	Pipe flow
3	>1.2	$<D$	Part full	Short	Any	Outlet	Orifice
Free entrance conditions							
4	<1.2	$<D$ >critical	Part full	Any	Mild	Outlet	Subcritical
5	<1.2	$<D$ <critical	Part full	Any	MIld	Outlet	Subcritical
6	<1.2	$<D$ <critical	Part full	Any	Steep	Inlet	Supercritical
		>critical	Formation of hydraulic jump in barrel				

The reader is referred to Ramsbottom *et al.* (1997), Chanson (1999) and Mays (1999) for additional information on culvert flows, establishment of stage–discharge relationships and culvert design in general. Charbeneau *et al.* (2006) use a two parameter model describing the hydraulic performance of highway culverts operating under inlet control for both unsubmerged and submerged conditions.

The hydraulic performance of a culvert can be improved by the adoption of the following guidelines.

(a) Culvert alignment

As a general rule, the barrel should follow the natural drainage alignment and its gradient, in order to minimize head losses and erosion. This may lead to a long skew culvert which will require more complex head and end walls. However, it is sometimes more economical to place the culvert perpendicular to the highway with certain acceptable changes in the channel alignment (see Linsley and Franzini, 1979).

(b) Culvert entrance structures

Properly designed entrance structures prevent bank erosion and improve the hydraulic characteristics of the culvert. The various types of entrance structures (end walls and wing walls) recommended are shown in Fig. 10.7.

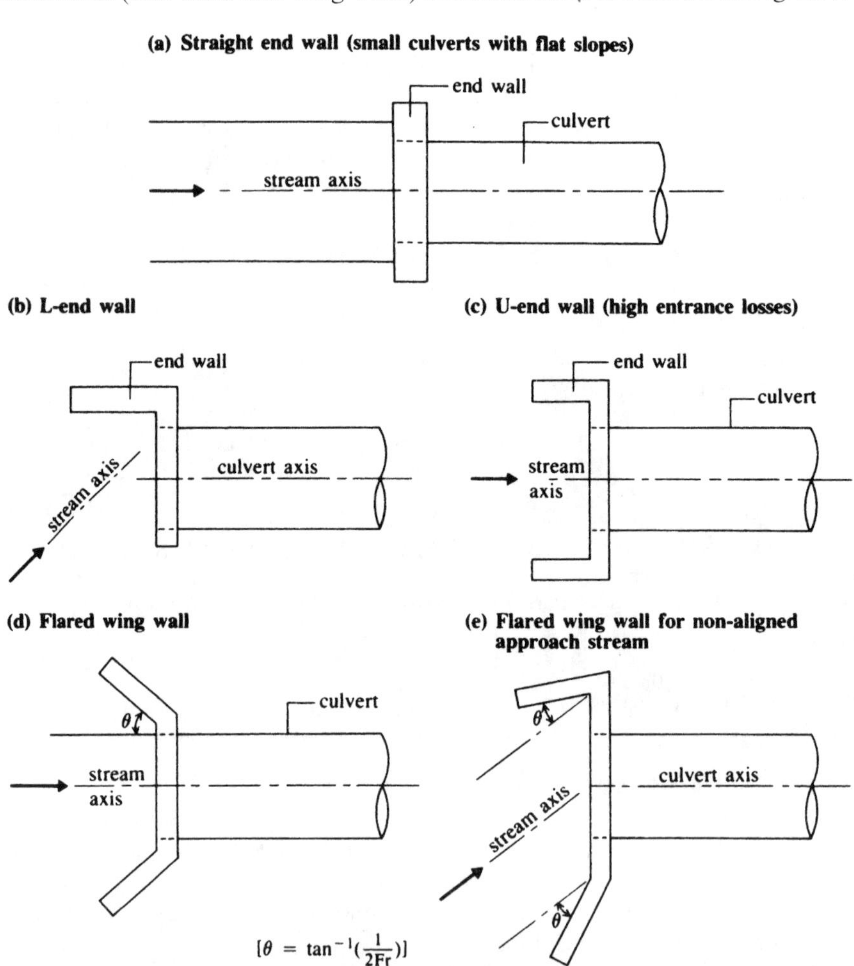

Fig. 10.7 Culvert entrance structures; plan views

A debris barrier (trash rack) must also be provided upstream of the culvert entrance to prevent the blockage of the barrel entrance.

In the case of a culvert with a submerged entrance, flaring the entrance will increase its capacity under a lower head for a given discharge. Such an arrangement for a box culvert (square or rectangular concrete barrel), the entrance area being double the barrel area over a length of $1.2D$, where D is the height of the barrel, is shown in Fig. 10.8.

A drop inlet structure with a necessary debris barrier (timber or concrete cribs) has to be provided whenever the culvert entrance is at the bed level (highway drainage facilities) of the drainage, requiring an abrupt break in the channel slope. Various arrangements of drop inlet culverts are shown in Fig. 10.9. The culvert sill length must be sufficient to discharge the design flow with a reasonably low-head water level. For high discharges, the entrance may be flared so as to increase the crest length. A flared entrance with a back wall (to prevent vortex action) considerably increases the inlet capacity. De-aeration chambers may have to be provided if a jump forms in the barrel of the culvert.

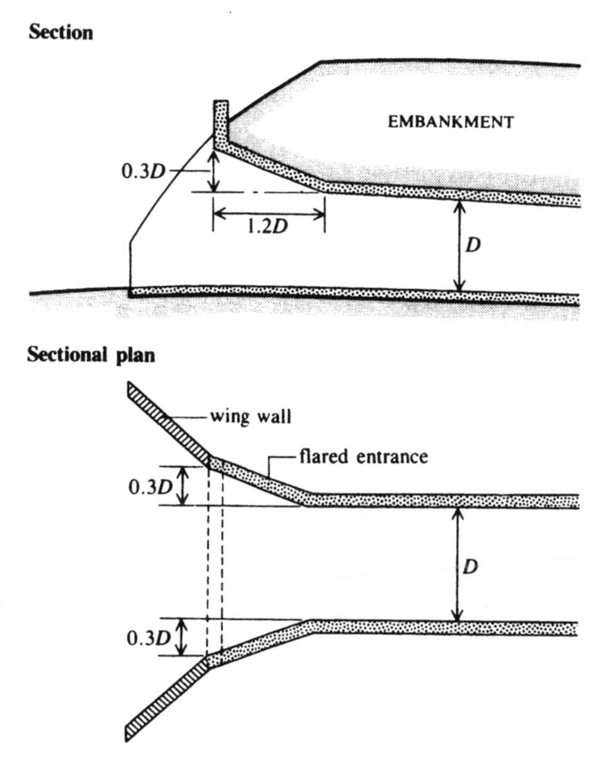

Fig. 10.8 Box culvert with flared entrance

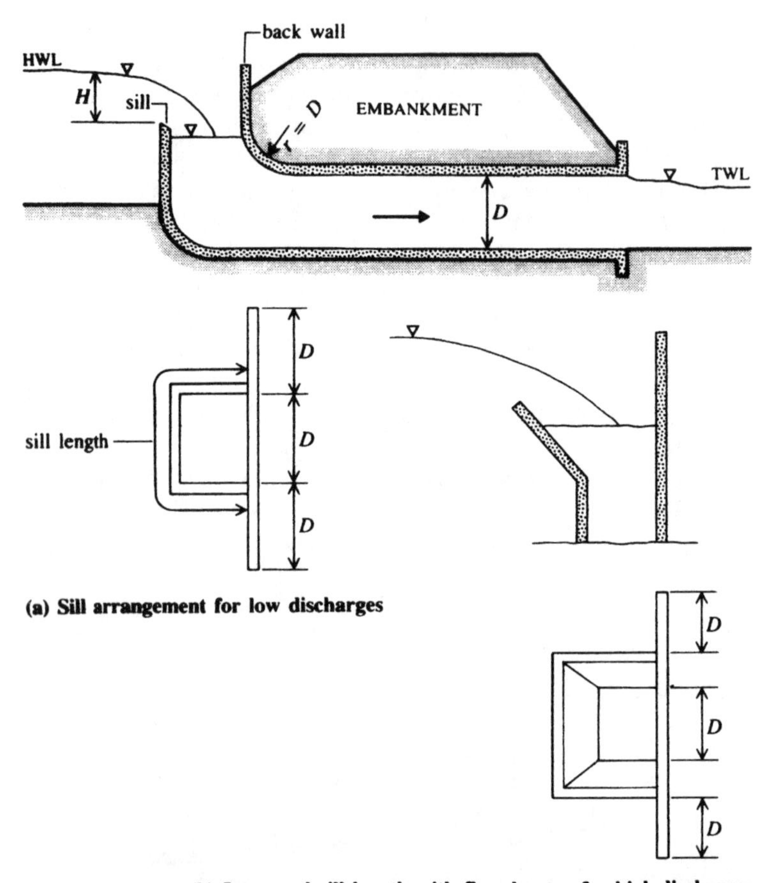

Fig. 10.9 Drop inlet culvert

(c) Culvert outlet structures

A proper device has to be provided at the outlet of a culvert to prevent the downstream erosion of the bed and the slopes of the embankment. For small discharges a straight or U-shaped end wall is sufficient. For moderate flows a flaring wing-walled outlet connecting the much wider downstream channel will reduce the scouring of the embankment and channel banks. The suggested flare angle for supercritical flows should be under 1 in 2, decreasing linearly with the flow Froude number. For subcritical flows it may be larger than 1 in 2.

(d) Scour below culvert outlets

The flow through a culvert may cause undesirable erosion (scour) at its unprotected outlet which can lead to undermining of the culvert structure.

Several researchers performed model tests on scour downstream of culvert structures and the combined results suggest the following design guidelines (see Breusers and Raudkivi, 1991):

$$\text{scour depth below bed level,} \quad y_s = 0.65D(U_o/U_{*c})^{1/3} \tag{10.1}$$

where U_o = flow velocity at exit and U_{*c} = Shields critical shear velocity $(=(\tau_c/\rho)^{1/2})$;

$$\text{scour width,} \quad B_s = 7.5DF_r^{2/3}, \tag{10.2}$$

$$\text{scour length,} \quad L_s = 15DF_r^{2/3} \tag{10.3}$$

where D = culvert height, $F_r = U_0/(gD)^{1/2}$, $0.27 < F_r < 2.7$ and $0.22 < d$ (mm) < 7.3. Equation (10.1) may be modified in the case of graded material as

$$y_s = 3.18DF_r^{0.57}\left(\frac{d_{50}}{D}\right)^{0.114}\sigma_g^{-0.4} \tag{10.4}$$

where $\sigma_g = (d_{84}/d_{16})^{1/2}$. In order to protect the channel bed against scouring, a minimum stone size is recommended as

$$d_s = 0.25DF_r \tag{10.5}$$

for low tail water levels. For high tail water levels ($>D/2$) the recommended stone size is reduced by $0.15D$.

Fletcher and Grace (1974) recommended a lining of trapezoidal cross-section downstream of the culvert exit extending to a length $= 5D$ with a bed slope of 1 in 10 followed by a curtain wall to a length $= D$ at a slope of 1 in 2; the side slopes of the trapezoidal lining are recommended to be 1 in 2. Alternatively, the design curves of Simons and Stevens (1972) may be used for non-scouring and scouring bed protection in rock basins (see Breusers and Raudkivi, 1991).

Blaisdell and Anderson (1988a, b) made a comprehensive study of scour at cantilevered pipe outlets and suggested the ultimate maximum scour hole depth, Z_{max}, below tailwater level (zero elevation) as

$$\frac{Z_{max}}{D} = -7.5[1 - e^{-0.6(F_{rd}-2)}] \tag{10.6}$$

for $Z_p/D \leq 1$ and

$$\frac{Z_{max}}{D} = -10.5[1 - e^{-0.35(F_{rd}-2)}] \tag{10.7}$$

for $Z_p/D>1$ where D = pipe diameter, Z_p = height of pipe outlet above tailwater level and $F_{rd}=V/(g\Delta d_{50})^{1/2}$, V being the jet plunge velocity at the tailwater

$$(F_{rd}>2;\ -2<Z_p(\mathrm{m})<8;\ \text{outlet slope},\ 0-0.782).$$

The usual energy-dissipating devices (sloping apron, cistern, stilling basin, plunge pool, etc.) may have to be provided if the culvert discharge velocities are very high (Chapter 5).

10.2.3 Bridges

The presence of a bridge across a stream creates constricted flow through its openings because of (a) the reduction in the width of the stream due to piers and their associated end contractions and (b) the fluming of the stream itself (in the case of wide streams with flood plains) to reduce the costs of the structure.

Apart from (local) scour around the piers and bridge abutments and possible bed erosion, there is a considerable backwater effect of the bridge. The corresponding afflux (rise in upstream water level) depends on the type of flow (subcritical or supercritical). As most bridges are designed for subcritical flow conditions in order to minimize scour and choking problems, further discussions here are mainly confined to subcritical flow.

The establishment of afflux levels is extremely important for the design of upstream dykes and other protective works and also for the location of safe bridge deck levels (to avoid the flooding of the deck and any consequent structural damage). It is equally important to determine the minimum clear length of span (economic considerations) which will not cause undesirable afflux levels. In order to establish permissible upstream stage levels, detailed investigations of the properties along the stream have to be investigated. Downstream of the bridge the water levels are only influenced by the nearest control section below the bridge. These levels can therefore be established by backwater computation (for further information see Hamill, 1999).

(a) Backwater levels

SHORT CONTRACTIONS

In flow through a relatively short contracted section (narrow bridge without approach fluming) with only a few piers, the backwater problem may be relatively less important. Referring to Fig. 10.10, the change in

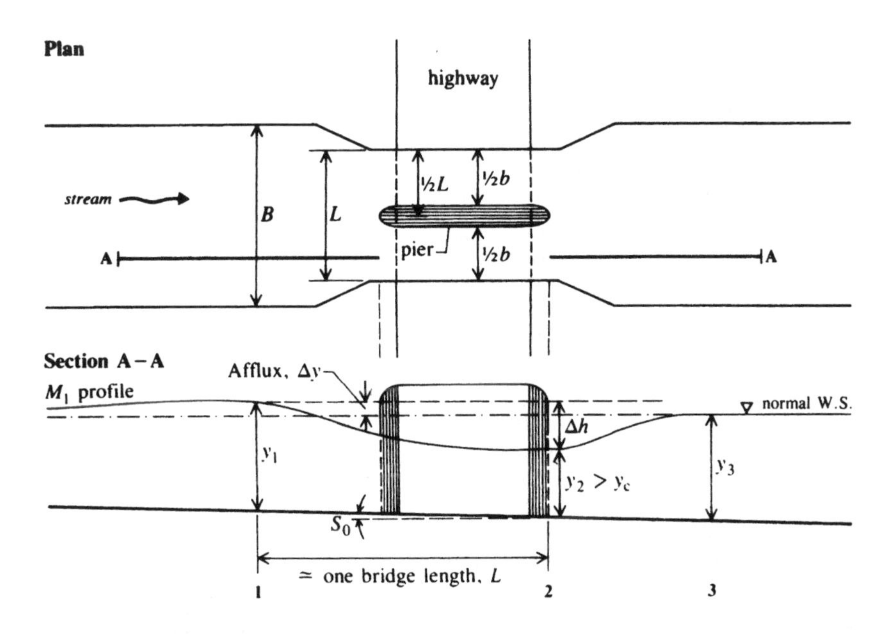

Fig. 10.10 Flow profile through bridge with contracted channel of relatively short length (subcritical flow)

water level, Δh, can be obtained by the energy equation between sections 1 and 2 (Kindsvater, Carter and Tracy, 1953) as

$$\Delta h = K_B V_2^2/2g + S_0 L/\sigma - \alpha_1 V_1^2/2g \tag{10.8}$$

where K_B is the bridge loss coefficient (Table 10.2), expressed as a function of the conveyance ratio,

$$\sigma = k_b/k_B, \tag{10.9}$$

Table 10.2 Bridge loss coefficient, K_B

σ	K_B
1.0	1.00
0.8	1.36
0.6	1.67
0.4	1.88
0.2	1.92

k_b being the conveyance of the gross contracted section with the same normal depth and roughness characteristics as the upstream approach section whose conveyance is k_B.

For rectangular unflumed sections the conveyance ratio (contraction ratio, $\alpha = 1 - \sigma$) becomes b/B, b being the clear width of the stream (of normal width, B) under the bridge (Fig. 10.10).

The bridge loss coefficient is also a function of the geometry of the bridge, its skew and eccentricity, and the submergence of the superstructure (i.e. the deck).

V_2 is the velocity just downstream of the piers, using the gross area under the bridge with the same upstream normal depth, and α_1 is the energy correction coefficient of the approach section. L is assumed to be equal to the bridge length (abutment to abutment), and S_0 is the normal bedslope of the unobstructed steam.

LONG CONTRACTIONS

In the case where the bridge has a number of large piers and/or long approach embankments contracting the water width, the backwater effect is considerable. Referring to the flow profile shown in Fig. 10.10, through such a long contracting section, Δy is the afflux entirely created by the presence of piers and channel contraction.

Momentum and continuity equations between sections 1 and 3 (assuming hydrostatic pressure distribution with a negligible bed slope and frictional resistance) result in

$$\Delta y / y_3 \simeq \{A + [A^2 + 12C_D(b/B)Fr_3^2]^{1/2}\}/6 \tag{10.10}$$

where

$$A = \{C_D(b/B) + 2\}Fr_3^2 - 2 \tag{10.11}$$

Fr_3 being the Froude number ($= V_3/(gy_3)^{1/2}$) at section 3.

Equation (10.10) should give good results if the drag coefficient C_D can be accurately estimated. The pier drag coefficient has been found to be a function of the velocity gradient of the approach flow, b/B, and the pier shape; however, owing to the non-availability of reliable drag coefficient values, the use of equation (10.10) is limited.

Yarnell's (1934) experimental data on the flow through bridge piers resulted in the following empirical equation:

$$\Delta y / y_3 = KFr_3^2(K + 5Fr_3^2 - 0.6)(\alpha + 15\alpha^4) \tag{10.12}$$

where

$$\alpha = 1 - \sigma = 1 - b/B \tag{10.13}$$

and K is a function of the pier shape according to Table 10.3.

Table 10.3 Values of *K* as a function of pier shape

Pier shape	*K*	*Remarks*
Semicircular nose and tail	0.9	All values applicable for piers with length to breadth ratio equal to 4; conservative estimates of Δy have been found for larger ratios;
Lens-shaped nose and tail	0.9	
Twin-cylinder piers with connecting diaphragm	0.95	
Twin-cylinder piers without diaphragm	1.05	Lens-shaped nose is formed from two circular curves, each of radius to twice the pier width and each tangential to a pier face
90° triangular nose and tail	1.05	
Square nose and tail	1.25	

Equation (10.12) is valid only if σ is large, i.e. the contraction cannot set up critical flow conditions between piers and choke the flow. If the flow becomes choked by excessive contraction the afflux increases substantially (Fig. 10.11). Referring to Fig. 10.11, the limiting values of σ (assuming uniform velocity at section 2) for critical flow at section 2 can be written as

$$\sigma = (2 + 1/\sigma)^3 Fr_3^4/(1 + 2Fr_3^2)^3. \tag{10.14}$$

In the case of choked flow the energy loss between sections 1 and 2 was given by Yarnell as

$$E_1 - E_2 = C_L V_1^2/2g \tag{10.15}$$

where C_L is a function of the pier shape (equal to 0.35 for square-edged piers and 0.18 for rounded ends, for a pier length:width ratio of 4). From equation (10.15) the upstream depth, y_1, can be calculated, from which the afflux Δy is obtained as $y_1 - y_3$.

Skewed bridges produce greater affluxes, and Yarnell found that a 10° skew bridge gave no appreciable changes, whereas a 20° skew produced about 250% more afflux values.

For backwater computations of arch bridges Martín-Vide and Prió (2005) recommend a head loss coefficient K for the sum of contraction and

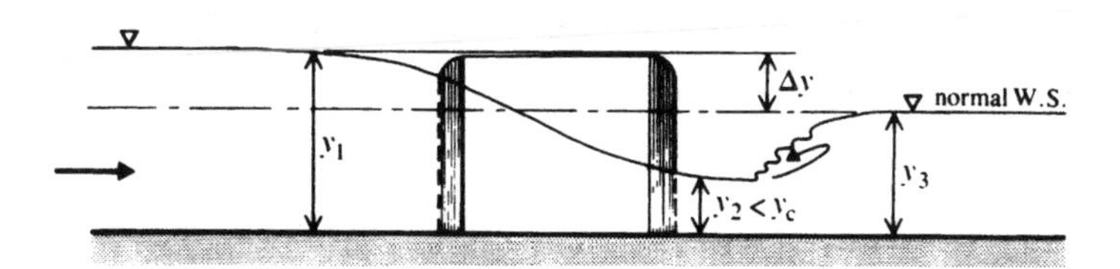

Fig. 10.11 Flow profile with choked flow conditions

expansion losses as $K = 2.3m - 0.345$, where m is the ratio of the obstructed and channel areas for $0.324 < m < 0.65$.

(b) Discharge computations at bridge piers

1. Nagler (1918) proposed a discharge formula for subcritical and near-critical flows as follows:

$$Q = K_N b(2g)^{1/2}(y_3 - \theta V_3^2/2g)(h_3 + \beta V_1^2/2g)^{1/2} \qquad (10.16)$$

the notation used in equation (10.16) being shown in Fig. 10.12(a). K_N is a coefficient depending on the degree of channel contraction and on the characteristics of the obstruction (Table 10.4); θ is a correction factor intended to reduce the depth y_3 to y_2 and β is the correction for the velocity of approach, depending on the conveyance ratio (Fig. 10.12(b)).

2. d'Aubuisson (1940) suggested the formula

$$Q = K_A b_2 y_3 (2gh_3 + V_1^2)^{1/2} \qquad (10.17)$$

where K_A is a function of the degree of channel contraction and of the shape and orientation of the obstruction (Table 10.4).

d'Aubuisson made no distinction between y_3 and y_2, and, although in many cases there is a small difference between them, equation (10.17) is recognized as an approximate formula.

3. Chow (1983) presents a comprehensive discussion of the discharge relationship between the flow through contracted openings and their shape, and other characteristics, together with a series of design charts produced by Kindsvater, Carter and Tracy (1953).

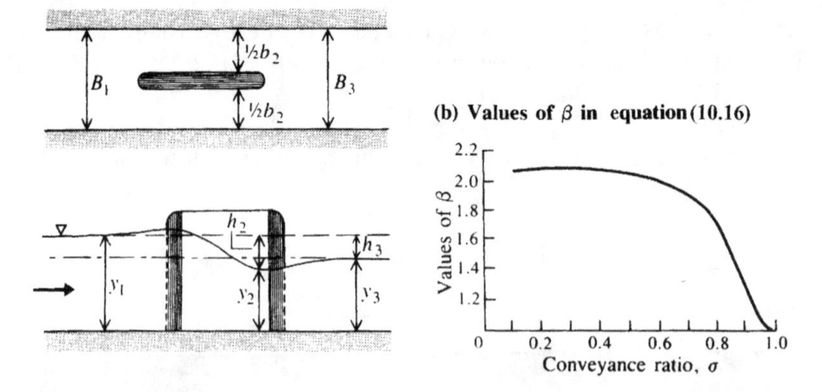

Fig. 10.12 Discharge computations through obstructions (definition sketch)

Table 10.4 Values of K_N and K_A

Type of pier	*Conveyance ratio, σ*									
	0.9		*0.8*		*0.7*		*0.6*		*0.5*	
	K_N	K_A	K_N	K_A	K_N	K_A	K_N	K_A	K_N	K_A
Square nose and tails	0.91	0.96	0.87	1.02	0.86	1.02	0.87	1.00	0.89	0.97
Semicircular nose and tails	0.94	0.99	0.92	1.13	0.95	1.20	1.03	1.26	1.11	1.31
90° triangular nose and tails	0.95		0.94		0.92					
Twin-cylinder piers with or without diaphragms	0.91		0.89		0.88					
Lens-shaped nose and tails	0.95	1.00	0.94	1.14	0.97	1.22				

(c) Scour depth under the bridge

If the contracted width (i.e. the bridge length, L) is less than the régime width, W (equation (9.9)), the normal scour depth, D_N, under the bridge is given by

$$D_N = R_s(W/L)^{0.61} \tag{10.18}$$

where R_s is the régime scour depth (equation (9.10)).

The maximum scour depth in a single-span bridge (no piers) with a straight approach (case 1) is about 25% more than the normal scour given by equation (10.18), whereas in the case of a multispan structure with a curved approach reach (case 2) it is 100% more than the normal scour. If the constriction is predominant, the maximum scour depth is the maximum of case 1 or case 2, or the value given by

$$D_{max} = R_s(W/L)^{1.56}. \tag{10.19}$$

(d) Scour around bridge piers

Several formulae based on experimental results have been proposed to predict the 'maximum' or 'equilibrium' scour depth (y_s, below general bed level) around bridge piers. In general, these assume the relationship

$$y_s/b' = \phi(y_0/b', Fr, d/b') \tag{10.20}$$

where b' is the pier width, y_0 is the upstream flow depth, d is the sediment size, and Fr is the flow Froude number.

Laursen's (1962) experimental results underestimate the scour depths, compared to many Indian experiments (Inglis, 1949) which suggest the formula (approach flow is normal to the bridge piers)

$$y_s/b' = 4.2(y_0/b')^{0.78}Fr^{0.52}. \qquad (10.21)$$

The Indian field data also suggest that the scour depth should be taken as twice the régime scour depth.

In the case of live beds (a stream with bedload transport) the formula

$$y_s/y_0 = (B/b')^{5/7} - 1 \qquad (10.22)$$

predicts the maximum equilibrium scour depth.

In a relatively deep flow a first-order estimate of (clear) local scour (around pier) may be obtained by

$$y_s = 2.3K_\alpha b' \qquad (10.23)$$

where K_α = angularity coefficient which is a function of the pier alignment, i.e. angle of attack of approach flow.

Once again the best estimate will be achieved with the appropriate coefficients for flow depth, alignment, etc. (see Breusers and Raudkivi (1991) for further information). The live bed, however, contributes to an appreciably reduced local scour depth. If the sediment bed is distinctly layered and the covering layer (normally coarse material) is of a thickness less than the local scour depth the overall scouring phenomenon is quite different (see Ettema, 1980).

The flow penetrates the covering layer, triggering its disintegration. The disintegration of the covering layer may at times take place only in the downstream direction, leaving a stepped scour just upstream of the pier followed by a further local pier scour at its bottom. The stepped scour depth in the covering layer, H, is given by

$$H = \eta(y_2 - y_1) \qquad (10.24)$$

where y_1 and y_2 are the uniform flow depths over a flat bed of grain roughness corresponding to the upstream surface particles (d_1) and the underlying surface fine particles (d_2) respectively; the coefficient for non-ripple-forming sediments $\eta = 2.6$ for design purposes. The total scour depth may lead to a gross underestimate if the lower layer is of very fine material (which may go into suspension).

The whole field of scour at bridge piers, piles and abutments is the subject of ongoing research focussing particularly on large scale experiments and development of scour with time. To proceed further with this topic is beyond the scope of this book and the reader is referred to e.g.

Breusers, Nicollet and Shen (1977), Clark and Novak (1983), Richardson and Richardson (1994), Melville and Chiu (1999), Melville and Coleman (2000), Oliveto and Hager (2002), Coleman, Lauchlan and Melville (2003), Sheppard, Odeh and Glasser (2004) and Dey and Barbhuiya (2005).

(e) Scour protection works around bridge piers

Although the presence of scour tends to reduce the backwater levels upstream of the bridge, the damage to the foundations of the structure may far outweigh the possible benefit. Hence protective measures, both to minimize the scour and to prevent undermining of the foundations, have to be taken. Piers with base diaphragms (horizontal rings) and multiple cylinder type piers have been found to minimize the scour considerably. The normal practice for protection of the foundation is to provide thick protective layers of stone or concrete aprons around the piers.

A riprap protection (Bonasoundas, 1973) in the shape of a longitudinal section of an egg with its broader end facing the flow is recommended for a cylindrical pier. The recommended overall width is $6b'$ and length $7b'$ of which $2.5b'$ is upstream of the pier. The thickness of riprap is $1/3b'$ with a maximum stone size, d, given by

$$d = 0.06 - 0.033U + 0.04U^2 \tag{10.25}$$

with U in metres per second and d in metres.

The mean critical flow velocity U_c (m s^{-1}) with a flow depth y_0 (m) is given by

$$U_c \approx 6d^{1/3}y_0^{1/6} \tag{10.26}$$

d being the armour stone size in metres (with $\rho_s = 2600\,\mathrm{kg\,m^{-3}}$).

For horizontal beds (US Army Coastal Engineering Research Center (1984); Chapter 14) the simplified empirical relationship is

$$U_c \approx 4.92d^{1/2}. \tag{10.27}$$

The riprap should be placed on a suitable inverted filter or a geotextile fabric (Fig. 9.11).

For further discussion of scour protection works refer to Zarrati *et al.* (2006) and Unger and Hager (2006). For a comprehensive treatment of bridge hydraulics (including hydraulic aspects of bridge construction and maintenance) refer to Neill (2004).

10.2.4 Dips

The dip is a shallow structure without excessive approach gradients. In arid regions, streams with infrequent flash floods and shallow depths (<0.3 m) may be allowed to flow through the dipped area. The upstream road edge should not be discontinuous with the stream bed in order to avoid scour, and at the downstream edge protection works such as a cut-off wall, concrete, or riprap paving must be provided. Also, the profile of the dip should, as far as possible, conform to the profile of the stream to minimize local disturbances to the flow.

The road surface has to withstand the expected flow velocities and debris. Bitumen-bound macadam may withstand velocities of up to $6\,m\,s^{-1}$, whereas up to $7\,m\,s^{-1}$ may be permitted on asphalted road surfaces. Low-level embankments, where occasional overtopping is permitted, must be protected against scour and bank-slope stability. The permissible mean velocities for a range of protective materials are suggested in Table 10.5 (Watkins and Fiddes, 1984).

The overflow discharge over an embankment may be predicted by using the weir formula of the type

$$Q = CbH^{3/2} \tag{9.29}$$

where C is the sill coefficient ($m^{1/2}s^{-1}$), b is the length of the flow section and H is the total head upstream of the sill.

The coefficient C is a function of h/L (h is the head over a sill of width L) for free flow conditions. For non-modular flow conditions a correction factor, f, as a function of $h_{d/s}/H$, may be incorporated in equation (9.29) (Tables 10.6 and 10.7).

Table 10.5 Permissible velocities to withstand erosion

Type of protection	*Velocity ($m\,s^{-1}$)*
Grass turfing:	
Bermuda grass	≃2.0
buffalo grass	1.5
Cobbles:	
≃100 mm	3.5
≃40 mm	2.5
Coarse gravel and cobbles (≃25 mm)	2.0
Gravel (≃10 mm)	1.5

Table 10.6 Range of values of *C* for free flow over the embankment

Type of surface	*Range of h/L*	*Range of C*
Paved surface	0.15	1.68
	0.20	1.69
	>0.25	1.70
Gravel surface	0.15	1.63
	0.20	1.66
	0.25	1.69
	0.30	1.70

Table 10.7 Correction factor, *f* (non-modular flows)

Type of surface	*Range of* $h_{d/s}/H$	*f*
Paved surface	≤0.8	1.0
	0.9	0.93
	0.95	0.8
	0.99	0.5
Gravel surface	≤0.75	1.0
	0.8	0.98
	0.9	0.88
	0.95	0.68
	0.98	0.50

Worked Example 10.3

Establish the stage (headwater level)–discharge relationship for a concrete rectangular box culvert, using the following data: width = 1.2 m; height = 0.6 m; length = 30 m; slope = 1 in 1000; Manning's $n = 0.013$; square-edged entrance conditions; free jet outlet flow; range of head water level for investigation = 0–3 m; neglect the velocity of approach.

Solution

1. $H/D \leq 1.2$. For $H < 0.6$ m, free flow open-channel conditions prevail. Referring to Fig. 10.6 and assuming that a steep slope entry gives entrance control, i.e. the depth at the inlet is critical, for $H = 0.2$ m, ignoring entry loss $y_c = (2/3) \times 0.2 = 0.133$ m and $V_c = 1.142\,\mathrm{m\,s^{-1}}$. This gives the critical slope $(Vn)^2/R^{4/3} = 0.00424$. Therefore the slope of the culvert is mild and hence subcritical flow analysis gives the following results:

$$Q = 1.2y_0[1.2y_0/(1.2 + 2y_0)]^{2/3}\,(0.001)^{1/2}/0.013$$

$$= 2.92y_0[1.2y_0/(1.2 + 2y_0)]^{2/3}; \qquad \text{(i)}$$

y_0 (m)	Q ($m^3 s^{-1}$)(equation (i))	y_c (m)
0.2	0.165	0.124
0.4	0.451	0.243
0.6 (=D)	0.785	0.352

At the inlet over a short reach,

$$H = y_0 + V^2/2g + K_e V^2/2g. \qquad \text{(ii)}$$

The entrance loss coefficient, K_e, is as follows:

for a square-edged entry, 0.5;

for a flared entry, 0.25;

for a rounded entry, 0.05;

y_0 (m)	H (m) (equation (ii))	Q ($m^3 s^{-1}$)
0.2	0.236	0.165
0.4	0.467	0.451
0.6	0.691	0.785
orifice ←>0.6←(1.2D=)	0.72 ⟶	0.817 (by interpolation)

2. $H/D \geq 1.2$.
   (a) For orifice flow

$$Q = C_d(1.2 \times 0.6)[2g(H - D/2)]^{1/2}. \qquad \text{(iii)}$$

With $C_d = 0.62$ the following results are obtained:

H(m)	Q ($m^3 s^{-1}$)	y_0 (m) (equation (i))
0.72	1.29	>0.6→no orifice flow exists

(b) For pipe flow the energy equation gives

$$H + S_0 L = D + h_L$$

where

$$h_L = K_e V^2/2g + (Vn)^2 L/R^{4/3} + V^2/2g.$$

Thus

$$Q = 2.08(H - 0.57)^{1/2}. \qquad \text{(iv)}$$

	H (m)	*Q (m^3s^{-1}) (equation (iv))*
$y_0 \simeq 0.6$ (equation (i) ←	0.691 ←	0.723
	0.72 ↓	0.805 ↑
	1.00	1.364
	2.00	2.487
	3.00	3.242

During rising stages the barrel flows full from $H = 0.72$ m and during falling stages the flow becomes free-surface flow when $H = 0.691$ m.

The following table summarizes the results:

H (m)	*Q (m^3s^{-1})*	*Type of flow*
Rising stages		
0.236	0.165	Open channel
0.467	0.451	Open channel
0.691	0.785	Open channel
0.720	0.805	Pipe flow
1.00	1.364	Pipe flow
2.00	2.487	Pipe flow
3.00	3.242	Pipe flow
Falling stages		
2.00	2.487	Pipe flow
1.00	1.364	Pipe flow
0.72	0.805	Pipe flow
0.691	0.723	Pipe flow
0.691	0.785	Open channel
0.467	0.451	Open channel
0.236	0.165	Open channel

Worked Example 10.4

Examine the stage–discharge relationship for the culvert in Worked example 10.3 if the bedslope is 1 in 100.

Solution

Rising stages are as follows.

1. For the open channel, preliminary calculations now indicate that the slope is steep and hence the entrance is the control, with the critical depth at the entry. The energy equation at the inlet gives

$$H = 1.5V^2/2g + y = 1.75y_c. \quad \text{(v)}$$

y_c (m)	H (m)	Type	Q ($m^3 s^{-1}$)
0.2	0.35	Free	0.336
0.4	0.70	Free	0.951
0.6	1.50 (>1.2D)	Submerged	–
0.411	0.72 (=1.2D)	Just free	0.990

2. For the orifice (equation (iii)),

H (m)	Type	Q ($m^3 s^{-1}$)	y_0 (m) (equation (i) with S_0 = 1/100)*
0.72	Orifice	1.29	0.36
1.00	Orifice	1.66	0.44
2.00	Orifice?	2.58	0.61 (>D)
1.95	Orifice	2.54	0.60

$*Q = 9.23y_0[1.2y_0/(1.2 + 2y_0)]^{2/3}$

3. For pipe flow (equation (iv) changes):

$Q = 2.08(H - 0.3)^{1/2}$ for $S_0 = 1/100$.

H (m)	Type	Q ($m^3 s^{-1}$)	y_0 (m) (equation (i))
1.95	Pipe flow	2.67	
2.00	Pipe flow	2.71	
3.00	Pipe flow	3.42	
Falling stages			
3.00	Pipe flow	3.42	
2.00	Pipe flow	2.71	
1.95	Pipe flow	2.67	
1.74	← Pipe flow	← 2.50	← 0.60
1.74 →	Orifice →	2.37	
1.00	Orifice	1.66	
0.72	Orifice	1.29	
0.72	Just free	0.99	
0.70	Free	0.951	
0.35	Free	0.336	

Worked Example 10.5

The design flood with a 20-year return period is 15 $m^3 s^{-1}$. Design the culvert-type cross-drainage structure with a high embankment with the following data: culvert length = 30 m; slope = 1.5%; available pipe barrel, corrugated pipes in multiples of 250 mm diameter; Manning's n = 0.024. The barrel protrudes from the embankment with no end walls, with an entry loss coefficient of 0.9. The maximum permissible head water level is 4 m above the invert with the barrel flowing full.

Solution

For full pipe flow the energy equation gives

$$H + 30 \times 0.015 = D + 0.9V^2/2g + (Vn)^2L/R^{4/3} + V^2/2g. \quad \text{(vi)}$$

Equation (vi) gives the following results:

D (m)	*H (m)*
1.500	12.61
2.000	4.74
2.500	3.25
2.250	3.70

Therefore provide a 2.25 m diameter barrel for $H \leq 4.0$ m.
Check for the flow conditions:

$$H/D = 3.70/2.25 = 1.65 > 1.2.$$

Hence the inlet is submerged. Using Manning's equation with the maximum discharge, the required diameter for the flow to be just free is 2.32 m, which is greater than the diameter provided. Hence the barrel flows full (under pressure).

Note that an improved entrance would considerably reduce the head loss and allow a smaller-diameter barrel to discharge the flood flow. For example a flare-edged entry (loss coefficient = 0.25) would produce a head of 3.93 m (<4.0 m) with a barrel of 2.00 m diameter.

Worked Example 10.6

A road bridge of seven equal span lengths crosses a 106 m wide river. The piers are 2.5 m thick, each with semicircular noses and tails, and their length:breadth ratio is 4. The streamflow data are given as follows: discharge = 500 $m^3 s^{-1}$; depth of flow downstream of the bridge = 2.50 m. Determine the afflux upstream of the bridge.

Solution

The velocity at the downstream section, $V_3 = 500/106 \times 2.5 = 1.887\,m\,s^{-1}$. Therefore the Froude number, $Fr_3 = 0.381$. Flow conditions within the piers are as follows: the limiting value of $\sigma \simeq 0.55$ (equation (10.14)), while the value of σ provided $= b/B = 13/15.5 = 0.839$. Since the value of σ provided is more than the limiting σ value, subcritical flow conditions exist between the piers. Using equation (10.12) with $K = 0.9$ (Table 10.3) and $\alpha = 1 - \sigma = 0.161$, the afflux, $\Delta y = 5.41 \times 10^{-2}$ m.

10.3 Drop structures

10.3.1 Introduction

A drop (or fall) structure is a regulating structure which lowers the water level along its course. The slope of a canal is usually milder than the terrain slope as a result of which the canal in a cutting at its headworks will soon outstrip the ground surface. In order to avoid excessive infilling the bed level of the downstream canal is lowered, the two reaches being connected by a suitable drop structure (Fig. 10.13).

The drop is located so that the fillings and cuttings of the canal are equalized as much as possible. Wherever possible, the drop structure may also be combined with a regulator or a bridge. The location of an offtake from the canal also influences the fall site, with offtakes located upstream of the fall structure.

Canal drops may also be utilized for hydropower development, using bulb- or propeller-type turbines. Large numbers of small and medium-sized drops are desirable, especially where the existing power grids are far removed from the farms. Such a network of micro-installations is extremely helpful in pumping ground water, the operation of agricultural equipment, village industries, etc. However, the relative economy of providing a large number of small falls versus a small number of large falls must be considered. A small number of large falls may result in unbalanced earthwork but, on the other hand, some savings in the overall cost of the drop structures can be achieved.

Drops are usually provided with a low crest wall and are subdivided into the following types: (i) the vertical drop, (ii) the inclined drop and (iii) the piped drop.

The above classification covers only a part of the broad spectrum of drops, particularly if structures used in sewer design are included; a comprehensive survey of various types of drops has been provided, e.g. by Merlein, Kleinschroth and Valentin (2002); Hager (1999) includes the treatment of drop structures in his comprehensive coverage of wastewater structures and hydraulics.

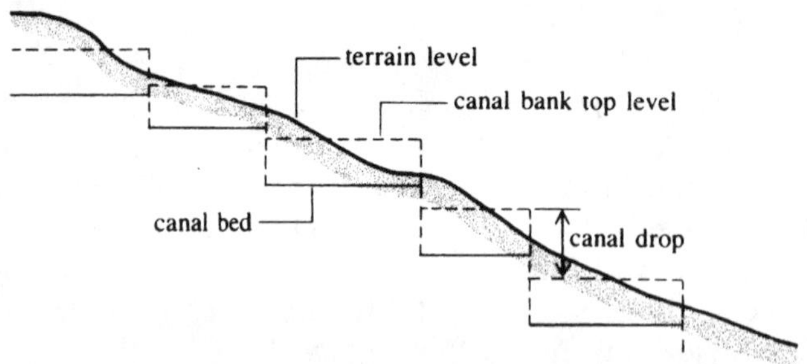

Fig. 10.13 Location of canal drops

10.3.2 Vertical drop structures

(a) Common (straight) drop

The common drop structure, in which the aerated free-falling nappe (modular flow) hits the downstream basin floor, and with turbulent circulation in the pool beneath the nappe contributing to energy dissipation, is shown in Fig. 10.14.

The following equations fix the geometry of the structure in a suitable form for steep slopes:

drop number, $D_r = q^2/g^{d3}$ (10.28)

where q is the discharge per metre width;

basin length, $L_B/d = 4.3D_r^{0.27} + L_j/d$; (10.29)

pool depth under nappe, $Y_P/d = D_r^{0.22}$; (10.30)

sequent depths, $y_1/d = 0.54D_r^{0.425}$; (10.31)

$y_2/d = 1.66D_r^{0.27}$; (10.32)

here d is the height of the drop crest above the basin floor and L_j the length of the jump.

A small upward step, h (around $0.5 < h/y_1 < 4$), at the end of the basin floor is desirable in order to localize the hydraulic jump formation. Forster and Skrinde (1950) developed design charts for the provision of such an abrupt rise.

The USBR (Kraatz and Mahajan, 1975) impact block type basin also provides good energy dissipation under low heads, and is suitable if the

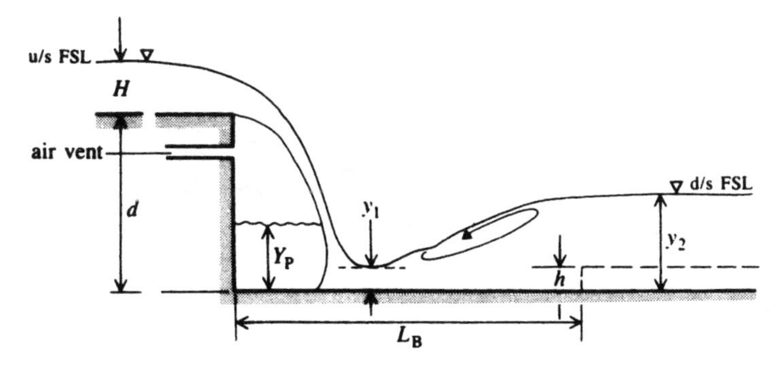

Fig. 10.14 Common drop structure (after Bos, 1976)

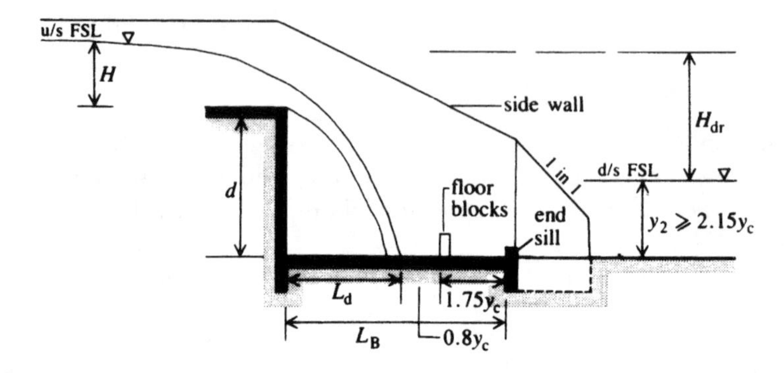

Fig. 10.15 Impact block type basin (after Bos, 1976)

tailwater level (TWL) is greater than the sequent depth, y_2. The following are the suggested dimensions of such a structure (Fig. 10.15):

basin length $L_B = L_d + 2.55y_c$; (10.33)

location of impact block, $L_d + 0.8y_c$; (10.34)

minimum TW depth, $y_2 \geq 2.15y_c$; (10.35)

impact block height, $0.8y_c$; (10.36)

width and spacing of impact block, $0.4y_c$; (10.37)

end sill height, $0.4y_c$; (10.38)

minimum side wall height, $y_2 + 0.85y_c$; (10.39)

here y_c is the critical depth.

The values of L_d can be obtained from Fig. 10.16.

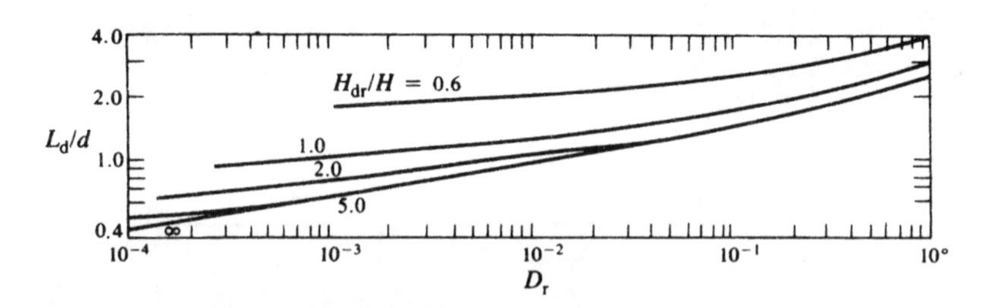

Fig. 10.16 Values of L_d/d (after Bos, 1976)

(b) Sarda-type fall (India)

This is a raised-crest fall with a vertical impact, consisting of a crest wall, upstream and downstream wing walls, an impervious floor and a cistern, and downstream bank and bed protection works (Fig. 10.17).

The crest design is carried out as follows. The crest length is normally kept equal to the bed width of the canal; however, an increase in length by an amount equal to the flow depth takes into account any future increase in discharge. Fluming may be provided to reduce the cost of construction of the fall. A flumed fall with a fluming ratio equal to $2F_1$, where F_1 is the approach flow Froude number, creates no choking upstream of the fall. A canal is not usually flumed beyond 50%. Whenever the canal is flumed, both upstream (contracting) and downstream (expanding) transitions have to be provided (Fig. 10.3).

The crest level must be so fixed that it does not create changes in upstream water levels (backwater or drawdown effects). If the reduced level (RL) of the full supply level (FSL) is Y, the RL of the total energy line (TEL) is

$$E = Y + V_a^2/2g \tag{10.40}$$

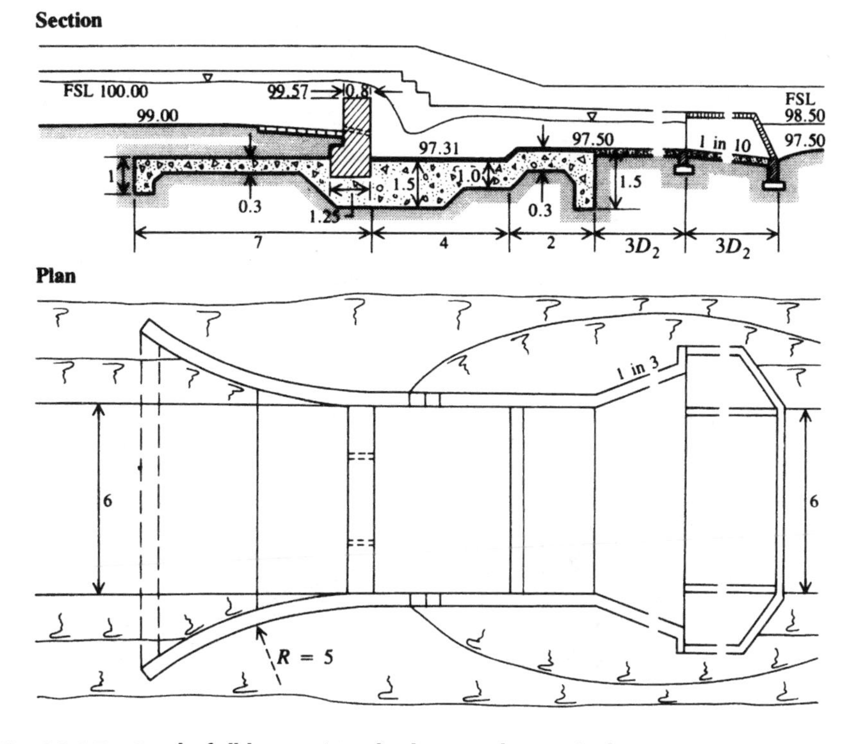

Fig. 10.17 Sarda fall layout (Worked example 10.7); dimensions in metres

where V_a is the approach velocity.

If L_e is the effective length of the crest, the head causing flow is given by the weir formula:

$$H = (Q/C_d L_e)^{2/3} \tag{10.41}$$

where Q is the discharge and C_d is the discharge coefficient of the crest. Therefore, the RL of the crest is $E - H$.

Two types of crest are used (Fig. 10.18); the rectangular one for discharges up to $10\,m^3 s^{-1}$ and the trapezoidal one for larger discharges (see Punmia and Lal, 1977).

The following are the design criteria established by extensive model studies at the Irrigation Research Institute in India.

1. For a rectangular crest,

$$\text{top width,} \quad B = 0.55d^{1/2}\ (\text{m}), \tag{10.42}$$

$$\text{base width,} \quad B_1 = (H + d)/S_s, \tag{10.43}$$

where S_s is the relative density of the crest material (for masonry, $S_s \simeq 2$). The discharge is given by the following formula:

$$Q = 1.835LH^{3/2}(H/B)^{1/6}. \tag{10.44}$$

2. For a trapezoidal crest,

$$\text{top width,} \quad B = 0.55(H + d)^{1/2}\ (\text{m}). \tag{10.45}$$

For the base width, B_1, upstream and downstream slopes of around 1

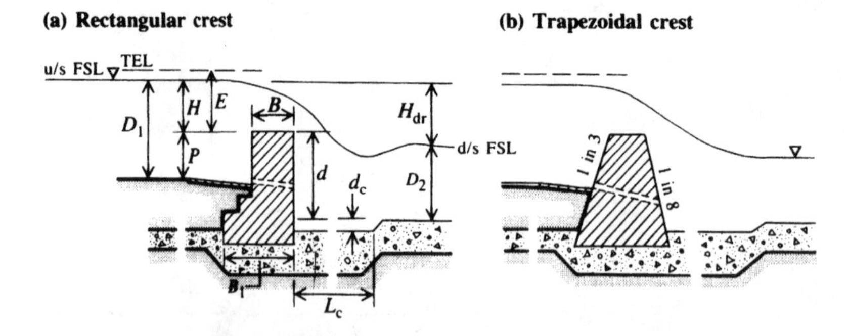

Fig. 10.18 Sarda fall crests

in 3 and 1 in 8 are usually recommended. The discharge is given by the following formula:

$$Q = 1.99LH^{3/2}(H/B)^{1/6}. \quad (10.46)$$

3. Design of cistern is as follows:

$$\text{length,} \quad L_c = 5(EH_{dr})^{1/2}; \quad (10.47)$$

$$\text{depth,} \quad d_c = \frac{1}{4}(EH_{dr})^{2/3}. \quad (10.48)$$

4. Minimum length of impervious floor downstream of the crest,

$$L_{bd} = 2(D_1 + 1.2) + H_{dr}. \quad (10.49)$$

(c) YMGT-type drop (Japan)

This type of drop is generally used in flumed sections suitable for small canals, field channels, etc., with discharges up to $1\,m^3 s^{-1}$ (Fig. 10.19). The following are the recommended design criteria:

1. sill height, P varies from 0.06 m to 0.14 m with the unit discharge q between 0.2 and $1.0\,m^3 s^{-1} m^{-1}$;
2. depth of cistern, $d_c = 1/2(E_c H_{dr})^{1/2}$; (10.50)
3. length of cistern, $L_c = 2.5L_d$, (10.51)

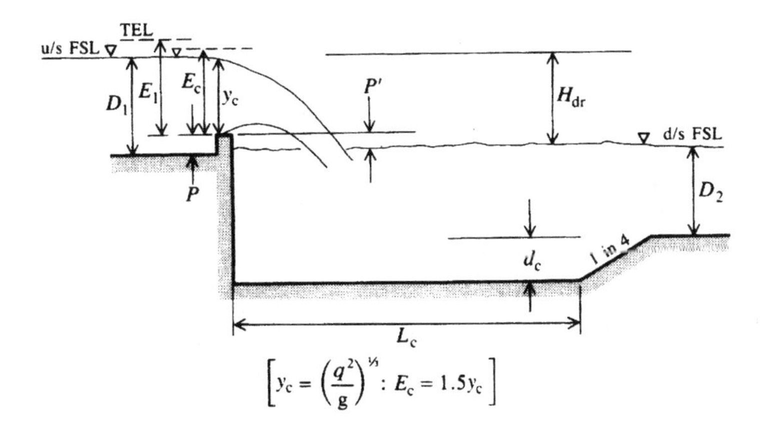

Fig. 10.19 YMGT-type drop, Japan (Kraatz and Mahajan, 1975)

where $L_d = L_{d1} + L_{d2}$ and

$$L_{d1}/E_c = 1.155[(P'/E_c) + 0.33]^{1/2}, \tag{10.52}$$

$$L_{d2} = (D_2 + d_c)\cot\alpha, \tag{10.53}$$

$$\cot\alpha = y_c/L_{d1}. \tag{10.54}$$

Alternatively, the recommendations of the IRI, India (previous section) may also be adopted.

(d) Rectangular weir drop with raised crest (France)

SOGREAH (Kraatz and Mahajan, 1975) have developed a simple structure suitable for vertical drops of up to 7 m (for channel bed widths of 0.2–1 m with flow depths (at FSL) of 0.1–0.7 m): Fig. 10.20 shows its design details.

1. For the design of crest,

$$\text{discharge,}\quad Q = CL(2g)^{1/2}H^{3/2}, \tag{9.29}$$

where $C = 0.36$ for the vertical upstream face of the crest wall and 0.40 for the rounded upstream face (5–10 cm radius). The crest length, $L = L_B - 0.10$ m for a trapezoidal channel and is B_1 (the bed width) for rectangular channels.

2. For the design of cistern,

$$\text{volume of basin,}\quad V = QH_{dr}/150\ (\text{m}^3), \tag{10.55}$$

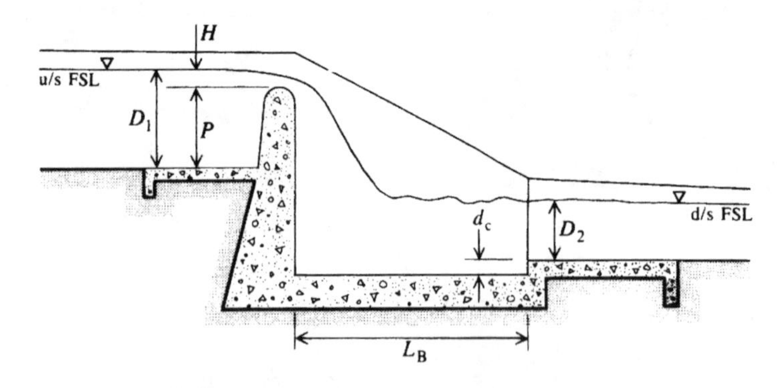

Fig. 10.20 Rectangular weir drop with raised crest, France (Kraatz and Mahajan, 1975)

$$\text{width of basin,} \quad W_B = V/[L_B(D_2 + d_c)], \tag{10.56}$$

where the depth of the basin, $d_c \simeq 0.1$–0.3 m.

10.3.3 Inclined drops or chutes

(a) Common chute

This type of drop has a sloping downstream face (between 1/4 and 1/6, called a glacis) followed by any conventional type of low-head stilling basin; e.g. SAF or USBR type III (Chapter 5). The schematic description of a glacis-type fall with a USBR type III stilling basin, recommended for a wide range of discharges and drop heights, is shown in Fig. 10.21.

(b) Rapid fall type inclined drop (India)

This type of fall is cheap in areas where stone is easily available, and is used for small discharges of up to $0.75\,m^3\,s^{-1}$ with falls of up to 1.5 m. It consists of a glacis sloping between 1 in 10 and 1 in 20. Such a long glacis assists in the formation of the hydrualic jump, and the gentle slope makes the uninterrupted navigation of small vessels (timber traffic, for example) possible.

(c) Stepped or cascade-type fall

This consists of stone-pitched floors between a series of weir blocks which act as check dams and are used in canals of small discharges; e.g. the tail of a main canal escape. A schematic diagram of this type of fall is shown in Fig. 10.22.

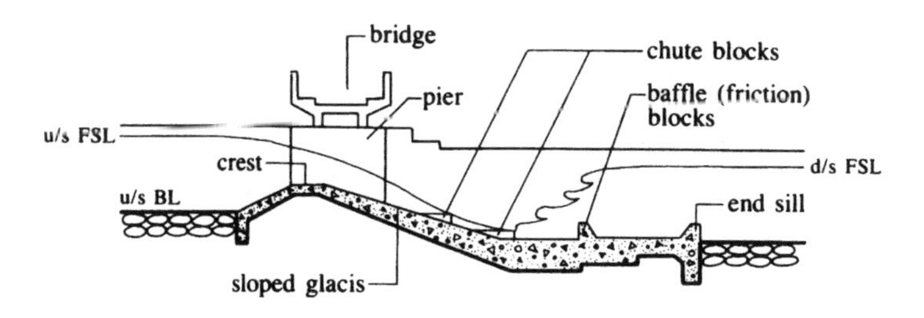

Fig. 10.21 Sloping glacis type fall with USBR type III stilling basin

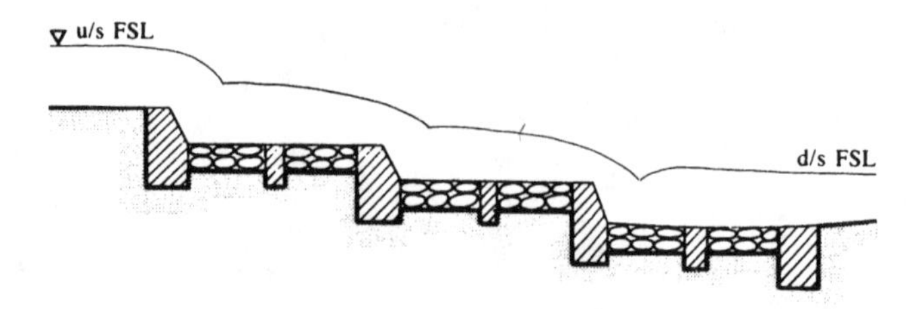

Fig. 10.22 Stepped or cascade-type fall

10.3.4 Piped drops

A piped drop is the most economical structure compared with an inclined drop for small discharges of up to $50 \mathrm{l\,s^{-1}}$. It is usually equipped with a check gate at its upstream end, and a screen (debris barrier) is installed to prevent the fouling of the entrance.

(a) Well drop structure

The well drop (Fig. 10.23) consists of a rectangular well and a pipeline followed by a downstream apron. Most of the energy is dissipated in the well, and this type of drop is suitable for low discharges (up to $50 \mathrm{l\,s^{-1}}$) and high drops (2–3 m), and is used in tail escapes of small channels.

(b) Pipe fall

This is an economical structure generally used in small channels. It consists of a pipeline (precast concrete) which may sometimes be inclined sharply downwards (USBR and USSR practice) to cope with large drops. However, an appropriate energy dissipator (e.g. a stilling basin with an end sill) must be provided at the downstream end of the pipeline.

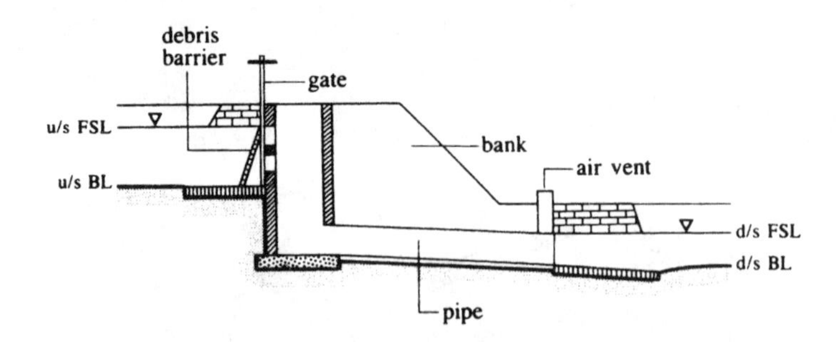

Fig. 10.23 Well drop structure

10.3.5 Farm drop structures

Farm channel drops are basically of the same type and function as those in distribution canals, the only differences being that they are smaller and their construction is simpler.

The notch fall type of farm drop structure (precast concrete or timber) consists of a (most commonly) trapezoidal notch in a crested wall across the canal, with the provision of appropriate energy-dissipation devices downstream of the fall. It can also be used as a discharge-measuring structure.

The details of a concrete check drop with a rectangular opening, widely used in the USA, are shown in Fig. 10.24. Up to discharges of about $0.5\,m^3\,s^{-1}$, the drop in the downstream floor level (C) is recommended to be around 0.2 m and the length of the apron (L) between 0.75 m and 1.8 m over a range of drop (D) values of 0.3–0.9 m.

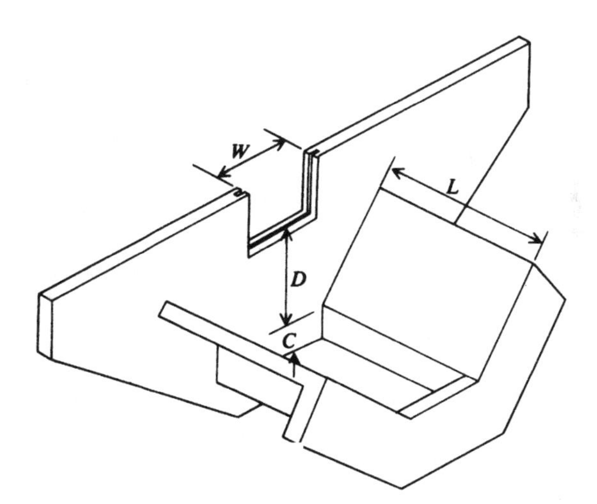

Fig. 10.24 Notch fall: concrete check drop (USA)

Worked Example 10.7

Design a Sarda-type fall using the following data: full supply discharge = $28\,m^3\,s^{-1}$; upstream FSL = 100.00 m AOD; downstream FSL = 98.50 m AOD; upstream bed level = 99.00 m AOD; downstream bed level = 97.50 m AOD; upstream bed width = 6.00 m; downstream bed width = 6.00 m; drop, H_{dr} = 1.50 m; safe exit gradient of the subsoil = 1 in 5.

Solution

CREST DESIGN

Adopt a rectangular crest (Fig. 10.18). $B = 0.55d^{1/2}$; for a trapezoidal crest, $B = 0.55\ (H + d)^{1/2}$ and $H + d = D_1 + 99.00 - 97.50 = 2.5$ m. Hence B (trapezoidal crest) = 0.87 m. Choose a rectangular crest, width $B = 0.80$ m. Adopting a crest length of $L = 6$ m, equation (10.51) gives the head over the crest,

$$H = 0.43\,\text{m}.$$

Check for B: the crest level = 100.00 − 0.43 = 99.57 m AOD. Therefore $d = 99.57 - 97.50 = 2.07$ m and $B = 0.55d = 0.79$ m. Therefore a crest width of 0.8 m is satisfactory.

The base width, $B_1 = 1.25$ m (equation (10.50) with $S_s = 2.0$). The velocity of approach (assuming a 1:1 trapezoidal channel) = 2.8/(6 + 1)1 = 0.4 m s^{-1}. Therefore the upstream total energy level (TEL) = 100.000 + 0.008 = 100.008 m AOD, and $E = 100.008 - 99.57 = 0.438$ m.

The depth of the cistern, $d_c = 0.19$ m (equation (10.55)), and the length of the cistern, $L_c \simeq 4$ m (equation (10.54)). The RL of the cistern bed = 97.50 − 0.19 = 97.31 m AOD.

IMPERVIOUS FLOOR DESIGN (WORKED EXAMPLE 9.1 PROVIDES DETAILED CALCULATIONS)

The maximum seepage head, $H_s = d$ (with no water downstream, and the upstream water level at crest level). Adopting nominal upstream and downstream cut-off depths of 1 m and 1.5 m respectively, the base length of the impervious floor for the exit gradient of 1 in 5 is approximately 13 m. The length of the impervious floor downstream of the crest is approximately 6 m (equation 10.56)). The upstream floor thickness (nominal thickness of 0.3 m) at the toe of the crest is approximately 1.5 m, and at 5 m from the toe it is approximately 0.14 m; adopt a minimum of 0.3 m.

Appropriate energy-dissipating devices (for large discharges) and upstream and downstream bed protection works must be provided. The detailed layout of the design is shown in Fig. 10.17.

References

d'Aubuisson, J.F. (1940) *Traité d'Hydraulique*, 2nd edn, Pitois, Levraut et Cie, Paris.

Blaisdell, F.W. and Anderson, CL. (1988a) A comprehensive generalized study of scour at cantilevered pipe outlets. *Journal of Hydraulic Research*, 26 (4): 357–76.

—— (1988b) A comprehensive study of scour of cantilevered pipe outlets. *Journal of Hydraulic Research*, 26 (5): 509–24.

Bonasoundas, M. (1973) *Strömungsvorgang und Kolkproblem*, Report 28, Oscar von Miller Institute, Technical University, Munich.

Bos, M.G. (ed.) (1976) *Discharge Measurement Structures*, Laboratorium voor Hydraulica en Afvoerhydrologia, Wageningen.

Breusers, H.N.C. and Raudkivi, A.J. (1991) *Scouring*, IAHR Hydraulic Structures Design Manual, Vol. 2, Balkema, Rotterdam.

Breusers, H.N.C., Nicollet, G. and Shen, H.W. (1977) Local scour around cylindrical piers. *Journal of Hydraulic Research*, 15 (3): 211–52.

Chanson, H. (1999) *Hydraulics of Open Channels – An Introduction*, Arnold, London.

Charbeneau, R.J., Henderson, A.D. and Sherman, L.C. (2006) Hydraulic performance curves for highway culverts, *J. of Hydraulic Engineering ASCE*, 132, No. 5: 474–81.

Chow, V.T. (1983) *Open Channel Hydraulics*, McGraw-Hill, New York.

Clark, A. and Novak, P. (1983) Local erosion at vertical piles by waves and currents, in *Proceedings of the IUTM Symposium on Seabed Mechanics*, Newcastle upon Tyne.

Coleman, S.E., Lauchlan, C.S. and Melville, B.W. (2003) Clear-water scour development at bridge abutments, *J. of Hydraulic Research*, IAHR, 41, No. 5: 521–31.

Dey, S. and Barbhuiya, A.K. (2005) Time variation of scour at abutments, *J. of Hydraulic Engineering*, ASCE, 13, No. 1: 11–23.

Ettema, R. (1980) *Scour at Bridge Piers*, Report 216, School of Civil Engineering, University of Auckland.

Fletcher, B.P. and Grace, J.L., Jr. (1974) *Practical Guidance for Design of Lined Channel Expansions at Culvert Outlets*, Technical Report H-74-9, US Army Engineers Waterways Experimental Station, Vicksburg.

Forster, J.W. and Skrinde, R.A. (1950) Control of the hydraulic jump by sills. *American Society of Civil Engineers Transactions*, 115: 973–87.

Hager, W.H. (1999) *Wastewater Hydraulics*, Springer Verlag, Berlin.

Hamill, L. (1999) *Bridge Hydraulics*, E & FN Spon, London.

Hinds, J. (1928) The hydraulic design of flume and siphon transitions. *Transactions of the American Society of Civil Engineers*, 92: 1423–59.

Inglis, C. (1949) *The Behaviour and Control of Rivers and Canals*, Research Publication 13, Central Water Power Irrigation and Navigation Report, Poona Research Station.

Kindsvater, C.E., Carter, R.W. and Tracy, H.J. (1953) *Computation of Peak Discharge at Contractions*, Circular 284, USGS.

Kraatz, D.B. and Mahajan, I.K. (1975) *Small Hydraulic Structures – Irrigation and Drainage*, FAO, Rome, Papers 26/1 and 26/2.

Laursen, E.M. (1962) Scour at bridge crossings. *Transactions of the American Society of Civil Engineers, Part 1*, 127: 166–209.

Linsley, R.K. and Franzini, J.B. (1979) *Water Resources Engineering*, 3rd edn, McGraw-Hill, New York.

Martín-Vide, J.P. and Prió, J.M. (2005) Backwater of arch bridges under free and submerged conditions, *J. of Hydraulic Research*, IAHR, 43, No. 5: 515–21.

Mays, L.W. (ed.) (1999) *Hydraulic Design Handbook*, McGraw-Hill, New York.

Melville, B.W. and Chiu, Yee-Meng (1999) Time scale for local scour at bridge piers, *J. of Hydraulic Engineering*, ASCE, 125, No. 1: 59–65.

Melville, B.W. and Coleman, S. (2000) *Bridge Scour*, Water Resources Publications, Colorado.

Merlein, J., Kleinschroth, A. and Valentine, F. (2002) *Systematisierung von Absturzbauwerken*, Mitteilung N. 69, Lehrstuhl und Laboratorium für Hydraulik und Gewässerkunde, Technische Universität, München.

Nagler, F.A. (1918) Obstruction of bridge piers to the flow of water. *Transactions of the American Society of Civil Engineers*, 82: 334–95.

Neil, C. (ed.) (2004) *Guide to Bridge Hydraulics*, 2nd edn, Thomas Telford, London, and Transportation Association of Canada, Ottawa.

Oliveto, G. and Hager, W.H. (2002) Temporal evolution of clear-water pier and abutment scour, *J. of Hydraulic Engineering*, ASCE, 128, No. 9: 811–20.

Punmia, B.C. and Lal, P.B.B. (1977) *Irrigation and Water Power Engineering*, 4th edn, Standard Publishers, New Delhi.

Ramsbottom, D., Day, R. and Rickard, C. (1997) *Culvert Design Manual*, CIRIA Report 168, Construction Industry Research and Information Association (CIRIA), London.

Ranga Raju, K.G. (1993) *Flow Through Open Channels*, 2nd edn, Tata McGraw-Hill, New Delhi.

Richardson, J.R. and Richardson, E.V. (1994) Practical method for scour prediction at bridge piers, in *Hydraulic Engineering '94* (eds J.V. Cotroneo and R.R. Rumer), *Proceedings of the ASCE Conference*, Buffalo, NY, 1994, American Society of Civil Engineers, New York.

Sheppard, D.M., Odeoh, M. and Glasser, T. (2004) Large scale clear-water local scour experiments, *J. of Hydraulic Engineering*, ASCE, 130, No. 10: 957–63.

Simons, D.G. and Stevens, M.A. (1972) Scour control in rock basins at culvert outlets, in *River Mechanics*, Vol. II (ed. H.W. Shen), Chapter 24.

Unger, J. and Hager, W.H. (2006) Riprap failure of circular bridge piers, *J. of Hydraulic Engineering, ASCE*, 132, No. 4: 354–62.

UPIRI (1940) Technical Memorandum 5, Uttar Pradesh Irrigation Research Institute, Roorkee.

US Army Coastal Engineering Research Center (1984) *Shore Protection Manual*, Washington, DC.

Vittal, N. and Chiranjeevi, V.V. (1983) Open channel transitions: rational method of design. *Journal of the Hydraulics Division, American Society of Civil Engineers*, 109 (1): 99–115.

Watkins, L.H. and Fiddes, D. (1984) *Highway and Urban Hydrology in the Tropics*, Pentech, London.

Yarnell, D.L. (1934) *Bridge Piers as Channel Obstructions*, Technical Bulletin 442, US Department of Agriculture.

Zarrati, A.R. Nazariha, M. and Mashahir, M.B. (2006) Reduction of local scour in the vicinity of bridge pier groups using collars and riprap, *J. of Hydraulic Engineering, ASCE*, 132, No. 2: 154–62.

Chapter 11

Inland waterways

11.1 Introduction

Navigation on inland waterways is the oldest mode of continental transport. Although during its long history it has passed through many stages of technological development and – in some countries – from prosperity to depression, there is no doubt that nowadays it forms an important and integral part of the transport infrastructure of many countries in the world.

In ancient civilizations, inland navigation flourished in the valleys of great rivers (the Nile, Euphrates, Ganges, Jang-c-tiang, etc.) and artificial waterways were known in ancient Egypt, Mesopotamia and China, where Emperor Yantei (Sui Dynasty AD 611) built the 'Great Canal' – a 2400 km waterway (linking the river systems of the north with the southern provinces).

In Europe in AD 793, the emperor Charles the Great had already started the building of a canal intended to link the Rhine and the Danube (Fossa Carolina), an attempt soon to be abandoned. The first clearly documented navigation lock dates from 1439 and was constructed on the Naviglio Grande canal in northern Italy.

Industrialization was the prime mover of modern waterways development in the 18th and 19th centuries with the network of navigable rivers and canals in England at the forefront of this type of development (e.g. the *Bridgewater canal* built by James Brindley and the *Ellesmere canal* built by Thomas Telford). *The Forth and Clyde canal* in Scotland completed in 1790, was the first sea-to-sea ship canal in the world.

The *Anderton boat lift* overcoming a head of 15 m between the Trent and the Mersey canal and the river Weaver in Cheshire, UK, was built in 1875 and is the first iron barge lift with a hydraulic lifting system.

The second half of the 19th and the beginning of the 20th century saw the construction of two great navigation canals of global importance. The 160 km long, 305–365 m wide and 19.5 m (minimum) deep *Suez canal* opened in 1869, shortening the sea routes between Europe and the Far East by 16 000 km; nowadays it is used by 15 000 ships/year including 150 000 t oil tankers. The 80 km long *Panama canal*, opened in 1914, links the Atlantic and the Pacific by a 13 km long and 153 m wide cut through the Continental Divide and a large artificial lake with three locks at the entrance and exit of the canal with a total lift of 26 m.

The present great European network of inland waterways is based on modernized and expanded 19th century navigation facilities. The same applies to the navigation facilities on the great American waterways, e.g. on the Mississippi and the Ohio River. Although in the 20th century inland waterways often could not compete with the railway and later motorway networks, they retained – and even increased – their rôle in the provision of a highly effective means of transport, particularly of bulk material.

The role of inland waterways in water resources management in the provision of modern recreational facilities and in the enhancement of the environment further contributed to this new perception.

In spite of the rapid development of other modes of transport there are some universally valid advantages in transport by inland navigation (Čábelka and Gabriel, 1985):

1. low energy requirements (the specific energy consumption for navigation is about 80% of that for rail and less than 30% of the consumption for road transport);
2. high productivity of labour per unit of transport output;
3. low material requirement per unit of transport volume (the corresponding values of rail and highway transport are two and four times higher respectively);
4. lowest interference with the environment (low noise, low exhaust fume generation);
5. very low land requirement (in the case of navigable rivers);
6. low accident incidence in comparison with other transport modes;
7. capability of easily transporting bulk cargo and large industrial products.

The detailed discussion of modes of transport on inland waterways and design and operation of associated hydraulic structures assumes that the reader is familiar with the concepts and equations of open-channel flow and at least some river engineering works, as discussed in Chapter 8.

11.2 Definitions, classification and some waterways

11.2.1 Definitions and classification of waterways

Waterways can be divided into three classes:

1. natural channels, i.e. rivers, or parts of rivers, the flow of which is not modified – the river channel may be improved by river training works;
2. canalized rivers, the flow of which is to a greater or smaller degree controlled by engineering works;
3. canals, entirely artificial waterways whose water is obtained by diversion from rivers, by pumping or from reservoirs.

In the endeavour to ensure gradual unification of European waterways and the standardization of their parameters, the Economic Commission for Europe adopted, in 1961, a uniform classification of inland waterways. This classification is based on the dimensions and the tonnage of traditional standard vessels, and classifies the waterways into six classes (Čábelka, 1976; Čábelka and Gabriel, 1985); I$\leq$400 t, II$\leq$650 t, III$\leq$1000 t, IV$\leq$1500 t, V$\leq$3000 t, and VI$>$3000 t. For every class the necessary parameters of the waterway and its structures were deduced from the parameters of standard vessels.

Simultaneously with the adoption of the above classification, it was agreed that the European waterways of international importance would be so built or reconstructed as to ensure that their parameters would correspond with the requirements of at least class IV and permit continuous passage to vessels of a tonnage between 1350 and 1500 tons. Waterways of classes I–III have a regional character.

Large European rivers, such as the Rhine, the Danube and others, are being made navigable at present mostly to the parameters of class V. Class VI includes primarily the Russian navigable rivers and canals, or the lowland stretches of the longest European rivers. Similar or even larger dimensions are found in navigable rivers of other continents, notably in North and South America.

The above-mentioned international classification of inland waterways was adopted in the period when – with the exception of motorboats – towing by tug boats was used almost exclusively. Since about 1970, however, this traditional navigation technology has been almost completely replaced in Europe (and much earlier in the USA) by the economically and operationally more advantageous pushing of the barges by push-boats (Section 11.4.2).

The introduction of this new technology resulted in proposals for the amendment of the existing waterway classification (Hilling, 1977). An economic Europe class II pushed barge type, with dimensions of 76.5 m $\times$ 11.4 m $\times$ 2.5 m and a tonnage of 1660 t, was proposed by Seiler

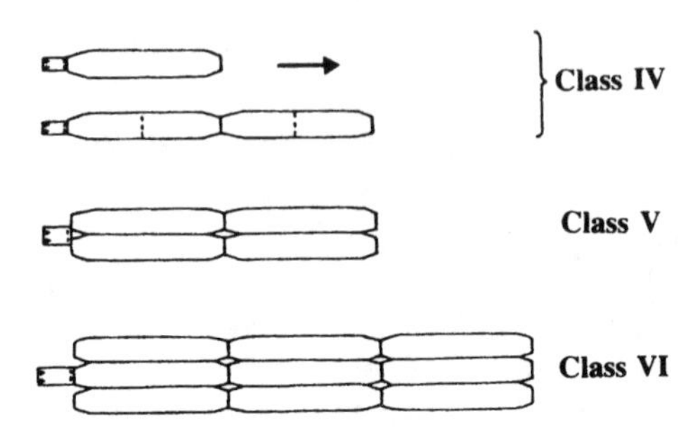

Fig. 11.1 Push trains used on European waterways (Čábelka and Gabriel, 1985)

(1972) as the standard barge for waterways of international importance, together with the recommendation that class IV to class VI waterways should be designated according to the number and arrangement of these barges in pushed trains (Fig. 11.1). Further modifications in the European classification took account also of the plan configuration in the push barge systems, their draught and bridge clearances (ECE, 1992).

11.2.2 Some waterways

(a) UK waterways

Although the building of inland waterways in the UK was in the forefront of European development in the late 18th and 19th centuries, the English canals fell into disuse towards the close of the 19th and in the first half of the 20th century, mainly because of the rapid development of rail and road links joining inland industrial centres to the coastal ports.

The past 30 years or so, however, have seen a renaissance in the use of inland waterways, as their role in water resources management, the provision of modern recreational facilities, and enhancement of the environment, quite apart from their commercial value, has become more widely appreciated.

The total length of usable inland waterways in the UK amounts to almost 4000 km. The more important inland waterways are those connected to the estuaries and the rivers Mersey, Severn, Thames and Humber.

The major UK canal is undoubtedly the Manchester Ship Canal, constructed between 1887 and 1893, which has parameters in class VI (Section 11.2.1). It is 58 km long, 36.5 m wide (bottom) and 8.5 m deep; its five locks can take vessels of up to 12 000 tons. The Avon is navigable to Bristol for a capacity of up to 5000 tons (also class VI). Examples of some other major inland waterways (other than estuaries) of class III are the Gloucester Ship Canal, the Weaver up to Northwick, the Ouse to Selby and the Tay to Perth. The Trent up to Newark falls into class II.

The famous Caledonian Canal linking the east and west coasts of Scotland (Inverness to Fort William) passes right through Loch Ness; it has a capacity of up to 600 t and its 29 locks were designed by Thomas Telford in the early 19th century.

The longest inland waterway is the Grand Union Canal which – including all branches – is almost 300 km long with over 200 locks. The longest tunnel in the UK network is the 2900 m long Dudley tunnel (Edwards, 1972).

(b) The European network

The European waterways of differing technical standards and slightly diverging parameters in the individual countries, form four more or less self-contained groups (Fig. 11.2):

1. French waterways;
2. Central European waterways between the Rhine in the west and the Vistula in the east, consisting of navigable rivers flowing to the north and the canals interconnecting them in an east–west direction;
3. South European waterways, comprising the Danube, the navigable sections of its tributaries, and accompanying canals;
4. East European waterways, consisting of the navigable rivers in the European part of Russia, and the Volga–Moskva canal, the Volga–Don canal, the Volga–Baltic Sea canal, the Baltic Sea–White Sea canal, etc.

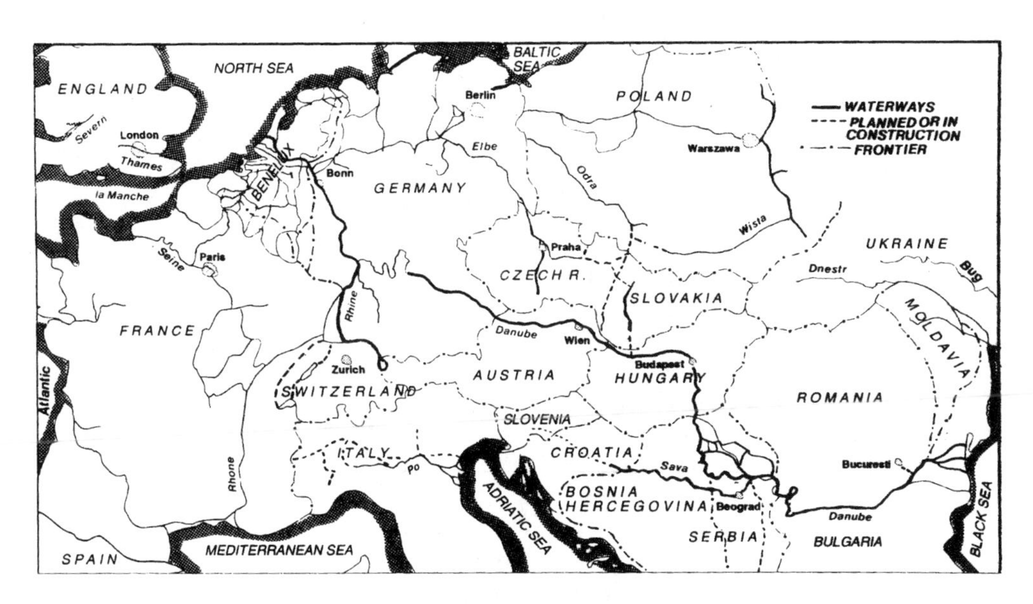

Fig. 11.2 European waterways (Novak, 1994)

There are also a large number of isolated waterways in the boundary regions of the European continent connected with the sea, especially in Italy, Portugal, Spain, Sweden, and Finland (see also UK above).

From the four above-mentioned European waterway groups, satisfactory navigation interconnection exists only between the French and the Central European waterways. The generation of an integrated network of European waterways necessitates primarily the link between the South and Central European waterways by means of two canal systems: the Rhine–Main–Danube canal, completed in 1992, and the Danube–Oder–Labe (Elbe) canal, the phased construction of which has already started. Of considerable importance also would be the connection with the East European waterways by means of the planned Oder–Vistula–Bug–Dniepr canal.

(c) US waterways

There is an important network of navigable rivers and canals, particularly in the central and eastern United States. Some of the most important waterways are shown schematically in Fig. 11.3 including the following: open river navigation on the Middle and Lower Mississippi from St Louis to the Gulf of Mexico, and on the Missouri River from Sioux City in Iowa downstream; major canalized rivers such as the Upper Mississippi, Ohio, Tennessee, Lower Columbia, and Arkansas Rivers and canals, e.g. the Chain of Rocks Canal on the Mississippi, the Arkansas Post Canal and the Tennessee–Tombigbee Waterway (opened in 1955) with a watershed section 65 km long (including the Bay Springs Lock – Fig. 11.13) and a canal section 70 km long, with 91 m bottom width, 3.65 m deep, with five locks and a total lift of 43 m.

11.3 Multipurpose utilization of waterways

Modern waterways practically always fulfil also other functions apart from inland navigation. The most common case is the utilization of water power in plants built next to navigation locks. Other uses of waterways are flood protection on trained rivers, provision of off-take facilities for water supply, drainage of adjacent land, waste water disposal, etc. The provision of recreational facilities and general improvement of the environment are some of the most important additional benefits provided by inland waterways.

Multipurpose use of waterways brings about also additional problems, none more so than the peak operation of power plants causing surges on canals and canalized rivers. The most serious are surges caused by sudden load rejection. In order not to affect navigation unfavourably, the effect of power plant operation must in the majority of cases be reduced by suitable measures (Čábelka and Gabriel, 1985).

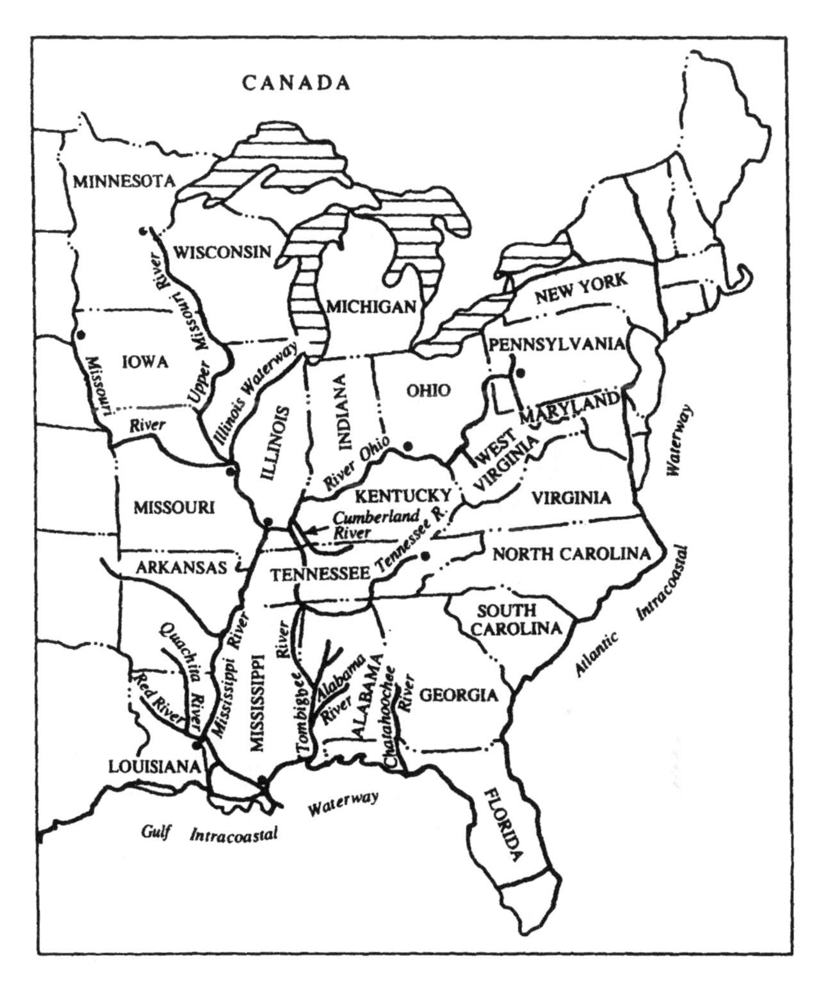

Fig. 11.3 Waterways in central and eastern USA (Petersen, 1986)

These measures may be electrical (switching of the generator outlet to water resistance), mechanical (disconnecting the linkage between guide and runner vanes of Kaplan turbines or providing automatic gate-controlled outlets connected directly to turbine spiral casings), or structural (provision of special auxiliary outlets designed as bypasses for the turbines).

In the case of power plants situated next to barrages the negative effect of surge waves on navigation can also be reduced by suitable (automatic) operation of the gates. Finally, the whole layout of the barrage and plant can be designed so as to minimize surges (widening of head and tail races, separation of the navigation channel from the power plant, etc.).

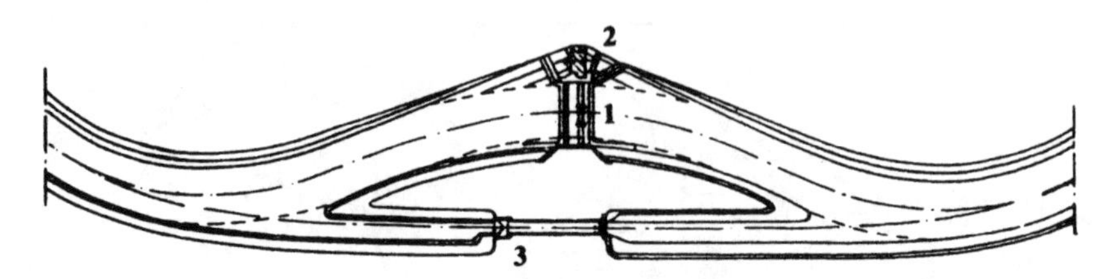

Fig. 11.4 Barrage (1) with power station (2) and navigation lock (3) (after Novak and Čábelka, 1981)

The above measures can be used singly or jointly. However, it must be emphasized that although they can reduce the surges from power plants to an acceptable level they cannot eliminate them completely. The optimum control of a whole cascade of plants is usually best developed by the use of mathematical modelling, combined with field measurements. The winter regime also has to be taken into account in these studies.

The layout of a barrage with a power plant, navigation lock(s) and their approaches requires careful consideration of the respective functions of the individual components of a complex hydraulic structure. This may result in different layouts on small and large navigable rivers and on navigation canals.

On smaller rivers and at older works the general layout usually followed a scheme with the navigation lock and power plant having parallel axes and being situated on either side of the weir or barrage, with the lock separated from the river by a long narrow dividing wall (Chapter 12). A better solution adopted in newer structures is one where the lock is separated from the river by an island, and the turbine axes (forebay and tail race) are set at an angle to the direction of flow in the river (Fig. 11.4). This arrangement prevents scour at the dividing walls of the power plant and lock and results in improved efficiency both of navigation and of the power plant. On large navigable rivers the considerations governing the layout are more complex and usually also have to take into account problems of sediment transport, ice, etc. (Kuhn, 1985 and Chapters 9 and 12).

The design of a barrage with power plant(s) and lock(s) on a river forming an international boundary may even call for a symmetrical arrangement, with two power plants and two sets of locks on either side of the river.

Details of a suitable layout of a complex water resources utilization, particularly one involving hydropower development and navigation, are best studied on scale models (Novak and Čábelka, 1981).

11.4 Transport on inland waterways

11.4.1 Utilization of inland waterways

The influence of inland navigation in various countries is very varied, depending primarily upon their geographical position and upon economic factors. Although it is difficult to obtain accurate data owing to different methods of registering freight on inland waterways (all freight or only that transported by the respective country), and the difficulty of distinguishing between coastal and inland navigation in some countries, Table 11.1 gives at least a general overview (relating to 1992) for some countries both in absolute (length and utilization of waterway) and relative terms when compared with rail and road transport (Savenije, 2000).

Table 11.1 Freight on inland waterways: annual throughput of shipping

Country	*Length of waterways[a] (km)*	*Utilization of waterways*		*Fraction of total transport capacity*	
		Volume ($\times 10^6$ t year^{-1})	*Output ($\times 10^9$ t km year^{-1})*	*Volume (%)*	*Output (%)*
Austria	358	7	1.5	1.7	1.2
Belgium	1513	90	5.1	17.7	10.7
Bulgaria	470	1	0.8	1.0	6.0
China	107800	312	57.0	14.4[b]	8.7[b]
Czech Republic	303	7	3.0	3.1	3.6
Finland	6245	2	2.0[b]	0.4	1.4
France	5817	71	8.6	3.6	3.4
Germany	6291	230	57.2	18.7	19.1
Hungary	1464	9	1.5	2.9	20.9
Italy	1366	1	0.1	1.0[b]	0.4[b]
Netherlands	5046	262	33.6	32.2	43.5
Poland	3805	8	0.7	0.6	0.7
Romania	1779	6	1.9	0.7	4.0
Russia[c]	34167	233	95.2	6.3	3.1
Slovakia	422	3	0.8	1.0	6.6
Switzerland	21	9	0.1	1.7	0.3
UK	1192	6	0.2	0.3	0.1
Ukraine	3647	41	8.2	0.8	1.8
USA	20573[d]	639	234.0[b]	?	10.6[b]

a Length regularly used.
b Not available for 1992, data for 1980–83.
c (European part).
d Over 40 000 km including intracoastal (Hilling, 1999).

11.4.2 Traction, push-tow and lighters

Barges may be self-propelled or towed or pushed by tugs. On some continental canals diesel or electric tractors, or engines on a track along the canal, have replaced the original form of traction by horses. On larger canals and navigable waterways the traditional method of using tug boats, pulling a number of barges, was replaced almost universally by 1970 by the control of a group of barges by push boats (Section 11.2). This development has been mainly due to the following advantages of the push-tow (Čábelka and Gabriel, 1985).

1. The resistance of a pushed barge train is lower than that of a towed train of the same tonnage; it is therefore possible either to increase the navigation speed or to reduce fuel consumption.
2. The crew required is fewer in number and thus the organization of labour and the living conditions of the crew are improved, and operating costs reduced.
3. Investment costs are up to 40% lower than for comparable tug boats.
4. The control and manoeuvrability of the whole train, and its safety, are improved.

The advantages of the push-boat technology increase with the size of the train which can navigate the waterways and pass through the locks without being disconnected. The size of the push trains varies according to the characteristics of the waterway (Section 11.2); on some larger ones (e.g. the lower Mississippi) 48-barge push-tows with towboats exerting a power over 5000 kW are not uncommon (Petersen, 1986).

With the development of international trade, intercontinental freight transport acquires an ever-increasing importance on the principal waterways. After an extraordinarily speedy development of container transport, the progressive method of international transport by means of floating containers, called lighters, has begun to assert itself (Kubec, 1981).

This system is intended above all for the transport of goods whose consigner and recipient are situated on navigable waterways of different continents. Lighters, grouped on inland waterways into pushed trains, are transported across the sea in special marine carriers provided with loading and unloading equipment of their own (lighter aboard ship – LASH – system).

A further development is the direct maritime link with cargo carried on a ship which combines full sea-going capacity with dimensions allowing maximum penetration of inland waterways – the river/sea ship – converting sometimes previous inland ports to mixed barge and sea-going traffic (e.g. Duisburg in Germany) (Hilling, 1999). The proposal for standardization of ships and inland waterways for river/sea navigation (PIANC, 1996) led in 1997 to the creation of the European River–Sea Transport Union (ERSTU).

11.5 Canalization and navigation canals

11.5.1 Canalization

A free-flowing river can be canalized by a series of barrages with navigation locks. Canalization becomes necessary from the navigation point of view if the free-flowing river has too shallow a depth and too high a velocity to permit navigation.

The advantages of canalization are as follows: the opportunity to develop multipurpose utilization of water resources; sufficient depth for navigation throughout the year, even at times of low river flows; reduced flow velocities; increased width of waterways; safer and cheaper navigation; often the reduced need for bank protection and its maintenance (compared with regulated rivers). The main disadvantages are the high capital cost, the need for protection of adjacent land, drainage problems, the delay of traffic passing through locks, the possible deposition of sediments at the upstream end and possible winter régime complications.

The upper reaches of most of the major navigable rivers are canalized or in the process of being canalized. The heights of individual steps in the cascade of barrages, i.e. the difference of water levels, varies greatly according to hydrological, morphological, and geological conditions, but is usually between 5 and 15 m.

11.5.2 Navigation canals

Navigation canals can be used to bypass a river section that is difficult to navigate and can be used in conjunction with a single barrage or several barrages spaced wider apart than in the case of river canalization. Furthermore, they are an essential part of inland navigation where they connect two watersheds. They require suitably shaped intakes, often a separate flow regulation structure, and navigation locks.

The position and layout of canals can – within the traffic and geological constraints – be adapted to general transport, land-use and industrial demands. The canal is usually appreciably shorter than a canalized river which, together with low (or zero) flow velocities, aids navigation in both directions. Their main disadvantage is use of land, and disruption of communications; thus when planning a canal, maximum use should be made of existing rivers, as far as their canalization is feasible.

Navigation canals can have a fall in one direction only or in both directions with a top water reservoir. They may connect two river systems or branch off a navigable waterway to give access to an industrial centre. The crossing of a canal with a navigable river, the branching of a canal from the river or the branching of canals may create special traffic and construction problems – see also Section 8.7.1.

Sections of canals which are either temporarily or permanently above the surrounding groundwater table need (apart from erosion protection) some means of protection against loss of water by seepage; proper underdrainage and protection of the impermeable or seepage-resistant layer (e.g. clay, concrete, plastics, etc.) against back pressure in the event of an increase of the groundwater level are essential. Bank protection on canalized rivers and canals is of the same type and variety as on trained rivers (Chapter 8).

Adequate canal depth and width are required, just as on regulated rivers and, because of the drift of tows passing through bends, a greater width is required there.

The minimum width, B, of a waterway in a straight section with simultaneous navigation in both directions is $B=3b$ or $B=2b+3\Delta b$, where b is the width of a barge (or a group of barges) and Δb is the side clearance, with $\Delta b \geq 5$ m. If navigation is in one direction only, $B=(1.5-2)b$.

The minimum radius, r, of a curved waterway is given by the length, L, of a typical barge multiplied by a constant which is about 3 for push boats and 4.5 for towed barges. The width of the waterway in a bend with a two-way traffic has to be increased to $B_0=B+\Delta B$ (Fig. 11.5(a)), where

$$\Delta B = \frac{L^2}{2r+B} \simeq \frac{L^2}{2r}. \tag{11.1}$$

The drift (deflection) angle, α, is the inclination of the tow to the tangent of the radius of curvature passing through the centre of the tow (Fig. 11.5(b)). The drift depends on the radius of the bend, the speed, power, and design of the tow (tug), loading of the tow, wind forces, and the flow pattern. The drift angle is larger for tows travelling in the downstream than the upstream direction.

The US Army Corps of Engineers (1980) extrapolated German drift angle data from the Rhine up to a tow length of 180 m and obtained, for the downstream direction, values of $2° < \alpha < 15°$ for radii of curves of $400\,\text{m} < r < 2500\,\text{m}$ (the larger the radius the smaller the value of α). For the upstream direction the values of α are halved.

According to the US Army Corps of Engineers, the following equations apply for the channel width B_0 in bends: for one-way traffic,

$$B_{01} = L_1 \sin\alpha_d + b_1 + 2c, \tag{11.2}$$

and for two-way traffic,

$$B_{02} = L_1 \sin\alpha_d + b_1 + L_2 \sin\alpha_u + b_2 + 2c + c', \tag{11.3}$$

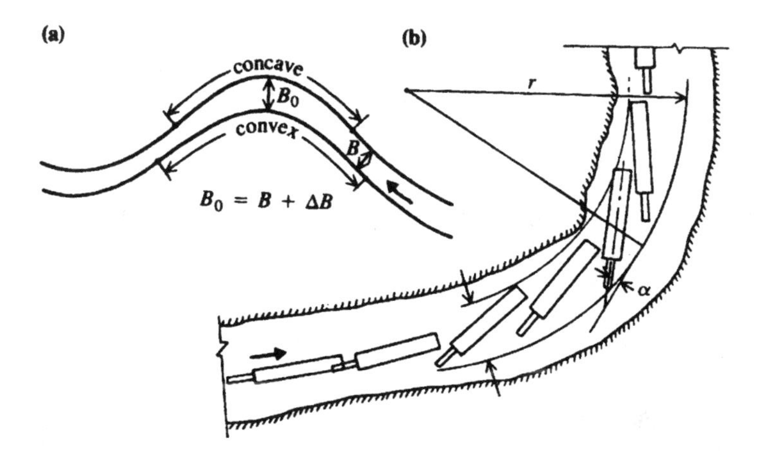

Fig. 11.5 Bends and tow drift on a waterway (Novak, 1994)

where L is the length of the tow, α the maximum drift angle, b the width of the tow, c the clearance between the tow and channel bank and c' the clearance between the passing tows; suffix 'd' refers to a downbound and 'u' to an upbound tow. The result of computations using equation (11.3) can be checked against equation (11.1).

11.6 Resistance of ships

The interaction of a ship with the surrounding body of water is a complicated one, particularly in a constricted waterway where, furthermore, the interaction of the flow and waves generated by navigation with the banks and bed of the waterway has to be taken into account. These factors, together with the power and speed requirements of the tow, determine the design of the navigation fairway, which also has to take into account the various bottlenecks encountered, e.g. fixed and movable bridges, off- and intakes, harbour entrances and exits, navigation locks, river and canal crossings, ferries, bypasses, crossing of lakes used for recreation, etc.

As a vessel moves along a waterway a backflow of water occurs, filling the space vacated by the submerged volume. In a restricted space the velocities of the backflow can be considerable. Furthermore, the water level along the vessel is depressed, with the greatest depression occurring near midship. The bank protection required on constricted waterways has to withstand these velocities, as well as the effect of waves generated by the movement of the vessels. Effective bank drainage is essential.

The maximum flow velocity for upstream navigation to be economically viable is about $2.5\,\mathrm{m\,s^{-1}}$. The speed of vessels and tows is limited by economic aspects depending on the type of vessel and the parameters of the waterway. It is usually up to 15–20 km/h on a deep and wide river and 11–15 km/h on canals and canalized navigable rivers (class IV).

The resistance of ships on restricted waters is influenced by many factors, the most important being speed, flow velocity, shape of bow and stern, length, squat and draught (both at bow and stern), keel clearance, and distance from canal banks. A general expression for the resistance, R, of a towed vessel was given by Kaa (1978), in a simplified form, as

$$R = C_{\mathrm{F}}\frac{1}{2}\rho(v+u)^2A' + \rho gBTz + \frac{1}{2}C_{\mathrm{p}}\rho v^2BT \tag{11.4}$$

where v is the ship's speed, u is the velocity of the return flow (u at the stern), z is the depression of the water level (equal to the squat), at stern or bow, C_{F} is the frictional resistance coefficient, A' is the wetted hull area, B is the width of the ship, T is the draught and C_{p} is a coefficient dependent on speed and draught.

The return flow velocity u and squat z in equation (11.4) can be computed from Bernoulli's equation and continuity:

$$2gz = (v+u)^2 - v^2 \tag{11.5}$$

$$vA = (v+u)(A_{\mathrm{c}} - A_{\mathrm{M}} - \Delta A_{\mathrm{c}}) \tag{11.6}$$

where A_{c} is the canal cross-section and A_{M} the midship sectional area. A good approximation for ΔA_{c} is given by

$$\Delta A_{\mathrm{c}} \simeq B_{\mathrm{c}}z \tag{11.7}$$

where B_{c} is the undisturbed canal width.

The resistance augmentation for push-tows over a single ship of the same dimensions and parameters is only moderate.

On a restricted waterway the resistance, R, increases as the ratio, n, of the canal cross-sectional area A_{c} and the immersed section of the barge(s) A (total immersed section of a train) decreases approximately in the ratio of $(n/(n-1))^2$ (i.e. assuming a variation of R with v^2). It follows that for values of n less than 4–5 the increase in the resistance becomes prohibitive (decreasing n from 6 to 5 increases R by 8%, 5 to 4 by 14% and 4 to 3 by 26%) and usually a value of n greater than 4 is used; (this limit is, of course, also speed dependent). Figure 11.6 shows the approximate variation of the resistance R for a 1350 t barge as a function of v, for three values of n and T. The curves have been computed from the simple equation by Gebers (Čábelka, 1976):

$$R = (\lambda A' + kBT)v^{2.25} \tag{11.8}$$

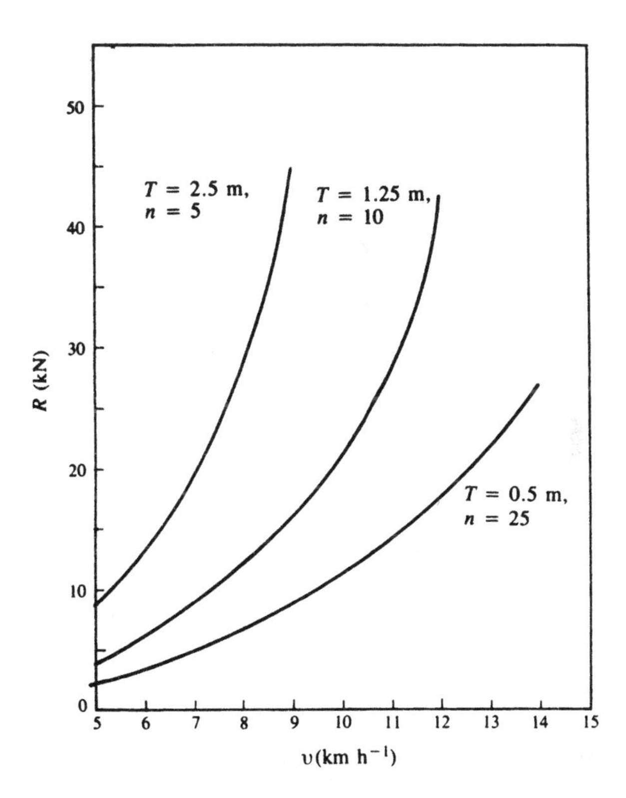

Fig. 11.6 Resistance as a function of velocity

(with R in $\mathrm{kN} \times 10^{-2}$ and v in $\mathrm{m\,s^{-1}}$), where $0.14 < \lambda < 0.28$ ($\lambda = 0.14$ for steel hulls and 0.28 for wooden ones) and $1.7 < k < 3.5$ ($k = 1.7$ for small boats and empty barges, $k = 3.5$ for full barges). The results of equation (11.8) agree quite well with measurements from models.

A ship navigating a bend of radius r encounters an increased resistance due to the centrifugal force acting on the side of the ship passing through the bend at a drift angle α (Section 11.5.2).

11.7 Wave action on banks

The ship-induced water motion causes waves that attack the banks of the waterway, which therefore requires suitable bank protection (Chapter 8). The whole complex interaction between boundary conditions, components of water motion, the forces acting, and the bank revêtment design for stability is shown schematically in Fig. 11.7. It is important (Bowmeester *et*

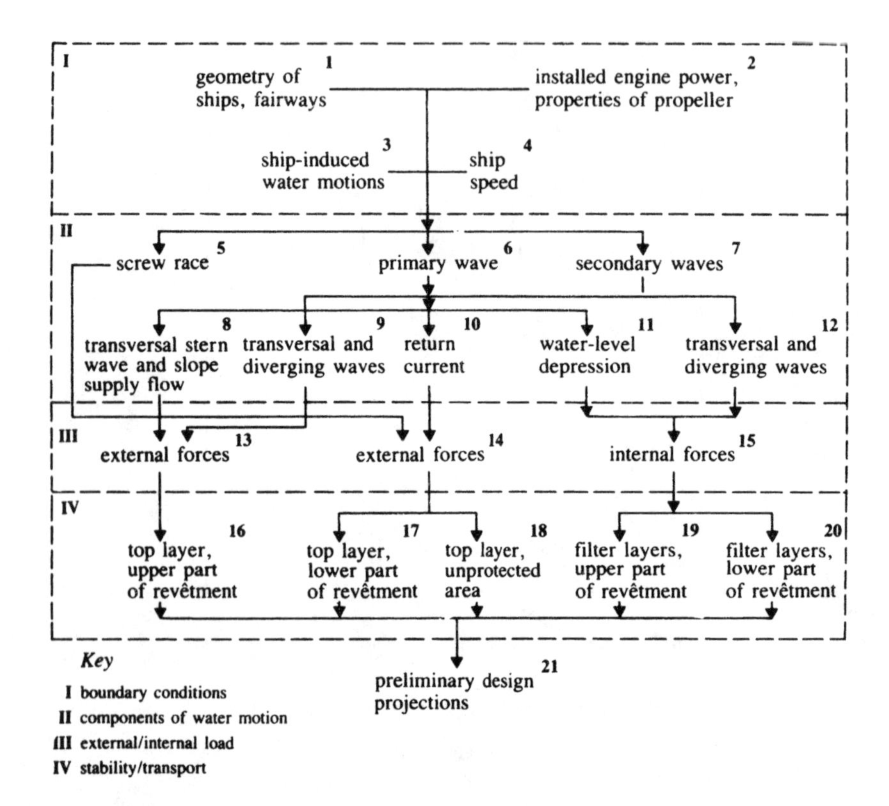

Fig. 11.7 Bank protection of navigable waterways (Hydro Delft, 1985)

al., 1977) to appreciate that the protective facing of banks must reach down below the water surface to a sufficiently low level, and its bottom edge must be flexible enough to ensure permanent contact with the subsoil. A good filter adapted to suit the subsoil is essential, as is drainage of sufficient capacity (Section 8.7.3). For further details see also PIANC (1987).

The height of the waves at the bank generated by navigation in constrained waterways depends primarily on the ship's speed and its relationship to the dynamic wave velocity, size, and form of the vessel, and its position relative to the bank. The effect of waterway cross-section is implicitly contained in the wave velocity. The boat speed is the most important factor, and the generated wave height rises quite steeply with the speed until a maximum is reached, corresponding to the speed at which the sailing motion becomes a gliding one with the bow above the water surface (on a medium-sized waterway fast boats moving downstream at speeds of 20–25 $\mathrm{km\,h^{-1}}$ typically generate waves 0.2–0.3 m high (Novak, 1994)). This speed corresponds to the surge velocity and is rarely attained by commercial shipping or larger pleasure boats. For further details on transients in constricted waterways, see also Blaauw and Verhey (1983), Fuehrer (1985), Kolkman (1978) and Schofield and Martin (1988).

11.8 Locks

11.8.1 General

Concentrated heads on canalized rivers and canals are usually overcome by navigation locks. The main components of locks are the lock gates, the lock chamber, and the lock valves and filling (emptying) systems (Novak, 1994).

Lock gates are of different types: mitre, hinged, sliding, vertical lift, submerged Tainter (horizontal axis), sector (vertical axis), reversed Tainter, etc. Valves in the lock-filling system are usually vertical lift, butterfly or cylindrical valves. Stop logs or vertical lift gates are used as emergency closure gates (for further details of gates and valves, see Chapter 6). Lock chambers must be designed with sufficient stability against surface- and groundwater and earth pressures, and must have sufficient resistance against ship impact.

Clearances of up to 1.0 m are usually allowed on either side of the largest vessel, with the effective length of a lock usually being 1–5 m greater than the longest vessel the lock is intended to accommodate. Because of the difference in elevation of gate sills, the upstream gate is nearly always smaller (lower) than the downstream one. The gate sill elevation controls the draft of tows that can use the lock; usually, 1–2 m is added to the required design depth as a provision for future development of the waterway. The lock sizes (length L, width B and particularly their head H), together with the selected system of lock filling and emptying, determine the design of the lock as well as the type and function of its gates.

In the course of the filling and emptying of the lock, a complicated unsteady flow occurs, not only in the lock itself but also in its approach basins. This flow exerts considerable forces on the barges; these forces must not exceed a permissible limit, and their effect must be eliminated by the tying of the vessels with mooring ropes in the lock or in its approach basin. During emptying of the locks the vessels are usually affected by smaller forces than during filling, because of the greater initial depth of water in the lock.

According to the size and type of filling we can divide locks into four categories:

1. locks with direct filling and emptying through their gates, this method being used mainly for small and medium lock sizes;
2. locks with indirect filling by means of short or long culverts situated either in the lateral lock walls or in its bottom, and connected with the lock chamber by means of suitably designed outlets;

3. locks of large dimensions in plan and high heads with more complex filling and emptying systems, designed to ensure uniform distribution of water during the filling and emptying along the whole lock area;
4. locks with combined direct and indirect filling.

11.8.2 Locks with direct filling

In the case of low- and medium-head locks ($H < 12\,m$), of small to medium dimensions in plan ($B = 12$–$24\,m$, $L = 190$–$230\,m$), the locks have the shape of a prismatic trough with vertical walls and a solid bottom, constructed of anchored steel sheet piles (Fig. 11.8(a)), suitably reinforced concrete precast components, as an *in situ* RC open frame (Fig. 11.8(b)), or concrete walls braced at the lock bottom.

Some older locks, and even some of the modern low-head locks, are fitted with mitre gates with vertical sluices built directly into the gates (Fig. 11.9(a)). On other low-head locks, particularly on the more recently built or renovated ones, upstream gates rotating about a horizontal axis or sliding gates are used (Fig. 11.10). These lock gates serve simultaneously as the direct filling mechanism, which considerably reduces the costs of the

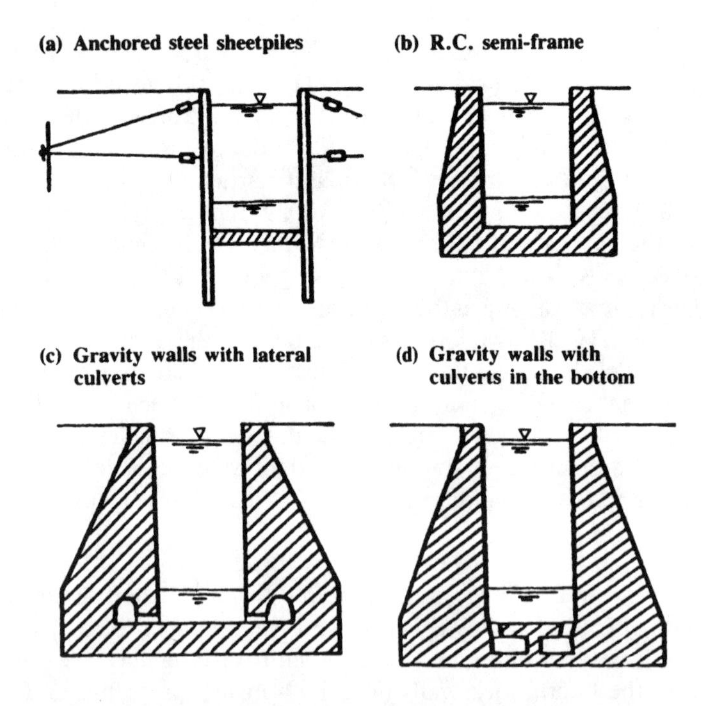

Fig. 11.8 Cross-section of locks

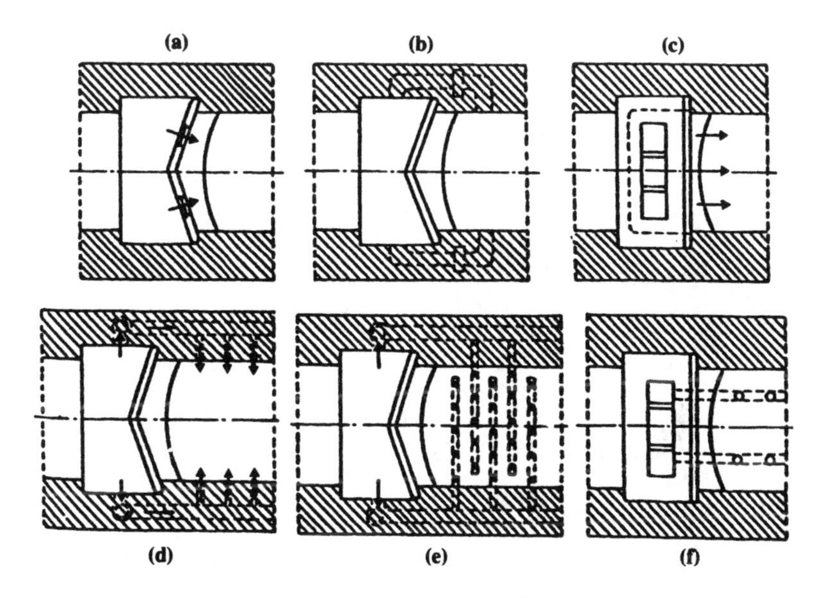

Fig. 11.9 Filling systems of locks (Novak and Čábelka, 1981)

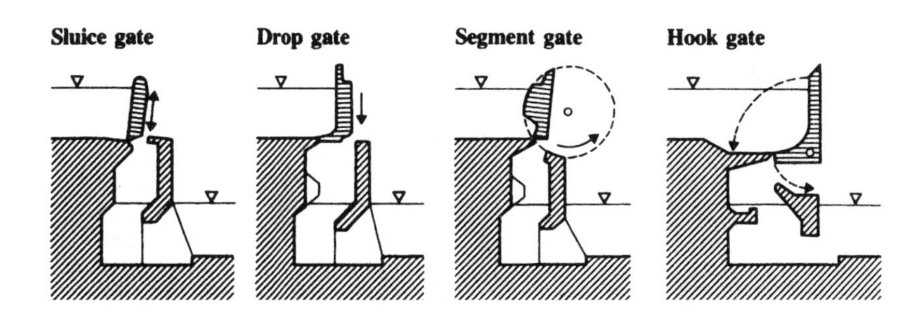

Fig. 11.10 Lock gates for direct lock filling (Čábelka and Gabriel, 1985)

construction as well as the maintenance of the locks; apart from that it accelerates the speedy passage of the vessels through the locks.

A rotating type of gate, according to a design by Čábelka, is shown in greater detail in Fig. 11.11. The lower edge of the upright gate is sufficiently submerged at the beginning of the lock filling below the lowest tailwater level so that the inflowing water is not aerated. For larger heads (up to 12 m) a similar gate design can be used, but it is necessary to separate the substantially deeper stilling basin below the gate from the lock chamber proper by a concrete screen. The installation of this type of gate is considerably eased by floating it into its position; this is made possible by the two horizontal tubes forming an integral part of its structure, which

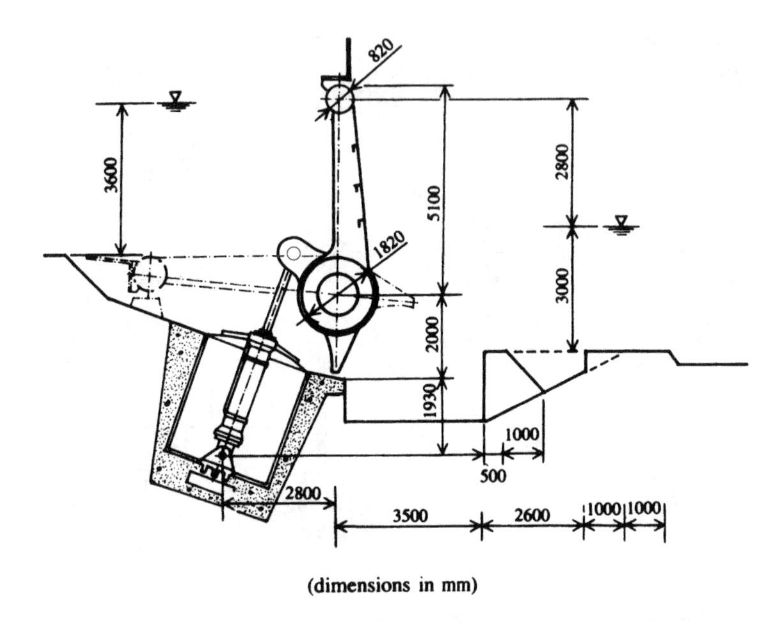

Fig. 11.11 Rotating gate of Čábelka type (after Čábelka and Gabriel, 1985)

are hermetically sealed and act as floats. The motion of the tilting gate is governed by a hydraulic motor, mounted at the bottom of the upper gate recess in a protective casing with sliding cover.

For direct filling of locks 12 m wide, with an initial water depth of 3.5–4 m, an incremental inflow below the gate at the beginning of its filling of $dQ/dt = 0.2\,m^3 s^{-2}$ is permissible without the forces in the mooring ropes of the handled ships exceeding their permissible values. The rate of rise of the water level varies within the limits of 0.8 and $1.2\,m\,min^{-1}$.

Directly filled locks may be emptied by means of short culverts or directly below the downstream lifting gates, or even through openings in the gates, closed with sluice, butterfly or flap valves.

Another type of a direct filling system suitable for low heads uses two sector gates turning along vertical axes, with the flow passage formed by their gradually increasing opening; the gates turn slowly through the initial stages of lock filling and increase their opening speed as the water level difference decreases. For larger heads a filling system utilizing a leaf gate, as shown in Fig. 11.12, can be used. The filling system is designed to prevent air entrainment in the filling passage below the gate, thus contributing to a calmer filling.

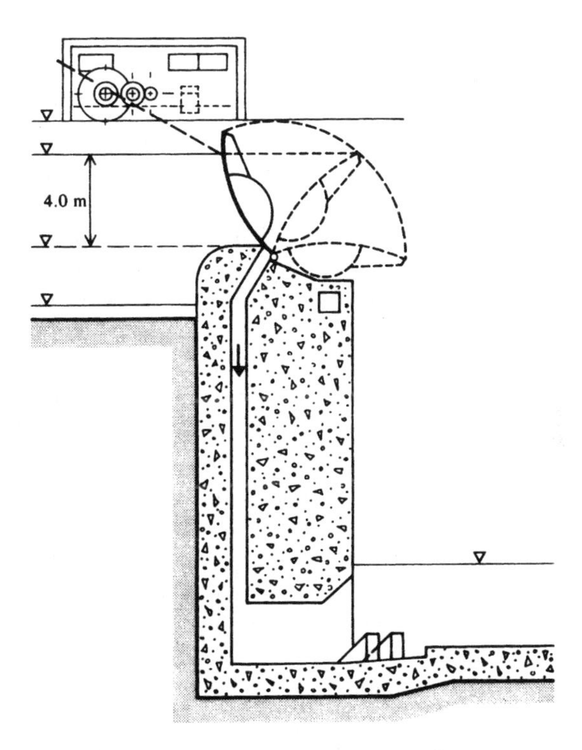

Fig. 11.12 Direct filling of high head locks (after Čábelka, 1976)

11.8.3 Locks with indirect filling and emptying

Indirect filling and emptying of low- and medium-head locks is usually carried out through short culverts in the lateral walls (Fig. 11.9(b)) or through culverts under the gate sill (Fig. 11.9(c)); for high-head locks ($H > 12$ m) or minor dimensions in plan ($B = 12$ m) it is best achieved by means of long culverts, situated either in the lateral walls (Figs 11.8(c), 11.9(d) and 11.9(e)) or in the bottom (Figs 11.8(d) and 11.9(f)). At the upstream and downstream ends of the lock, the culverts are provided with gates (sluice or segment), which must be situated below the lowest possible water level. To reduce outlet losses the overall cross-sectional area of all outlet ports should be 1.3–1.5 times as large as the cross-sectional area of the fully opened culvert gate.

The outlet ports are distributed to ensure, as far as possible, equal outflow along the culvert length, and are usually staggered and positioned and shaped to ensure that the outflow into the lock is directed below the bottom of the vessels in order not to exert lateral forces on them. For the same reason, the outlet ports of long culverts situated in the lock bottom are directed towards its walls (Fig. 11.8(d)). The permissible rate of increase of the inflow of water into the locks filled by a long culvert should not exceed $dQ/dt < 0.6\,m^3\,s^{-2}$ at the beginning of the filling. The mean rate of rise of the water level in these locks varies between $1.5\,m\,min^{-1}$ and $2.0\,m\,min^{-1}$.

For locks of large dimensions in plan and/or very high heads a more complicated filling and emptying system, usually designed on the basis of model studies (Novak and Čábelka, 1981), is necessary. There are many locks in existence with various designs of filling and emptying systems. An example of such a system used in the lock at Bay Springs on the Tennessee–Tombigbee waterway, USA ($L = 183$ m, $B = 33.50$ m and $H = 26$ m (Petersen, 1986)) is shown in Fig. 11.13.

The maximum head for which single-stage locks can be used is basically determined by the limit for which a rational filling and emptying system – technically feasible and viable from the economic and water resources point of view – can be designed. Surface flow and translation waves, which would generate unacceptable forces in the mooring ropes of the handled vessels, must be eliminated as far as possible. The filling velocity is usually limited by the danger of cavitation in the filling system.

Examples of very high heads used for a single-stage lock are $H = 42.5$ m on the Ust–Kamenogorsk scheme on the Irtysh river, Russia (lock dimensions $100\,m \times 18\,m$, minimum water depth 2.5 m), $H = 35$ m at the Carrapatelo dam on the Duero river, Portugal, and $H = 34.5$ m at the John Day Dam on the Columbia River, USA.

On canalized rivers also used for power-generating purposes, where water requirements for navigation must be minimized, high heads are overcome by means of either coupled locks (e.g. on the Gabčíkovo scheme on the Danube), two-stage locks (e.g. on the Djerdab scheme on the Danube), three-stage locks (e.g. in the Dnieprogress scheme on the Dniepr river) or a whole cascade of locks (e.g. the Three Gorges Project on the Yangtze River uses a cascade of 5 locks next to a 183 m high dam with a total head of about 110 m each lock chamber of dimensions $280 \times 34 \times 5$ (m) and capable of passing a 10000 ton barge fleet) sometimes

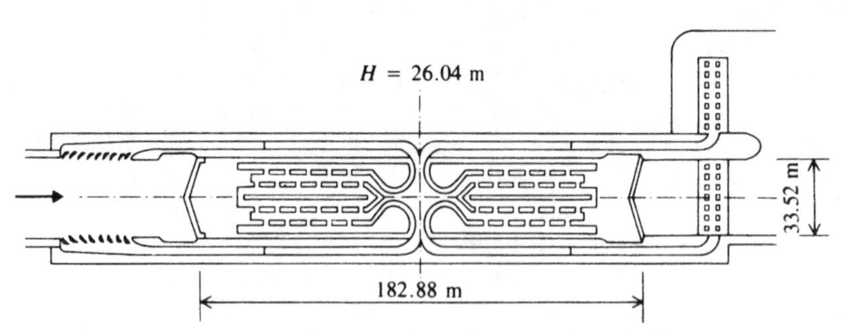

Fig. 11.13 Filling and emptying system of the lock at Bay Springs on the Tennessee–Tombigbee Waterway, USA (after Ables, 1978)

with intermediate reservoirs. The economy of handling water is achieved at the cost of increased capital outlay and a longer time for the passage of vessels through the given step. Another way of saving water are thrift locks (see Section 11.9) and lifts (see Section 11.10).

11.8.4 Hydraulics of locks

The design parameters of major interest are the time of filling (emptying) T, the maximum discharge Q, and the maximum forces acting on a vessel during lockage. If we assume that the lock is filled from a large forebay with a constant water level, we can write

$$Q\,\mathrm{d}t = ca(2gh)^{1/2}\mathrm{d}t = -A\,\mathrm{d}h \tag{11.9a}$$

where A is the lock area in plan, h is the instantaneous head (the difference between the forebay and lock water levels), c is a coefficient and a is the filling system area (valve area); both c and a are functions of time, but c is usually taken as a constant.

For an instantaneous complete opening of the filling system, equation (11.9a) yields, for the time of filling a lock with a total head H,

$$T = \int_0^T \mathrm{d}t = -\frac{A}{ca(2g)^{1/2}}\int_H^0 \frac{\mathrm{d}h}{h^{1/2}} = \frac{2A(H)^{1/2}}{ca(2g)^{1/2}}. \tag{11.10}$$

For a linear opening of the system in time $T_1(a = a_1T_1/t)$,

$$\int_0^{T_1} \mathrm{d}t = -\frac{A}{c(2g)^{1/2}}\int_H^{h_{T_1}} \frac{\mathrm{d}h}{a_t h^{1/2}} = -\frac{AT_1}{cat(2g)^{1/2}}\int_H^{h_{T_1}} \frac{\mathrm{d}h}{h^{1/2}}. \tag{11.11}$$

Thus

$$\int_0^{T_1} t\,\mathrm{d}t = \frac{2AT_1}{ca(2g)^{1/2}}(H^{1/2} - h_{T_1}^{1/2}) \tag{11.12}$$

and

$$T_1 = \frac{4A(H^{1/2} - h_{T_1}{}^{1/2})}{ca(2g)^{1/2}}. \tag{11.13}$$

The total filling time from equations (11.10) and (11.13) is then

$$T = T_1 + \frac{2Ah_{T1}{}^{1/2}}{ca(2g)^{1/2}} = \frac{T_1}{2} + \frac{2AH^{1/2}}{ca(2g)^{1/2}} \tag{11.14}$$

(generally, the opening of the filling system is non-linear). In the same way we could derive the equation for the time to equalize the water levels between two locks of areas A_1 and A_2. For an instantaneous full opening of the filling system,

$$T = \frac{2A_1A_2H^{1/2}}{(A_1 + A_2)ca(2g)^{1/2}}. \tag{11.15}$$

If $A_1 = A_2$,

$$T = \frac{AH^{1/2}}{ca(2g)^{1/2}} \tag{11.16}$$

and for $A_1 = \infty$ we again obtain equation (11.10).

The time of filling of a lock by an overfall height h_1 over a gate width B is given approximately by

$$T = \frac{AH}{2/3C_dB(2g)^{1/2}h_1^{3/2}}. \tag{11.17}$$

(Equation (11.17) neglects the change from a modular to non-modular overflow at the end of the filling.)

If the opening of the filling system is gradual but not linear, we have to compute the filling time by a step method, e.g. from equation (11.10) it follows that

$$\Delta t = \frac{2A}{ca(2g)^{1/2}}(h_{i-1}^{1/2} - h_i^{1/2}); \tag{11.18}$$

thus

$$h_i = \left(h_{i-1}^{1/2} - \frac{c(2g)^{1/2}a}{2A}\Delta t\right)^2. \tag{11.19}$$

Equation (11.19) gives also the rate of change of the depth in the lock $(H - h_i)$, and it permits us to compute the change of discharge with time from equation (11.9a):

$$Q_i = ca_i(2gh_i)^{1/2} \tag{11.9b}$$

(in equations (11.18) and (11.19), a is a function of time).

Of particular interest, of course, is the maximum discharge Q_{max}. In the special case of a linear opening of the filling (emptying) system we can determine Q_{max} and the head at which it occurs analytically from the two equations

$$t^2 = \frac{4AT_1}{ca(2g)^{1/2}}(H^{1/2} - h_t^{1/2})$$

and

$$Q = \frac{cat(2gh_t)^{1/2}}{T_1}.$$

For Q_{max}, $dQ/dh_t = 0$, giving

$$h_t = \frac{4}{9}H. \tag{11.20}$$

The maximum discharge occurs at $4/9H$ if the filling system is not fully open yet by the time this level is reached, i.e. the criterion is the value of h_{T_1} computed from equation (11.13). If $h_{T_1} < 4/9H$ the maximum discharge occurs at $4/9H$; if $h_{T_1} > 4/9H$ then the maximum discharge occurs at the head of h_{T_1} corresponding to the end of opening of the filling (emptying) system. The real time of filling can be up to 12% shorter than the computed one owing to inertia effects in the filling system.

The coefficient of discharge usually varies between 0.6 and 0.9 and is a function of the geometry of the system. Although also a function of time, an average value of c, best determined by field or model experiments, is usually used in the computation. Typical shapes of the Q, a and h values as functions of time (Novak, 1994) are shown in Fig. 11.14.

During lockage the vessel is tied by hawsers to bollards, with the ropes at an angle between 20° and 40° to the longitudinal axis of the lock (vessel). Because of inertia, during the small movements of the vessel the force in the ropes, R, is about 35% larger than the force, P, acting on the vessel. The resultant tension in the rope is then

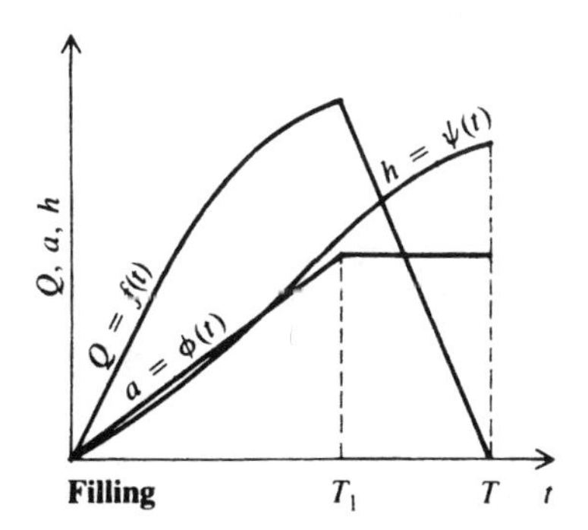

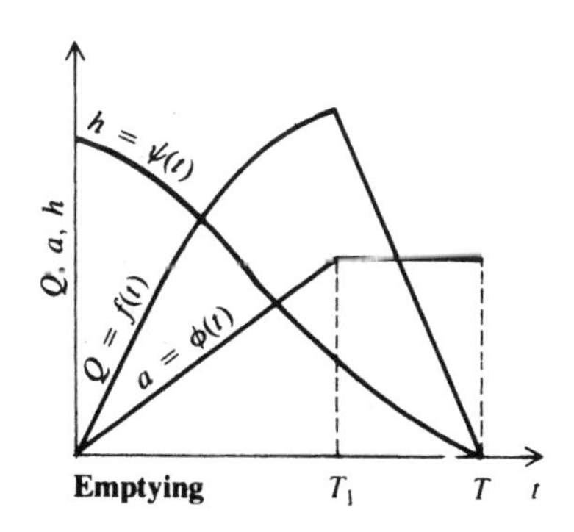

Fig. 11.14 Variation of *Q*, *a* and *h* with time

$$R = 1.35P/\cos 40° = 1.75P. \tag{11.21}$$

For safety, the permissible tension is limited by

$$R_{per} = D/600 \tag{11.22}$$

where D is the displacement of the vessel (in tons). In Eastern Europe the relationship

$$R_{per} = D^{3/5}/20 \tag{11.23}$$

is used (Čábelka, 1976). The force P has three main components: P_1, the resistance of the vessel; P_2, the force due to the longitudinal slope of water surface in the lock; P_3, the force due to translatory wave action caused mainly by 'sudden' changes of inflow into (outflow from) the lock. P_1 and P_2 act against each other and, to a great extent, cancel each other out (for further details, see also Jong and Vrijer (1981)).

The value of P_3, which is about 80% of P, is mainly influenced by the value of dQ/dt. Thus the force acting on the vessel during lockage can be limited by imposing a permissible limit on dQ/dt during the filling (emptying) operation (Sections 11.8.2 and 11.8.3). This is best achieved by controlling the rate of opening of the valves of the filling system and/or by the shape of the culverts at the valves. The limiting value of the transverse component of the horizontal forces acting on the vessel should not exceed 50% of the longitudinal component P.

11.9 Thrift locks

On canals where there is water shortage, high heads may be overcome by lifts (Section 11.10) or by locks with thrift basins; the latter have the advantage that they permit simultaneous handling of large tows but require greater land use than lifts. In thrift locks, considerable reduction of water consumption is achieved by conveying, by gravity, part of the water during lock emptying to thrift basins, to be returned again by gravity to the lock during its subsequent filling (Fig. 11.15).

Thrift basins are usually constructed next to the lateral lock wall, either as open or as superimposed closed reservoirs. Every basin is connected with the lock by its own conduit, provided with two-way gates. By increasing the number of basins to more than four, only a very small additional reduction of water consumption can be achieved, and the handling time increases disproportionately in comparison with simple locks. Locks with thrift basins are used for heads of up to 30 m. For these heads they can still be designed so that their efficiency is comparable with standard types of boat lifts.

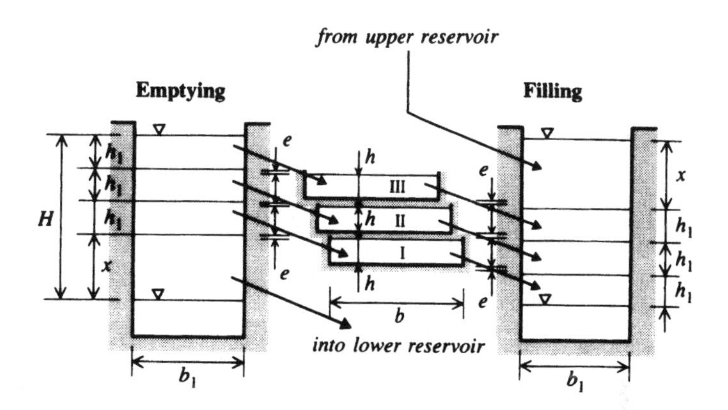

Fig. 11.15 Thrift basins

If b and b_1 are the thrift basin and lock widths, h and h_1 are the basin and lock depth increments, n is the number of thrift basins, e is the excess head allowed during the operation (it would take too long to wait for complete level equalization) and x is the residual depth to be filled from (during filling) or discharged to (during emptying) the canal, then, assuming the lock and basin to have equal lengths:

$$b_1 h_1 = bh \quad \text{or} \quad K = b_1/b = h/h_1 (<1),$$

$$x = e + h + e + h_1 = 2e + h_1 + h = 2e + (K+1)h_1. \tag{11.24}$$

For a total lift, H, and the lift provided by the basins, nh_1,

$$H = x + nh_1,$$

$$h_1 = (H - x)/n. \tag{11.25}$$

Thus, from equations (11.24) and (11.25), the head loss

$$x = \frac{2en + (K+1)H}{n + K + 1} \tag{11.26}$$

or the relative head loss

$$\frac{x}{H} = h_r = \frac{K + 1 + 2en/H}{n + K + 1} \simeq \frac{K+1}{n+K+1}. \tag{11.27}$$

The efficiency, η, of the thrift lock is

$$\eta = 1 - h_r = \frac{n - 2en/H}{K + n + 1} \simeq \frac{n}{K+n+1}. \tag{11.28}$$

η and h_r are not strongly dependent on H as, usually, $2en \ll H$. The value of K is usually 0.5–0.7. Little is gained by increasing n beyond 4 or 5, or by decreasing K. A larger number of narrow basins is often cheaper than a smaller number of broader ones.

11.10 Lifts and inclined planes

If the provision of water for the operation of high-head locks causes major problems, it is possible to use boat lifts, for the operation of which the water requirements are almost zero. To overcome very great heads (as much as 100m) only boat lifts are really feasible. As a rule, boat lifts consist of horizontal water-filled troughs provided at both ends with gates. The troughs of the boat lifts have a maximum length of about 100m. Therefore, they are suitable for the handling of barges and short (1 + 1) push trains; major push trains have to be disconnected. However, because of the great travelling speed of the trough, the capacity of the boat lifts is relatively high.

According to the direction of motion of the trough, lifts can be either vertical or inclined. For balancing and moving the boat trough filled with water vertical lifts use pistons, floats (Fig. 11.16(a)), counterweight balances (Fig. 11.16(b)), or other special mechanisms.

Inclined boat lifts usually have the boat trough mounted on a special undercarriage which travels on a track on an inclined plane, either in the direction of the longitudinal trough axis or normal to it (Fig. 11.17). As a rule, the trough is counterbalanced by a suspended weight travelling on an inclined track below the undercarriage of the boat trough. The acceleration of the trough during starting and the deceleration during braking must be small enough to maintain the variations of the water level in the trough within permissible limits, in order to reduce the forces in the mooring ropes of the boat to an acceptable magnitude. To reduce these forces, part of the water is sometimes let out of the trough before its lifting to settle the boat at the bottom of the trough and thus stabilize it. Boat lifts are more sensitive in operation than locks and more prone to damage.

A special type of inclined boat lift is a design developed by J. Aubert; it consists of an inclined trough with a mobile water-retaining wall (*pente d'eau*), forming a water wedge on which the boat floats. The 'wall' is moved by two coupled electric locomotives with the floating boat moored to them. Difficulties in operation can be caused by the circumferential sealing of the travelling water-retaining wall. This type of boat lift has been built at Montech on the Garonne River (Aubert, Chaussin and Cancelloni, 1973).

The first *rotating boat lift* – the 'Falkirk wheel' – was built in 2002 in Scotland replacing a flight of 19th century locks connecting the Forth and Clyde (see Section 11.1) and the Union canals. Although the concept of a rotating lift has been proposed before (in 1902 a rotating drum with two internal rotating drum-troughs was proposed for the Danube–Oder canal

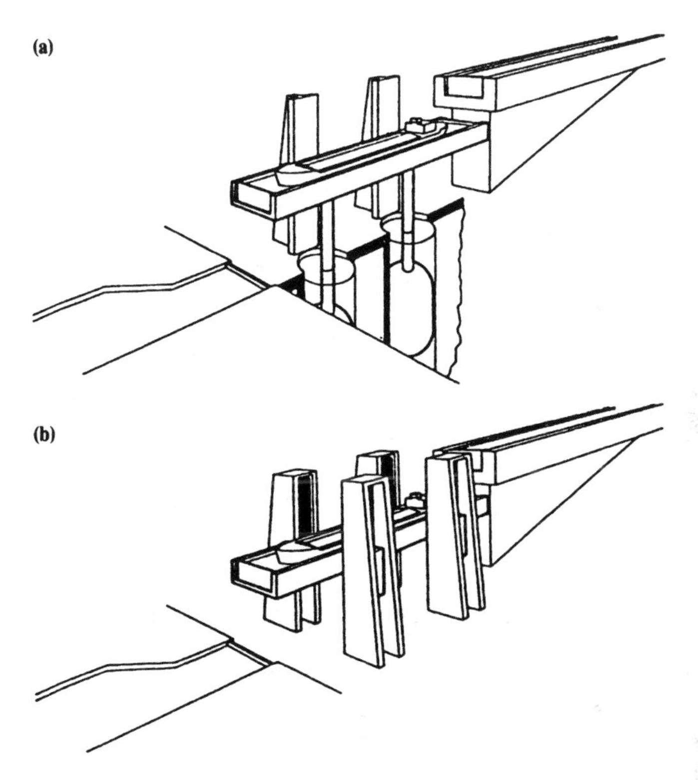

Fig. 11.16 Vertical boat lifts with (a) floats or (b) counterweights (Čábelka and Gabriel, 1985)

(never built)) the 35 m high 'Falkirk wheel' originally proposed as a Ferris type wheel has some unique features. The lift consists of two 30 m long gondolas, which sit within two propeller shaped arms, each capable of transporting four average sized canal boats. The arms rotating 180 degrees as a wheel around a central axle are driven by hydraulic motors, which also, by a set of gears, turn the two gondolas so that they remain horizontal. The whole structure was assembled on site and parts bolted (rather than welded) to deal with fatigue stresses.

Examples of exceptionally high and large vertical boat lifts are the lift at Strepy-Thieu on the Canal de Centre in Belgium and the proposed lift at the Three Gorges Project in China. The former, replacing four older lifts of about 17 m height each and capacity 1350 tons overcomes a rise of about 73 m and has two (balanced) troughs each 112 m long, 12 m wide and 8 m high weighing 2200 tons. The 113 m high ship lift at the Three Gorges Project is designed with a container 120 m long, 18 m wide, with water depth 3.5 m, capable of carrying 3000 ton passenger or cargo boats. An example of a large high inclined lift is the Krasnojarsk (Russia) 101 m high lift, capacity 1500 tons with a 90 m long and 18 m wide container.

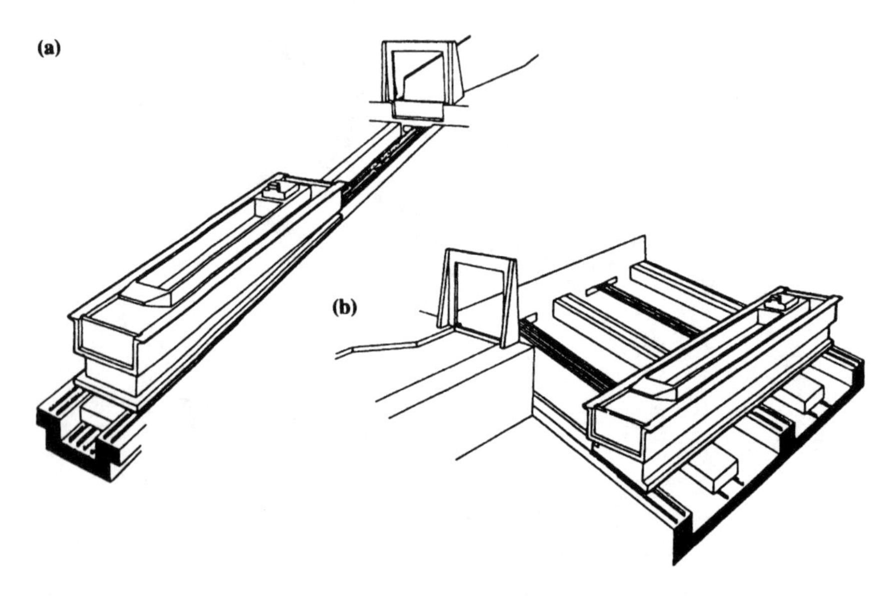

Fig. 11.17 Inclined boat lifts (Čábelka and Gabriel, 1985)

11.11 Lock approaches

Lock approaches provide the transition between the navigable river and the lock and must be designed both to ensure a safe and speedy entry into the approach basin and the lock and to permit the mooring of boats waiting to enter the lock while this is operating to pass other vessels down or upstream. The approach basin width will thus depend on the above factors as well as on the number of locks (single, twin, etc.) and the likely number of push trains waiting for handling.

On a waterway with flowing water the approach basin is divided into three parts (Fig. 11.18). The first part l_a is intended for the braking of the vessels entering the lock or for their accelerating at departure. The next part l_b is intended for passing and overtaking of vessels or push trains, and possibly for their mooring. The third part l_c with jetties or guide walls (at inclinations of 1 in 4 or 1 in 5) represents a transition between the wider approach basin and the narrower lock head.

The approach basins of locks on canalized rivers are usually separated from the power plant or weir by a long dividing wall or an island. A sudden change of the width and cross-section here results in critical flow regions with lateral contraction, and transverse or even reverse flow; these conditions are very unfavourable for navigation and may cause accidents. To reduce the transverse velocity below the maximum permissible value of about $0.35\,\mathrm{m\,s^{-1}}$ it is advisable to provide a water passage in the dividing wall near its head (Fig. 11.18).

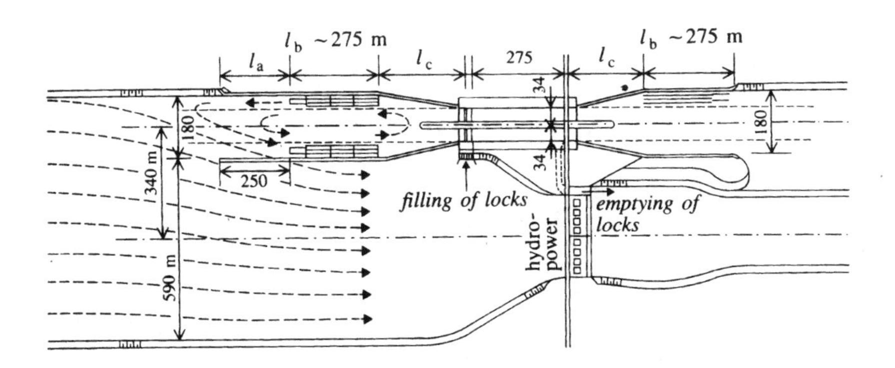

Fig. 11.18 Lock with approaches on the Danube (after Čábelka and Gabriel, 1985); dimensions in metres

The approach basins of navigation locks on still-water canals may be symmetrical or asymmetrical in plan and are usually relatively short, since no braking length is required, as vessels can reduce their speed before entering the approach basin. The same applies to the downstream approach basins of locks on waterways with flowing water, since the vessels enter them against the flow direction.

However, downstream lock approaches have to be protected against undesirable currents arising from spillway and/or power-plant discharges, and from the lock emptying system.

When designing fender structures, dolphins, jetties, etc., it is necessary to account for the forces (and displacements) which are likely to occur owing to the impact of ships during mooring or collisions. The resultant force will depend on the velocity of the ship and its angle with the fender structure as well as, of course, on the ship displacement. Generally, loaded push-tows have smaller velocities and angles of collision than unloaded ones. The theoretical computation is fairly complicated, but measurements in prototype indicate that the collision can be schematized as a linear damped mass–spring system. The contribution of the dolphin and ground to damping is considerable (Vrijer, 1983).

11.12 Inland ports

Inland ports serving the loading and unloading of vessels, transfer of goods, and their further handling are connected with their hinterland by water, highway and/or pipeline transport routes. The ports can have specialized zones or basins intended for the transfer of certain cargoes (ore, coal, aggregates, sand, individual shipments, containers, etc. (Porteous, 1977)).

The extent, location, and layout of an inland port are determined by its transfer capacity. For a small capacity a port can be built directly on the bank of a navigable river or canal by widening it by at least two or three standard boat widths, or by the width required for barge turning. For a medium transfer capacity it is more advantageous to build one or two port basins outside the waterway, connected with the fairway by a suitably designed entry.

A large transfer capacity port should be as compact as possible; it has several basins connected with the waterway by means of an approach canal, extending beyond the entry into a port approach basin intended for the formation of push trains or for vessels waiting for unloading. A turning basin is usually situated in the proximity of the port approach basin. The port layout depends above all on local conditions and the purpose which the port should serve.

Manual or semimechanical transfer of freight in ports has been almost fully replaced by discontinuous or continuous mechanical handling (conveyor belts, pneumatic conveyors and pumps), particularly for the conveyance of liquid substances. Automated continuous transfer suitable for large quantities of freight, above all for bulk and liquid cargo, is the most productive form.

Gantry cranes have their rail tracks laid along the waterfront, as close to the water as possible, to ensure that at least two barges are within reach of their jibs. In the interests of efficiency the cranes should not travel over excessively long distances but they must have a large action radius. Some ports have large-capacity stationary gantry cranes with travelling trolleys, intended for the transfer of very heavy and bulky goods. Concrete assembly surfaces are provided in their vicinity, intended for the assembly of large-size products which could not be transported to the port on the highway or railway owing to their large dimensions. For products of excessive weight and dimensions, which cannot be handled otherwise, the roll-on–roll-off transfer system has to be used.

For goods sensitive to moisture roofed berths provided with overhead travelling cranes are used. Additionally, ports are provided with modern storage capacities for packaged goods, dumps for temporary bulk storage, and grain silos. Separated from the main port area are large-capacity tanks for the storage of inflammable substances, situated in the proximity of tanker berths.

The ever-increasing intensity of utilization of inland waterways and the mechanization of transfer operations necessitates automated control of ports and transfer operations. This is particularly so in the rapidly developing container transport system. Automatic container terminal control systems are based on a suitable combination of computer data processing and remote control of man-operated transfer of goods (Bourrieres and Chamreroy, 1977).

For an overall survey of planning and design of ports see e.g. Agerschou *et al.* (2004).

Worked Example 11.1

A navigation lock, 200 m × 12 m in plan and with a 9.00 m head, is filled through two longitudinal conduits with rectangular gates 3.00 m wide controlling the flow. The overall coefficient of discharge of the filling system is 0.65 and the gates open 8.5 mm s^{-1} at a uniform speed in 4.5 min. Determine the maximum discharge entering the lock and the total time of filling.

Solution

The time of opening of the gates is $T = 4.5 \times 60 = 270$ s. The flow area of a fully open gate is $3 \times 0.0085 \times 270 = 6.885\,\text{m}^2$. The flow area of the filling system is $2 \times 6.885 = 13.77\,\text{m}^2$. From equation (11.14), for a linear opening of the filling system the total time of filling is

$$T = \frac{T_1}{2} + \frac{2AH^{1/2}}{ca(2g)^{1/2}} = \frac{270}{2} + \frac{2 \times 200 \times 12\sqrt{9}}{0.65 \times 13.77\sqrt{19.62}} = 496\,\text{s}.$$

The head on the lock at the end of the opening of the gates, h_{T_1}, is from equation (11.13)

$$270 = \frac{4A(H^{1/2} - h_{T_1}^{1/2})}{ca(2g)^{1/2}} = \frac{4 \times 200 \times 12(3 - h_{T_1}^{1/2})}{0.65 \times 13.77\sqrt{19.62}}.$$

Thus $h_{T_1} = 3.55$ m. Q_{max} occurs at either h_{T_1} or $4/9H$, whichever is greater; in this case $4/9H = 4/9 \times 9 = 4\,\text{m} > 3.55\,\text{m}$.

The time at which $h = 4$ m can be obtained from the equation

$$\int_0^t t\,dt = -\frac{AT_1}{ca(2g)^{1/2}}\int_H^{h_t}\frac{dh}{h^{1/2}} = \frac{2AT_1}{ca(2g)^{1/2}}(H^{1/2} - h_t^{1/2}) = t^2/2;$$

for $h_t = 4$ m

$$t^2 = \frac{4 \times 200 \times 12 \times 270}{0.65 \times 13.77\sqrt{19.62}}(\sqrt{9} - \sqrt{4}) = 65\,370\,\text{s}^2.$$

Thus $t = 256$ s (<270 s) and the maximum discharge occurs before the filling system is fully open. At time $t = 256$ s, the area of opening of the system is $(13.77 \times 256)/270 = 13.056\,\text{m}^2$. Therefore (from equation (11.9b)) $Q_{max} = 0.65 \times 13.056 \times (4 \times 19.6)^{1/2} = 75.19\,\text{m}^3\,\text{s}^{-1}$.

Worked Example 11.2

A thrift lock has four basins of the same length (225 m) as the lock, and each of width 20 m. If the width of the lock chamber is 12 m, its total lift is 24 m, and the residual head on each thrift basin is 0.20 m, determine the efficiency of the thrift lock and the daily saving in water consumption if there are 18 lockings per day.

Solution

From equation (11.26) with $K = 12/20 = 0.6$,

$$x = \frac{2en + (K+1)H}{n+K+1} = \frac{2 \times 0.2 \times 4 + 1.6 \times 24}{4 + 0.6 + 1} = 7.143\,\text{m}.$$

Normal water consumption for one lockage is $12 \times 24 \times 225\,\text{m}^3$, and thrift lock consumption for one lockage is $12 \times 7.143 \times 225\,\text{m}^3$. The efficiency of the thrift lock is $(24 - 7.43)/24 = 70\%$. The saving of water per day is $18 \times 225 \times 12(24 - 7.143) = 819\,250\,\text{m}^3$.

References

Ables, J.H., Jr (1978) *Filling and Emptying System for Bay Springs Lock, Tennessee–Tombigbee Waterway, Mississippi*, Technical Report h-78-19, US Army Corps of Engineers Waterways Experimental Station.

Agerschou, H. *et al.* (2004) *Planning and Design of Ports and Marine Terminals*, 2nd edn., Thomas Telford Publishing, London.

Aubert, J., Chaussin, P. and Cancelloni, M. (1973) La pente d'eau à Montech, in *Navigation, Ports et Industries*, Paris: 291–6.

Blaauw, H.G. and Verhey, H.J. (1983) Design of inland navigation fairways. *American Society of Civil Engineers Journal of Waterway, Port, Coastal, and Ocean Engineering*, 109 (1): 18–30.

Bourrieres, P. and Chameroy, J. (1977) *Ports et Navigation Modernes*, Eyrolles, Paris.

Bouwmeester, J., Kaa, E.J. van de, Nuhoff, H.A. and van Orden, R.G.J. (1977) Recent studies on push towing as a base for dimensioning waterways, in *Proceedings of the 24th International Navigation Congress*, PIANC, Leningrad, Paper SI-C3, 34 pp.

Čábelka, J. (1976) *Inland Waterways and Inland Navigation*, SNTL, Prague (in Czech).

Čábelka, J. and Gabriel, P. (1985) Inland waterways, in *Developments in Hydraulic Engineering*, Vol. 3 (ed. P. Novak), Elsevier Applied Science, London.

ECE (1992) Document TRANS/SC3/R No. 153, Brussels.

Edwards, L.A. (1972) *Inland Waterways of Great Britain*, Imray, Laurie, Norie and Wilson, London.

Fuehrer, M. (1985) Wechselbeziehungen zwischen Schiff und beschraenktem Fahrwasser. *Mitteilungen der Forschungsanstalt für Schiffahrt, Wasser und Grundbau* (49), 125 pp.

Hilling, D. (1977) *Barge Carrier Systems – Inventory and Prospects*, Benn, London.

—— (1999) Inland shipping and the maritime link, *Proceedings of Institution of Civil Engineers, Water Maritime and Energy*, 136, No. 4: 193–7.

Hydro Delft (1985) Report No. 71, Delft Hydraulics Laboratory (special issue on navigation in restricted waterways).

Jong, R.J. de and Vrijer, A. (1981) Mathematical and hydraulic model investigation of longitudinal forces on ships in locks with door filling systems, in *Proceedings of the 19th IAHR Congress*, New Delhi.

Kaa, E.J. van de (1978) Power and speed of push-tows in canals, in *Proceedings of the Symposium on Aspects of Navigability of Constraint Waterways Including Harbour Entrances*, Delft Hydraulic Laboratory, 16 pp.

Kolkman, P.A. (1978) Ships meeting and generating currents (general lecture), in *Proceedings of the Symposium on Aspects of Navigability of Constraint Waterways Including Harbour Entrances*, Delft Hydraulic Laboratory, 26 pp.

Kubec, J. (1981) Improvement of the integration of ocean and inland navigation by means of a unified system of dimensions of barges and lighters, in *Proceedings of the 25th International Navigation Conference*, PIANC, Edinburgh, Section S1, Vol. 4: 577–85.

Kuhn, R. (1985) *Binnenverkehrswasserbau*, Ernst, Berlin.

Novak, P. (1994) Inland waterways, in *Kempe's Engineering Yearbook* (ed. C. Sharpe), Benn, Tonbridge, Chapter L2.

Novak, P. and Čábelka, J. (1981) *Models in Hydraulic Engineering: Physical Principles and Design Application*, Pitman, London.

Petersen, M.S. (1986) *River Engineering*, Prentice-Hall, Englewood Cliffs, NJ.

PIANC (1987) *Guidelines for the Design and Construction of Flexible Revêtments Incorporating Geotextiles for Inland Waterways*, Report of Working Group 4 of the Permanent Technical Committee 1, Permanent International Association of Navigation Congresses, Brussels.

—— (1996) *Standardisation of Ships and Inland Waterways for River/Sea Navigation*, Permanent International Association of Navigation Congresses, Brussels.

Porteous, J.D. (1977) *Canal Ports: the Urban Achievement of the Canal Age*, Academic Press, London.

Savenije, R.Ph.A.C. (2000) (Rijkswaterstaat Adviesdienst), Personal communication.

Schofield, R.B. and Martin, C.A. (1988) Movement of ships in restricted navigation channels. *Proceedings of the Institution of Civil Engineers*, 85 (March): 105–20.

Seiler, E. (1972) Die Schubschiffahrt als Integrationsfaktor zwischen Rhein und Donau. *Zeitschrift für Binnenschiffahrt und Wasserstrassen* (8).

Vrijer, A. (1983) Fender forces caused by ship impacts, in *Proceedings of the 8th International Harbour Congress*, Antwerp.

US Army Corps of Engineers (1980) *Layout and Design of Shallow Draft Waterways*, Engineering 1110-2-1611.

Chapter 12

Hydroelectric power development

12.1 Introduction

Hydropower is extracted from the natural potential of usable water resources, and about 20% of the world's power requirement is at present derived in this way. Any water resources development of which a hydropower scheme may form part has environmental and social impacts, which must be taken into consideration at the initial planning stage. Also, legal and political implications must be carefully considered.

One of the most important factors affecting any hydropower development is the cost of the scheme. With the rising costs and shortage of resources, economic comparisons with various energy sources – oil, coal, nuclear, gas and other renewable energy resources have to be made. In this, the fact that, unlike the 'fuel' costs of the hydropower plant, the costs of fuel for power generation by a conventional thermal plant rise at least with inflation makes hydroelectric plants economically advantageous, particularly in the long term.

The latest technological advancements in hydroelectric power generation permit the selection of proper designs incorporating environmental and social requirements. Some innovations in the development of special turbine–generator units, such as bulb and slant-axis units, also suggest considerable reductions in construction costs, thus augmenting the benefits-costing of a project. Therefore, the selection of the final design requires the comparison of many alternative proposals from several sites, which may incorporate expansion facilities to meet future demands. Land use is one example of environmental considerations in hydropower development. Over 100 kW/km^2 is exceptionally good, whereas 1 kW/km^2 is not efficient land use (Vladut, 1977).

12.2 Worldwide hydroelectric power development in perspective

The total installed hydropower capacity worldwide is about 740 GW and the energy produced annually is about 2770 TWh i.e. approximately 19% of the world's electricity supply (Aqua-Media International, 2004).

The gross theoretical annual potential worldwide for primary hydropower (i.e. excluding pumped storage schemes) is over 40 000 TWh of which about 14 000 TWh is considered to be technically possible and about 8000 TWh economically feasible. (Bartle and Hallowes, 2005). The above figures hide, of course, the large variability of the potential and installed hydropower across continents and countries due to morphology, climate and economic development. Thus, e.g. in Norway over 99% of the total power supply is from hydropower whereas in the UK it is only about 3%.

Some very large hydropower schemes are under construction or have been recently completed. The Three Gorges project when completed (in 2009) will have 26 turbine generator units of 700 MW each, totalling 18 200 MW installed capacity producing 84.7 TWh annually; there is also provision on site for a future expansion by 6 further units. The Ghazi Barotha project in Pakistan, completed in 2004 (which features a 52 km long lined channel), has a generating capacity 1450 MW with an average output of 6600 GWh. However, in many countries the major emphasis in new hydropower development is on pumped storage schemes (see Section 12.5) and particularly on small hydro plants (see Section 12.11).

12.3 Power supply and demand

Demand for electrical power varies from hour to hour during the day, from day to day, and from year to year; the power demand is defined as the total load which consumers choose, at any instant, to connect to the supplying power system. The system should have enough capacity to meet the expected demand, in addition to unexpected breakdowns and maintenance shutdowns. A daily load (demand) curve for a typical domestic area is shown in Fig. 12.1.

Base load is the load continuously exceeded, whereas the average load is the area under the curve divided by the time. The load factor over a certain period is the ratio of the average load to the peak load and is expressed as a daily, weekly, monthly or yearly value. A single station connected to an industrial plant may have a load factor of, for example 80%. In a country where the supply is distributed through a national grid system for a diversity of uses, the annual load factor may be of the order of 40%.

It must be appreciated that low load factors represent a degree of inefficiency, as sufficient capacity in the form of generating machinery

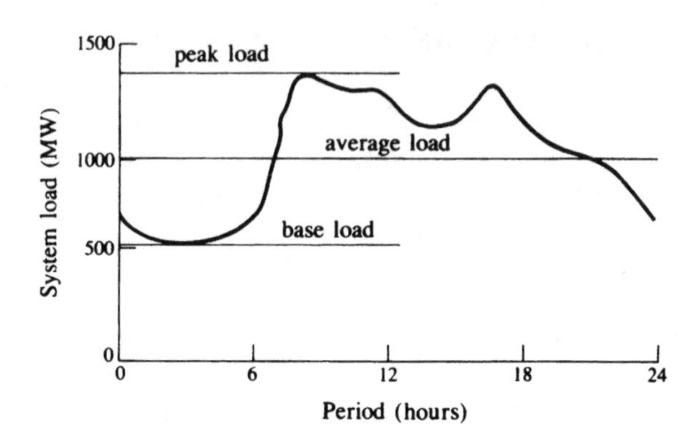

Fig. 12.1 A typical demand curve

must be installed to meet the peak demand while, on average, a considerable part of this machinery is standing idle.

The choice of type of power system depends on the kind of fuel available, its costs, availability of suitable sites, etc. While the fuel costs of a hydroelectric plant are virtually nil, the construction costs of civil engineering works are usually much greater than those of a thermal plant owing to the additional costs of the impounding structures, penstocks, long transmission lines, etc.

Thermal plants are more efficient to run at full load and hence are suitable for generating continuously near capacity to carry the base load. A hydroelectric plant can be put into operation in a time of a few seconds to 4–5 min, whereas at least 30 min are required to start up and load a thermal system. Hydroelectric plants are therefore very suitable for meeting variations in load with little waste of power. In an ideal interconnected system the thermal stations would be used to generate up to the maximum base load, with the hydrostation providing peak power.

Nuclear power plants are equally unsuitable for variable load operation, as the reactors cannot be easily controlled to respond quickly to load changes, and hence are used as baseload plants at a load factor of at least 80%. Because of lack of bulk, nuclear fuel transport costs are negligible and hence nuclear power plants are advantageous in locations where conventional fossil fuels and hydropower are unavailable. Although nuclear power plants, unlike fossil-fuel thermal plants, do not need expensive air-pollution control systems, safety and waste disposal remain a serious consideration.

12.4 Some fundamental definitions

The gross head, H_0, at a hydroelectric plant is the difference in water level between the reservoir behind the dam and the water level in the tail race.

Because of the variable inflow and operating conditions of the plant these levels vary. The effective or net head, H, is the head available for energy production after the deduction of losses in the conveying system of the plant (Figs 12.8 and 12.15).

The water falling from a high-level source drives turbines, which in turn drive generators that produce the electricity. The hydraulic power is given by

$$P = \eta \rho g Q H / 1000 \text{ (kW)} \tag{12.1}$$

where η is the turbine efficiency and Q is the flow rate (in $m^3 s^{-1}$) under a head of H (m). The hydraulic efficiency of the plant is the ratio of net head to gross head (i.e. H/H_0), and the overall efficiency is equal to the hydraulic efficiency times the efficiency of turbine and generator. The installed capacity of a hydroplant is the maximum power which can be developed by the generators at normal head with full flow. The unit of electrical power is the kilowatt, and that of the electrical energy, defined as the power delivered per unit time, is the kilowatt-hour (kW h).

Primary, or 'firm', power is the power which is always available, and which corresponds to the minimum streamflow without consideration of storage. Secondary, or surplus, power is the remainder and is not available all the time. Secondary power is useful only if it can be absorbed by relieving some other station, thus affecting a fuel (thermal) saving or water saving (in the case of another hydrostation with storage).

12.5 Types of water power development

(a) Run-of-river plant (local development)

A weir or barrage is built across a river and the low head created is used to generate power, the power station often being an integral part of the weir structure. It has very limited storage capacity and can only use water when available. Its firm capacity is low, because the water supply is not uniform throughout the year, but it can serve as a baseload plant. Some plants may have enough upstream storage to meet the peak demands of the day. Two typical layouts of the run-of-river plant are shown in Fig. 12.2.

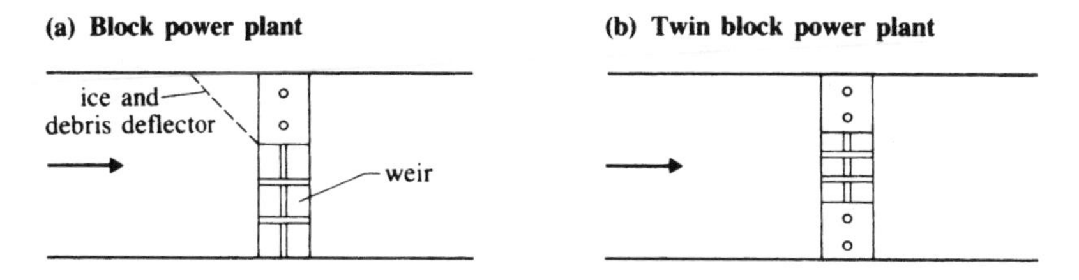

Fig. 12.2 Run-of-river plants

(b) Diversion canal plant

Sometimes topographic, geological and hydrological conditions and environmental and economic considerations may favour diversion-type power development schemes. The flow from the impounded water in the river upstream of the barrage is diverted into a power canal which rejoins the river further downstream (Fig. 12.3), with the power station located either next to the intake, or within the canal, or at the outlet. A rocky stretch of the river containing rapids, where regulation could be difficult, may be avoided by this type of layout. A fall of considerable height can be developed by means of a diversion canal in a river valley with a relatively steep slope.

(c) Storage plant (remote development)

The dam structure is separated from the power station by a considerable distance over which the water is conveyed, generally by a tunnel and pipeline, so as to achieve medium and high heads at the plants (Fig. 12.4). The reservoir storage upstream of the dam increases the firm capacity of the plant substantially, and depending on the annual run-off and power requirements, the plant may be used as a baseload and/or peak-load installation.

(d) Pumped storage plant

Where the natural annual run-off is insufficient to justify a conventional hydroelectric installation, and where it is possible to have reservoirs at the head- and tailwater locations, the water is pumped back from the lower to the headwater reservoir. This kind of plant generates energy for peak load, and at off-peak periods water is pumped back for future use. During off-peak periods excess power available from some other plants in the system (often a run-of-river, thermal, or tidal plant) is used in pumping the water from the lower reservoir. A typical layout of a pumped storage plant is shown in Fig. 12.5.

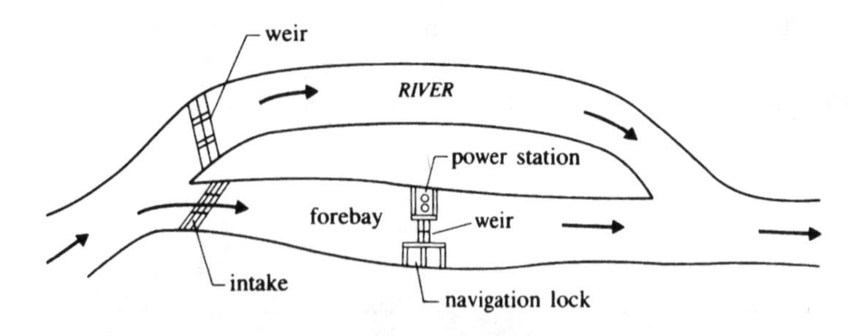

Fig. 12.3 Diversion canal plant

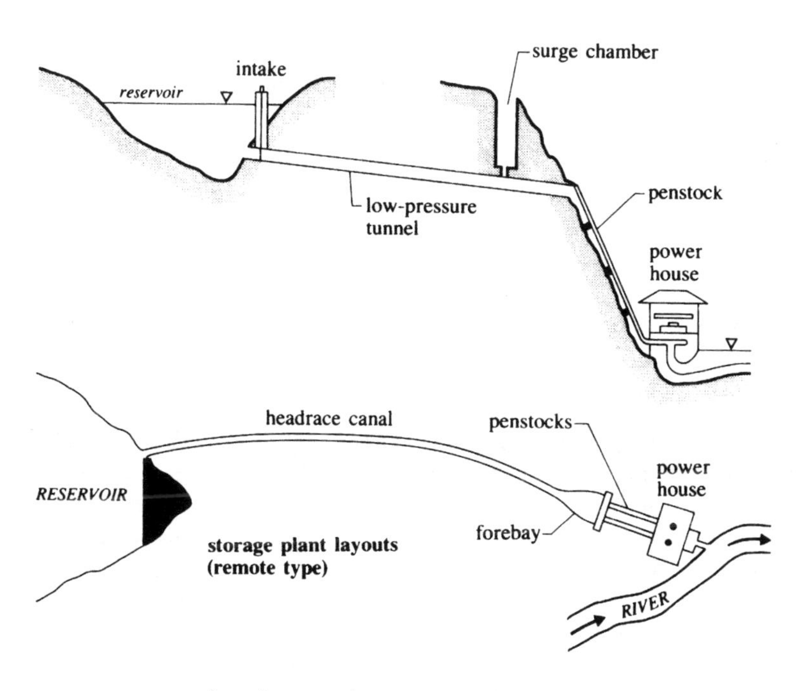

Fig. 12.4 Storage plant layouts (remote type)

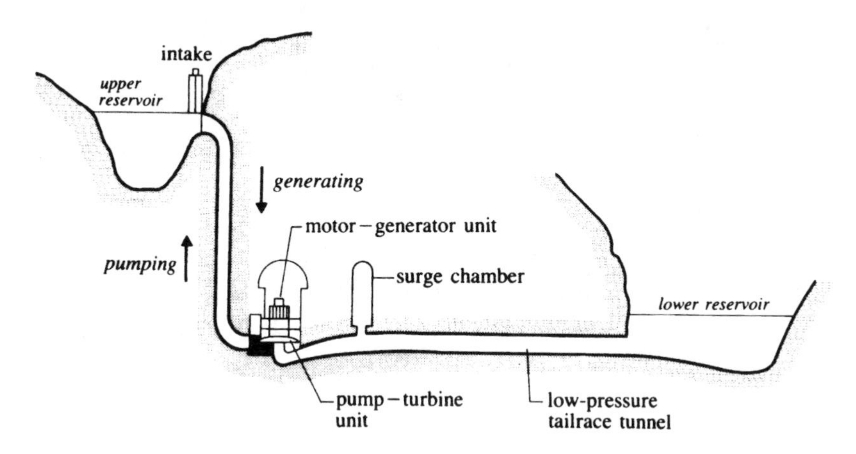

Fig. 12.5 Pumped storage plant

A pumped storage plant is an economical addition to a system which increases the load factor of other systems and also provides additional capacity to meet the peak loads. These have been used widely in Europe, and about 50 plants are in the US with a total installed capacity of about 6700 MW. In the UK large plants are in operation at Ffestiniog (300 MW) and Dinorwic (1680 MW) in North Wales, and at Foyers (300 MW) and Cruachan (450 MW) in Scotland. A major recently completed 1060 MW pumped storage scheme is in Germany (Goldisthal).

Reversible pump–turbine and generator–motor machines are in use nowadays, thus reducing the cost of a pumped storage plant installation by eliminating the additional pumping equipment and pump house.

12.6 Head classification of hydropower plants

(a) Low-head plants

These plants have a gross head of less than about 50 m and are usually of the run-of-river type, with or without pondage, the power house being an integral part of the dam or barrage. Tidal power plants (Section 12.12.1) are also low-head plants. The discharges are usually large in low-head plants. The following are some typical installations of low-head plants: Pitlochry, Scotland (17 m); Owen Falls, Uganda (21 m); St Lawrence, Canada (22 m); La Rance Tidal plant, France (3–11 m).

(b) Medium-head plants

These plants may be either locally or remotely controlled, with a head of about 50–300 m, and some of the well known large medium-head installations are: Castello de Bode; Portugal (97 m), Hoover, USA (185 m); Three Gorges, China (113 m); Itaipu, Brazil (126 m); Guri, Venezuela (146 m); Kasnojarsk, Russia (100 m).

(c) High-head plants

Most of the high head plants (head $\geqslant$ 300 m) are of the remote-controlled type. The following are some of the installations of this type: Laures, Italy (2030 m); Reisseck-Kreuzeck, Austria (1771 m); Chandoline, Switzerland (1750 m); Mar, Norway (780 m); Cruachan, Scotland (401 m); Dinorwic, North Wales (440 m).

12.7 Streamflow data essential for the assessment of water-power potential

The gross head of any proposed scheme can be assessed by simple surveying techniques, whereas hydrological data on rainfall and run-off are essential in order to assess the available water quantities (see Shaw, 1994). The following hydrological data are necessary: (a) the daily, weekly or monthly flow over a period of several years, to determine the plant capacity and estimated output which are dependent on the average flow of the stream and its distribution during the year; (b) low flows, to assess the primary, firm, or dependable power.

(a) Streamflow data analysis

A typical streamflow hydrograph, including a dry period from which the frequency of occurrence of a certain flow during the period can be calculated (Worked example 12.1), is shown in Fig. 12.6.

The flow duration curve (Fig. 12.7) is a plot of the streamflow in ascending or descending order (as ordinate) and its frequency of occurrence as a percentage of the time covered by the record (as abscissa). Also shown on this figure is the flow duration curve of a flow regulated by storage. Losses due to evaporation and leakage from the proposed reservoir, and flow rates relating to low-water management downstream of the reservoir and to any other water demand, have to be taken into account in arriving at the regulated flow duration curve.

(b) Power duration curve

If the available head and efficiency of the power plant are known, the flow duration curve in Fig. 12.7 may be converted into a power duration curve by changing the ordinate to the available power (i.e. $\eta\rho gQH$).

The power which is available for 95% to 97% of the time on the reservoir regulated schemes is usually considered to be the primary or firm power, and the area of the power duration curve under the minimum amount of flow available for 95% to 97% of the time thus gives the total amount of the primary power. Primary power is not necessarily produced continuously. If pondage and interconnection facilities are available, the

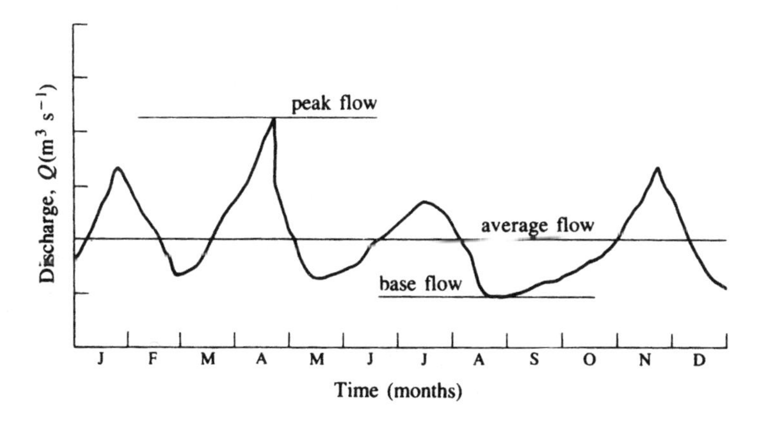

Fig. 12.6 Streamflow hydrograph

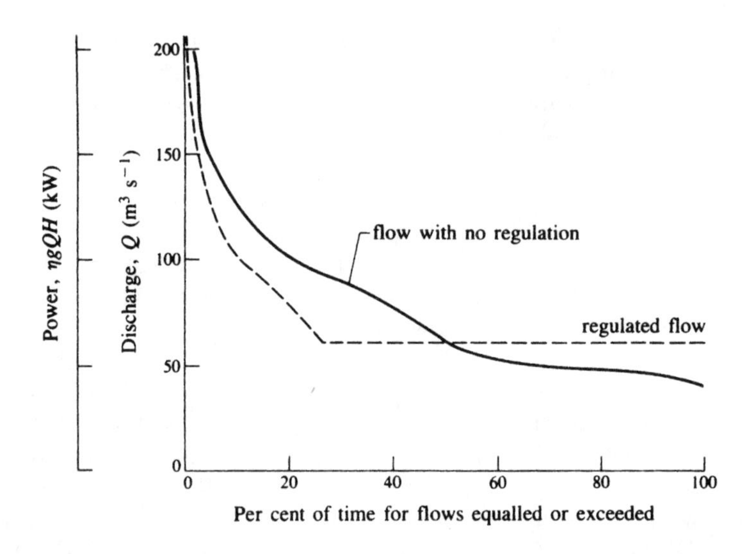

Fig. 12.7 Flow–power duration curve (power scale multiplying factor = ηgH)

plant may be operated on the peak load only. The surplus or secondary power is all the available power in excess of the primary power, and is given by the area under the power duration curve between the firm power line and the total installed capacity of the power plant.

(c) Mass curve

The mass curve (Fig. 12.24) is a plot of cumulative flow against time throughout the period of record. It is used to estimate storage requirements and usable flow for power production (Linsley and Franzini, 1979; Twort, Ratnayaka and Brandt, 2000). The slope of the curve at any point indicates the rate of inflow at that particular time.

The gradient of the line OA represents the average rate of inflow over the total period for which the mass curve has been plotted. If the rate of withdrawal (draw-off or demand) of water for power production is uniform, and is given by the gradient of OA, the gradient of the mass curve at any instant indicates either the reservoir being filled or emptied according to whether the slope of the mass curve is greater or smaller than that of the demand line OA. Thus, in order to permit the release of water at this uniform rate for the entire period, the reservoir should have a minimum capacity given by the sum of the vertical intercepts BC and DE. Assuming the reservoir to contain an initial storage of water equal to DE, it would be full at C and empty at D. Mass curves may also be used to determine the spill of water from the reservoir of a given capacity from which water is drawn off to meet the power demand.

12.8 Hydraulic turbines and their selection

12.8.1 Types of hydraulic turbines

Hydraulic turbines may be considered as hydraulic motors or prime movers of a water-power development, which convert water energy (hydropower) into mechanical energy (shaft power). The shaft power developed is used in running electricity generators directly coupled to the shaft of the turbine, thus producing electrical power.

Turbines may be classified as impulse- and reaction-type machines (Nechleba, 1957). In the former category, all of the available potential energy (head) of the water is converted into kinetic energy with the help of a contracting nozzle (flow rate controlled by spear-type valve – Fig. 12.8(a)) provided at the delivery end of the pipeline (penstock). After impinging on the curved buckets the water discharges freely (at atmospheric pressure) into the downstream channel (called the tail race). The most commonly used impulse turbine is the Pelton wheel (Fig. 12.8(b)). Large units may have two or more jets impinging at different locations around the wheel.

In reaction turbines only a part of the available energy of the water is converted into kinetic energy at the entrance to the runner, and a substantial part remains in the form of pressure energy. The runner casing (called the scroll case) has to be completely airtight and filled with water throughout the operation of the turbine. The water enters the scroll case and moves into the runner through a series of guide vanes, called wicket gates. The flow rate and

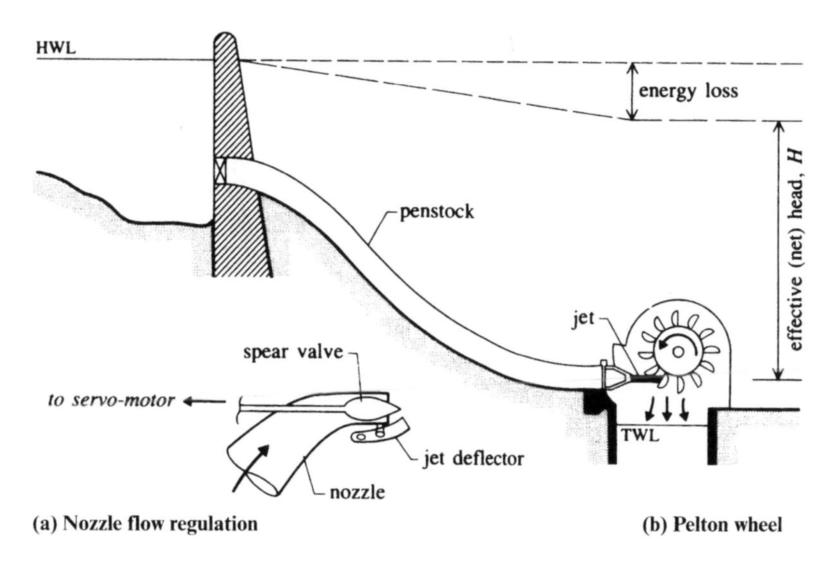

Fig. 12.8 Impulse turbine (Pelton wheel)

its direction can be controlled by these adjustable gates. After leaving the runner, the water enters a draft tube which delivers the flow to the tail race. There are two main types of reaction turbine, the Francis and the propeller (also known as Kaplan or bulb) turbine (Fig. 12.9).

Turbines may also be classified according to the main direction of flow of water in the runner as: (a) tangential flow turbine (Pelton wheel); (b) radial flow turbine (Francis, Thompson, Girard); (c) mixed-flow turbine (modern Francis type); (d) axial-flow turbine of fixed-blade (propeller) type or movable blade (Kaplan or bulb) type.

For history of the development of hydraulic turbines see Viollet (2005).

12.8.2 Specific speed, N_s, speed factor, ϕ, and turbine classification

The specific speed, N_s, of a turbine is its most important characteristic, and is of paramount importance in design. It is defined as the speed at which a geometrically similar runner would rotate if it were so proportioned that it would develop 1 kW when operating under a head of 1 m, and is given by

$$N_s = NP^{1/2}/H^{5/4} \qquad (12.2)$$

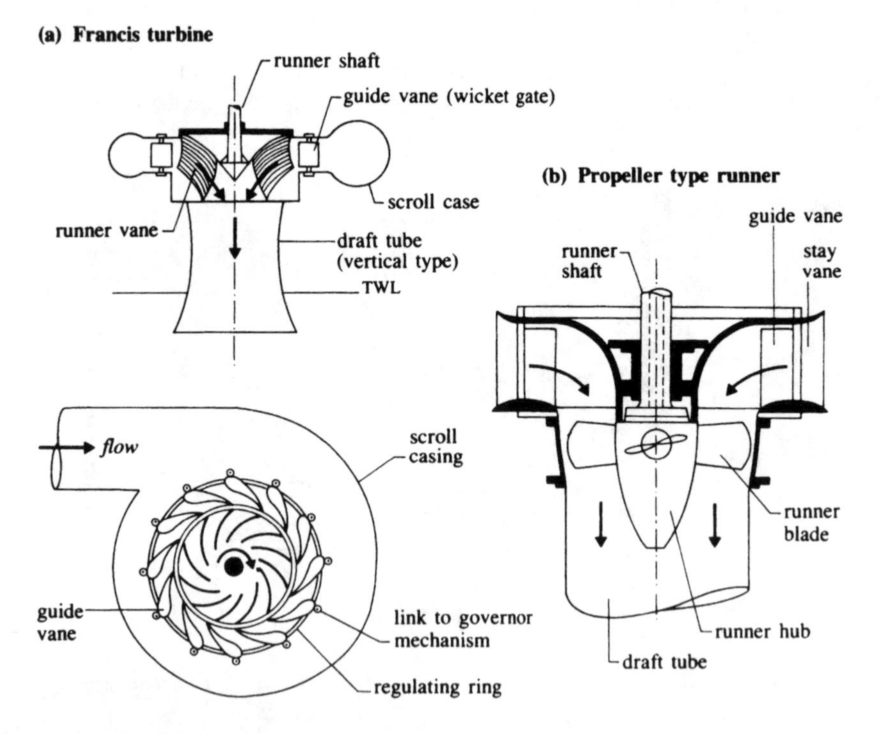

Fig. 12.9 Reaction turbines

where N is the rotational speed in revolutions per minute (rev min^{-1}), P is the power developed (kW), and H is the effective head (m).

The speed factor, ϕ, is the ratio of the peripheral speed, v, of the buckets or vanes at the nominal diameter, D, to the theoretical velocity of water under the effective head, H, acting on the turbine:

$$\phi = v/(2gH)^{1/2} = DN/(84.6H^{1/2}). \quad (12.3)$$

Table 12.1 suggests appropriate values of ϕ which give the highest efficiencies for any turbine. Also shown in the table are the head and specific speed ranges and the efficiencies of the three main types of turbine.

12.8.3 Turbine rating and performance

Hydraulic turbines are generally rated under maximum, minimum, normal, and design head. The runner is designed for optimum speed and maximum efficiency at design head, which is usually selected as the head above and below which the average annual generation of power is approximately equal. However, in reality, head and load conditions change during operation and it is extremely important to know the performance of the unit at all other heads. Such information is usually furnished by the manufacturer in the form of plots of efficiency versus various part-loading conditions (Fig. 12.10(a)). The change of efficiency with specific speed is shown in Fig. 12.10(b).

For a given total plant capacity, total costs will generally increase with an increase in the number of units. As efficiencies of large units (turbine and generators) are higher than those of smaller ones of the same type, if the power demand is reasonably uniform, it is good practice to install a small number of large units.

On the other hand as the efficiency of the hydraulic turbine decreases with the decreasing flow rate (e.g. when running under part-full-

Table 12.1 Range of ϕ values, specific speeds and heads

Type of runner	ϕ	N_s	*H (m)*	*Efficiency (%)*
Impulse	0.43–0.48	8–17		85–90
		17	>250	90
		17–30		90–82
Francis	0.60–0.90	40–130		90–94
		130–350	25–450	94
		350–452		94–93
Propeller	1.4–2.0	380–600	<60	94
		600–902		94–85

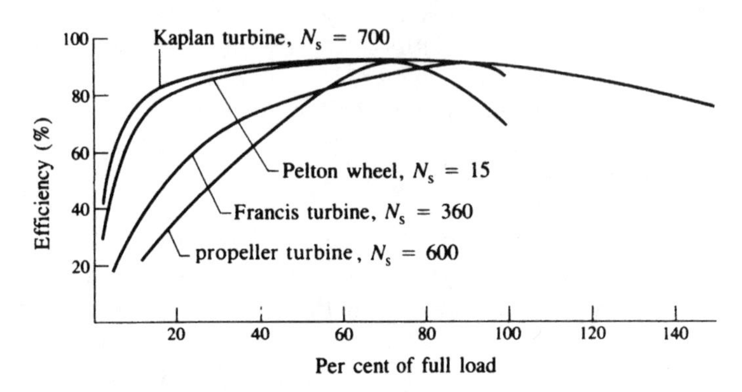

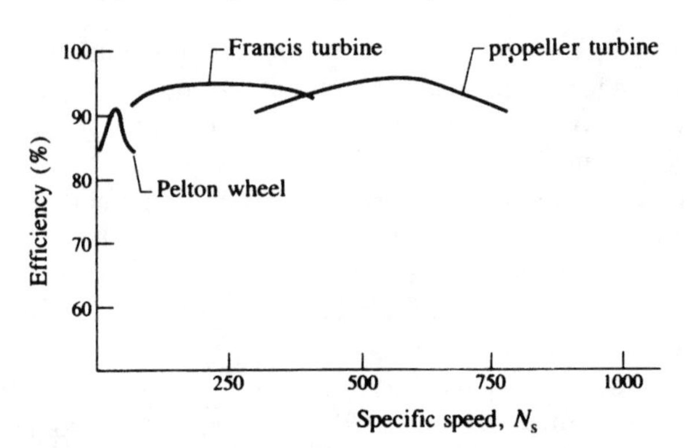

Fig. 12.10 Turbine performance curves

load condition) it is better to use a greater number of smaller machines for widely variable operating conditions. The load variation is then met by putting machines into and out of service, so that individual sets do not have too wide a variation of load and therefore always operate as efficiently as possible.

12.8.4 Operating or normal speed, *N* and runaway speed

All runners having the same geometrical shape, regardless of size, have the same specific speed; specific speed is a useful parameter for the selection of a runner for particular duties. Charts or tables (Table 12.2) can be used for this selection. If the head and flow rate are known, the speed of the unit N can be calculated from equations (12.1) and (12.2) (see worked example 12.3).

Table 12.2 Q–H–N_s data (after Raabe, 1985): Q($m^3 s^{-1}$); H (m); P (kW); D (m); N (rev min^{-1})

Type of runner	*Specific speed,* $N_s = NP^{1/2}/H^{5/4}$	*Maximum head, H (m)*	*Unit discharge,* $Q_u = Q/D^2H^{1/2}$	*Unit speed* $N_u = ND/H^{1/2}$
Impulse (Pelton)	7–11	1650–1800	0.011–0.007	39.4–39.8
	11–17	700–1650	0.024–0.011	38.9–39.4
	17–26	350–700	0.055–0.024	37.6–38.9
Francis:				
slow	51–107	410–700	0.35–0.1	63.6–60.8
medium	107–150	240–410	0.59–0.35	67.5–63.6
	150–190	150–240	0.83–0.59	72.6–67.5
fast	190–250	90–150	1.13–0.83	81–72.6
	250–300	64–90	1.28–1.13	92.2–81
Kaplan	240–450	50	1.22–0.93	145–85
	330–560	35	1.61–1.29	155–100
	390–690	20	2.0–1.6	170–110
	490–760	15	2.35–2.0	180–120
	570–920	6	2.45–2.35	200–135

If the turbines are to drive electricity generators their speeds must correspond to the nearest synchronous speed for a.c. machines and the correct speed physically possible for d.c. machines. For synchronous running, the speed N is given by

$$N = 120f/p \tag{12.4}$$

where f is the frequency of the a.c. supply in Hz (50–60 Hz) and p is the number of poles of the generator (divisible by 4 for heads up to 200 m or by 2 for heads above 200 m). The recommended (Mosonyi, 1987) normal speeds for 50 Hz a.c. supply machines are 3000, 1500, 1000, 750, 600, 500, 375, 300, 250, 214, 188, 167, 150, 125, 107, 94, 83, 75 and 60 rev min^{-1}.

If the external load on the machine suddenly drops to zero (sudden rejection) and the governing mechanism fails at the same time, the turbine will tend to race up to the maximum possible speed, known as *runaway speed*. This limiting speed under no-load conditions with maximum flow rate must be considered for the safe design of the various rotating components of the turbogenerator unit. The suggested runaway speeds of the various runners for their appropriate design considerations are given in Table 12.3.

Table 12.3 Runaway speeds and acceptable head variations

Type of runner	*Runaway speed (% of normal speed)*	*Acceptable head variation (% of design head)*	
		Minimum	*Maximum*
Impulse (Pelton)	170–190	65	125
Francis	200–220	50	150
Propeller	250–300	50	150

12.8.5 Cavitation in turbines and turbine setting, Y_s

Cavitation results in pitting, vibration and reduction in efficiency and is certainly undesirable. Runners most seriously affected by cavitation are of the reaction type, in which the pressures at the discharge ends of the blades are negative and can approach the vapour pressure limits. Cavitation may be avoided by suitably designing, installing and operating the turbine in such a way that the pressures within the unit are above the vapour pressure of the water. Turbine setting or draft head, Y_s (Figs 12.15 and 12.16), is the most critical factor in the installation of the reaction turbines. The recommended limits of safe specific speeds for various heads, based on experience with existing power plants, are shown in Fig. 12.11.

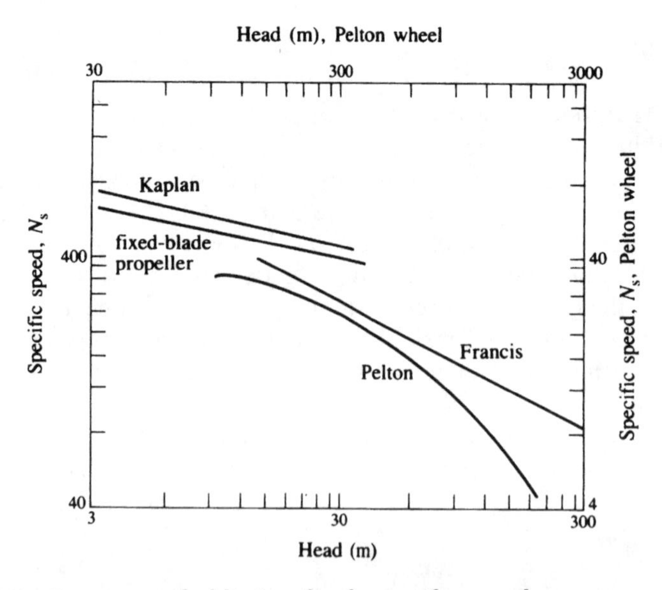

Fig. 12.11 Recommended limits of safe specific speeds

The cavitation characteristic of a hydraulic machine is defined as the cavitation coefficient or plant sigma (σ), given by

$$\sigma = (H_a - H_v - Y_s)/H \tag{12.5}$$

where $H_a - H_v = H_b$, is the barometric pressure head (at sea level and 20 °C, $H_b = 10.1$ m), and H is the effective head on the runner.

From equation (12.5) the maximum permissible turbine setting $Y_{s,\max}$ (elevation above tailwater to the centreline of the propeller runners, or to the bottom of the Francis runners) can be written as

$$Y_{s,\max} = H_b - \sigma_c H \quad \text{(Thoma's formula)} \tag{12.6}$$

where σ_c is the minimum (critical) value of σ at which cavitation occurs (usually determined by experiments). If Y_s is negative the runner must be set below the tailwater.

Typical values of σ_c for reaction turbines, versus their specific speeds, are shown in Table 12.4.

Table 12.4 Critical plant sigma values, σ_c

	Francis runners					*Propeller runners*			
Specific speed (N_s)	75	150	225	300	375	375	600	750	900
σ_c	0.025	0.10	0.23	0.40	0.64	0.43	0.8	1.5	3.5

The above recommended limiting values of σ may also be approximated by

$$\sigma_c = 0.0432(N_s/100)^2 \quad \text{for Francis runners} \tag{12.7}$$

and

$$\sigma_c = 0.28 + 0.0024(N_s/100)^3 \quad \text{for propeller runners} \tag{12.8}$$

with an increase of σ_c by 10% for Kaplan turbines (Mosonyi, 1987).

The preliminary calculations of the elevation of the distributor above the tailwater level (Y_t, Fig. 12.15) suggest the following empirical relationships (based on knowledge of the existing plants (Doland, 1957)):

$$Y_t = Y_s + 0.025DN_s^{0.34} \quad \text{for Francis runners} \tag{12.9}$$

and

$$Y_t = Y_s + 0.41D \quad \text{for propeller runners} \tag{12.10}$$

where D is the nominal diameter of the runner.

12.8.6 Runner diameter, D

For the approximate calculations of the runner diameter, the following empirical formula (Mosonyi, 1987) may be used:

$$D = a(Q/N)^{1/3} \tag{12.11}$$

where $a = 4.4$ for Francis- and propeller-type runners and 4.57 for Kaplan-type turbines (D in m, Q in $m^3 s^{-1}$, N in rev min^{-1}).

The equation (Mosonyi, 1988)

$$D = 7.1Q^{1/2}/(N_s + 100)^{1/3}H^{1/4} \tag{12.12}$$

may also be used to fix the propeller-type runner diameter (H in m).

The impulse wheels are fed by contracting nozzles and, in the case of the Pelton wheel turbine, the hydraulic efficiency is at its maximum when the speed factor ϕ is around 0.45 and the smallest diameter of the jet,

$$d_j = 0.542(Q/H)^{1/2}. \tag{12.13}$$

The nominal diameter, D, of the Pelton wheel (also known as mean or pitch diameter measured to the centreline of the jet) is thus given by

$$D = 38H^{1/2}/N. \tag{12.14}$$

The jet ratio m, defined as D/d_j, is an important parameter in the design of Pelton wheels, and for maximum efficiency a jet ratio of about 12 is adopted in practice. The number of buckets for a Pelton wheel is at an optimum if the jet is always intercepted by the buckets, and is usually more than 15. The following empirical formula gives the number of buckets, n_b, approximately as:

$$n_b = 0.5m + 15. \tag{12.15}$$

This holds good for $6 < m < 35$. It is not uncommon to use a number of multiple jet wheels mounted on the same shaft so as to develop the required power.

12.8.7 Turbine scroll case

A scroll case is the conduit directing the water from the intake or penstock to the runner in reaction-type turbine installations (in the case of impulse wheels a casing is usually provided only to prevent splashing of water and to lead water to the tail race). A spiral-shaped scroll case of the correct geometry ensures even distribution of water around the periphery of the runner with the minimum possible eddy formations. The shape and internal dimensions are closely related to the design of the turbine.

(a) Full spiral case

A full spiral case (Fig. 12.12(a)) entirely enclosing the turbine with a nose angle, φ, of 360° ensures most perfect flow conditions. However, in practice spiral cases with $320° < \varphi < 340°$ are also called full spiral cases.

This kind of spiral case will generally be used in medium- and high-head installations where discharge requirements are smaller.

(b) Partial spiral case

For low-head plants the entrance area should be large so as to allow large flows. This is achieved by choosing nose angles that are less than 320°. The spacing of the units is therefore governed by these large entry widths of the partial spiral scroll cases (Fig. 12.2(b)).

For high-head plants, a circular scroll case cross-section is normally adopted; a metal casing is more suited to this shape. For low-head plants where the water quantities are large a rectangular section with rounded corners may be constructed *in situ* in concrete. The approximate dimensions of a Francis turbine layout with steel scroll case and a propeller turbine layout with concrete scroll, respectively, are shown in Figs 12.12(a) and (b). The design of the shape of the spiral case is governed by the flow requirements. Initial investigations should be based on the following

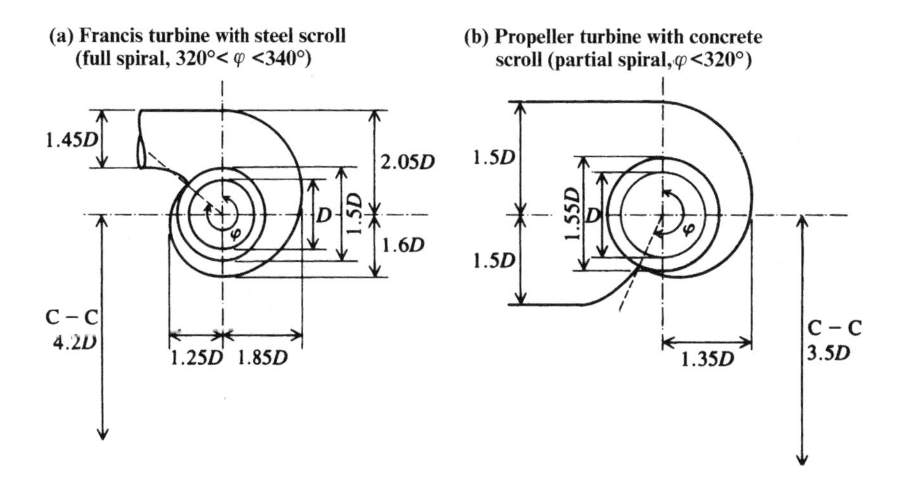

Fig. 12.12 Recommended dimensions of scroll casings

assumptions: (a) a spiral case of constant height; (b) an evenly distributed flow into the turbine; (c) no friction losses.

Referring to Fig. 12.13(a), the discharge in the section of the spiral case defined by an angle θ is given by $q = Q\theta/2\pi$, where Q is the total discharge to the runner.

The velocity at any point within the spiral case can be divided into radial (V_r) and tangential (V_t) components.

The tangential component, $V_t = K/r$, where $K = 30\eta gH/N\pi$ (from the basic Euler equation for the power absorbed by the machine) and the discharge through the strip dq is given by

$$dq = V_t h_0 dr = Kh_0 dr/r.$$

Therefore

$$q = \int_{r_0}^{R} Kh_0 dr/r = Q\theta/2\pi \quad \text{or} \quad \ln R/r_0 = Q\theta/2\pi Kh_0. \tag{12.16}$$

Equation (12.16) shows that for a given vortex strength, K, a definite relationship exists between θ and R.

The most economical design of a power station substructure and the narrowest spiral case can be obtained by choosing a rectangular section adjoining the guide vanes (entrance ring) by steep transition (symmetrical or asymmetrical), as shown in Fig. 12.13(b).

We can write

$$h = h_0 + \alpha(r - r_0) \tag{12.17}$$

where $\alpha = \cot\beta_1 + \cot\beta_2$.

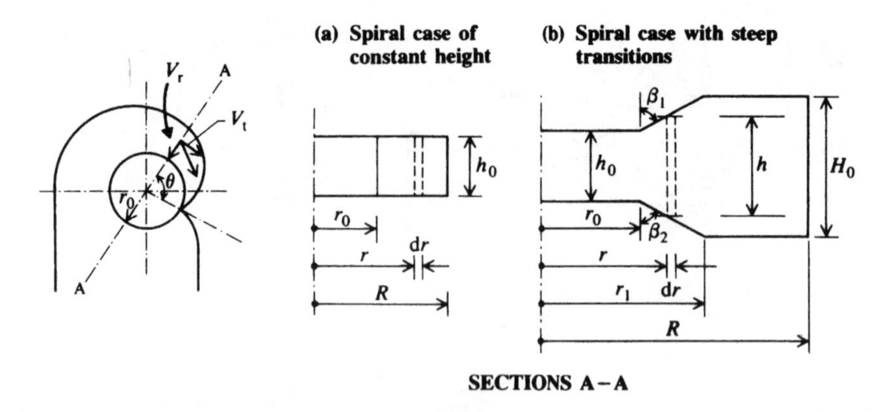

Fig. 12.13 Typical cross-sections of a spiral case

Equation (12.16) now becomes

$$Q\theta/2\pi K = \int_{r_0}^{r_1} h\,\mathrm{d}r/r + \int_{r_1}^{R} H_0\,\mathrm{d}r/r \tag{12.18}$$

which, on integration, after replacing h by equation (12.17) gives

$$Q\theta/2\pi K = (h_0 - \alpha r_0)\ln(r_1/r_0) + (H_0 - h_0) + H_0 \ln(R/r_1). \tag{12.19}$$

Knowing r_1 from

$$r_1 = (H_0 - h_0)/\alpha + r_0 \tag{12.20}$$

the value of R defining the shape of the spiral casing can be determined. The height H_0 at any angle θ may be assumed to be linearly increasing from h_0 at the nose towards the entrance.

The shape of the cross-section is determined at various values of θ by assuming the existence of uniform velocity for the entire spiral case (Mosonyi, 1987) to be equal to the entrance velocity, $V_0 \simeq 0.2\,(2gH)^{1/2}$. Thus, knowing $q_i = Q\theta_i/2\pi$, the area of cross-section at an angle θ_i is given by

$$A_i = q_i/V_0 = 0.18Q\theta_i/H^{1/2}. \tag{12.21}$$

This approximation results in larger cross-sections towards the nose, desirable in order to minimize the friction losses (ignored in the theoretical design development) which are more pronounced in the proximity of the nose.

It is desirable to provide streamlined dividing piers in entrance flumes of large widths to ensure as even a flow distribution as possible. The final design of the scroll case should preferably be checked by model tests, especially in the cases of unconventional arrangements of large turbine units. An example of the inner shaping of a concrete spiral casing is shown in Fig. 12.14.

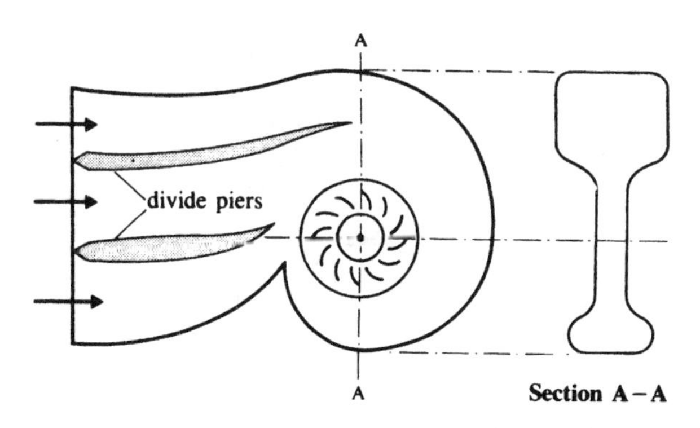

Fig. 12.14 Inner shape of a concrete scroll case with divide piers

12.8.8 Draft tubes

The draft tube is a conduit discharging water from the runner to the tail race, and has a twofold purpose: (a) to recover as much as possible of the velocity energy of the water leaving the runner, thus increasing the dynamic draft head; (b) to utilize the vertical distance between the turbine exit and the tailwater level, called the static draft head (Fig. 12.15). The most common draft tube is the elbow type (Fig. 12.15), which minimizes the depth of the substructure. Compared with the vertical type (Fig. 12.16)

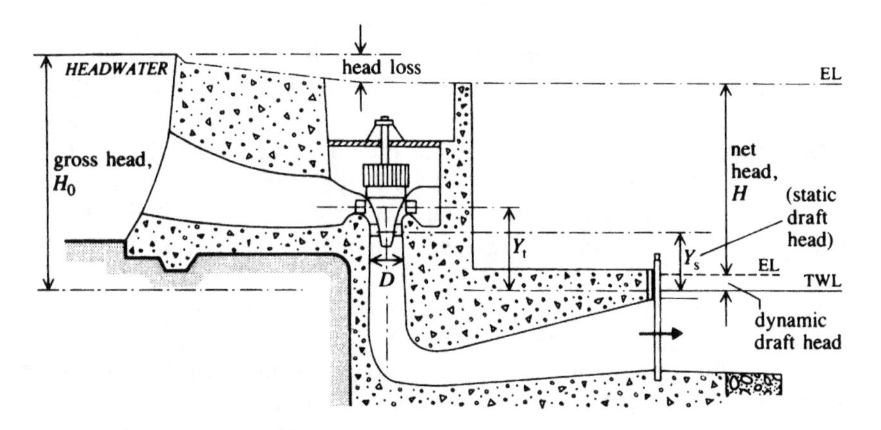

Fig. 12.15 Elbow-type draft tube

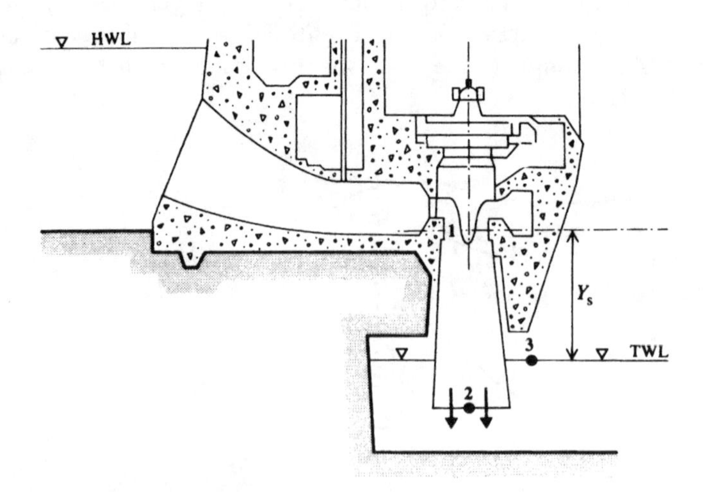

Fig. 12.16 Vertical draft tube

it also has the desirable effect of directing the flow in the direction of the tailwater flow.

The elbow-type draft tube is divided into three parts, all three sections gradually expanding like diffusers:

1. a vertical (entrance) part in circular cross-section, gradually expanding;
2. a bend part (its aim being to minimize losses due to changes in the direction of flow) in gradual transition from the circular section into a rectangular section;
3. an almost horizontal part in rectangular section, gradually expanding to direct the flow into the tail race with minimum losses.

Draft tubes of large discharge capacities (units operating under low heads) are usually designed by model tests, investigated jointly with the runner models. Considering the layout in Fig. 12.16, for example, the energy equation between 1 and 3 gives:

$$Y_s + p_1/\rho g + v_1^2/2g = p_a/\rho g + v_2^2/2g + H_L \tag{12.22}$$

where p_a is the atmospheric pressure and H_L is the friction and eddy losses in the draft tube.

Therefore the pressure head at the runner exit, $p_1/\rho g$, is given by

$$p_1/\rho g = p_a/\rho g - Y_s - (v_1^2/2g - v_2^2/2g - H_L). \tag{12.23}$$

Denoting $v_1^2/2g - v_2^2/2g - H_L = H_d$, the head regained

$$H_d = \eta_d(v_1^2/2g - v_2^2/2g) \tag{12.24}$$

where η_d is the efficiency of the draft tube. By a proper design of the draft tube, the exit velocity v_2 can be reduced to $1-2\,\mathrm{m\,s^{-1}}$, with η_d as high as 85%.

In order to avoid cavitation at the exit from the runner, the condition $p_1/\rho g > p_v/\rho g$, where p_v is the saturated vapour pressure (around 0.3 m of water absolute), must be satisfied with a sufficient safety factor, since the flow over parts of the runner will be at lower pressures.

The suggested dimensions of the draft tube used for high specific speed turbines are shown in Fig. 12.17 (Mosonyi, 1988).

12.9 Other components of hydropower plants

The various appurtenances connecting the construction and operation of hydroelectric power plants are as follows: dams (storage or diversion type control works); gates; valves; intakes; waterways (open channel or low-pressure tunnel head race); high-pressure pipes (penstocks); fish passes;

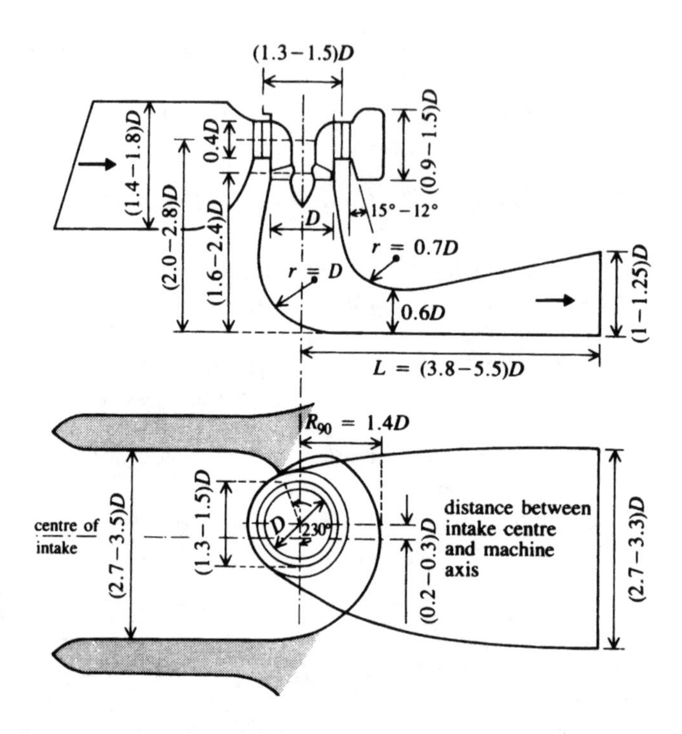

Fig. 12.17 Recommended dimensions of an elbow-type draft tube (Mosonyi, 1988)

pressure relief valves; surge tanks; turbine governors; generators; the power house superstructure; cranes; a switπch yard for transformers and switching equipment; transmission lines. The only accessories briefly described here are those that are of direct concern to the civil engineer; dams, control works, intakes, fish passes, canals, gates and valves are discussed in previous chapters.

For more detailed information of the components of hydropower plants including economic analyses see e.g. Mosonyi (1987), Gulliver and Arndt (1991), or Zipparo and Hasen (1993).

12.9.1 Head race

The head race is a conveyance for water from the source (reservoir or river) to the power plant in the form of a canal (open waterway), tunnel (low-pressure conveyance) or penstock (high-pressure conveyance) (Fig. 12.5). The open waterway usually terminates in a forebay which is an enlarged body of water from which the penstocks convey the water to the turbines in the power house.

The forebay functions as a small balancing reservoir (with spilling arrangements) upstream of the power house which accommodates the sudden rejections and increased demands of the load. This may be developed by enlarging the channel just upstream of the intake for the penstocks. The usual components of an intake structure, such as trash racks, gates, etc., must be included. The provision of air inlets at the back of the gates to the penstocks is essential in order to prevent the development of excessive negative pressures in the penstocks as a result of their sudden draining, i.e. when the head gates are closed and the turbine gates opened.

(See also Section 9.2 for details of intake structures).

12.9.2 Penstocks

The penstocks are pipes of large diameter, usually of steel or concrete, used for conveying water from the source (reservoir or forebay) to the power house. They are usually high-pressure pipelines designed to withstand stresses developed because of static and waterhammer pressures created by sudden changes in power demands (i.e. valve closures and openings according to power rejection and demand). The provision of such a high-pressure line is very uneconomical if it is too long, in which case it can be divided into two parts, a long low-pressure conveyance (tunnel) followed by short high-pressure pipeline (penstock) close to the turbine unit, separated by a surge chamber which absorbs the waterhammer pressure rises and converts them into mass oscillations.

(a) Design criteria

The hoop stress, p_t, can be obtained as

$$p_t = pD/2e \qquad (12.25)$$

where D is the internal diameter, e is the wall thickness of the penstock and p is the internal pressure including all the waterhammer effects. Equation (12.25) gives the wall thickness as

$$e = pD/2p_t\eta_j \qquad (12.26)$$

where η_j is the joint efficiency (0.9–1.0 for welded joints).

For a chosen penstock pipeline, the wall thickness can be calculated by equation (12.26) assuming an allowable tensile stress, p_t, of the material (for steel, $p_t = 150\,N\,mm^{-2}$). Further, the pipe thickness should be such as to withstand additional stresses due to free spanning or cantilevering between the support and expansion joints, and longitudinal movements over the supports due to temperature changes.

The energy of the flow through a penstock is inevitably reduced owing to entry and friction losses. Although the friction losses can be minimized by careful selection of the pipe diameter, and its entrance losses can be minimized by a bell-mouthed entrance, an economical penstock diameter may be determined from a study of the annual charges on the cost of the installed pipe compared with the lost revenue due to this power loss. As can be seen from Fig. 12.18, energy losses decrease with the increasing diameters while construction costs increase. A diameter which minimizes the total annual costs can be determined from the sum of the two costs.

Fahlbusch (1982) reformulated the objective of the economic analysis in terms of the amount of the invested capital and the capitalized value of the lost energy, and arrived at the conclusion that the most economical diameter can be computed within an accuracy of about ±10% from

$$D = 0.52H^{-0.17}(P/H)^{0.43} \tag{12.27}$$

where P is the rated capacity of the plant (kW), H is the rated head (m), and D is the diameter (m).

The design of the anchorages and support rings for the penstocks (additional momentum forces) has to be carefully considered wherever there is a change of gradient and direction at branch outlets. The treatment of the subject is beyond the scope of this text; further information can be found in structural design books. Surge tanks are dealt with in Section 12.10.

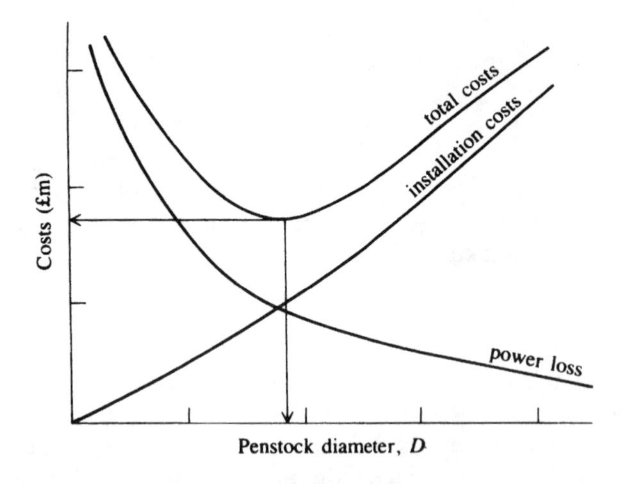

Fig. 12.18 Economic diameter of penstock

12.9.3 Turbine governors

The governor is a mechanism controlling the rotational speed of the turbo-generator unit; constant speed must be maintained in order to obtain the a.c. supply with constant frequency. As the turbine and hence its interconnected generator tend to decrease or increase speed as the load varies, the maintenance of an almost constant speed requires regulation of the amount of water allowed to flow through the turbine by closing or opening the gates (or nozzles) of the turbine automatically, through the action of a governor. A simple governing mechanism for turbines is shown in Fig. 12.19. Increase in the rotor speed raises piston A, permitting oil to enter chamber B, thus closing the gates slightly. The operation is reversed if the speed drops.

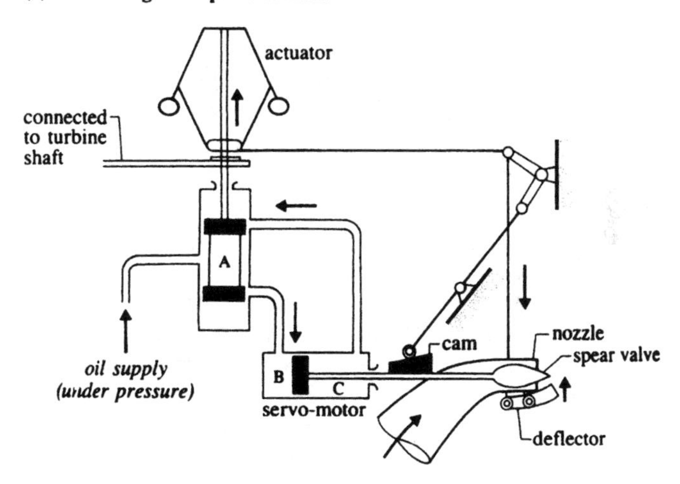

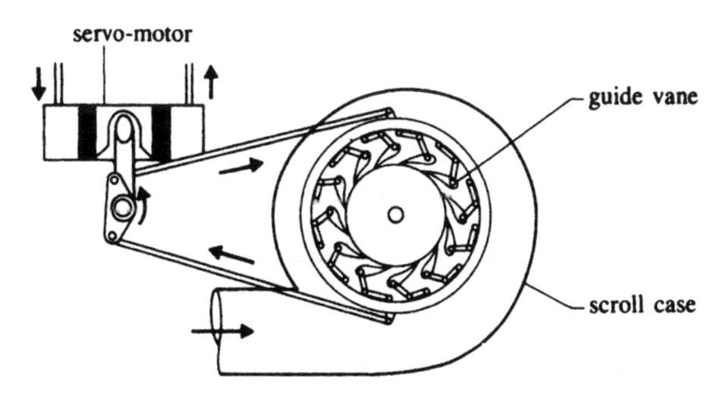

Fig. 12.19 Turbine governors

A rapid closing or opening of the nozzle or guide vanes (gates) is undesirable, as serious waterhammer problems may result in the penstocks. Sudden changes may be avoided in the case of a Pelton turbine if a deflector is activated in front of the jet, thus diverting part of the flow away from the turbine. Similarly, in the case of a reaction-type turbine a relief valve may allow a part of the discharge to flow directly to the tail race without entering the runner.

12.9.4 Generators and transformers

The generator is an electrical machine coupled to the turbine shaft (either horizontally or vertically). The alternating current (a.c.) synchronous generator is widely used in hydroelectric power production practice. It has two elements, a magnetic field consisting of an assembly of electromagnets (poles) which rotates (hence rotor), within a stator (stationary unit) which is a system of conductors (armature windings). The relative displacement between the rotor and the stator induces an alternating electromotive force. The a.c. supply in the UK is produced with a standard frequency of 50 Hz. The turbine is governed to operate at constant speed and the generator is designed with the appropriate number of poles to produce the designed frequency at the selected speed (equation (12.4)).

Owing to the physical limitations imposed by the mechanical properties of the materials, the rotational speeds of a hydroelectric machine are limited, e.g. for an output of 15 MW a limiting speed of 1500 rev min^{-1}, 250 MW at 600 rev min^{-1}, 1000 MW at 120 rev min^{-1} etc. At the sites with larger potential power, the turbine speed is generally chosen at 100–375 rev min^{-1}, thus reducing the number of sets for the same installed capacity.

The sizes of the generators vary depending on their rating and on their shaft arrangements (either vertical or horizontal). A.c. generators are rated in kilovolt-amperes (kVA). The apparent or nominal rating (output) differs from the actual output, P_a, the difference P_m being used to magnetize the rotor field. Thus the rated kVA is given by

$$\text{kVA} = (P_a^2 + P_m^2)^{1/2} \tag{12.28}$$

and the power factor is given by

$$PF = \cos\phi = P_a/\text{kVA}. \tag{12.29}$$

Exciters (direct-current generators) mounted on the generator shaft energize the rotor field of the main generator.

A.c. generators with a rating of 300 kVA vary in weight from about 160 kN for an operating speed of 900 rev min^{-1} to 500 kN for a speed of 100 rev min^{-1}, their sizes being about 3 m high and 2.5–5 m in diameter. Horizontal shaft units are generally suitable for low-speed plants and are heavy, large and costly, whereas vertical shaft units used with high-speed plants are rather small and less expensive. For speeds larger than 1000 rev min^{-1} additional stresses will be induced, warranting special materials and designs. Special structural problems often arise in providing proper support for generators. The generators must also be adequately ventilated to keep them from overheating, and this is achieved by air and/or water cooling.

For high-head (up to 300 m) storage schemes, reversible Francis-type pump–turbines (Raabe, 1985) have been developed to operate at a relatively high efficiency as either a pump or a turbine. The same electrical unit works as a generator or motor by reversing the poles. A plant equipped with reversible facilities may reduce the costs of a pumped storage scheme through the elimination of the additional pumping equipment and pumping house. There may be abnormal wear on all components of such units due to their frequent operation in starting and stopping modes several times a day. The hydraulic pump–turbine characteristics, dual rotation and method of starting the units in pumping mode are some of the additional factors which will affect the generator design.

Exceptionally large units – e.g. the turbines at the Three Gorges Project with runner diameter 10 m and generator diameter 18.5 m and weight 1800 tons – require ingenuity in their design to avoid vibration problems.

The transformers connecting the power source (generators) and the receiving circuit (transmission lines) step up the voltage for transmission, thus reducing the power loss and permitting the use of smaller conductors (cables) in the transmission line. The transformers are usually located in outside switchyards adjacent to the power house, as a necessary precaution to avoid high voltage and other hazards.

12.9.5 Power house

The power house structure can be divided in two sections, a substructure supporting the hydraulic and electrical equipment and a superstructure housing the equipment. The substructure is usually a concrete block with all the necessary waterways formed within it. The scroll case and draft tube are usually cast integrally (especially in large low-head plants) with the substructure with steel linings.

The superstructure usually houses the generating units and exciters, the switch board and operating room. Vertical-axis units (whose turbines are placed just below the floor level) generally require less floor space than those mounted on horizontal axes. The cost of the superstructure can be reduced considerably by housing individual generators only (outdoor power house), although it has the disadvantage that maintenance works have to be restricted to good weather conditions only. Under certain topographic conditions, particularly when the power plant is situated in narrow canyons with no convenient site for a conventional type of power house, this may be located underground. Many examples of this type exist in Europe and elsewhere (e.g. the Cruachan and Dinorwic plants in the UK).

It is essential to equip a power house with a crane to lift and move equipment for installation and maintenance purposes. Travelling cranes spanning the width of the building and capable of traversing its entire length are normally used. The crane rail elevation depends on the maximum clearance required when the crane is in operation which, in turn, determines the overall height of the superstructure (Fig. 12.26).

12.9.7 Tail race

The tail race is the waterway into which the water from the turbine units is discharged (through draft tubes if reaction-type units are used). It may be very short and if the power house is close to the stream the outflow may be discharged directly into it. On the other hand, if the power house is situated at a distance from the stream the tail race may be of considerable length. Proper tail race design ensures, especially in low-head plants, that more of the plant gross head is available for power development.

The tail race in the vicinity of the draft tube exit (head of tail race) must be properly lined, as it may otherwise degrade and cause lowering of the tailwater elevation of scouring of the channel bottom. Should this be allowed to progress the designed turbine setting level would alter, thus causing reduced efficiency of the turbine (cavitation in the turbine runner), and remedial measures (artificial raising of the water level) would have to be taken. The tail race channel may sometimes aggrade, in which case the gross head at the plant decreases, with a resulting reduction in power output. This situation may arise if the main spillway outflow is close to the tail race without an adequate separating wall.

Gates, with an appropriate hoisting mechanism, must be provided at the draft tube outlet (between the piers and tail race) to isolate the draft tube for maintenance works.

The tail race of the underground power house is invariably a horizontal tunnel into which the turbine units discharge the water. Such tunnel flow could sometimes take place under pressure, calling for the necessity of a surge tank close to the turbine units (i.e. at the head of the tail race tunnel).

12.10 Surge tanks

12.10.1 General

Surge tanks may be considered essentially as a forebay close to the machine. Their primary purpose is the protection of the long pressure tunnel in medium- and high-head plants against high waterhammer pressures caused by sudden rejection or acceptance of load. The surge tank converts these fast (waterhammer) pressure oscillations into much slower – and lower – pressure fluctuations due to mass oscillation in the surge chamber; the detailed treatment of waterhammer analysis is beyond the scope of this textbook and the reader is referred to Jaeger (1997), Novak (1983), Chaudry (1987), Fox (1989) and Wylie and Streeter (1993).

The surge chamber (Fig. 12.4) dividing the pressure tunnel into a short high-pressure penstock downstream and a long low-pressure tunnel upstream thus functions as a reservoir for the absorption or delivery of water to meet the requirements of load changes. It quickly establishes the equilibrium of the flow conditions, which greatly assists the speed regulation of the turbine.

12.10.2 Surges in surge chambers

Sudden changes in load conditions of the turbine cause mass oscillations in surge tanks which are eventually damped out by the hydraulic friction losses of the conveyance. The amplitude of these oscillations is inversely proportional to the area of the surge tank, and if the area provided were very large dead beats would be set in the tank. Although these conditions would be favourable to achieving the new equilibrium state very quickly, the design would be uneconomical. On the other hand, if too small an area is provided the oscillations become unstable; this is unacceptable. It is therefore essential to choose a section in which the oscillations become stable within a short period of time.

The critical section for stability is given by the equation

$$A_{sc} = V_0^2 A_t L_t / 2gP_0 H_0 \quad \text{(Thoma criterion)} \tag{12.30}$$

where A_t and L_t are the tunnel cross-sectional area and length respectively, and the suffix '0' defines the steady state conditions prior to the load variation (Fig. 12.20). A stable tank area is usually chosen with a safety factor of about 1.5.

Using Manning's equation with $1/n = 85$ (smooth concrete finish) gives, from equation (12.30),

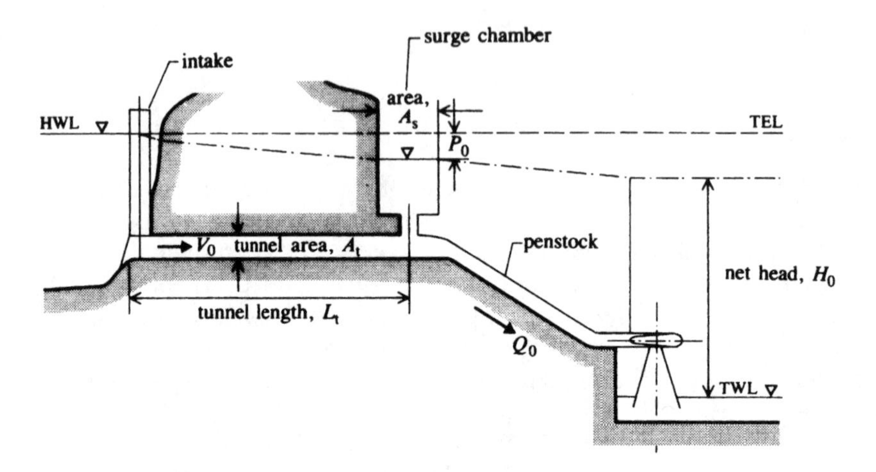

Fig. 12.20 A typical power plant layout: steady state conditions

$$A_{sc} = 45D^{10/3}/H_0 \tag{12.31}$$

where D is the diameter of the tunnel (in m).

The maximum upsurge and downsurge should be contained within the chamber. For simple surge tanks the following equations may be used to calculate these maximum surges. For a sudden 100% load rejection, maximum upsurge

$$Z_{*max} = 1 - 2K_{*0}/3 + K_{*0}^2/9 \quad (\text{for } K_{*0} < 0.7), \tag{12.32}$$

where $Z_* = Z/Z_{max}$, $K_{*0} = P_0/Z_{max}$, $Z_{max} = Q_0/A_s r$ and $r = (gA_t/L_t A_s)^{1/2}$, and maximum downsurge

$$Z_{*min} = -1/(1 + 7K_{*0}/3). \tag{12.33}$$

For a sudden 100% load demand, maximum downsurge

$$Z_{*max} = -1 - 0.125K_{*0} \quad (\text{for } K_{*0} < 0.8) \tag{12.34}$$

where Z is the surge amplitude with respect to the reservoir level, A_s is the cross-sectional area of the surge tank and P_0 is the head loss in the tunnel. The range of surge levels (amplitudes) must not be too large to minimize the governing difficulties. The maximum upsurge and downsurge are computed for extreme conditions, i.e. the top level of the surge chamber is governed by the maximum upsurge level when the reservoir level is at its maximum and the bottom level of the chamber is controlled by the maximum downsurge level when the reservoir is at its lowest drawdown level.

Instantaneous 100% demand conditions result in too large a maximum downsurge, as the normal practice is to allow for 0–10% of full load demand quickly but, thereafter, the unit is brought to full load only gradually. Maximum downsurges are normally calculated against 75–100% of full load and, once again, the bottom level of the chamber is controlled by the reservoir at its lowest drawdown level condition. This condition is invariably more critical than the one governed by the maximum downsurge after a load rejection.

Excessive surges may occur if several quick load variations are imposed on the unit (overlapping surges). These may create additional governing difficulties and the top and bottom levels of the surge chamber may have to be modified to accommodate these excessive surges.

In order to achieve conservative designs of the surge chamber it is usual to assume a lower conduit friction factor than average for calculating the maximum upsurge and a higher friction factor for the maximum downsurge. In all cases due consideration of the effect of ageing of the tunnel must be given. Head losses of all types are assumed to be proportional to V^2. Turbine efficiency is assumed to be constant throughout its operational range; it may be recalled that the Kaplan-type turbine satisfies this assumption over a very wide range of its loading conditions.

12.10.3 Types of surge tanks

1. *Simple surge tanks.* The simple surge tank (Fig. 12.21(a)) is of uniform cross-section and is open to the atmosphere, acting as a reservoir. It is directly connected to the penstock so that water flows in and out with small head losses when load variations occur. It is usually large in size with expensive proportions and sluggish in responding to damping surges. These are very rarely used in modern practice except in installations where load changes are either small or very gradual.
2. *Throttled tank.* In the throttled tank (restricted orifice type tank) the restricted entry (Fig. 12.21(b)) to the surge tank creates retardation and acceleration conditions of flow in the tunnel upstream of it, thus reducing the storage requirement and minimizing the maximum up- and downsurges. Although this type of surge tank is economical (because of its smaller size) compared with the simple tank section, the rapid creation of retarding and accelerating heads complicates the governing mechanism, requiring additional inertia in the turbo-generator units.
3. *Surge tank with expansion chambers.* This type of surge tank (Fig. 12.21(c)) consists of a narrow riser (main surge shaft); attached to it at either end are large expansion chambers. The narrow riser

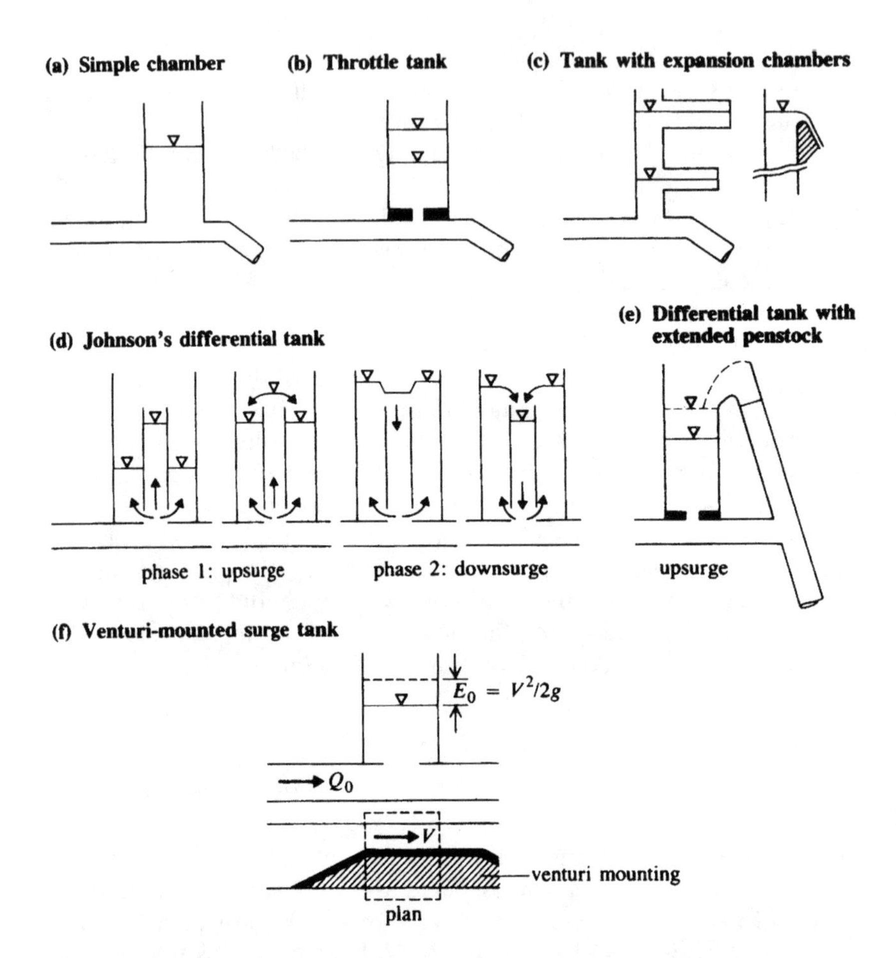

Fig. 12.21 Types of surge tanks

reacts quickly, creating accelerating or decelerating heads, and at the same time the expansion chambers minimize the maximum up- and downsurge levels, thus limiting the range of surge levels (i.e. easier governing). In order to reduce the costs of the structure, spilling arrangements may sometimes be provided either to wastage (if water is not scarce) or back to the penstock.

4. *Differential surge tank.* This type (also known as Johnson's differential tank – Fig. 12.21(d)) consists of an internal narrow riser shaft with an orifice entry to the larger outer shaft at the bottom. As the central riser is narrow it responds instantaneously during the upward phase; at the same time the maximum amplitude is restricted to its top level, any excess water spilling back into the outer chamber. Similarly, during the downward phase water spills into the narrow riser

while the riser itself responds quickly to maintaining the desired level. The differential tank with an extended penstock, which acts as a central riser, is shown in Fig. 12.21(e).

5. *Surge tanks with venturi mounting.* Considering the velocity energy under the surge tank ($V^2/2g = E_0$), Thoma's critical section can be written as

$$A_{sc} = V_0^2 A_t L_t / 2g(P_0 + E_0)H_0. \tag{12.35}$$

More economical sections may result by providing a venturi contraction (Fig. 12.21(f)) under the surge tank (thus increasing the velocity head, E_0 (Escande, Dat and Nalluri, 1962)). For the detailed theory and design of surge tanks, see Jaeger (1977), Novak (1983), Wylie and Streeter (1993) and Popescu *et al.* (2003).

12.11 Small hydraulic power plant development

There are some inconsistencies in the use of the term 'small' hydraulic power, but the following classification adopted by the United Nations International Development Organisation is widely used: *micropower* <100 kW, *minipower* 100 kW to 1 MW, *small power* 1 MW to 10 MW. Another term which could be added to the above classification is '*Picohydro*' used for developments of less than 5 kW.

The potential for the development of small hydraulic power plants worldwide is substantial and its use particularly in rural or isolated areas is growing at such a rate that any statistics become quickly obsolete. As an example in UK the technically feasible potential for small hydraulic power plants (<5 MW) is about 4.6 TWh/year, the economically feasible potential about 1.1 TWh/year, of which about 12% has been developed (Bartle and Hallowes, 2005). In China small plants owned by municipalities and local authorities account for about 16 000 MW; Indonesia's micro-potential (between 250 kW and 500 kW) is estimated at about 500 MW (Hayes, 2004).

Very small plants (pico – micro) require the development of special suitable machinery (e.g. fixed blade turbines) and many can be sited at disused mills, which used water wheels. Although the installation of new *water wheels* disappeared virtually early in the 20th century, their potential use as a cost effective low head low flow hydraulic energy converter (contrary to turbines operating under atmospheric conditions) is making them attractive again to a certain degree. The *overshot* water wheels were employed for head differences 2.5–10 m and flow rates of 0.1–0.2 m^3/s, the *breast* wheels for 1.5–4 m and 0.5–0.95 m^3/s and the *undershot* wheels for 0.5–2.5 m and 0.5–0.95 m^3/s. 'Modern' water wheels employ the potential energy of the water and are built of steel; their efficiencies are about 85% (overshot) and 75% (undershot) for a wide range of flow conditions $0.2 < Q/Q_{max} < 1$. For further details of water wheels see Viollet (2005) and Müller and Kauppert (2004).

The main advantage of small hydropower schemes which are – like all hydropower – a valuable renewable energy source are their flexibility of siting, relatively short periods of design and construction, modest capital costs, possible multiple development, automation of control, conjunctive use with other energy sources and flexibility in meeting demand. However, they are not suitable for meeting major concentrated demands and require careful considerations of suitable plant siting, capacity and type.

For further discussion of small hydroelectric plants see e.g. Monition, Le Nir and Roux (1984) and Gosschalk (2002).

12.12 Other energy resources

Hydropower is of course only one of many resources for generating electricity. Electricity from coal, oil, gas and nuclear fission power stations and other renewable resources like wind power, tidal and wave energy, geothermal, solar and depression energy, combustion of biomass as well as the nuclear fusion power (currently at the feasibility stage) all provide the components for an energy policy which is constantly evolving particularly at the time when the need for the reduction of carbon emissions is paramount. Even a superficial discussion of energy policies as well as of the above mentioned alternatives is beyond the scope of this text and only a brief mention of marine based power development is included.

In principle three forms of marine energy can be considered for power generation (for a more detailed discussion see Kerr 2005):

- tidal barrage power using head difference;
- tidal stream power using kinetic energy;
- wave power using the energy of wave motion.

In all cases economic, technical and environmental issues have to be considered.

(a) Tidal power

The use of tidal energy dates back to the 12th century, when tidal mills worked along the coast of Brittany, France. The basic principle of their operation was to form a storage basin by constructing a dyke closing off a cove; the basin filled through gates during the flood tide, and during the ebb tide it emptied through an undershot wheel, thus producing a driving force. The operational principle of a tidal power plant remains the same as that of the tidal mills, but the turbine units may produce power during filling as well as emptying of the basin (Fig. 12.22(a)).

Tidal amplitudes attain considerable magnitudes along certain coastal stretches (Canadian Atlantic coast 13.5 m, Bristol Channel, UK, 10 m, French Atlantic coast 8 m; figures relate to mean annual ranges). In the Pacific region, e.g. the coasts of China and Russia, mean amplitudes of 6–9 m have also been recorded.

In spite of the fact that tidal power production costs are slightly higher (around 3p per kW h compared with 2–2.5p. for nuclear power, at 1980 prices) and that it fluctuates both daily and seasonally, it could displace oil- or coal-fired stations for peak generation and also provide power for pumped storage schemes.

The first tidal power station at La Rance on the west coast of France has an installed capacity of 240 MW (24 bulb-type units of 10 MW each) with a load factor of 25%, the peaks in the generation of power being directed to displace the peak generation of oil-fired plants and to supply the pumped storage schemes in the Pyrenees. The bulb turbines also permit pumping, thus superelevating the basin level (Fig. 12.22(b)) with respect to the sea level at the end of filling. This installation has now been accepted as successful since its energy production costs are competitive with those of other resources in the French power system. Furthermore,

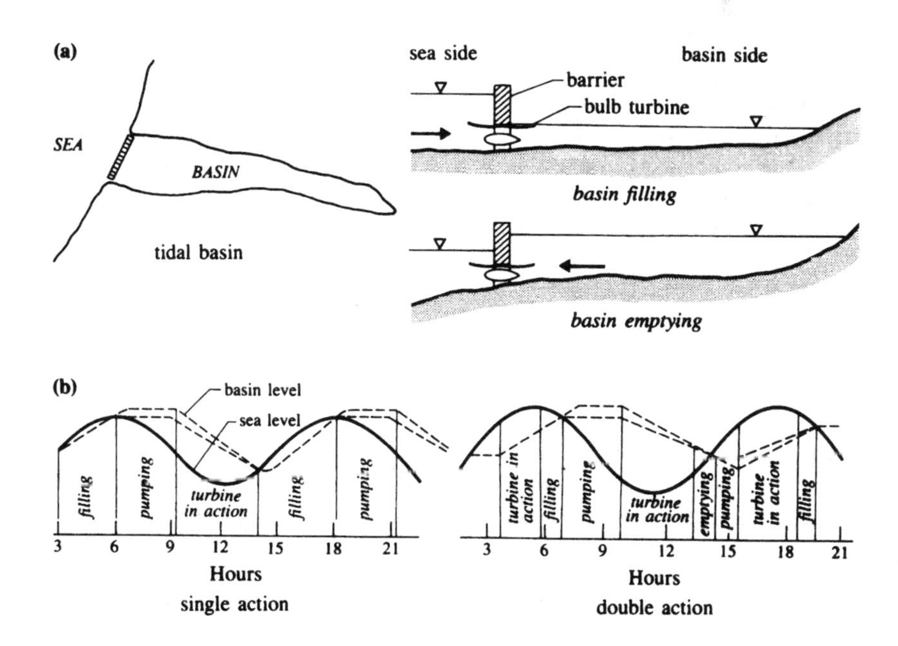

Fig. 12.22 Tidal power development

technological developments since the completion of the La Rance power scheme have stimulated the construction of smaller tidal plants in Canada, China and Russia and the planning of major schemes in Argentina, Australia, Canada, China, India, Korea, Russia and the UK with a total capacity of 43 GW and annual power of 63 TWh. For further details of tidal power generation see e.g. Wilson and Balls (1987) and Haws (1997).

(b) Tidal stream power

Marine currents caused by tides or/and oceanic circulation have a large potential energy even if most are too slow for exploitation. Some development projects are under way.

(c) Wave power

Using the sinusoidal wave equations and the expressions for its energy (see Sections 14.2.3 and 14.2.4) and applying some approximations Mosonyi (1988) derived a very simple equation for the specific wave power:

$$P = H^2 T \text{ (kW/}m\text{)} \qquad (12.36)$$

(H is the wave height (m) and T the wave period(s)). Mosonyi assumes that about 50% of this energy can be captured and utilized with a further 50% efficiency (generation and transmission).

Worldwide the wave energy is a very large renewable resource even if only a small part is realistically recoverable. The UK has the largest potential of European countries with locations in Australia and South and North America having substantial resources.

Wave energy devices are either shoreline, near-shore (mounted on the sea bed) or offshore (floating) and convert energy by using an oscillating water column, hinged contour devices (power is extracted from the motion of the joints) or the motion of the device relative to the sea bed. The on-shore and near-shore plants have the additional advantage of the technology being integrated into harbour walls and coastal protection structures.

A cylindrical device seated on the sea bed with its upper part oscillating vertically as water pressure changes with passing waves was as a prototype installed in 2004 off the coast of Portugal; the first commercial wave energy installation (hinged contour device) is likely also to be in Portugal (Kerr, 2005). Devices using an oscillatory water column as shoreline development were tested in UK.

For further details of wave power development and devices see Taylor (1983) or Thorpe (1999).

Worked Example 12.1

The average monthly flows of a stream in a dry year are as follows:

Month	*J*	*F*	*M*	*A*	*M*	*J*	*J*	*A*	*S*	*O*	*N*	*D*
$Q(m^3 s^{-1})$	117	150	203	117	80	118	82	79	58	45	57	152

It is intended to design a hydroelectric power plant across the stream, using the following data: net head at the plant site = 20 m; efficiency of the turbine = 90%.

1. Plot the flow and power duration curves and calculate the firm and secondary power available from this source if the maximum usable flow is limited to $150\,m^3 s^{-1}$.
2. If it is intended to develop the power at a firm rate of 15 MW, either by providing a storage or by providing a standby diesel plant with no storage, determine the minimum capacity of the reservoir and of the diesel unit.

Solution

Flow in descending order $(m^3 s^{-1})$	*Frequency (months)*	*Frequency equalled/ exceeded (%)*
203	1	8.3
152	2	16.7
150	3	25.0
118	4	33.3
117	5	
117	6	50.0
82	7	58.3
80	8	66.7
79	9	75.0
58	10	83.3
57	11	91.7
45	12	100.0

1. The flow duration curve (flow versus frequency equalled or exceeded) is plotted in Fig. 12.23. The same plot can be used as a power duration curve by multiplying the ordinates by a factor of 0.176 ($=\eta\rho g H/10^6$) to obtain the power in MW with $\eta = 90\%$ and $H = 20$ m. The firm power available (equal to the area of the power duration curve under the $45\,m^3 s^{-1}$ line) is 7.95 MW. The secondary power (equal to the area under the power duration curve between the 150 and $45\,m^3 s^{-1}$ lines) is 10 MW.

2. The power to be supplemented by the storage or standby unit to obtain a firm power of 15 MW is the area abc (Fig. 12.23) = 17.76 MW month. Therefore the storage required is

$$17.76 \times 10^6 \times 30 \times 24 \times 60/1000 \times 9.81 \times 20 \times 0.9 = 2.6 \times 10^8\,\text{m}^3.$$

With no storage, the firm power available is 7.95 MW. Therefore the capacity of the standby unit is $15 - 7.95 = 7.05$ MW.

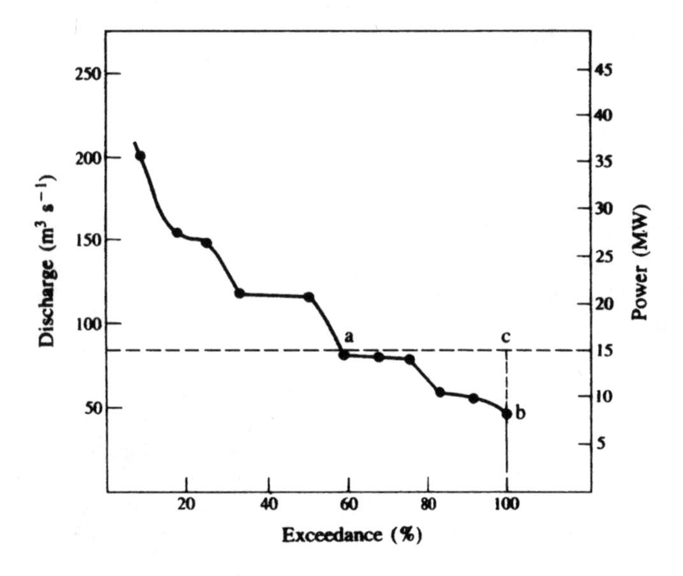

Fig. 12.23 Flow–power duration curve

Worked Example 12.2

The monthly flows of a stream over the period of the driest year on record are as shown below:

Month	*J*	*F*	*M*	*A*	*M*	*J*	*J*	*A*	*S*	*O*	*N*	*D*
Flow ($\times 10^6\,\text{m}^3$)	4.0	2.25	5.0	1.25	0.5	0.75	0.5	0.75	1.25	1.25	5.0	6.25

1. Estimate the maximum possible uniform draw-off from this stream, and determine the reservoir capacity to achieve the uniform draw-off and the minimum initial storage to maintain the demand.
2. If the reservoir has only a total capacity of $8 \times 10^6\,\text{m}^3$ with an initial storage of $4 \times 10^6\,\text{m}^3$, determine (a) the maximum possible uniform draw-off and (b) the spillage.

Solution

Referring to the mass curve plot between cumulative flows and months in Fig. 12.24 we obtain the following.

1. The uniform draw-off is the gradient of line OA

$$= 28.75 \times 10^6/12 \times 30 \times 24 \times 60 \times 60$$

$$= 0.924\,\mathrm{m^3\,s^{-1}}.$$

The reservoir capacity $= \mathrm{BC} + \mathrm{DE} = 10.5 \times 10^6\,\mathrm{m^3}$.
The initial storage $= \mathrm{DE} = 6.5 \times 10^6\,\mathrm{m^3}$.

2. (a) The new demand with an initial storage of $4 \times 10^6\,\mathrm{m^3}$ (DG) (Fig. 12.24) is the gradient of new demand line OF

$$= 25.75 \times 10^6/12 \times 30 \times 24 \times 60 \times 60 = 0.827\,\mathrm{m^3\,s^{-1}}.$$

(b) The spillage $= \mathrm{HC} = 1 \times 10^6\,\mathrm{m^3}$.

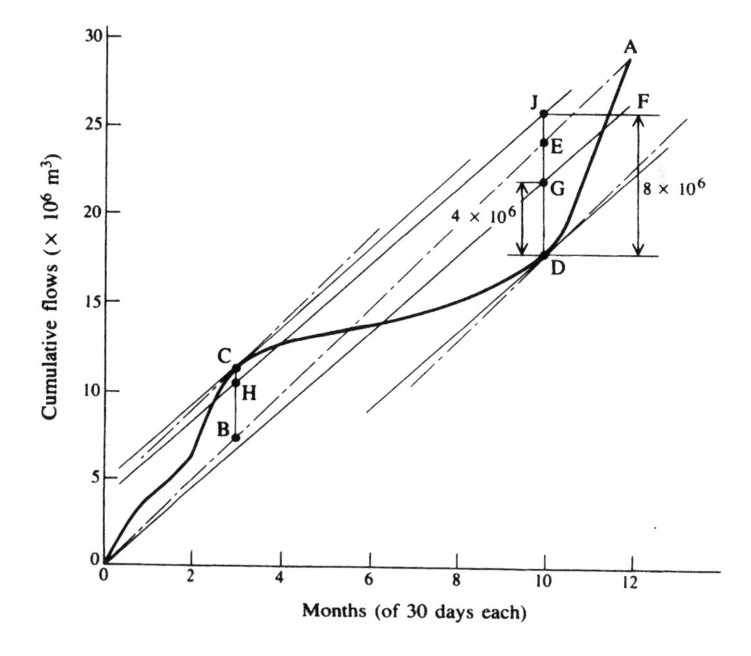

Fig. 12.24 Mass curve

Worked Example 12.3

The monthly flow of a river supplying water to a power plant in 14 successive 4-week periods of the driest year is as follows:

Period (4 weeks)	*1*	*2*	*3*	*4*	*5*	*6*	*7*	*8*	*9*	*10*	*11*	*12*	*13*	*14*
Flow (m^3s^{-1})	1.14	0.58	0.68	0.78	0.94	1.16	0.56	0.56	0.5	0.67	1.36	1.65	1.49	0.83

1. Calculate the volume of reservoir storage to be provided to maintain the highest uniform output throughout the year, and calculate the continuous power if available head is 40 m.
2. If a reservoir of only half the required capacity can be provided, how much power is to be supplied by an auxiliary plant to produce the continuous power in 1?
3. How many turbines and what type and specific speed units would you install for the power in 1?

Solution

To plot the mass curve the cumulative volumes of flow need to be calculated:

Period (4 week)	*Flow rate (m^3s^{-1})*	*ΣFlow rates (m^3s^{-1})*	*ΣVolumes ($\times 10^6 m^3$)*	*ΣOutflows ($\times 10^6 m^3$)*	*ΣBalance volumes ($\times 10^6 m^3$)*
1	1.14	1.14	2.76	2.23	+0.53
2	0.58	1.72	4.16	4.46	−0.30
3	0.68	2.40	5.80	6.69	−0.89
4	0.78	3.18	7.69	8.92	−1.23
5	0.94	4.12	9.97	11.15	−1.18
6	1.16	5.28	12.77	13.38	−0.61
7	0.56	5.84	14.13	15.61	−1.48
8	0.56	6.40	15.48	17.84	−2.36
9	0.50	6.90	16.69	20.07	−3.38
10	0.67	7.57	18.31	22.30	−3.99
11	1.36	8.93	21.60	24.53	−2.93
12	1.65	10.58	25.60	26.76	−1.16
13	1.49	12.07	29.20	28.99	+0.21
14	0.83	12.90	31.21	31.21	0

1. The uniform draw-off rate

$$= 31.21 \times 10^6/14 \times 28 \times 24 \times 60 \times 60$$

$= 0.92\,\mathrm{m^3\,s^{-1}}$

$= 2.23 \times 10^6\,\mathrm{m^3}$ per 4-week period.

Improved accuracy in calculating the required reservoir capacity can be achieved by plotting the net cumulative mass curve (Fig. 12.25(b)), which gives the capacity as $4.52 \times 10^6\,\mathrm{m^3}$. The continuous

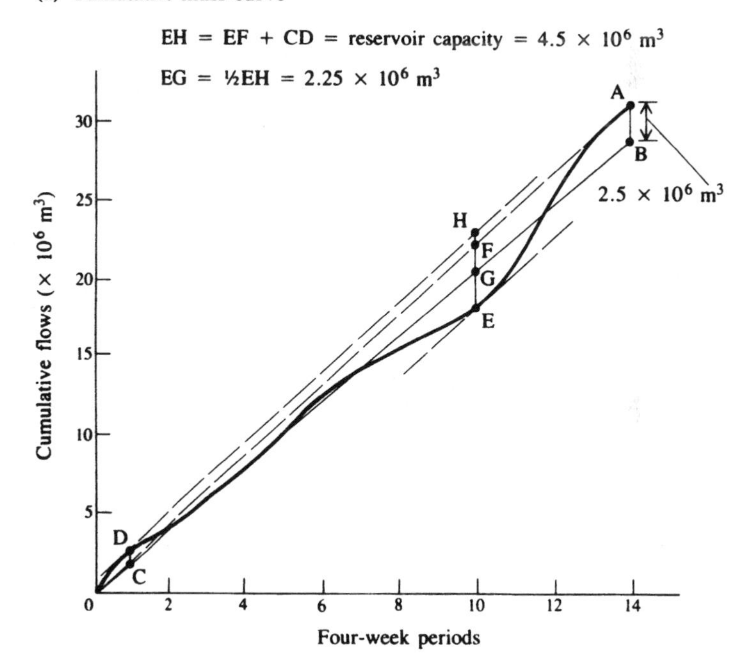

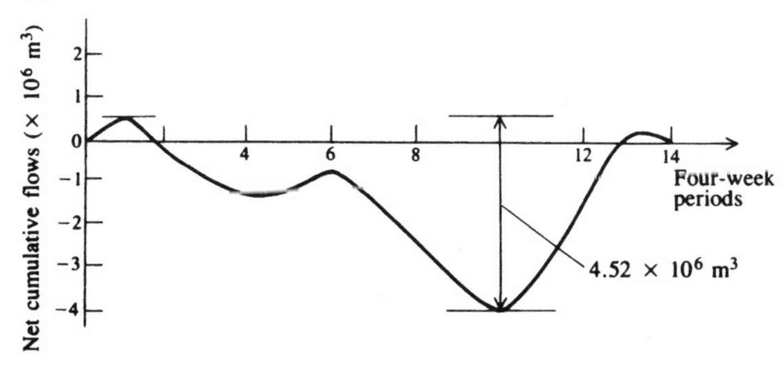

Fig. 12.25 Mass curves

power output $= \eta\rho gQH = 332\,\text{kW}$, assuming a Francis-type turbine (low head, small discharge and efficiency of around 92%).

2. The reservoir capacity $= 0.5 \times 4.52 \times 10^6 = 2.26 \times 10^6\,\text{m}^3$. The reduced demand line with this storage is the line OB on Fig. 12.25(a), giving a drop of $2.5 \times 10^6\,\text{m}^3$ in 14 4-week periods. Therefore, the drop in power

$$= 0.92 \times 1000 \times 9.81 \times 2.5 \times 10^6 \times 40/14 \times 4 \times 7 \times 24 \times 60 \times 60 \times 10^3$$

$$= 26.6\,\text{kW}.$$

Hence 26.6 kW must be supplied by an auxiliary plant.

3. For constant-output, low-head, small discharges, a Francis-type machine is appropriate. The power output is rather small and uniform, and hence one machine will normally do. However, a minimum of two units may be provided in case of breakdowns/maintenance. The following options exist:

(a) designing and operating two units sharing the power output equally (too small a power output per machine and hence inefficient);
(b) designing one unit only and the second, of similar size, as a standby unit (uneconomical);
(c) in view of the above, it is desirable to design and operate one unit only and to provide a diesel unit as a standby.

The specific speed,

$$N_s = NP^{1/2}/H^{5/4}.$$

Adopting $\phi = 0.7$ (Table 12.1) and runner diameter $D = 4.43\,(Q/N)^{1/3}$ we can obtain the operating speed (from equation (12.3)) as $N = 810\,\text{rev min}^{-1}$. Choosing the nearest synchronous speed of $750\,\text{rev min}^{-1}$ (page 482),

$$D = 4.43(0.92/750)^{1/3} = 475\,\text{mm}$$

and the specific speed

$$N_s = 750 \times 332^{1/2}/40^{5/4} = 136.$$

Worked Example 12.4

In a pumped-storage hydropower project, water is delivered from the upper impounding reservoir through a low-pressure tunnel and four high-pressure penstocks to the four pump–turbine units. The elevation of the impounding reservoir water level is 500 m AOD, and the elevation of the downstream reservoir water level is 200 m AOD. The maximum reservoir storage which can be utilized continuously for a period of 48 h is $15 \times 10^6\,m^3$.

The low-pressure tunnel is constructed as follows: length = 4 km; diameter = 8 m; friction factor, $\lambda = 0.028$.

The high-pressure penstocks (4 nos.) are constructed as follows: length of each penstock = 500 m; diameter = 2 m, friction factor, $\lambda = 0.016$; turbine efficiency when generating = 90%; generator efficiency (16 poles, 50 Hz) = 90%, turbine efficiency when pumping = 80%; barometric pressure = 10.3 m of water; Thoma's cavitation coefficient, $\sigma = 0.043(N_s/100)^2$.

1. Determine the maximum power output from the installation.
2. Estimate the specific speed and specify the type of turbine.
3. Determine the safe turbine setting relative to the downstream reservoir water level.
4. If a simple surge chamber 6 m in diameter is provided at the end of the low-pressure tunnel, estimate

   (a) the maximum upsurge and downsurge in the surge chamber for a sudden rejection of one unit and
   (b) the maximum downsurge for a sudden demand of one unit.

Solution

The discharge available $= 15 \times 10^6/(48 \times 60 \times 60) = 86.8\,m^3\,s^{-1}$. The power output is calculated as follows:

$$\text{velocity in tunnel} = 86.8/((\pi/4)8 \times 8) = 1.73\,m\,s^{-1}.$$

Therefore

$$\text{head loss in tunnel} = \lambda L V^2/2gD = 2.13\,m,$$

$$\text{discharge per penstock} = 86.8/4 = 21.7\,m^3\,s^{-1},$$

$$\text{velocity in penstock} = 21.7/((\pi/4)2 \times 2) = 6.91\,m\,s^{-1}.$$

Therefore

$$\text{head loss in penstock} = \lambda L V^2/2gD = 9.73\,m,$$

$$\text{gross head at turbine} = 500 - 200 = 300\,m,$$

and so

$$\text{net head} = 300 - 2.13 - 9.73 = 288.14\,\text{m},$$

$$\text{output/turbine} = \eta\rho g Q H/10^6 = 55\,\text{MW},$$

$$\text{total output} = 4 \times 55 = 220\,\text{MW}.$$

The net output of the generators is $0.9 \times 220 = 198$ MW. The generator speed, $N = 120f/p = 375$ rev min^{-1} (acceptable synchronous speed). Therefore the specific speed, $N_s = NP^{1/2}/H^{5/4} = 76$. A Francis-type turbine is suitable (efficiency, specific speed, and head match this type). The turbine setting

$$Y_s = B - \sigma H,$$

$$\sigma = 0.043 \times (76/100)^2 = 0.0248,$$

and therefore

$$Y_s = 10.3 - 0.0248 \times 288 = 3.16\,\text{m} \quad \text{or} \quad 203.16\,\text{m AOD}.$$

The distributor elevation

$$Y_t = Y_s + 0.025DN_s^{0.34}.$$

The approximate runner diameter

$$D = 4.43(Q/N)^{1/3} = 1.72\,\text{m}.$$

Therefore

$$Y_t = 3.16 + 0.187 = 3.347\,\text{m} \quad \text{or} \quad 203.35\,\text{m AOD}.$$

The surge chamber calculations are as follows:

$$\text{area of surge chamber,} \quad A_s = 28.27\,\text{m}^2;$$

$$\text{area of tunnel,} \quad A_t = 50.26\,\text{m}^2;$$

$$\text{length of tunnel,} \quad L_t = 4000\,\text{m}.$$

Therefore

$$r = (gA_t/L_tA_s)^{1/2} = 0.066.$$

For one unit rejection or demand, $Q_0 = 21.7\,m^3 s^{-1}$ and $P_0 = 2.13\,m$. Therefore

$$Z_{max} = Q_0/A_s r = 11.63\,m$$

and

$$K_0^* = P_0/Z_{max} = 0.183.$$

Upon sudden rejection, the maximum upsurge (equation (12.32)), $Z_{max}^* = 0.88$. Therefore

$$Z_{max} = 0.88 \times 11.63 = 10.23\,m.$$

The maximum downsurge (equation (12.33)), $Z_{min}^* = -0.7$. Therefore

$$Z_{min} = -0.7 \times 11.63 = 8.14\,m.$$

Upon sudden demand (equation (12.34)), $Z_{min}^* = -1.023$. Therefore

$$Z_{min} = -1.023 \times 11.63 = -11.9\,m.$$

Worked Example 12.5

In a pumped-storage scheme, water is supplied to a single turbine through a 200 m long, 2 m diameter penstock line. The power plant uses two centrifugal pumps to pump water back to the upper reservoir, using the same penstock pipeline. Basic data are as follows. Pumps (2 nos.), rated capacity = 6 MW each, efficiency of pump = 87%, efficiency of pump motor = 85%; turbine (1 no), installed capacity = 25 MW, efficiency of turbine = 91%, efficiency of generator = 89%; Penstock pipeline, length = 200 m, diameter = 2 m, Manning's $n = 0.015$, gross head above tailwater level = 115 m.

Determine

1. the maximum discharge rate per pump,
2. the maximum discharge to the turbine and
3. the overall efficiency of the system if all the water used by the turbine is pumped back to the upper reservoir.

Solution

PUMP DISCHARGE

The power output from the pumps

$$= \rho g Q(115 + h_f)/1000\,kW$$

where h_f is the system losses. Ignoring minor losses, Manning's formula gives

$$h_f = Q^2n^2L/A^2R^{4/3} = 0.0155Q^2.$$

Therefore

$$2 \times 6000 \times 0.87 = 9.81Q(115 + 0.0115Q^2)$$

giving $Q \simeq 9.2\,m^3\,s^{-1}$. Therefore the discharge per pump $= 9.2/2 = 4.6\,m^3\,s^{-1}$. The input to the motor $= 2 \times 6/0.85 = 14.12$ MW. A 1 h operation of the pump (i.e. an input of 14.12 MW h units) stores $9.2 \times 60 \times 60 = 3.31 \times 10^4\,m^3$ of water.

TURBINE OPERATION

The input to the turbine

$$= \eta\rho g Q(115 - h_f)/1000\,kW.$$

Therefore

$$25\,000 = 0.91 \times 9.81Q(115 - 0.0115Q^2)$$

giving $Q \simeq 26.1\,m^3\,s^{-1}$. The generator output $= 25 \times 0.89 = 22.25$ MW. Therefore the maximum duration of turbine operation from $3.31 \times 10^4\,m^3$ of stored water is $3.31 \times 10^4/26.1 \times 60 \times 60 = 0.352$ h. Therefore the total generated units (i.e. the output) $= 22.25 \times 0.352 = 7.84$ MW h, and the overall efficiency of the system is $7.84/14.12 \simeq 55.5\%$.

Worked Example 12.6

The following data refer to a proposed hydroelectric power plant: turbines, total power to be produced = 30 MW, normal operating speed = 150 rev min^{-1}, net head available = 16 m; draft tube, maximum kinetic energy at exit of draft tube = 1.5% of H, efficiency of draft tube = 85%, vapour pressure $\leq$ 3 m of water, atmospheric pressure = 10.3 m of water.

1. What size, type, and number of units would you select for the proposed plant?
2. Starting from first principles, determine the turbine setting relative to the tailrace water level.

Solution

For a low-head, high-discharge plant, Kaplan-type units are suitable.

Assuming a specific speed of, say, 500, the power per machine

$$= (N_s H^{5/4}/N)^2 = 11\,377\,\text{kW}.$$

Therefore the number of units is 30 000/11 377 = 2.64. Therefore, choose three units, each having an installed capacity of 10 MW. Note that the number of units depends on other factors such as the variability of power demand, breakdown–maintenance works, the availability of national grid power supply in case of emergencies, etc.

The specific speed

$$N_s = 150 \times \sqrt{10\,000}/16^{5/4} = 468.$$

The discharge per unit is 10 000/0.94 × 9.81 × 16 (assuming an efficiency of 94% – Table 12.1) = 67.75 $m^3 s^{-1}$. Therefore the runner diameter, $D = 4.57(Q/N)^{1/3} = 3.50$ m, and the inlet velocity (i.e. the exit velocity at the runner) is 67.75/(π/4) × 3.5 × 3.5 = 7.04 $m s^{-1}$; hence the inlet velocity head is 2.53 m. The exit velocity head is 1.5 × 16/100 = 0.24 m.

Applying Bernoulli's equation between the inlet of the draft tube and the tailwater level (Fig. 12.16)

$$Y_s = p_a/\rho g - p_v/\rho g - (V_1^2/2g - V_2^2/2g - h_{fd})$$

$$= 10.3 - 3.0 - 0.85(2.53 - 0.24) = 5.35\,\text{m above TWL}.$$

From Thoma's cavitation limiting conditions,

$$Y_s = B - \sigma H,$$

$$\sigma = 0.58 \text{ (from Table 12.4)},$$

giving $Y_s = 1.02$ m above TWL. In the absence of further data, Thoma's criterion may be adopted.

Worked Example 12.7

A run-of-river plant uses a mean head of 10 m and generates approximately 30 MW. The load factor of the installation is 40%.

1. Determine the number, type, and specific speed of the turbines.
2. Design the scroll case for one of the units.

Solution

1. For low heads and large discharges the use of Kaplan-type units is suggested. Also, these units can cope with the variable demand very efficiently.

The average output of the plant is $30 \times 0.4 = 12$ MW. The units will be designed to produce this power most efficiently and any variation in demand will be provided by additional units. There are several alternatives: one 12 MW unit and three units of 6 MW each, five units of 6 MW each, three units of 10 MW each, or two units of 15 MW each, as it is better to have fewer larger units than many smaller units, provided that the units are capable of working efficiently under varying load conditions. More details on the nature of demand variability would provide better answers.

2. As an example, let us design the scroll casing for a 6 MW unit. Assuming a specific speed of 800, the operational speed of the unit

$$N = N_s H^{5/4} / \sqrt{6000} = 183 \text{ rev min}^{-1}.$$

Adopt a synchronous speed of 150 rev min^{-1}, as lower specific speeds are desirable in order to minimize the cavitational problems. This gives the specific speed, $N_s = 653$. The discharge Q per unit is $6000/0.9 \times 9.81 \times 10 = 68 \text{ m}^3 \text{s}^{-1}$. Note that an efficiency of 90% is used, which corresponds to high specific speeds. Therefore the runner diameter, $D = 4.57(68/150)^{1/3} = 3.50$ m.

Adopting a maximum entrance velocity of 1.7 m s^{-1} (Zipparo and Hasen, 1993), the inlet area of the flume entrance to the scroll is $68/1.7 = 40 \text{ m}^2$. Assuming a height of 5 m, the width of the entrance flume is $40/5 = 8$ m.

SCROLL CASE DESIGN

A concrete scroll with a contracting section is adopted. Equation (12.19) gives R values as a function of the position angle θ of the spiral (Fig. 12.13). The vortex strength

$$K = 30\eta g H / \pi N = 5.61 \text{ m}^2 \text{s}^{-1}.$$

The outer diameter of the guide vane assembly $(=2r_0) = 1.5D = 5.25$ m. Therefore $r_0 = 5.25/2 = 2.625$ m. The height of the inflow section at guide vane assembly, $h_0 = 0.4D = 1.40$ m. $\alpha = 7.47$ with $\beta_1 = \beta_2 = 15°$. For the nose angle, a partial spiral is suitable because of large discharges and low heads. Adopting a 240° spiral, the heights H_0 (Fig. 12.13(b)) can be calculated for different values of θ.

Assuming the maximum spiral height to be 5 m (to conform to the entrance height adopted) and assuming a linear variation between h_0 and H_{0max}, i.e. between 1.4 m and 5 m, the following table gives the values of $H_0(\theta)$:

$\theta°$ from nose	0	30	60	90	120	180	240
H_0 (m)	1.40	1.85	2.30	2.75	3.20	4.10	5.00

Knowing that $r_1 = (H_0 - h_0)/\alpha + r_0$, equation (12.19) gives the spiral radius for angle θ, and the results are tabulated below:

	$\theta°$ from nose						
	0	30	60	90	120	180	240
H_0 (m)	1.40	1.85	2.30	2.75	3.20	4.10	5.00
r_1 (m)	2.62	2.68	2.71	2.81	2.87	2.99	3.10
R (m)	2.62	4.57	6.55	8.25	9.75	12.52	14.52

The maximum width of the spiral is $R_{60} + R_{240} = 6.55 + 14.52 = 21.07$ m. Practical widths are within the range $(2.7 - 3)D$ and hence let us adopt a maximum width of $2.7D = 9.45$ m. Therefore the reduced maximum radius R'_{240} is obtained as

$$R'_{240} = 9.45 \times 14.52/21.07 = 6.52 \text{ m}.$$

This reduction increases the vorticity of the flow which can be obtained through equation (12.19). Thus the new vortex strength, $K' = 10.55 \text{ m}^2 \text{s}^{-1}$. Using this new K' value the spiral radii for all angles can be recomputed and the results shown as in the following table:

$\theta°$ from nose	0	30	60	90	120	180	240
R (m)	2.62	3.53	4.18	4.82	5.22	6.02	6.52

The maximum scroll case width is now $4.18 + 6.52 = 10.70$ m.

DRAFT TUBE SETTING

Using Thoma's criterion,

$$Y_s = B - \sigma H.$$

Using a value $\sigma_c = 1.15$

$$Y_s = 10 - 1.15 \times 10$$

$$= -1.5 \text{ m below TWL},$$

i.e. the unit is to be submerged.

The sectional details through the unit are shown in Fig. 12.26.

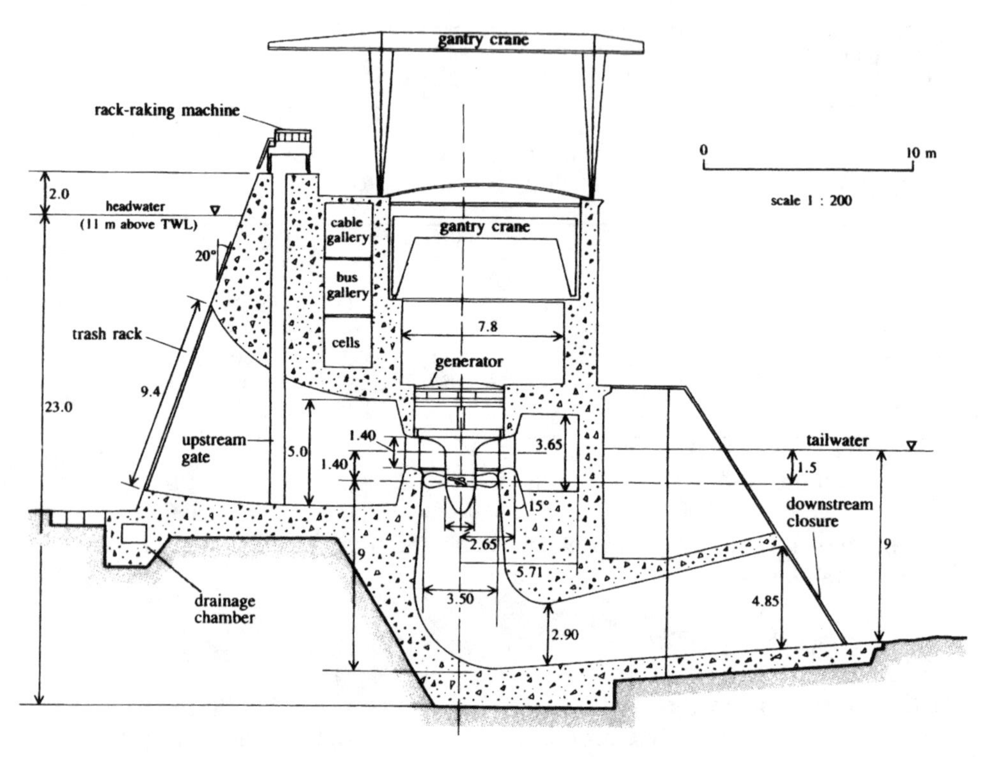

Fig. 12.26 Section through power station; dimensions in metres

References

Aqua-Media International (2004) *Hydropower and Dams World Atlas and Industry Guide*, Aqua-Media International, Sutton.

Bartle, A. and Hallowes, G. (2005) Hydroelectric power: present role and future prospects, *Civil Engineering*, Proc. of the Inst. of Civil Engrs., 158, November: 28–31.

Chaudry, M.H. (1987) *Applied Hydraulic Transients*, 2nd edn, Van Nostrand Reinhold Co, New York.

Doland, J.J. (1957) *Hydropower Engineering*, Ronald, New York.

Escande, L., Dat, J. and Nalluri, C. (1962) *Stabilité de Cheminée d'Equilibre Placée sur Canal de Fuite*. Académie des Sciences, Paris.

Fahlbusch, F. (1982) Power tunnels and penstocks – economics re-examined. *Water Power & Dam Construction*, 34 (June): 13–15.

Fox, J.A. (1989) Transient flow in *Pipes, Open Channels and Sewers*. Ellis Horwood Ltd, Chichester.

Gosschalk, E.M. (2002) *Reservoir Engineering: Guidelines for Practice*, Thomas Telford, London.

Gulliver, J.S. and Arndt, R.E.A. (eds) (1991) *Hydropower Engineering Handbook*, McGraw Hill, New York.

Haws, E.T. (1997) Tidal Power – a major prospect for the 21st century, *Water, Maritime and Energy*, Proc. ICE, 124, No. 1: 1–24.

Hayes, D. (2004) Small plants, *Int. Water Power and Dam Construction*, February: 12–13.

Jaeger, C. (1977) *Fluid Transients in Hydro-Electric Engineering Practice*, Blackie, Glasgow.

Kerr, D. (2005) Marine energy: getting power from tides and waves, *Civil Engineering – Special Issue*, Proc. ICE, 158, November: 32–9.

Linsley, R.K. and Franzini, J.B. (1979) *Water Resources Engineering*, 3rd edn, McGraw-Hill, New York.

Monition, L., Le Nir, M. and Roux, J. (1984) *Micro Hydropower Stations*, Wiley, Chichester.

Mosonyi, E. (1987) *Water Power Development*, 3rd edn, Vols I and II, Hungarian Academy of Sciences, Budapest.

—— (1988) Water power developments (low head plants), in *Developments in Hydraulic Engineering*, Vol.5 (ed. P. Novak), Elsevier Applied Science, London.

Müller, G, and Kauppert, K. (2004) Performance characteristics of water wheels, *Journal of Hydraulic Research*, 42, No. 5: 451–60.

Nechleba, M. (1957) *Hydraulic Turbines*, Artia, Prague.

Novak, P. (1983) *Waterhammer and Surge Tanks*, 3rd revised edn, International Institute for Hydraulic and Environmental Engineering, Delft.

Popescu, M., Arsenic, D. and Vlase, P. (2003) *Applied Hydraulic Transients for Hydropower Plants and Pumping Stations*, Balkema, Rotterdam.

Raabe, J. (1985) *Hydro Power*, VDI, Düsseldorf.

Shaw, E.M. (1994) *Hydrology in Practice*, 3rd edn, Chapman & Hall, London.

Taylor, R.H. (1983) *Alternative Energy Sources*, Adam Hilger, Bristol.

Thorpe, T.W. (1999) *A Brief Review of Wave Energy*, UK Dept. of Trade and Industry, ETSU-R120, London.

Twort, A.C., Ratnayaka, D.D. and Brandt, M.J. (2000) *Water Supply*, 5th edn, Edward Arnold, London.

Viollet, P.-L. (2005) *Histoire de l'Energie Hydraulie*, Presses de l'Ecole Nationale de Ponts et Chaussées, Paris

Vlaudut, T. (1997) Reservoirs and environment, *Int. Water Power and Dam Construction*, March: 28–30.

Wilson, E.M. and Balls, M. (1987) Tidal power generation, in *Developments in Hydraulic Engineering*, Vol.4 (ed. P. Novak), Elsevier Applied Science, London.

Wylie, E.B. and Streeter, V.L. (1993) *Fluid Transients in Systems*, Prentice Hall, Englewood Cliffs, New York.

Zipparo, V.J. and Hasen, H. (1993) *Davis' Handbook of Applied Hydraulics*, 4th edn, McGraw Hill, New York.

Chapter 13

Pumping stations

13.1 Introduction

Pumping facilities have to be provided for water supply if economical gravity systems cannot be constructed. Most large pumping stations abstract water from surface sources such as rivers, canals, lakes, etc., whereas groundwater abstraction is usually provided by smaller (usually submerged) pumping units. Sometimes pumping installations may have to be provided to pump surface water (low-lift drainage installations) behind a dyke (e.g. cofferdam enclosure) or from a shallow sump. High-lift drainage pumps may sometimes be arranged in a grid pattern covering a large area which needs to be drained (e.g. lowering a water table). Pumping installations may be needed to pump sewage or storm sewer flows from low-level networks to high-level screening–treatment plants. Booster pumps may be needed in water supply networks to boost pressure heads. Reversible pump–turbine units are utilized in pumped storage hydroelectric schemes. In all of these cases, different types of pumps with appropriate sump and intake arrangements at the abstraction point are used to transfer the liquid from low to high levels.

13.2 Pumps and their classification

13.2.1 General

Pumps are hydraulic machines which convert mechanical energy (imparted by rotation) into water energy used in lifting (pumping) water/sewage to higher elevations. The mechanical energy is provided by electrical power (motor) or diesel, gas or steam prime movers using either vertical or horizontal spindles.

13.2.2 Types of pumps

Pumps may be classified as rotodynamic (radial (centrifugal), axial or mixed-flow type)) and reciprocating (positive displacement) pumps. Also, there exist several other types such as air-lift (pneumatic) and jet pumps, screw and helical rotor pumps, etc.

In the rotodynamic-type pump water, while passing through the rotating element (impeller), gains energy which is converted into pressure energy by an appropriate impeller casing (Fig. 13.1(a) and Fig. 13.1(b)). The reciprocating pump (Fig. 13.1(c)) utilizes the energy transmitted by a moving element (piston) in a tightly fitting case (cylinder). The applications of positive displacement pumps, air-lift, and jet pumps are very much limited nowadays and most pumping stations utilize centrifugal-type

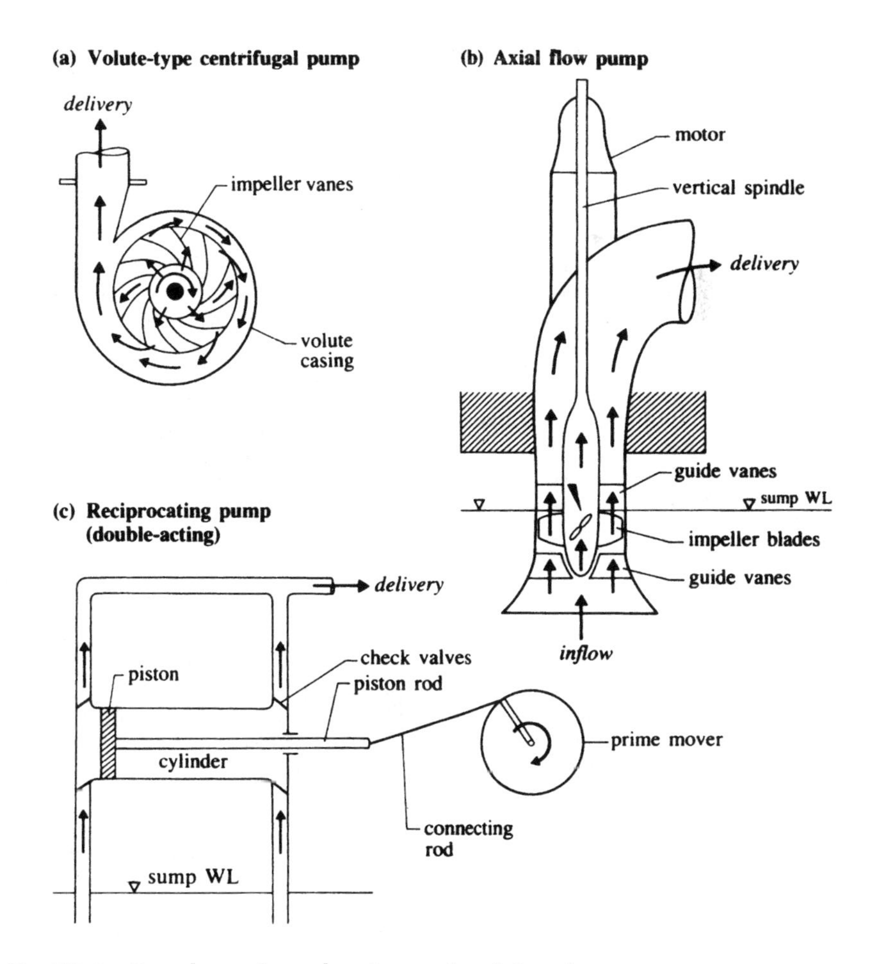

Fig. 13.1 Rotodynamic and reciprocating (piston) type pumps

pumps. Anderson (1994) and Wislicenus (1965) give details of construction and performance characteristics of pumps. For a detailed discussion of rotodynamic pump design see Turton (2006).

The types of pumps and their applications are listed in Table 13.1 as a guideline for their selection.

13.2.3 Specific speed N_s

Specific speed is the parameter which characterizes the rotodynamic pumps more explicitly and is given by

$$N_s = NQ^{1/2}/H_m^{3/4} \tag{13.1}$$

where Q is the discharge ($1\,s^{-1}$), H_m the total (manometric) head (m), and N the rotational speed (rev min^{-1}). The ranges of the specific speed for the various types of pumps are shown in Table 13.2.

Table 13.1 Types of pumps and their applications

Type	*Discharge*	*Head*	*Applications and remarks*
Rotodynamic pump:			To pump water and sewage; higher
radial-flow type	Low	High (>30 m)	efficiencies in pumping clean water; sewage pumps are slow speed,
axial-flow type	High	Low (up to ≃ 15 m)	unchokable pumps (capable of disintegrating solids);
mixed-flow type	Medium	Medium (25–30 m)	usually axial flow (propeller) type (Fig. 13.1(b))
Reciprocating pump	Low	Medium	Very viscous fluid pumping; well and borehole pumping; leakages unavoidable; inefficient
Air-lift pump	Low	Low	Groundwater recovery from wells with large quantities of sand and silt; inefficient
Jet pump	Low	Medium	Combined with centrifugal pump; borehole abstraction; inefficient
Screw pump	High	Low	Archimedes screw principle; low speed; to pump activated sludge or liquids with flocs
Helical rotor pump	Low	Low	Positive displacement pump with helical rotor and stator elements; used for sewage or liquids with suspended matter pumping; borehole pump for water supply

Table 13.2 Specific speeds for rotodynamic pumps

Type		*Specific speed*
Radial flow	(i) slow speed	300–900
	(ii) medium speed	900–1500
	(iii) high speed	1500–2400
Mixed flow		2400–5000
Axial flow		3400–15 000

13.2.4 Pump operation

A typical installation of a centrifugal pump, transferring water from a sump (low level) to a higher elevation, is shown in Fig. 13.2. V_s and V_d are the velocities in suction and delivery pipes; h_{fs} and h_{fd} are the head losses in suction and delivery pipes (friction, valves, bends, etc.); h_s and h_d are the suction and delivery heads, the sum of which is the total static lift, H_s. H_m is the total (manometric) head the pump must develop to lift water through the static lift.

Referring to Fig. 13.2, the manometric head can be written as

$$H_m = H_s + h_{fs} + h_{fd} + V_d^2/2g. \quad (13.2)$$

Equation (13.2) can be rewritten with negligible velocity head as

$$H_m = H_s + h_{fs} + h_{fd}. \quad (13.3)$$

If the impeller losses are considered, the efficiency (manometric efficiency) of the pump is affected. The total head the pump must develop to overcome the impeller and system losses (in pipelines and pipe fittings) in lifting the water through a given static lift is given by

$$H = H_m/\eta \quad (13.4)$$

where η is the manometric efficiency of the pump.

The pump characteristics, such as discharge versus total head, efficiency and power for the design speed of rotation are obtained from the manufacturers. Pumps are sometimes required to operate under varying speeds (due to varying demands) and their corresponding characteristics change considerably. Within a reasonable range of operational speeds the performance may be assessed with the help of hill-diagrams (curves of equal efficiencies).

From dynamic similarity criteria, the corresponding discharge and head that a pump can develop with varying speed are written as

$$Q_2 = Q_1(N_2/N_1) \tag{13.5}$$

and

$$H_2 = H_1(N_2/N_1)^2. \tag{13.6}$$

Where the demand for pumping varies (e.g. variation in the inflow rate during the day at a sewage pumping station) several pumps are operated in a 'parallel' arrangement with any number of pumps brought into operation according to the level in the suction well (sump). For pumping against large heads (deep borewell pumping) pumps are operated in 'series', i.e. water from the first pump (impeller) is delivered to the inlet of the second pump, and so on. A pump with impellers in series is also called a multistage or booster pump. All pumps in the series system must operate simultaneously.

13.2.5 Pump setting (suction lift)

Referring to Fig. 13.2 application of Bernoulli's equation between the sump water level and the pump (impeller) inlet gives

$$p_s/\rho g = p_a/\rho g - (h_s + h_{fs} + V_s^2/2g). \tag{13.7}$$

Equation (13.7) indicates that the pressure at the pump inlet, p_s, is below atmospheric pressure, p_a, and if this negative pressure exceeds the vapour pressure limits cavitation sets in. To avoid cavitation (otherwise the efficiency drops and the impeller becomes damaged) the suction head, h_s, is limited so that the pressure at the inlet is equal to the allowable vapour pressure p_v. Other measures such as minimizing head losses (by choosing large diameters and shorter lengths of suction pipes without regulating valves) are also taken.

The maximum suction head (pump setting) can thus be given as

$$h_s \leq (p_a - p_v)\rho g - (h_{fs} + V_s^2/2g). \tag{13.8}$$

From equation (13.7) the net positive suction head ($NPSH$) can be written as

$$(p_s - p_v)/\rho g = NPSH = (p_a - p_v)/\rho g - (h_s + h_{fs} + V_s^2/2g) \tag{13.9}$$

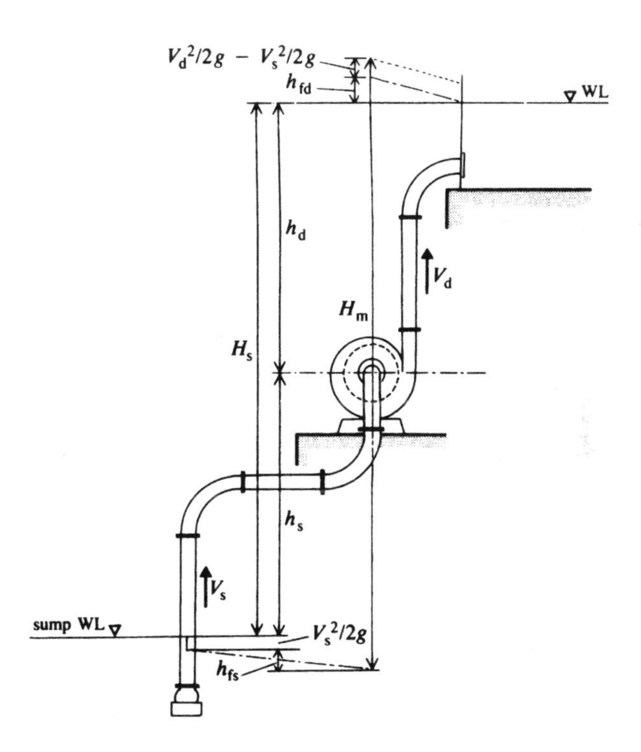

Fig. 13.2 Definition sketch of pump installation

and Thoma's cavitation number, σ, given by (equation (4.14))

$$\sigma = NPSH/H_m, \tag{13.10}$$

is found to be related to specific speed. The *NPSH* values are supplied by the pump manufacturers.

Defining the suction specific speed, N_{ss}, as

$$N_{ss} = NQ^{1/2}/(NPSH)^{3/4} \tag{13.11}$$

the cavitation number, σ, is given by

$$\sigma = (N_s/N_{ss})^{4/3}. \tag{13.12}$$

The suction specific speeds for most centrifugal and axial (propeller) type units range between 4700 and 6700. The critical cavitation number, σ_c, to avoid cavitation is suggested (from model tests) to be

$$\sigma_c = 0.103(N_s/1000)^{4/3}. \tag{13.13}$$

Electrosubmersible units, which are in widespread use nowadays, eliminate the need for suction pipes, and the problems of cavitation and cooling are avoided.

13.2.6 Priming of the pump

Prior to the starting-up of a centrifugal-type pump, both the pump and its suction pipe must be full of water. If the pump is located below the water level in the sump this condition will always exist; otherwise the air in the pump and its suction pipe must be expelled and replaced by water, i.e. ‘primed’. A foot valve (non-return type) at the suction inlet end, together with an air vent cock in the volute casing of the pump, assist the priming process. Self-priming pump units which do not need any external priming are also commercially available.

13.3 Design of pumping mains

13.3.1 System characteristic, duty point, and operational range

The efficient design of a pumping station largely depends on the piping system used to convey the fluid. Friction and other losses in the system, which are a function of discharge, have to be overcome by the pump, the performance of which is interrelated with the external pipe (system,) characteristic.

The pump characteristic at a given speed, N, is a function of discharge and is written as

$$H_m = AQ^2 + BQ + C \tag{13.14}$$

where A, B and C are the coefficients which can be evaluated from its $Q(H)$ curve.

The system curve is written as

$$H_m = H_s + h_m + h_f \tag{13.15}$$

where h_m and h_f are the minor and major system losses respectively; both these losses can be expressed as KQ^2, K being the appropriate loss coefficient. The solution of equations (13.14) and (13.15) (either analytical or graphical – Fig. 13.3) gives the duty point of the installation at which the pump delivers the required discharge. It is important that the duty point coincides with the peak efficiency of the pump for its economical opera-

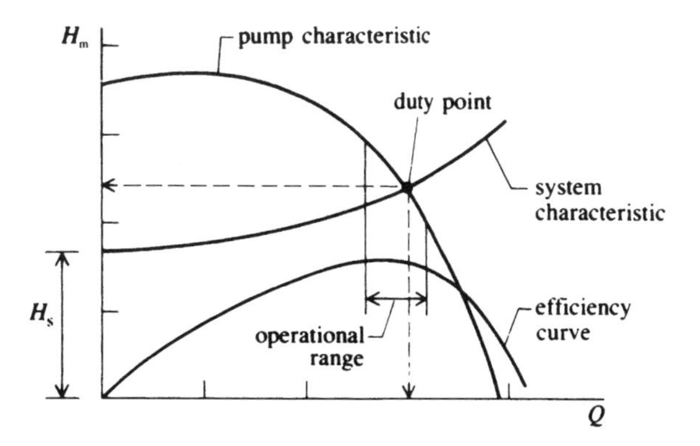

Fig. 13.3 Pump–pipeline characteristics

tion. For varying discharges the unit may be throttled over an operational range at the expense of the head. However, the extent of the operational range may have to be limited to one giving reasonably high pump efficiencies.

13.3.2 Selection of pipeline diameter

The best possible diameter of a pipeline system for a pumping unit depends on the system characteristic. The various pipeline systems which could be considered as matching the pump characteristic are shown in Fig. 13.4. It must be noted that each operational point corresponds to a

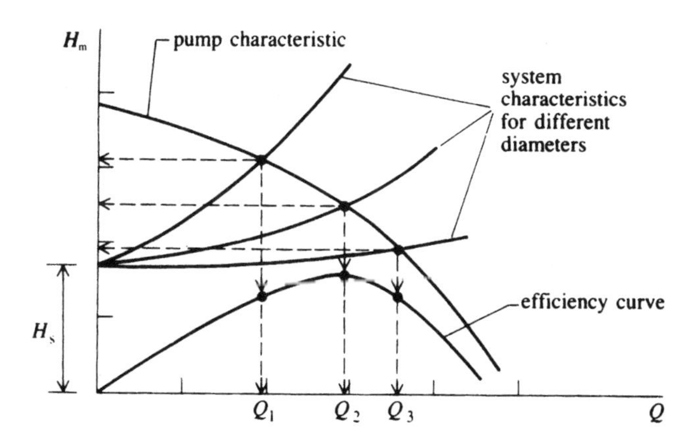

Fig. 13.4 Selection of system pipeline diameter

particular efficiency of the pump, and the system selection largely depends on the discharge–head requirements and on pump efficiency.

13.3.3 Variations in sump level, pumping demand, and friction losses

During the operational period of a pumping system the demand may increase or decrease, the water level in the sump well may change (e.g. variation in drawdown of a well) and the wall roughness of the pipeline may increase with age; even partial clogging of the suction pipe (e.g. clogging of the well screen) may occur.

A system initially designed with its characteristic as (a) corresponding to a discharge Q_1 is shown in Fig. 13.5. If the water table in the well, for example, decreases by an amount ΔS (dry period) the system curve changes to (b), delivering a reduced discharge of Q_2. A larger pump with a different $Q(H_m)$ characteristic may be necessary to maintain the original discharge Q_1 with the new system characteristic. If the conditions return to normal (i.e. $\Delta S = 0$; wet period) the original system curve (a) controls the operation, giving a discharge $Q_3 > Q_1$. If no storage facilities are available

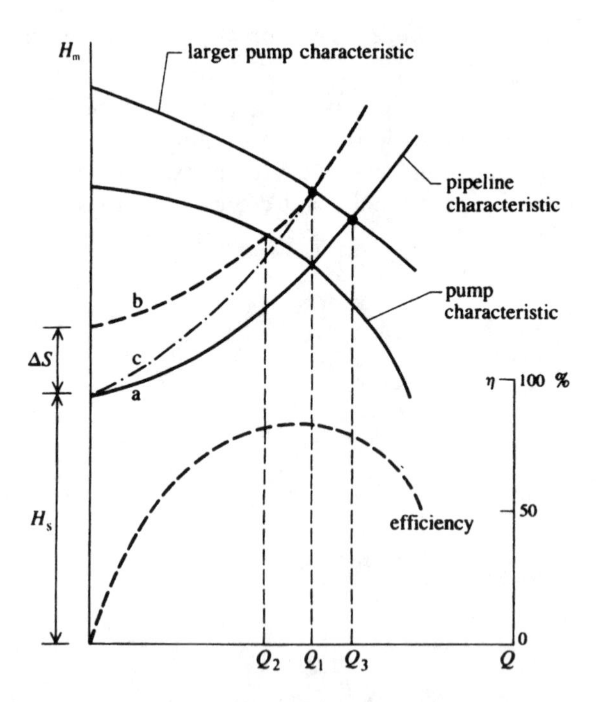

Fig. 13.5 Pipeline and pump characteristics with varying sump levels

to store the extra water, the pump may be throttled, changing the system curve to (c), thus reducing the discharge to Q_1 and incurring certain loss of energy which, if possible, should be avoided. For continuous operation of a pumping station the units must operate with as high an efficiency as possible, to minimize the running costs of the station.

13.4 Classification of pumping stations and intakes

(a) Abstraction from surface sources

The pumping station is fed from an open-surface source such as a canal, a river, or a reservoir, often through a sump and an intake. With water levels varying over a large range, sediment may enter the sump and intake. Sediment traps and screens (to trap floating debris) are therefore usually provided (Fig. 13.6). The station will probably also have multiple pumps (including standby units) which cater for the changes in sump levels.

(b) Water supply from treatment plants

In the absence of gravity flow, treated water is supplied to a distribution network or a storage tower-reservoir through a pumping station. Silt- and debris-free water is directed to either a wet or a dry sump (without screens or traps) from which it is pumped to the network (booster pumps) or to another storage tank. The possible arrangements of the sump layout are shown in Fig. 13.7.

The wet well arrangement (Fig. 13.7(a)) is simple, economical and most widely used. Pumps installed below water level (submersible) are preferable (reduction in suction lift and no priming needed) but involve maintenance problems. The dry well arrangement (Fig. 13.7(b)) is more reliable because of easy access for pump maintenance at all times.

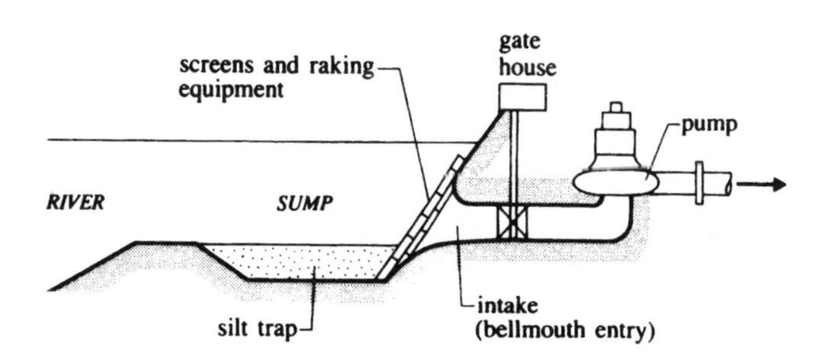

Fig. 13.6 River intake (after Prosser, 1977)

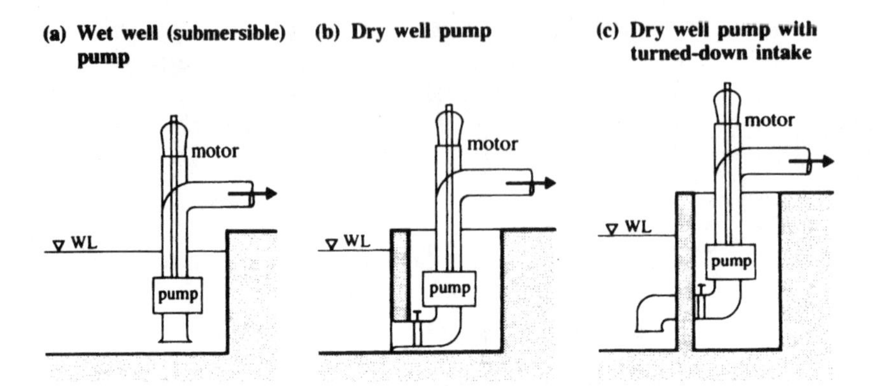

Fig. 13.7 Wet and dry well (sump) pump installations. Note that the best location for the pump in a dry well is below WL

The bellmouth entry to the pump suction pipe suppresses flow separation and ensures uniform flow throughout the intake cross-section. The turned-down bellmouth (Fig. 13.7(c)) allows a lower sump water level than a horizontal intake (Fig. 13.7(b)) and is less prone to vortex formation (Section 13.5).

(c) Stormwater pumping installations

Provision of coarse screens with wet well arrangements is recommended in this type of pumping installation. Since the pump duty is intermittent, maintenance is possible during dry periods. Similar arrangements apply to de-watering and land-drainage installations.

(d) Sewage (untreated) pumping to treatment works

The pump sump size is so designed as to keep the approach flow velocities sufficient for the solid matter to remain in suspension. Stagnant areas or corners must be avoided. A dry well with a turned-down bellmouth is the most suitable type of sump, and a typical section of such an arrangement is shown in Fig. 13.8.

Screening of the sewage prior to pumping should be avoided if possible as the screening equipment requires regular maintenance with additional costs for the disposal of the screened wastes. Normally small pumping stations (pump sizes up to 200 mm in diameter), preferably with two pump units, are desirable to cope with the pumping of untreated sewage. Pumps with appropriate impeller design are available to pass almost all solids that can enter the pump suction bellmouth (Prosser, 1992)

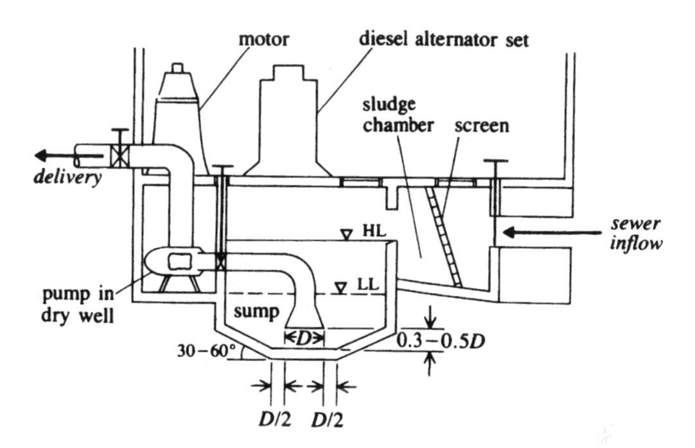

Fig. 13.8 Pump sump and intake for sewage pumping

(Section 13.6). Archimedes screw pumps are used to lift wastewater with large floating bodies into treatment works.

(e) Abstraction from boreholes

The installations are normally of the wet well type with pumps located within the wells. In deep wells special multistage pumps are used, whereas for shallow wells the pumps may be located at ground level. Well screens are essential to prevent sand from entering into the system. The deep well pumps are normally less bulky (around 100–400 mm diameter) to fit into well diameters of 150–600 mm.

13.5 Sump design

The most important aspect of pumping installation design is the suction well (sump), with a good intake (inlet to suction pipe) arrangement. In the case of shallow submergence of the inlet, the formation of a local vortex at the water surface is often observed. Such a situation may also arise if the volume of water in the sump is inadequate, in which case the sudden starting of the pump lowers the water surface. Equally sudden steep velocity gradients, high velocities, sudden changes in the flow direction as water enters the bellmouth inlet in the sump and several pump units set in a long narrow sump with inflow from one end also aggravate the formation of vortices.

The vortex entrains large quantities of air into the pump suction pipe. This causes a drop in efficiency and structural vibrations, and increases the corrosion damage of the pump and its accessories.

A good sump design must therefore avoid the formation of vortices; this can be achieved by directing the flow uniformly across the sump width with an approach velocity of around $0.3\,m\,s^{-1}$ and by avoiding abrupt expansion of the side walls and large stagnation zones in the sump (minimizing large-scale swirling flow). For larger expansion ratios (area at outlet/area at inlet >2) vanes may have to be provided for uniform flow distribution. The pump intake must be located in the direction of the approach flow if possible. Multiple intakes from the same sump should be separated by dividing walls (to minimize interference). The most effective free-surface vortex-suppressing device is a horizontal floor grating, installed about 100 to 150 mm below the water level. Other devices such as floating rafts, grating cages, and curtain walls are also used as vortex suppressors. Provision of a bellmouth entry (Fig. 13.9(a)) at the inlet to the suction pipe minimizes entrance losses and facilitates smooth axial flow. Some typical sump layouts of good design (Prosser, 1977) are shown in Figs 13.9(b)–13.9(e).

For intakes with proper approach flow conditions (without any vortex-suppression devices) the minimum required submergence could be written as

$$h/d = a + bF_{rd} \tag{13.16}$$

where h is the depth of submergence to the centre of intake pipe of diameter, d, F_{rd} $(=V/(gd)^{1/2}$, V being the velocity in the intake pipe) is the Froude number and $a=0.5$–1.5 and $b=2$–2.5 (Knauss, 1987). At low Froude numbers (≤ 0.3, i.e. large-size intakes) a value of a of at least 1 is recommended.

The minimum sump volume for good flow conditions also depends on the maximum allowable number of pump starts in a given time. Start-up of an electric motor generates considerable heat energy in the motor, and hence the number of starts must be limited.

The minimum sump volume, V_{min}, between stop and start of a single pump unit is given by

$$V_{min} = Q_P T/4 \tag{13.17}$$

where Q_P is the pumping rate and T is the time between starts. Equation (13.17) then suggests that for a frequency of 10 starts per hour ($T = 6\,min$) the minimum volume is 1.5 times the pumping rate per minute. When two or more pumps are used the start levels are normally staggered and equation (13.17) is applied to the largest pump.

For a multisystem an additional volume of 0.15 times the plan area of the sump (in m^3) should be provided. However, there are certain restraints in the maximum volume of the sump, e.g. a septicity problem of sewage in the sewage sump (BS, 1987).

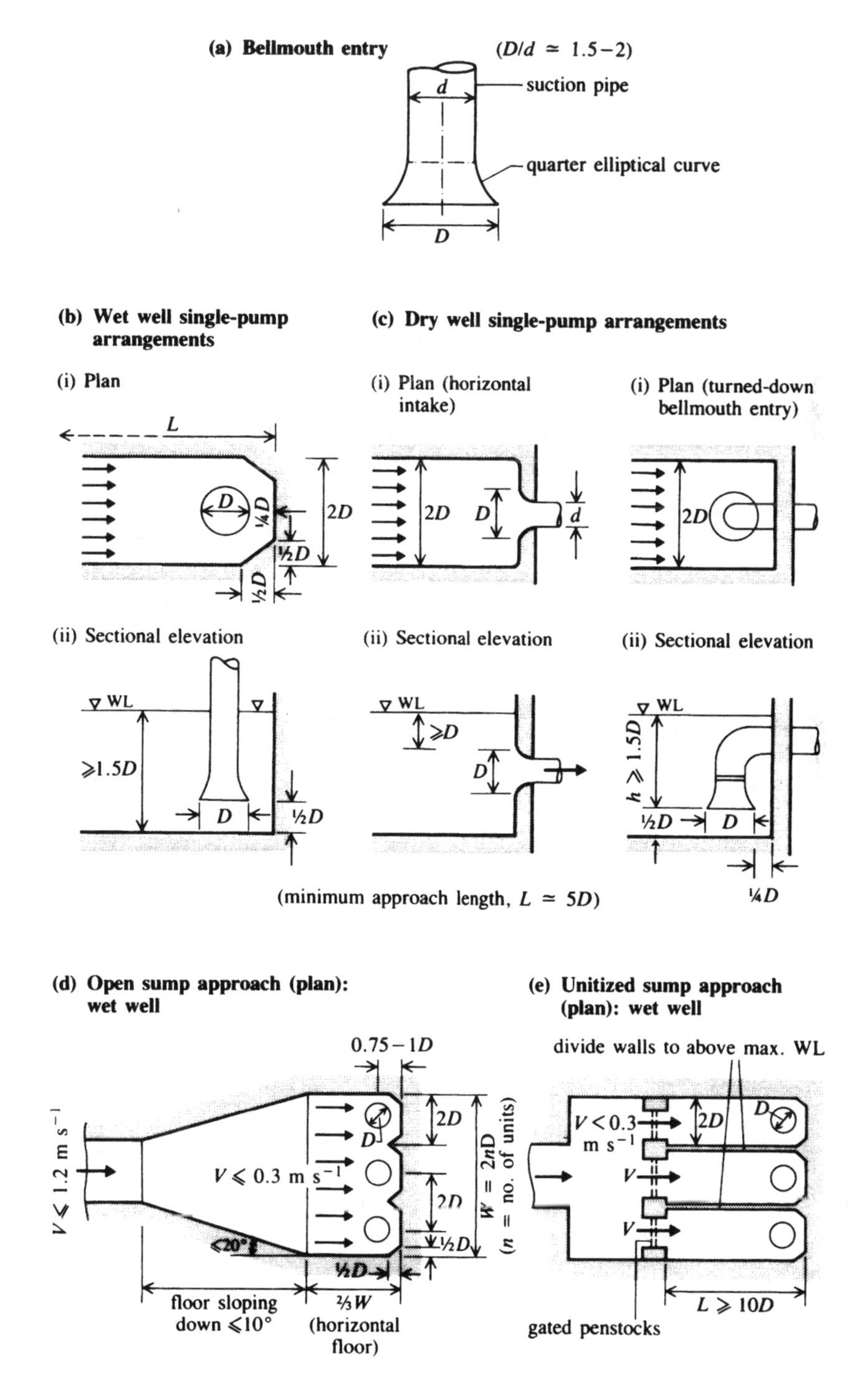

Fig. 13.9 Typical recommended sump designs (after Prosser, 1977)

13.6 Screening devices

All major pumping stations (especially sewage and stormwater pumps) have to be provided with bar screens to prevent large objects from entering the wet well. They consist of steel bars spaced at 20–40 mm with a blockage ratio (area of bars/total area) between 0.3 and 0.5. The spacings could be larger if the pumps are able to pump solids (sewage pumps). Appropriate mechanical cleaning (raking) devices for the screens may also be provided. The screens are usually laid at an angle of 60°–90° to the horizontal and a head loss of at least around 150 mm (Section 9.2) must be incorporated in the hydraulic design of bar screen devices.

The presence of screens increases turbulence (proportional to bar size) which would cause undesirable flow conditions if the screens were situated very close to the pump. Any swirling in the flow may be reduced by a screen or grid, or even by a honeycomb device. Alternatively, splitters may be used under the bellmouth intake as an antiswirl device.

13.7 Benching

As it is a desirable practice to pump all the solids, their deposition on the floor or walls of the wet well sump must be minimal. This is achieved by providing a suitable benching for the (self-cleansing) pump sumps.

The benching should be steep with a minimum slope of 45°; if possible, 60°–75° is preferable. It should extend right up to the pump intake where it terminates in a small pit in which the intake bellmouth of the pump must be installed. It is advisable to locate stop levels–override controls in such a way that the sump level is drawn down to the lowest possible level at regular intervals (e.g. once a day).

13.8 Surges

It is essential to protect the pipeline and the pumping station against pressure surges (waterhammer) in the system due to the closure or opening of valves, or sudden power failure causing pump stoppage. The undesirable effects of surges in pumping station systems can be controlled by antiwaterhammer devices (flywheels, air vessels, etc.). For the treatment of waterhammer analysis and surge protection see e.g. Jaeger (1977), Novak (1983), Sharp and Sharp (1996) and Popescu *et al.* (2003).

13.8.1 Flywheels

A flywheel coupled with the pump unit provides additional inertia so that the pump continues to rotate even after a power cut. This reduces the pressure transients (analogous to slow valve closure).

The rate of change of speed of a pump is given by

$$dN/dt = 900\rho gQH/\pi^2 IN\eta \qquad (13.18)$$

where I is the moment of inertia of the rotating parts of the pump and motor and η is the pump efficiency at discharge Q, head H and speed N. This method of providing protection without any ancillary equipment is recommended for small installations.

13.8.2 Bypasses and pressure relief valves

When a pump stops suddenly a low-pressure wave in the delivery pipe and a high-pressure wave in the suction pipe develop, resulting in a discharge from the suction side to the delivery side through a non-return valve (Fig. 13.10) in the bypass. In addition (or alternatively) pressure relief valves and air inlet valves may be provided in the pipeline system.

13.8.3 Surge tanks and air vessels

In contrast to a hydropower plant, it is usually not practicable to construct an open surge tank in pumping installations to suppress the surges. Instead, a closed air vessel with an air compressor is provided close to the pump. When a negative transient occurs due to pump ‘tripping’ the air vessel supplies water under pressure into the pipeline, thus reducing the

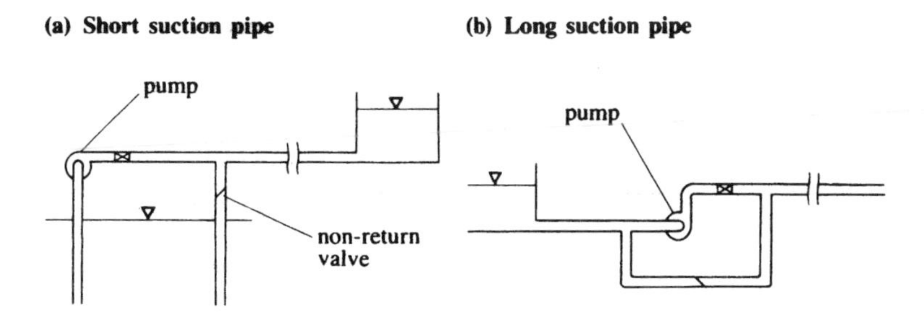

Fig. 13.10 Bypass arrangements

pressure drop. On the other hand, it accepts water when the pump opens, i.e. under positive transient conditions.

Between the pump and air vessel (Fig. 13.11) a check valve has to be provided, which closes quickly after a power failure as the pump, during its slowing down, cannot deliver against the head maintained by the air vessel. The compressed air delivery is triggered by the predetermined maximum and minimum water levels in the air vessel.

Analogous to surge tank operations, the simplified solution of dynamic and continuity equations of an air vessel for a sudden closure ($Q=0$; check valve closure) without entrance losses and slow isothermic pressure variations yields

$$H = H_0 \pm Q_0(LH_0/gAV_0)^{1/2} \qquad (13.19)$$

where the positive sign is taken for H_{max} and the negative sign for H_{min}, and suffix '0' refers to steady state conditions.

Design charts based on the assumption of complete closure, a ratio of head losses at entrance to air vessel in both directions of 2.5 at the air vessel base and a gas law of $HV^{1.2}$ = constant are available. The dimensionless surges for the total head loss coefficient K_s for a flow into the air vessel of 0.3 are shown in Fig. 13.12. Similar design charts are available for other values of K_s.

H_0 and v_0 are the steady-state pressure head in the air vessel and the velocity in the pipeline respectively, and V_0 is the corresponding air volume in the air vessel.

For a given pipeline (diameter D, length L and wall thickness e) and discharge Q_0 the pressure wave celerity a is calculated by

$$a = 1/[\rho(1/K + D/eE)]^{1/2} \qquad (13.20)$$

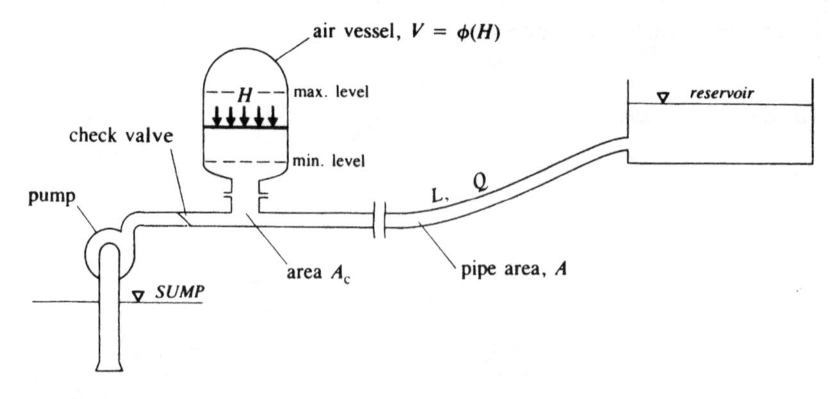

Fig. 13.11 Layout of an air vessel

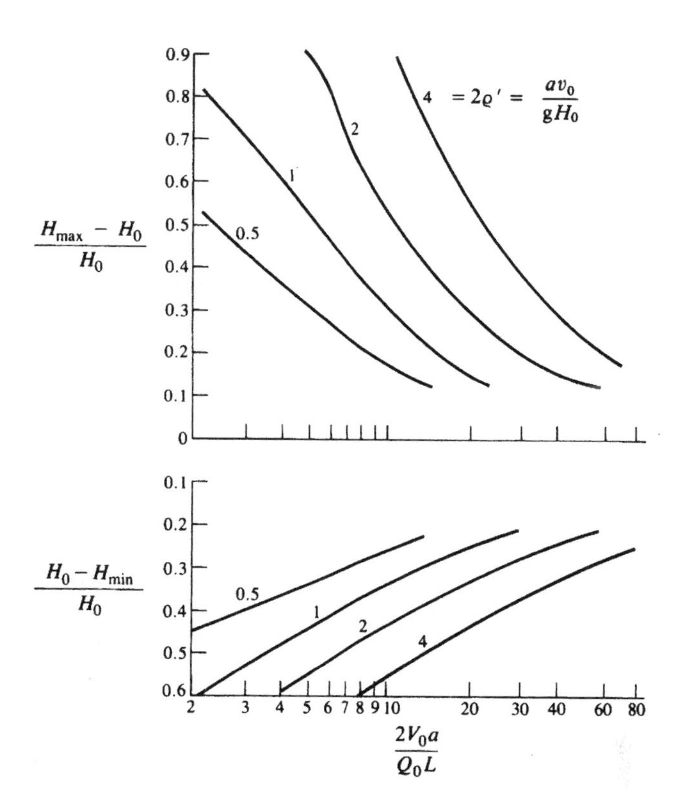

Fig. 13.12 Design charts for an air vessel with $K_s = 0.3$

where ρ is the density of water ($1000\,\text{kg}\,\text{m}^{-3}$), K is the bulk modulus of water ($\simeq 2 \times 10^9\,\text{N}\,\text{m}^{-2}$) and E is the Young's modulus of the pipe material (for steel, $E \simeq 2 \times 10^{11}\,\text{N}\,\text{m}^{-2}$).

The parameter $2\rho'$ is given by

$$2\rho' = av_0/gH_0 \tag{13.21}$$

and the head loss coefficient K_s by

$$K_s = (P + R')v_0^2/H_0 \tag{13.22}$$

where P is the friction loss term in $P_0 = Pv_0^2$ (Fig. 12.20) and R' is the head loss term at the entrance (throttle) to the air vessel, given by

$$R' = k(A/c_cA_c)^2/2g, \tag{13.23}$$

in which k is the head loss coefficient at entrance, A is the area of pipeline, A_c is the throttled (at entrance) area and c_c is the contraction coefficient.

Selecting H_{max} and H_{min}, suitable values of K_s, R' and V_0 (from $2V_0a/Q_0L$) can be found, and the required volume of the air vessel $V > V_{max}$ $(=V_0(H_0/H_{min})^{1/1.2})$ can be determined. It is more practical to choose H_{max} and K_s and for known H_0, a, v_0 (i.e. ρ') to find H_{min} and V_0 (Worked example 13.4).

13.9 General design considerations of pumping stations and mains

Rising mains (pipeline systems) from a pumping station normally follow the ground contours, with their carrying capacities dependent on their hydraulic gradients. The hydraulic gradient of a rising main is given by the pressure and friction heads determined by the characteristics of the pump–pipeline system.

When boosting the flow through an existing pipeline by installing a booster pump the elimination of suction troubles may have to be considered. A simple arrangement of a booster pump installed in a gravity pipeline is shown in Fig. 13.13. The increased hydraulic gradient (CB) to augment the flow may result in negative pressures upstream of the pump (between pump and E). A possible solution to avoid undue negative pressures is to place the pump upstream of E.

Booster pumps are normally commissioned whenever there is an increased demand (e.g. peak daytime consumption). Automatic cut-in and cut-out arrangements of the booster must be provided so that the pipeline is not subjected to high pressures during off-peak (night-time) periods when the demand reduces considerably. A simple and economical device to control the operation of the booster pump is to provide a balancing tank in the system at an appropriate point with the water levels in the tank triggering the pump on and off according to demand (see Twort, Ratnayaka and Brandt (2000).

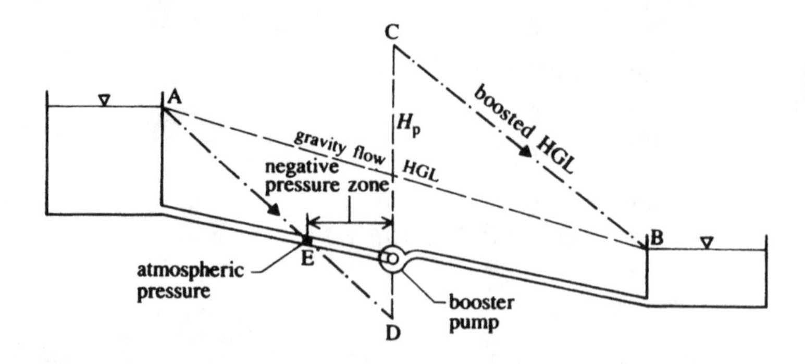

Fig. 13.13 Effect of booster pump on gravity line

Pumping stations for water supply from river intakes or boreholes are normally designed to discharge continuously over a period of 20–22 h a day. Boosters may also need storage facilities.

It is advisable to install more than one pump so that maintenance may be carried out without interrupting the supply. Often, several pumps of differing capacities are installed in a water supply pumping station, and the units are switched on or off as demand variations dictate. If a very fine controlled output is needed, variable-speed motors (expensive) must be used.

Sewage pumps must be so designed as to be capable of pumping the peak flows as they occur in sewers, with a maximum delay of no more than 12 h. Otherwise, the sewage may become septic. The flows may vary considerably throughout the year, month, week or day. The daily variations may be considered to be between zero (at night) to about six times the dry weather flow (d.w.f.) at times during the day, and it is preferable to install pumping capacity up to 6 d.w.f. using at least two or three pumps, each capable of about 2 d.w.f.

Pumping mains are usually designed for velocities of around 0.9–1 $m s^{-1}$ when supplying water at a constant rate throughout the day. This may be doubled for short-period pumping. Prior to the final selection of the pipeline diameter, economical analyses balancing the costs of large-diameter pipelines and their maintenance, and the savings in pumping costs due to the reduction in friction losses, must be carried out.

The minimum diameter suitable for pumping unscreened crude sewage is 100 mm and velocities around 0.75–2 $m s^{-1}$ are normally recommended. However, in stormwater–sewage pumping higher velocities (even 3–4 $m s^{-1}$) may be more economical as their operation is intermittent. The friction head losses in the case of sludge flows are much larger than for clear water flows.

The suggested (Bartlett, 1978) multiplying factors, F, for the calculation of head losses in sludge flows, given by

$$h_f = F\lambda(L/D)V^2/2g \tag{13.24}$$

(where λ is the friction factor for clear water flows), are given in Table 13.3.

A pumping station normally consists of a substructure below ground level in two compartments (if a dry well sump is used; Fig. 13.8). The roof of the dry well forms the floor of the superstructure of the pumping station, with appropriate access to the pumping units. The access to the wet well (sump) is normally provided from outside. The functional requirements, with the necessary dimensions, must be carefully assessed and incorporated in the design.

Table 13.3 Sludge flow head losses

Moisture content of sludge (%)	*Factor F (equation (13.24))*
100	1.00
98	1.25
96	1.75
94	2.75
92	4.75
90	7.00

Worked Example 13.1

The gravity flow between an impounding reservoir and a service reservoir is to be augmented by 25% owing to immediate future demand. Gravity flow data are as follows: head available = 100 m; pipeline length = 1000 m; pipe diameter = 250 mm; pipe roughness = 0.15 mm. A pump with an operating speed of 1450 rev min^{-1} will be used to boost the flow in the existing main.

1. What type of pump would you recommend?
2. Assuming the pump manometric efficiency to be 70% and the electrical motor efficiency to be 90%, calculate the continuous power consumption in kW.
3. Where on the pipeline would you install the pump?

Solution

GRAVITY FLOW

As given above, $L = 1000$ m, $D = 0.25$ m, $k = 0.15$ mm, and $H = 100$ m. Combining the Darcy–Weisbach and Colebrook–White formulae (Featherstone and Nalluri, 1995) gives

$$V = -2(2gDS_f)^{1/2}\log[k/3.7D + 2.51\nu/D(2gDS_f)^{1/2}] \quad \text{(i)}$$

where S_f is the friction gradient.

The first iteration with $S_f = 100/1000 = 0.1$ gives $V = 5.256\,\text{m}\,\text{s}^{-1}$. Therefore minor losses, $h_m = 1.5V^2/2g = 2.11$ m. Hence the new friction gradient, $S_f = (100.00 - 2.11)/1000 = 0.0979$.

A better estimate of V (from equation (i)) is $5.2\,\text{m}\,\text{s}^{-1}$, giving $Q = 255\,\text{l}\,\text{s}^{-1}$. The boosted flow is $1.25 \times 255 = 319\,\text{l}\,\text{s}^{-1}$.

Equation (i) gives (by iteration) the corresponding head loss $S_fL = 153$ m. Hence the total head needed to give $319\,\text{l}\,\text{s}^{-1} = 153 + 1.5V^2/2g = 156.23$ m. Therefore the increase in the head needed for the new discharge is 56.23 m. This head must be supplied by the pump and hence its output, ρgQH, is $9.81 \times 0.319 \times 56.23 = 176$ kW.

Therefore the power input = power output/overall efficiency = $176/0.9 \times 0.7 = 280$ kW.

TYPE OF PUMP

The specific speed, $N_s = NQ^{1/2}/H^{3/4} = 1450(319)^{1/2}/(56.23)^{3/4} = 1261$. A radial-flow type (medium-speed) pump is recommended (Table 13.2).

LOCATION OF THE PUMP

The pump is located so that the suction pressure at the inlet is zero. The increased friction gradient to obtain $319 \, \mathrm{l\,s^{-1}} = 156.23/1000 = 0.15623$. The available gravity head of 100 m is lost over a length L, beyond which the pump supplies the required additional head. Therefore, $L = 100/0.15623 = 640$ m. Hence the pump must be located at a distance of 640 m or less from the impounding reservoir inlet.

Worked Example 13.2

Two identical pumps are installed in a pumping station. One of them is mainly used as a standby, but due to increased future flow requirements it is proposed to bring this also into continuous operation. The following are the pump–pipeline characteristics:

Pump characteristics:				
discharge ($\mathrm{l\,s^{-1}}$)	0	20	40	60
head (m)	50	45	33	19
efficiency (%)	–	65	69	40
System characteristics:				
head losses (m)	–	1.0	4.0	10.0

The static head is 20 m.

1. Determine the present maximum flow.
2. Investigate the relative suitability of arranging the two pumps in series and in parallel to provide the increased flow. Justify your choice of arrangement.

Solution

The plots of the pump–pipeline characteristic curves with one pump operating, and two pumps operating in series and in parallel, are shown in Fig. 13.14.

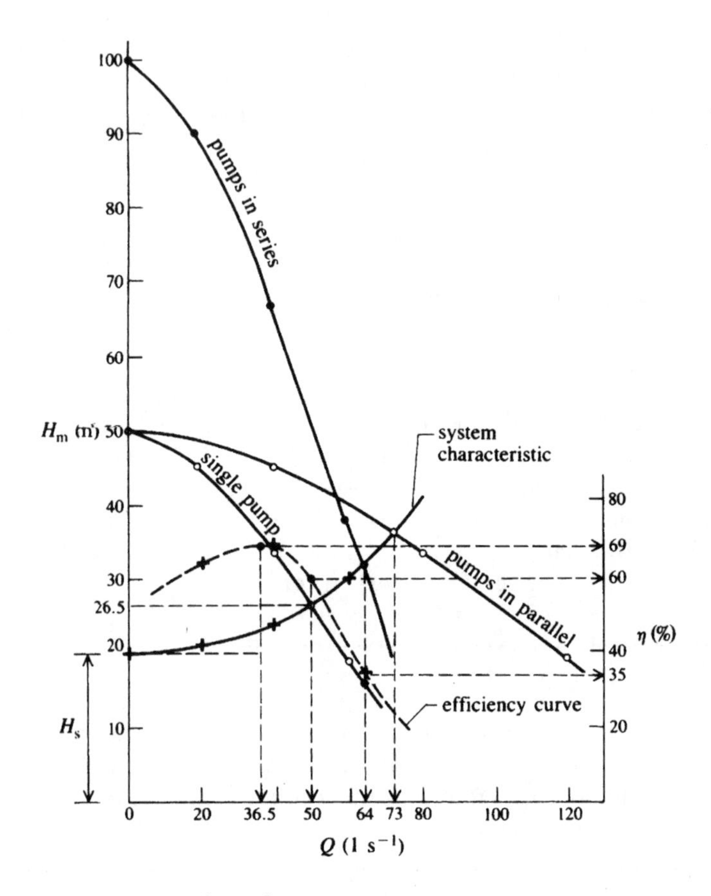

Fig. 13.14 Pump–pipeline characteristics

1. The present maximum flow is $50 \mathrm{l s^{-1}}$, with the pump working against a head of 26.5 m at an efficiency of 60%.
2. *Two pumps in parallel.* Pump characteristics for in-parallel operations are as follows:

Discharge ($\mathrm{l s^{-1}}$)	0	40	80	120
Head (m)	50	45	33	19

The total discharge is $73 \mathrm{l s^{-1}}$ at a head of 36 m. Therefore, the discharge per pump is $73/2 = 36.5 \mathrm{l s^{-1}}$, and the corresponding efficiency is 69%. Hence the input power per pump (when in parallel operation) is $9.81 \times 0.0365 \times 36/0.69 = 18.68$ kW. Therefore the total power consumption $= 2 \times 18.68 = 37.36$ kW.

3. *Two pumps in series.* Pump characteristics for in-series operation are as follows:

Discharge ($l\,s^{-1}$)	0	20	40	60
Head (m)	100	90	66	38

The discharge is $64\,l\,s^{-1}$ at a total head of 31.5 m. Hence the head per pump is $31.5/2 = 15.75$ m. The corresponding efficiency is 35%. Therefore the total power consumption in series operation is $2 \times 9.81 \times 0.064 \times 15.75/0.35 = 56.5$ kW.

Two pumps in parallel deliver $73\,l\,s^{-1}$ ($>Q$ in series) to a head of 36 m ($>H$ in series) with a power consumption of 37.36 kW ($<$power in series). Hence a parallel system is the best option to provide the increased flow.

Worked Example 13.3

Water from an abstraction well in a confined aquifer is pumped to the ground level by a submersible borehole pump; an in-line booster pump delivers the water to a reservoir, with the level 20 m above ground level at the well site. The system's pipework has the following characteristics:

System characteristics:					
discharge ($l\,s^{-1}$)	20	30	40	50	60
head losses (m)	1.38	3.14	5.54	8.56	12.21
Pump discharge ($l\,s-1$)	0	10	20	30	40
Borehole pump head (m)	10.0	9.6	8.7	7.4	5.6
Booster pump head (m)	22.0	21.5	20.4	19.0	17.4

Aquifer and well data are as follows: the coefficient of permeability of the aquifer, $K = 50$ m per day; the aquifer thickness, $b = 20$ m; the radius of the well, $r_w = 0.15$ m.

A pumping test suggested that the drawdown was 3 m when the abstraction rate was $30\,l\,s^{-1}$. The water table is 2 m below ground level. The drawdown is as follows:

$$z_w = (Q/2\pi Kb)\ln(R_0/r_w) \qquad \text{(ii)}$$

where Q is the abstraction rate in m^3 per day, and R_0 is the radius of influence of the well, which may be assumed to be linearly related to the abstraction rate.

Determine the maximum discharge which the combined pumps would deliver to the reservoir.

Solution

Referring to Fig. 13.15 and to equation (ii), the pump test results give

$$R_0 = 215\,\text{m} \quad \text{or} \quad R_0/Q = 7.15. \qquad \text{(iii)}$$

Q (l s^{-1})	20	30	40	50	60
R_0 (m) (from equation (iii))	143	215	286	357	427
z_w (m) (from equation (ii))	1.88	3.0	4.15	5.35	6.57
System losses (m)	1.38	3.14	5.54	8.56	12.21
Static head (m) (20 + 2)	22.00	22.00	22.00	22.00	22.00
Total system head (m)	25.26	28.14	31.69	35.91	40.78

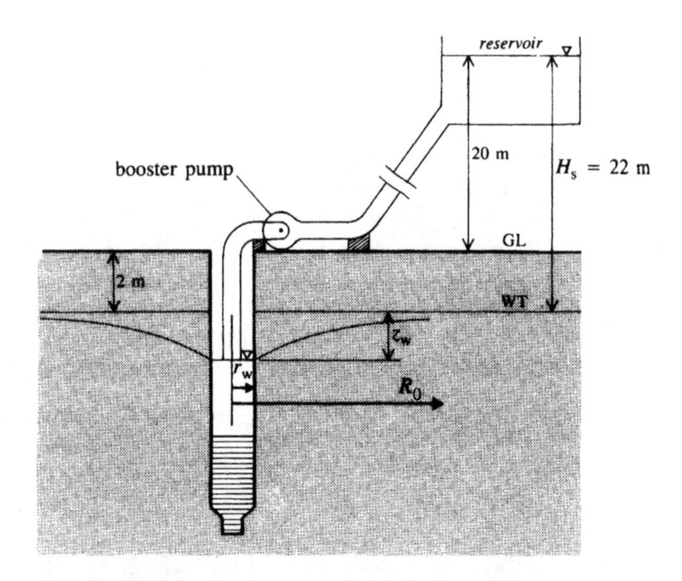

Fig. 13.15 Borehole pumping arrangement

For pump operation in series,

Q (l s^{-1})	0	10	20	30	40
Head (m)	32.0	31.1	29.1	26.4	23.0

Pump–pipeline characteristic curves (Fig. 13.16) show that the combined pumps deliver a discharge of $27.5\,\text{l s}^{-1}$, working against a total head of 26 m.

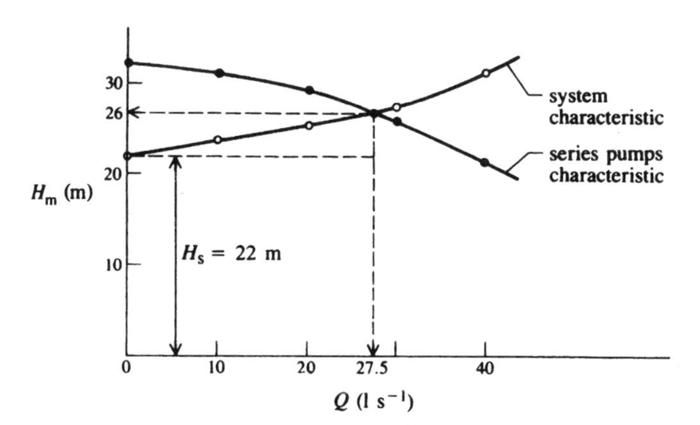

Fig. 13.16 Pump–pipeline characteristics

Worked Example 13.4

A pumping plant delivers $1.3\,m^3\,s^{-1}$ of water into a reservoir with a water level 75 m above steady state water level in the air vessel. The length of the supply line from the air vessel to the reservoir is 1500 m, its diameter is 800 mm and the friction coefficient is 0.019. Assuming a head loss coefficient $K_s = 0.3$ for flow into the vessel, calculate the minimum necessary volume of an air vessel required for the protection of the plant if the maximum permissible head rise at the pump is 40% of the pumping head. The pressure wave celerity may be assumed as $1000\,m\,s^{-1}$. Compare the result with that of the appropriate simplified method of solution.

Solution

As given above, $a = 1000\,m\,s^{-1}$, $\lambda = 0.019$, $L = 1500\,m$ and $Q = 1.3\,m^3\,s^{-1}$. The steady flow velocity, $v_0 = Q/A = 1.3/0.503 = 2.59\,m\,s^{-1}$. The head loss, $H_f = \lambda L v_0^2/2gD = 12\,m$. Therefore $H_0 = 75 + 12 + 10 = 97\,m$ (in absolute units). $(H_{max} - H_0)/H_0 < 0.4$ (given), so

$$2\rho' = av_0/gH_0 = 2.7 \quad \text{(equation (13.21))}.$$

From the design graph (Fig. 13.12), $2V_0a/Q_0L \simeq 20$, giving $V_0 \simeq 20\,m^3$. From the adiabatic gas law it follows that

$$V_{max} = V_0(H_0/H_{min})^{1/1.2}. \qquad \text{(iv)}$$

From the design graph (Fig. 13.12), $(H_0 - H_{min})/H_0 \simeq 0.37$, therefore $H_{min} = 97 - 0.37 \times 97 = 61\,m$, and from equation (iv), $V_{max} = 29.6 \simeq 30\,m^3$. Therefore, the minimum volume of the vessel $> 30\,m^3$.

The simplified solution (no losses; isothermal gas law) is given as follows: $H_{max} = 1.4H_0 = 1.4(75 + 10) = 119\,m$ (absolute units). From equation (13.19),

$$H_{max} - H_0 = Q_0(LH_0/gAV_0)^{1/2}$$

giving $V_0 \simeq 38\,m^3$, and

$$H_{min} = H_0 - Q_0(LH_0/gAV_0)^{1/2}$$

$$= 85 - 34 = 51\,m.$$

Therefore $V_{max} = H_0V_0/H_{min}$ (isothermal) $= 63\,m^3$, and the minimum volume $> 63\,m^3$.

References

Anderson, H.H. (1994) Liquid pumps, in *Kemp's Engineers Yearbook* (ed. C. Sharpe), Vol. 1, Benn, Tonbridge, Chapter F7.

Bartlett, R.E. (1978) *Pumping Stations for Water and Sewage*, Applied Science, London.

BS (1987) *Sewerage*, CP 8005, British Standards Publications, London.

Featherstone, R.E. and Nalluri, C. (1995) *Civil Engineering Hydraulics*, 3rd edn, Blackwell Scientific, Oxford.

Jaeger, C. (1977) *Fluid Transients in Hydraulic Engineering Practice*, Blackie, Glasgow.

Knauss, J. (ed.) (1987) *Swirling Flow Problems at Intakes*, IAHR Hydraulic Structures Design Manual, Vol. 1, Balkema, Rotterdam.

Novak, P. (1983) *Waterhammer and Surge Tanks*, 3rd revised edn, International Institute for Hydraulic and Environmental Engineering, Delft.

Popescu, M., Arsenic, D. and Vlaso, P (2003) *Applied Hydraulic Transients for Hydroelectric Plants and Pumping Stations*, Balkema, Rotterdam.

Prosser, M.J. (1977) *The Hydraulic Design of Pump Sumps and Intakes*, BHRA Fluid Engineering and CIRIA, Cranfield.

—— (1992) *Design of Low-lift Pumping Station*, CIRIA Report 121, Construction Industry Research and Information Association, London.

Sharp, B.B. and Sharp, D.B. (1996) *Water Hammer: Practical Solutions*, Arnold, London.

Turton, R.K. (2003) *Rotodynamic Pump Design*, Cambridge University Press.

Twort, A.C., Ratnayaka, D.D. and Brandt, M.J. (2000) *Water Supply*, 5th edn, Edward Arnold, London.

Wislicenus, G.F. (1965) *Fluid Mechanics of Turbomachinery*, Vols 1 and 2, Dover Publications, London.

Chapter 14

Waves and offshore engineering

14.1 Introduction

A large number of offshore structures of various types and sizes have been built to extract fossil fuels from wells at considerable depths below seabed. Offshore engineering is a multi-disciplinary activity with inputs from civil, mechanical, marine and chemical engineering and geology.

Offshore structures may be either fixed or floating and they support platforms which in turn support various components such as drilling derricks, processors, flare stacks, radio masts, accommodation and a helideck. Steel or concrete or a combination of both is used for the construction. The majority of structures are fixed jacket platforms that are tubular steel members welded together to form a three-dimensional truss and held in place by means of piles driven through the jacket legs into the sea floor. Another common fixed structure is one that has a massive concrete base that rests on the seabed and the stability of this installation is ensured by its weight. Once the structure is in position, the base can be used as a storage tank for crude oil. A floating structure on the other hand is held in position by either cables or vertical tethers anchored to the sea bed or it may simply be a floating rig connected to a wellhead located at the sea bed and kept in position by thrusters. Chakrabarti (1987) describes the different types of offshore platforms and Mathur (1995) gives an overview of the various offshore activities concerning the production of the oil and gas.

Oil is usually transported from the offshore platform by means of tankers or submarine pipelines. These pipelines are usually but not always buried in the seabed but they will be exposed to waves and currents during pipe laying and in deep waters. The common geometrical shape of the various structural elements forming an offshore platform and obviously of submarine pipelines is cylindrical. They can be subjected to severe environmental forces due to wind, currents and waves.

Under unfavourable conditions flow-induced vibrations could occur leading to structural failure. The design of offshore structures must therefore not only take account of the forces exerted on them by currents and waves, but also guard against fluid-induced vibration. Another important mechanism to be wary of is the fatigue failure of structures which can occur due to stress reversals from incessant exposure to wave-induced forces. Information on forces exerted on the elements of offshore structures in the presence of irregular waves is thus important for design of structures for fatigue. However, the treatment of such forces arising from irregular waves is beyond the scope of this book.

The most noticeable feature of the sea is the oscillations of the free surface. The nature of these oscillations depends on the way they have been generated, the most important being the wind-generated waves that are of a relatively short period. The understanding of wave motion and its interaction with structure is of paramount importance in the design of offshore and coastal structures.

14.2 Wave motion

14.2.1 Solution for velocity potential

Wind waves generated on the oceans are random and short crested. Although most wave theories have been developed for long-crested periodic waves, they have been applied with satisfactory results to waves which have travelled out of the generating areas.

Periodic waves are those in which a phase, for example the crest, is observed by a stationary observer at equal intervals of time, T, called the wave period. The still-water level (SWL) in Fig. 14.1 is the mean water level about which the oscillations of the surface due to waves take place. A sinusoidal wave and some of its properties are shown in Fig. 14.1. The

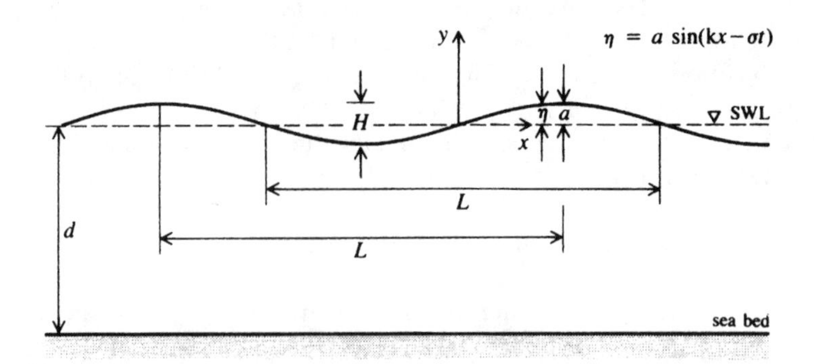

Fig. 14.1 Definition sketch for a sinusoidal wave

vertical distance between the crest and the trough is the height of the wave H, which is twice the amplitude, a. The wavelength, L, is the horizontal distance between the crests. The phase velocity, or celerity, of the wave, c, is:

$$c = L/T. \tag{14.1}$$

The steepness of the wave is H/L. If the height of the wave is extremely small compared with the wavelength and the depth of water, the governing equations are linear and the waveform is usually referred to as a linear or Airy wave. Some other waves that a coastal engineer may find as a better approximation of waves on shorelines are shown in Fig. 14.2. These are non-linear waves which occur for large wave heights. In non-linear theory it is usual to classify waves in terms of the wavelength relative to the water depth. In deep water or for short waves, a finite-height wave known as a Stokes wave is applicable (Fig. 14.2(a)). In shallow water or for long waves, the cnoidal wave theory is applied as an approximation. Both Stokes and cnoidal waves (Fig. 14.2(b)) are asymmetrical with respect to the still-water level and have sharp crests and elongated troughs. A solitary wave characterized by a single hump above still water, moving in shallow water, is shown in Fig. 14.2(c). Linear wave theory is widely used in engineering applications because of its simplicity, but, for cases where better evaluation of wave properties is required, complex non-linear wave theories have to be applied. However, if the waves are not large in relation to the depth, or steep enough to break, linear theory is sufficiently accurate.

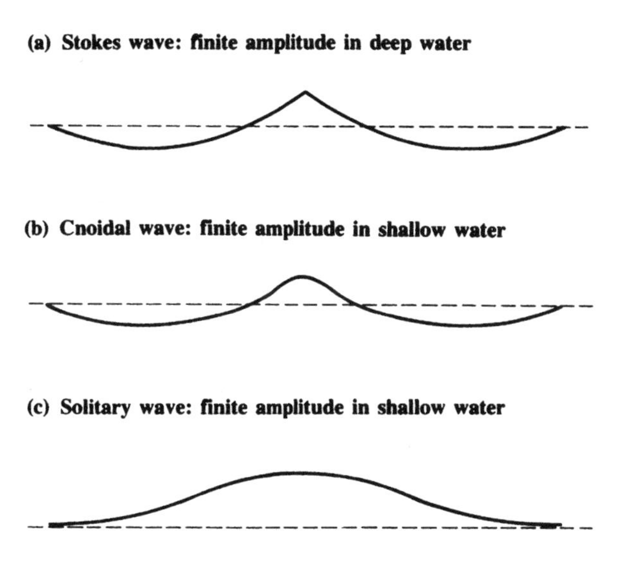

Fig. 14.2 Some non-linear waves

If the waves are also long crested, the fluid motion is two dimensional. Since waves can travel over large distances without significant decay of energy, it can be assumed that wave motion is irrotational. Thus, in linear wave analysis, the study of wave motion makes use of inviscid fluid theory.

The equation of continuity for an incompressible flow is

$$\frac{\partial u}{\partial x}+\frac{\partial v}{\partial y}=0 \tag{14.2}$$

where u is the horizontal velocity, v is the vertical velocity and x, y are the coordinate axes, as shown in Fig. 14.1. Note that the positive y-axis is upwards from the still water level. A velocity potential ϕ, is defined as

$$u=\partial\phi/\partial x, \quad v=\partial\phi/\partial y. \tag{14.3}$$

After substitution for u and v from equation (14.3), equation (14.2) becomes

$$\frac{\partial^2\phi}{\partial x^2}+\frac{\partial^2\phi}{\partial y^2}=0. \tag{14.4}$$

Equation (14.4) is the Laplace equation for the velocity potential. The advantage of introducing the potential is that ϕ is the only property of the field to be determined (instead of the two velocities); the penalty is a second-order partial differential equation.

Bernoulli's equation for unsteady flow, expressing the conservation of energy, can be written as

$$\frac{\partial\phi}{\partial t}+\frac{u^2+v^2}{2}+\frac{p}{\rho}+gy=0 \tag{14.5}$$

where p is the pressure, ρ is the density of the fluid, g is the acceleration due to gravity and t is the time. For waves of very small height relative to the wavelength and depth, the velocity-squared terms in equation (14.5) are only of second-order importance and hence can be neglected. Consequently, equation (14.5) becomes

$$\frac{\partial\phi}{\partial t}+\frac{p}{\rho}+gy=0. \tag{14.6}$$

At the horizontal bed, with d the water depth in the undisturbed state,

$$v=\left.\frac{\partial\phi}{\partial y}\right|_{y=-d}=0, \tag{14.7}$$

and at the free surface, the vertical component of velocity v must be such that

$$v = \left.\frac{\partial \phi}{\partial y}\right|_{y=\eta} = \frac{d\eta}{dt} \tag{14.8}$$

where η is the surface elevation (Fig. 14.1). At the crest $\eta = a$, the amplitude. Because of the small-amplitude assumption, equation (14.8) leads to

$$v = \left.\frac{\partial \phi}{\partial y}\right|_{y=0} = \frac{\partial \eta}{\partial t}. \tag{14.9}$$

Only the temporal derivative of η is retained since convective terms are negligible.

At the surface, $y = \eta \simeq 0$, $p = 0$, and hence equation (14.6) becomes

$$\frac{\partial \phi}{\partial t} + g\eta = 0 \tag{14.10}$$

The wave profile as shown in Fig. 14.1 is given by

$$\eta = a \sin\left[2\pi\left(\frac{x}{L} - \frac{t}{T}\right)\right]. \tag{14.11a}$$

This wave moves along the positive x-direction with a celerity c; the frequency of the wave, f, is $1/T$. Denoting $2\pi/T$ as the circular frequency, σ, and $2\pi/L$ as the wavenumber k, equation (14.11a) may be written as

$$\eta = a \sin(kx - \sigma t). \tag{14.11b}$$

For the wave profile of equation (14.11b), the solution for the velocity potential satisfying equation (14.4) along with boundary conditions given by equations (14.7) and (14.9) is

$$\phi = \frac{-ac \cosh[k(y+d)]}{\sinh(kd)} \cos(kx - \sigma t). \tag{14.12}$$

14.2.2 Wave celerity

Substituting for ϕ from equation (14.12) into equation (14.10), the following expression is obtained for the celerity of the wave:

$$c^2 = \frac{g}{k}\tanh(kd) = \frac{gL}{2\pi}\tanh\left(\frac{2\pi d}{L}\right). \tag{14.13}$$

Figures 14.3 and 14.4 show, respectively, the variations of the linear wave celerity and period as functions of wavelength and depth. In deep water, for which d/L is large, $\tanh(kd)$ tends to unity. Hence, equation (14.13) approximates to

$$c^2 = g/k = gL/2\pi. \qquad (14.14)$$

On the other hand, for waves in shallow water, the wavelength is large in relation to the depth. Thus in shallow water, for which $\tanh(kd) \rightarrow kd$, the celerity of the linear wave is given as

$$c^2 = gd. \qquad (14.15)$$

In deep water the longer the wave, the greater is the celerity. This phenomenon is usually called the normal dispersion. Equation (14.14) is a very close approximation for celerity for values of d/L greater than 0.5. On the other hand, the shallow-water result of equation (14.15) is a good approximation for values of L/d greater than 20.

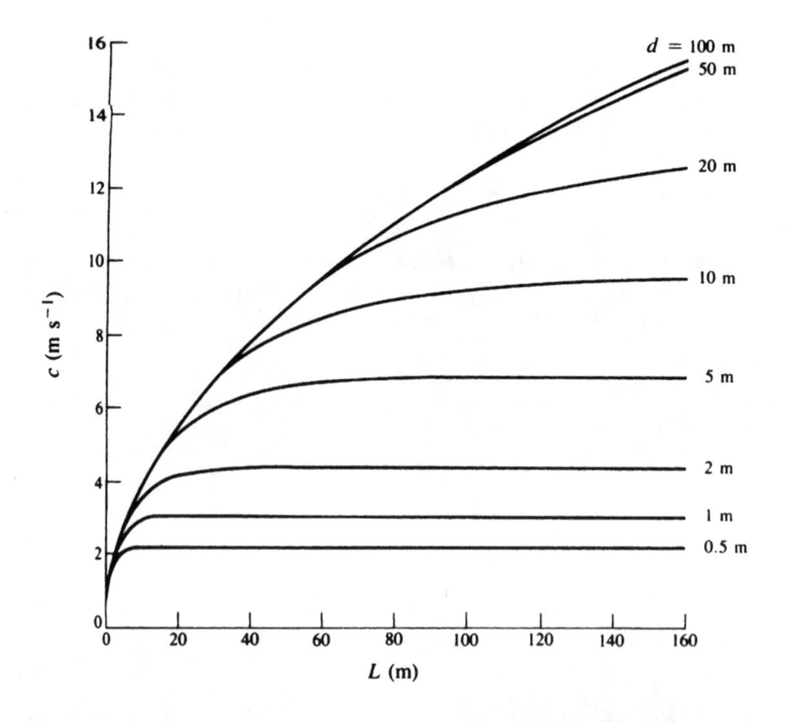

Fig. 14.3 Wave celerity as a function of wavelength and depth

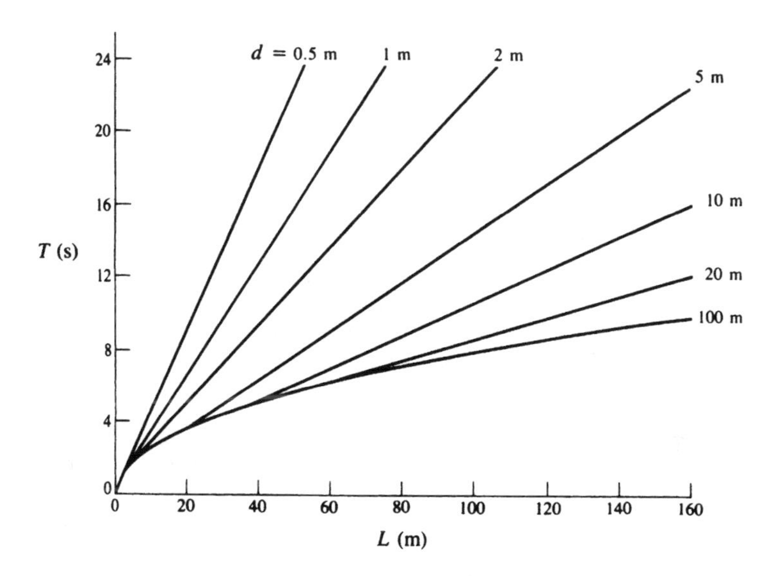

Fig. 14.4 Wave period as a function of wavelength and depth

14.2.3 Particle kinematics

The velocities u and v are obtained from their definitions in terms of the potential (equation (14.3)). They are

$$u = \frac{a\sigma\cosh[k(y+d)]}{\sinh(kd)}\sin(kx - \sigma t), \tag{14.16}$$

$$v = \frac{-a\sigma\sinh[k(y+d)]}{\sinh(kd)}\cos(kx - \sigma t).$$

For the Airy wave the fluid particles do not make large excursions from the mean position, the coordinates of which are the same as the point occupied by the particle before the start of the wave motion. The horizontal and vertical displacements of the particle from the mean position respectively, X, Y, are

$$X = \int_0^t u\,\mathrm{d}t, \quad Y = \int_0^t v\,\mathrm{d}t. \tag{14.17}$$

Substituting for velocities u and v from equation (14.16), the temporal variations of the x-wise and y-wise displacements X and Y respectively are

$$X = \frac{a \cosh[(k(y+d)]}{\sinh(kd)} \cos(kx - \sigma t),$$

$$Y = \frac{a \sinh[k(y+d)]}{\sinh(kd)} \sin(kx - \sigma t). \qquad (14.18)$$

In the above equations the mean particle position is given by (x,y), i.e. by its coordinates before it is perturbed by the wave. It may be easily shown that, in general, the particles perform elliptical orbits, the horizontal displacements being greater than the vertical displacement (Fig. 14.5). In deep water, the orbit becomes circular, while in shallow water the particles tend to move back and forth in a straight line.

Since the particles move in closed orbits, there is no net transport of mass due to a linear wave. This is not the case with the non-linear waves shown in Fig. 14.2.

14.2.4 Energy of waves

The kinetic energy (KE) in a wave is obtained by the following integration:

$$KE = \int_{-d}^{0} \int_{0}^{L} \frac{\rho}{2}(u^2 + v^2) \mathrm{d}x \, \mathrm{d}y.$$

The integrand is the kinetic energy of a particle of volume dx dy per unit length at an instant of time. Substituting for u and v from equation (14.16) at an instant, say $t=0$, and performing the double integration, the kinetic energy is

$$KE = \frac{1}{4} \rho g a^2 L \qquad (14.19)$$

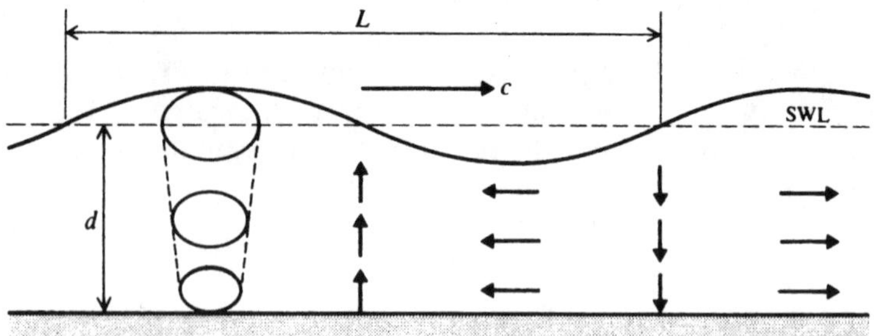

Fig. 14.5 Elliptical orbits of particles and directions of particle motions under various phases

over one wavelength per unit length of the wave crest. The potential energy (PE) in a wave is obtained by finding the work done by the wave to displace water vertically from the still-water level. It is simply

$$PE = \int_0^L \frac{1}{2}\rho g \eta^2 \, dx$$

with η given by equation (14.11a); the potential energy (PE) in a wavelength per unit length of wave crest is

$$PE = \frac{1}{4}\rho g a^2 L. \tag{14.20}$$

The total energy in a wavelength per unit length parallel to the crest is therefore

$$KE + PE = \frac{1}{2}\rho g a^2 L. \tag{14.21}$$

The equal division of energy between kinetic and potential energies, i.e. $PE = KE$, is an essential requirement of free vibrations. The total energy E per unit plan area for unit length parallel to the crest is

$$E = \frac{1}{2}\rho g a^2 \quad \text{or} \quad = \frac{1}{8}\rho g H^2. \tag{14.22}$$

14.2.5 Radiated energy

Radiated energy, R, or energy flux is the rate at which the wave energy moves in the direction of wave propagation, and it is the rate at which work is done by the pressure forces: R is also the wave power. It is given by

$$R = \int_{-d}^{0} pu \, dy$$

where p is the pressure given by the linearized Bernoulli equation (14.6) without the hydrostatic term gy. Hence

$$R = -\int_{-d}^{0} \rho \frac{\partial \phi}{\partial t} u \, dy. \tag{14.23}$$

Integration of the above equation after substituting for u from equation (14.16), and for ϕ and hence for $\partial\phi/\partial t$ from equation (14.12) results in

$$R = \frac{1}{2}\rho g a^2 \frac{c}{2}\left[1 + \frac{2kd}{\sinh(2kd)}\right] \tag{14.24}$$

or

$$R = EC_g$$

where

$$C_g = \frac{c}{2}\left[1 + \frac{2kd}{\sinh(2kd)}\right]. \tag{14.25}$$

C_g is called the group velocity. It can be seen from equation (14.25) that in deep water ($kd \to \infty$), the group velocity $C_g \to c/2$, and that in shallow water ($kd \to 0$), $C_g \to c$.

14.2.6 Waves on currents

The properties of waves in the presence of currents are different from those in still water. A frame of reference moving with the component of current velocity V in the direction of wave propagation is chosen; the problem is then reduced to wave propagation appropriate to still water. In the following equations, subscripts a and r denote the absolute and moving frames of reference.

$$c_a = c_r + V$$

$$Lf_a = Lf_r + V.$$

From equation (14.13)

$$c_r^2 = \frac{g}{k}\tanh(kd)$$

14.3 Range of validity of linear theory

For engineering purposes it is important to establish when linear theory ceases to be applicable, and when it becomes necessary to apply non-linear deep- and shallow-water wave theories, as forces acting on structures calculated from linear wave theory can be underestimated.

The ranges over which linear and non-linear theories are applicable are presented in terms of H/d and d/L (Komar, 1976). The lines that delin-

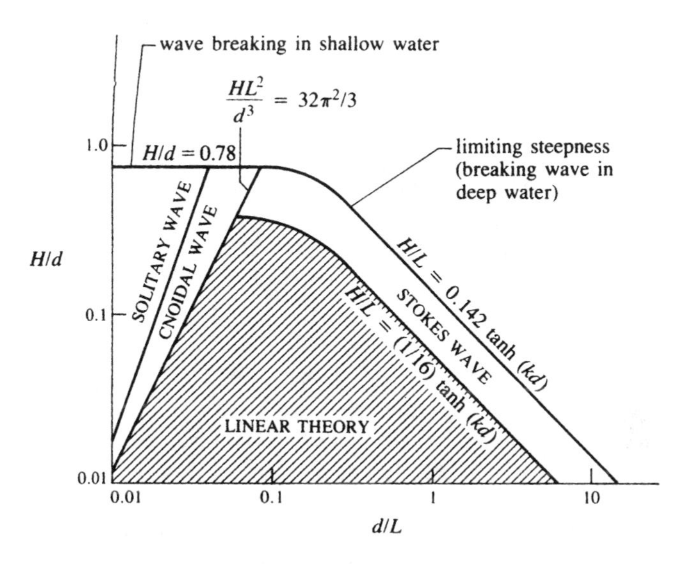

Fig. 14.6 Regions of validity of wave theories (Komar, 1976)

eate one theory from another are shown in Fig. 14.6. In deep water, linear theory is applicable as long as the steepness of the wave

$$H/L < \frac{1}{16}\tanh(kd). \tag{14.26}$$

The upper limit at which Stokes waves exist is determined by wave breaking, the criterion of which in deep water is

$$H/L = \frac{1}{7}\tanh(kd). \tag{14.27}$$

Thus Stokes non-linear theory for deep water is applicable for wave steepness in the range given by equations (14.26) and (14.27).

In shallow water, the limit of linear theory is given as (Fig. 14.6)

$$HL^2/d^3 = 32\pi^2/3. \tag{14.28}$$

Non-linear wave theories, in particular the cnoidal wave theory, are complex and difficult to apply. For further information on non-linear waves, refer to Dean and Dalrymple (1991). However, Skjelbreia and Hendrickson (1961) provided tabulated solutions to fifth-order Stokes waves, and Wiegel (1964) presented charts relevant to cnoidal waves for engineering applications. Tables of results pertaining to wave properties for a wide range of conditions produced by Williams (1985) are extremely useful in applications.

14.4 Waves approaching a shore

Waves generated in the deep water of the ocean travel without change of form for a considerable distance. As they approach the shore, the decreasing depth begins to have an effect. The waves undergo a decrease in wavelength and increase in height, and hence become steeper as they travel towards the shore. Waves at an oblique angle of approach are refracted (so that the crests are turned to become nearly parallel to the shoreline) before they break. As the waves break (Section 14.5), with the attendant air entrainment, virtually all of the energy is lost and a swash (the onrush of water towards the beach followed by the downrush away from the beach) is established. When the line of breaking is at an angle to the beach, currents parallel to the beach, known as longshore currents, are established. The swash and the longshore currents are important causes of sediment motion that shapes the coastline.

Apart from refraction, waves at the shore will be affected by reflection (Section 14.6), and diffraction (Section 14.8) depending on the bed features. For a gently sloping beach, the wave behaviour is mainly determined by refraction. The analysis of refraction is usually carried out using linear theory, even though its basic assumptions are violated in the shallow-water region where the wave breaks. In fact, steep waves assume a non-linear form in the same way as do cnoidal waves, changing almost to solitary waves before they break on the beach.

The refraction diagram showing wave crests and rays, drawn orthogonal to the crests, is extremely useful in coastal engineering practice. Generally, a ridge in the bed topography causes the rays to converge while a valley makes them diverge. Conservation of radiated energy within adjacent rays means that for converging rays the wave height will increase, and a structure in the region of converging rays may be subjected to potentially damaging wave forces. A study of wave refraction is also useful in the design of sea outfalls and the location of a harbour entrance.

The continuity principle requires that the period of the wave approaching the shore should be the same everywhere, and in refraction analysis it is assumed that the radiated energy is conserved between two adjacent rays.

For a constant wave period, T,

$$T = L_0/c_0 = L/c = \text{constant}$$

or

$$c/c_0 = L/L_0 \tag{14.29}$$

where the subscript ‘0’ denotes deep-water conditions.

From equations (14.13) and (14.14), respectively, for c and c_0,

$$c^2 = c_0^2 \frac{L}{L_0} \tanh\left(\frac{2\pi d}{L}\right).$$

After substitution for L/L_0 from equation (14.29) and rearranging,

$$\frac{2\pi d}{L_0} = \frac{c}{c_0} \ln\left(\frac{1 + c/c_0}{1 - c/c_0}\right)^{1/2}. \tag{14.30}$$

A wave approaching a shore of uniform slope is shown in Fig. 14.7. In deep water, the wave crests form an angle α_0 with the bed contours. The crests are stretched and swung around so that they make an angle α as they approach the shore, as shown in Fig. 14.7. The celerity c depends on the local depth and wavelength; it is possible to relate it to c_0 by applying Snell's law of refraction:

$$c/c_0 = \sin \alpha / \sin \alpha_0. \tag{14.31}$$

Consider a crest length b_0 between two adjacent rays in deep water. If the normal distance between the chosen rays locally is b, then

$$b/b_0 = \cos \alpha / \cos \alpha_0. \tag{14.32}$$

For shoaling water, Snell's law shows that $\alpha < \alpha_0$; the rays tend to diverge as the wave moves shorewards. $(b_0/b)^{1/2}$ is called the refraction coefficient, K_r.

Consider the energy flux, R, normal to the wave crests:

$$\frac{1}{8}\rho g H^2 b C_g = \frac{1}{8}\rho g H_0^2 b_0 C_{g0}. \tag{14.33}$$

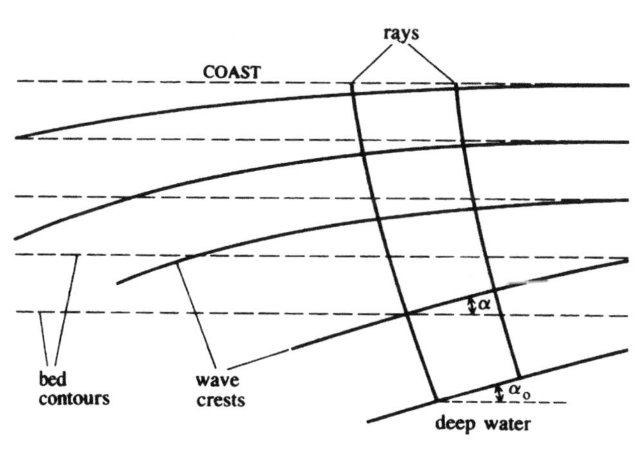

Fig. 14.7 Waves approaching the shore

H may be expressed as

$$H = K_s K_r H_0 \tag{14.34}$$

where $K_s = (C_{g0}/C_g)^{1/2}$ is the shoaling coefficient. To facilitate drawing of refraction patterns, use can be made of Fig. 14.8, which shows graphically the various wave properties locally at depth d, expressed in relation to the deep-water conditions.

As the bed topography can be highly irregular, the numerical approach to the refraction diagram involves a step-by-step determination of the advancement of the wave crests. The local celerity and wavelength are taken to be those of small-amplitude, long-crested waves over uniform depth equal to the local depth. A method of tracing rays from deep water is illustrated in a worked example at the end of this chapter.

14.5 Wave breaking

Miche's criterion for breaking of regular waves at constant depth is

$$H/L = 0.142 \tanh(5.5d/L) \tag{14.35}$$

Waves in deep water break when their steepness H_0/L_0 exceeds 1/7 (equation (14.27)). In shallow water, the wave profile approximates that of a solitary wave before breaking. A solitary wave breaks when

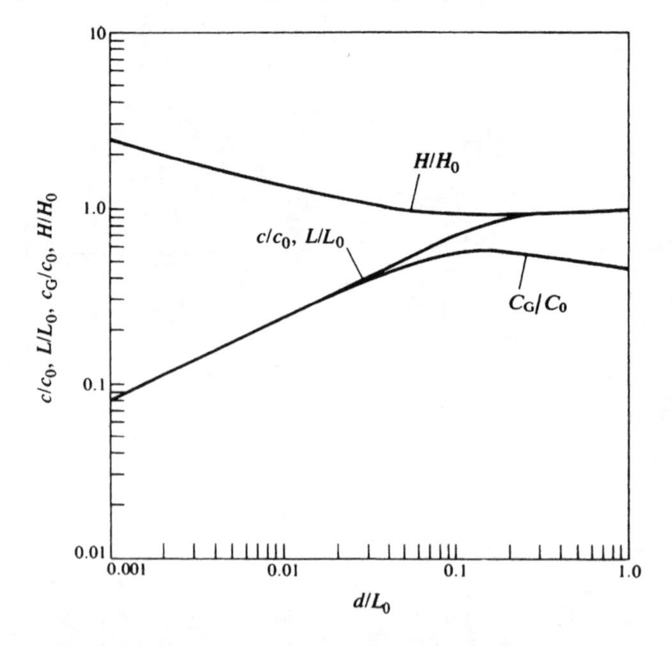

Fig. 14.8 The shore transforms for an Airy wave

$$H/d = 0.78. \tag{14.36}$$

The *Shore Protection Manual* (US Army, 1984) recommends a formula in SI units for wave-breaking in shallow water as

$$H/d = b - aH_0/gT^2 \tag{14.37}$$

where $a = 4.46g(1 - e^{-19s})$ and $b = 1.561/(1 + e^{-19.5s})$, s being the beach slope.

Three kinds of breakers, spilling, plunging and surging, as shown in Fig. 14.9, are distinguished. As a guide, a dimensionless parameter

$$\xi = s/(H_0/L_0)^{0.5} \tag{14.38}$$

is related to the type of breaker (Fig. 14.9).

It is possible for normal approach to relate the wave height and the depth of water at the location of wave breaking to the conditions upstream without making detailed refraction calculations. As before, the waveform at the instant of breaking is considered to be that of a solitary wave

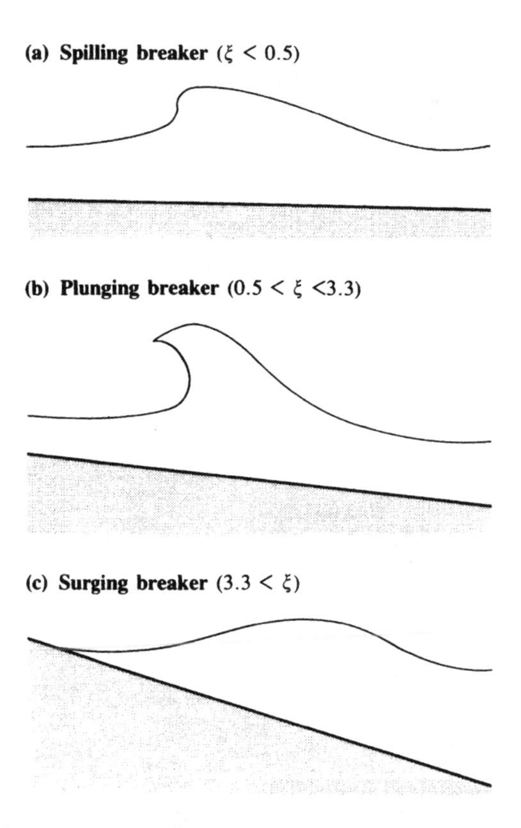

Fig. 14.9 Breaking waves at the beach

(Fig. 14.2(c)). The profile of the solitary wave relative to the axis moving with the celerity of the wave is

$$\eta/H = \text{sech}^2[(3H/d)^{1/2}(x/2d)]. \tag{14.39}$$

As a result of energy conservation, the energy contained in the solitary wave is equal to that in the wave under deep-water conditions and to twice the potential energy (consistent with the earlier finding that the kinetic and potential energies in a wave are equal). The energy in a solitary wave per unit length along the crest is therefore (Morris, 1963)

$$E_s = 2\int_{-\infty}^{\infty} \frac{1}{2}\rho g \eta^2 \, dx$$

$$= 2 \times 2 \frac{\rho g H^2}{2} \int_0^{\infty} \text{sech}^4\left[\left(\frac{3H}{d}\right)^{1/2} \frac{x}{2d}\right] dx$$

$$= \frac{8\rho g}{3\sqrt{3}} (Hd)^{3/2}.$$

Using the criterion of equation (14.36), the above equation becomes

$$E_s = \frac{11.6}{3\sqrt{3}} \rho g H^3. \tag{14.40}$$

As the energy of the oncoming waves at deep water within a single wave-length is

$$E_s = (1/8)\rho g H_0^2 L_0$$

this results in

$$(H_b/H_0) = 0.38(H_0/L_0)^{-1/3} \tag{14.41}$$

and

$$(d_b/H_0) = 0.49(H_0/L_0)^{-1/3}. \tag{14.42}$$

Equations (14.41) and (14.42) are useful in determining the wave height H_b and the shallow-water depth d_b at which the wave breaks, given the properties H_0 and L_0 in deep water.

Coastal Engineering Manual (US Army, 2002) suggests a criterion for breaking of irregular waves on horizontal bed in shallow water as $H_{sb} \approx 0.6 d_b$. Alternate expressions similar to Miche's criterion using H_{rms} at breaking point have also been proposed. For the definition of H_{rms}, refer to equation (14.47).

14.6 Wave reflection

A wall in a wave field interacts with a progressive wave by reflecting it; the reflected wave moves in a direction depending on the angle of the incident wave. If the crests of the incident wave are parallel to the wall, the crests of the reflected wave will also be parallel to it. The coefficient of reflection, which is the ratio of the amplitude of the reflected wave to the amplitude of the incident wave, depends on the angle of the incident wave and the energy absorption capability of the wall which, in turn, depends on its geometry, porosity, and roughness.

In linear theory, linear superposition of the surface profiles due to the incident and reflected waves is permissible. The surface profile of the incoming wave is given by equation (14.11b) as

$$\eta_i = a\sin(kx - \sigma t)$$

and that due to the reflected wave is given by

$$\eta_r = a\sin(kx + \sigma t). \tag{14.43}$$

(Note that the direction of the reflected wave is in the negative x-direction and that total reflection is assumed.)

The resulting wave surface is described by

$$\eta = \eta_i + \eta_r = 2a\sin kx\cos\sigma t. \tag{14.44}$$

Equation (14.44) represents a standing wave, as shown in Fig. 14.10. The positions of no vertical motion are called nodes, and those of maximum amplitudes are called antinodes.

Standing waves forming adjacent to a vertical wall can cause deep erosion because the particle velocity increases near the bed owing to the increase in wave height.

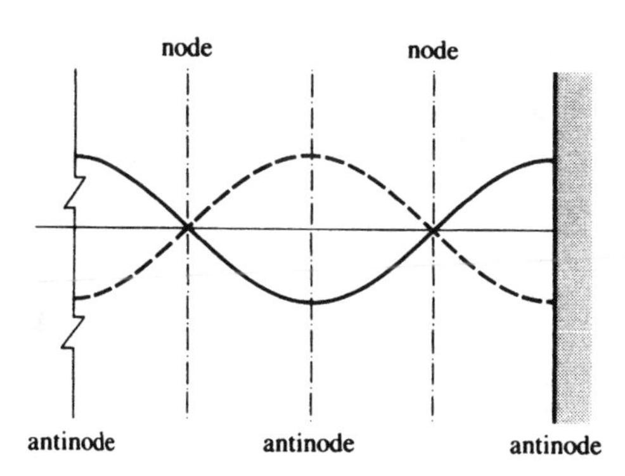

Fig. 14.10 Standing wave formed by reflection at a vertical wall

14.7 Basin oscillations

In a basin the width of which is uniform and much smaller than the length, oscillations occur in the longitudinal direction. The basins may be idealized with both ends closed or one end closed. The former is analogous to a harbour with an entrance at the seaward end, while the latter may be taken to approximate a bay or an estuary. The oscillations in the harbour and the estuary will be triggered by the waves or tides at the seaward end.

The fundamental, the first and the second modes of basin oscillations are shown in Fig. 14.11. The water surface forms antinodes at the vertical walls and nodes at the open end of the basin. The wavelength in the n-th node of oscillation L_n is related to length of the basin L_b as follows:

$$\text{both ends closed, } L_n = \frac{2L_b}{n+1}$$

$$\text{one end closed and one end open, } L_n = \frac{4L_b}{2n+1}.$$

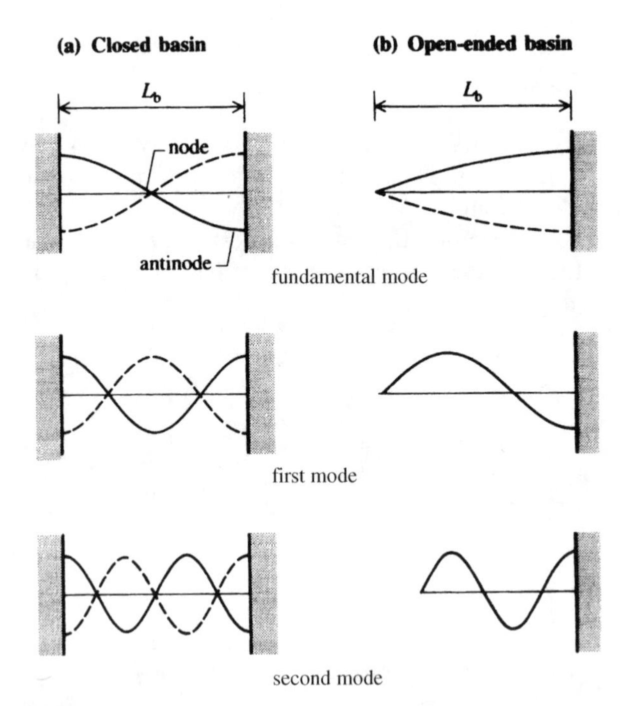

Fig. 14.11 Basin oscillations

If the basin is long, the wavelength can be large in relation to the depth of water, and the shallow-water result for the period of oscillations can be applied, i.e.

$$T_n = L_n/c = L_n/(gd)^{1/2}.$$

If the width of a harbour L_w is comparable to its length L_b, more complex patterns of resonant oscillations may occur. The harbour may be assumed as both ends closed. The period of oscillation is

$$T_{nm} = 2\left[gd\left(\frac{n^2}{L_b^{\,2}} + \frac{m^2}{L_w^{\,2}}\right)\right]^{-0.5}$$

where n and m are various modes assuming values of 0, 1, 2, ...

14.8 Wave diffraction

Waves impinging on a structure of finite length bend in the lee of the structure due to radiation of wave energy. This is the process of diffraction: wave diffraction at a semi-infinite breakwater is shown in Fig. 14.12.

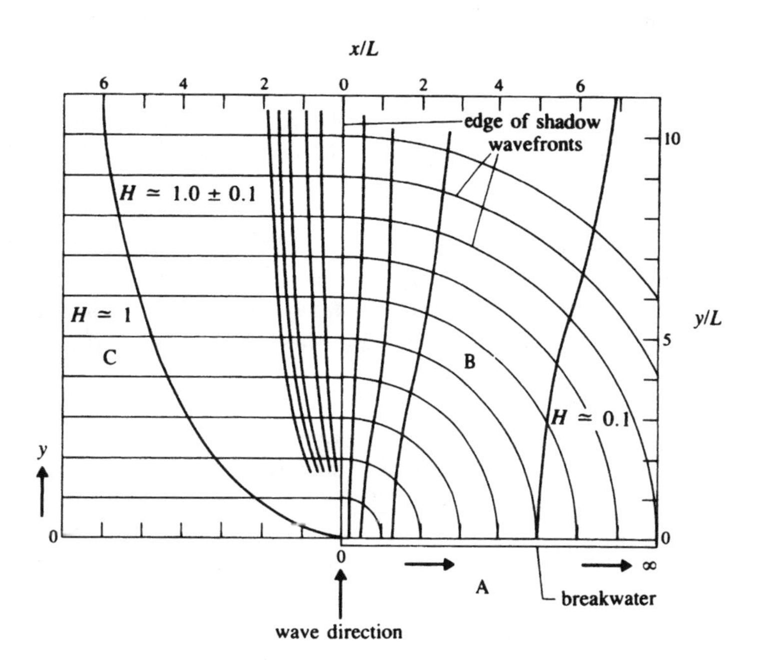

Fig. 14.12 Diffraction of waves in the presence of a semi-infinite breakwater (Muir Wood and Fleming, 1969)

Three regions may be distinguished in the figure: (1) the region consisting of incident and reflected waves (A), (2) a region in the shadow of the breakwater in which the crests form circular arcs (B) and (3) a region in which the incident waves progress undisturbed (C). As the waves spread behind the structure, their amplitudes decrease. The ratio of the diffracted wave height to the height of the incident wave is termed the diffraction coefficient. For the specific case of diffraction with incident waves at various angles at a long breakwater, values of diffraction coefficients have been tabulated by Muir Wood and Fleming (1969).

The diffraction pattern through a breakwater gap is usually obtained assuming the gaps to be small or large. The solution for a large gap is obtained by a superposition of the two solutions for the two breakwaters for normal incidence of the waves. For graphical solutions for a number of cases, the reader is referred to Wiegel (1964) and the *Shore Protection Manual* (US Army, 1984).

14.9 Wave prediction

14.9.1 General

The energy of waves is provided by the wind blowing over the ocean. The onset of wave formation is influenced by surface tension; although important in model experiments this is of no consequence in engineering practice. Waves are generated on the initially flat water surface by the motion of the pressure-producing patterns of turbulence of the air stream; in the later stages of development of the waves, the energy required for the wave growth comes directly from the mean motion of the wind.

Waves generated by wind have a wide range of frequencies and wavelengths. The longest waves are those the celerity of which is equal to the wind speed. The steepest waves are determined by the breaking condition in deep water. The heights and periods generated depend on the wind speed, U, the distance or fetch, F, over which the wind blows and the duration, T_w, of the wind.

There is a certain interaction between the wind and the isobar spacing as given in the meteorological charts. In meteorological practice, isobars are spaced at 4 mbar in the UK and at 3 mbar in the USA. The wind direction is parallel to the lines of isobars but is modified by friction over the water surface. The pressure distribution normal to the isobars is determined by the Coriolis force resulting from the Earth's rotation and the centripetal force due to the curvature of the moving air masses. The resultant wind is called the gradient wind. When the isobars are parallel and straight, only the Coriolis force is important and the wind is called geostrophic wind. The equation governing the motion of the geostrophic wind at a particular point is

$$\frac{1}{\rho_a}\frac{\partial p}{\partial n} = 2U\Omega \sin \lambda \tag{14.45}$$

where $\partial p/\partial n$ is the pressure gradient normal to the isobar, U is the wind speed, ρ_a is the density of air, Ω is the angular velocity of the Earth's rotation, and λ is the latitude.

In order to estimate the characteristic height of waves generated by wind, the fetch F has to be estimated from the weather map (US Army, 1984). An approximate estimate of the fetch is the distance along the isobar curve from the location to a position upwind at which the tangent to the isobar deviates by about 15°. If the wind blows seaward from the coast, then the fetch length is limited by the coast. Meteorological data provide help in estimating the wind duration.

14.9.2 Significant height and period of the waves

Significant wave height, H_s, and the wave period, T_s, are the mean height and period of the highest third of waves in a sample. Other characteristic wave periods are the mean period of the waves whose troughs and crests are above SWL and the wave period corresponding to that at which the spectrum exhibits a peak.

Another measure of the wave period, called the zero-crossing period, is obtained from the output of a wave record. It is determined by finding the number of times, N, that the record trace crosses the still-water level as the water level changes from the trough to the crest. The zero-crossing period, T_z, is equal to the duration of the record divided by N. For engineering applications (Tucker, 1963), T_z is nearly equal to the significant period T_s.

The significant wave height, H_s, is used as a design wave height in coastal engineering practice. In the earliest method of wave forecasting, known as the Sverdrup, Munk and Breitshneider (SMB) method (King, 1972) and applicable to deep water, H_s and T_s are related to the fetch F, duration of the wind T_w, and acceleration due to gravity g. The relationship is shown in Fig. 14.13 in non-dimensional form. The $T_{min}U/F$ versus gF/U^2 curve gives the minimum time, T_{min}, required for the generation of waves of maximum energy for the given fetch and wind speed. If the duration of wind, T_w, is greater than T_{min} the generation of the wave is fetch limited. In that case, H_s is found for the given fetch.

On the other hand, if $T_w < T_{min}$ the production of waves is duration limited. From the $T_{min}U/F$ versus gF/U^2 curve (Fig. 14.13) we can find F corresponding to $T_w = T_{min}$ and from the value of F thus determined, we find H_s. Figure 14.13 is also used to find T_s.

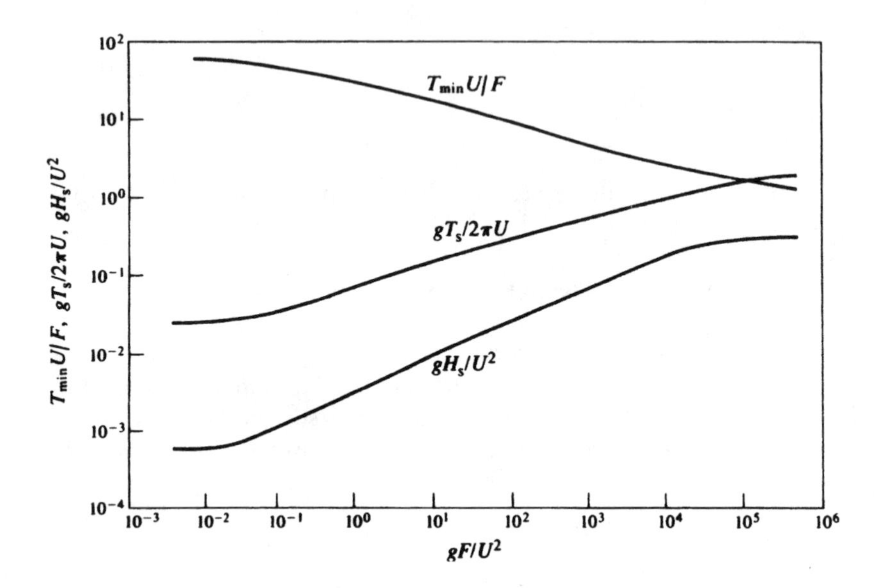

Fig. 14.13 Fetch graph for deep water (King, 1972)

Darbyshire and Draper (1963) presented widely used charts (Fig. 14.14) for wave prediction, based on observations around Britain. Figure 14.14 permits the determination of H_{max} and the period of waves in deep water as functions of wind speed (km h^{-1}), duration (h) and fetch (km). H_{max} is the maximum wave height in a wave record of 10 min length. As with the SMB method, wave height is found to be either fetch limited or duration limited. Darbyshire and Draper found that the records collected in coastal areas, where depth varied from 30 m to 45 m, showed significant differences from those collected in oceanic waters. Figure 14.15 facilitates the determination of H_{max} and the period for coastal waters as functions of wind speed (km h^{-1}), duration (h) and fetch (km).

An alternative method of wave forecasting is based on the frequency spectrum of the waves. In this method the irregular waves of the ocean are assumed to consist of frequencies, f, ranging from 0 to infinity. If the mean-square value of the wave amplitudes in the range $f-\Delta f/2$ and $f+\Delta f/2$ is $\overline{a^2(f)}$, the energy per unit frequency centred at f is

$$E(f) = \frac{1}{2}\overline{a^2(f)}/\Delta f. \tag{14.46}$$

$E(f)$ gives the energy spectrum showing the energy content at various frequency components. The mean-square value of the wave amplitudes is given by

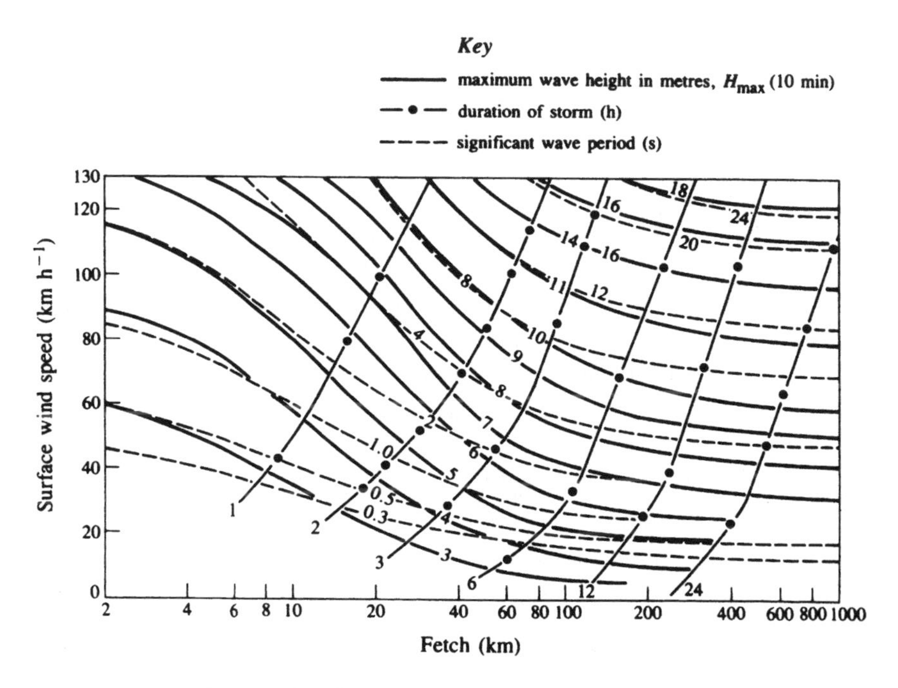

Fig. 14.14 Fetch graphs for oceanic waters (Darbyshire and Draper, 1963)

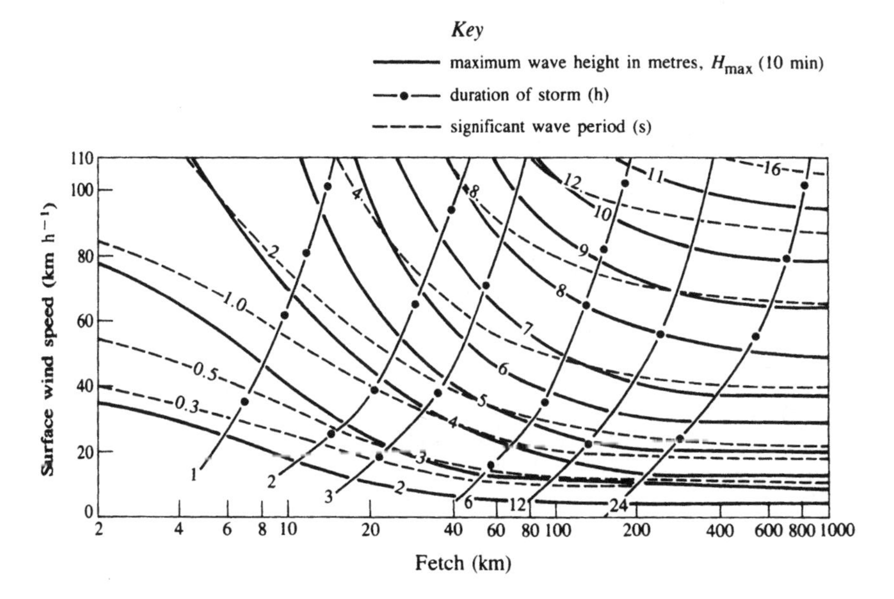

Fig. 14.15 Fetch graphs for coastal waters (Darbyshire and Draper, 1963)

$$\frac{1}{2}\overline{a^2} = \int_0^\infty E(f)\mathrm{d}f$$

or

$$= \frac{1}{8}H_{\mathrm{rms}}^2. \tag{14.47}$$

A fully developed sea is one for which the fetch and the duration of wind are unlimited, so that the waves do not grow any further for a given wind speed and the wave field has attained a steady state. The Pierson–Moskowitz spectrum for the fully developed sea (Pierson and Moskowitz, 1964) is

$$E_{\mathrm{PM}}(f) = \alpha g^2 (2\pi)^{-4} f^{-5} \exp\left[-\frac{5}{4}\left(\frac{f}{f_{\mathrm{m}}}\right)^{-4}\right] \tag{14.48}$$

where α is the Phillips constant (equal to 0.0081) and $f_{\mathrm{m}} = 0.8772$ $(g/2\pi U_{19.5})$. $U_{19.5}$ is the wind speed measured at 19.5 m above the sea surface.

Substituting equation (14.48) into equation (14.47) and performing the integration, the rms value of the wave heights is

$$H_{\mathrm{rms}} = 0.316\frac{g\alpha^{1/2}}{\pi^2 f_{\mathrm{m}}^2}, \tag{14.49}$$

$$T_{\mathrm{s}} \simeq T_{\mathrm{z}} = 0.7104/f_{\mathrm{m}}.$$

The Joint North Sea Wave Project (JONSWAP) produced a frequency spectrum for the developing sea (Hasselmann *et al.*, 1976). The wind speeds for the JONSWAP wave measurements were taken at 10 m above the still-water level: $U_{10} = 0.93U_{19.5}$. The spectrum is

$$E_{\mathrm{J}}(f) = \alpha g^2 (2\pi)^{-4} f^{-5} \exp\left[-\frac{5}{4}\left(\frac{f}{f_{\mathrm{m}}}\right)^{-4}\right]\gamma^q \tag{14.50}$$

where

$$\alpha = 0.076(gF/U_{10}^2)^{-0.22}, \quad q = \exp[-(f - f_{\mathrm{m}})^2/2\sigma_1^2 f_{\mathrm{m}}^2],$$

$$U_{10}f_{\mathrm{m}}/g = 3.5(gF/U_{10}^2)^{-0.33},$$

$$\sigma_1 = 0.07 \quad \text{for } f < f_{\mathrm{m}},$$

$$\sigma_1 = 0.09 \quad \text{for } f \geq f_{\mathrm{m}},$$

$$\gamma = 3.3, \quad T_{\mathrm{s}} \simeq T_{\mathrm{z}} = 0.777/f_{\mathrm{m}}.$$

The wave energy in the JONSWAP spectrum is concentrated within a narrower band of frequencies and is more peaked than that given by the Pierson–Moskowitz spectrum. Significant wave height H_s is related to H_{rms} in the next section.

JONSWAP spectrum given by equation (14.50) is for fetch-limited conditions of deep seas. If waves generated are duration-limited, then an effective fetch F_{eff}, is calculated as

$$F_{eff} = \left(\frac{gT_w}{68.8U}\right)^{1.5} \tag{14.51}$$

If a given fetch $F < F_{eff}$, then the waves are fetch-limited. If $F > F_{eff}$, the waves are duration-limited; then F_{eff} is substituted for F in the calculation of wave height and peak frequency.

Coastal Engineering Manual (US Army, 2002) recommends slightly modified forms for JONSWAP spectrum.

In shallow water, both friction and percolation act to modify wave growth. *Coastal Engineering Manual* (US Army, 2002) suggests that wave growth formulas of deep water may be used with the restriction that no wave with period exceeding $9.78(d/g)^{0.5}$ exists.

14.10 Wave statistics

Wave records at a particular site are usually collected over a time period ranging from 15 min to 1 h, spaced at about 3 h intervals. Each record is a sample that provides the short-term statistics. For a narrow band of frequencies over which the wave energy is concentrated, the record may be described by the Rayleigh distribution. If $P(H)$ is the probability that a wave height will exceed H, then

$$P(H) = \exp[-(H/H_{rms})^2]. \tag{14.52}$$

H_{rms} is the root-mean-square value of the wave heights, defined as

$$H_{rms}^2 = \frac{1}{N-1}\sum_{i=1}^{N} H_i^2 \tag{14.53}$$

where N is the number of values of H in the data. Note that $P(H) = 1$ for $H = 0$ and 0 for H tending to infinity. Using the Rayleigh distribution, it is now possible to estimate the average of the highest nth of the waves. For example, for $n = 3$, the average is the significant wave height H_s

$$H_s = \sqrt{2}H_{rms}. \tag{14.54}$$

From equation (14.49), for the Pierson–Moskowitz spectrum,

$$H_s = 0.447 g\alpha^{1/2}/\pi^2 f_m^2. \tag{14.55}$$

Numerical integration of the JONSWAP spectrum carried out by Carter (1982) leads to an expression for H_s as

$$H_s = 0.552 g\alpha^{1/2}/\pi^2 f_m^2. \tag{14.56}$$

The estimate of a design wave on the basis of recorded data may also be made using the long-term wave statistics. H_s is often selected with a certain return period T_R, i.e. the design (significant) wave height is expected to be exceeded in one year over T_R years. Thus

$$T_R = 1/P(H_s). \tag{14.57}$$

It is common practice to use a 50- or 100-year return period as a level of protection for most coastal structures. Alternatively the design may be based on the encounter probability E. It is the probability that the design wave is equalled or exceeded during the life of L years of the structure. The encounter probability E is

$$E = 1 - (1 - 1/T_R)^L \tag{14.58}$$

For example, $E = 0.33$ for $T_R = 50$ and $L = 20$ or for $E = 0.1$ and $L = 20$, $T_R = 190$.

Choice of larger return period for the design of a structure means smaller probability E that the design wave will be exceeded during the life of the structure.

Several probability distributions have been proposed to describe the long-term statistics. Among them is the log-normal distribution commonly used in coastal engineering. It is a normal or Gaussian distribution for the variate $\ln(H_s)$ instead of H_s. The probability $P(H_s)$ for the log-normal distribution is

$$P(H_s) = 1 - \frac{1}{(2\pi)^{1/2}} \int_0^{H_s} \frac{1}{\sigma_s H_s} \exp\left[-\frac{1}{2} \left(\frac{z_s - \mu_s}{\sigma_s} \right)^2 \right] dH_s \tag{14.59}$$

where $z_s = \ln(H_s)$. The two parameters μ_s and σ_s are the mean and standard deviation of the variate $\ln(H_s)$. For a sample of N values of H_s, they are defined as

$$\mu_s = \frac{1}{N} \sum_{i=1}^{N} \ln(H_s)$$

and

$$\sigma_s^2 = \frac{1}{N-1}\sum_{i=1}^{N}[\ln(H_s) - \mu_s]^2.$$

Before attempting to estimate the design wave height for a return period, we must first ensure that the collected data are closely approximated by the chosen probability distribution. The following steps may be adopted.

1. The values of H_s for each of the record are arranged in the ascending order of magnitude. Let the number of H_s be N.
2. A plotting formula is used to relate $P(H_s)$ with H_s and to plot this on the probability paper is

$$P(H_s) = \frac{\text{number of waves exceeding } H_s}{N+1}.$$

3. The data points $(H_s, P(H_s))$ are plotted on a probability paper corresponding to the chosen probability distribution. The data points will collapse on a straight line if they follow the chosen distribution (see Worked example).
4. The straight line is extrapolated to determine the design wave height for the chosen return period T_R. If the time interval of the records is τ, then

$$T_R = \tau/P(H_s)$$

in which τ is expressed in units of years. For example, if $\tau = 3\,\text{h}$, it is equivalent to 1/2920 year.

If the log-normal distribution does not fit the data, the data may be approximated by the Weibull distribution:

$$P(H_s) = \exp\left[-\left(\frac{H_s - H_c}{H_0}\right)^{k_1}\right]. \quad (14.60)$$

H_c is the minimum wave height at the site; it is found by examining the long-term records by trial and error in the search for the best fit to the Weibull distribution. The parameters H_0 and k_1 are obtained after arriving at the best fit. Note that the Weibull distribution uses three parameters, as against two in the log-normal distribution.

Extreme value distributions may be used for estimating the design wave height for a specified return period. They deal with the largest value of the wave height in each year over a period of years. Extreme value statistics require data for a considerable number of years. If they are not available over at least 30 years, long-term statistics are preferred. Extreme value statistics are dealt with comprehensively by Goda (1979), Isaacson and Mackenzie (1981) and Sarpkaya and Isaacson (1981).

14.11 Forces on cylindrical structures

14.11.1 Forces due to currents

Offshore structures constructed of cylindrical members and submarine pipelines are exposed to hydrodynamic forces due to waves and currents. A steady current approximated, for instance, by the tidal flow will exert steady and fluctuating forces on the cylindrical structures. If the cylinder is well away from solid boundaries, the forces are a steady in-line force together with the fluctuating in-line and lift forces. On the other hand, a pipeline resting on the sea bed will be additionally subjected to a steady lift force acting away from the bed owing to the asymmetry of the flow structure. The effects of the fluctuating forces are discussed in Section 14.12.

Consider an isolated, smooth cylinder of diameter D, the axis of which is normal to uniform flow. The steady in-line or drag force per unit length of the cylinder F_{D}, is customarily expressed as

$$F_D = \frac{1}{2} C_{\mathrm{D}} \rho V^2 D \tag{14.61}$$

where C_{D} is the drag coefficient. C_{D} is a function of the Reynolds number $R_{\mathrm{e}} = VD/\nu$, where V is the velocity upstream of the cylinder and ν is the kinematic viscosity of the fluid (Schlichting, 1960). C_{D} decreases with R_{e} for laminar flow in the boundary layer along the periphery of the cylinder. At $R_{\mathrm{e}} \approx 10^5$ (the precise value depending on the turbulence level of the stream and the roughness of the cylinder), transition to turbulent flow in the boundary layer takes place with an abrupt drop in the value of C_{D} (Schlichting, 1960).

The lift force or the force transverse to the direction of the current per unit length of the cylinder is expressed similarly to equation (14.61) as

$$F_{\mathrm{L}} = \frac{1}{2} C_{\mathrm{L}} \rho V^2 D \tag{14.62}$$

where C_{L} is the lift coefficient. Isolated cylinder experiences no steady lift but a considerable fluctuating lift with a frequency equal to the shedding frequency of vortices from the cylinder.

The forces exerted on a pipeline in the proximity of the sea bed are dependent on the Reynolds number of the flow, the relative roughness of the bed and the cylinder, and the gap between the bed and the pipe, expressed as a ratio of the pipe diameter. The boundary layer on the sea bed will have an influence too. Experiments carried out by Littlejohns (1974) in the Severn Estuary (UK) show that for cylinders on the sea bed the steady drag and lift coefficients are 1.15 and 1.27 respectively.

14.11.2 Wave forces

In estimating the wave forces, the size of the cylinder in relation to the wavelength is important. If the ratio of the diameter of the cylinder to the wavelength D/L is less than about 0.2, the viscosity of the fluid and hence the separation effects become important. In this case the perturbations due to the presence of the cylinder are local and the wave forces are determined using the Morison equation (14.63). On the other hand, for $D/L > 0.2$, the excursions of the fluid particles are small relative to the diameter of the cylinder and the flow around the body experiences no separation. The waves are scattered and a diffraction analysis is used to find the pressure distribution on the cylinder and hence the force. The subject of wave forces is exhaustively treated by Sarpkaya and Isaacson (1981) and Chakrabarti (1987).

(a) In-line forces on small vertical cylinders

The Morison equation for the wave force exerted on a submerged cylinder of small diameter considers that the force is simply a sum of the drag and inertia forces. The latter arise as a consequence of the unsteady nature of the wave field. The Morison equation for the in-line force, F_i, per unit length of the cylinder is

$$\frac{\mathrm{d}F_i}{\mathrm{d}y} = C_D \frac{\rho}{2} |u| uD + C_M \frac{\rho \pi D^2}{4} \frac{\mathrm{d}u}{\mathrm{d}t} \tag{14.63}$$

where C_D is the drag coefficient, C_M is the inertia coefficient, u and $\mathrm{d}u/\mathrm{d}t$ are respectively the particle velocity and the acceleration normal to the axis of the cylinder and D is the cylinder diameter; they are determined along the axis of the cylinder as though it were absent. The first term on the right-hand side of equation (14.63) is the drag term, which contains the modulus of velocity in order for the direction of force to be along the instantaneous particle velocity vector. The second term is the inertia force that will arise from the unsteadiness of the flow field even if the fluid is inviscid (C_M is often computed using the inviscid theory).

Placing $x = 0$ at the axis of the cylinder, u and $\mathrm{d}u/\mathrm{d}t$ may be expressed as

$$u = u_m \sin \sigma t, \quad \mathrm{d}u/\mathrm{d}t = u_m \sigma \cos \sigma t,$$

where u_m is the maximum velocity. Using the above expressions, equation (14.63) results (in non-dimensional form) in

$$\frac{d(2F_i/\rho u_m^2 D^2)}{d(y/D)} = C_D |\sin \sigma t| \sin \sigma t + C_M \frac{\pi^2 D}{u_m T} \cos \sigma t. \tag{14.64}$$

The term $u_m T/D$ is known as the Keulegan–Carpenter number, K_c. It may be shown from equations (14.16) and (14.18) that $K_c = 2\pi X/D$. Thus K_c specifies the relative importance of the distance of travel of the fluid particles to the diameter of the cylinder. It is clear from equation (14.64) that for small values of K_c (in practice, less than about 1.0), inertia force dominates drag force. If K_c is large, separation becomes important so the drag outweighs the inertia force.

The in-line force per unit length, dF_i/dy, given by equation (14.63) is obtained by substituting for u from equation (14.16) and du/dt, the time derivative of u.

Assuming constant values of C_D and C_M across the depth, F_i after integration of equation (14.63) is

$$F_i = -C_D \frac{\rho}{32k} H^2 \sigma^2 D \frac{\sinh(2kd) + 2kd}{\sinh^2(kd)} |\sin \sigma t| \sin \sigma t - C_M \frac{\rho \pi D^2 a \sigma^2}{4k} \cos \sigma t. \tag{14.65}$$

Let

$$A_D = \frac{\sigma^2}{k} \frac{\sinh(2kd) + 2kd}{\sinh^2(kd)}$$

and $A_1 = \sigma^2/k$. For the maximum force, F_m, on the cylinder, the time derivative of F_i must be zero, i.e. $dF_i/dt = 0$. If $t = t_m$, at which F_i is equal to the maximum value F_{im}, then

$$\sigma t_m = \arccos\left[2\pi\left(\frac{C_M}{C_D}\right)\frac{A_1 D}{A_D H}\right]. \tag{14.66}$$

From equation (14.65)

$$F_{im} = C_D \frac{\rho}{32} H^2 D A_D \sin^2(\sigma t_m) + C_M \frac{\rho \pi D^2}{8} H A_1 \cos(\sigma t_m). \tag{14.67}$$

For an isolated cylinder in inviscid fluid, $C_M = 2.0$. However, both C_D and C_M are functions of the Reynolds number, the Keulegen–Carpenter number and the surface roughness. The growth of marine plants and organisms on the surface of the cylinder will undoubtedly affect the roughness and also the effective diameter of the cylinder. The presence of currents also affects the coefficients. The choice of suitable values of C_D and C_M is therefore not easy. Muir Wood and Fleming (1969) suggest values of C_M and C_D as functions of R_e and K_c. Figures 14.16 and 14.17 show the experimental results of Sarpkaya (1976) concerning the coefficients C_M and C_D with respect to R_e and K_c. Different certifying agencies such as the British Standards Institution, Det Norske Veritas and the American Petroleum Institute suggest values of C_D and C_M for design.

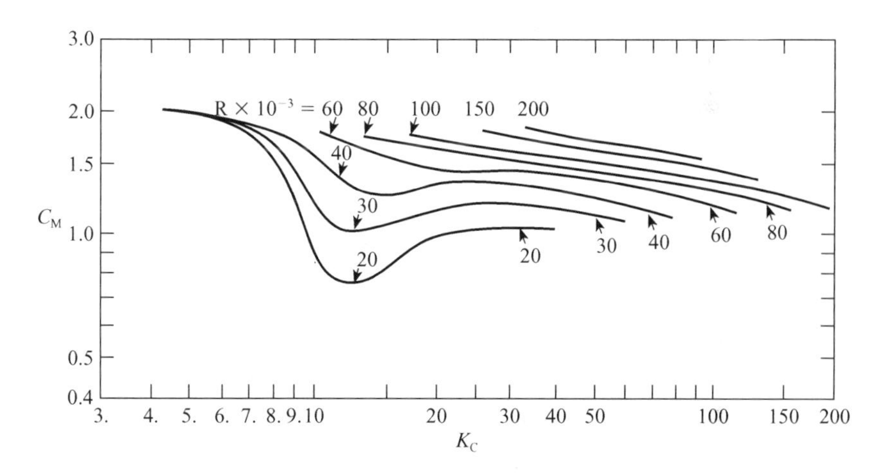

Fig. 14.16 C_M vs K_C number for various values of Reynolds number (Re) for smooth circular cylinder (Sarpkaya, 1976a)

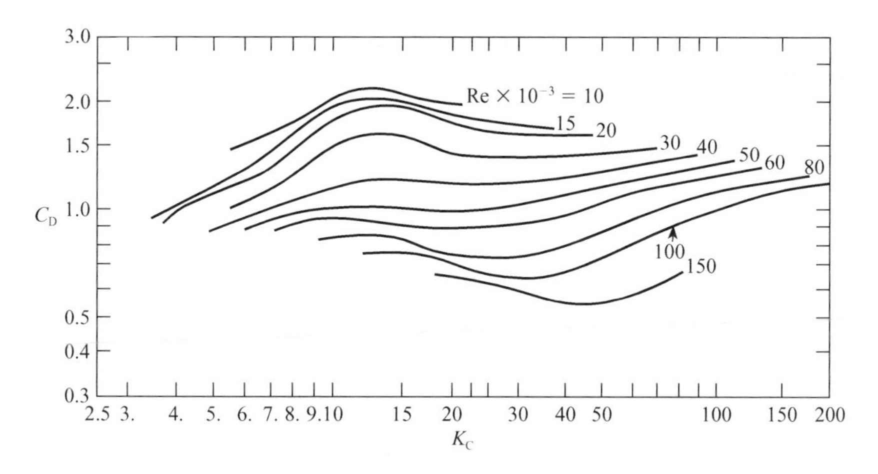

Fig. 14.17 C_D vs K_C number for various values of Reynolds number (Re) for smooth circular cylinder (Sarpkaya, 1976a)

It is common practice to use Morison equation (14.63) for the in-line force for co-existing waves and currents, by replacing the velocity component u in equation (14.63) by the total velocity which is the algebraic sum of the current velocity and the particle velocity due to the waves (see Worked example 14.2).

If the axis of the cylinder is inclined at an angle to the direction of wave propagation, then the velocity and the acceleration normal to the axis are used in the Morison equation to obtain the force. The tangential force exerted by the component of the velocity along the cylinder is relatively small.

Vertical cylinders can experience the force of breaking waves. If the height of the wave at breaking is H_b, the force exerted on the cylinder may be expressed as

$$F_B = C_B \rho g D H_b^2. \tag{14.68}$$

The coefficient C_B is found to vary between 1.2 and 3.0.

(b) In-line forces on small horizontal cylinders

For a horizontal cylinder, the axis of which is parallel to the crest of the wave, the force per unit length of the cylinder is, according to the Morison equation,

$$F_i = C_D \frac{\rho}{2} |u| u D + C_M \frac{\rho \pi D^2}{4} \frac{du}{dt}. \tag{14.69}$$

u and du/dt are perpendicular to the axis of the cylinder. In equation (14.69), C_D and C_M again depend on the Reynolds number, the Keulegan–Carpenter number and the roughness, and also on the gap ratio G/D (G is the gap between the cylinder and the sea bed). Sarpkaya (1976b, 1977) provides C_M and C_D against R_e for various K_c and $\frac{G}{D}$. C_M for frictionless flow is 3.3 for a smooth cylinder resting on the sea bed ($G/D = 0$). Experiments have been conducted by Littlejohns at Perrin Bay, Cornwall, UK, to determine the values of the coefficients under field conditions (Littlejohns, 1982). Allowing for the scatter of the results, which is natural for the test conditions, the appropriate values are $C_D = 1.0$ and $C_M = 2.4$ for cylinders resting on the sea bed, with K_c up to 13.0.

(c) Transverse forces on small horizontal cylinders

In oscillatory flow the flow separates from the surface of the cylinder for $K_c > 3$. For $K_c > 5$, vortices are shed asymmetrically from the top and bottom of the cylinder causing alternate transverse or lift forces normal to the oscillatory flow direction. Transverse forces can be of similar magnitude to the in-line forces. The oscillatory flow reverses its direction in every half cycle and the flow has to adjust to this reversal. The flow structure is complex with some vortices interacting with the cylinder when they are swept back and forth. The dominant frequency of the lift force increases with the Keulegan–Carpenter number. It is the same as the wave frequency for $K_c<5$, the second harmonic of the wave frequency for $5<K_c<15$ and the third harmonic for $15<K_c<20$ and so on. At K_c numbers greater than about 25, the fluctuating lift coefficient tends to attain a value typical of the steady flow and the wake resembles a steady flow Karman Vortex street.

The lift force per unit length of the cylinder F_L is usually presented as:

$$F_L = \frac{1}{2} C_L \rho u^2 D$$

where C_L is the lift coefficient. Sarpkaya (1976a) presents C_L in terms of K_c and R_e for smooth cylinders exposed to periodically oscillated flow in a U-tube (Fig. 14.18). Further experiments of Sarpkaya (1976b) provide values of the lift coefficient for cylinders above a plane bed.

No accurate representation of the lift force variation with time in oscillatory flows has been achieved. Chakrabarti (1987) represented the lift force for vertical cylinders by a Fourier series having frequencies equal to the wave frequency and its multiples as:

$$f_y(t) = \frac{1}{2} \rho u^2 D \sum_{n=1}^{N} C_{Ln} \cos(n\sigma t + \epsilon_n) \qquad (14.70)$$

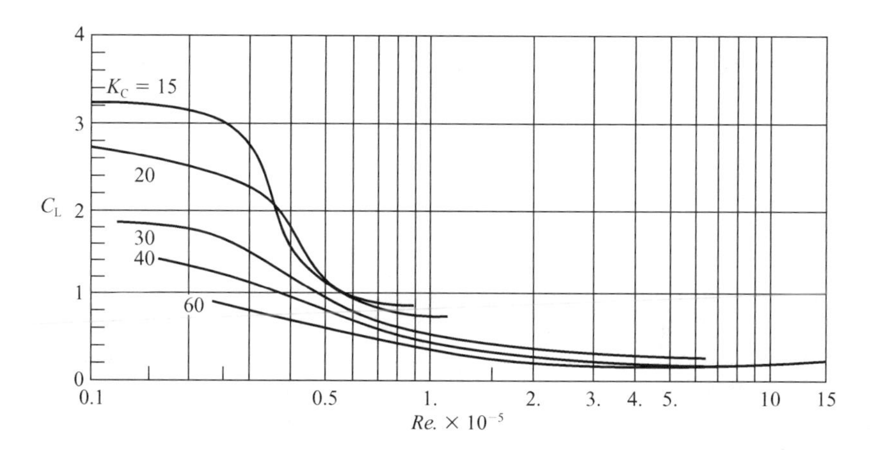

Fig. 14.18 Lift coefficient as a function of Reynolds number for various values of K_C for smooth circular cylinder (Sarpkaya, 1976a)

where N is the number of Fourier components, C_{Ln} and ϵ_n are the nth lift coefficient and corresponding phase angle respectively.

For Keulegan–Carpenter numbers above 25 Bearman *et al.* (1984) developed an equation for the lift force assuming quasi-steady vortex shedding at a constant Strouhal number of 0.2 based on the instantaneous flow velocity. The lift force for the half cycle on a unit length of the cylinder is:

$$f_y(t) = \frac{1}{2} C_L \rho u_m^2 D \sin^2 \sigma t \cos[\alpha(1 - \cos \sigma t) + \psi] \tag{14.71}$$

where $\alpha = K_c S$ and ψ is the phase angle for the lift force depending on the history of the reverse flow.

For $K_c < 20$, an empirical expression of Kao *et al.* (1984) for the lift force on horizontal cylinder is:

$$F_y(t) = \rho \frac{\pi D^2}{4} L C_{my} \frac{\partial v}{\partial t} + \frac{1}{2} \rho D L C_L u^2 + C_w \rho \frac{D}{2} L \left[u^2 \left(t - \frac{T\theta}{2\pi} \right) - \epsilon u_m^{\ 2} \right] \tag{14.72}$$

where C_w, θ and ϵ are empirical coefficients. The first two terms on the right of the above equation are the transverse inertia and lift terms predicted by the potential theory, the third term is due to vortex influence. Chioukh and Narayanan (1997) present values of C_{my} and C_L for potential flow as functions of G/D. Empirical values of the coefficients C_w, θ and ϵ expressed as functions of G/D and K_c are presented by Kao *et al.* (1984).

(d) Wave forces on large cylinders

Linear diffraction theory is used for the determination of wave forces on large cylinders, the diameters of which are greater than 0.2 times the wavelength: it is only briefly described here. The velocity potential, ϕ_T, of the field in the presence of the cylinder is expressed as

$$\phi_T = \phi_i + \phi_s$$

where ϕ_i is the incident wave potential and ϕ_s is the scattered wave potential. ϕ_s must satisfy the linearized energy condition on the free surface, the no-penetration condition on the sea bed and the condition that the velocity on the surface of the cylinder due to the scattered wave must be equal and opposite to that of the incident wave. At large distances from the cylinder, a boundary condition called Sommerfeld's radiation condition must be imposed for the scattered wave. The Laplace equation (equation (14.4)) for ϕ_s is solved satisfying the kinematic boundary conditions on the sea bed and on the cylinder, dynamic boundary conditions on the free surface and the radiation conditions at infinity. Numerical approaches using Green functions (Garrison and Chow, 1972), the finite element method (Bai, 1975) and boundary element methods (Brebbia, 1978) are used to calculate ϕ_s.

Once ϕ_s and hence ϕ_T are known, the pressures on the cylinder are obtained from equation (14.6) without the hydrostatic pressure term. Integration of the pressure distribution leads to the force on the cylinder. The variation of the dimensionless maximum in-line force $F_{im}/[\rho gHDd \tanh(kd)/kd]$ as a function of kD for a surface-piercing vertical cylinder is shown in Fig. 14.19.

14.11.3 Wave forces on pipelines in the shoaling region

As the waves move progressively towards the shore, the non-linear effects have a significant effect on the wave height. The wave height of parallel waves predicted by the finite amplitude wave theory can be larger than that calculated by equation (14.33) from linear theory. Iwagaki, Shiota and Doi (1982) propose a simple approximate expression for the wave height in shoaling waters avoiding complex calculations of cnoidal wave theory. Swift and Dixon (1987) present shoaling curves based on a stream function series solution.

Wave forces exerted on horizontal cylinders parallel to wave crests are calculated using the Morison equation (14.69) in the shoaling region before wave breaking. The kinematics at the position of the cylinder for the undisturbed wave should be determined by a finite amplitude theory such as the streamfunction theory (Huang and Hudspeth, 1982). Laboratory experiments show that the cnoidal wave theory for wave kinematics

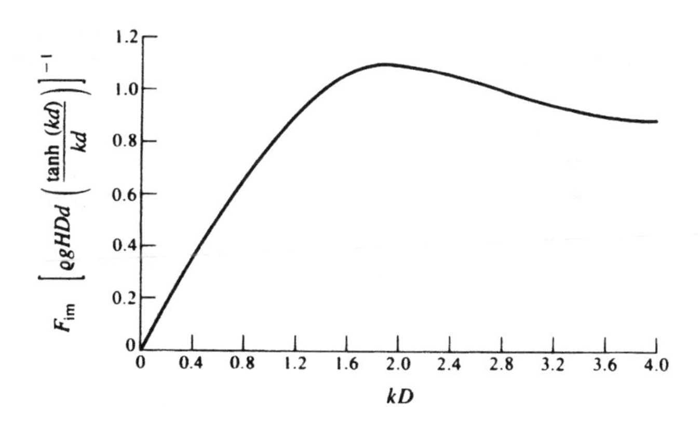

Fig. 14.19 Maximum wave force on a large vertical cylinder

(Yuksel and Narayanan, 1994a) is reasonably satisfactory for the wave forces. However, the difficulty remains in the choice of appropriate drag and inertia coefficients. When the waves approach the cylinder obliquely, then the component of the velocity normal to the cylinder is applied in the Morison equation to determine the wave force.

The incipient wave breaking for which the criteria are given by equations (14.41) and (14.42) coincides with the occurrence of the maximum wave height showing the initiation of a bubble with attendant foam formation. At the later stage of plunging, the crest of the wave falls onto the forward moving water. The wave height at the plunge point is about 60% of that at the breaking point. Experimental studies concerning breaking wave forces are available only for waves approaching normally to the cylinder (Yuksel and Narayanan, 1994b). For horizontal cylinders fully submerged in water and resting on a rigid beach, the maximum forces occur when the cylinder is placed at the plunge point. The impingement force on the cylinder placed at the plunge point is not a constant but exhibits randomness; therefore a statistical measure of the breaking wave force for certain probability of exceedence is used.

14.11.4 Pipeline stability

For the design of submarine pipelines in shallow waters the refraction pattern of the wave has to be obtained. The direction of the waves with respect to the pipeline is determined along the pipeline before the waves break. In order to evaluate the wave forces on the pipeline the components of the particle velocity and acceleration normal to the pipe axis are used in the Morison equation. Very little is known about the forces on the pipe once the waves break. Submarine pipelines, especially those in shallow water, are normally buried. However, they may be allowed to rest on the sea bed before burial. Sometimes the characteristics of the sea bed may be such that neither burial nor anchoring is possible. In this case the stability of the pipeline against rolling must be considered.

Referring to Fig. 14.20, the pipeline is subjected to the in-line force due to the combined action of currents and waves, F_i, and the lift force F_L (both F_i and F_L are per unit length of the pipeline). Let the submerged weight of the pipeline per unit length be W. For stability against rolling

$$F_i < \mu(W - F_L)$$

where μ is the coefficient of friction for the seabed–pipeline interface.

So far in this section, linear theory has been used for fluid velocity and acceleration. In deep water Stokes' higher order theory should be

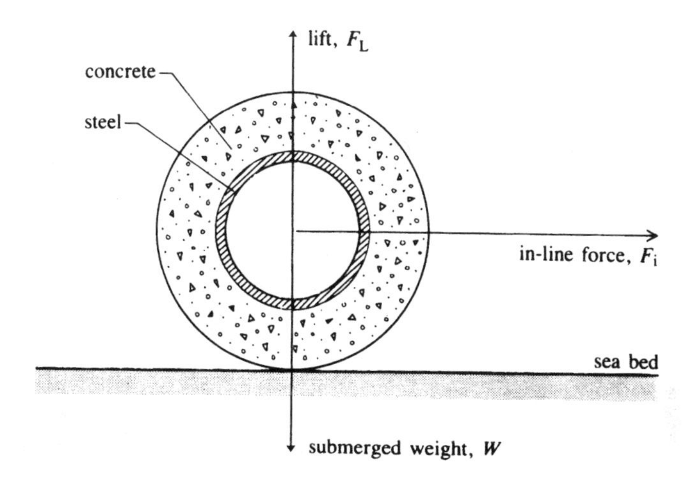

Fig. 14.20 Forces exerted on a submarine pipeline

used to describe the particle kinematics. Near the coast the linear theory may not be accurate enough, so that the forces are underestimated.

When pipelines are placed on erodible seabed, they sag and bury themselves due to action of waves and currents. This self-burial process is advantageous as it saves costly artificial trenching or burial of the pipes. During the erosion process or in the presence of scour beneath the pipelines, spanning conditions of pipelines may prevail. The pipes can experience large static deflection or fluid-induced vibrations endangering their structural integrity. Generally the local scour in the vicinity of pipelines is three-dimensional. The equilibrium depth of scour S beneath fixed pipes in laboratory conditions is typically 0.6 times the diameter of the pipelines for steady currents and in the presence of waves it is $S/D = 0.1\sqrt{K_c}$ (Sumer and Fredsoe, 1990; Sumer *et al.* 2001).

If the pipes are laid partially buried, the relative scour depth S/D is also a function of the relative embedment depth e/D (Hoffmanns and Verheij (1997), *Coastal Engineering Manual* (US Army, 2002)).

Pipelines generally are covered using layers of materials such as gravel, concrete blocks, and mattresses of concrete, asphalt or polypropylene to protect them against erosion and mechanical damage and to reduce their weight (Hoffmanns and Verheij, 1997); they are completely buried to protect them from fishing trawlers.

14.12 Vortex-induced oscillations

14.12.1 Vertical cylinders in currents

The fluctuating forces exerted on the piles in the presence of currents can excite oscillations and lead to failure of the structure; oscillations of piles during construction are also possible. This subject of flow-induced vibrations is only briefly dealt with in this section. For extensive treatment, the reader is referred to Hallam *et al.* (1978), Blevins (1990), Chakrabarti (1987) and Sumer and Fredsoe (1997).

Consider a rigidly mounted cylinder exposed to a steady two-dimensional flow normal to its axis. The flow pattern in the wake of the cylinder is dependent on the Reynolds number of the flow. When the Reynolds number exceeds about 70, separation of the boundary layer takes place. The separated shear layers roll up into vortices which shed alternatively from the cylinder at the extremities of a line perpendicular to the flow. The frequency of vortex shedding, f_v, is expressed in terms of the Strouhal number, S, defined as f_vD/U. For the Reynolds number greater than about 1000, $S = f_vD/U \simeq 0.2$.

The alternate shedding of vortices is responsible for the periodic in-line and cross-flow components of force exerted on the cylinder. The cross-flow excitation is at the same frequency as the vortex shedding frequency f_v. On the other hand, the in-line fluctuations of force are at twice the frequency f_v. The in-line fluctuations of forces are not significant with respect to flow-induced oscillations in air, but in water they could be significant. Hence vertical piles in water are susceptible under unfavourable circumstances to both in-line and cross-flow oscillations.

The behaviour of a vertical pile is represented by a cantilever system with the fixed end at the sea bed. Such a single-degree-of-freedom system under the action of the exciting force and linear damping is represented by the differential equation

$$M\ddot{x} + C\dot{x} + Kx = F(t) \tag{14.73}$$

where M is the mass, C the damping coefficient, K is the stiffness, and $F(t)$ is the exciting force, which is a function of time. $\ddot{x}$ and $\dot{x}$ are the second and first derivatives of the displacement x with respect to time t respectively.

For free oscillations, $C = 0$ and $F(t) = 0$, and the system has a natural frequency $f_n = (1/2\pi)(K/M)^{1/2}$. With only $F(t) = 0$ in equation (14.73), the system performs damped oscillations if $C < C_c$ (the critical damping):

$$C_c = 4\pi M f_n. \tag{14.74a}$$

The decay of the amplitude of damped oscillations is logarithmic. A convenient measure of the damping is the logarithmic decrement, δ, which is defined as the natural logarithm of the ratio of any two successive amplitudes of oscillations. It may be expressed as

$$\delta = 2\pi\zeta/(1-\zeta^2)^{1/2} \tag{14.74b}$$

where $\zeta = C/C_c$. When the frequency of the exciting force $F(t)$ coincides with the natural frequency f_n, resonance takes place.

In a real structure, the mass distribution, m_s, along the pile may be non-uniform; the mass of entrained water, m_w, in a hollow pile and the added mass, m_a, resulting from the motion of the pile in water should also be accounted for. The mass per unit length is then

$$m = m_s + m_w + m_a. \tag{14.75}$$

The added mass may be expressed as $m_a = C_a \rho \pi D^2/4$. For an isolated cylinder, $C_a = 1$. C_a is related to C_M, the inertia coefficient, as $C_a = C_M - 1$. The pile may not be completely immersed; in this case, the mass of entrained water and the added mass are to be considered only for the immersed part of the pile.

In the analysis of dynamic response, the real structure is replaced by an equivalent cylinder of the same cross-section but of a length equal to the depth of water (Fig. 14.21). Both the real structure and the equivalent cylinder have the same mode shape, natural frequency, and inertial properties. By this artifice, the experimental results pertaining to a cantilever with the fixed end at the sea bed can be used to analyse the real structure. In this section only the simple case of the vertical pile with a uniform distribution of mass and without any end mass or constraints is considered. Piles with lengths greater than the water depth and with end constraints and end masses are treated fully in Hallam, Heaf and Wootton (1978), with worked examples.

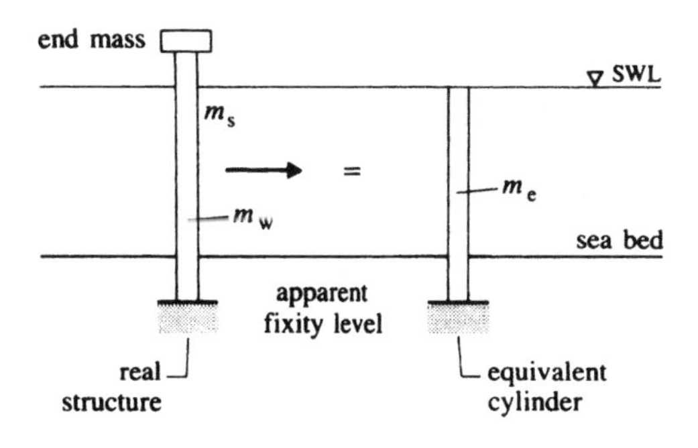

Fig. 14.21 Equating a real structure to an equivalent structure

The natural frequency of a cantilever with effective mass m_e per unit length is

$$f_n = 0.56(EI/m_e l^4)^{1/2} \tag{14.76}$$

where E is the modulus of elasticity, and I is the second moment of area: l is the 'effective length' of the pile, equal to the water depth plus the apparent fixity depth (Fig. 14.21) to allow for the stiffness of the pile–soil interaction. For stiff clay, the apparent fixity depth is $3.5D$ to $4.5D$, where D is the diameter of the pile. For very soft silts it is $7D$ to $8.5D$.

The resonance condition in the cross-flow direction occurs when the forcing frequency $f_v = f_n$. For the cylinder, with the vortex shedding frequency given by $S = 0.2$, resonance takes place when the flow speed is as follows:

$$\text{for cross-flow motion,} \qquad V = 5 f_n D; \tag{14.77}$$

$$\text{for in-line motion,} \qquad V = 2.5 f_n D. \tag{14.78}$$

The amplitude, η, of oscillations of the vertical pile in water may be expressed functionally as

$$\eta \sim m_e, f_n, \delta, V, \rho, \mu, D.$$

Dimensional considerations show that

$$\frac{\eta}{D} \sim \frac{V}{f_n D}, \frac{m_e}{\rho D^2}, \delta, \frac{\rho V D}{\mu}.$$

The effect of the Reynolds number may be considered insignificant. Therefore,

$$\frac{\eta}{D} \sim \frac{V}{f_n D}, \frac{m_e}{\rho D^2}, \delta.$$

The first term on the right-hand side of the above equation is usually called the reduced velocity. The last two terms are combined to form a stability parameter K_s' defined as

$$K'_s = 2 m_e \delta / \rho D^2.$$

Thus

$$\frac{\eta}{D} \sim K_s', \frac{V}{f_n D}. \tag{14.79}$$

14.12.2 In-line oscillations

Experimental studies on the in-line response of flexible cylinders exposed to steady flow exhibit two peaks, as shown in Fig. 14.22. The peak at the higher value of the reduced velocity V/f_nD is associated with the alternate shedding of the vortices and the one at the lower reduced velocity with the vortices shed symmetrically from the cylinder. The critical value of V/f_nD at which oscillations are initiated is found to be a function of K_s' (Fig. 14.23). It has also been found that no excitation in the in-line direction occurs for $K_s'>1.8$.

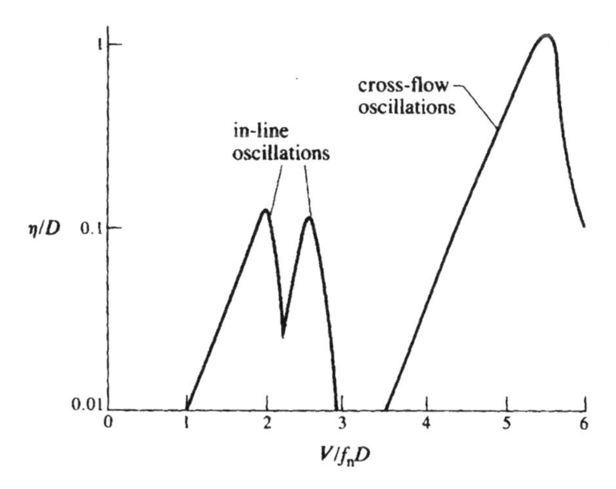

Fig. 14.22 Typical response of a vertical pile in a steady current

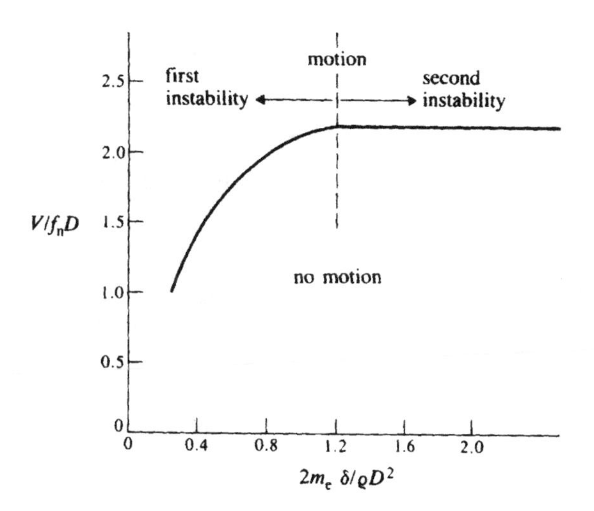

Fig. 14.23 Criterion for onset of instability with respect to in-line oscillations (Hallam, Heaf and Wootton, 1978)

14.12.3 Cross-flow oscillations

It is shown in Fig. 14.22 that the peak response of the cylinder in the cross-flow direction occurs at values of V/f_nD greater than those for in-line oscillations. The onset of cross-flow oscillations is expressed in terms of the critical value of V/f_nD as a function of the Reynolds number of the flow (Fig. 14.24). No cross-flow oscillations have been observed for $K_s' > 10$.

The damping associated with marine structures is the sum of the structural damping and fluid dynamic drag. Both of these are difficult to evaluate. Structural damping, which is expressed in terms of the logarithmic decrement δ_s, is usually estimated from full-scale tests. Typically values of δ_s, as given in Hallam, Heaf and Wootton (1978) are as follows:

for structural steel, $\delta_s = 0.02$;

for concrete, $\delta_s = 0.05$ (uncracked) or $\delta_s = 0.16 - 0.36$ (cracked);

for wood, $\delta_s = 0.05 - 0.2$;

for marine steel structures, $\delta_s = 0.08 - 0.20$;

for marine concrete structures, $\delta_s \simeq 0.06$.

The other source of damping is the drag force arising from the oscillatory motion in the fluid. Equation (14.61) gives the drag force F_D as

$$F_D = \frac{1}{2} C_D \rho |u| u D = \frac{1}{2} C_D \rho |\dot{x}| \dot{x} D.$$

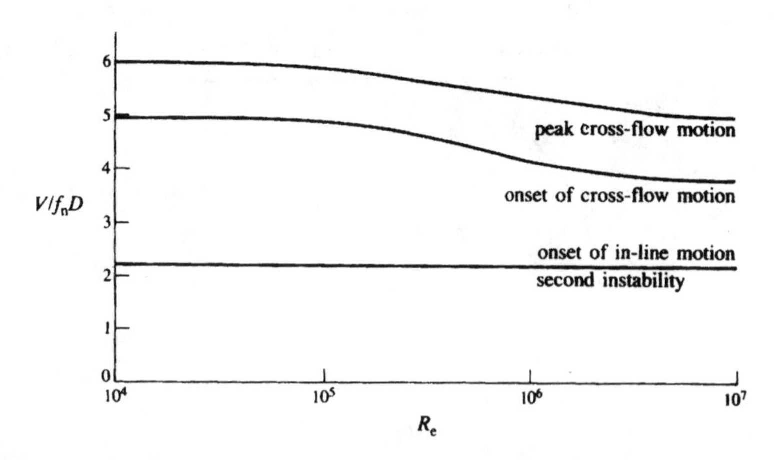

Fig. 14.24 Criterion for onset of instability with respect to cross-flow oscillations (Hallam, Heaf and Wootton, 1978)

If the above non-linear term for the damping force is used, it is impossible to solve equation (14.73) in closed form. Hence the drag term is linearized as

$$F_{\mathrm{D}} = \frac{1}{2} c_{\mathrm{w}} \rho \dot{x} D \tag{14.80}$$

where c_{w} is the value of $C_{\mathrm{D}} | \dot{x} |$, averaged over a whole cycle.

14.12.4 Onset of instability

In order to determine whether a vertical pile will be subjected to flow-induced vibrations in a steady current, the pile equivalent to the structure under investigation is first established. For the equivalent pile, the stability parameter K_{s}', the reduced velocity and the Reynolds number for the current are found. It is shown in Figs 14.23 and 14.24 whether onset of instability in the in-line or cross-flow can occur. The amplitude of oscillations is then given as a function of K_{s}' in Hallam, Heaf and Wootton (1978).

Vortex-induced oscillations may be prevented at the design stage by a suitable choice of the values of reduced velocity and the stability parameter. Devices that modify the flow and reduce excitation may be fitted to the circular structure in the field. The ones that are commonly used are strakes (fins wound around the cylinder) and a shroud (a tube with a number of small holes placed over the cylinder and separated from it by a small gap).

14.13 Oscillations of cylinders in waves

In time-dependent flows, the vibratory behaviour of a cylinder is different from that in steady flow. For such a cylinder, dimensional considerations show that the amplitude of motion may be expressed as:

$$\frac{\eta}{D} \approx \frac{u_{\mathrm{m}}}{f_{\mathrm{n}} D}, K_{\mathrm{s}}', K_{\mathrm{c}}, R_{\mathrm{e}}, \frac{k}{D}.$$

For a smooth cylinder without Reynolds number dependence the above expression becomes

$$\frac{\eta}{D} \approx \frac{u_{\mathrm{m}}}{f_{\mathrm{n}} D}, K_{\mathrm{s}}', K_{\mathrm{c}}. \tag{14.81}$$

When a small rigid cylinder is free to move in waves, the response equation is similar to equation (14.73). For the inline oscillations a simple expression for the time-dependent force is the Morison equation:

$$F_x(t) = C_D\rho\frac{D}{2}(u-\dot{x})\,|u-\dot{x}| + C_M\rho\frac{\pi D^2}{4}\frac{\partial u}{\partial t}. \tag{14.82}$$

The drag term accounts for the hydrodynamic damping through the use of the relative velocity between the fluid and structure. For the random wave fields, it is usual to linearize the drag term to analyse the response of the structure.

With regard to vertical oscillation of horizontal cylinders due to waves, equation (14.73) should consider the displacements in the vertical direction. For smooth cylinders, the lift force $F_y(t)$ on the cylinder may be expressed by equation (14.71) for $K_c>25$ due to Bearman *et al., 1984* or by equation for $K_c<20$ due to Kao *et al.*, 1984 (see Section 14.11.2(c)). The peak oscillation of cylinders is found to occur when the ratio of the flow frequency to the natural frequency of the cylinder assumes integer values (Bearman and Hall, 1987; Chioukh and Narayanan, 1997). The amplitude of displacement of the cylinder is of the order of one cylinder diameter and is affected by the proximity of the cylinder to the channel bed.

Instead of the empirical expressions for the forces $F_x(t)$ and $F_y(t)$, attempts are being made to obtain these forces by solving hydrodynamic equations or by discrete modelling (see Sumer and Fredsoe, 1997).

Worked Example 14.1

1. At a site off a coast, measurements of wave heights have been made for 15 min durations at 3 h intervals. For a sample, the wave heights are as below. Check whether the data fit the Rayleigh distribution.

   The number of waves is given within an interval of wave height. The interval is the wave height below which the same is given minus the preceding height in the table:

H (m)	0.25	0.5	0.75	1.0	1.25	1.50	1.75	2.0	2.25	2.5	2.75	3.0	3.25	3.5
Number of waves	3	7	7	7	12	11	10	8	9	13	7	2	3	1

   The significant wave height for the sample is 2.5 m. The total number of waves in the sample, $N=100$.

2. The highest significant wave on each day has been recorded, to obtain 365 values for a year. Their distribution is given below:

H_s (m)	0.5	1.0	1.5	2.0	2.5	3.0	3.5	4.0	4.5	5.0	5.4
Number of waves ($>H_s$)	340	128	84	73	48	37	18	15	6	2	1

Estimate the significant wave height for a return period of 20 years.

Solution

1. $P(H) = \dfrac{\text{number of waves exceeding } H}{N+1}$, $H_s = 2.5\,\text{m}$.

H/H_s	0.1	0.2	0.3	0.4	0.5	0.6	0.7	0.8	0.9	1.0	1.1	1.2	1.3	1.4
Number of waves exceeding H	97	90	83	76	64	53	43	35	26	13	6	4	1	0
$P(H)(\%)$	96	89	82	75	63	52	43	35	26	13	6	4	1	

$P(H)$ versus H/H_s is plotted on probability paper for the Rayleigh distribution, which fits the data reasonably well (Fig. 14.25).

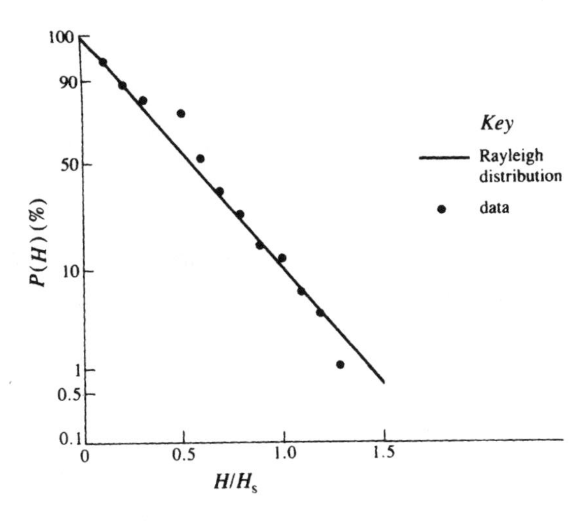

Fig. 14.25 Probability distribution (Worked example 14.1, part 1)

2.

H_s (m)	0.5	1.0	1.5	2.0	2.5	3.0	3.5	4.0	4.5	5.0	5.4
$P(H_s)$	0.93	0.35	0.23	0.2	0.13	0.1	0.05	0.04	0.016	0.005	0.003

$P(H)$ versus H_s is plotted on probability paper for the log-normal distribution (Fig. 14.26). The interval of measurement is 1 day = 1/365 yr. $P(H_s)$ for a return period of 20 years is

$$P(H) = \frac{1}{20 \times 365} = 1.37 \times 10^{-4}.$$

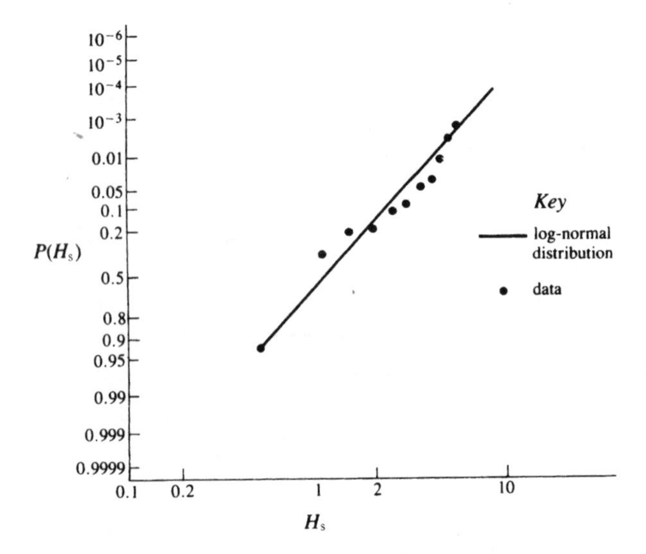

Fig. 14.26 Log-normal distribution (Worked example 14.1, part 2)

Extrapolating the straight line on the probability paper, H_s corresponding to $P(H) = 1.37 \times 10^{-4}$ is the value $H_s \simeq 9$ m.

Worked Example 14.2

A submarine pipeline of internal diameter 275 mm rests on a rocky sea bed at a depth of 20 m. The pipe is made of 16 mm thick steel with a 100 mm thick concrete coating. Waves of 6 m height and 9 s period approach the shore with the crests at an angle of 60° with the pipeline. The prevailing current of 0.5 m s^{-1} is along the same direction as that of the waves. Check whether the pipe is stable against rolling.

Assume the following: $C_D = 1.1$; $C_M = 3.1$; $C_L = 1.2$; density of sea water = 1030 kg m^{-3}; density of concrete = 2700 kg m^{-3}; density of steel = 7800 kg m^{-3}. The coefficient of friction for the rock–pipe interface is 0.3.

Solution

The external diameter of the pipe is $275+32+200=507\,\text{mm}$. The weight of concrete is $1/4\pi(0.507^2-0.307^2)\times 2700\times 9.81=3.39\times 10^3\,\text{N m}^{-1}$. The weight of steel is $1/4\pi(0.307^2-0.275^2)\times 7800\times 9.81=1.13\times 10^3\,\text{N m}^{-1}$. The weight of the pipe in air is $4.52\times 10^3\,\text{N m}^{-1}$. The buoyancy is $1/4\pi\times 0.507^2\times 1030\times 9.81=2.04\times 10^3\,\text{N m}^{-1}$. The submerged weight of the pipe is $(4.52-2.04)\times 10^3=2.48\times 10^3\,\text{N m}^{-1}$.

From Fig. 14.4, for $d=20\,\text{m}$, $T=9\,\text{s}$ and $L=105\,\text{m}$,

$$H=6\,\text{m},\quad k=2\pi/105=0.06\,\text{m}^{-1},\quad \sigma=2\pi/9=0.7\,\text{s}^{-1}.$$

From equation (14.16)

$$\begin{aligned} u &= -\frac{6}{2}\times\frac{2\pi}{9}\,\frac{\cosh[0.06(-19.746+20)]}{\sinh(0.06\times 20)}\sin\sigma t \\ &= -1.4\sin\sigma t. \end{aligned}$$

MORISON EQUATION (14.69)

In determining the wave force, the component of velocity normal to the pipeline is considered. The algebraic sum of the velocity of the current and the particle velocity due to the wave is

$$u=(0.5-1.4\sin\sigma t)\cos 60°.$$

DRAG TERM

$$\begin{aligned} C_D\frac{\rho}{2}|u|uD &= 1.1\times\frac{1030}{2}|0.5-1.4\sin\sigma t|(0.5-1.4\sin\sigma t) \\ &\qquad \times 0.507\times\cos^2 60° \\ &= 71.8|0.5-1.4\sin\sigma t|(0.5-1.4\sin\sigma t)(\text{N m}^{-1}). \end{aligned}$$

INERTIA TERM

$$\begin{aligned} \frac{\partial u}{\partial t} &= -\frac{6}{2}\times\left(\frac{2\pi}{9}\right)^2\cosh\frac{[0.06(-19.746+20)]}{\sinh(0.06\times 20)}\cos\sigma t\cos 60° \\ &= -\cos\sigma t\cos 60°(\text{m s}^{-2}). \end{aligned}$$

The inertia force is

$$-3.1\times 1030\times\frac{\pi}{4}\times 0.507^2\cos 60°\cos\sigma t=-322.3\cos\sigma t\ (\text{N m}^{-1}).$$

The in-line force on the pipe per unit length is

$$71.8|0.5 - 1.4\sin\sigma t|(0.5 - 1.4\sin\sigma t) - 322.3\cos\sigma t \ (\mathrm{N\,m^{-1}}).$$

The lift force per unit length is

$$1.2 \times \frac{1030}{2} \times (0.5 - 1.4\sin\sigma t)^2 \times 0.507\cos^2 60^\circ$$

$$= 78.3(0.5 - 1.4\sin\sigma t)^2 \ (\mathrm{N\,m^{-1}}).$$

t	*Drag*	*Inertia force*	*In line*	*Lift*	*Frictional force*
0	18.0	−322.3	−304.3	19.6	738
1	−11.5	−246.9	−258.4	12.5	740
2	−55.4	−56.0	−111.4	60.5	726
3	−36.4	161.1	124.7	39.8	−732
4	0.0	302.9	302.9	0.0	−744
5	68.8	302.9	371.7	75.0	−722
6	210.5	161.2	371.7	229.7	−675
6.5	253.4	56.0	309.4	276.5	−661
7	253.4	−56.0	197.5	276.5	−661
8	140.7	−246.9	−106.2	153.2	698
9	18.0	−322.3	−304.3	19.6	738

The frictional force = 0.3 × (submerged weight − lift). For stability, the frictional force must be greater than the in-line force. The pipeline is found to be stable.

Worked Example 14.3

Waves of period 8 s in deep water approach the shore from the south-west, as shown in Fig. 14.27. In the figure, bed contours are shown as broken lines. Draw the wave refraction pattern.

Solution

Referring to Fig. 14.27, the steps for drawing the refraction diagram are as follows.

1. Sketch the wave crest at deep water and divide the crest into equal strips AB, BC, CD, etc. Usually ABCD... are drawn at deep water, which may be taken to be $d \simeq 0.5L_0$. From Fig. 14.4, $L_0 = 100$ m for $T = 8$ s.
2. Compute celerities at A, B, C, D... as c_A, c_B, c_C, c_D... using Fig. 14.3.

3. Choose a convenient time of travel t, which is 10 s in this example. Obviously t should be small for closely packed bed contours in order to achieve better accuracy. Draw the orthogonals or rays at A, B, C, D... and advance these rays by $c_A t$, $c_B t$, $c_C t$, $c_D t$,... to A′, B′, C′, D′....
4. Repeat steps 2 and 3 at A′, B′, C′, D′ ... to advance the rays for time t.
5. The lines drawn normal to the rays are the wave crests. Rays and crests are sketched until wave breaking is reached. Refraction coefficients may be determined from the spacing of the rays and wave heights calculated if required.

The refraction pattern for the example is shown in Fig. 14.27.

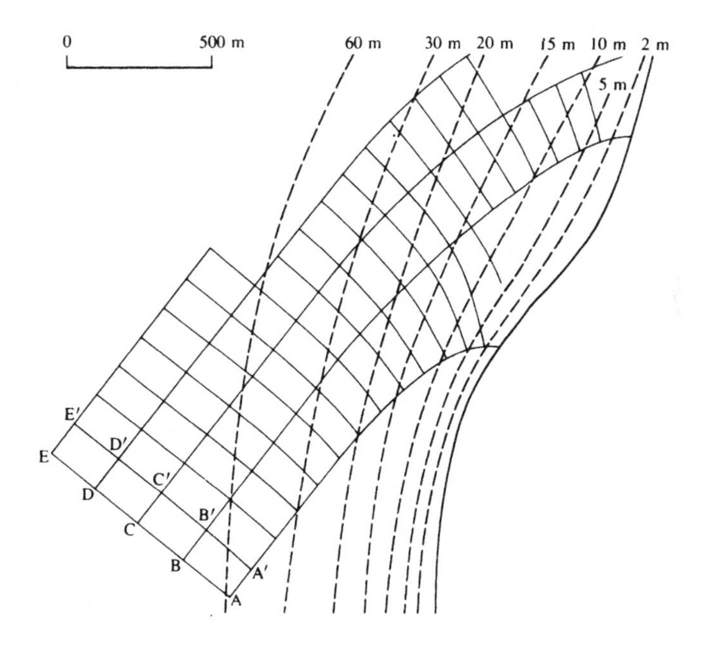

Fig. 14.27 Refraction diagram (Worked example 14.3)

Worked Example 14.4

During construction of a pier, one of the vertical steel piles, diameter 0.508 m, thickness 12.5 mm, and length equal to the depth of water of 10 m, is subjected to a current of $0.5\,\mathrm{m\,s^{-1}}$. The mass of the pile is $195\,\mathrm{kg\,m^{-1}}$. The modulus of elasticity is $200 \times 10^9\,\mathrm{N\,m^{-2}}$. The second moment of area of the section is $6.0 \times 10^{-4}\,\mathrm{m^4}$. The density of sea water is $1030\,\mathrm{kg\,m^{-3}}$.

Will this pile be subjected to flow-induced oscillations?

Solution

The inner diameter of the pile is $0.508 - 2 \times 0.0125 = 0.483\,\text{m}$. The mass of entrained water per unit length is

$$1030 \times \frac{\pi}{4} \times 0.483^2 = 189\,\text{kg}\,\text{m}^{-1}.$$

The added mass per unit length is

$$1030 \times \frac{\pi}{4} \times 0.508^2 = 209\,\text{kg}\,\text{m}^{-1}.$$

The effective mass per unit length, $m_e = 195 + 189 + 209 = 593\,\text{kg}\,\text{m}^{-1}$. The apparent fixity depth is $6 \times 0.508 = 3.05\,\text{m}$. The effective length of the pile $= 10 + 3.05 = 13.05\,\text{m}$. The resonant frequency, f_n, is

$$0.56\left(\frac{200 \times 10^9 \times 6.0 \times 10^{-4}}{593 \times 13.05^4}\right)^{1/2} = 1.48\,\text{Hz}.$$

The reduced velocity, $V' = V/f_n D = 0.5/(1.48 \times 0.508) = 0.66$. Assuming that the damping factor for marine structure steel is 0.08,

$$K_s = \frac{2 \times 593 \times 0.08}{1030 \times 0.508^2} = 0.357.$$

From Fig. 14.23, for $K_s = 0.357$, the critical reduced velocity $V'_c = 1.33$, which is larger than $V'(= 0.66)$. Therefore in-line oscillations will not be initiated.

The kinematic viscosity is $10^{-6}\,\text{m}^2\,\text{s}^{-1}$, and the Reynolds number is $0.5 \times 0.508 \times 10^6 = 2.4 \times 10^5$. From Fig. 14.24, the critical value of $V/f_n D = V'_c = 4.7$. As V'_c is greater than V', cross-flow oscillations will not occur.

References

Bai, K.J. (1975) Diffraction of oblique waves by an infinite cylinder. *Journal of Fluid Mechanics*, 68 (3): 513–45.

Bearman, P.W., Graham, J.M.R. and Obasaju, E.D. (1984) A model equation for the transverse forces on cylinders in oscillatory flows. *Applied Ocean Research*, 6 (3): 166–72.

Bearman, P.W. and Hall, P.F. (1987) Dynamic response of circular cylinders in oscillatory flow and waves, *Proceedings of the International Conference on*

Flow-induced Vibrations, Bowness-on-Windermere, UK, BHRA, 183–90, BHRA, Cranfield.

Blevins, R.D. (1990) *Flow-induced Vibration*, Van Nostrand Reinhold, London.

Brebbia, C.A. (1978) *The Boundary Element Method*, Pentech, Plymouth.

Carter, D.J. (1982) Prediction of wave height and period for a constant wind velocity using JONSWAP results. *Ocean Engineering*, 9 (1): 17–33.

Chakrabarti, S.K. (1987) *Hydrodynamics of Offshore Structures*, Computational Mechanics Publications, Southampton.

Chioukh, N. and Narayanan, R. (1997) Oscillations of elastically-mounted cylinders over plane beds in waves. *Journal of Fluids and Structures*, 11: 447–63.

Darbyshire, M. and Draper, L. (1963). Forecasting wind generated sea waves. *Engineering*, 195 (April): 482–4.

Dean, R.G. and Dalrymple, R.A. (1991) *Water Wave Mechanics for Engineers and Scientists*, World Scientific, Singapore.

Garrison, C.J. and Chow, P.Y. (1972) Wave forces on submerged bodies. *Journal of Waterways, Harbors and Coastal Engineering, Proceedings of the American Society of Civil Engineers*, 98 (WW3): 375–92.

Goda, Y. (1979) A review of statistical interpretation of wave data. *Report of the Port and Harbour Instititute*, Japan, 18: 5–32.

Hallam, M.G., Heaf, N.S. and Wootton, L.R. (1978) *Dynamics of Marine Structures*, Report UR8, CIRIA Underwater Engineering Group, October.

Hasselmann, K.D., Ross, D.B., Muller, P. and Sell, W. (1976) A parametric wave prediction model. *Journal of Physical Oceanography*, 6: 200–28.

Hoffmans, G.J.C.M. and Verheij, H.J. (1997) *Scour Manual*, Balkema, Rotterdam, The Netherlands.

Huang, M.C. and Hudspeth, R.T. (1982) Pipeline stability under finite amplitude waves. *Journal of the Waterways and Harbors Division, Proceedings of the American Society of Civil Engineers*, 108 (WW2): 125–45.

Isaacson, M. and Mackenzie, N.G. (1981) Long-term distributions of ocean waves: a review. *Journal of the Waterways, Port, Coastal Ocean Division, Proceedings of the American Society of Civil Engineers*, 107 (WW2): 93–109.

Iwagaki, Y., Shiota, K. and Doi, H. (1982) Shoaling and refraction coefficient of finite amplitude waves. *Coastal Engineering, Japan*, 25: 25–35.

Kao, C.C., Daemrich, K.F., Kohlbase, S. and Partenscky, H.W. (1984) Transverse force due to wave on cylinder near bottom. Symposium on Ocean Structures Dynamics, Corvallis, Oregon, USA: 356–68.

King, C.A.M. (1972) *Beaches and Coasts*, 2nd edn, Arnold, London.

Komar, P.D. (1998) *Beach Processes and Sedimentation*, Second Edition, Prentice-Hall, Englewood Cliffs, NJ.

Littlejohns, P.S.G. (1974) *Current-induced Forces on Submarine Pipelines*, Report No. INT 1238, Hydraulics Research Station, Wallingford, England.

—— (1982) *Wave forces on pipelines*, Report No. Ex 106, Hydraulics Research Station, Wallingford, England.

Mathur, A. (1995) *Offshore Engineering: An Introduction*, Witherby Publishers, London.

Morris, H.M. (1963) *Applied Hydraulics in Engineering*, Ronald Press, New York.

Muir Wood, A.M. and Fleming, C.A. (1969) *Coastal Hydraulics*, Macmillan, London.

Pierson, W.J. and Moskowitz, L. (1964) A proposed spectral form for fully developed wind seas. *Journal of Geoophysical Research*, 69: 5181–90.

Sarpkaya, T. (1976a) In-line and transverse forces on cylinders in oscillatory flow at high Reynolds numbers, *Proceedings of the 8th Annual Offshore Technology Conference*, Paper OTC 2533: 95–100.

—— (1976b) Forces on cylinders near a plane boundary in a sinusoidally oscillating fluid. *Journal of Fluids Engineering*, ASME, 98: 499–505.

—— (1977) In-line and transverse forces on cylinders near a wall in oscillatory flow at high Reynolds number, *Proceedings of the 9th Annual Offshore Technology Conference*, Houston, Paper OTC 2989, 3: 161–6.

Sarpkaya, T. and Isaacson, M. (1981) *Mechanics of Wave Forces on Offshore Structures*, Van Nostrand, London.

Schlichting, H. (1960) *Boundary Layer Theory*, McGraw-Hill, New York.

Skjelbreia, L. and Hendrickson, A. (1961) Fifth order gravity wave theory, in *Proceedings of the 7th Conference on Coastal Engineering*, Vol. 1, American Society of Civil Engineers, Chapter 10.

Sumer, B.M. and Fredsoe, J. (1990) *Scour below Pipelines*, *Journal of Waterway, Port, Coastal and Ocean Engineering*, ASCE, 116 (3): 307–23.

Sumer, B. and Fredsoe, J. (1997) *Hydrodynamics Around Cylindrical Structures*, World Scientific, Singapore.

Sumer, B.M, Whitehouse, R.J.S. and Torum, A. (2001) Scour around Coastal Structures: A Summary of Recent Research, *Coastal Engineering*, 44, pp. 153–90.

Swift, R.H. and Dixon, J.C. (1987) Transformation of regular waves. *Proceedings of the Institution of Civil Engineers, Part 2*, 83 (June): 359–80.

Tucker, M.J. (1963) Analysis of records of sea waves. *Proceedings of the Institution of Civil Engineers*, 26: 305–16.

US Army (1984) *Shore Protection Manual*, US Army Coastal Engineering Research Center, Washington, DC.

—— (2002) *Coastal Engineering Manual*, Corps of Engineers, Washington, DC.

Wiegel, R.L. (1964) *Oceanographical Engineering*, Prentice-Hall, Englewood Cliffs, NJ.

Williams, J.G. (1985) *Tables of Progressive Gravity Waves*, Pitman, London.

Yuksel, Y. and Narayanan, R. (1994a) Wave forces on horizontal cylinder resting on sloping bottom, in *Proceedings of the 4th International Offshore and Polar Engineering Conference*, Tokyo.

—— (1994b) Breaking wave forces on horizontal cylinders close to the sea bed. *Journal of Coastal Engineering*, 23: 115–23.

Chapter 15

Coastal engineering

15.1 Introduction

Coastal engineering encompasses a variety of problems of practical importance, e.g. provision of harbours and their protection against sedimentation, provision for discharge of effluents into the sea, design and construction of works to protect coastal areas from flooding, defence against erosion, etc. A coastal structure should not only satisfy the functions it is intended for but also structurally withstand the hostile environment. Most coastal problems are difficult to tackle owing to the complexity of the processes involved. A solution to one problem may very well cause others, and so particular attention should be given to the interaction between the various elements that determine the coastal régime. Over the years coastal processes have been better understood and designs have been rationalized with the aid of laboratory studies, theoretical methods and field observations. For extensive treatment of the subject of coastal engineering the reader is referred to Ippen (1966), Muir Wood and Fleming (1969), Silvester (1974), Horikawa (1978), Fredsoe and Deigaard (1992), Herbich (2000), Kamphuis (2000) and Reeve *et al.* (2004).

Waves play the most dominant role in the coastal processes causing erosion, movement of sediments, development of beach profile, overtopping of coastal defence structures, harbour resonance, etc. The properties of wind-generated waves of short period and their refraction, diffraction and breaking are treated in Chapter 14.

The very slow rise and fall of water level is due to the astronomical tides produced by the gravitational field in the presence of rotating Earth, Moon and Sun. The timescale of tidal oscillations is very much larger than that of the wind-generated waves. Water level variation due to astronomical tides is generated by the interaction of the gravitational fields of the rotating Earth, Moon and Sun during their orbital motions. These motions cause periodic forces on water on the surface of the Earth introducing a large number of periodicities in the tidal motion. The time periods of these periodic constituents are known. Further, tidal flow is strongly influenced by the Coriolis acceleration, by the sea bed topography and features of coastlines. It may be amplified by the local resonance effects in bays and estuaries. Accurate prediction of tidal behaviour at a particular site by theoretical means is thus not possible. In practice, the measured data at a site are expressed as a summation of periodic components, the number of which depends on the duration of the record. Once the periodic components have been computed, they are then used in the summation to predict the water level at the site in the future. In the USA, tide predictions are made by the National Oceanic and Atmospheric Administration; in the UK, the Admiralty Tide Tables provide information about astronomical tides around the coasts.

Seawater levels are also raised significantly by strong onshore winds and by suction due to low atmospheric pressure. Apart from generating waves the wind causes the water surface to assume a slope due to wind shear, thus forming a storm surge which is further affected by the topography of the coastline. The *Shore Protection Manual* (US Army, 1984) presents a method of determining the storm surge based on wind shear.

Water level at a particular site depends on tidal and wind conditions. The design of structures must take into account the tide levels to which should be added the depth due to storm surge. The combined effects of high water level and large waves are the cause of coastal flooding or damage to coastal structures. Observations made along the coast over a period of time could form a useful basis for statistical analysis to determine the individual probabilities of wave heights and water levels (Will, Willis and Smith, 1985). However, in recent times the joint probability of a given water level with a wave of given height occurring at the same time, forms the basis for specifying the design conditions for coastal structures.

Seiches are oscillations of water levels in a body of water such as lakes and reservoirs; the disturbance may be initially caused by sudden gusts of wind (see also Section 4.4). Tsunamis are caused by submarine earthquakes. Seiches, harbour oscillations and tsunamis are long period waves and hence are treated as shallow water waves. Tsunamis generated in deep sea can grow to a considerable height as they approach the shore due to shoaling effects (equation (14.34)) and they can cause extensive flooding, costly damage to property and heavy loss of life along coastal regions.

Coastal currents may be generated by the density differences in the seas, wind stresses, tidal flow, rivers entering the sea and waves. Currents produced by breaking waves are important as far as sediment transport along the beaches is concerned. River flows discharging into the sea can cause near shore currents and circulation. Currents caused by the flood and ebb of the tide and by wind stresses, particularly in shallow water, can lead to vortex shedding in the wake of immersed structures. As with offshore structures, flow-induced vibrations of structures such as piles may also occur, threatening the structural integrity (see Sections 14.12 and 14.13).

Global warming causes rise of sea water temperatures and melting of the polar ice both of which result in rising sea levels making the efficacy of some of the existing structures as measures against flooding questionable in the future. Engineers are therefore beginning to introduce future water levels due to global warming into design calculations.

15.2 Coastal defence

15.2.1 General

Coastal defence is used to describe natural features or man-made measures that prevent coastal flooding and prevent the existing shoreline from erosion. Natural coastal defence is provided by shorelines in the form of cliffs, sand dunes and beaches. Cliffs can be either erodible or non-erodible. The former is a good source of beach material and the latter protects the shoreline from the severity of the sea possibly causing depletion of coastal sediments in their vicinity. Wind action is responsible for transporting and depositing sand to form sand dunes along the coast. They are very effective as sea defence against flooding during extreme stormy conditions. Vegetation, particularly grass, protects sand dunes from eroding due to wind and traps sand transported by wind. Coastal processes affecting the sediment motion and physical and geological features of the coastline help form a healthy natural beach which is the best form of coastal defence.

In planning engineering works, the factors that must be taken into account are the causes of erosion of the existing coastline, the direction and the magnitude of the waves and surges, the tidal range, the transport of sediments, the effect of the planned structure on the coastal régime and, of course, economic, social, and environmental factors. Before proceeding with a coastal defence scheme, cost–benefit and environmental impact studies are thus essential.

Coastline protection may involve schemes such as seawalls, breakwaters parallel to the shoreline, groynes and provision of beach nourishment. The interaction of waves with these structures is clearly very important in their design.

Beaches adjust themselves to absorb the energy of the waves; if the wave climate exhibits large variability, the beach profiles become complex. When the waves break, the flow is intensely turbulent and large quantities of sediment may be transported in suspension. The subsequent onrush of water can carry even coarse sediments towards the beach. The downrush is slow and can transport relatively small sediments offshore. If the energy of oncoming waves is small, only finer sediments are moved onshore. The sediment transported over a period of time will depend on the variations in the wave climate, including the direction of wave attack. In addition to the onrush and downrush of water on the beach, two important types of currents are set up by the wave motion: rip currents and longshore currents.

The rip current is the narrow and strong flow seaward as a result of bed features and the formation of opposing along-shore currents. For example, when water is trapped behind obstacles such as sandbars, it forms a current along the coast which can drain with a high velocity current towards the sea through an opening in the sandbars (Fig. 15.1). Thus the transport of water over the sandbar (e.g. by a large breaker) feeds the rip current to form a nearshore circulation.

When waves break at an angle to the coastline, a longshore current parallel to the coastline is established (Fig. 15.2), responsible for the net

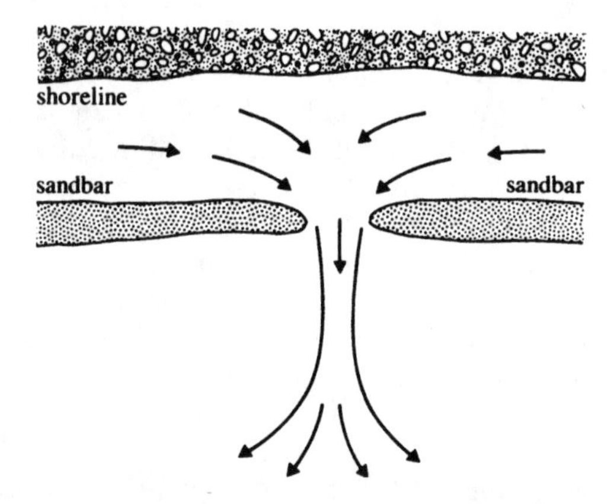

Fig. 15.1 Rip currents

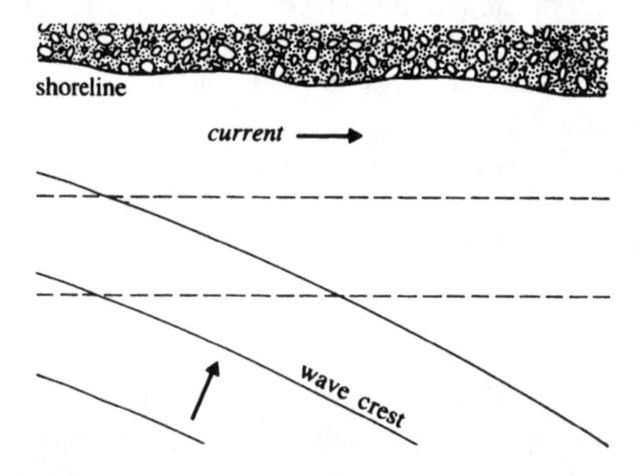

Fig. 15.2 Longshore currents

transport of sediment along the shore. Knowledge of the sediment transport capacity of the longshore currents is important in preventing erosion of the beach or in siting a harbour entrance.

The sediment transport capacity is related to the wave power at the instant of wave breaking:

$$R_b = \frac{\rho g}{8} H_{s_b}^2 C_{g_b} \sin \alpha_b \cos \alpha_b \tag{15.1}$$

where the subscript ‘b’ refers to the quantities at the instant of wave breaking, and α is the angle of the crests with the bed contour. The submerged weight of the sediments transported per unit time is

$$G_s = (\rho_s - \rho) g a' Q_s \tag{15.2}$$

where ρ_s is the density of the sediments, Q_s is the volume rate of sediment plus water transported and a' is the ratio of the volume of sediments to the total volume (typically 0.6).

An empirical relationship between G_s and R_b is

$$G_s = kR_b. \tag{15.3}$$

Shore Protection Manual (1984) recommends a value of 0.39 for the coefficient k. Some authors use r.m.s value of wave height H_{rmsb} at breaking point instead of H_{sb} for wave power R_b in equation (15.1). By virtue of equation (14.54), $k = 0.78$ in equation (15.3) when H_{rmsb} is used.

The coefficient k is not a constant. Its dependence on the sediment diameter is discussed in detail in *Coastal Engineering Manual* (US Army, 2002) to suggest Bailard’s formula

$$k = 0.05 + 2.6 \sin^2(\alpha_b) + 0.007\, u_b/w_s \tag{15.4}$$

in which u_b is the maximum velocity of particle at breaking point and w_s is the fall velocity of sand. Using shallow water approximation

$$u_b = 0.5(H_b/d_b)(gd_b)^{0.5} \tag{15.5}$$

The value of k obtained from equation (15.4) is applicable when Equation (15.3) is expressed in H_{rmsb} instead of H_{sb}.

Most of the drift occurs in the breaker zone. The direction of littoral drift or transport is determined by the direction of the longshore current. A temporary reversal of its direction may take place as a result of the variation of the approach angle of the waves. As the wave attack usually predominates in one direction over a period of time, there is a net transport of sediment in that direction. Wave climate at a particular site on the coastline can be expressed in terms of bands of wave heights in sectors of wave direction. It is possible to use this statistical information along with equation (15.1) to determine the net sediment discharge and its direction and then evaluate whether erosion or deposition takes place at the site taking account of the tidal currents which will have an influence on the sediment transported.

Structures such as detached breakwaters and groynes interfere with the littoral drift. A detached breakwater is normally built parallel to the shore to provide an area sheltered from the wave action (Fig. 15.3(a)). The refraction and diffraction effects produce a decrease in energy behind the breakwater; hence deposition takes place in the protected area. In the presence of a structure, the wave height varies in the direction alongshore. Such variation influences the longshore sediment transport in addition to the contribution from wave breaking.

Shore-connected breakwaters (Fig. 15.3(b)), which extend across the littoral zone, act as a barrier. If the littoral drift was originally in equilibrium, the breakwater will initially cause deposition on the updrift side, and because of the reduced sediment concentration in the longshore current, erosion on the downdrift side of the breakwater. As time proceeds, the

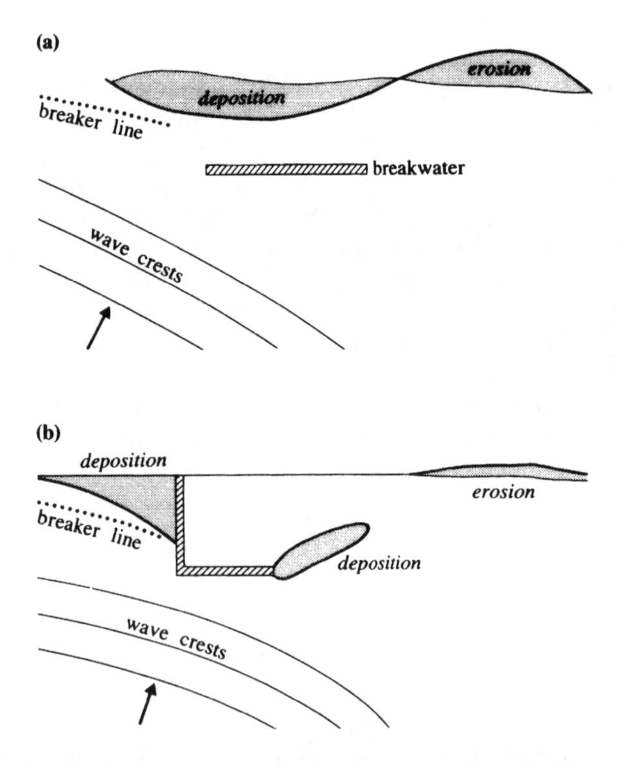

Fig. 15.3 Effects of breakwaters on the littoral drift

sediment motion will tend again towards equilibrium, with sediments moving along the face of the breakwater and depositing at its lee side.

Waves approaching the shore can cause dominant movement of sediments cross-shore to shape the beach profile. The shape of the beach will also be affected by the presence of coastal structures. An approximate criterion for the direction of movement of sediments due to Dean (1973) is: if $H_0/(w_s T) < 1$, sediment movement is onshore and if $H_0/(w_s T) > 1$, it is offshore. The method of Bailard (1982) is widely used to estimate the volume of cross-shore sediment transport.

It is assumed that over a long period of time, the beach profile tends to a state of equilibrium. The equilibrium profile proposed by Bruun and Dean (Dean, 2002) is

$$d = Ax^{2/3} \tag{15.6}$$

in which d is the depth and x is the seaward distance, and $A = 0.067\, w_s^{0.44}$.

The limit of the profile is taken to extend closure depth $d^* = 1.57\, H_{s12}$ (Birkemeier, 1985). H_{s12} is the significant wave height with the frequency of occurrence of 12 hours per year.

Beach profile and closure depth are useful to compute sediment volume in beach nourishment schemes. For further details about the shore evolution and beach nourishment, reference is made to Pilarczyk and Zeidler (1996), Kamphuis (2000), Dean (2002) and Reeve *et al.* (2004).

15.2.2 Coastal models

Many coastal numerical models couple hydrodynamic equations with equations for transport of sediments, pollutants etc. The hydrodynamic equations represent the presence of waves, currents and water levels. They also include forces arising from sources such as Coriolis acceleration, wind stresses and density effects. Some form of turbulence structure and empirical expression for boundary friction are assumed. The three dimensional equations are usually difficult to solve and hence are simplified into two dimensional ones. Kamphuis (2000) discusses the suitability of these models such as 2DH (depth-integrated) and 2DW (width-integrated) in predicting short term, medium term and long term morphological changes based on time and spatial scales. DELFT3D of Delft Hydraulics Laboratory and MIKE 21 of Danish Hydraulics Institute are commercial softwares that are popular in the numerical study of the hydrodynamics, sediment transport, pollutant dispersion, morphological changes etc., over extensive coastal regions.

In the numerical models uncertainties remain with respect to bed forms, bed shear and turbulent stresses, and the interaction between the flow and the sediment motion is complicated. As coastal morphology depends on sediment transport, many developed models are deficient in these areas.

Numerical models have to be calibrated against field data which should be of sufficient accuracy. Calibration is used to fine tune the model by adjusting some of the parameters such as the eddy viscosity, roughness coefficient etc. The calibrated model is then verified against another independent set of field data which have not been used in the calibration.

In coastal engineering, continuity of sediment volume is normally used to model erosion or accretion of a beach. The so-called one-line model assumes a typical beach profile like equation (15.6) assuming to be unchanging up to the closure depth. But the beach itself is assumed to move on a horizontal plane onshore or offshore depending on erosion or accretion. A formula like equation (15.3) is used for the quantity of sediment motion alongshore. The model should also incorporate computation of wave transformation along the beach. The one-line model predicts the movement of the beach cross-shore. On the other hand, N-line coastal models introduce refinements by taking account of the cross-shore sediment transport. Kamphuis (2000) lists some available computer programs based on one-line or N-line models.

Roelvink and Broeker (1993) compared the performance of five commonly used profile change models. Cross-shore sediment transport is important in these models. Comparison of the computed results with the results of laboratory results showed that these models can only predict short-term changes. They also find wide differences between the predictive capabilities of the various models used.

Numerical models are also used to investigate the impact of structures such as offshore or shore-connected breakwaters on the sediment motion and hence the coastal behaviour. Although significant progress has been made so far in the development of these models, research is actively pursued to improve their predictions. Physical models are also used to investigate coastal behaviour under controlled conditions. However, it may not be possible to simulate in the physical model all the processes present in the field. Currently design of coastal structures makes use of results obtained from numerical and physical models, and field measurements.

For a general discussion of models in hydraulic engineering see Chapter 16.

15.2.3 Groynes

One of the methods of saving a beach or a shoreline from erosion is to construct a series of groynes, usually built perpendicular to the shoreline. They trap sand transported by the longshore currents, but downdrift erosion may take place due to starvation of beach material.

Groynes are constructed from timber, concrete, stone or steel; riprap or concrete blocks are also used, the choice of material depending on availability and cost. The length of the groynes is related to the littoral drift; the smaller it is, the shorter the groynes. The height of the groynes is related to the depth of deposition required. In the initial stages after the erection of the groynes, beach material may have to be fed at the down-drift side of the groynes to avoid undesirable recession of the beach there. As time progresses, the beach material tops over the groynes to fill successively the adjoining compartment until the beach encroaches into the sea (Fig. 15.4).

The length and spacing of the groynes depend on the amount of sediment to be trapped, the size of the beach material and the angle of wave attack. Shingle beaches require short groynes, while sand beaches require long groynes. At the seaward end the groynes should extend to the low-water level. One American recommendation is that they should end at the 2 m depth contour. The ratio of spacing to the length of groynes varies between 1 and 4. The *Shore Protection Manual* (US Army, 1984) recommends 2 to 3. If the wave crests are almost parallel to the beach, larger ratios may be used. High groynes which reduce the drift significantly are often not the best solution for stabilizing the beach. The recommended height of groynes is 0.5–1 m above beach level. Initially, groynes may be constructed with their tops above the foreshore level and, as accretion takes place, the groynes are raised to ensure a uniform build-up of the foreshore. Low groynes reduce local erosion.

Straight groynes are the most common, although zig-zag groynes and breakwaters parallel to the beach at the end of the groynes have also been used. If the prevailing angle of wave attack is 30° or more, groynes are inclined at an angle towards the downdrift side. Groynes may either be impermeable or permeable. Permeable groynes are intended to allow the continuation of some littoral drift. They may trap larger sediments but allow smaller sediments to be transported, thus reducing erosion downdrift. Experience has shown, however, that on the whole they are no better than the impermeable ones (Berkeley, Thorn and Roberts, 1981).

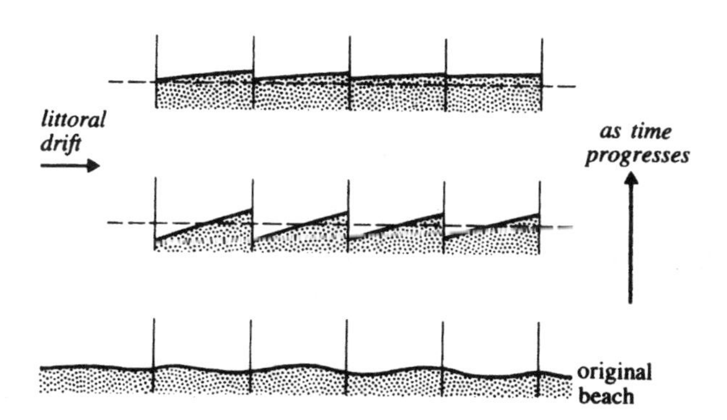

Fig. 15.4 Operation of a successful groyne system

No well-defined procedure exists at present for the design of a groyne system. Experience gained from field observations, model studies and mathematical models is all used in the design of groynes to avoid their failure and undesirable side-effects such as recession at the downdrift side (Summers and Fleming, 1983).

15.2.4 Beach nourishment

Beach nourishment may be used as an alternative to the installation of, or in conjunction with, groynes for the protection of shoreline. It is deployed where the coastal areas experience loss of sediments without being replenished by littoral drift or where the purpose is to create a wider beach for recreation or land reclamation. Beach nourishment involves the supply of suitable materials from quarries, mines or from offshore by dredging to the shore and dumping at suitable places on the beach so that the sea action will distribute them to shape the required beach profile. The amount of material for recharge will depend on the rate of coastal erosion, environmental requirements, required beach shape and onshore–offshore movement of bed material. It is advisable to use material for beach nourishment with properties like sizes and grading similar to the native material at the beach. Usually flat average slope of beaches is made of finer sediments. The quantity of material and hence the annual cost including transport from the material source and dumping along the beach will dictate the cost of recharging the coast. Computational and physical modelling can give better understanding of the processes involved and hence better estimates of the recharge volume especially for larger schemes. Details of design of beach profiles with beach nourishment are clearly set out in Simm *et al.* (1996). Reference is also made to *Shore Protection Manual* (US Army, 1984), Davison *et al.*, (1992) and Dean (2002).

15.3 Wave forces on coastal structures

Coastal structures are defined as rigid, semirigid or flexible, according to their rate of failure. A rigid structure exposed to a high wave might collapse completely. The design wave height for rigid structures is the average of the highest one-tenth of the waves. For flexible structures, which very rarely collapse in their entirety, the design wave is the significant wave. Some damage to flexible structures is tolerated if the functional requirements are not seriously lost. These design waves are chosen for a specified return period (Goda, 1979).

Sea walls, breakwaters and other engineering structures have to withstand the hydrostatic, wave and impact forces of the breaking waves.

If the depth in the vicinity of the wall is such that the waves break before or at the wall, it will be exposed to large impact pressures owing to the compression of a trapped cushion of air between the advancing wave and the wall. If water near the wall is deep enough, a standing wave or clapotis is established adjacent to the structure and the force exerted on the wall is given by the pressure distribution due to the clapotis.

Sainflou's analysis is used to estimate the pressures developed in the presence of the clapotis (Morris, 1963). It is for fully reflected standing wave and cannot be applied when wave breaking or overtopping occurs. According to this theory, the horizontal plane of the oscillations of the standing wave is raised by an amount h_0 above the still-water level:

$$h_0 = \frac{\pi H^2}{L} \operatorname{cotanh}\left(\frac{2\pi d}{L}\right) \tag{15.7}$$

where H and L are, respectively, the height and wavelength of the incident wave and d is the undisturbed water depth (Fig. 15.5). The clapotis has oscillations with a maximum height of $2H$. From the linear theory, the velocity potential, ϕ, for the standing wave of height $2H$ is

$$\phi = -\frac{Hc \cosh[k(y+d)]}{\sinh(kd)} \cos kx \cos \sigma t$$

and

$$\frac{\partial \phi}{\partial t} = \frac{Hc\sigma \cosh[k(y+d)]}{\sinh(kd)} \cos kx \sin \sigma t.$$

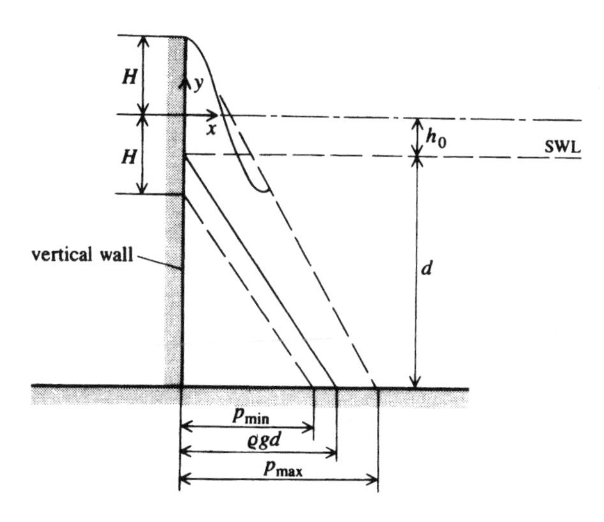

Fig. 15.5 Pressure distribution on vertical wall

Note that the velocity potential satisfies $u = 0$ at the wall for which $x = 0$. At the wall

$$\frac{\partial \phi}{\partial t} = \frac{Hc\sigma \cosh[(k(y+d)]}{\sinh(kd)} \sin \sigma t. \qquad (15.8)$$

Using equation (15.8) in the linearized energy equation (14.6), the pressure distribution on the wall is given by

$$p = \rho\left[-Hc\sigma \frac{\cosh[k(y+d)]}{\sinh(kd)} \sin \sigma t - gy\right].$$

(The second term in the bracket on the right-hand side of the above equation is the contribution from the hydrostatic pressure distribution.) With $c = \sigma/k$, and c^2 given by equation (14.13), the pressure can be written as

$$p = \rho\left[-gH \frac{\cosh[k(y+d)]}{\cosh(kd)} \sin \sigma t - gy\right]. \qquad (15.9)$$

At the bed ($y = -d$), the maximum and minimum pressures are

$$p = \rho g\left[\pm \frac{H}{\cosh(kd)} + d\right]. \qquad (15.10)$$

The maximum and the minimum forces exerted on the wall may be estimated assuming that the pressure varies linearly from 0 at the surface to the value on the bed given by equation (15.10). The pressure profiles are shown in Fig. 15.5. The maximum and minimum pressures at any depth, y, below the still-water level are

$$p_{\max} = \rho g \frac{-y + h_0 + H}{d + h_0 + H}\left[\frac{H}{\cosh(kd)} + d\right] \qquad (15.11)$$

and

$$p_{\min} = \rho g \frac{-y + h_0 - H}{d + h_0 - H}\left[-\frac{H}{\cosh(kd)} + d\right]. \qquad (15.12)$$

The force on the wall per unit width due to the clapotis at the crest position is

$$F_{\max} = \frac{\rho g}{2}(d + h_0 + H)\left[\frac{H}{\cosh(kd)} + d\right]. \qquad (15.13)$$

With the trough of the clapotis at the wall, the force per unit width is

$$F_{\min} = \frac{\rho g}{2}(d + h_0 - H)\left[-\frac{H}{\cosh(kd)} + d\right]. \tag{15.14}$$

The maximum and the minimum bending moments about the base per unit width of the wall are

$$M_{\max} = \frac{\rho g}{6}(d + h_0 + H)^2\left[\frac{H}{\cosh(kd)} + d\right] \tag{15.15}$$

and

$$M_{\min} = \frac{\rho g}{6}(d + h_0 - H)^2\left[-\frac{H}{\cosh(kd)} + d\right]. \tag{15.16}$$

In equations (15.13)–(15.16) the effects of the hydrostatic pressure distribution due to the still water have been included. In designing the structure, account must be taken of any still water behind the structure.

Significant pressures may be developed by the breaking waves on the wall. The *Shore Protection Manual* (US Army, 1984) recommends the method of Minikin (1963); it results in the pressure distribution shown in Fig. 15.6. It consists of a dynamic pressure peaking at the still-water level and a hydrostatic pressure due to the wave amplitude. The pressures are in addition to the hydrostatic pressures of the still water.

The peak dynamic pressure at the still-water level is

$$p_{\mathrm{m}} = p_{\mathrm{b}} + \frac{1}{2}\rho g H. \tag{15.17}$$

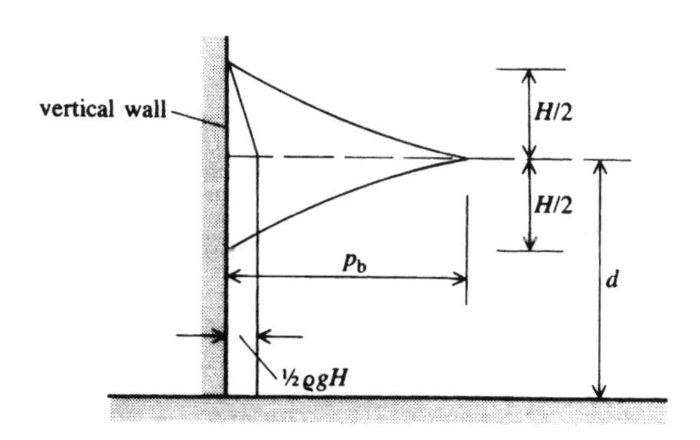

Fig. 15.6 Pressure distribution on vertical wall due to breaking wave

The force on the wall is obtained by integrating the pressure variation. The variation of the dynamic pressure component with respect to the depth is assumed as parabolic with zero at $\pm H/2$. Hence the force on the wall per unit width due to breaking waves is

$$F_b = \frac{1}{3}Hp_b + \frac{1}{2}\rho gH\left(d + \frac{H}{4}\right)$$

$$= \frac{\rho gH}{6}\left[4\pi H\frac{d}{L} + 3\left(d + \frac{H}{4}\right)\right]. \quad (15.18)$$

If the wall slopes at an angle θ to the vertical, then the dynamic pressure p_b must be multiplied by $\cos^2\theta$.

Goda's method (Goda, 1974, 2000) is used widely to calculate the horizontal force on a vertical breakwater resting on a rubble mound or foundation. The formulae cover both breaking and broken conditions of waves but not aeration. The wall pressure distribution on the vertical wall resting on rubble mound without overtopping is shown in Fig. 15.7. The pressure is atmospheric at a vertical distance R which is the run up, given as

$$R = 0.75\ (1 + \cos\beta)\, H_d$$

where H_d is the design wave height. If the waves do not break before the structure $H_d = 1.8\ H_s$ and for broken waves H_d is the highest of the random waves at a distance $5H_s$ seaward of the structure. β is the angle between wave crests and the wall. Referring to Figure 15.7,

$$P_1 = 0.5\ (1 + \cos\beta)(\alpha_1 + \alpha_2\cos^2\beta)\ \rho gH_d$$

$$P_3 = \alpha_1 P_1$$

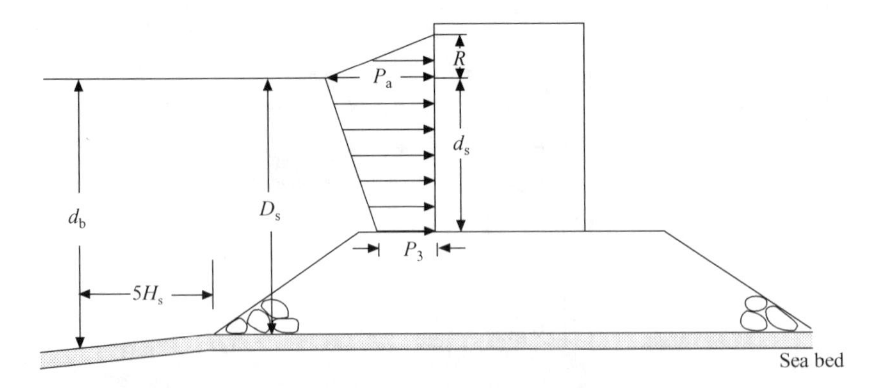

Fig. 15.7 Pressure distribution in Goda's method

The coefficients α_1, α_2 and α_3 are

$$\alpha_1 = 0.6 + 0.5\left[\frac{4\pi D_s/L}{\sinh(4\pi D_s/L)}\right]^2$$

$$\alpha_2 = \min\left[\frac{1}{3}(1 - d_s/d_b)(H_d/d_s)^2, 2d_s/H_d\right]$$

$$\alpha_3 = 1 - (d_s/D_b)[1 - 1/\cosh(2\pi D_s/\mathrm{L})]$$

D_s is the water depth measured from SWL to the seabed, d_s is the depth from SWL to the top of the rubble mound, d_b is the water depth at breaking which is assumed at a distance of $5\,H_s$ in front of the vertical wall, min (a, b) is the smaller of the two values a and b.

Goda (1974, 2000) also presents formulae for uplift pressures on the wall placed into the foundation. Wave forces exerted on vertical walls resting on caissons are extensively treated in *Coastal Engineering Manual* (US Army, 2002).

Sea walls and other coastal structures are susceptible to wave and current induced erosion. At sloping front structures, scour undermines the toe leading to their failure. The local scour can be reduced by reducing wave reflection, achieved by decreasing the front slope of the structure, making it permeable, increasing the roughness of the face or introducing a horizontal berm. The extent of the scour is also controlled by placing large stones in the affected areas or by other toe protection measures (Hoffmans and Verheij (1994)).

15.4 Wave run-up

Sea walls, many of which are sloping with berms, provide an almost impervious surface over which wave uprush and backrush occur. As the severity of the waves increases, the protection of the seaward face of the wall changes from turf to heavy stone or concrete construction. Some examples of sea walls are shown in Fig. 15.8: they are (a) stepped, (b) smooth sloping and (c) vertical. The sea wall shown in Fig. 15.8(d) is similar in construction to a rubble-mound breakwater, with the cover layer consisting of armour units designed to withstand large wave forces. A gently sloping wall reduces reflection and thus erosion at the toe of the wall. Modern designs frequently incorporate a curved section at the top of the sea wall to deflect the waves downwards (Fig. 15.8(e)). Sea walls are expensive to construct and to maintain. Their design must prevent undue overtopping (see Section 15.5); the toe of the wall must be protected against erosion

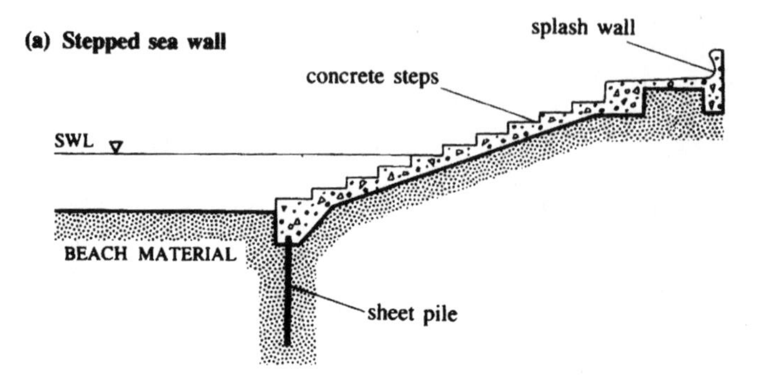

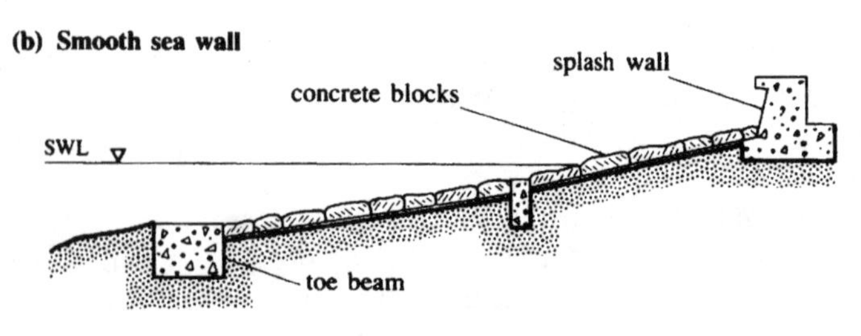

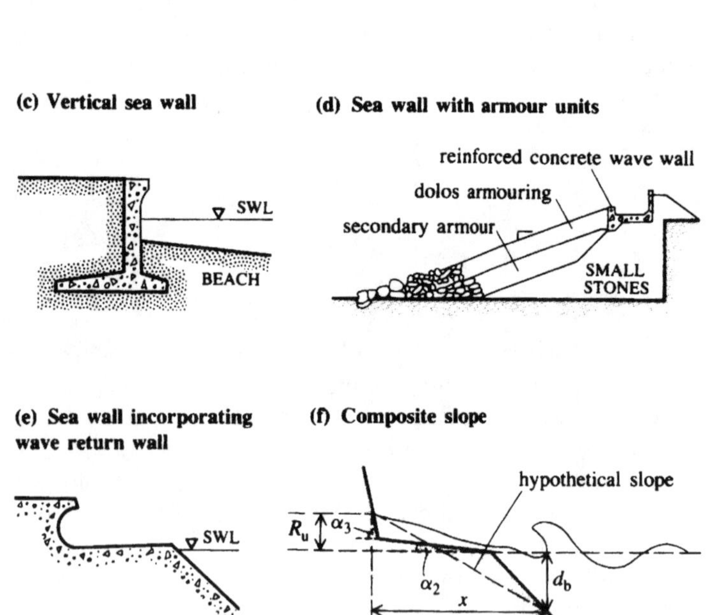

Fig. 15.8 Examples of sea walls

and adequate drainage must be provided behind the wall to reduce slope failure and undermining.

When a wave breaks on a structure such as a sea wall or breakwater, some of its energy is dissipated in turbulence and the rest is used to run up on the wall, gaining potential energy. The height of the run-up determines whether there will be overtopping. The slope, β, of the face of an impermeable wall to the horizontal to ensure breaking of the wave is given by Iribarren and Nogales (1949) as

$$\tan\beta = \frac{8}{T}\left(\frac{H_{\mathrm{I}}}{2g}\right)^{1/2}. \tag{15.19}$$

Slopes steeper than β given by the above expression cause surging and reflection; for flatter slopes, the waves break on the slope. H_{I} is the incident wave height.

For surging waves on steep slopes, the run-up R_{u} is

$$R_{\mathrm{u}}/H_{\mathrm{I}} = (\pi/2\beta)^{1/2}. \tag{15.20}$$

Laboratory studies show that for surging waves $R_{\mathrm{u}}/H_{\mathrm{I}}$ is no greater than about 3.

For small slopes, the run-up R_{u} of the breaking wave is the vertical height above the still-water level and is related to the wave height in non-dimensional form as

$$R_{\mathrm{u}}/H_0 = 1.016\tan\beta(H_0/L_0)^{-0.5}. \tag{15.21}$$

The subscript ‘0’ denotes the deep-water conditions. With $L_0 = gT^2/2\pi$, the above equation may be expressed as

$$R_{\mathrm{u}}/H_0 = 0.405\tan\beta(H_0/gT^2)^{-0.5}. \tag{15.22}$$

The *Shore Protection Manual* (US Army, 1984) presents charts for the estimation of the wave run-up of regular waves on rubble-mound breakwaters and riprap slopes. R_{u} is reduced by the porosity and roughness of the wall but materials used under permeable conditions are exposed to uplift pressures due to the receding wave. Equation (15.21) for an impervious surface can be modified for a porous surface, as

$$R_{\mathrm{u}}/H_0 = 1.016\tan\beta(H_0/L_0)^{-0.5}r \tag{15.23}$$

where $r < 1$ is an empirically determined factor (Table 15.1).

Equation (15.23) gives the mean run-up for regular waves; this, however, will vary from wave to wave under the action of irregular waves. The run-ups due to irregular waves on coastal structures with smooth slope have been determined from model studies for two measures of wave

Table 15.1 Factor *r* for various armour units

Armour unit	*r*
Smooth, impervious	1.0
Concrete slabs	0.9
Concrete block	0.85–0.9
Grass on clay	0.85–0.9
One layer of quarrystone (impervious)	0.8
Rubble stone placed at random	0.5–0.8
Two or more layers of rockfill	0.5
Tetrapods	0.5

Those values may be considerably exceeded if $\tan\beta(H_0/L_0)^{-0.5}>2$ ($\beta=$ slope of cover layer).

height (Allsop *et al.*, 1985). The average run-up R_s reached by significant wave of height H_s is

$$\frac{R_s}{H_s}=2.11-0.09\xi_p. \tag{15.24}$$

The average run-up R_2 for waves whose height is exceeded by 2% of the incoming waves is

$$\frac{R_2}{H_s}=3.39-0.21\xi_p \tag{15.25}$$

where ξ_p is the Irribarren number defined as $\tan\alpha/(H_s/L_p)^{0.5}$ in which L_p is the deep water wavelength corresponding to the peak period (see worked example).

Over rough slopes the values of run-up are obtained by multiplying the values determined for smooth slopes (equations (15.24) and (15.25)) by correction factors r as given in Table 15.1. For slopes with tetropods as armour units the equation for run-up due to Allsop *et al.* (1985) is

$$\frac{R_s}{H_s}=1.32(1-\exp(-0.31\xi_p)) \tag{15.26}$$

$$\frac{R_2}{H_s}=1.83(1-\exp-(0.31\xi_p)). \tag{15.27}$$

Coastal Engineering Manual (US Army, 2002) expresses run-up R_2 for smooth slopes as

$$\frac{R_2}{H_s}=1.5\xi_p \quad for \quad 0.5<\xi_p<2.0$$

$$=3.0 \quad for \quad 2<\xi_p<3-4 \tag{15.28}$$

The correction factors r for rough slopes have been updated but are only slightly different from the values given in Table 15.1.

Waves breaking on a sloping structure causes run-down in addition to run-up. Run-down is the vertical distance that the water level reaches below the still water level. The average run-down R_{d2} for waves whose height is exceeded by 2% of the incoming waves is estimated as (*Coastal Engineering Manual* (US Army, 2002))

$$\frac{R_{d2}}{H_s} = 0.33\xi_p \quad \text{for} \quad 0 < \xi_p < 4$$

$$= 1.5 \quad \text{for} \quad \xi_p > 4 \qquad (15.29)$$

Sea walls are often designed with composite slopes including a berm which reduces the wave run-up. Hunt (1959) recommends a berm width of at least one-fifth of the wavelength. For the effect of berm on run-up, refer to *Coastal Engineering Manual* (US Army, 2002). Run-up on a composite slope is estimated using Saville's method (1957) where a single slope extending from the point of wave breaking to the point of maximum run-up is assumed and the wave run-up is estimated for this hypothetical slope (Fig. 15.8(f)). Since the wave run-up is not known initially, it is determined by trial and error. First, the maximum run-up on a slope is assumed. The depth d_b at which the wave breaks is $0.78H_b$, where H_b is the height of the wave at the instant of breaking. The run-up for the hypothetical slope is found from equation (15.21) and compared with the assumed one (worked example). As wall profiles have large effects on the reflection of the wave and on the run-up on the wall, the design of large projects should involve model studies (Novak and Čábelka, 1981). For fuller treatment of sea wall design, the reader is referred to Thomas and Hall (1992).

15.5 Wave overtopping

Overtopping of the sea wall can take place as 'solid water' or as 'sprays'. Sea walls should be designed to satisfy their functional requirements without causing undesirable impact on the surroundings. If the freeboard is generous, the scheme can be too expensive. On the other hand, if the freeboard is too low, the land that it is designed to protect may be subjected to flooding and erosion. Under certain conditions occasional overtopping is not unusual with irregular waves. It is important to estimate the quantity of water overtopping a sea wall and to assess the consequences.

The mean overtopping discharge Q is usually expressed functionally as

$$Q = \textit{function of}\,(H_s, T, \beta, R, d, g, d, s\ldots) \qquad (15.30)$$

in which H_s is offshore significant wave height, T is the mean wave period, α is the approach angle, d is the depth at the toe of the wall, R is freeboard of the wall from SWL and s is the beach slope.

Overtopping of non-porous vertical walls due to regular waves may be determined from the model tests reported in *Shore Protection Manual*, (US Army, 1984) (see also Thomas and Hall, 1992).

For simple slopes, the test results of Owen (1980) with random waves are applied for the prediction of discharge due to overtopping. The results are expressed in terms of a relationship between two non-dimensional quantities Q^* and R^*.

$$Q^* = A\exp\left(-\frac{BR^*}{r}\right) \tag{15.31}$$

in which $Q^* = \dfrac{Q}{TgH_s}$ and $R^* = \dfrac{R}{T\sqrt{gH_s}}$. Q is the mean discharge in m^3s^{-1} per m run of the wall overtopping the crest of the sea wall, T is the mean wave period, R is the freeboard of the wall from the still water level to the crest and r is the roughness factor as given in Table 15.1. The results are valid for the range of slopes from 1:1 to 1:4 and wave steepness from 0.035 to 0.055.

Typical values of A and B are given in Table 15.2

For values of the coefficients A and B of equation (15.31) to estimate the mean overtopping discharge to include bermed sea wall and the effects of angle of wave attack, refer to Owen (1980).

Tolerable overtopping discharges

For vehicles Besley *et al.* (1998) suggest mean values of overtopping discharge <0.001 l/s/m as safe at all speeds, between 0.001 and 0.02 l/s/m as unsafe at high speed and >0.021 l/s/m as unsafe at any speed. For pedestri-

Table 15.2 Values of coefficients *A* and *B* for simple sea walls

Seawall slope	*A*	*B*
1:1	7.9×10^{-3}	20.12
1:1.5	1.02×10^{-2}	20.12
1:2	1.25×10^{-2}	22.06
1:2.5	1.45×10^{-2}	26.1
1:3	1.63×10^{-2}	31.9
1:3.5	1.78×10^{-2}	38.9
1:4	1.92×10^{-2}	46.96
1:4.5	2.15×10^{-2}	55.7
1:5	2.50×10^{-2}	65.2

ans they suggest mean values <0.004 as not uncomfortable but wet, between 0.004 and 0.03 l/s/m as uncomfortable but not dangerous, and >0.03 l/s/m as dangerous.

15.6 Rubble-mound breakwaters

15.6.1 General

Most breakwaters are constructed for the protection of harbours. In some cases they may become part of a jetty or support a roadway. A rubble-mound breakwater and a vertical breakwater built on rubble-mound (Quinn, 1972) are shown in Fig. 15.9. Rubble-mound breakwaters usually consist of a core of small quarry-run rock, protected by one or more intermediate layers or underlayers separating the core from the cover layers of large armour units. Failure of the rubble-mound breakwaters may be due to removal of or damage to armour units, overtopping causing scour, toe erosion, loss of core material, or foundation problems (Institution of Civil Engineers, 1983). The armour units of the rubble-mound breakwaters are

(a) Rubble mound breakwater (Quinn, 1972)

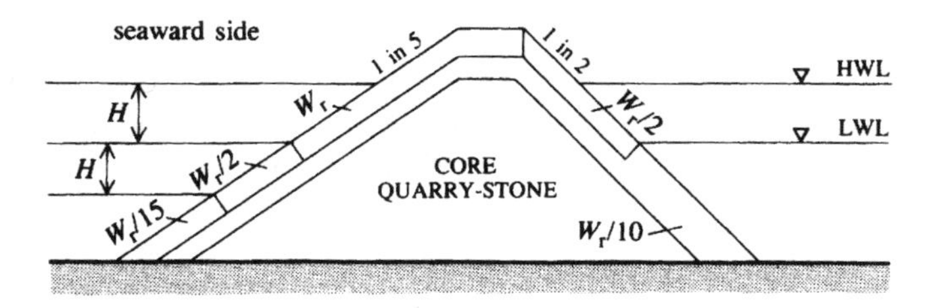

(b) Vertical-walled breakwater

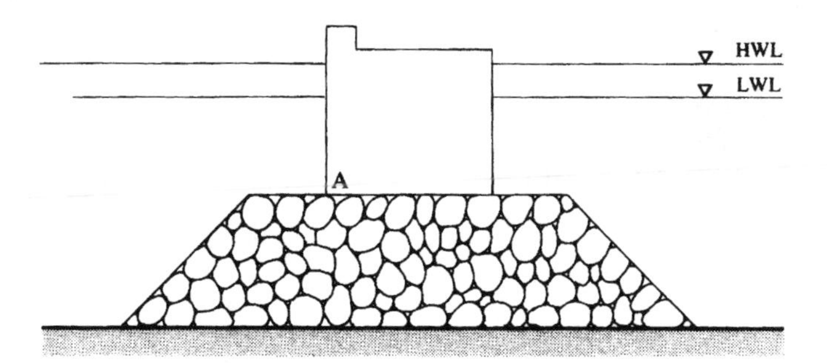

Fig. 15.9 Typical breakwaters

made of large quarry-stones or concrete prefabricated units such as tetrapods, cobs and dolos (Fig. 15.10), the latter specially formed to reduce wave reflection and produce a good degree of interlocking.

15.6.2 Stability of breakwaters

(a) Shore Protection Manual *formulae*

Hudson (1961) presented a simpler formula for determining the stability of the rubble-mound breakwater. It gives the required weight of the armour units in the cover layer in the form

$$W_r = \frac{\rho_s g H^3}{K_D(\rho_s/\rho - 1)^3 \cot\beta} \qquad (15.32)$$

where W_r is the weight of the individual armour unit, ρ_s is the density of the armour units, β is the inclination of the structure to the horizontal and K_D is a dimensionless coefficient. Typically, breakwater slopes on the seaward side vary from 1 in 1.5 to 1 in 3 (V:H). Hudson's empirical formula is applicable for slopes of the cover layer ranging from 1 in 1.5 to 1 in 5. The coefficients K_D as recommended by the *Shore Protection Manual* (US Army, 1984), are given in Table 15.3 for various armour units. For the armour units used at the structure head, K_D is found to depend on the slope of the cover layer. The use of a single quarrystone cover layer is not recommended unless special precautions are taken in placing. Of the specially formed units, tetrapods have probably been the most popular. From Table 15.3, it is clear that large values of K_D are associated with the specially formed units.

It should be noted that Hudson's equation does not take account of the wave period, the oblique approach of the waves, or the irregularity of

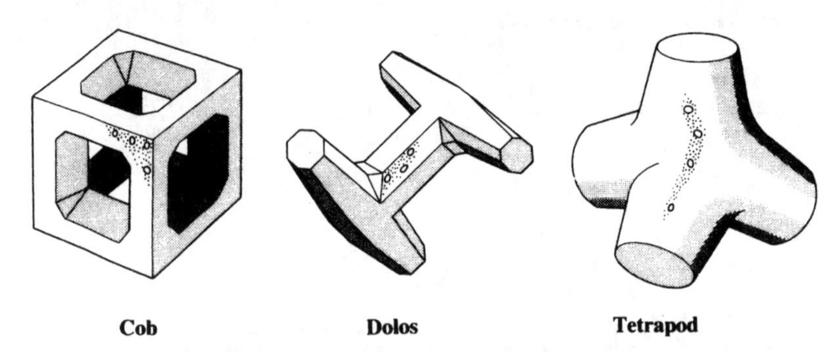

Fig. 15.10 Examples of specially formed armour units (Chadwick, Morfett and Borthwick, 2004)

Table 15.3 Values of K_D in Hudson's formula (SPM): no damage and minor overtopping

Armour unit	*Number of units in cover layer*	*Structure trunk*		*Structure head*		*Slope*
		Breaking wave	*Non-breaking wave*	*Breaking wave*	*Non-breaking wave*	
Smooth quarrystone	2	1.2	2.4	1.1	1.9	1.5–3.0
Smooth quarrystone	>3	1.6	3.2	1.4	2.3	1.5–3.0
Rough angular quarrystone	2	2.0	4.0	1.9	3.2	1.5
				1.6	2.8	2.0
				1.3	2.3	3.0
Rough angular quarrystone	>3	2.2	4.5	2.1	4.2	1.5–3.0
Tribar	2	9.0	10.0	8.3	9.0	1.5
				7.8	8.5	2.0
				6.0	6.5	3.0
Dolos	2	15.8	31.8	8.0	16.0	2.0
				7.0	14.0	3.0
Tetrapod	2	7.0	8.0	5.0	6.0	1.5
				4.5	5.5	2.0
				3.5	4.0	3.0

the waves approaching the coast. There is some evidence that the stability of the armour units is affected by the wave groups, so named because of the tendency of large waves to travel together.

The thickness, t_1, of the cover layer consisting of n layers of armour units is calculated from

$$t_1 = nK'_D(W_r/\rho_s g)^{1/3} \tag{15.33}$$

where K'_D is the dimensionless coefficient of the layer. The required number, N, of units for a given surface area A is

$$N = An\left(1 - \frac{p_r}{100}\right)\left(\frac{\rho_s g}{W_r}\right)^{2/3} \tag{15.34}$$

where p_r (%) is the porosity. Porosity and layer coefficients are given, for various armour units, in Table 15.4.

The crest width of the breakwater depends on the amount of overtopping that might occur. It is also dictated by the construction method. The recommended minimum width is three times the thickness of the cover layer, i.e.

$$b_{min} = 3K'_D(W_r/\rho_s g)^{1/3}. \tag{15.35}$$

There are many variations in the size and proportion of materials used in rubble-mound breakwaters. A breakwater, with the grading of stones used, is shown in Fig. 15.9(a). The forces exerted on the vertical part of the breakwater shown in Fig. 15.9(b) may be determined by the methods described in Section 15.3. The vertical structure of a rubble mound may fail owing to its sliding or overturning, toe erosion, pressures generated in the mound, and previously mentioned problems with the rubble mound itself.

In the design of a cover layer, a certain amount of damage of armour units in the region of wave attack may be allowed for by reducing their size (and cost). Damage is defined as a percentage of the volume of

Table 15.4 Layer coefficient K'_D and porosity for various armour units

Armour unit	*Number of layers, n*	*Layer coefficient, K'_D*	*Porosity, P_r (%)*
Smooth quarrystone	2	1.02	38
Rough quarrystone	2	1.0	37
Rough quarrystone	>3	1.0	40
Tetrapod	2	1.10	50
Tribar	2	1.02	54
dolos	2	0.94	56

armour units displaced in the zone of wave attack. Damage is permissible if the intermediate layers and the core of the rubble-mound breakwater are not exposed to wave attack. Also, a structure designed to resist waves of moderate severity may suffer damage without complete destruction. There must be a trade-off between the initial cost of a damage-free breakwater for most severe waves and the maintenance cost of the breakwater designed on the basis of permissible damage. The *Shore Protection Manual* (US Army, 1984) gives the ratio of wave height causing per cent damage to the wave height responsible for 0–5% damage. The wave run-up on smooth slopes is given by equations (15.23) to (15.28). The reduction factors due to the porosity and roughness for some units are given in Table 15.1 (Bruun, 1972).

(b) Van der Meer formulae

Van der Meer (1987) presents formulae for the stability of rock armour from results of model tests with random waves taking into account the effects of a number of variables not included in the Hudson equation (15.32). The formulae are:

(i) for breakers with Irribarren number $\xi < 2.5$ on smooth slopes and $\xi < 2.0$ on rough slopes

$$\frac{H_s}{\Delta D_{50}} = 6.2 P^{0.18} \left(\frac{S}{\sqrt{N}} \right)^{0.2} \xi^{-0.5}; \tag{15.36}$$

(ii) for breakers with $\xi > 2.5$ on smooth slopes and $\xi > 2.0$ on rough slopes

$$\frac{H_s}{\Delta D_{50}} = 1.0 P^{-0.13} \left(\frac{S}{\sqrt{N}} \right)^{0.2} \sqrt{\cot \alpha}\, \xi^{P}. \tag{15.37}$$

ξ is the Irribarren number $= \tan\beta/(H/L_0)^{0.5}$ where β is the slope of the structure and H is the wave height at the structure. Δ is $(\rho_s - \rho)/\rho$, N is the number of waves in design wave conditions, S is the damage number defined as A_d/D_{50}^2, P is the notional permeability factor and D_{50} is the nominal diameter of the rocks. A_d is the cross-sectional area of erosion.

Van der Meer (1987) relates the mean mass of the rocks M_{50} to the size D_{50} as

$$D_{50} = \left(\frac{M_{50}}{\rho_s} \right)^{1/3}. \tag{15.38}$$

The recommended values of the design damage number S equivalent to the number of stones of size D_{50} from a D_{50} width of the structure are given in Table 15.5.

Table 15.5 Variation in damage number for failure conditions

Slope	*Initial*	*Damage (S) intermediate*	*Failure*
1:1.5	2	–	8
1:2	2	5	8
1:3	2	8	12
1:1.4	3	8	17

Typical permeability factors P for different types of rock construction are presented in Fig. 15.11 (Van der Meer, 1987). The equations (15.36) and (15.37) have been arrived at from model testing with the layer thickness of about $2.0D_{50}$. Van der Meer (1988) has also investigated the stability of specially formed units such as cubes and tetrapods.

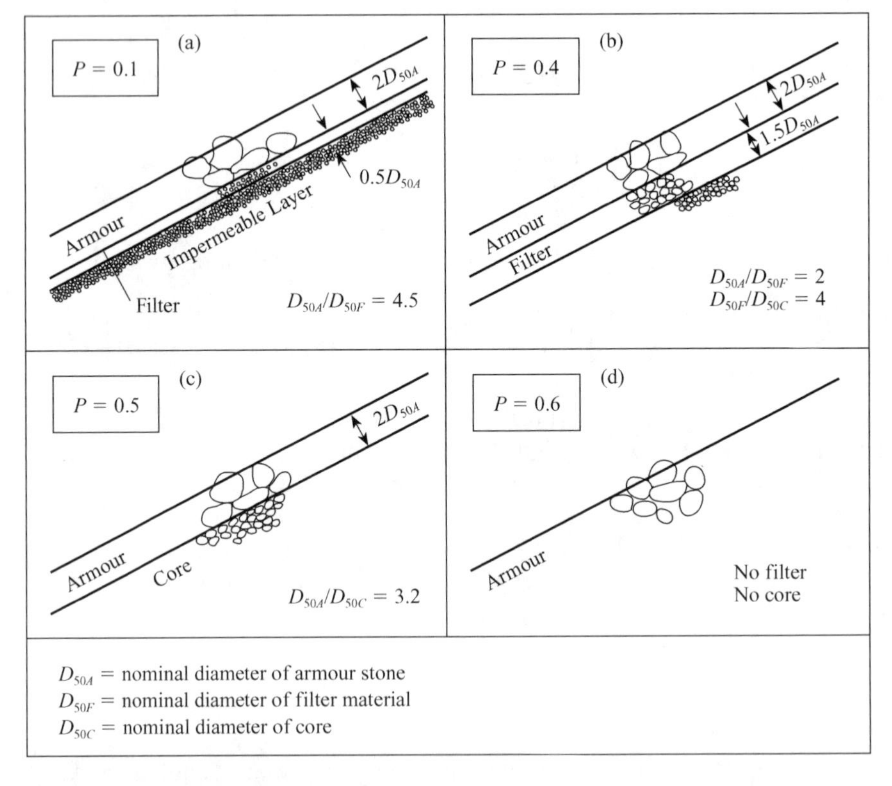

Fig. 15.11 Typical permeability factors (P) for different types of rock construction

15.7 Sea outfalls

15.7.1 General

The oceans have a great capacity to assimilate wastes and to make them harmless and they are being exploited as receivers of wastes from outfalls. A typical outfall consists of a pipe discharging effluents from the land to the sea and laid on or buried in the sea bed (or housed in a tunnel). It is extremely important that the design of outfall should take account not only of its hydraulic function but also of the environmental impact. Some of the short outfalls constructed have failed from the environmental standpoint since effluents have found their way back polluting the beaches. Even long outfalls may sometimes be unsatisfactory since effluents can return to the beach depending upon the direction of currents that are produced by the tidal flow, waves and wind.

There are essentially two outfall types. One consists of a pipe with a diffuser at its seaward end, both above the sea bed, and the wall ports, through which the effluent discharges, are just holes through the wall of the pipe; the second type, buried under the sea bed, has a number of risers from the diffuser. The outfall pipe and the risers must be protected from ship anchors and fishing lines.

In large outfall schemes and in schemes involving confined waters, the concentrations of the pollutants discharged should be checked periodically by sampling the quality of water which is usually characterized by the biological oxygen demand (BOD), bacterial content, suspended matter, turbidity, pH, temperature, toxic chemicals, minerals and organic and inorganic matter. The impact of these various constituents on the environment is often complex and not necessarily immediate.

When the effluent is discharged into the sea, it is diluted rapidly by initial dilution and subsequently by the secondary dispersion. The BOD and the suspended solids are reduced quickly by these processes. It is, however, necessary that the ambient currents are favourable and that the site is at a sufficient distance from the shore to reduce the bacterial content still further through dilution, mortality and sedimentation before the diluted sewage may reach the shore.

Apart from the above environmental aspects, the design of the outfall system will depend on the characteristics of the sewage system, storm overflows, the outfall site, the headworks and the outfall itself. The storm overflows on combined systems (collecting both sewage and rainwater run-off) are designed to reduce the flow in the sewers to about 6 times the dry weather flow and may also be a cause of pollution whose impact on the environment has to be assessed.

The preliminary treatment of the wastewater on land and the subsequent marine treatment should complement each other in order to achieve environmentally acceptable standards. Before the effluent is discharged into the sea, the grit is removed to avoid its deposition in the outfall pipe and the suspended matter is reduced to small sizes (the screens should pass a maximum particle size of 5–6 mm). If the suspended load in the effluent is such that the marine treatment is not very effective environmentally, then primary sedimentation may be incorporated before discharging the effluent into the sea. If the effectiveness of the outfall falls short of the acceptable bacterial content, then disinfection may be recommended. It is worth noting that the sludge produced in primary sedimentation may be difficult to dispose of and that disinfection of the effluent can be detrimental to the aquatic life.

In this concise text many aspects of sea outfalls could be touched on only briefly; for a fuller treatment of the subject, the reader is referred to Charlton (1985), Neville-Jones and Dorling (1986), Wood *et al.* (1995) and Roberts (1996).

15.7.2 Site surveys

For effective design of outfalls, site surveys should be undertaken to assess the quality of receiving water, tides and currents, and environmental conditions (e.g. seaweed, animal life, bacteria, etc.). Charlton (1985) describes the methods used to obtain a form of tidal atlas in which the current variation throughout the tidal cycle is presented on an hourly basis. The overall performance of the outfall is predicted by tracer tests. These tests are carried out usually in normal conditions of the sea by introducing tracers at the site and the dispersion and dilution of the tracer are continuously monitored mainly at a depth less than 1 m.

15.7.3 Initial dilution and secondary dispersion

The effluent passing through a port is lighter than the surrounding sea water; the density of sea water is typically $1.026\,\mathrm{kg\,m^{-3}}$ and depends on the ambient temperature and salt content. Thus a density gradient across the depth may exist due to variations of temperature and salinity with the density increasing in depth. The density stratification is likely to be more pronounced in summer than in winter.

Consider that the effluent from the port exhausts in the vertical direction into the sea of uniform density without ambient currents. The resulting buoyant jet spreads owing to its initial momentum rate and the buoyancy effects. Without the momentum rate, the effluent from the port will spread as pure plume driven only by the buoyancy. As the jet rises to the surface, entrainment of the surrounding water takes place with the maximum velocity and concentration along the centreline of the jet decaying in the streamwise direction. Instead of concentration, the term dilution is commonly used. If, say, 99 ml of pure salt water mixes with 1 ml of the effluent, then the dilution is 100:1 or simply 100. On reaching the surface of the sea the effluent tends to spread horizontally.

The density of the effluent increases through dilution. If the receiving water is density stratified, then the effluent may not even reach the surface as, when it reaches a value at some depth equal to the density of the surrounding water, the plume is arrested and spreads horizontally.

The next phase of the effluent transport is the advection by ocean currents and turbulent diffusion. The phase, called the secondary dispersion, depends on the current and the turbulence structure of the ocean.

Ports generally discharge effluents horizontally to achieve the best dilution (Fig. 15.12). Figure 15.12 also shows the spread of the jet to be discussed further in Section 15.7.6. For horizontally discharged effluent, about 20% to 50% greater dilutions are obtained at the water surface than for vertical discharge.

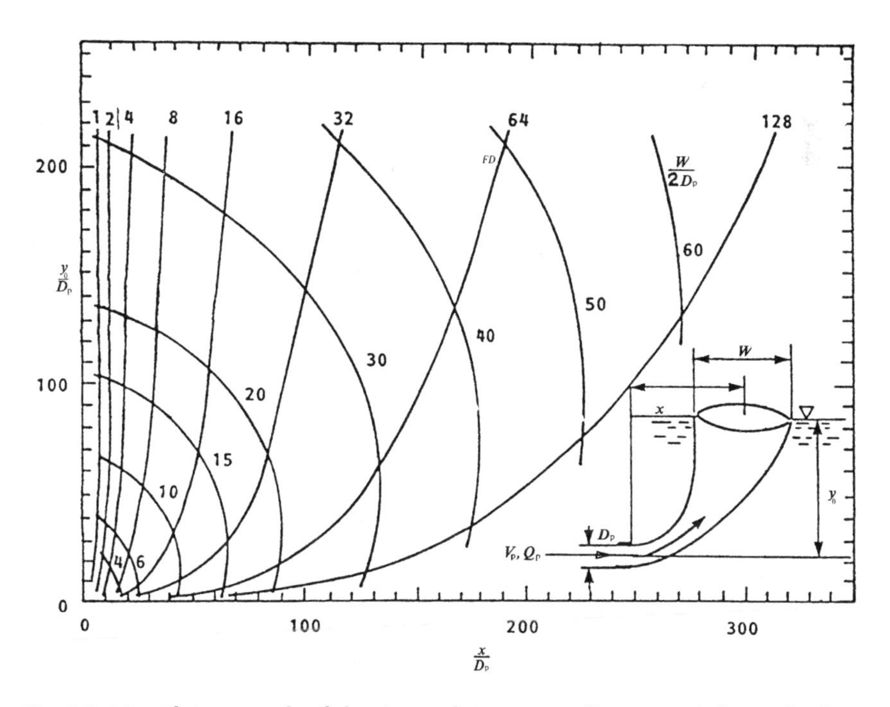

Fig. 15.12 The spread of horizontal jets in still water (after Charlton, 1985)

15.7.4 Still ambient fluid

The initial dilution of the effluent S_m along the centreline of the jet issuing from the port may be expressed in still water conditions in a functional form as

$$S_m = f_0\left[\frac{V_p}{(g'D_p)^{1/2}}, \frac{y_0}{D_p}\right] \tag{15.39}$$

where V_p is the velocity at the exit of the port, y_0 is the depth of the port exit below the surface and D_p is the diameter of the port. $g' = g(\rho_a - \rho_0)/\rho_0$ where ρ_a is the density of the ambient fluid and ρ_0 is the density of the effluent. The first term in the brackets on the right-hand side of equation (15.39) is the densimetric Froude number denoted further as FD (see also equation (8.20)).

The spread of the buoyant jet horizontally discharged in still surrounding fluid has been studied extensively and the results are presented in the form of equation (15.39) as charts or equations. Figure 15.13 shows the graphical expression of the functional form due to Fan and Brooks (1966). Cederwall (1968) expresses the dilution in the form of equations as

$$S_m = 0.54\,FD\left(\frac{y_0}{D_p FD}\right)^{7/16} \quad \text{for} \quad \frac{y_0}{D_p} < 0.5\,FD \tag{15.40}$$

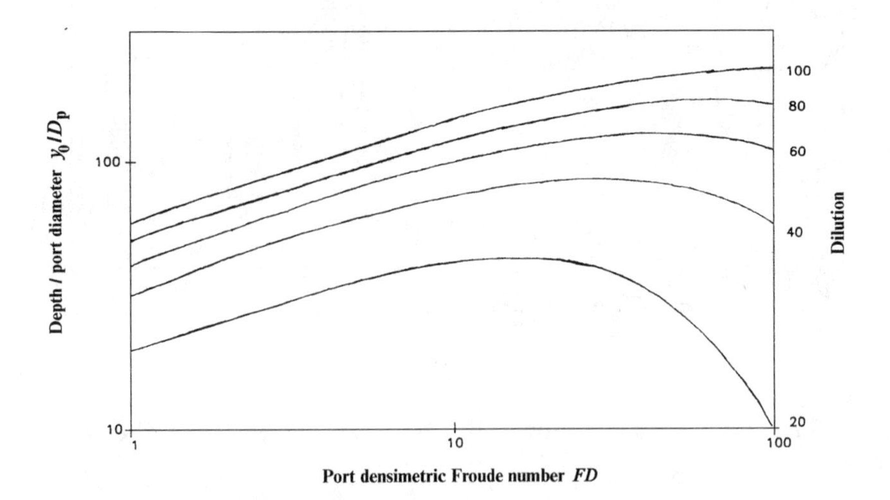

Fig. 15.13 Centreline dilution as a function of y_0/D_p and FD

and

$$S_m = 0.54\,FD\left(\frac{0.38y_0}{D_p FD} + 0.66\right)^{5/3} \quad \text{for} \quad \frac{y_0}{D_p} \geq 0.5\,FD. \tag{15.41}$$

S_m is the minimum dilution occurring along the centreline of the plume. For vertical port discharges, the average plume dilutions S_a in the initial mixing region can be determined by the equations due to Muellnoff *et al.* (1985) with Q_p being the port discharge:

$$S_a = 0.13\,g^{1/3} Q_p^{-2/3} y_0^{5/3}$$

which can be written in non-dimensional form as

$$S_a = 0.13\left(\frac{\pi}{4}\right)^{-2/3} \frac{1}{FD^{2/3}} \left(\frac{y_0}{D_p}\right)^{5/3}. \tag{15.42}$$

The average plume dilution S_a is approximately 1.8 times the minimum dilution S_m at the centreline.

Figure 15.13 shows that the same dilution can be obtained for either low or high exit densimetric Froude numbers. Low velocities mean low driving heads while large velocities need large heads. The choice of low velocities can lead to malfunctions of the outfall if deposits of sediments on the bed occur or if the densimetric Froude number is below the critical limit of unity when salt water intrusion reducing the effluent flow will develop.

15.7.5 Moving receiving water

When the effluent is discharged into receiving water moving with velocity V_a the dilution is significantly modified from that in calm water. Field studies (Agg and Wakeford, 1972) give an equation for an additional dilution factor α which is a ratio of the dilution in moving water to the dilution in still water. The empirical equation for α applicable in the range $0.1 < V_a/V_p < 2.0$ is

$$\log \alpha = 1.107 + 0.938 \log(V_a/V_p). \tag{15.43}$$

(The dilution in moving water is determined by multiplying the dilution in still water of Fig. 15.13 by the factor α.) Lee and Neville-Jones (1987) present the following equations for dilution in moving water:

$$S_m = \frac{0.31 B^{1/3} y_0^{5/3}}{Q_p} \quad \text{for} \quad y_0 < \frac{5B}{V_a^3}, \tag{15.44}$$

$$S_{\mathrm{m}} = \frac{0.32 V_{\mathrm{a}} y_0^2}{Q_{\mathrm{p}}} \quad \text{for} \quad y_0 \geq \frac{5B}{V_{\mathrm{a}}^3} \tag{15.45}$$

and

$$B = Q_{\mathrm{p}} g \frac{\rho_{\mathrm{a}} - \rho_0}{\rho_0}.$$

For small values of y_0/D_{p}, the results of Agg and Wakeford are comparable with equations (15.44) and (15.45) but, for y_0/D_{p} between 50 and 100, they underpredict the initial dilution (Neville Jones and Dorling, 1986).

For vertical single ports discharging into moving receiving water, Muellenhoff *et al.* (1985) present the following equation for average dilution S_{a}:

$$S_{\mathrm{a}} = \frac{0.49\, V_{\mathrm{a}} y_0^2}{Q_{\mathrm{p}}}. \tag{15.46}$$

For further information about dilution in moving fluid, either stratified or unstratified, refer to Wood *et al.* (1993) and Roberts (1996).

15.7.6 Spacing of ports

The spacing of the ports depends on the geometry of the plume as it spreads. Under still water conditions the buoyant rising plumes should not overlap till they reach the surface. The geometry of the buoyant plume in still water according to Brooks (1970) is shown in Fig. 15.12. The dimensionless radial extent of the plume W/D_{p} at the free surface is presented in Fig. 15.12 as a function of the densimetric Froude number FD and y_0/D_{p}. The spacing of the ports should be greater than W. In moving receiving water, the size of the plume is increased with the increased degree of dilution. Neville-Jones and Dorling (1986) suggest that for $V_{\mathrm{a}}/V_{\mathrm{p}} < 0.2$, $W = 0.5 y_0$ for $y_0 > 2B/V_{\mathrm{a}}^3$. The width $W = 0.9 y_0$ for $V_{\mathrm{a}}/V_{\mathrm{p}} > 0.2$.

15.7.7 Diffuser design

The flow through the outfall may be due to gravity depending on the head available between the headworks and the highest water level at the discharge location in the sea. If enough head is not available, then the wastewater is pumped through the outfall.

The frictional loss in the straight pipe of the outfall leading to the diffuser is calculated using the Darcy–Weisbach equation (equation (8.4)) with friction factor computed by equation (8.5) or the Colebrook–White equation (Worked example 13.1). During the operation of the outfall, slime builds up on the wall of the pipe increasing the roughness height and reducing the effective diameter of the pipe; this should be taken into account in headloss calculations.

Figure 15.14 shows typically a diffuser with three ports with vertical risers. An expression for the discharge through a port is obtained by applying the energy equation between the exit of the port and the centreline of the diffuser at the junction. Consider for example, port 2:

$$\frac{p_2}{\rho g} + \frac{V_2^2}{2g} = \frac{p_{a2}}{\rho g} + \frac{V_{p2}^2}{2g} + l_{p2} + \frac{V_{p2}^2}{2g}(k_e + k_{st} + k_b) \tag{15.47}$$

where p_2 and V_2 are the pressure and velocity respectively just before the junction at 2. l_{p2} is the vertical distance between the port exit and the centreline of the diffuser and p_{a2} is the ambient pressure at the port exit. k_e, k_{st} and k_b are respectively the coefficients for the headlosses at entry to the riser, along the straight riser and in the bend (Miller, 1994). The discharge through the second port Q_{p2} is then $Q_{p2} = a_{p2}V_{p2}$.

Wall ports are simply holes through the wall of the diffuser and may have a rounded or sharp-edged entrance. Equation (15.47) applies with $l_{p2} = k_{st} = k_b = 0$. Therefore for V_{p2}

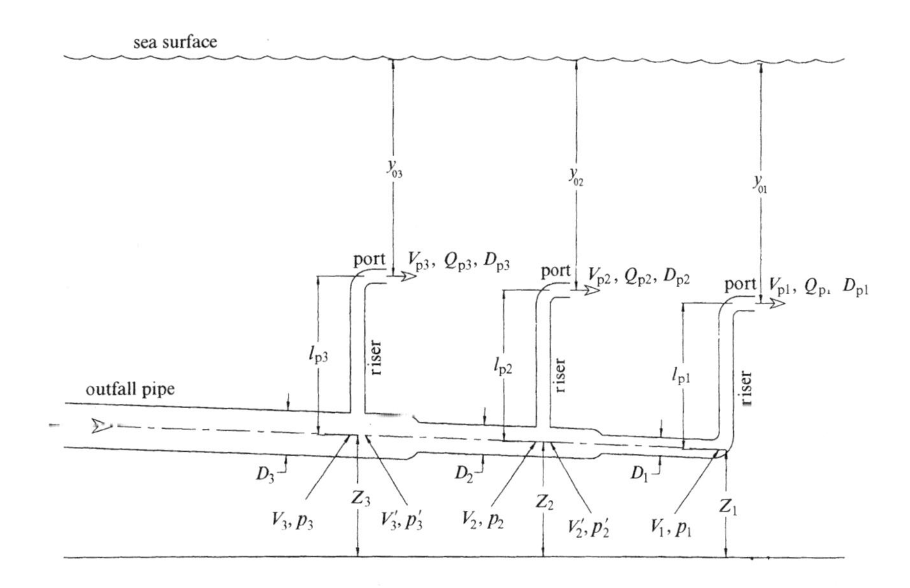

Fig. 15.14 Diffuser with risers

$$V_{p2} = \left(\frac{1}{1+k_e} 2gE_2 \right)^{1/2} \tag{15.48}$$

where

$$E_2 = \frac{p_2}{\rho g} + \frac{V_2^2}{2g} - \frac{p_{a2}}{\rho g} \tag{15.49}$$

and the flow rate is

$$Q_{p2} = c' a_{p2} V_{p2} = c_d a_{p2} (2gE_2)^{1/2} \tag{15.50}$$

with

$$c_d = c' \left(\frac{1}{1+k_e} \right)^{1/2}.$$

An empirical equation (Koh and Brooks, 1975) based on laboratory experiments for port diameters less than one-tenth of the diameter of the diffuser for a bell-mouth entrance is

$$c_d = 0.975 \left(1 - \frac{V_j^2}{2gE_j} \right)^{3/8} \tag{15.51}$$

and for a sharp-edged entrance is

$$c_d = 0.63 - 0.58 \frac{V_j^2}{2gE_j}. \tag{15.52}$$

E_j is as given by equation (15.49). (V_j is the velocity in the diffuser just before the junction j and V_{pj} is the velocity through the port at junction j.) For the riser ports the experimental discharge coefficients vary from 0.648 to 0.848 (Grace, 1978). These coefficients may be used for the preliminary design; for detailed calculations equation (15.47) taking account of the losses in the individual components of the riser assembly has to be followed.

(a) Analysis of flow in the diffuser

Referring to Fig. 15.14, the energy equation for the segment of the diffuser between ports 2 and 3 is

$$\frac{p_3'}{\rho g} + \frac{V_3'^2}{2g} + \Delta Z_{23} = \frac{p_2}{\rho g} + \frac{V_2^2}{2g} + h_{f23} + h_c. \tag{15.53}$$

p'_3 and V'_3 are respectively the pressure and velocity in the diffuser at the section immediately downstream of junction 3 (Fig. 15.14). p_2 and V_2 refer respectively to pressure and velocity occurring upstream of junction 2 (Fig. 15.14). ΔZ_{23} is the change in elevation . h_{f23} is the frictional head loss in the straight pipe and h_c is the head loss in the contraction which is usually negligible. Z_2 and Z_3 are the elevations of junctions 1 and 2 respectively (Fig. 15.14).

The pressure across the junction (Fig. 15.14) is determined using the energy equation which, as an example, for junction 2 is

$$\frac{p_2}{\rho g} + \frac{V_2^2}{2g} = \frac{p'_2}{\rho g} + \frac{V_2'^2}{\rho g} + k_2\frac{V_2'^2}{2g}. \tag{15.54}$$

Note that

$$Q_2 = Q'_2 + Q_{p2} \tag{15.55}$$

where Q_2, Q'_2 and Q_{p2} are respectively the discharges upstream and downstream of the junction and through the port.

The analysis of flow in the diffuser is a step-by-step procedure starting from the extreme seaward end. The calculation procedure for the wall ports (or ports with risers) is as follows.

1. Estimate E_1 relevant to port 1 in the diffuser at the junction with the first port for the discharge Q_{p1} through the first port, using equation (15.49); determine p_1.
2. Determine the pressure p_2 just downstream of junction 2 from equation (15.53).
3. Solve equations (15.54) and (15.55) or (15.50) to find the pressure p_2 just upstream of the junction and the flow Q_{p2} through the second port.
4. Continue the procedure to port 3 and so on.
5. Check whether the sum of the discharges through all the ports is the same as the design discharge.

The calculation procedure can be carried out using computer programs.

It is difficult to avoid at all times deposition of sediments in the outfall pipe and the diffuser (at low flows) as a small pipe diameter increasing the flow velocity would result in large energy losses and greater pump power requirements. Usually some sediments are allowed to deposit at low flows to be flushed out at high flows. Generally velocities in excess of about $0.9\,\mathrm{m\,s^{-1}}$ will be required at high flows. For further information on sediment transport in sewers and outfall see Novak and Nalluri (1987) and Ackers (1991). In order that the velocities along the diffuser are sufficiently large to dislodge the sediments from the bed and transport them, the diffuser diameter is reduced progressively as shown in Fig. 15.14.

An outline of a preliminary outfall design is given in Worked example 15.4.

15.8 Coastal management

A large percentage of the world population lives along the coasts. A coastal zone is the interface between land and sea including beaches, coastal marshes, mudflats, mangroves, estuaries, etc. The principal demands on coastal zone development include property (buildings, highways, land), leisure activities (tourism, recreation), commercial interests (shipping, fishing), minerals and disposal of wastes (industrial and domestic wastes).

Coastal planners are interested in the use of coastal zones to maintain and develop some of the existing facilities and to initiate new schemes adding fresh demands on the coasts. Thoughtless planning can upset the very character of the coasts forever and seriously reduce coastal resources. Engineers design, construct and maintain breakwaters for harbours, structures for coastal defence, etc. In many instances construction of harbours or prevention of erosion along a particular reach of coastline has interrupted littoral drift causing serious downdrift problems. Site-specific solution can transfer the problem to adjoining sites. Hence traditional solutions of defending a particular site along the shoreline against flooding or erosion are now being challenged in the light of better understanding of the coastal processes.

Conservationists are concerned about the impact of growing demands on the coastal environment and the loss or decline of natural habitats, marine life etc., by coastal activities and/or pollution. The environmental assessment of its impact on the various features of the coast such as flora and fauna forms therefore an essential part of any project.

The objective of coastal management is to preserve the coastal resources at the same time balancing the sometimes conflicting requirements of development, protection, usage and protection (see Barrett, 1989). In a broader sense social, economic, cultural, national and international considerations play an important rôle in the development of coastal management strategies (see Kay and Alder, 1999).

In the past administrators, planners and engineers who have the responsibility for providing and monitoring coastal defences were usually only interested in the length of coast falling within their administrative boundary. But it is the physical features of the coast and the coastal processes, particularly the sediment motion, that should define the boundaries for integrated shoreline management plans.

Shoreline management plans should have the following objectives

(MAFF, 1995): to study the coastal processes in a cell defined by the physical features and sediment transport, to predict the likely shape of the coasts in the future, to identify all the resources within the area covered by the plan, to assess various coastal defence options including 'doing nothing', to carry out regional and site-specific research and monitoring, to facilitate consultation between interested parties and to develop future objectives for shoreline management.

Worked Example 15.1

A storm in deep water generates waves which travel towards the shore, impinging on a breakwater. The breakwater is a vertical wall erected on a rubble mound (Fig. 15.9b). The fetch-limited storm has a wind speed of $10\,\mathrm{m\,s^{-1}}$ and a fetch of 100 km. The base of the vertical wall and the sea bed are respectively 3 m and 8 m below HWL.

Find the maximum force on the vertical wall and the bending moment about A using Sainflou's method.

Solution

As given below, fetch = 100 km and wind speed = $10\,\mathrm{m\,s^{-1}}$.

JONSWAP spectrum (SEE SECTION 14.9.2)

$$f_m = 3.5 \times \frac{9.81}{10}\left(\frac{9.81 \times 100 \times 10^3}{10^2}\right)^{-0.33}$$

$$= 0.165\,\mathrm{Hz};$$

$$T_s = 0.777/f_m = 4.7\,\mathrm{s};$$

$$\alpha = 0.076\left(\frac{9.81 \times 100 \times 10^3}{10^2}\right)^{-0.22}$$

$$= 0.01;$$

$$H_s = \frac{0.552 \times 9.81}{\pi^2 \times 0.165^2}\sqrt{0.01} = 2.01\,\mathrm{m};$$

$$L_0 = \frac{9.81 \times 4.7^2}{2\pi} = 34.5\,\mathrm{m}.$$

For $d/L_0 = 8/34.5 = 0.23$, $H/H_0 = 0.93$ and $L/L_0 = 0.93$ (Fig. 14.8), $H = 0.93 \times 2.01 = 1.87\,\mathrm{m}$ and $L = 0.93 \times 34.5 = 32\,\mathrm{m}$,

$$\frac{2\pi d}{L} = \frac{2\pi \times 8}{32} = 1.57.$$

p_{max} on the bed is given by

$$\rho g\left[d + \frac{H}{\cosh(2\pi d/L)}\right] = 1030 \times 9.81\left(8 + \frac{1.87}{2.51}\right) = 88.4 \times 10^3\,\mathrm{N\,m^{-2}},$$

$$h_0 = \frac{\pi \times 1.87^2}{32}\,\mathrm{cotanh}(1.57) = 0.37\,\mathrm{m}.$$

Assume linear distribution:

$$d + H + h_0 = 8 + 1.87 + 0.37 = 10.24\,\mathrm{m}.$$

Take y positive downwards from SWL. The pressure on the base of the vertical wall, with the crest of clapotis at the wall, is

$$\frac{88.4 \times 10^3}{10.24}(y + h_0 + H) = 8.63 \times 10^3(3 + 0.37 + 1.87)$$

$$= 8.63 \times 10^3 \times 5.24\,\mathrm{N\,m^{-2}}.$$

The force on the vertical face is

$$8.63 \times 10^3 \times \frac{5.24^2}{2} = 118.5 \times 10^3\,\mathrm{N}$$

per unit length of the wall.

MOMENT ABOUT *A*

Because of the linear pressure distribution,

$$M_{\mathrm{A}} = 0.5 \times 8.63 \times 10^3 \times 5.24^2 \times \frac{5.24}{3}$$

$$= 206.9\,\mathrm{N\,m}$$

per unit length of the vertical wall.

Worked Example 15.2

At a depth of 20 m, waves of height 1.31 m and period 7 s are observed to travel inshore. Estimate the run-up of the waves on the composite slope as shown in Fig. 15.15.

Solution

At the place of observation, $H = 1.31\,\text{m}$, $d = 20\,\text{m}$ and $T = 7\,\text{s}$. From Fig. 14.4, $L = 70\,\text{m}$, and from Fig. 14.3, $c = 10.2\,\text{m s}^{-1}$.

$$L_0 = \frac{gT^2}{2\pi} = 76.5\,\text{m}.$$

For $d/L_0 = 20/76.5 = 0.26$, $H/H_0 = 0.95$ (Fig. 14.8) and so $H_0 = 1.31/0.95 = 1.38\,\text{m}$.

WAVE BREAKING

$$\frac{H_b}{1.38} = 0.38\left(\frac{1.38}{76.5}\right)^{-1/3} \qquad \text{(equation (14.41))}$$

$$H_b = 2.0\,\text{m}; \quad d_b = 2.0/0.78 = 2.56\,\text{m}.$$

RUN-UP

1. *First trial.* Assume a run-up of 2 m above SWL. The horizontal distance from the wave breaking section to the run-up, $s = 10.0 + 5 + 6 + 5.6 = 26.6\,\text{m}$. The vertical distance is $2.0 + 2.56 = 4.56\,\text{m}$. The average slope is $4.56/26.6 = 0.17$.

$$\frac{R_u}{H_0} = \frac{R_u}{1.38} = 1.016 \times 0.17 \times \left(\frac{1.38}{76.5}\right)^{-0.5} \qquad \text{(equation (15.21))}$$

$$= 1.28$$

$$R_u = 1.28 \times 1.38 = 1.8\,\text{m}.$$

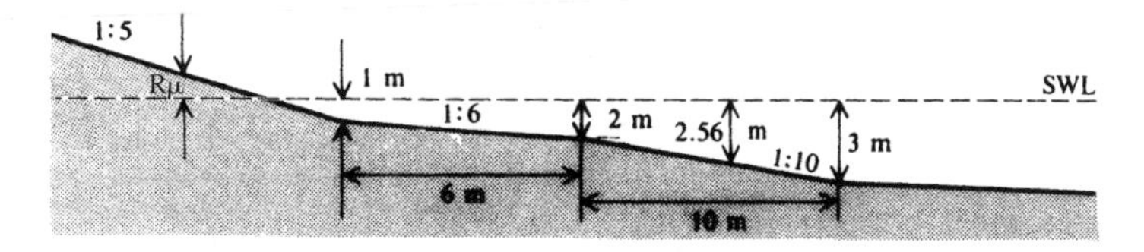

Fig. 15.15 Sea wall of composite slope (Worked example 15.2)

2. *Second trial.* Assume that $R_u = 1.8\,\text{m}$. The average slope is

$$(1.8 + 2.56)/(9.0 + 5 + 6 + 5.6) = 4.36/25.6 \approx 0.17.$$

$$\frac{R_u}{1.38} = 1.016 \times 0.17(1.38/76.5)^{-1/2} = 1.78\,\text{m}. \quad \text{(equation (15.21))}$$

Worked Example 15.3

An embankment at 10m OD along an east-facing bay is to be protected against erosion and flooding. The table below gives fetches, wind speeds, and refraction coefficients for obtaining design wave heights:

	Direction				
	N	*NE*	*E*	*SE*	*S*
Fetch (km)	400	250	100	80	60
Design wind speed (km h^{-1})	75	75	50	50	75
Refraction coefficient, K_r	0.2	0.35	0.8	0.7	0.3

(a) Provide a hydraulic design of sea wall protecting the embankment. The maximum still-water level is 3.5m OD, the depth of water being 3.5m.

Solution

The wave heights and periods are calculated using the JONSWAP spectrum for fetch-limited sea (equation (14.55)), given in the table on p. 667.

Choose 2.17m as the significant wave height for the design, the significant wave period being 8.16s.

Consider the sea wall profile shown in Fig. 15.16. Two layers of rough quarry stone will form the cover layer. The density of quarry stone is

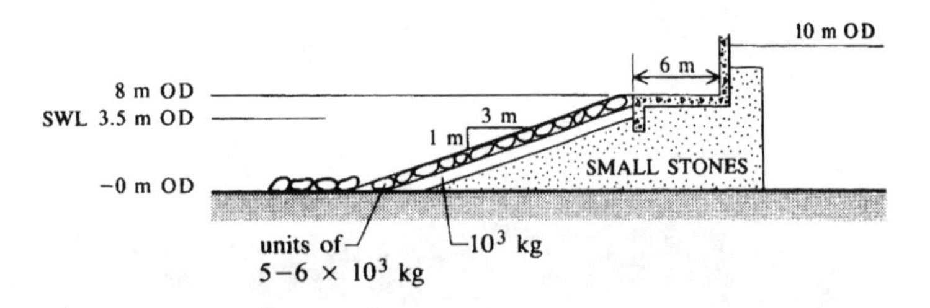

Fig. 15.16 Sea wall profile adopted (Worked example 15.3)

2650 kg m^{-3}. The design wave does not break before reaching the wall, because $H/d = 2.17/3.5 = 0.62 < 0.78$, the criterion for wave breaking. For minimum wave reflection, the slope of the sea wall (equation (15.19)) should be less than

$$\tan\beta = \frac{8}{8.16}\left(\frac{2.17}{2.0 \times 9.81}\right)^{1/2}$$

$$= 0.33.$$

	Direction				
	N	*NE*	*E*	*SE*	*S*
Wind speed, $U(\mathrm{m\,s^{-1}})$	20.8	20.8	13.9	13.9	20.8
gF/U^2	9070	5669	5077	4062	1360
Uf_m/g	0.173	0.202	0.21	0.226	0.324
α	0.01	0.0114	0.0116	0.0122	0.0155
$1/f_m$ or T_m	12.26	10.5	6.75	6.27	6.54
$T_z = T_s(\mathrm{s})$	9.53	8.16	5.24	4.87	5.08
$H_s(\mathrm{m})$	8.25	6.46	2.69	2.38	2.92 (equation (14.53))
$L_0(\mathrm{m})$	141.8	104.0	42.9	37.0	40.3 ($L_0 = gT_s^2/2\pi$)
d/L_0	0.025	0.034	0.082	0.095	0.087
C_g/c_0	0.36	0.41	0.55	0.58	0.56
C_g/Cg_0	0.72	0.82	1.1	1.16	1.12 ($C_{g0} = c_0/2$)
$K_s = (Cg_0/C_g)^{1/2}$	1.18	1.12	0.95	0.93	0.94
K_r	0.2	0.3	0.8	0.7	0.3
Design wave height (m)	1.94	2.17	2.04	1.55	0.82

The slope of the upstream face is chosen to be $1V{:}3H$. From Table 15.3, K_D for rough quarry stone is 1.0, the individual weight being given as (equation (15.32))

$$W_r = \frac{2650 \times 9.81 \times 2.17^3}{1.0 \times (2.65 - 1)^3 \times 3} \simeq 20\,\mathrm{kN},$$

$$\text{mass of units} = \frac{20 \times 10^3}{9.81} \simeq 2 \times 10^3\,\mathrm{kg}.$$

The deep-water design wave height $H_0 = H(C_g/C_{g0})^{1/2} = 2.17 \times \sqrt{0.82} = 1.97$ m. r in equation (15.23) is assumed to be 0.8 (Table 15.1). The wave run-up, R_u (equation (15.23)) is

$$R_u = 1.97 \times 1.016 \times \frac{1}{3}\left(\frac{1.97}{104.0}\right)^{-0.5} \times 0.8 = 3.88\,\mathrm{m}.$$

The run-up corresponds to $3.5 + 3.88 = 7.38$ m OD.

The horizontal part at 8.0 m OD is assumed to be 6 m, which is greater than the crest width given by equation (15.35). The thickness of the cover layer (equation (15.33)) is

$$t_1 = 2 \times 1.0 \times \left(\frac{20 \times 10^3}{2.65 \times 10^3 \times 9.81}\right)^{1/3}$$

$$= 1.83\,\text{m}.$$

The area of the cover layer per unit width of sea wall, $A = [12^2(3^2 + 1^2)]^{1/2} \simeq 38\,\text{m}^2$. From Table 15.4, the porosity of the cover layer is 37%.

The number of units per unit width of cover layer (equation (15.34)) is

$$N = 38 \times 2 \times \left(1 - \frac{37}{100}\right)\left(\frac{2650 \times 9.81}{20 \times 10^3}\right)^{2/3} = 57.$$

Figure 15.16 gives the profile of the sea wall, which is to be developed further from model studies and other considerations such as site and economic constraints.

(b) For the wave conditions and slope of the breakwater of Worked example 15.3, using Van der Meer's formula determine the mass of the armour units for rough quarry stone and the wave run-up if the design storm duration is 6.0 hours. Assume damage number for failure conditions as low.

Solution

$$\xi = \frac{1}{3}\left(\frac{2.17}{104.0}\right)^{-0.5}$$

$$= 2.3$$

From the above table on the peak period $T_z = 8.16\,\text{s}$.

$$N_z = 6.0 \times 3600/8.16 \cong 2700.$$

From Table 15.5, $S = 2.0$. $P = 0.4$ from Fig. 15.11.
For $\xi = 2.3$, using equation (15.37) for run-up on rough slopes

$$\frac{2.17}{1.65 D_{50}} = 1.0 \times 0.4^{-0.13} \times \left(\frac{2}{\sqrt{2700}}\right)^{0.2} \times \sqrt{3} \times 2.3^{0.4}$$

$$D_{50} = 0.93\,\text{m}.$$

Mean mass of rocks $M_{50} = 0.93^3 \times 2.65 = 2.13 \times 10^3\,\text{kg}$.

RUN-UP

Peak Irribarren number at the structure $\xi_p = (1/3) \times (2.17/172.0)^{-0.5}$
$= 2.7.$

The correction factor r from Table 15.1 is 0.8.

$$R_u = 2.17 \times (2.11 - 0.09 \times 2.7) \times 0.8$$

$$= 3.24\,\text{m}.$$

Worked Example 15.4

Produce a preliminary design for a sea outfall. The dry weather flow through the outfall is $0.1\,\text{m}^3\text{s}^{-1}$ and the peak flow is $0.5\,\text{m}^3\text{s}^{-1}$. Site investigations show that a sea outfall extending to 2 km seaward would not cause pollution of foreshore and recreational waters. The still water depth there is 15 m. The density of sea water is $1026\,\text{kg}\,\text{m}^{-3}$.

If a minimum dilution of 50 is required, estimate the diameter of the ports, number of ports and the length of the diffuser for still water.

Solution

Cederwall equations (15.40) and (15.41) are used to obtain the following table (choose FD and compute y_0/D_p for $S_m = 50$).

FD	1	2	5	10	13	15	20	50
y_0/D_p	38	49	67	82	88	91	96	100
D_p (m)	0.39	0.31	0.22	0.18	0.17	0.165	0.156	0.15
V_p (m s^{-1})	0.32	0.56	1.2	2.16	2.7	3.1	4.0	9.8
Q_p (m^{3}s^{-1})	0.038	0.041	0.047	0.057	0.062	0.066	0.076	0.17

The diameter of the ports is chosen as 175 mm which is large enough to avoid blockage. The exit velocity is about $2.4\,\text{m}\,\text{s}^{-1}$ which is less than the highest acceptable velocity of $3\,\text{m}\,\text{s}^{-1}$ to avoid undue headloss at the exit.

The discharge through the port of 175 mm diameter $= 0.06\,\text{m}^3\text{s}^{-1}$. Number of ports is $0.5/0.06 = 10$ (say). The flow through the ports at DWF $= 0.1/10 = 0.01\,\text{m}^3\text{s}^{-1}$. The velocity through the ports at DWF

$$= 0.01/(\pi 0.175^2/4)$$

$$= 0.4\,\text{m}\,\text{s}^{-1}.$$

Discharge through the ports at peak flow $= 0.5/10 = 0.05\,\text{m}^3\text{s}^{-1}$. Velocity through the ports at peak flow

$$= 0.05/(\pi 0.175^2/4)$$

$$= 2.1\,\mathrm{m\,s^{-1}}.$$

A velocity of $2.1\,\mathrm{m\,s^{-1}}$ is expected to be sufficient to avoid blockage of the ports. Densimetric Froude number

$$FD = 2.1/(9.81 \times 0.026 \times 0.175)^{0.5}$$

$$= 9.9,$$

$$y_0/D_p = 15/0.175 = 85.7.$$

From Fig. 15.12, for $y_0/D_p = 85.7$ and $FD = 9.9$,

$$w/2D_p = 15,$$

$$w = 2 \times 15 \times 0.175 \simeq 6\,\mathrm{m}.$$

The diffuser length $= 6 \times 10 = 60\,\mathrm{m}$.

References

Ackers, P. (1991) Sediment aspects of drainage and outfall design, in *Environmental Hydraulics* (eds J.H.W. Lee and Y.K. Cheung), Balkema, Rotterdam, pp. 10–29.

Agg, A.R. and Wakeford, A.C. (1972) Field studies of jet dilution of sewage at sea outfalls. *Institute of Public Health Engineers*, 71: 126–49.

Allsop, N.W.H., Hawkes, P.J., Jackson, F.A. and Franco, L. (1985) *Wave run-up on steep slopes – model tests under random waves*, Report SR2, Hydraulic Research, Wallingford, UK.

Bailard, J.A. (1981) An energetics total load sediment transport model for a plane sloping beach. *Journal of Geophysical Research*, 86 (C11): 10938–54.

Barrett, M.G. (1989) What is coastal management, *Proceedings of the Conference organised by the Maritime Engineering Board*, ICE, Bournemouth, UK.

Berkeley, R., Thorn, R. and Roberts, A.G. (1981) *Sea Defence and Coast Protection Works*, Thomas Telford, London.

Besley, P., Stewart, T. and Allsop, N.W.H. (1998) *Overtopping of vertical structures: New prediction methods to account for shallow water conditions*. Proceedings of International Conference on Coastlines, Structures and Breakwaters, ICE, Thomas Telford, London.

Birkemeier, W.A. (1985) Field data on seaward limit of profile change. *Journal of the Waterway, Port, Coastal and Ocean Engineering*, American Society of Civil Engineers, 111 (3): 598–602.

Brooks, N.H. (1970) Conceptual design of submarine outfalls I. Jet diffusion. California Institute of Technology, W.M. Keck Laboratory Technical Memorandum 70–1.

Bruun, P. (1985) *Design and Construction of Mounds for Breakwaters and Coastal Protection*, Elsevier, Amsterdam.

Cederwall, K. (1968) *Hydraulics of Marine Waste Disposal*, Report No. 42, Hydraulics Division, Chalmers Institute of Technology, Gotenburg, January.

Chadwick, A, Morfett, J.C. and Borthwick, M. (2004) *Hydraulics in Civil and Environmental Engineering*, 4th edn, E. & F.N. Spon, London.

Charlton, J.A. (1985) Sea outfalls, in *Developments in Hydraulic Engineering*, Vol. 3 (ed. P. Novak), Applied Science, London.

Davison, A.T., Nicholls, R.J. and Leatherman, S.P. (1992) Beach nourishment as a coastal management tool: an annotated bibliography on development associated with the artificial nourishment of beaches. *Journal of Coastal Research*, 8 (4): 984–1.

Dean, R.G. (2002) *Beach Nourishment*: Theory and Practice, World Scientific, Singapore.

Fan, L.H. and Brooks, N.H. (1966) Discussion of 'Horizontal jets in stagnant fluid of other density' by G. Abraham. *Journal of the Hydraulics Division, Proceedings of the American Society of Civil Engineers*, 92 (HY2): 423–9.

Fredsoe, J. and Deigaard, R. (1992) *Mechanics of Coastal Sediment Transport*, World Scientific, Singapore.

Goda, Y. (1974) New wave pressure formulae for composite breakers. *Proceedings of the 14th Conference on Coastal Engineering*, ASCE, New York: 1702–20.

—— (1979) A review of statistical interpretation of wave data. *Report of the Port and Harbour Institute, Japan*, 18: 5–32.

—— (2000) *Random Seas and Design of Maritime Structures*, World Scientific, Singapore.

Grace, R.A. (1978) *Marine Outfall Systems*, Prentice-Hall, Englewood Cliffs, NJ.

Herbich, J.B. (2000) *Handbook of Coastal Engineering*, McGraw-Hill, New York.

Hoffmans, G.J.C.M. and Verheij, H.J. (1997) *Scour Manual*, Balkema, Rotterdam, The Netherlands.

Horikawa, K. (1978) *Coastal Engineering – An Introduction to Ocean Engineering*, University of Tokyo Press, Tokyo, Japan.

Hudson, R.Y. (1961) Laboratory investigation of rubble-mound breakwaters. *Transactions of the American Society of Civil Engineers*, 126 (lv): 492–541.

Hunt, J.R., Jr (1959) Design of sea walls and breakwaters. *Journal of the Waterways and Harbours Division, Proceedings of the American Society of Civil Engineers*, 85 (WW3): 123–52.

Institution of Civil Engineers (1983) Breakwaters – Design and Construction, *Proceedings of Conference*, London: May: 4–6.

Ippen, A.T. (1966) *Estuary and Coastline Hydrodynamics*, McGraw-Hill, New York.

Iribarren, C.R. and Nogales, S. (1949) Protection of Ports, *Proceedings of the 17th Congress, International Association of Navigation Congresses*, Lisbon.

Kamphuis, J.W. (2000) *Introduction to Coastal Engineering and Management*, World Scientific, Singapore.

Kay, R. and Alder, J. (1999) *Coastal Planning and Management*, E. & F.N. Spon, London.

King, C.A.M. (1972) *Beaches and Coasts*, 2nd edn, Arnold, London.

Koh, R.C.Y. and Brooks, N.H. (1975) Fluid mechanics of waste-water disposal in the ocean. *Annual Review of Fluid Mechanics*, 7: 187–212.
Lee, J.H.W. and Neville-Jones, P. (1987) Initial dilution of horizontal jet in crossflow. *Journal of Hydraulic Engineering*, 113 (5): 615–30.
MAFF (1993) *Shoreline Management Plans – A Guide for Coastal Defence Authorities*, Ministry of Agriculture, Fisheries and Food, UK.
Miller, D.S. (ed.) (1994) *Discharge Characteristics*, IAHR Hydraulic Structures Design Manual, Vol. 8, Balkema, Rotterdam.
Minikin, R.R. (1963) *Wind, Waves and Marine Structures*, Griffin, London.
Morris, H.M. (1963) *Applied Hydraulics in Engineering*, Ronald Press, New York.
Muellenhoff, W.P., *et al.* (1985) *Initial Mixing Characteristics of Municipal Ocean Discharges*, Report EPA/600/3-85/073, US Environmental Protection Agency, Washington, D.C.
Muir Wood, A.M. and Fleming, C.A. (1969) *Coastal Hydraulics*, Macmillan, London.
Neville-Jones, P. and Dorling, C. (1986) *Outfall Design for Environmental Protection: a Discussion Document*, ER 209E, Water Research Centre, Marlow.
Novak, P. and Čábelka, J. (1981) *Models in Hydraulic Engineering – Physical Principles and Design Applications*, Pitman, London.
Novak, P. and Nalluri, C. (1987) Sediment transport in sewers and their sea outfalls, in *Proceedings of the 22nd Congress*, IAHR, Lausanne, Vol. D: 337–42.
Owen, M.W. (1980) *Design of Seawalls Allowing for Overtopping*, Report EX924, Hydraulics Research, Wallingford, UK.
Quinn, A.D. (1972) *Design and Construction of Ports and Marine Structures*, McGraw-Hill, New York.
Pilarczyk, K.W. and Zeidler, R.B. (1996) *Offshore Breakwaters and Shore Evolution Control*, A.A. Balkema, Rotterdam.
Reeve, D., Chadwick, A. and Fleming, C. (2004) *Coastal Engineering: Processes*, Theory and Practice, Spon Press, London.
Roberts, P.J.W. (1996) *Sea Outfalls, in Environmental Hydraulics*, edited by V.P. Singh and W.H. Hager, Water Science Technology Library, Kluwer Academic Publishers, Dordrecht.
Roelvink, J.A. and Broeker, H. (1993) Cross-shore profile models, *Coastal Engineering*, 21: 163–191.
Saville, T. Jr. (1957) Wave run-up on composite slopes, in *Proceedings of the 6th International Conference on Coastal Engineering*, University of Florida, USA: 691–9.
Silvester, R. (1974) *Coastal Engineering*, 2 volumes, Elsevier Publishing.
Simm, J.D., Brampton, A.H., Beech, N.W. and Brooke, J.S. (1996) *Beach Management Manual*, Report 153, Construction Industry Research and Information Association, UK.
Summers, L. and Fleming, C.A. (1983) *Groynes in Coastal Engineering*, Technical note 111, CIRIA, London.
Thomas, R.S. and Hall, B. (1992) *Seawall Design*, Butterworth–Heinemann, CIRIA, Oxford.
US Army (1984) *Shore Protection Manual*, US Army Coastal Engineering Research Center, Washington, DC.
—— (2002) *Coastal Engineering Manual*, Corps of Engineers, Washington, DC.

Van der Meer, J.W. (1987) Stability of breakwater armour layers – Design formulae, *Coastal Engineering*, 11: 219–39.

—— (1988) Stability of cubes, tetrapods and accropode, *Proceedings of the Conference Breakwaters 1988*, ICE, Eastbourne, Thomas Telford, London: 71–80.

Will, A.L., Willis, T.A.F. and Smith, D.D.S. (1985) Design and construction of sea wall and breakwater at Tornes power station. *Proceedings of the Institution of Civil Engineers, Part 1*, 78: 1165–89.

Wood, I.R., Bell, R.G. and Wilkinson, D.L. (1993) *Ocean Disposal of Wastewater*. World Scientific, Singapore.

Chapter 16

Models in hydraulic engineering

16.1 Hydraulic models

16.1.1 General

Hydraulics and hydraulic engineering have been marked during the 20th century by an extraordinary development of experimental methods leading to the widespread use of *scale models* and by the application of *computational techniques*.

Although many engineers use the terms mathematical, numerical and computational model as synonyms, there is a clear distinction between them. A *mathematical model* is a set of algebraic and differential equations representing the flow and based on assumptions about the physics of the prototype flow and environmental processes. A *numerical model* is an approximation of the mathematical model in the form of a computable set of parameters describing the flow at a set of discrete points. The *computational model* is the implementation of a general numerical model for a specific situation. There are many computational systems available and the user has to choose carefully among them; this choice requires, or at least is helped by, the understanding of the underlying mathematical model.

Computational models are often cheaper than the equivalent physical scale model (see below) and do not suffer from scale effects. On the other hand they can be applied only where the physics is known and included in the model and where sufficient topographical and other relevant data are available. Furthermore, their accuracy may be limited – sometimes severely – by the schematization and discretization procedure and lack of calibration.

From data handling the discipline of computational hydraulics has grown to *hydroinformatics*, which uses simulation modelling and information and communication technologies (ICT) to help solve problems in hydraulics, hydrology and environmental engineering for further management of water based systems (Abbott, 1991). Although hydraulics is a central component of hydroinformatics it itself is being influenced by this new area (Abbott, Babovic and Cunge, 2000).

In a further development *artificial neural networks* attempt to simulate the working of a human brain by passing information from one 'neuron' to all others connected with it; although a large set of data is required to 'train' the network, the method has been applied successfully to a number of flow situations.

The use of *experiments in the solution of hydraulic problems* can be traced back over many centuries, but it was not until the second half of the 19th century that the idea of using scale models to solve engineering problems was evolved and gradually put on a sound basis. In 1869 W. Froude constructed the first water basin for model testing of ships, and in 1885 O. Reynolds designed a tidal model of the Upper Mersey. The turn of the century saw the establishment of two pioneering river and hydraulic structures laboratories by Hubert Engels in Dresden (1898) and Theodor Rehbock in Karlsruhe (1901). These were followed by many new laboratories all over the world, with the major expansion occurring during the first half of the 20th century.

The increasing use of mathematical techniques and computers since about 1960 may have led to a reorientation, but not necessarily diminished use, of the hydraulic laboratory in solving hydraulic engineering problems. The reasons for this are manifold: the continuously increasing size and complexity of some schemes requires new, untested designs which defy mathematical solution; hydraulic scale models are increasingly being used as aids in the solution of environmental problems by studies both of basic physical phenomena of, for example, sediment and pollutant transport and of design applications; the development of the theory of similarity led to the realization of the inevitability of scale effects in the use of scale models which, in turn, became the impetus both of wider use of field studies and of specially designed laboratory investigations; mathematical models require data which are often derived from physical models – leading also to an increasing use of *hybrid models* combining the advantages of both modelling techniques.

In this text only a very brief overview of hydraulic models can be given; for further details see references quoted in the following sections.

16.1.2 Mathematical, numerical and computational models

As stated above the basis of numerical and computational models is a mathematical model, i.e. a set of algebraic and differential equations representing the flow. These are e.g. continuity and Navier–Stokes equations for turbulent flow (including terms for turbulent stresses) and/or shallow water and Saint Venant equations (see Section 8.2.2) involving variation of velocity and/or concentration in space and time.

The simplest partial differential equations (PDEs) frequently occurring in engineering are linear second order equations which in fluid dynamics often take the form of hyperbolic (wave), parabolic (diffusion) or elliptical (Laplace) equations. For the solution of PDEs initial and /or boundary conditions must be given.

The solution of PDEs requires in most cases recourse to numerical solutions based on finite difference, finite element or finite volume methods with explicit or implicit solutions; numerical stability, convergence, dissipation and dispersion are important aspects of numerical models. The method of characteristics transforms the PDEs into ordinary differential equations along certain characteristics resulting in simplified step solutions.

There are many computer packages available for the solution of hydraulic engineering problems, but their successful use requires a good background knowledge of underlying hydraulic physical and mathematical principles. Many of these packages relate to specific areas, e.g. Mike 11 (open channel flow), Flowmaster (pipelines and pressure surges) DAMBRK (dam-break) (see also Section 7.5.2), etc.

For a general discussion of the mathematical background of PDEs and their solution see, e.g. Jeffrey (2003); for an overview of the development of software (and hardware) in computational hydraulics see, e.g. Abbott (1991), Anderson (1995) and Abbott and Minns (1997); for the background to turbulence modelling see, e.g. ASCE (1988) and Rodi (1996). Zinkiewicz and Taylor (2000) provide a detailed account of the finite element method, Vreugdenhill (1994) gives details of numerical modelling of shallow water flows and Chadwick *et al.* (2004) give examples of applications of computational hydraulics.

Computational fluid dynamics (CFD) has an important role to play in hydraulic engineering design – see, e.g. Verwey (1983). There have been rapid developments e.g. in the application of numerical techniques to overfall spillway design – see Spaliviero and Seed (1998), Song and Zhou (1999) and Assy (2001). The use of computational hydraulics is particularly widespread in river engineering – see, e.g. Cunge, Holly and Verwey (1980) – including computation of the flow field round groynes and bridge abutments (Biglari and Sturm, 1998). Bürgisser (1999) gives a general overview of the solution of the free water surface at hydraulic structures using numerical modelling (including an extensive bibliography of the subject). The application of numerical techniques in the computation of pressure transients is covered e.g. in Wilie and Streeter (1993). For a brief discussion of numerical modelling in coastal engineering see Chapter 15.

16.1.3 Scale models

A *scale model* in hydraulic engineering (as opposed to analogue and mathematical models) (ASCE, 1982) uses the method of direct (physical) simulation of (hydraulic) phenomena, (usually) in the same medium as in the prototype. Models are designed and operated according to scaling laws, i.e. conditions that must be satisfied to achieve the desired similarity between model and prototype. The ratio of a variable in prototype to the corresponding variable in the model is the scale factor (scale); in the literature the reciprocal of this ratio is sometimes used. In the following text the scale factor is denoted by M_x.

Distortion is a conscious departure from a scaling law (e.g. geometric distortion – discussed later). Non-similarity between model and prototype, resulting from the fact that not all pertinent dimensionless numbers (physically meaningful ratios of parameters used in determining scaling laws) are the same in the model and prototype, is called the scale effect. In other words, the scale effect is the error caused by using the model according to the main determining law and neglecting others, e.g. errors resulting from modelling the prototype on the basis of scales chosen to suit the dominant force action and allowing the other forces to be out of scale.

The analysis of flow often leads to the use of geometrically distorted models; indeed, even a geometrically similar model almost inevitably introduces some degree of distortion of flow and some scale effects. The modeller has to be aware of these effects and, in relation to hydraulic structures design, of model–prototype conformity. He or she has to relate this particularly to the required precision of the answer and, most importantly, must be aware of whether the model answer enhances or reduces the safety of the prototype structure. For example, a model of a ski-jump spillway will produce a jet with less air entrainment than would be the case in prototype; because of this, and because of the reduced velocity of the jet on the model and hence the reduced air resistance, when interpreting the distance of the jet impact on the river bed downstream of the structure one obtains from the model a distance (when scaled up according to model scale) that is greater than is likely to be the case in prototype but possibly with a deeper scour hole. The scale effect in this case means that the actual scour is likely to be somewhat nearer the dam but shallower than indicated by the model tests. Does this matter? Does it increase or decrease the safety of the structure? The answers will depend, amongst other factors, on geological conditions, operational rules, etc.; the interpretation of the scale-model result thus requires knowledge, and also intuition and experience, both of the modeller and the design engineer.

Only a few aspects of physical scale models as used in hydraulic engineering can be touched upon here, and for more detailed treatment the reader is referred to specialized texts (e.g. Kobus, 1980; Novak and Čábelka, 1981; Novak, 1984).

(a) Geometric, kinematic, dynamic and mechanical similarity

Geometric similarity is similarity in form, i.e. the length scale M_l is the same in all directions.

Kinematic similarity denotes similarity of motion, i.e. similarity of velocity and acceleration components along the x, y, z axes; M_u, M_v, M_w are all constants (not necessarily equal); the same applies for the acceleration scales.

Dynamic similarity denotes similarity of forces; thus with M_m as the scale for mass and M_t for time we can write

$$M_{Px} = M_m M_{ax} = M_m M_{lx} M_t^{-2} = M_m M_u^2 M_{lx}^{-1} = \text{constant} \tag{16.1}$$

(etc., in the y and z directions).

Mechanical similarity is an all-embracing term including geometric, kinematic and dynamic similarity, i.e. M_l, M_v and M_P are all constants, the same in all directions. Mechanical similarity can be defined as follows: two formations are (mechanically) similar if they are geometrically similar and if, for proportional masses of homologous points, their paths described in proportional times are also geometrically similar. This definition, based on Newton's law, thus includes geometric similarity of the two formations, the proportionality of times and the geometric similarity of the paths travelled (kinematic similarity) as well as the proportionality of masses and thus also of forces (dynamic similarity).

The *theory of similarity*, leading to dimensionless numbers and scaling laws, can be elaborated in three ways. The first determines the criteria of similarity from a system of basic homogeneous (differential) equations which mathematically express the investigated physical phenomena. The second path leads to the conditions of similarity through dimensional analysis carried out after a careful appraisal of the physical basis of each phenomenon and of the parameters which influence it. The combined use of physical and dimensional analyses is often the best route to a successful formulation of similarity criteria. The third way could be denoted as the method of synthesis (Barr, 1983; Sharp, 1981).

An example of the first route – the use of physical laws and governing equations – is the formulation of the scaling laws (criteria) by writing, for example, the Navier–Stokes equations, both for the model and the prototype, and inserting the scales of the various parameters onto one set of the equations. In this way we arrive from

$$\frac{\partial v}{\partial t} + u\frac{\partial v}{\partial x} + v\frac{\partial v}{\partial y} + w\frac{\partial v}{\partial z} = Y - \frac{1}{\rho}\frac{\partial p}{\partial y} + \frac{\mu}{\rho}\left(\frac{\partial^2 v}{\partial x^2} + \frac{\partial^2 v}{\partial y^2} + \frac{\partial^2 v}{\partial z^2}\right)\dots \tag{16.2}$$

with $Y = g$ at the condition

$$\frac{M_v}{M_t} = \frac{M_v^2}{M_l} = M_g = \frac{M_p}{M_\rho M_l} = \frac{M_\mu M_v}{M_\rho M_l^2}$$

or

$$\frac{M_l}{M_v M_t} = \frac{M_g M_l}{M_v^2} = \frac{M_p}{M_\rho M_v^2} = \frac{M_\mu}{M_\rho M_v M_l} = 1 \qquad (16.3a–d)$$

i.e. the condition that the scales of the Strouhal (*Sh*), Froude (*Fr*), Euler (*Eu*) and Reynolds (*Re*) numbers must be 1. The dimensionless numbers thus derived can, however, be criteria of similarity only if the initial equations have an unambiguous solution. This can only be attained if the equations are limited by certain boundary conditions which assume the character of conditions of unambiguity of the solution.

An inspection of equations (16.3a–d) shows that with $M_g = 1$ and using the same liquid for the model as in prototype ($M_\rho = M_\mu = 1$) they can be satisfied only if $M_l = 1$, i.e. in a model the same size as the prototype. Therefore we have to design and operate our models almost always with *approximate mechanical similarity*, choosing a dominant force component (e.g. gravity) and neglecting the effects of the others. To minimize the resulting scale effects we have to impose limiting (boundary) conditions on our scales, e.g. choose a scale where in the model flow the effect of viscosity will be negligible.

Taking gravity as our dominant force (correct in most models of hydraulic structures) results from equations (16.3) with $M_g = 1$ in

$$\frac{M_v}{(M_g M_l)^{1/2}} = \frac{M_v}{M_l^{1/2}} = M_{Fr} = 1. \qquad (16.4)$$

The same result can be obtained form equation (16.1) by writing

$$M_a M_m = M_g M_\rho M_l^3 = M_\rho M_l^2 M_v^2.$$

Equation (16.4) represents the Froude law of similarity. From it we can obtain all the other required scale factors expressed in terms of the length scale M_l (Table 16.1).

(b) Scale models of hydraulic structures

In the vast majority of cases, design problems associated with hydraulic structures as described in the previous chapters are investigated on geometrically similar scale models, operated according to the Froude law of

Table 16.1 Scale factors

Parameter	*Scale factor*	*Parameter*	*Scale factor*
Velocity	$M_v = M_l^{1/2}$	Area	$M_A = M_l^2$
Volume	$M_V = M_l^3$	Mass	$M_m = (M_\rho)\, M_l^3 = M_l^3$
Time	$M_t = M_l^{1/2}$	Discharge	$M_Q = M_l^{5/2}$
Force	$M_P = (M_\rho) M_l^3 = M_l^3$	Specific discharge	$M_q = M_l^{3/2}$
Pressure (intensity)	$M_p = (M_\rho) M_l = M_l$	Energy	$M_E = (M_\rho) M_l^4 = M_l^4$
		Momentum	$M_M = (M_\rho)\, M_l^{7/2} = M_l^{7/2}$

similarity. The main exceptions are distorted models of rivers, river training and coastal engineering projects (Section 16.1.4).

The main causes of scale effects are model roughness and model approach conditions associated with turbulent boundary layer development, surface tension effects and associated aeration and vortex-formation problems and cavitation phenomena. Some of these scale effects can be overcome, or at least minimized, by using model scales giving sufficiently high model Reynolds numbers (which are reduced against the prototype when using the Froude scaling law by $M_l^{3/2}$ for the same model viscosity as in prototype) and Weber numbers (reduced on the model for the same liquid as in prototype by M_l^2).

For example, for correct extrapolation of the shape of a nappe the head on a sharpcrested notch should be at least 60 mm; for a head below 20 mm the overflow parabola of the free jet (see Section 4.7) is deformed almost in to a straight line. Hence it is advisable to choose a spillway model scale so that the head on the spillway crest in the model is at least 60 mm (for the maximum discharge). Similar conditions apply for the shape of the outflow under a gate where the minimum gate opening should also be about 60 mm. The diameter of model bottom outlets should preferably be larger than 50 mm to avoid scale effects in the entry loss coefficients. The Reynolds number for smooth models should be such that it corresponds to the fully turbulent hydraulically rough prototype value to obtain the correct friction losses. Thus, for a pipe with a relative roughness of $k/D = 0.001$, the friction factor λ is independent of the Reynolds number for $Re = VD/\nu > 10^6$. To achieve the same value of λ for a very smooth model pipe we need a Reynolds number of about 70 000 (see Moody diagram). Thus if the prototype Reynolds number is 10^7, from $M_l^{3/2} = 10^7/(7 \times 10^4)$, the model scale should be $M_l = 27.5$ – a condition which may be difficult to fulfil. On models of outlet works with spillway and bottom outlet(s) we may thus have to accept some minor scale effects in the reproduction of the bottom outlet(s) performance. If flow in open channels is involved, the Reynolds number on the model ($Re = VR/\nu$) should be greater than

$$Re = \frac{126R}{k\lambda^{1/2}} \tag{16.5}$$

(with $\lambda = h_f 8gR/lV^2$) to avoid viscous effects; this usually requires $10^{3.5} < Re < 10^{4.5}$. For models of intakes the Weber number (defined as $V(\rho D/\sigma)^{1/2}$) should be greater than 11 and the Reynolds number (VD/ν) greater than 3×10^4 to avoid surface tension and viscous effects. The above are only approximate guidelines; for further details, see, for example, Kobus (1980), Novak and Čábelka (1981) and Knauss (1987).

It is extremely unlikely that on a standard reduced model problems associated with the aeration and bulking of flow over spillways can be fully investigated, although useful results can be obtained for the initiation of aeration on chutes, the design of artificial aerators, aeration in shaft spillways, etc. The scale effects in, for example, energy dissipation also have to be carefully analysed (Kobus, 1984).

An ordinary scale model of, say, a spillway will not cavitate where cavitation would occur in prototype, because the ambient atmospheric pressure has not been reduced to the model scale. Pressure measurements taken on the model and converted to prototype would, however, indicate whether cavitation in prototype is likely. For example, if a model scale $M_l = 25$ exhibits a negative pressure of 0.5 m, cavitation would occur in prototype because $25 \times 0.5 = 12.5$ m below atmospheric pressure, which is physically not possible, as cavitation would start close to the minimum possible value of 10 m. If we want to investigate the behaviour of a spillway with cavitation occurring on the model, this has to be placed in a cavitation tunnel where we can control and reduce the ambient pressure (Novak and Čábelka, 1981); even in this case we have to take into account water quality and other scale effects (Kobus, 1984; Burgi, 1988; Eickmann, 1992).

Flow-induced vibrations on model gates with the model operated according to the Froude law reproducing correctly the flow field are subject to scale effects which can be overcome by a hybrid rigid model (simulating the low frequency interaction of the gate with the flow) or by use of models with an overall reproduction of elasticity (see e.g. Haszpra, 1979 or ICOLD, 1996; refer also to Section 16.2).

Investigations on models of hydraulic structures often require sophisticated instrumentation for the measurement of discharge, velocity and pressure fluctuations, air concentration, etc. Models of navigable waterways and of locks with model barges in particular need specialized equipment. On some waterway models the steering and propulsion of the model barge(s) are computer controlled and model tests are used to calibrate mathematical simulation models (M_l for such models is likely to be smaller than 15). Forces acting on model barges during lock operation are recorded by special dynamometers eliminating, or at least minimizing, the inertial forces.

(c) River and coastal engineering models

From Bernoulli's equation using the same procedure as used for the derivation of equation (16.3), we can obtain

$$M_z = M_h = M_v^2 = M_\lambda M_v^2 M_l M_R^{-1} = M_\xi M_v^2 \qquad (16.6a–d)$$

where M_z, M_h, M_ξ are scale ratios for height above datum, depth and local loss coefficient. Equation (16.6a) indicates that for open-channel models with non-uniform flow the height and depth scales must be identical, and a tilting of the model about one of its ends is permissible for uniform flow conditions only. As we frequently require a different vertical and horizontal (length and width) scale to achieve a sufficiently high Reynolds number on the model, and to ensure a fully turbulent flow régime, i.e. $M_h < M_l$, the discharge scale from equation (16.6b) (which again represents the Froude law) will be

$$M_Q = M_A M_v = M_h^{3/2} M_l. \qquad (16.7)$$

Equation (16.6c) results in

$$M_R = M_l M_\lambda \qquad (16.8)$$

and equation (16.6d) in

$$M_\xi = 1, \qquad (16.9)$$

i.e. local loss coefficients should be the same on the model as in prototype. This last condition is practically impossible to achieve in distorted river models for every local loss but can be achieved overall, taking all local losses (at changes of section and direction) together.

The above equations contain seven variables (M_z, M_h, M_l, M_v, M_λ, M_R, M_ξ); furthermore, M_R must be a function of M_h and M_l, and M_λ of M_R and M_k, where M_k is the roughness size scale:

$$M_R = f_1(M_h, M_l), \qquad (16.10)$$

$$M_\lambda = f_2(M_R, M_k). \qquad (16.11)$$

Thus we have the six equations (16.6a), (16.6b), (16.8), (16.9), (16.10) and (16.11) for eight variables, giving two degrees of freedom. In the design of the model we can therefore choose only two variables, usually M_l and M_h (or M_Q).

In the case of a movable bed model, in the first approach we can substitute for M_k the variable M_d, i.e. the sediment size scale, and if we want to achieve similarity of incipient sediment motion from the Shields criterion (equations (8.19) or (8.20)) we obtain another equation for M_Δ, ($\Delta = (\rho_s - \rho)/\rho$):

$$M_\Delta = \frac{M_{U*}^2}{M_d} = \frac{M_R M_S}{M_d} = \frac{M_R M_h}{M_l M_d}. \tag{16.12}$$

We have now nine variables and seven equations, and again only two degrees of freedom for choice of scales. For similarity of sediment transport, additional boundary conditions may be required and similarity of bedforms also has to be investigated.

To model tidal motion on estuary and coastal engineering models similar considerations apply; however, we must remember that for the vertical motion of the water surface (or the sediment fall velocity) the corresponding velocity scale is given by

$$M_w = M_h/M_t = M_h^{3/2}/M_l \tag{16.13}$$

(as $M_t = M_l/M_v = M_l M_h^{-1/2}$). For studies of wave refraction, which depends only on depth, the wave celerity scale $M_c = M_h^{1/2}$; we can also obtain the same result from the long shallow-water wave equation (equation (14.15)). As for short deep-water waves, $M_c = M_L^{1/2}$ (equation (14.14)) and $M_L = M_c M_T$ (equation (14.1)), where M_T is the wave period scale and M_L the wavelength scale, this results in

$$M_c (= M_L^{1/2} = M_h^{1/2}) = M_T. \tag{16.14}$$

Thus we can have in this case a distorted model $M_h \neq M_l$. For reproducing wave diffraction, the wave height along the obstacle must be correctly reproduced and, therefore, $M_L = M_l$, i.e. an undistorted model is required if major scale effects are to be avoided.

Again, only the outline approach to river and coastal engineering models has been discussed; for further detailed treatment of the subject the reader is referred to, for example, Allen (1947), Yalin (1971), Kobus (1980) or Novak and Čábelka (1981).

16.2 Structural models

16.2.1 General

Structural models of hydraulic structures are mainly models of dams and their foundations – but see also models of gates (Section 16.1.3). Conceived as a technique for verifying and developing the theoretical analysis of more complex structures, physical modelling offers the advantage of a tangible rather than a solely mathematical representation of structural response. Modelling techniques were developed and perfected in the period 1950–65, essentially for multicurvature and complex concrete dams, against limitations imposed on the application of sophisticated mathematical analyses by non-availability of the necessary computing power.

Provided that a physical model is constructed in strict accord with the appropriate laws of similitude (Section 16.1.3) it will function as a structural analogue, yielding a valid prediction of prototype deformation and stress distribution. In practice, as with hydraulic models, limitations are imposed by conflicting requirements for compliance with the different laws of similitude, most notably those relating to model material characteristics and load response. Structural models are also relatively inflexible in practice, and investigation of the effects of a change in geometry or structural detail, e.g. the presence of joints in a concrete dam, or a change in any major parameter, may require construction of a completely new model. For these and other reasons computational methods (Chapters 2 and 3 and Section 16.1.2) have largely displaced physical modelling techniques. Consideration of the latter is therefore restricted here to the essential principles only.

Structural modelling of dams almost invariably relates to a static loading condition. The relevant relationships are, therefore, those governing the stress (σ) and force (P) ratios respectively (Table 16.1), i.e. $M_\sigma = M_\rho M_l$ and $M_P = M_\rho M_l^3$.

For any material, the Poisson ratio, ν, and the linear strain, ϵ, are dimensionless parameters, and thus for structural similitude of model (subscript m) and prototype (subscript p) $\epsilon_p = \epsilon_m$ and $\nu_p = \nu_m$. From Hooke's law the consequence of these statements is

$$M_\sigma = M_E M_\epsilon \tag{16.15a}$$

or, for $M_\epsilon \simeq 1$,

$$M_\sigma \simeq M_E (= M_\rho M_l). \tag{16.15b}$$

The dominant stresses in a dam are those generated by hydrostatic loads, including seepage and uplift, by structural self-weight and, if secondary stresses are considered, the effects of temperature or of foundation deformation. Difficulties inherent in simultaneously satisfying requirements for material and structural similitude and reconciling scale ratios for geometry, stress and weight have been alluded to and are readily apparent, given that the objective is for the model to correctly predict prototype deformation and stress.

16.2.2 Modelling concrete dams

Concrete dam models are almost invariably of arch or cupola structures. Simulation of the external hydrostatic load in compliance with the appropriate scaling laws requires a liquid of very high density. In practice a mercury-filled flexible bag is most commonly employed; the implication of a predetermined density ratio of 13.6 with respect to other scale ratios will be noted (see Table 16.1).

Desirable characteristics of the model material are a correct representation of stress–strain response and a relatively low modulus of elasticity, E, to enhance model deformations and permit the use of mercury loading on a model of manageable dimensions. The material should have appreciable tensile and compressive strength to reduce the risk of cracking, and must be readily cast or machined to the correct profile for the dam. Materials which have been successfully employed include microconcrete, plastics and plaster-based compounds. Mixtures of high-grade plaster of Paris with a chemically inert filler (such as diatomite or sodium montmorillonite) and water have proved particularly suitable. Such mixtures have the following additional desirable characteristics:

1. isotropy and homogeneity;
2. they are uniformly elastic within model stress range;
3. the Poisson ratio, ν, is approximately equal to that of concrete and rock (0.15–0.22);
4. elastic properties are not time dependent;
5. modulus of elasticity, E, and strength characteristics can be accurately controlled.

The purpose of the filler is to permit a high water:plaster ratio and hence a low E value. The plaster–filler mixtures require careful curing over a period, the final product having the texture of a soft chalk and being readily machined to profile. It will be noted that appropriately scaled representation of E and ν, rather than strength, is the most important parameter in relation to correct simulation of structural response.

Representative values of the principal engineering characteristics of plaster–filler mixtures include E ranging from 0.7×10^3 to $5.0 \times 10^3\,\mathrm{MN\,m^{-2}}$ and $\nu = 0.20$. Compressive strengths lie in the range 2.0–20.0 $\mathrm{MN\,m^{-2}}$, with tensile strengths typically 12% to 25% as great. Linear scales for concrete dam models are commonly in the range 1:50 to 1:200, care being taken to model an appreciable portion of the abutments and underlying foundations to correctly simulate interaction.

Major structural features of the prototype abutment and/or foundation geology, e.g. faults or crush zones etc., must be reproduced on the model. Plaster–filler mixes with appropriate stiffness characteristics may be employed to model significant changes in rock deformability associated with such features or with differing rock types.

Deformations under hydrostatic and other loadings are determined by suitably mounted transducers or dial gauges. Stresses are determined from strains recorded by surface-bonded strain gauges or strain rosettes at strategic locations on upstream and downstream faces of the model dam.

Self-weight loads are the most difficult to simulate at model scale. One technique involves progressively cutting the model down in stages following completion of all hydrostatic load tests. At each stage or level in turn, the superincumbent self-weight load is represented through a system of vertical springs acting on spreader plates on the model. An alternative technique involves inversion of the model and its immersion in mercury.

Construction details, e.g. joints, or 'defects' such as cracks can be represented in a sophisticated model, and temperature effects can also be studied if the scaling laws are further developed. Comprehensive reviews of structural modelling techniques and their application to specific studies have been presented by Rydzewski (1963) and Rocha, Serafim and Azeveda (1961).

It may be noted that plaster–filler mixtures also lend themselves to simulation of geomechanical problems involving rocks and jointed rock masses. They have been applied to investigations of this nature associated with major dam projects, as discussed by Oberti and Fumagalli (1963) and by Fumagalli (1966).

16.2.3 Modelling of embankments

Application of physical modelling to the study of geotechnical problems and embankments is severely constrained by the dominance of self-weight loading and by the complexity and non-uniformity of the range of prototype materials, i.e. foundation soils, compacted earthfills and rockfills. The approach to design for fill dams is in any event very different from that for concrete dams, focusing on seepage, deformation and stability rather than upon stress. Physical modelling has therefore been confined to limited investigations of embankment slopes and of embankments on soft foundations in terms of pore pressure changes, deformation and stability, self-weight scaling being achieved by mounting the model on the rotating arm of a large centrifuge. In the case of the Cambridge geotechnical centrifuge as described by Schofield (1980), model packages of the order of 1000 kg, could be subjected to accelerations of up to 125g at a radius (rotor arm length) of 4 m.

Geotechnical models offer many attractions in principle, but in practice the problems are almost intractable. Natural soils must be used to construct the model, and a valid simulation of zoning in the prototype dam is impossible. Even for simple homogeneous model embankments on a homogeneous foundation difficulties arise in attempting to translate model data to prototype scale as most problems are stress-path dependent, i.e. they relate to short- and longer-term loading histories. The problems are also stress-level dependent, i.e. they are a function of self-weight for the embankment and foundation complex. Bassett (1981) and Ko (1988) provide a comprehensive introduction to contemporary views on the use and limitations of physical models in geotechnical design and a further perspective is provided in Taylor (1994). The utility of such models is effectively restricted to study of deformation modes and failure mechanisms for simplified profiles.

16.2.4 Modelling of seismic response

Studies of the seismic response of models of concrete gravity dams using the 'shaking table' technique have been reported by Mir and Taylor (1994). The requirement for specialist facilities and the problems with valid physical simulation of the prototype dam constrain any application of this technique outside the research laboratory. Comparable and valid model studies of the behaviour of embankment dams are generally considered not to be possible, but physical model tests conducted for comparison with numerical analyses have been reported by Finn (1990).

The application of centrifugal modelling techniques to specific aspects of seismic stability for embankments is discussed in Pilgrim and Zeng (1994) and in Dewoolkar, Ko and Pak (1999).

Worked Example 16.1

A river, transporting sediment, flows into a tidal estuary. The maximum freshwater flow into the estuary is $4000\,m^3\,s^{-1}$. It is required to make a preliminary design of a scale model in a laboratory where the space available dictates the horizontal scale, $M_l = 250$; the pumping capacity available for the model is $27\,l\,s^{-1}$ and it is desirable to use it reasonably fully to avoid viscous effects on the model. Establish

1. a suitable vertical scale for the model,
2. the discharge rate,
3. the tidal period scale,
4. the scale of the fall velocity of suspended sediment,
5. the probable scale of the bed material, and
6. the scale for the density of the bed load.

Solution

1. Utilizing fully the discharge capacity gives a discharge scale $4 \times 10^6/27 = 148\,150$. From equation (16.7), $M_h = (M_Q/M_l)^{2/3} = (148\,150/250)^{2/3} = 70.57$. Therefore let us choose $M_h = 75$, giving a distortion of 3.33, which is probably quite acceptable in this case.
2. $M_Q = M_h^{3/2} M_l = 75^{3/2} \times 250 = 162\,380$. The maximum model discharge will be $4 \times 10^6/162\,380 = 24.6\,\mathrm{l\,s^{-1}}$.
3. $M_t = M_l/M_v = M_l M_h^{-1/2} = 250/75^{1/2} = 28.87$ (equation (16.3)).
4. $M_{w_s} = M_h^{3/2}/M_l = 75^{3/2}/250 = 2.6$ (equation (16.13)).
5. From the Manning–Strickler equation for a wide channel with $R \simeq y$,

$$M_v = M_n^{-1} M_h^{2/3} M_s^{1/2} = M_d^{-1/6} M_h^{2/3} M_h^{1/2} M_l^{-1/2} = M_d^{-1/6} M_h^{7/6} M_l^{-1/2} = M_h^{1/2},$$

$$M_d = M_h^4 M_l^{-3} = 75^4/250^3 = 2.025 \simeq 2.$$

6.

$$M_\Delta = \frac{M_h^2}{M_l M_d} = \frac{M_h^2 M_l^3}{M_h^4 M_l} = \frac{M_l^2}{M_h^2} = \left(\frac{250}{75}\right)^2 = 11.11 \text{ (equation (16.12))}.$$

Note that the answer to 5 assumes that there is no effect of bedforms (i.e. that the bed is flat) which may be unrealistic; this has to be checked by further computation, and the final value of M_d may influence the whole model design.

References

Abbott, M.B. (1991) *Hydroinformatics: Information Technology and the Aquatic Environment*, Avebury Technical, Aldershot.

Abbott, M.B. and Minns, A.W. (1998) *Computational Hydraulics*, 2nd edn, Ashgate, Brookfield, USA.

Abbott, M.B., Babovic, V.M. and Cunge, J.A. (2001) Towards the hydraulics of the hydroinformation era. *Journal of Hydraulic Engineering*, ASCE, 39, No. 4: 339–49.

Allen, J. (1947) *Scale Models in Hydraulic Engineering*, Longman, London.

Anderson, J.D. (1995) *Computational Fluid Dynamics*, McGraw Hill, New York.

ASCE (1982) American Society of Civil Engineers, Task Committee on Glossary of Hydraulic Modeling Terms, Modeling hydraulic phenomena – a glossary of terms. *Journal of the Hydraulics Division, Proceedings of the American Society of Civil Engineers*, 108 (NY7): 45–852.

—— (1988) Turbulence modelling of surface water flow and transport, Part I–V; Task Committee on Turbulence Models in Hydraulic Computations. *Journal of Hydraulic Engineering*, ASCE, 114, No. 9: 970–1073.

Assy, T.M. (2001) Solution for spillway flow by finite difference method. *Journal of Hydraulic Research*, IAHR, 39, No. 3: 241–7.

Barr, D.I.H. (1983) A survey of procedures for dimensional analysis. *International Journal of Mechanical Engineering in Education*, 11 (3): 147–59.

Bassett, R.H. (1981) The use of physical models in design, in *Proceedings of the 7th European Conference on Soil Mechanics and Foundation Engineering*, Brighton, Vol. 2, British Geotechnical Society, London.

Biglari, B. and Sturm, T.W. (1998) Numerical modelling of flow around bridge abutments in compound channel. *Journal of Hydraulic Engineering*, ASCE, 124, No. 2: 156–64.

Burgi, P.H. (ed.) (1988) Model–prototype correlation of hydraulic structures, in *Proceedings of the International Symposium*, ASCE, Colorado Springs, American Society of Civil Engineers, New York.

Bürgisser, M. (1999) *Numerische Simulation der freien Wasseroberfläche bei Ingenieurbauten*, Mitteilungen No. 162, Versuchsanstalt fur Wasserbau, Hydrologie and Glaziologie, ETH, Zürich.

Chadwick, A., Morfett, J. and Borthwick, M. (2004) *Hydraulics in Civil and Environmental Engineering*, 4th edition, Spon Press, London.

Cunge, J.A., Holly Jr, F.M. and Verwey A. (1980) *Practical Aspects of Computational River Hydraulics*, Pitman, London.

Dewoolkar, M.M., Ko, H.Y. and Pak, R.Y.S. (1999) Centrifuge modelling of models of seismic effects on saturated earth structures. *Geotechnique*, 49, 2: 247–66.

Eickman, G. (1992) *Massstabseffekte bei der beginnenden Kavitation*, Berichte der Versuchsanstalt Obernach, No. 69, TU, Munchen.

Finn, W.D.L. (1990) Seismic analysis of embankment dams. *Dam Engineering*, 1 (1): 59–75.

Fumagalli, E. (1966) Stability of arch dam rock abutments, in *Proceedings of the 1st International Congress of Rock Mechanics*, Lisbon, Vol. II, Laboratorio Nacional de Engenharia Civil, Lisbon.

Haszpra, O. (1979) *Modelling Hydroelastic Vibrations*, Pitman, London.

ICOLD (1996) *Vibration of Hydraulic Equipment for Dams*, Bulletin 102, International Commission on Large Dams, Paris.

Jeffrey, A. (2003) *Applied Partial Differential Equations: An Introduction*, Academic Press/Elsevier, New York.

Knauss, J. (1987) *Swirling Flow Problems at Intakes*, IAHR Hydraulic Structures Design Manual, Balkema, Rotterdam.

Ko, H.Y. (1988) Summary of the state-of-the-art in centrifuge model testing, in *Centrifuge in Soil Mechanics*, 11–18, Balkema, Rotterdam.

Kobus, H. (ed.) (1980) *Hydraulic Modelling*, Bulletin 7, German Association for Water Resources and Land Development.

—— (ed.) (1984) *Proceedings of the Symposium on Scale Effects in Modelling Hydraulic Structures*, Technische Akademie, Esslingen.

Mir, R. and Taylor, C.A. (1994) *Shaking Table Studies of the Performance of Gravity Dam Models*, Report No. UBCE-EE-94, Earthquake Engineering Research Centre, University of Bristol.

Novak, P. (1984) Scaling factors and scale effects in modelling hydraulic structures. General lecture, in *Proceedings of the Symposium on Scale Effects in Modelling Hydraulic Structures*, Technische Akademie, Esslingen, Paper 03: 1–6.

Novak, P. and Čábelka, J. (1981) *Models in Hydraulic Engineering – Physical Principles and Design Applications*, Pitman, London.

Oberti, G. and Fumagalli, E. (1963) Results obtained in geomechanical model studies, in *Proceedings of the Symposium on Concrete Dam Models*, Laboratorio Nacional de Engenharia Civil, Lisbon.
Pilgrim, N.K. and Zeng, X. (1994) Slope stability with seepage in centrifuge model earthquakes, in *Centrifuge '94*, 233–8, Balkema, Rotterdam.
Rocha, M., Serafim, J.L. and Azeveda, M.C. (1961) *Special Problems of Concrete Dams Studied by Models*, Bulletin No. 12, RILEM.
Rodi, W. (1996) *Numerische Berechnung Turbulenter Strömungen in Forschung und Praxis*, Institut für Hydromechanik, Universität Karlsruhe, Karlsruhe.
Rydzewski, J.R. (1963) The place of models in the study of arch dams under hydrostatic and gravity loading, in *Proceedings of the Symposium on Concrete Dam Models*, Laboratorio Nacional de Engenharia Civil, Lisbon.
Taylor, R.N. (ed.) (1994) *Geotechnical Centrifuge Technology*, Spon, London.
Schofield, A.N. (1980) Cambridge geotechnical centrifuge operations (20th Rankine lecture). *Géotechnique*, 30 (3): 225–67.
Sharp, J.J. (1981) *Hydraulic Modelling*, Butterworth, London.
Spaliviero, F. and Seed, D. (1998) *Modelling of Hydraulic Structures*, Report SR 545, HR, Wallingford.
Song, C.C.S. and Zhou, F. (1999) Simulation of free surface flow over spillway. *Journal of Hydraulic Engineering*, ASCE, 125, No. 9: 959–67.
Verwey, A. (1983) The rôle of computational hydraulics in the hydraulic design of structures, in *Developments in Hydraulic Engineering*, Vol. 1 (ed. P. Novak), Applied Science, London.
Vreugdenhill, C.B. (1994) *Numerical Methods for Shallow-water Flows*, Kluwer Academic Publishers, Dordrecht.
Wylie, E.B. and Streeter, V.L. (1993) *Fluid Transients in Systems*, Prentice Hall, Englewood Cliffs, N.J.
Yalin, M.S. (1971) *Theory of Hydraulic Models*, Macmillan, London.
Zinkiewicz, O. and Taylor, R.L. (2000) *The Finite Element Method*, 5th edition, Butterworth-Heinemann, London.

Author Index

Page references in **bold** refer to tables and page references in *italic* refer to figures.

AUTHOR INDEX

AUTHOR INDEX

Subject index

Page references in **bold** refer to tables and page references in *italic* refer to figures.

SUBJECT INDEX

SUBJECT INDEX

ted in the United States